Fundamentals of Structural Mechanics

Second Edition

Fundamentals of Structural Mechanics

Second Edition

Keith D. Hjelmstad
University of Illinois at Urbana-Champaign
Urbana-Champaign, Illinois

ISBN 978-1-4419-3609-7
e-ISBN 978-0-387-23331-4

Printed in the United States of America. (BS/DH)

9 8 7 6 5 4 3 2 1

springeronline.com

To the memory of
Juan Carlos Simo
(1952–1994)
who taught me
the joy of mechanics

and

To my family
Kara, David, Kirsten, and Annika
who taught me
the mechanics of joy

Contents

Preface

The last few decades have witnessed a dramatic increase in the application of numerical computation to problems in solid and structural mechanics. The burgeoning of computational mechanics opened a pedagogical gap between traditional courses in elementary strength of materials and the finite element method that classical courses on advanced strength of materials and elasticity do not adequately fill. In the past, our ability to formulate theory exceeded our ability to compute. In those days, solid mechanics was for virtuosos. With the advent of the finite element method, our ability to compute has surpassed our ability to formulate theory. As a result, continuum mechanics is no longer the province of the specialist.

What an engineer needs to know about mechanics has been forever changed by our capacity to compute. This book attempts to capitalize on the pedagogical opportunities implicit in this shift of perspective. It now seems more appropriate to focus on fundamental principles and formulations than on classical solution techniques.

* * * *

The term *structural mechanics* probably means different things to different people. To me it brings to mind the specialized theories of beams, plates, and shells that provide the building blocks of common structures (if it involves bending moment then it is probably structural mechanics). Structural elements are often slender, so structural stability is also a key part of structural mechanics. This book covers the fundamentals of structural mechanics. The treatment here is guided and confined by the strong philosophical framework of continuum mechanics and is given wings to fly by the powerful tools of numerical analysis.

In essence, this book is an introduction to computational structural mechanics. The emphasis on computation has both practical and pedagogical roots. The computational methods developed here are representative of the methods prevalent in the modern tools of the trade. As such, the lessons in computation are practical. An equally important outcome of the computational framework is the great pedagogical boost that the student can get from the notion that most problems are amenable to the numerical methods advocated herein. A theory is ever-so-much more interesting if you really believe you can crunch numbers with it. This optimistic outlook is a pedagogical boon to learning mechanics and the mathematics that goes along with it.

This book is by no means a comprehensive treatment of structural mechanics. It is a simple template to help the novice learn how to think about structural mechanics and how to express those thoughts in the language of mathematics. The book is meant to be a preamble to further study on a variety of topics from continuum mechanics to finite element methods. The book is aimed at advanced undergraduates and first-year graduate students in any of the mechanical sciences (e.g., civil, mechanical, and aerospace engineering).

* * * *

The book starts with a brief account of the algebra and calculus of vectors and tensors (chapter 1). One of the main goals of the first chapter is to introduce some requisite mathematics and to establish notation that is used throughout the book. The next three chapters lay down the fundamental principles of continuum mechanics, including the geometric aspects of deformation and motion (chapter 2), the laws governing the transmission of force (chapter 3), and elements of constitutive theory (chapter 4).

Chapters 5 and 6 concern boundary value problems in elasticity and their solution. We introduce the classical (strong form) and the variational (weak form) of the governing differential equations. Many of the ideas are motivated with the one-dimensional *"little boundary value problem."* The Ritz method is offered as a general approach to numerical computations, based upon the principle of virtual work. Although we do not pursue it in detail, we show how the Ritz method can be specialized to form the popular and powerful *finite element method.* The Ritz method provides a natural tool for all of the structural mechanics computations needed for the rest of the book.

Chapters 7 and 8 cover the linear theories of beams and plates, respectively. These structural mechanics theories are developed within the context of three-dimensional continuum mechanics with the dual benefit of lending a deeper understanding of beams and plates and, at the same time, of providing two relevant applications of the general equations of continuum mechanics presented in the first part of the book. The classical constrained theories of beams (Bernoulli-Euler) and plates (Kirchhoff-Love) are examined in detail. Each theory is cast both as a classical boundary value problem and as a variational problem.

Chapters 9 through 11 concern structural stability. Chapter 9 explores the concept of energy principles, observing that if an energy functional exists we can deduce it from a virtual-work functional by a theorem of Vainberg. The relationship between virtual work and energy provides an opportunity for further exploration of the calculus of variations. This chapter ends with the observation that one can use an energy criterion to explore the stability of static equilibrium if the system possesses an energy functional. Chapter 10 gives a brief, illustrative invitation to static stability theory. Through the examination of some simple discrete systems, we encounter many of the interesting phenomena associated with nonlinear systems. We distinguish limit points from bifurcation points, explore the effects of imperfections, and examine the role of linearized buckling analysis. Chapter 11 extends the ideas of chapter 10 to continuous systems, applying the machinery developed in chapter 9 to nonlinear planar beam theory.

Structural stability problems create a strong need for a general approach to nonlinear computations. Chapter 12 provides an introduction to nonlinear computations in mechanics. Newton's method serves as the unifying framework for organizing the nonlinear computations. The arc-length method is offered as a general strategy for numerically tracing equilibrium paths of nonlinear mechanical systems. We illuminate the curve-tracing algorithm by numerically solving Euler's elastica and subsequently apply the algorithm to the solution of the fully nonlinear beam problem. Computer programs are presented at each level of the development to help cement the understanding of the algorithms.

* * * *

Juan C. Simo, to whose memory this book is dedicated, was my closest and dearest friend. We were graduate students at the University of California at Berkeley in the early 1980s. We spent countless hours in the coffee shops near campus discussing mechanics. I learned to appreciate mechanics by watching his deep and clear insight flow from his pen onto his "pad yellow," as he called it in his inimitable Spanish accent. In his hands, the equations of mechanics came to life. Juan's love for mechanics, his tireless pursuit of knowledge, and his gift for developing and expressing theory made an indelible mark on me. His influence is clearly written on these pages.

Juan Simo passed away on September 26, 1994, at the age of 42, after an eight-month battle with cancer. In his short career, Juan made tremendous contributions to the field of computational mechanics, many of them in the area of nonlinear structural mechanics. Unfortunately, a classical education in structural mechanics leaves the student ill-equipped to appreciate Juan's contributions (not to mention the contributions of many others). The approach I have taken in this book was inspired by the hope of narrowing the gap between classical structural mechanics and some of the modern innovations in the field.

I owe a great debt to Juan that I can repay only by passing on what he taught me to the next generation of scholars and engineers. I hope this book defrays some of that debt.

The first edition of the book was born as a brief set of class notes for my course *Applied Structural Mechanics* at the University of Illinois. I am indebted to Bill Hall, Narbey Khachaturian, and Arthur Robinson for enabling the teaching opportunity that led to this book. I have loved every minute of the 24 times I have taught this course. I am indebted to the many students who have taken my course, first for inspiring me to write the book and then for gamely trying to learn from it. I appreciate the help of my former students and post-docs—Parvis and Bijan Banan (a.k.a. "the bros"), Jiwon Kim, Ertugrul Taciroglu, Eric Williamson, and Ken Zuo—for their assistance with the first edition and later the completion of the solutions to *all* of the problems in that edition. I also appreciate the support of my colleagues—especially Bob Dodds, Dennis Parsons, and Glaucio Paulino—for believing in the course enough to make it a cornerstone of our graduate curriculum in structural engineering at Illinois.

This second edition of the book is informed by nearly a decade of using the first edition in my class. I have refined the story and added some important topics. I have tightened up some of the things that were a little loose and loosened a few that were a bit tight. I even rewrote the computer programs in MATLAB. I have expanded the number of examples in the text and I have augmented the problems at the ends of the chapters—tapping into my extensive collection of problems that have grown from my proclivity to facilitate the learning of mechanics through a diet of fortnightly "quizzes." The revisions for the second edition were largely made during the fall semester of 2003. The students in that class endured last minute delivery of the new chapters and did yeoman's work in tracking down typographical errors. I am especially appreciative of my own research assistants—Steve Ball, Kristine Cochran, Ghadir Haikal, Kalyanababu Nakshatrala, and Arun Prakash—for proofreading the text and making suggestions for its improvement. Special thanks to Kalyan for providing a tidy proof of Vainberg's theorem.

Finally, I am grateful to my wife, Kara, and my children, David, Kirsten, and Annika for being cheerful and supportive while the book robbed them of my time and attention. While it was Juan Simo who taught me the joy of mechanics, my family has taught me the mechanics of joy.

Keith D. Hjelmstad

1
Vectors and Tensors

The mechanics of solids is a story told in the language of vectors and tensors. These abstract mathematical objects provide the basic building blocks of our analysis of the behavior of solid bodies as they deform and resist force. Anyone who stands poised to undertake the study of structural mechanics has undoubtedly encountered vectors at some point. However, in an effort to establish a least common denominator among readers, we shall do a quick review of vectors and how they operate. This review serves the auxiliary purpose of setting up some of the notational conventions that will be used throughout the book.

Our study of mechanics will naturally lead us to the concept of the tensor, which is a subject that may be less familiar (possibly completely unknown) to the reader who has the expected background knowledge in elementary mechanics of materials. We shall build the idea of the tensor from the ground up in this chapter with the intent of developing a facility for tensor operations equal to the facility that most readers will already have for vector operations. In this book we shall be content to stick with a Cartesian view of tensors in rectangular coordinate systems. General tensor analysis is a mathematical subject with great beauty and deep significance. However, the novice can be blinded by its beauty to the point of missing the simple physical principles that are the true subject of mechanics. So we shall cling to the simplest possible rendition of the story that still respects the tensorial nature of solid mechanics.

Mathematics is the natural language of mechanics. This chapter presents a fairly brief treatment of the mathematics we need to start our exploration of solid mechanics. In particular, it covers some basic algebra and calculus of vectors and tensors. Plenty more math awaits us in our study of structural mechanics, but the rest of the math we will develop on the fly as we need it, complete with physical context and motivation.

This chapter lays the foundation of the mathematical notation that we will use throughout the book. As such, it is both a starting place and a refuge to regain one's footing when the going gets tough.

The Geometry of Three-dimensional Space

We live in three-dimensional space, and all physical objects that we are familiar with have a three-dimensional nature to their geometry. In addition to solid bodies, there are basically three primitive geometric objects in three-dimensional space: the *point*, the *curve*, and the *surface*. Figure 1 illustrates these objects by taking a slice through the three-dimensional solid body $\mathcal{B}$ (a cube, in this case). A point describes position in space, and has no dimension or size. The point $\mathcal{P}$ in the figure is an example. The most convenient way to describe the location of a point is with a *coordinate system* like the one shown in the figure. A coordinate system has an origin $\mathcal{O}$ (a point whose location we understand in a deeper sense than any other point in space) and a set of three coordinate directions that we use to measure distance. Here we shall confine our attention to Cartesian coordinates, wherein the coordinate directions are mutually perpendicular. The location of a point is then given by its coordinates $\mathbf{x} = (x_1, x_2, x_3)$. A point has a location independent of any particular coordinate system. The coordinate system is generally introduced for the convenience of description or numerical computation.

A curve is a one-dimensional geometric object whose size is characterized by its arc length. In a sense, a curve can be viewed as a sequence of points. A curve has some other interesting properties. At each point along a curve, the curve seems to be heading in a certain direction. Thus, a curve has an orientation in space that can be characterized at any point along the curve by the line tangent to the curve at that point. Another property of a curve is the rate at which this orientation changes as we move along the curve. A straight line is

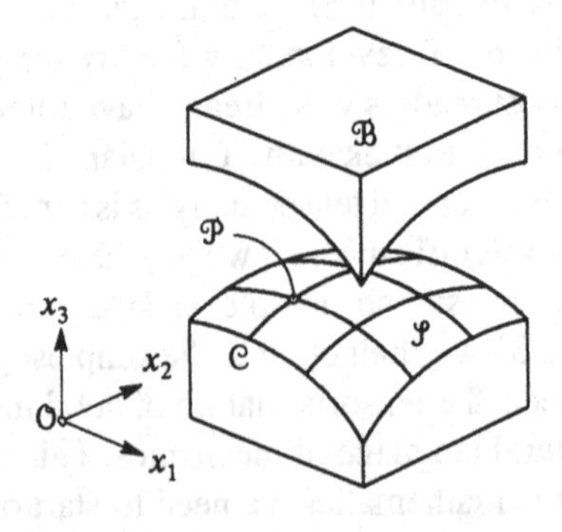

Figure 1 The elements of the geometry of three-dimensional space

a curve whose orientation never changes. The curve $\mathcal{C}$ exemplifies the geometric notion of curves in space.

A surface is a two-dimensional geometric object whose size is characterized by its surface area. In a certain sense, a surface can be viewed as a family of curves. For example, the collection of lines parallel and perpendicular to the curve $\mathcal{C}$ constitute a family of curves that characterize the surface $\mathcal{S}$. A surface can also be viewed as a collection of points. Like a curve, a surface also has properties related to its orientation and the rate of change of this orientation as we move to adjacent points on the surface. The orientation of a surface is completely characterized by the single line that is perpendicular to the tangent lines of all curves that pass through a particular point. This line is called the normal direction to the surface at the point. A flat surface is usually called a plane, and is a surface whose orientation is constant.

A three-dimensional solid body is a collection of points. At each point, we ascribe some physical properties (e.g., mass density, elasticity, and heat capacity) to the body. The mathematical laws that describe how these physical properties affect the interaction of the body with the forces of nature summarize our understanding of the behavior of that body. The heart of the concept of continuum mechanics is that the body is continuous, that is, there are no finite gaps between points. Clearly, this idealization is at odds with particle physics, but, in the main, it leads to a workable and useful model of how solids behave. The primary purpose of hanging our whole theory on the concept of the continuum is that it allows us to do calculus without worrying about the details of material constitution as we pass to infinitesimal limits. We will sometimes find it useful to think of a solid body as a collection of lines, or a collection of surfaces, since each of these geometric concepts builds from the notion of a point in space.

Vectors

A *vector* is a directed line segment and provides one of the most useful geometric constructs in mechanics. A vector can be used for a variety of purposes. For example, in Fig. 2 the vector **v** records the position of point *b* relative to point *a*. We often refer to such a vector as a *position vector*, particularly when *a* is the origin of coordinates. Close relatives of the position vector are *displacement* (the difference between the position vectors of some point at different times), *velocity* (the rate of change of displacement), and *acceleration* (the rate of change of velocity). The other common use of the notion of a vector, to which we shall appeal in this book, is the concept of *force*. We generally think

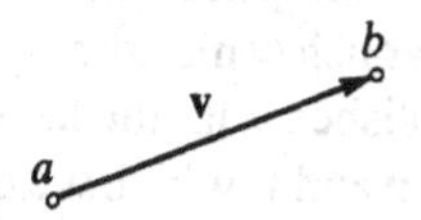

Figure 2 A vector is a directed line segment

of force as an action that has a magnitude and a direction. Likewise, displacements are completely characterized by their magnitude and direction. Because a vector possesses only the properties of magnitude (length of the line) and direction (orientation of the line in space), it is perfectly suited to the mathematical modeling of things like forces and displacements. Vectors have many other uses, but these two are the most important in the present context.

Graphically, we represent a vector as an arrow. The shaft of the arrow gives the orientation and the head of the arrow distinguishes the direction of the vector from the two possibilities inherent in the line segment that describes the shaft (i.e., line segments *ab* and *ba* in Fig. 2 are both oriented the same way in space). The length, or magnitude, of a vector $\mathbf{v}$ is represented graphically by the length of the shaft of the arrow and will be denoted symbolically as $\|\mathbf{v}\|$ throughout the book.

The magnitude and direction of a vector do not depend upon any coordinate system. However, for computation it is most convenient to describe a vector in relation to a coordinate system. For that purpose, we endow our coordinate system with unit *base vectors* $\{\mathbf{e}_1, \mathbf{e}_2, \mathbf{e}_3\}$ pointing in the direction of the coordinate axes. The base vectors are geometric primitives that are introduced purely for the purpose of establishing the notion of direction. Like the origin of coordinates, we view the base vectors as vectors that we understand more deeply and intuitively than any other vector in space. Basically, we assume that we know what it means to be pointing in the $\mathbf{e}_1$ direction, for example. Any collection of three vectors that point in different directions makes a suitable basis (in the language of linear algebra we would say that three such vectors *span* three-dimensional space). Because we have introduced the notion of base vectors for convenience, we shall adopt the most convenient choice. Throughout this book, we will generally employ *orthogonal unit vectors* in conjunction with a Cartesian coordinate system.

Any vector can be described in terms of its components relative to a set of base vectors. A vector $\mathbf{v}$ can be written in terms of base vectors $\{\mathbf{e}_1, \mathbf{e}_2, \mathbf{e}_3\}$ as

$$\mathbf{v} = v_1\mathbf{e}_1 + v_2\mathbf{e}_2 + v_3\mathbf{e}_3 \tag{1}$$

where v_1, v_2, and v_3 are called the *components* of the vector relative to the basis. The component v_i measures how far the vector extends in the $\mathbf{e}_i$ direction, as shown in Fig. 3. A component of a vector is a scalar.

Vector operations. An abstract mathematical construct is not really useful until you know how to operate with it. The most elementary operations in mathematics are *addition* and *multiplication*. We know how to do these operations for scalars; we must establish some corresponding operations for vectors.

Vector addition is accomplished with the head-to-tail rule or parallelogram rule. The sum of two vectors $\mathbf{u}$ and $\mathbf{v}$, which we denote $\mathbf{u}+\mathbf{v}$, is the vector connecting the tail of $\mathbf{u}$ with the head of $\mathbf{v}$ when the tail of $\mathbf{v}$ lies at the head of $\mathbf{u}$, as shown in Fig. 4. If the vectors $\mathbf{u}$ and $\mathbf{v}$ are replicated to form the sides of a

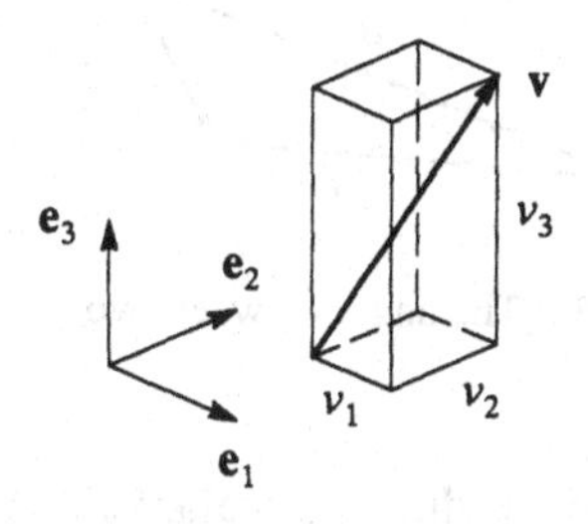

Figure 3 The components of a vector relative to a basis

parallelogram *abcd*, then $\mathbf{u}+\mathbf{v}$ is the diagonal *ac* of the parallelogram. Subtraction of vectors can be accomplished by introducing the negative of a vector, $-\mathbf{v}$ (segment *bf* in Fig. 4), as a vector with the same magnitude that points in exactly the opposite direction of $\mathbf{v}$. Then, $\mathbf{u}-\mathbf{v}$ is simply realized as $\mathbf{u}+(-\mathbf{v})$. If we construct another parallelogram *abfe*, then $\mathbf{u}-\mathbf{v}$ is the diagonal *af*. It is evident from the figure that segment *af* is identical in length and direction to segment *db*. A vector can be added to another vector, but a vector and a scalar cannot be added (the well-worn analogy of the impossibility of adding apples and oranges applies here).

We can multiply a vector $\mathbf{v}$ by a scalar α to get a vector $\alpha\mathbf{v}$ having the same direction but a length equal to the original length $\|\mathbf{v}\|$ multiplied by α. If the scalar α has a negative value, then the sense of the vector is reversed (i.e., it puts the arrow head on the other end). With these definitions, we can make sense of Eqn. (1). The components v_i multiply the base vectors $\mathbf{e}_i$ to give three new vectors $v_1\mathbf{e}_1$, $v_2\mathbf{e}_2$, and $v_3\mathbf{e}_3$. The resulting vectors are added together by the head-to-tail rule to give the final vector $\mathbf{v}$.

The operation of multiplication of two vectors, say $\mathbf{u}$ and $\mathbf{v}$, comes in three varieties: The *dot product* (often called the scalar product) is denoted $\mathbf{u}\cdot\mathbf{v}$; the *cross product* (often called the vector product) is denoted $\mathbf{u}\times\mathbf{v}$; and the *tensor product* is denoted $\mathbf{u}\otimes\mathbf{v}$. Each of these products has its own physical significance. In the following sections we review the definitions of these terms, and examine the meaning behind carrying out such operations.

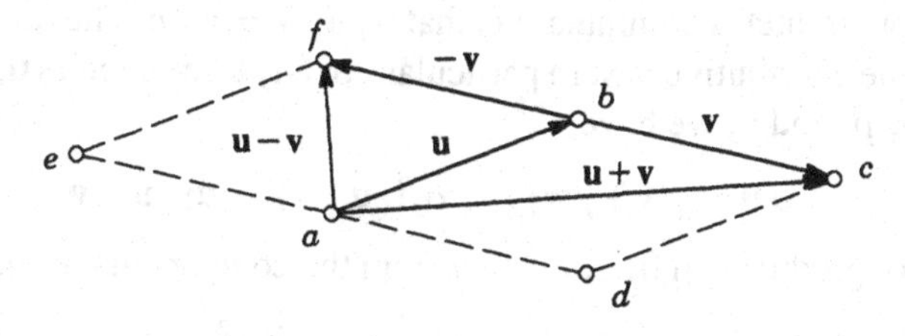

Figure 4 Vector addition and subtraction by the head-to-tail or parallelogram rule

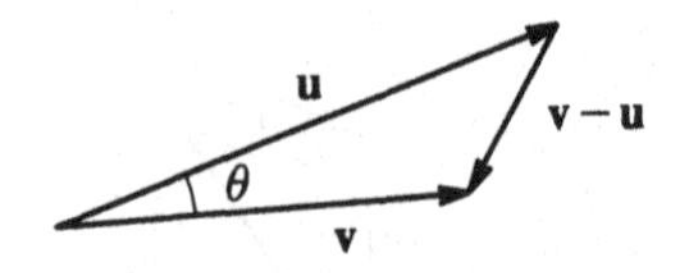

Figure 5 The angle between two vectors

The dot product. The dot product is a scalar value that is related to not only the lengths of the vectors, but also the angle between them. In fact, the dot product can be defined through the formula

$$\mathbf{u} \cdot \mathbf{v} \equiv \| \mathbf{u} \| \, \| \mathbf{v} \| \cos\theta(\mathbf{u}, \mathbf{v}) \tag{2}$$

where $\cos\theta(\mathbf{u}, \mathbf{v})$ is the cosine of the angle θ between the vectors $\mathbf{u}$ and $\mathbf{v}$, shown in Fig. 5. The definition of the dot product can be expressed directly in terms of the vectors $\mathbf{u}$ and $\mathbf{v}$ by using the law of cosines, which states that

$$\| \mathbf{u} \|^2 + \| \mathbf{v} \|^2 = \| \mathbf{v} - \mathbf{u} \|^2 + 2 \| \mathbf{u} \| \, \| \mathbf{v} \| \cos\theta(\mathbf{u}, \mathbf{v})$$

Using this result to eliminate θ from Eqn. (2), we obtain the equivalent definition of the dot product

$$\mathbf{u} \cdot \mathbf{v} \equiv \tfrac{1}{2}\left(\| \mathbf{u} \|^2 + \| \mathbf{v} \|^2 - \| \mathbf{v} - \mathbf{u} \|^2 \right) \tag{3}$$

We can think of the dot product as measuring the relative orientation between two vectors. The dot product gives us a means of defining orthogonality of two vectors. Two vectors are orthogonal if they have an angle of $\pi/2$ radians between them. According to Eqn. (2), any two nonzero vectors $\mathbf{u}$ and $\mathbf{v}$ are orthogonal if $\mathbf{u} \cdot \mathbf{v} = 0$. If $\mathbf{u}$ and $\mathbf{v}$ are orthogonal, then they are the legs of a right triangle with the vector $\mathbf{v} - \mathbf{u}$ forming the hypotenuse. In this case, we can see that the Pythagorean theorem makes the right-hand side of Eqn. (3) equal to zero. Thus, $\mathbf{u} \cdot \mathbf{v} = 0$, as before.

Equation (3) suggests a means of computing the length of a vector. The dot product of a vector $\mathbf{v}$ with itself is $\mathbf{v} \cdot \mathbf{v} = \| \mathbf{v} \|^2$. With this observation Eqn. (2) verifies that the cosine of zero (the angle between a vector and itself) is one.

The dot product is commutative, that is, $\mathbf{u} \cdot \mathbf{v} = \mathbf{v} \cdot \mathbf{u}$. The dot product also satisfies the distributive law. In particular, for any three vectors $\mathbf{u}$, $\mathbf{v}$, and $\mathbf{w}$ and scalars α, β, and γ, we have

$$\alpha \mathbf{u} \cdot (\beta \mathbf{v} + \gamma \mathbf{w}) = \alpha\beta(\mathbf{u} \cdot \mathbf{v}) + \alpha\gamma(\mathbf{u} \cdot \mathbf{w}) \tag{4}$$

The dot product can be computed from the components of the vectors as

$$\mathbf{u} \cdot \mathbf{v} = \sum_{i=1}^{3} u_i \mathbf{e}_i \cdot \sum_{j=1}^{3} v_j \mathbf{e}_j = \sum_{i=1}^{3} \sum_{j=1}^{3} u_i v_j \left(\mathbf{e}_i \cdot \mathbf{e}_j \right)$$

In the first step we merely rewrote the vectors **u** and **v** in component form. In the second step we simply distributed the sums. If the last step puzzles you then you should write out the sums in longhand to demonstrate that the mathematical maneuver was legal. Because the base vectors are orthogonal and of unit length, the products $\mathbf{e}_i \cdot \mathbf{e}_j$ are all either zero or one. Hence, the component form of the dot product reduces to the expression

$$\boxed{\mathbf{u} \cdot \mathbf{v} = \sum_{i=1}^{3} u_i v_i} \tag{5}$$

The dot product of the base vectors arises so frequently that it is worth introducing a shorthand notation. Let the symbol δ_{ij} be defined such that

$$\boxed{\delta_{ij} \equiv \begin{cases} 1 & \text{if } i = j \\ 0 & \text{if } i \neq j \end{cases}} \tag{6}$$

The symbol δ_{ij} is often referred to as the *Kronecker delta*. Clearly, we can write $\mathbf{e}_i \cdot \mathbf{e}_j = \delta_{ij}$. When the Kronecker delta appears in a double summation, that part of the summation can be carried out explicitly (even without knowing the values of the other quantities involved in the sum!). This operation has the effect of contraction from a double sum to a single summation, as follows

$$\sum_{i=1}^{3} \sum_{j=1}^{3} u_i v_j \delta_{ij} = \sum_{i=1}^{3} u_i v_i$$

A simple way to see how this contraction comes about is to write out the sum of nine terms and observe that six of them are multiplied by zero because of the definition of the Kronecker delta. The remaining three terms always share a common value of the indices and can, therefore, be written as a single sum, as indicated above.

One of the most important geometric uses of the dot product is the computation of the projection of one vector onto another. Consider a vector **v** and a unit vector **n**, as shown in Fig. 6. The dot product $\mathbf{v} \cdot \mathbf{n}$ gives the amount of the vector **v** that points in the direction **n**. The proof is quite simple. Note that *abc* is a right triangle. Define a second unit vector **m** that points in the direction *bc*. By construction $\mathbf{m} \cdot \mathbf{n} = 0$. Now let the length of side *ab* be γ and the length

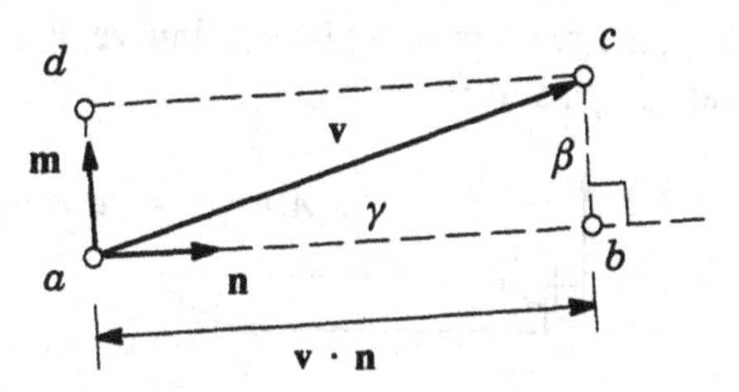

Figure 6 The dot product gives the amount of **v** pointing in the direction **n**

of side bc be β. The vector ab is then $\gamma\mathbf{n}$ and the vector bc is $\beta\mathbf{m}$. By the head-to-tail rule we have $\mathbf{v} = \gamma\mathbf{n}+\beta\mathbf{m}$. Taking the dot product of both sides of this expression with $\mathbf{n}$ we arrive at the result

$$\mathbf{v} \cdot \mathbf{n} = (\gamma\mathbf{n}+\beta\mathbf{m}) \cdot \mathbf{n} = \gamma$$

since $\mathbf{n} \cdot \mathbf{n} = 1$. But γ is the length of the side ab, proving the original assertion. This observation can be used to show that the dot product of a vector with one of the base vectors has the effect of picking out the component of the vector associated with the base vector used in the dot product. To wit,

$$\mathbf{e}_m \cdot \mathbf{v} = \mathbf{e}_m \cdot \sum_{i=1}^{3} v_i \mathbf{e}_i = \sum_{i=1}^{3} v_i \delta_{im} = v_m \tag{7}$$

We can summarize the geometric significance of the vector components as

$$\boxed{v_m = \mathbf{e}_m \cdot \mathbf{v}} \tag{8}$$

That is, v_m is the amount of $\mathbf{v}$ pointing in the direction $\mathbf{e}_m$.

The cross product. The cross product of two vectors $\mathbf{u}$ and $\mathbf{v}$ results in a vector $\mathbf{u} \times \mathbf{v}$ that is orthogonal to both $\mathbf{u}$ and $\mathbf{v}$. The length of $\mathbf{u} \times \mathbf{v}$ is defined as being equal to the area of a parallelogram, two sides of which are described by the vectors $\mathbf{u}$ and $\mathbf{v}$. To wit

$$\boxed{A(\mathbf{u},\mathbf{v}) \equiv \|\mathbf{u} \times \mathbf{v}\|} \tag{9}$$

as shown in Fig. 7. The direction of the resulting vector is defined according to the right-hand rule. The cross product is not commutative, but it satisfies the condition of skew symmetry $\mathbf{u} \times \mathbf{v} = -\mathbf{v} \times \mathbf{u}$. In other words, reversing the order of the product only changes the direction of the resulting vector. The base vectors satisfy the following identities

$$\begin{aligned} \mathbf{e}_1 \times \mathbf{e}_2 &= \mathbf{e}_3 & \qquad \mathbf{e}_2 \times \mathbf{e}_1 &= -\mathbf{e}_3 \\ \mathbf{e}_2 \times \mathbf{e}_3 &= \mathbf{e}_1 & \qquad \mathbf{e}_3 \times \mathbf{e}_2 &= -\mathbf{e}_1 \\ \mathbf{e}_3 \times \mathbf{e}_1 &= \mathbf{e}_2 & \qquad \mathbf{e}_1 \times \mathbf{e}_3 &= -\mathbf{e}_2 \end{aligned} \tag{10}$$

Like the dot product, the cross product is distributive. For any three vectors $\mathbf{u}$, $\mathbf{v}$, and $\mathbf{w}$ and scalars α, β, and γ, we have

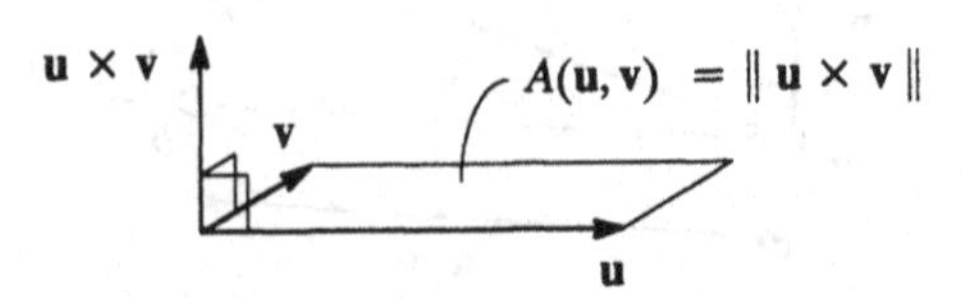

Figure 7 Area and the cross product of vectors

$$\alpha\mathbf{u} \times (\beta\mathbf{v}+\gamma\mathbf{w}) = \alpha\beta(\mathbf{u} \times \mathbf{v}) + \alpha\gamma(\mathbf{u} \times \mathbf{w}) \tag{11}$$

The component form of the cross product of vectors **u** and **v** is

$$\mathbf{u} \times \mathbf{v} = \sum_{i=1}^{3} u_i \mathbf{e}_i \times \sum_{j=1}^{3} v_j \mathbf{e}_j = \sum_{i=1}^{3}\sum_{j=1}^{3} u_i v_j (\mathbf{e}_i \times \mathbf{e}_j)$$

where, again, we have first represented the vectors in component form and then distributed the product. Carrying out the summations, substituting the appropriate incidences of Eqn. (10) for each term of the sum, the component form of the cross product reduces to the expression

$$\mathbf{u} \times \mathbf{v} = (u_2 v_3 - u_3 v_2)\mathbf{e}_1 + (u_3 v_1 - u_1 v_3)\mathbf{e}_2 + (u_1 v_2 - u_2 v_1)\mathbf{e}_3 \tag{12}$$

The triple scalar product. The *triple scalar product* of three vectors **u**, **v**, and **w** is denoted as $(\mathbf{u} \times \mathbf{v}) \cdot \mathbf{w}$. Since the dot product results in a scalar and the cross product results in a vector, the order of multiplication is important (and is shown with parentheses). The triple scalar product has an important geometric interpretation. Consider the parallelepiped defined by the three vectors **u**, **v**, and **w** shown in Fig. 8. The cross product of **u** and **v** results in a vector that is normal to both **u** and **v**. Let us normalize this vector by its length to define the unit vector $\mathbf{n} \equiv \mathbf{u} \times \mathbf{v} / \| \mathbf{u} \times \mathbf{v} \|$. The height of the parallelepiped perpendicular to its base is the length of the component of **w** that lies along the unit vector **n**. This height is simply $h = \mathbf{w} \cdot \mathbf{n}$. Thus, the volume of the parallelepiped is the base area times the height

$$V(\mathbf{u}, \mathbf{v}, \mathbf{w}) = hA(\mathbf{u}, \mathbf{v}) = \left(\mathbf{w} \cdot \frac{\mathbf{u} \times \mathbf{v}}{\| \mathbf{u} \times \mathbf{v} \|}\right) \| \mathbf{u} \times \mathbf{v} \|$$

Upon simplification, we get the following formula for the volume of the parallelepiped as the triple scalar product of the three vectors **u**, **v**, and **w**

$$\boxed{V(\mathbf{u}, \mathbf{v}, \mathbf{w}) = (\mathbf{u} \times \mathbf{v}) \cdot \mathbf{w}} \tag{13}$$

The triple scalar product can be computed in terms of components. Taking the dot product of **w** with $\mathbf{u} \times \mathbf{v}$, as already given in Eqn. (12), we find

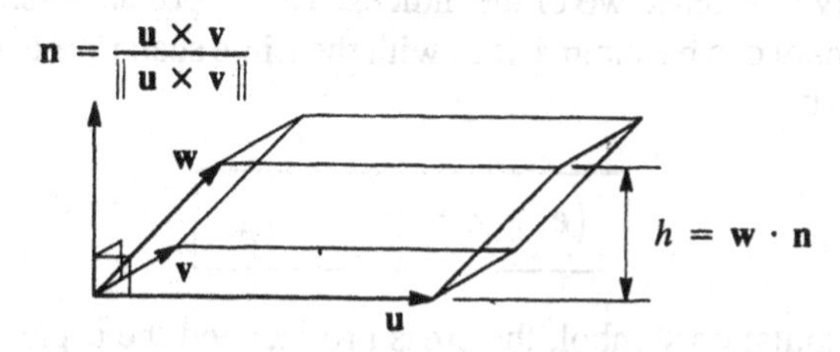

Figure 8 Volume and the triple scalar product

$$(\mathbf{u}\times\mathbf{v})\cdot\mathbf{w} = w_1(u_2v_3-u_3v_2) + w_2(u_3v_1-u_1v_3) + w_3(u_1v_2-u_2v_1)$$
$$= (u_1v_2w_3+u_2v_3w_1+u_3v_1w_2) - (u_3v_2w_1+u_2v_1w_3+u_1v_3w_2)$$

where the second form shows quite clearly that the indices are distinct for each term and that the indices on the positive terms are in *cyclic* order while the indices on the negative terms are in *acyclic* order. Cyclic and acyclic order can be easily visualized, as shown in Fig. 9. If the numbers 1, 2, and 3 appear on a circle in clockwise order, then a cyclic permutation is the order in which you encounter these numbers when you move clockwise from any starting point, and an acyclic permutation is the order in which you encounter them when you move anticlockwise. The indices are in cyclic order when they take the values (1, 2, 3), (2, 3, 1), or (3, 1, 2). The indices are in acyclic order when they take the values (3, 2, 1), (1, 3, 2), or (2, 1, 3).

Figure 9 Cyclic and acyclic permutations of the numbers 1, 2, and 3

The triple scalar product of base vectors represents a fundamental geometric quantity. It will be used in Chapter 2 to describe the volume of a solid body and the changes in that volume. Let us introduce a shorthand notation that is related to the triple scalar product. Let the (permutation) symbol ϵ_{ijk} be

$$\epsilon_{ijk} \equiv \begin{cases} 1 & \text{if } (i,j,k) \text{ are in cyclic order} \\ 0 & \text{if any of } (i,j,k) \text{ are equal} \\ -1 & \text{if } (i,j,k) \text{ are in acyclic order} \end{cases} \tag{14}$$

The scalars ϵ_{ijk} are sometimes referred to as the components of the *permutation tensor*. There are 27 possible permutations of three indices that can each take on three values. Of these 27, only three have (distinct) cyclic values and only three have (distinct) acyclic values. All other permutations of the indices involve equality of at least two of the indices. The 27 possible values of the permutation symbol can be summarized with the triple scalar products of the base vectors. To wit,

$$(\mathbf{e}_i \times \mathbf{e}_j)\cdot\mathbf{e}_k = \epsilon_{ijk} \tag{15}$$

With the permutation symbol, the cross product and the triple scalar product can be expressed neatly in component form as

$$\begin{aligned} \mathbf{u} \times \mathbf{v} &= \sum_{i=1}^{3} \sum_{j=1}^{3} \sum_{k=1}^{3} u_i v_j \epsilon_{ijk} \mathbf{e}_k \\ (\mathbf{u} \times \mathbf{v}) \cdot \mathbf{w} &= \sum_{i=1}^{3} \sum_{j=1}^{3} \sum_{k=1}^{3} u_i v_j w_k \epsilon_{ijk} \end{aligned} \tag{16}$$

You should verify that these formulas involving ϵ_{ijk} give the same results as found previously.

Tensors

The cross product is an example of a vector operation that has as its outcome a new vector. It is a very special operator in the sense that it produces a vector orthogonal to the plane containing the two original vectors. There is a much broader class of operations that produce vectors as the result. The *second-order tensor* is the mathematical object that provides the appropriate generalization. (If the context is not ambiguous, we will often refer to a second-order tensor simply as a tensor.)

> **Definition.** A *tensor* is an object that operates on a vector to produce another vector. (17)

Schematically, this operation is shown in Fig. 10, wherein a tensor **T** operates on the vector **v** to produce the new vector **Tv**. Unlike a vector, there is no easy graphical representation of the tensor **T** itself. In abstract we shall understand a tensor by observing what it does to a vector. The example shown in Fig. 10 is illustrative of all tensor actions. The vector **v** is stretched and rotated to give the new vector **Tv**. In essence, tensors stretch and rotate vectors.

A tensor is a linear operator that satisfies

$$\mathbf{T}(\alpha\mathbf{u}+\beta\mathbf{v}+\gamma\mathbf{w}) = \alpha\mathbf{T}\mathbf{u} + \beta\mathbf{T}\mathbf{v} + \gamma\mathbf{T}\mathbf{w} \tag{18}$$

for any three scalars α, β, γ, and any three vectors **u**, **v**, **w**. Because any vector in three-dimensional space can be expressed as a linear combination of three vectors that span the space, it is sufficient to consider the action of the tensor on three independent vectors. The action of the tensor **T** on the base vectors, for example, completely characterizes the action of the tensor on any other vector. Thus, it is evident that a tensor can be completely characterized by nine

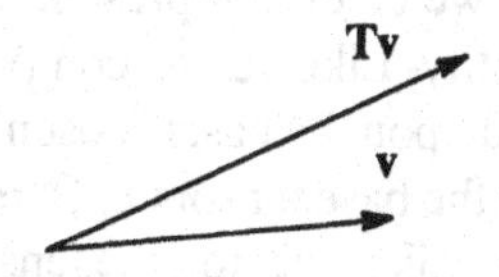

Figure 10 A tensor operates on a vector to produce another vector

scalar quantities: the three components of the vector $\mathbf{T}\mathbf{e}_1$, the three components of the vector $\mathbf{T}\mathbf{e}_2$, and the three components of the vector $\mathbf{T}\mathbf{e}_3$. We shall refer to these nine scalar quantities as the components of the tensor. Like a vector, which can be expressed as the sum of scalar components times base vectors, we shall represent a tensor as the sum of scalar components times base tensors. We introduce the tensor product of vectors as the building block to define a natural basis for a second-order tensor.

The tensor product of vectors. The tensor product of two vectors $\mathbf{u}$ and $\mathbf{v}$ is a special second-order tensor which we shall denote $\left[\mathbf{u} \otimes \mathbf{v}\right]$. The action of this tensor is embodied in how it operates on a vector $\mathbf{w}$, which is

$$\boxed{\left[\mathbf{u} \otimes \mathbf{v}\right]\mathbf{w} \equiv \left(\mathbf{v} \cdot \mathbf{w}\right)\mathbf{u}} \tag{19}$$

In other words, when the tensor $\mathbf{u} \otimes \mathbf{v}$ operates on $\mathbf{w}$ the result is a vector that points in the direction $\mathbf{u}$ and has the length equal to $\left(\mathbf{v} \cdot \mathbf{w}\right) \| \mathbf{u} \|$, the original length of $\mathbf{u}$ multiplied by the scalar product of $\mathbf{v}$ and $\mathbf{w}$. The tensor product of vectors appears to be a rather curious object, and it certainly takes some getting used to. It will, however, prove to be highly useful in developing a coordinate representation of a general tensor $\mathbf{T}$.

The tensor products of the base vectors $\mathbf{e}_i \otimes \mathbf{e}_j$ comprise a set of second-order tensors. Since there are three base vectors, there are nine distinct tensor product combinations among them. These nine tensors provide a suitable basis for expressing the components of a tensor, much like the base vectors themselves provided a basis for expressing the components of a vector. Like the base vectors, we presume to understand these base tensors better than any other tensors in the space. We can confirm that by noting that their action is given simply by Eqn. (19). In fact, we can observe from Eqn. (19) that

$$\left[\mathbf{e}_i \otimes \mathbf{e}_j\right]\mathbf{e}_k = \left(\mathbf{e}_j \cdot \mathbf{e}_k\right)\mathbf{e}_i = \delta_{jk}\,\mathbf{e}_i \tag{20}$$

We will use this knowledge of the tensor product of base vectors to help us with the manipulation of tensor components.

The second-order tensor $\mathbf{T}$ can be expressed in terms of its components T_{ij} relative to the base tensors $\mathbf{e}_i \otimes \mathbf{e}_j$ as

$$\mathbf{T} = \sum_{i=1}^{3}\sum_{j=1}^{3} T_{ij}\left[\mathbf{e}_i \otimes \mathbf{e}_j\right] \tag{21}$$

It will soon be evident why we elect to represent the nine scalar components with a double indexed quantity. Like vector components, the components T_{ij} are scalar values that depend upon the basis chosen for the representation. The tensor part of $\mathbf{T}$ comes from the base tensors $\mathbf{e}_i \otimes \mathbf{e}_j$. The tensor, then, is a sum of scalars times base tensors. Like a vector, the tensor $\mathbf{T}$ itself does not depend upon the coordinate system; only the components do.

A tensor is completely characterized by its action on the three base vectors. Let us compute the action of $\mathbf{T}$ on the base vector $\mathbf{e}_n$

$$\mathbf{T}\mathbf{e}_n = \sum_{i=1}^{3}\sum_{j=1}^{3} T_{ij}[\mathbf{e}_i \otimes \mathbf{e}_j]\mathbf{e}_n = \sum_{i=1}^{3}\sum_{j=1}^{3} T_{ij}\delta_{jn}\mathbf{e}_i = \sum_{i=1}^{3} T_{in}\mathbf{e}_i \tag{22}$$

The first step simply introduces the coordinate form of $\mathbf{T}$. The second step carries out the tensor product of vectors as in Eqn. (20). The final step recognizes that the sum of nine terms reduces to a sum of three terms because six of the nine terms are equal to zero.

We can get some insight into the physical significance of the components by taking the dot product of $\mathbf{e}_m$ and $\mathbf{T}\mathbf{e}_n$. Recall from Eqn. (8) that dotting a vector with $\mathbf{e}_m$ simply extracts the mth component of the vector. Starting from the result of Eqn. (22) we compute

$$\mathbf{e}_m \cdot \mathbf{T}\mathbf{e}_n = \mathbf{e}_m \cdot \sum_{i=1}^{3} T_{in}\mathbf{e}_i = \sum_{i=1}^{3} T_{in}\delta_{im} = T_{mn} \tag{23}$$

Thus, we can see that T_{mn} is the mth component of the vector $\mathbf{T}\mathbf{e}_n$. We can summarize the physical significance of the tensor components as follows

$$\boxed{T_{mn} = \mathbf{e}_m \cdot \mathbf{T}\mathbf{e}_n} \tag{24}$$

The identity tensor. The identity tensor is the tensor that has the property of leaving a vector unchanged. We shall denote the identity tensor as $\mathbf{I}$, and endow it with the property that $\mathbf{I}\mathbf{v} = \mathbf{v}$, for all vectors $\mathbf{v}$. The identity tensor can be expressed in terms of orthonormal (i.e., orthogonal and unit) base vectors

$$\boxed{\mathbf{I} \equiv \sum_{i=1}^{3} \mathbf{e}_i \otimes \mathbf{e}_i} \tag{25}$$

Of course, this definition holds for any orthonormal basis. To prove that Eqn. (25), we need only consider the action of $\mathbf{I}$ on a base vector $\mathbf{e}_j$. To wit

$$\mathbf{I}\mathbf{e}_j \equiv \sum_{i=1}^{3}[\mathbf{e}_i \otimes \mathbf{e}_i]\,\mathbf{e}_j = \sum_{i=1}^{3}\left(\mathbf{e}_i \cdot \mathbf{e}_j\right)\mathbf{e}_i = \sum_{i=1}^{3}\delta_{ij}\,\mathbf{e}_i = \mathbf{e}_j$$

Since the base vectors span three-dimensional space, it is apparent that $\mathbf{I}\mathbf{v} = \mathbf{v}$ for any vector. Observe that Eqn. (25) can br expressed in terms of the Kronecker delta as

$$\mathbf{I} = \sum_{i=1}^{3}\sum_{j=1}^{3}\delta_{ij}[\mathbf{e}_i \otimes \mathbf{e}_j]$$

Hence, δ_{ij} can be interpreted as the ijth component of the identity tensor.

The tensor inverse. Let us assume that we have a tensor $\mathbf{T}$ and that it acts on a vector $\mathbf{v}$ to produce another vector $\mathbf{Tv}$. A tensor stretches and rotates a vector. It seems reasonable to imagine a tensor that undoes the action of another tensor. Such a tensor is called the *inverse* of the tensor $\mathbf{T}$, and we denote it as $\mathbf{T}^{-1}$. Thus, $\mathbf{T}^{-1}$ is the tensor that exactly undoes what the tensor $\mathbf{T}$ does. To be more specific, the tensor $\mathbf{T}^{-1}$ can be applied to the vector $\mathbf{Tv}$ to give back $\mathbf{v}$. Conversely, if the tensor $\mathbf{T}^{-1}$ is applied to the vector $\mathbf{v}$ to give the vector $\mathbf{T}^{-1}\mathbf{v}$, then the tensor $\mathbf{T}$ can be applied to $\mathbf{T}^{-1}\mathbf{v}$ to give back the vector $\mathbf{v}$. These operations define the inverse of a tensor and are summarized as follows

$$\mathbf{T}^{-1}(\mathbf{Tv}) = \mathbf{v}, \quad \mathbf{T}(\mathbf{T}^{-1}\mathbf{v}) = \mathbf{v} \tag{26}$$

The above relations hold for any vector $\mathbf{v}$. As we will soon see, the composition of tensors (a tensor operating on a tensor) can be viewed as a tensor itself. Thus, we can say that $\mathbf{T}^{-1}\mathbf{T} = \mathbf{I}$ and $\mathbf{T}\mathbf{T}^{-1} = \mathbf{I}$.

Example 1. As a simple example of a tensor and its operation on vectors, consider the *projection* tensor $\mathbf{P}$ that generates the image of a vector $\mathbf{v}$ projected onto the plane with normal $\mathbf{n}$, as shown in Fig. 11.

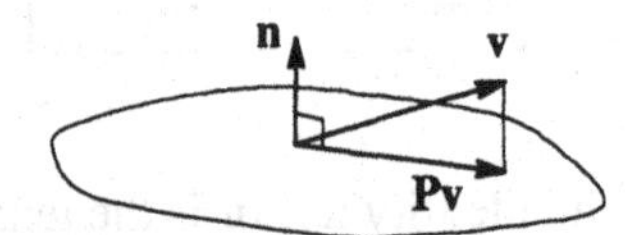

Figure 11 The action of the projection tensor

The explicit expression for the tensor is given by

$$\mathbf{P} \equiv \mathbf{I} - \mathbf{n} \otimes \mathbf{n} \tag{27}$$

where $\mathbf{I}$ is the identity tensor. The action of $\mathbf{P}$ on $\mathbf{v}$ gives the result

$$\begin{aligned} \mathbf{Pv} &= [\mathbf{I} - \mathbf{n} \otimes \mathbf{n}]\mathbf{v} \\ &= \mathbf{Iv} - [\mathbf{n} \otimes \mathbf{n}]\mathbf{v} \\ &= \mathbf{v} - (\mathbf{n} \cdot \mathbf{v})\mathbf{n} \end{aligned}$$

To see that the vector $\mathbf{Pv}$ lies in the plane we need only to show that its dot product with the normal vector $\mathbf{n}$ is zero. Accordingly, we can make the computation $\mathbf{Pv} \cdot \mathbf{n} = (\mathbf{v} \cdot \mathbf{n}) - (\mathbf{v} \cdot \mathbf{n})(\mathbf{n} \cdot \mathbf{n}) = 0$, since $\mathbf{n}$ is a unit vector.

It is interesting to note that we can derive the tensor $\mathbf{P}$ from geometric considerations. From Fig. 11 we can see that, by vector addition, $\mathbf{Pv} + \beta\mathbf{n} = \mathbf{v}$ for some, as yet unknown, value of the scalar β. To determine β we simply take the dot product of the previous vector equation with the vector $\mathbf{n}$, noting that $\mathbf{n}$ has unit length and is perpendicular to $\mathbf{Pv}$. Hence, $\beta = \mathbf{v} \cdot \mathbf{n}$. Now, we substitute back to get

$$\mathbf{P}\mathbf{v} = \mathbf{v} - \beta\mathbf{n} = \mathbf{v} - (\mathbf{v}\cdot\mathbf{n})\mathbf{n} = [\mathbf{I} - \mathbf{n}\otimes\mathbf{n}]\mathbf{v} \tag{28}$$

thereby determining the tensor **P**.

Component expression for operation of a tensor on a vector. Equipped with the component representation of a tensor we can now take another look at how a tensor **T** operates on a vector **v**. In particular, let us examine the components of the resulting vector **Tv**.

$$\mathbf{T}\mathbf{v} = \sum_{i=1}^{3}\sum_{j=1}^{3} T_{ij}[\mathbf{e}_i \otimes \mathbf{e}_j]\sum_{k=1}^{3} v_k\mathbf{e}_k = \sum_{i=1}^{3}\sum_{j=1}^{3}\sum_{k=1}^{3} T_{ij}v_k[\mathbf{e}_i \otimes \mathbf{e}_j]\mathbf{e}_k \tag{29}$$

Carrying out the summations in Eqn. (29), noting the properties expressed in Eqn. (20), we finally obtain the result

$$\mathbf{T}\mathbf{v} = \sum_{i=1}^{3}\sum_{j=1}^{3} T_{ij}v_j\mathbf{e}_i \tag{30}$$

From this expression, we can see that the result is a vector (anything expressed in a vector basis is a vector). Furthermore, we can observe from Eqn. (30) that the *i*th component of the vector **Tv** is given by

$$(\mathbf{T}\mathbf{v})_i = \sum_{j=1}^{3} T_{ij}v_j \tag{31}$$

That is, we compute the *i*th component of the resulting vector from the components of the tensor and the components of the original vector. The similarity between the operation of a tensor and that of a matrix in linear algebra should be apparent.

The summation convention. General relativity is a theory based on tensors. While Einstein was working on this theory, he apparently got rather tired of writing the summation symbol with its range of summation decorating the bottom and top of the Greek letter sigma. What he observed was that, most of the time, the range of the summation was equal to the dimension of space (three dimensions for us, four for him) and that when the summation involved a product of two terms, the summation was over a repeated index. For example, in Eqn. (31) the index j is the index of summation, and it appears exactly twice in the summand $T_{ij}v_j$. Einstein decided that, with a little care, summations could be expressed without laboriously writing the summation symbol. The summation symbol would be understood to apply to repeated indices.

The summation convention, then,.means that any repeated index, also called a *dummy index*, is understood to be summed over the range 1 to 3. With the summation convention, then, Eqn. (30) can be written as

$$\mathbf{T}\mathbf{v} = T_{ij}v_j\mathbf{e}_i$$

with the summation on the indices i and j implied because both are repeated. All we have done is to eliminate the summation symbol, a pretty significant economy of notation. The triple scalar product of vectors can now be written

$$(\mathbf{u} \times \mathbf{v}) \cdot \mathbf{w} = u_i v_j w_k \epsilon_{ijk}$$

Indices that are not repeated in a product are called *free indices.* These indices are not summed and must appear on both sides of the equation. For example, the index i in the equation

$$(\mathbf{T}\mathbf{v})_i = T_{ij}v_j$$

is a free index. The presence of free indices really indicate multiple equations. The index equation must hold for all values of the free index. The equation above is really three equations,

$$(\mathbf{T}\mathbf{v})_1 = T_{1j}v_j, \quad (\mathbf{T}\mathbf{v})_2 = T_{2j}v_j, \quad (\mathbf{T}\mathbf{v})_3 = T_{3j}v_j$$

That is, the free index i takes on values 1, 2, and 3, successively.

The letter used for a dummy index can be changed at will without changing the value of the expression. For example,

$$(\mathbf{T}\mathbf{v})_i = T_{ij}v_j = T_{ik}v_k$$

A free index can be renamed if it is renamed on both sides of the equation. The previous equation is identical to

$$(\mathbf{T}\mathbf{v})_m = T_{mj}v_j = T_{mk}v_k$$

The beauty of this shorthand notation should be apparent. But, like any notational device it should be used with great attention to detail. The mere slip of an index can ruin a derivation or computation.

Perhaps the greatest pitfall of the novice index manipulator is to use an index too many times. An expression with an index appearing more than twice is ambiguous and, therefore, meaningless. For example, the term $T_{ii}v_i$ has no meaning because the summation is ambiguous. The summation convention applies only to terms involved in the same product; to indices of the same tensor, as in the case $T_{ii} = T_{11}+T_{22}+T_{33}$; and to indices in a quotient, as in the expression for divergence, i.e., $\partial v_i/\partial x_i = \partial v_1/\partial x_1+\partial v_2/\partial x_2+\partial v_3/\partial x_3$. Terms separated by a + operation are not subject to the summation convention, and in such a case an index can be reused, as in the expression $T_{ij}v_j+S_{ij}w_j$. Whenever the Kronecker delta appears in a summation, it has the net effect of contracting indices. For example

$$T_{ij}\delta_{jk} = T_{ik}$$

Observe how the summed index j on the tensor component T_{ij} is simply replaced by the free index k on δ_{jk} in the process of contraction of indices.

In this book the summation convention will be in force unless specifically indicated otherwise.

Generating tensors from other tensors. We can define sums and products of tensors using only the geometric and operational notions of vector addition and multiplication. For example, we know how to add two vectors so that the operation $\mathbf{Tv}+\mathbf{Sv}$ makes sense (by the head-to-tail rule). The question is: Does the operation $\mathbf{T}+\mathbf{S}$ make sense? In other words, can you add two tensors together? It makes sense if we define it to make sense. So we will.

Let us define the sum of two tensors $\mathbf{T}$ and $\mathbf{S}$ through the following operation

$$\boxed{[\mathbf{T}+\mathbf{S}]\mathbf{v} \equiv \mathbf{Tv} + \mathbf{Sv}} \tag{32}$$

In other words, the tensor $[\mathbf{T}+\mathbf{S}]$ operating on a vector $\mathbf{v}$ is equivalent to the sum of the vectors created by $\mathbf{T}$ and $\mathbf{S}$ individually operating on the vector $\mathbf{v}$.

An expression for the components of the tensor $[\mathbf{T}+\mathbf{S}]$ can then be constructed simply using the component expressions for Eqn. (32). Let us use Eqn. (30), which gives the formula for computing the components of a tensor operating on a vector, as the starting point (no need to reinvent the wheel). We can write each term of Eqn. (32) in component form and then gather terms on the right side of the equation to yield

$$\begin{aligned} [\mathbf{T}+\mathbf{S}]_{ij}v_j\mathbf{e}_i &= T_{ij}v_j\mathbf{e}_i + S_{ij}v_j\mathbf{e}_i \\ &= (T_{ij}+S_{ij})v_j\mathbf{e}_i \end{aligned}$$

From simple identification of terms on both sides of the equation, we get

$$[\mathbf{T}+\mathbf{S}]_{ij} = T_{ij} + S_{ij}$$

In other words, the ijth component of the sum of two tensors is the sum of the ijth components of the two original tensors.

We can follow the same approach to define multiplication of a tensor by a scalar, as in $a\mathbf{T}$. The scaled tensor $a\mathbf{T}$ is defined through the operation

$$\boxed{[a\mathbf{T}]\mathbf{v} \equiv a(\mathbf{Tv})} \tag{33}$$

Again, the component expression can be deduced by applying Eqn. (30) to get

$$\begin{aligned} [a\mathbf{T}]_{ij}v_j\mathbf{e}_i &= a(T_{ij}v_j\mathbf{e}_i) \\ &= (aT_{ij})v_j\mathbf{e}_i \end{aligned}$$

Thus, the components of the scaled tensor are $[a\mathbf{T}]_{ij} = aT_{ij}$. That is, each component of the original tensor is scaled by a.

The definition of the *transpose* of a tensor can be constructed as follows. The dot product $\mathbf{u} \cdot \mathbf{T}\mathbf{v}$ is a scalar. One might wonder if there is a tensor for which we could reverse the order of operation on $\mathbf{u}$ and $\mathbf{v}$ and get exactly the same scalar value. There is and the tensor is called the *transpose* of $\mathbf{T}$. We shall use the symbol $\mathbf{T}^T$ to denote the transpose. The transpose of $\mathbf{T}$ is defined through the identity

$$\boxed{\mathbf{v} \cdot \mathbf{T}^T\mathbf{u} \equiv \mathbf{u} \cdot \mathbf{T}\mathbf{v}} \tag{34}$$

The components of the transpose $\mathbf{T}^T$ can be shown to be $[\mathbf{T}^T]_{ij} = [\mathbf{T}]_{ji}$ (see Problem 10). That is, the first and second index (row and column in matrix notation) of the tensor components are simply swapped. A tensor is called *symmetric* if the operation of the tensor and its transpose give identical results, i.e., $\mathbf{u} \cdot \mathbf{T}\mathbf{v} = \mathbf{v} \cdot \mathbf{T}\mathbf{u}$. The components of a symmetric tensor satisfy $T_{ij} = T_{ji}$.

We can define a new tensor through the composition of two tensors $[\mathbf{ST}]$. Let the tensor $\mathbf{S}$ operate on the vector $\mathbf{T}\mathbf{v}$. We can define the tensor $[\mathbf{ST}]$ as

$$\boxed{[\mathbf{ST}]\mathbf{v} \equiv \mathbf{S}(\mathbf{T}\mathbf{v})} \tag{35}$$

The components of the tensor $\mathbf{ST}$ can be computed as follows

$$\begin{aligned} [\mathbf{ST}]_{ij} v_j \mathbf{e}_i &= S_{ik}[\mathbf{e}_i \otimes \mathbf{e}_k]\left(T_{mj} v_j \mathbf{e}_m\right) \\ &= \left(S_{ik} T_{mj} \delta_{km}\right) v_j \mathbf{e}_i \end{aligned}$$

Contracting the index m in the above expression leads to the formula for the components of the composite tensor

$$[\mathbf{ST}]_{ij} = S_{ik} T_{kj} \tag{36}$$

Notice how close is the resemblance between this formula and the formula for the product of two square matrices.

An alternative composition of two second-order tensors can also be defined using the dot product of vectors. Consider two tensors $\mathbf{S}$ and $\mathbf{T}$. Let the two tensors operate on the vectors $\mathbf{u}$ and $\mathbf{v}$ to give two new vectors $\mathbf{S}\mathbf{u}$ and $\mathbf{T}\mathbf{v}$. Now we can take the dot product of the new vectors. According to Eqn. (34), this product is equal to

$$\mathbf{S}\mathbf{u} \cdot \mathbf{T}\mathbf{v} = \mathbf{u} \cdot \mathbf{S}^T(\mathbf{T}\mathbf{v}) = \mathbf{u} \cdot [\mathbf{S}^T\mathbf{T}]\mathbf{v}$$

We can view the tensor $\mathbf{S}^T\mathbf{T}$ as a second-order tensor in its own right, operating on the vector $\mathbf{v}$ and then dotted with $\mathbf{u}$. The tensor $\mathbf{S}^T\mathbf{T}$ has components

$$[\mathbf{S}^T\mathbf{T}]_{ij} = S_{ki} T_{kj} \tag{37}$$

Notice the subtle difference between Eqns. (36) and (37). The tensor $\mathbf{T}^T\mathbf{T}$ is always symmetric, even if $\mathbf{T}$ is not (see Problem 11).

It should be clear that we could go on defining new tensor objects ad infinitum. Any such definition will emanate from the same basic considerations, and the computation of the components of the resulting tensors follows exactly along the lines given above. We shall have the opportunity to make such definitions throughout this book, and thus defer further discussion until needed.

Tensors, tensor components, and matrices. A tensor is not a matrix. However, if the foregoing discussion of tensors has left you thinking of matrices, you are not far off the mark. The way we have chosen to denote the components of a second-order tensor (with two indices, that is) makes the temptation to think of tensors as matrices quite compelling. We can list the components of a tensor in a matrix; all of the formulas for tensor-index manipulation are then exactly the same as standard matrix algebra. To some extent, matrix algebra can be an aid to understanding formulas like Eqn. (36). On the other hand, a second-order tensor is no more a three by three matrix than a vector is a three by one matrix.

Matrices are for keeping books, for organizing computations. A tensor or a vector exists independent of a particular manifestation of its components; a matrix *is* a particular manifestation of its components. So take the analogy between tensors and matrices for what it is worth, but try not to confuse a tensor with its components. To do so is rather like being unable to feel cold because you don't know the value of the temperature in degrees Celsius. The fundamental property of "cold" exists independent of what scale you choose to measure temperature.

That said, let us back off from this purist view a little and introduce a notational shorthand that will be useful in stating and solving problems in tensor analysis. When we solve a particular problem, we will select a coordinate system having a particular set of base vectors. The components of any tensor will be expressed relative to those base vectors. For expedience, we will often collect those components in a matrix as

$$\mathbf{T} \sim \begin{bmatrix} T_{11} & T_{12} & T_{13} \\ T_{21} & T_{22} & T_{23} \\ T_{31} & T_{32} & T_{33} \end{bmatrix}$$

where the notation $\mathbf{T} \sim [\ ]$ should be read as "the components of the tensor $\mathbf{T}$, relative to the understood basis, are stored in the matrix $[\ ]$ with the convention that the first index i on the tensor component T_{ij} is the row index of the matrix and the second index j on the tensor component is the column index of the matrix." We avoid the temptation to use the notation $\mathbf{T} = [\ ]$ because we do not want to give the impression that we are setting a tensor equal to a matrix of its components. If there is any question as to what the basis is, then this abbreviated notation does not make sense, and should not be used. The reason this

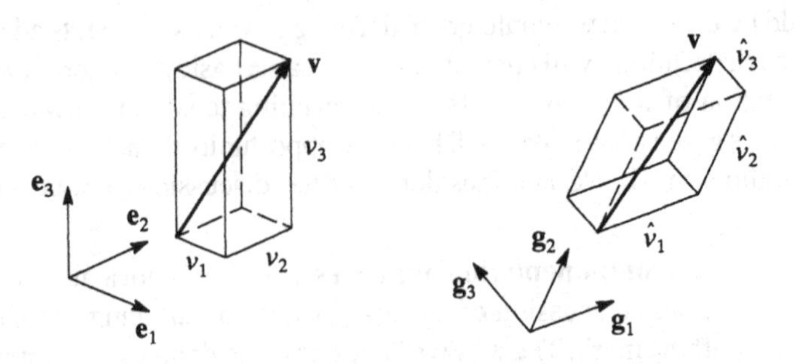

Figure 12 Components of a vector in different coordinate systems

notation is useful is that tensor multiplication is the same as matrix multiplication if the components are stored in the manner shown.

Change of basis. Consider two different coordinate systems, the first with unit base vectors $\{\mathbf{e}_1, \mathbf{e}_2, \mathbf{e}_3\}$ and the second with unit base vectors $\{\mathbf{g}_1, \mathbf{g}_2, \mathbf{g}_3\}$. Any vector $\mathbf{v}$ can be expressed in terms of its components along the base vectors of a coordinate system, as shown in Fig. 12. Clearly, the components of a vector depend upon the coordinate system even though the vector itself does not. It seems reasonable that the components of the vector with respect to one basis should be related somehow to the components of the vector with respect to the other basis. In this section we shall derive that relationship.

A vector can be expressed equivalently in the two bases as

$$\mathbf{v} = v_j \mathbf{e}_j = \hat{v}_j \mathbf{g}_j \tag{38}$$

We can derive the relationship between the two sets of components by taking the dot product of the vector $\mathbf{v}$ with one of the base vectors, say $\mathbf{g}_i$. From Eqn. (38) we obtain

$$\mathbf{g}_i \cdot \mathbf{v} = \hat{v}_i = v_j(\mathbf{g}_i \cdot \mathbf{e}_j)$$

since $\hat{v}_j(\mathbf{g}_j \cdot \mathbf{g}_i) = \hat{v}_j \delta_{ij} = \hat{v}_i$. Let us define the nine scalar values

$$\boxed{Q_{ij} \equiv \mathbf{g}_i \cdot \mathbf{e}_j} \tag{39}$$

that arise from the dot products of the base vectors. The nine values record the cosines of the angles between the nine pairings of the base vectors. Note that the first index of Q is associated with the $\mathbf{g}$ base vector and the second index of Q is associated with the $\mathbf{e}$ base vector. Be careful. The dot product is commutative so $Q_{ij} = \mathbf{e}_j \cdot \mathbf{g}_i$ (the first index of Q is still associated with $\mathbf{g}$ and the second index is still associated with $\mathbf{e}$!).

The formula giving one set of vector components in terms of the other is then

$$\boxed{\hat{v}_i = Q_{ij} v_j} \tag{40}$$

We can find the reverse relationship by dotting Eqn. (38) with $\mathbf{e}_i$ instead of $\mathbf{g}_i$. Carrying out a similar calculation we find that

$$\boxed{v_i = Q_{ji}\hat{v}_j} \tag{41}$$

The components of a second-order tensor $\mathbf{T}$ transform in a manner similar to vectors. A tensor can be expressed in terms of components relative to two different bases in the following manner

$$\mathbf{T} = \hat{T}_{ij}[\mathbf{g}_i \otimes \mathbf{g}_j] = T_{ij}[\mathbf{e}_i \otimes \mathbf{e}_j]$$

where $\hat{T}_{ij}$ is the ijth component of $\mathbf{T}$ with respect to the base tensor $[\mathbf{g}_i \otimes \mathbf{g}_j]$ and T_{ij} is the ijth component of $\mathbf{T}$ with respect to the base tensor $[\mathbf{e}_i \otimes \mathbf{e}_j]$. The relationship between the components in the two coordinate systems can be found by computing the product $\mathbf{g}_m \cdot \mathbf{T}\mathbf{g}_n$, as follows

$$\mathbf{g}_m \cdot \mathbf{T}\mathbf{g}_n = \hat{T}_{mn} = T_{ij}(\mathbf{g}_m \cdot \mathbf{e}_i)(\mathbf{g}_n \cdot \mathbf{e}_j)$$

Computing instead $\mathbf{e}_m \cdot \mathbf{T}\mathbf{e}_n$ we can find the inverse relationship. Once again noting that $Q_{ij} \equiv \mathbf{g}_i \cdot \mathbf{e}_j$, we can write the formulas for the transformation of second-order tensor components as

$$\boxed{\hat{T}_{mn} = Q_{mi}Q_{nj}T_{ij} \qquad T_{mn} = Q_{im}Q_{jn}\hat{T}_{ij}} \tag{42}$$

The main difference between transforming the components of a tensor and those of a vector is that it took two Q terms to accomplish the task for a tensor, one for each index, but only one Q term for a vector. It should be evident that higher-order tensors, i.e., those with more indices, will transform analogously with the appropriate number of Q terms present.

As you might expect, the components of the coordinate transformation $Q_{ij} \equiv \mathbf{g}_i \cdot \mathbf{e}_j$ have some interesting properties. These components make up what is called an *orthogonal transformation*. The orthogonal transformation components have the following property

$$\boxed{Q_{ki}Q_{kj} = \delta_{ij} \qquad Q_{ik}Q_{jk} = \delta_{ij}} \tag{43}$$

The proof of each equation relies on the expression for the identity tensor:

$$(\mathbf{g}_k \cdot \mathbf{e}_i)(\mathbf{g}_k \cdot \mathbf{e}_j) = \mathbf{e}_i \cdot [\mathbf{g}_k \otimes \mathbf{g}_k]\mathbf{e}_j = \mathbf{e}_i \cdot \mathbf{e}_j = \delta_{ij}$$
$$(\mathbf{g}_i \cdot \mathbf{e}_k)(\mathbf{g}_j \cdot \mathbf{e}_k) = \mathbf{g}_i \cdot [\mathbf{e}_k \otimes \mathbf{e}_k]\mathbf{g}_j = \mathbf{g}_i \cdot \mathbf{g}_j = \delta_{ij}$$

Problem 13 asks you to explore further the relationship between the two bases, and clarifies the notion of the Q_{ij} being components of a tensor $\mathbf{Q}$.

Example 2. There is a relationship between the permutation symbol and the Kronecker delta that is often referred to as the $\epsilon - \delta$ identity. The identity is

$$\epsilon_{ijk}\epsilon_{imn} = \delta_{jm}\delta_{kn} - \delta_{jn}\delta_{km}$$

Let us prove this identity.

First note that the cross product is equivalent to operation by a skew-symmetric tensor $[\mathbf{u}\times]$ defined to have components as follows

$$[\mathbf{u}\times] \sim \begin{bmatrix} 0 & -u_3 & u_2 \\ u_3 & 0 & -u_1 \\ -u_2 & u_1 & 0 \end{bmatrix}$$

One can easily verify that $[\mathbf{u}\times]\mathbf{v} = \mathbf{u}\times\mathbf{v}$. By matrix multiplication one can also verify that $[\mathbf{u}\times]^T[\mathbf{v}\times] = (\mathbf{u}\cdot\mathbf{v})\mathbf{I} - \mathbf{v}\otimes\mathbf{u}$. Now,

$$\begin{aligned}
\epsilon_{ijk}\epsilon_{imn} &= \big((\mathbf{e}_i\times\mathbf{e}_j)\cdot\mathbf{e}_k\big)\big((\mathbf{e}_i\times\mathbf{e}_m)\cdot\mathbf{e}_n\big) \\
&= \big((\mathbf{e}_k\times\mathbf{e}_j)\cdot\mathbf{e}_i\big)\big(\mathbf{e}_i\cdot(\mathbf{e}_n\times\mathbf{e}_m)\big) \\
&= (\mathbf{e}_k\times\mathbf{e}_j)\cdot[\mathbf{e}_i\otimes\mathbf{e}_i](\mathbf{e}_n\times\mathbf{e}_m) \\
&= [\mathbf{e}_k\times]\mathbf{e}_j\cdot[\mathbf{e}_n\times]\mathbf{e}_m \\
&= \mathbf{e}_j\cdot[\mathbf{e}_k\times]^T[\mathbf{e}_n\times]\mathbf{e}_m \\
&= \mathbf{e}_j\cdot\big[(\mathbf{e}_k\cdot\mathbf{e}_n)\mathbf{I} - \mathbf{e}_n\otimes\mathbf{e}_k\big]\mathbf{e}_m \\
&= (\mathbf{e}_j\cdot\mathbf{e}_m)(\mathbf{e}_k\cdot\mathbf{e}_n) - (\mathbf{e}_j\cdot\mathbf{e}_n)(\mathbf{e}_k\cdot\mathbf{e}_m) \\
&= \delta_{jm}\delta_{kn} - \delta_{jn}\delta_{km}
\end{aligned}$$

There are other, possibly simpler proofs of the ϵ-δ identity. For example, one can recognize that the identity is simply 81 equations. You can verify them one by one. This example has the additional merit of illustrating various vector and tensor manipulation techniques.

Tensor invariants. In subsequent chapters we will have occasions to wonder whether there are properties of the tensor components that do not depend upon the choice of basis. These properties will be called *tensor invariants*. The identities of Eqn. (43) will be useful in proving the invariance of these properties. The argument will go something like this: Let $f(T_{ij})$ be a function of the components of the tensor $\mathbf{T}$. Under a change of basis, we can write this function in the form $f(Q_{ik}Q_{jl}T_{kl})$. If the function has the property that

$$f(Q_{ik}Q_{jl}T_{kl}) = f(T_{ij})$$

then the function f is a tensor invariant. Since it does not depend upon the coordinate system, we can say that it is an intrinsic function of the tensor $\mathbf{T}$, and write $f(\mathbf{T})$. Three fundamental tensor invariants are given by

$$\boxed{f_1(\mathbf{T}) \equiv T_{ii} \qquad f_2(\mathbf{T}) \equiv T_{ij}T_{ji} \qquad f_3(\mathbf{T}) \equiv T_{ij}T_{jk}T_{ki}} \tag{44}$$

The proof that $f_1(\mathbf{T})$ is invariant is straightforward

$$f_1(\mathbf{T}) = \hat{T}_{ii} = Q_{ik}Q_{il}T_{kl} = \delta_{kl}T_{kl} = T_{kk}$$

by the formula for change of basis, contracted to give $\hat{T}_{ii}$, and Eqn. (43). The invariance of the other two functions can be proved in a similar manner (see Problem 18). Any function of tensor invariants is itself a tensor invariant. We shall sometimes refer to the invariant functions $f_1(\mathbf{T})$, $f_2(\mathbf{T})$, and $f_3(\mathbf{T})$ as the *primary invariants* to distinguish them from other invariant functional forms.

The *trace* of a tensor is simply the sum of its diagonal components. We use the operator "tr" to designate the trace. Thus, $\text{tr}(\mathbf{T}) = T_{ii}$ is the first invariant of the tensor $\mathbf{T}$. The second and third invariants can also be expressed in terms of the trace operator. Let us introduce the notation of a tensor raised to a power as $\mathbf{T}^2 \equiv \mathbf{TT}$ and $\mathbf{T}^3 \equiv \mathbf{TTT}$, where the components are given by the formula for products of tensors, Eqn. (36), as

$$\left[\mathbf{T}^2\right]_{ij} = T_{im}T_{mj} \qquad \left[\mathbf{T}^3\right]_{ij} = T_{im}T_{mn}T_{nj} \tag{45}$$

It should be evident that a tensor can be raised to any (integer) power. Taking the trace of $\mathbf{T}^2$ and $\mathbf{T}^3$ gives $\text{tr}(\mathbf{T}^2) = \left[\mathbf{T}^2\right]_{ii}$ and $\text{tr}(\mathbf{T}^3) = \left[\mathbf{T}^3\right]_{ii}$. Using these expressions in Eqn. (45) we find that the three invariants can be equivalently cast in terms of traces of powers of the tensor $\mathbf{T}$ as

$$f_1(\mathbf{T}) = \text{tr}(\mathbf{T}), \quad f_2(\mathbf{T}) = \text{tr}(\mathbf{T}^2), \quad f_3(\mathbf{T}) = \text{tr}(\mathbf{T}^3) \tag{46}$$

By extension, one can establish that $f_n(\mathbf{T}) \equiv \text{tr}(\mathbf{T}^n)$ is an invariant of the tensor $\mathbf{T}$ for any value of n (see Problem 18). One can prove that the invariants for $n \geq 4$ can all be computed from the first three invariants (see Problem 19).

Eigenvalues and eigenvectors of symmetric tensors. A tensor has properties independent of any basis used to characterize its components. As we have just seen, the components themselves have mysterious properties called invariants that are independent of the basis that defines them. It seems reasonable to expect that we might be able to find a representation of a tensor that is canonical. Indeed, this canonical form is the spectral representation of the tensor that can be built from its eigenvalues and eigenvectors. In this section we shall build the mathematics behind the spectral representation of tensors.

Recall that the action of a tensor is to stretch and rotate a vector. Let us consider a symmetric tensor $\mathbf{T}$ acting on a unit vector $\mathbf{n}$.† If the action of the tensor is simply to stretch the vector but not to rotate it then we can express it as

$$\boxed{\mathbf{Tn} = \mu\mathbf{n}} \tag{47}$$

where μ is the amount of the stretch. This equation, by itself, begs the question of existence of such a vector $\mathbf{n}$. Is there any vector that has the special property that action by $\mathbf{T}$ is identical to multiplication by a scalar? Is it possible that more than one vector has this property?

Equation (47) is called an *eigenvalue problem.* Eigenvalue problems show up all over the place in mathematical physics and engineering. The tensor in three dimensional space is a great context in which to explore the eigenvalue problem because the computations are quite manageable (as opposed to, say, solving the vibration eigenvalue problem of structural dynamics on a structure with a million degrees of freedom).

A vector $\mathbf{n}$ that satisfies the eigenvalue problem is a special vector (an *eigenvector*) that has the property that operation by the second-order tensor $\mathbf{T}$ is the same as operation by the scalar μ (the *eigenvalue*). Equation (47) can be written as $\left[\mathbf{T}-\mu\mathbf{I}\right]\mathbf{n} = \mathbf{0}$, which is a linear homogeneous system of equations. (Note that $\mathbf{0}$ is the zero vector). In order for this system to have a nontrivial solution (i.e., $\mathbf{n} \neq \mathbf{0}$), the determinant of the coefficient matrix must be equal to zero. That is,

$$\det\left[\mathbf{T}-\mu\mathbf{I}\right] = \det\begin{bmatrix} T_{11}-\mu & T_{12} & T_{13} \\ T_{21} & T_{22}-\mu & T_{23} \\ T_{31} & T_{32} & T_{33}-\mu \end{bmatrix} = 0 \tag{48}$$

If we carry out the computation of the determinant, we get the *characteristic equation* (a cubic equation in the case of a three by three matrix) for the eigenvalues μ. The characteristic equation can be written in the form

$$\boxed{-\mu^3 + I_T\mu^2 - II_T\mu + III_T = 0} \tag{49}$$

where the coefficients of the characteristic polynomial

$$I_T = \mathrm{tr}(\mathbf{T}), \qquad II_T = \tfrac{1}{2}\left[I_T^2 - \mathrm{tr}(\mathbf{T}^2)\right], \qquad III_T = \det(\mathbf{T}) \tag{50}$$

are *invariants* of the tensor $\mathbf{T}$. We shall refer to I_T, II_T, and III_T as the *principal invariants* to distinguish these functions from the primary invariants. The determinant of a tensor can be expressed in terms of the primary invariants $f_1(\mathbf{T})$, $f_2(\mathbf{T})$, and $f_3(\mathbf{T})$ (see Problem 23), so all three of the principal invariants are functions of the primary invariants (and vice versa). The principal invariants can be expressed in component form as

† The definition of the eigenvalue problem does not require that $\mathbf{n}$ be a unit vector. In fact, it should be obvious that if $\mathbf{n}$ satisfies Eqn. (47) then so does any scalar multiple of $\mathbf{n}$. Setting the length of the eigenvector is usually considered arbitrary with many choices available. However, in many applications there is an auxiliary condition that determines the length of the vector. For the two most important cases that we will consider in solid mechanics (principal values of stress and strain tensors) the vector $\mathbf{n}$ must be unit length. Assuming unit length from the outset removes some ambiguity without loss of generality.

$$I_T = T_{ii}, \quad II_T = \tfrac{1}{2}\left(T_{ii}T_{jj} - T_{ij}T_{ij}\right), \quad III_T = \tfrac{1}{6}\epsilon_{ijk}\epsilon_{lmn}T_{il}T_{jm}T_{kn} \tag{51}$$

Because the coefficients of the characteristic equation are invariants of the tensor **T** it follows that the roots μ do not depend upon the basis chosen to describe the components and hence are intrinsic properties of **T**.

Finding the roots of the characteristic equation. The cubic equation has three roots (not necessarily distinct) that correspond to three (not necessarily unique) directions. If the cubic equation cannot be factored, then the roots can be found iteratively. For example, we can use Newton's method to solve the nonlinear equation $g(x) = 0$. Given a starting value x_o, we can compute successive estimates of a root of $g(x) = 0$ (see Chapter 12) as

$$x_{i+1} = x_i - \frac{g(x_i)}{g'(x_i)} \tag{52}$$

where $g'(x_i)$ is the derivative of $g(x)$ evaluated at the current iterate x_i. The starting value determines the root to which the iteration converges if there are multiple roots. In the present context, let x_i be the estimate of the eigenvalue μ at the ith iteration. The next estimate can be computed from Newton's formula as

$$x_{i+1} = \frac{2x_i^3 - I_T x_i^2 + III_T}{3x_i^2 - 2I_T x_i + II_T} \tag{53}$$

The iteration continues until $|x_n - x_{n-1}|$ is less than some acceptable tolerance. Then the eigenvalue is $\mu \approx x_n$.

We can always take, as a starting value, $x_o = 0$. However, Gershgorin's theorem might be of some help in estimating a good starting point for the Newton iteration. Gershgorin's theorem simply states that the diagonal element T_{ii} of the tensor **T** might be a good estimate of the eigenvalue μ_i. The quality of the estimate depends upon the size of the off-diagonal elements of **T**. In fact, the theorem states that if you draw a circle centered at T_{ii} with radius

$$r_i = \sum_{\substack{j=1 \\ j\neq i}}^{3} |T_{ij}| \tag{54}$$

i.e., the sum of the absolute values of the off-diagonal elements, then μ_i lies somewhere in that circle, as shown in Fig. 13. (For symmetric matrices the eigenvalues are always real, so that they lie on the real axis. Nonsymmetric matrices can have complex eigenvalues, and in such cases the extra dimension implied by the circle is important.) There is a catch. If two circles overlap, then the only thing we can conclude is that both of the two associated eigenvalues lie somewhere in the union of those two circles. For the case illustrated in Fig. 13, we know that $T_{33} - r_3 \leq \mu_3 \leq T_{33} + r_3$. We also know that the other two

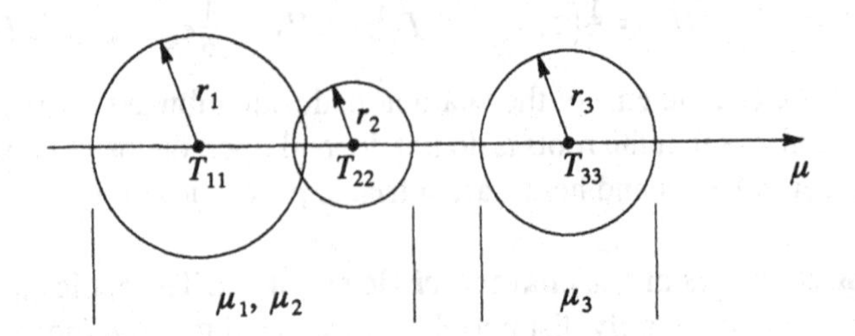

Figure 13 Graphical representation of Gershgorin's theorem

eigenvalues satisfy $T_{11} - r_1 \leq \mu_1, \mu_2 \leq T_{22} + r_2$, i.e., they lie somewhere between extremes of the two circles. Clearly, if the off-diagonal elements of the tensor are small, the diagonal elements are very good estimates of the eigenvalues. In any case, the diagonal elements should be good starting points for the Newton iteration. It also provides a means of checking our eigenvalues once we have found them. If they do not lie within the proper bounds, they cannot be correct. This theorem applies to matrices of any dimension.

Once one root is determined, one can use synthetic division to factor the root out of the cubic, leaving a quadratic that can be solved by the quadratic formula. Alternatively, we could simply use Eqn. (53) from another starting point in the hope that it would converge to one of the other roots (there is no guarantee that the iteration will converge to a root different from one already found).

Determination of the eigenvectors. The cubic equation has three roots, which we call μ_1, μ_2, and μ_3. Each of these roots corresponds to an eigenvector. Let the eigenvectors corresponding to μ_1, μ_2, and μ_3 be called $\mathbf{n}_1$, $\mathbf{n}_2$, and $\mathbf{n}_3$, respectively. These eigenvectors can be determined by solving the system of equations $[\mathbf{T} - \mu_i \mathbf{I}]\, \mathbf{n}_i = \mathbf{0}$ (no implied sum on i). However, by the very definition of the eigenvalues, the coefficient matrix $[\mathbf{T} - \mu_i \mathbf{I}]$ is singular, so we must exercise some care in solving these equations.

Let us try to find the eigenvector $\mathbf{n}_i$ associated with μ_i, (any one of the eigenvalues). Let us assume that the eigenvector has the form

$$\mathbf{n}_i = n_1^{(i)} \mathbf{e}_1 + n_2^{(i)} \mathbf{e}_2 + n_3^{(i)} \mathbf{e}_3$$

Our aim is to determine the, as yet unknown, values of $n_1^{(i)}$, $n_2^{(i)}$, and $n_3^{(i)}$. To aid the discussion let us define three vectors that have components equal to the columns of the coefficient matrix $\left[\mathbf{T} - \mu_i \mathbf{I}\right]$

$$\mathbf{t}_1^{(i)} \sim \begin{bmatrix} T_{11} - \mu_i \\ T_{21} \\ T_{31} \end{bmatrix} \qquad \mathbf{t}_2^{(i)} \sim \begin{bmatrix} T_{12} \\ T_{22} - \mu_i \\ T_{32} \end{bmatrix} \qquad \mathbf{t}_3^{(i)} \sim \begin{bmatrix} T_{13} \\ T_{23} \\ T_{33} - \mu_i \end{bmatrix}$$

The equation $\left[\mathbf{T} - \mu_i \mathbf{I}\right] \mathbf{n}_i = \mathbf{0}$ can be written as (dropping the superscript "(i)" just to simplify the notation)

$$n_1\mathbf{t}_1 + n_2\mathbf{t}_2 + n_3\mathbf{t}_3 = \mathbf{0} \tag{55}$$

It should first be obvious that the vectors $\{\mathbf{t}_1, \mathbf{t}_2, \mathbf{t}_3\}$ are not linearly independent. In fact, we selected μ_i precisely to create this linear dependence. Besides, if these vectors were linearly independent then, by a theorem of linear algebra, the only possible solution to Eqn. (55) would be $n_1 = n_2 = n_3 = 0$, which is clearly at odds with our original aim.

Consider the case where the eigenvalue μ_i is distinct (i.e., neither of the other two eigenvalues is equal to it). In this case at least two of the three vectors $\{\mathbf{t}_1, \mathbf{t}_2, \mathbf{t}_3\}$ are linearly independent. The trouble is we do not know in advance which two. There are three possibilities: $\{\mathbf{t}_1, \mathbf{t}_2\}$, $\{\mathbf{t}_1, \mathbf{t}_3\}$, and $\{\mathbf{t}_2, \mathbf{t}_3\}$. We can write Eqn. (55) as

$$n_\alpha \mathbf{t}_\alpha + n_\beta \mathbf{t}_\beta = -n_\gamma \mathbf{t}_\gamma \tag{56}$$

where no summation is implied and the integers $\{\alpha,\beta,\gamma\}$ take on distinct values of 1, 2, or 3 (i.e., no two can be the same). Our three choices are then $\{\alpha,\beta,\gamma\} = \{1,2,3\}$, $\{2,3,1\}$, or $\{3,1,2\}$. Equation (56) is overdetermined. There are more equations (3) than unknowns (2). However, by construction these equations should be consistent with each other. Hence, any two of the equations should be sufficient to determine n_α and n_β. To remove the ambiguity we can replace Eqn. (56) with its normal form by taking the dot product first with respect to $\mathbf{t}_\alpha$ and then with respect to $\mathbf{t}_\beta$ to give two equations in two unknowns:

$$\begin{bmatrix} \mathbf{t}_\alpha \cdot \mathbf{t}_\alpha & \mathbf{t}_\alpha \cdot \mathbf{t}_\beta \\ \mathbf{t}_\beta \cdot \mathbf{t}_\alpha & \mathbf{t}_\beta \cdot \mathbf{t}_\beta \end{bmatrix} \begin{bmatrix} n_\alpha \\ n_\beta \end{bmatrix} = -n_\gamma \begin{bmatrix} \mathbf{t}_\alpha \cdot \mathbf{t}_\gamma \\ \mathbf{t}_\beta \cdot \mathbf{t}_\gamma \end{bmatrix} \tag{57}$$

Among the three choices of $\{\alpha,\beta,\gamma\}$ at least one must work. Equation (57) will not be solvable if the coefficient matrix is singular. That would be true if its determinant was zero, i.e., if $(\mathbf{t}_\alpha \cdot \mathbf{t}_\alpha)(\mathbf{t}_\beta \cdot \mathbf{t}_\beta) = (\mathbf{t}_\alpha \cdot \mathbf{t}_\beta)^2$. If this is the case then it is also true that $n_\gamma = 0$, which can certainly be verified once you have successfully solved the problem. If your first choice of $\{\alpha,\beta,\gamma\}$ did not work out, then try another one.

One of the important things to notice from Eqn. (57) is that n_α and n_β can only be determined up to an arbitrary multiplier n_γ. To solve the equations one can simply specify a value of n_γ ($n_\gamma = 1$ will work just fine). The vector can be scaled by a constant ϱ to give the final vector $\mathbf{n} = \varrho\left(n_\alpha \mathbf{e}_\alpha + n_\beta \mathbf{e}_\beta + n_\gamma \mathbf{e}_\gamma\right)$. The condition of unit length of $\mathbf{n}$ establishes the value of ϱ as

$$\varrho = (n_\alpha^2 + n_\beta^2 + n_\gamma^2)^{-1/2} \tag{58}$$

Orthogonality of the eigenvectors. One interesting feature of the eigenvalue problem is that the eigenvectors for distinct eigenvalues are orthogonal, as suggested in the following lemma.

Lemma. Let $\mathbf{n}_i$ and $\mathbf{n}_j$ be eigenvectors of the symmetric tensor $\mathbf{T}$ corresponding to distinct eigenvalues μ_i and μ_j, respectively (that is, they satisfy $\mathbf{Tn} = \mu\mathbf{n}$). Then $\mathbf{n}_i$ is orthogonal to $\mathbf{n}_j$, i.e., $\mathbf{n}_i \cdot \mathbf{n}_j = 0$.

Proof. The proof is based on taking the difference of the products of the eigenvectors with $\mathbf{T}$ in different orders (no summation on repeated indices)

$$\begin{aligned} 0 &= \mathbf{n}_j \cdot \mathbf{T}\mathbf{n}_i - \mathbf{n}_i \cdot \mathbf{T}\mathbf{n}_j \\ &= \mathbf{n}_j \cdot (\mu_i \mathbf{n}_i) - \mathbf{n}_i \cdot (\mu_j \mathbf{n}_j) \\ &= (\mu_i - \mu_j)\, \mathbf{n}_i \cdot \mathbf{n}_j \end{aligned} \tag{59}$$

The first line of the proof is true by definition of symmetry of $\mathbf{T}$. The second line substitutes the eigenvalue property $\mathbf{Tn} = \mu\mathbf{n}$. The last line reflects that the dot product of vectors is commutative. Since we assumed that the eigenvalues were distinct, Eqn. $(59)_C$ can be true only if $\mathbf{n}_i \cdot \mathbf{n}_j = 0$, that is, if they are orthogonal. ❑

Notice that orthogonality does not hold if the eigenvalues are repeated because Eqn. $(59)_C$ is satisfied even if $\mathbf{n}_i \cdot \mathbf{n}_j \neq 0$. We will see the ramification of this observation in the following examination of the special cases.

Special cases. There are two special cases that deserve mention. Both correspond to repeated roots of the characteristic equation. The main concern is how to find the eigenvectors associated with repeated roots.

If $\mu_\alpha = \mu_\beta \neq \mu_\gamma$ we have the case that two of the roots are equal, but the third is distinct. For the distinct root μ_γ we can follow the above procedure and find the unique eigenvector $\mathbf{n}_\gamma$. The vectors corresponding to the double eigenvalue are not unique. If we have two eigenvectors $\mathbf{n}_\alpha$ and $\mathbf{n}_\beta$ corresponding to $\mu_\alpha = \mu_\beta = \mu$, then any vector that is a linear combination of those two vectors, $\mathbf{n} = a\mathbf{n}_\alpha + b\mathbf{n}_\beta$, is also an eigenvector. The proof is simple

$$\begin{aligned} \mathbf{Tn} &= \mathbf{T}(a\mathbf{n}_\alpha + b\mathbf{n}_\beta) \\ &= a\mathbf{T}\mathbf{n}_\alpha + b\mathbf{T}\mathbf{n}_\beta \\ &= a\mu\,\mathbf{n}_\alpha + b\mu\,\mathbf{n}_\beta \\ &= \mu(a\mathbf{n}_\alpha + b\mathbf{n}_\beta) \quad = \mu\mathbf{n} \end{aligned}$$

Since the eigenvectors are orthogonal for distinct eigenvalues, the physical interpretation of an eigenvector $\mathbf{n}$ corresponding to the double eigenvalue μ is that it is any vector that lies in the plane normal to $\mathbf{n}_\gamma$, as shown in Fig. 14.

There is a clever way of finding such a vector. The tensor $[\mathbf{I} - \mathbf{n} \otimes \mathbf{n}]$ is a projection tensor. When applied to any vector $\mathbf{m}$, it will produce a new vector that is orthogonal to $\mathbf{n}$. Specifically

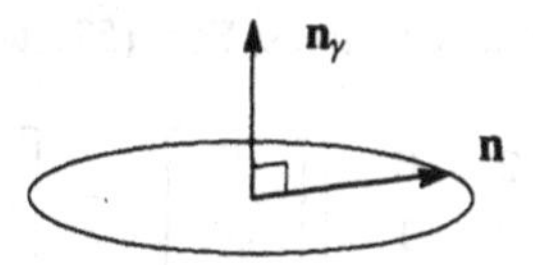

Figure 14 Physical interpretation of eigenvectors for repeated eigenvalues

$$\bar{\mathbf{m}} = [\mathbf{I} - \mathbf{n} \otimes \mathbf{n}]\mathbf{m} = \mathbf{m} - (\mathbf{n} \cdot \mathbf{m})\mathbf{n} \tag{60}$$

is orthogonal to $\mathbf{n}$ (prove it by computing the value of the dot product of vectors $\mathbf{n}$ and $\bar{\mathbf{m}}$). Thus, to compute the eigenvectors corresponding to the double root, we need only take any vector $\mathbf{m}$ in the space (not collinear with $\mathbf{n}_\gamma$) and compute

$$\bar{\mathbf{n}}_\beta = \mathbf{m} - (\mathbf{n}_\gamma \cdot \mathbf{m})\mathbf{n}_\gamma \tag{61}$$

then normalize as $\mathbf{n}_\beta = \bar{\mathbf{n}}_\beta / \| \bar{\mathbf{n}}_\beta \|$. To get a third eigenvector that is orthogonal to the other two, we can simply compute the cross product $\mathbf{n}_\alpha = \mathbf{n}_\beta \times \mathbf{n}_\gamma$.

The second special case has all three of the eigenvalues equal, $\mu_1 = \mu_2 = \mu_3 = \mu$. In this case, any vector in the space is an eigenvector. If we need an orthonormal set of three specific vectors, we can apply the same procedure as before, starting with any two (noncollinear) vectors.

Example 3. *Distinct roots.* Consider that the components of the tensor $\mathbf{T}$ are given by the matrix of values

$$\mathbf{T} \sim \begin{bmatrix} 3 & -1 & 0 \\ -1 & 3 & 0 \\ 0 & 0 & 3 \end{bmatrix}$$

The invariants are $I_T = 9$, $II_T = 26$, and $III_T = 24$. The characteristic equation for the eigenvalues is $-\mu^3 + 9\mu^2 - 26\mu + 24 = 0$. This equation can be factored (not many real problems have integer roots!) as

$$-(\mu - 2)(\mu - 3)(\mu - 4) = 0$$

showing that the roots are $\mu_1 = 2$, $\mu_2 = 3$, and $\mu_3 = 4$. (Note that Gershgorin's theorem holds!) The eigenvector associated with the first eigenvalue can be found by solving the equation $[\mathbf{T} - \mu_1 \mathbf{I}]\mathbf{n}_1 = \mathbf{0}$. We can observe that

$$[\mathbf{T} - \mu_1 \mathbf{I}] \sim \begin{bmatrix} 1 & -1 & 0 \\ -1 & 1 & 0 \\ 0 & 0 & 1 \end{bmatrix} \quad \Rightarrow \quad \begin{aligned} \mathbf{t}_1^{(1)} &= \mathbf{e}_1 - \mathbf{e}_2 \\ \mathbf{t}_2^{(1)} &= -\mathbf{e}_1 + \mathbf{e}_2 \\ \mathbf{t}_3^{(1)} &= \mathbf{e}_3 \end{aligned}$$

Taking the choice $\{\alpha, \beta, \gamma\} = \{2, 3, 1\}$, Eqn. (56) gives

$$n_2^{(1)} \mathbf{t}_2^{(1)} + n_3^{(1)} \mathbf{t}_3^{(1)} = -n_1^{(1)} \mathbf{t}_1^{(1)}$$

Letting $n_1^{(1)} = 1$, the normal equations, Eqn. (57), take the form

$$\begin{bmatrix} 2 & 0 \\ 0 & 1 \end{bmatrix} \begin{bmatrix} n_2^{(1)} \\ n_3^{(1)} \end{bmatrix} = \begin{bmatrix} 2 \\ 0 \end{bmatrix}$$

which gives $n_2^{(1)} = 1$ and $n_3^{(1)} = 0$. Thus, the eigenvector for $\mu_1 = 2$ is

$$\mathbf{n}_1 = \mathbf{e}_1 + \mathbf{e}_2$$

The remaining two eigenvectors can be found in exactly the same way, and are

$$\mathbf{n}_2 = \mathbf{e}_3, \quad \mathbf{n}_3 = \mathbf{e}_1 - \mathbf{e}_2$$

These vectors can, of course, be normalized to unit length.

It is interesting to note what happens for other choices of the normal equations in the preceding example. In particular, it is evident that $\mathbf{t}_2^{(1)} = -\mathbf{t}_1^{(1)}$. If we were to make the choice $\{\alpha, \beta, \gamma\} = \{1, 2, 3\}$ then the coefficient matrix for the normal equations would be singular. This observation is also consistent with the fact that $n_3^{(1)} = 0$.

Example 4. *Repeated roots.* Consider that the components of the tensor **T** are given by the matrix of values

$$\mathbf{T} \sim \begin{bmatrix} 5 & -1 & -1 \\ -1 & 5 & -1 \\ -1 & -1 & 5 \end{bmatrix}$$

The invariants are $I_T = 15$, $II_T = 72$, and $III_T = 108$. The characteristic equation for the eigenvalues is

$$-\mu^3 + 15\mu^2 - 72\mu + 108 = 0$$
$$\text{or} \quad -(\mu - 3)(\mu - 6)(\mu - 6) = 0$$

showing that the roots are $\mu_1 = \mu_2 = 6$, and $\mu_3 = 3$. The eigenvector associated with the distinct eigenvalue μ_3 can be found by solving the equation $[\mathbf{T} - \mu_3 \mathbf{I}]\mathbf{n}_3 = \mathbf{0}$ as in the previous example. The result is

$$\mathbf{n}_3 = \frac{1}{\sqrt{3}}(\mathbf{e}_1 + \mathbf{e}_2 + \mathbf{e}_3)$$

The eigenvectors corresponding to the repeated root must lie in a plane orthogonal to $\mathbf{n}_3$. We can select any vector in the space and project out the component along $\mathbf{n}_3$. Let us use $\mathbf{m} = \mathbf{e}_1$. Project out the part of the vector along $\mathbf{n}_3$ (see Example 1)

$$\begin{aligned}\mathbf{n}_2 = \varrho\mathbf{P}\mathbf{m} &= \varrho\left[\mathbf{I}-\mathbf{n}_3\otimes\mathbf{n}_3\right]\mathbf{e}_1\\ &= \varrho\left[\mathbf{e}_1-(\mathbf{n}_3\cdot\mathbf{e}_1)\mathbf{n}_3\right]\\ &= \varrho\left[\mathbf{e}_1-\tfrac{1}{3}(\mathbf{e}_1+\mathbf{e}_2+\mathbf{e}_3)\right]\\ &= \varrho\left(\tfrac{2}{3}\mathbf{e}_1-\tfrac{1}{3}\mathbf{e}_2-\tfrac{1}{3}\mathbf{e}_3\right)\\ &= \tfrac{1}{\sqrt{6}}(2\mathbf{e}_1-\mathbf{e}_2-\mathbf{e}_3)\end{aligned}$$

where the constant ϱ was selected to give the vector unit length. Finally, $\mathbf{n}_1$ can be computed as $\mathbf{n}_1 = \mathbf{n}_2\times\mathbf{n}_3$ to give

$$\mathbf{n}_1 = \tfrac{1}{\sqrt{2}}(-\mathbf{e}_2+\mathbf{e}_3)$$

The spectral decomposition. If the eigenvalues and eigenvectors are known, we can express the original tensor in terms of those objects in the following manner

$$\boxed{\mathbf{T} = \sum_{i=1}^{3}\mu_i\,\mathbf{n}_i\otimes\mathbf{n}_i} \tag{62}$$

Note that we need to suspend the summation convention because of the number of times that the index i appears in the expression. This form of expression of the tensor $\mathbf{T}$ is called the *spectral decomposition* of the tensor. How do we know that the tensor $\mathbf{T}$ is equivalent to its spectral decomposition? As we indicated earlier, the operation of a second-order tensor is completely defined by its operation on three independent vectors. Let us assume that the eigenvectors $\{\mathbf{n}_1, \mathbf{n}_2, \mathbf{n}_3\}$ are orthogonal (which means that any eigenvectors associated with repeated eigenvalues were orthogonalized). Let us examine how the tensor and its spectral decomposition operate on $\mathbf{n}_j$

$$\mathbf{T}\mathbf{n}_j = \sum_{i=1}^{3}\mu_i\left[\mathbf{n}_i\otimes\mathbf{n}_i\right]\mathbf{n}_j = \sum_{i=1}^{3}\mu_i\left(\mathbf{n}_j\cdot\mathbf{n}_i\right)\mathbf{n}_i = \sum_{i=1}^{3}\mu_i\delta_{ij}\mathbf{n}_i = \mu_j\mathbf{n}_j$$

Thus, we have concluded that both tensors operate the same way on the three eigenvectors. Therefore, the spectral representation must be equivalent to the original tensor. A corollary of the preceding construction is that any two tensors with exactly the same eigenvalues and eigenvectors are equivalent.

The spectral decomposition affords us another remarkable observation. We know that we are free to select any basis vectors to describe the components of a tensor. What happens if we select the eigenvectors $\{\mathbf{n}_1, \mathbf{n}_2, \mathbf{n}_3\}$ as the basis? According to Eqn. (62), in this basis the off-diagonal components of the tensor $\mathbf{T}$ are all zero, while the diagonal elements are exactly the eigenvalues

$$\mathbf{T} \sim \begin{bmatrix} \mu_1 & 0 & 0 \\ 0 & \mu_2 & 0 \\ 0 & 0 & \mu_3 \end{bmatrix}$$

The invariants of **T** also take a special form when expressed in terms of the eigenvalues. The invariants are, by their very nature, independent of the basis chosen to represent the tensor. As such, one must get the same value of the invariants in *all* bases. Those values will, of course, be the values computed in any specific basis. The simplest basis, often referred to as the *canonical basis*, is the one given by the eigenvectors. In this basis, the invariants can be represented as

$$\boxed{\begin{aligned} I_T &= \mu_1 + \mu_2 + \mu_3 \\ II_T &= \mu_1\mu_2 + \mu_1\mu_3 + \mu_2\mu_3 \\ III_T &= \mu_1\mu_2\mu_3 \end{aligned}} \tag{63}$$

Example 5. Consider a tensor **T** that has one distinct eigenvalue μ_1 and a repeated eigenvalue $\mu_2 = \mu_3$. Use the spectral decomposition to show that the tensor **T** can be represented as

$$\mathbf{T} = \mu_1[\mathbf{n} \otimes \mathbf{n}] + \mu_2[\mathbf{I} - \mathbf{n} \otimes \mathbf{n}]$$

where **n** is the unit eigenvector associated with the distinct eigenvalue μ_1.

Let $\mathbf{n}_1 \equiv \mathbf{n}$, $\mathbf{n}_2$, and $\mathbf{n}_3$ be eigenvectors of **T**. Further assume that these vectors are orthogonal (remember, if they are not orthogonal due to a repeated root, they can always be orthogonalized). The sum of outer products of orthonormal vectors is the identity. Thus,

$$\mathbf{I} = \sum_{i=1}^{3} \mathbf{n}_i \otimes \mathbf{n}_i = \mathbf{n} \otimes \mathbf{n} + \sum_{i=2}^{3} \mathbf{n}_i \otimes \mathbf{n}_i$$

Write **T** in terms of its spectral decomposition as

$$\begin{aligned} \mathbf{T} = \sum_{i=1}^{3} \mu_i[\mathbf{n}_i \otimes \mathbf{n}_i] &= \mu_1 \mathbf{n} \otimes \mathbf{n} + \sum_{i=2}^{3} \mu_i[\mathbf{n}_i \otimes \mathbf{n}_i] \\ &= \mu_1 \mathbf{n} \otimes \mathbf{n} + \mu_2 \sum_{i=2}^{3} \mathbf{n}_i \otimes \mathbf{n}_i \\ &= \mu_1 \mathbf{n} \otimes \mathbf{n} + \mu_2[\mathbf{I} - \mathbf{n} \otimes \mathbf{n}] \end{aligned}$$

There is great significance to this result. Notice that the final spectral representation does not refer to $\mathbf{n}_2$ and $\mathbf{n}_3$ at all. Since these vectors are arbitrarily chosen from the plane orthogonal to **n** these vectors have no intrinsic significance (other than that they faithfully represent the plane). In this case there are only three intrinsic bits of information: μ_1, μ_2, and $\mathbf{n}$. Hence, this representation of $\mathbf{T}$ is canonical.

The Cayley-Hamilton theorem. The spectral decomposition and the characteristic equation for the eigenvalues of a tensor can be used to prove the *Cayley-Hamilton theorem*, which states that

$$\boxed{\mathbf{T}^3 - I_T\mathbf{T}^2 + II_T\mathbf{T} - III_T\mathbf{I} = \mathbf{0}} \tag{64}$$

where $\mathbf{T}^2 = \mathbf{TT}$ and $\mathbf{T}^3 = \mathbf{TTT}$ are products of the tensor $\mathbf{T}$ with itself. Using the spectral decomposition, one can show that (Problem 22)

$$\mathbf{T}^m = \sum_{i=1}^{3} (\mu_i)^m \mathbf{n}_i \otimes \mathbf{n}_i$$

Using this result, and noting that $\mathbf{I} = \mathbf{n}_i \otimes \mathbf{n}_i$ (sum implied), we can compute

$$\mathbf{T}^3 - I_T\mathbf{T}^2 + II_T\mathbf{T} - III_T\mathbf{I} = \sum_{i=1}^{3} \left(\mu_i^3 - I_T\mu_i^2 + II_T\mu_i - III_T\right)\mathbf{n}_i \otimes \mathbf{n}_i$$

All of the eigenvalues satisfy the characteristic equation. Thus, the term in parentheses is always zero, thereby proving the theorem.

Vector and Tensor Calculus

A *field* is a function of position defined on a particular region. In our study of mechanics we shall have need of scalar, vector, and tensor fields, in which the output of the function is a scalar, vector, or tensor, respectively. For problems defined on a region of three-dimensional space, the input is the position vector $\mathbf{x}$. A function defined on a three-dimensional domain, then, is a function of three independent variables (the components x_1, x_2, and x_3 of the position vector $\mathbf{x}$). In certain specialized theories (e.g., beam theory, plate theory, and plane stress) position will be described by one or two independent variables.

A field theory is a physical theory built within the framework of fields. The primary advantage of using field theories to describe physical phenomena is that the tools of differential and integral calculus are available to carry out the analysis. For example, we can appeal to concepts like infinitesimal neighborhoods and limits. And we can compute rates of change by differentiation and accumulations and averages by integration.

Figure 15 shows the simplest possible manifestation of a field: a scalar function of a scalar variable, $g(x)$. A scalar field can, of course, be represented as a graph with x as the abscissa and $g(x)$ as the ordinate. For each value of position x the function produces as output $g(x)$. The derivative of the function is defined through the limiting process as

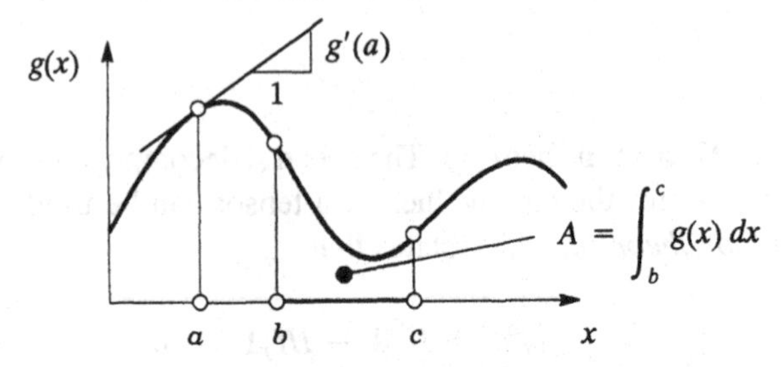

Figure 15 A scalar field $g(x)$ is a scalar-valued function of a scalar variable. Differentiation gives the slope of the curve at a point and integration gives the area A under the curve between two points.

$$\frac{dg}{dx} \equiv \lim_{\Delta x \to 0} \left(\frac{g(x+\Delta x) - g(x)}{\Delta x} \right) \equiv g'(x) \tag{65}$$

The derivative has the familiar geometrical interpretation of the slope of the curve at a point and gives the rate of change of g with respect to change in position x. Many of the graphical constructs that serve so well for scalar functions of scalar variables do not generalize well to vector and tensor fields. However, the concept of the derivative as the limit of the ratio of flux, $g(x+\Delta x) - g(x)$ in the present case, to size of the region, Δx in the present case, will generalize for all cases.

Figure 16 illustrates that a segment $[x, x+\Delta x]$ has a left end and a right end. If we ascribe a directionality to the segment by imagining the positive direction to be in the direction of the $+x$ axis, then the left end is the "inflow" boundary and the right end is the "outflow" boundary of the segment. We can think of the *flux* of g as being the difference between the outflow and the inflow. For a scalar function of a scalar variable that is simply $g(x+\Delta x) - g(x)$. According to Eqn. (65), the derivative dg/dx is the limit of the ratio of flux to size of the region.

In three-dimensional space we shall generalize our concept of derivative (rate of change) using an arbitrary region $\mathfrak{B}$ having volume $\mathcal{V}(\mathfrak{B})$ with surface Ω having unit normal vector field $\mathbf{n}$, as shown in Fig. 17. We will define various

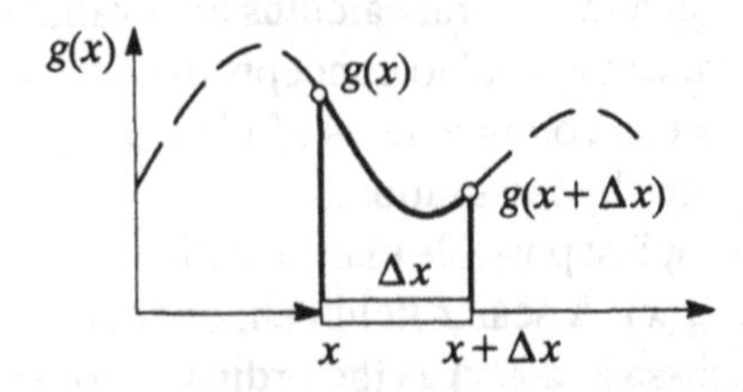

Figure 16 The "flux" of a scalar field $g(x)$ on the region $[x, x+\Delta x]$ is the difference in the function values at the ends of the segment.

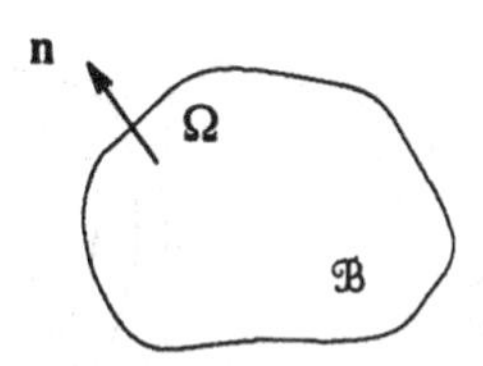

Figure 17 A region $\mathfrak{B}$ in three-dimensional space with volume $\mathcal{V}(\mathfrak{B})$ and surface Ω with outward unit normal vector field **n**.

types of derivatives of various types of fields in the following sections, but all of these derivatives will be the limit of the ratio of some sort of flux (outflow minus inflow) to the volume of the region as the volume shrinks to zero. In these definitions the flux will involve an integral over the surface area and the normal vector **n** will help to distinguish "inflow" from "outflow" for the situation at hand. For each definition of derivative we will develop a coordinate expression that will tell us how to formally "take the derivative" of the field. The coordinate expressions will all involve partial derivatives of the vector or tensor components.

The integral of the function between the limits b and c gives the area between the graph of the function $g(x)$ and the x axis (see Fig. 15). For any scalar function of a scalar variable one can think of the integral as the "area under the curve." Integration is the limit of a sum of infinitesimal strips with area $g(x)\,dx$. The total area is the accumulated sum of the infinitesimal areas. The geometric notion of integration is quite independent of techniques of integration based upon anti-derivatives of functions because there are methods of integration (e.g., numerical quadrature) that do not rely upon the anti-derivative. In our developments here we need to think of integrals both in the sense of executing integrals (mostly later in the book) and in the more generic sense of accumulating the limit of a sum.

In three dimensional space we will encounter surface integrals and volume integrals. Most of the time we will not use the notation of "double integrals" for surface integrals and "triple integration" for volume integrals, but rather understand that

$$\int_{\Omega} (\cdot)\, dA = \int\!\!\int (\cdot)\, dx\, dy, \qquad \int_{\mathfrak{B}} (\cdot)\, dV = \int\!\!\int\!\!\int (\cdot)\, dx\, dy\, dz \tag{66}$$

where the variables and infinitesimals must be established for the coordinate system that is being used to characterize the problem at hand. Again, techniques of integration are important only in particular problems to carry out computations.

The second aspect of integration that we will introduce in this chapter is the idea of integral theorems that provide an equivalence between a surface inte-

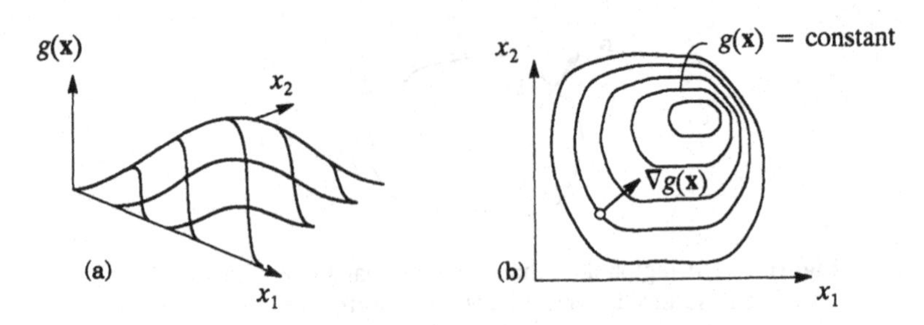

Figure 18 (a) A graph and (b) a contour map of a scalar field in two dimensions

gral and a volume integral. We shall see that the divergence theorem (in any of its many specific forms) is the multivariate counterpart to the one-dimensional fundamental theorem of calculus

$$\int_a^b \left(\frac{dg}{dx}\right) dx = g(b) - g(a) \tag{67}$$

The remainder of this chapter is devoted to reviewing of some of the basic ideas from vector calculus and the extension of those ideas to tensor fields.

Scalar fields of vector variables. A *scalar field* is a function $g(\mathbf{x})$ that assigns a scalar value to each point $\mathbf{x}$ in a particular domain. The temperature in a solid body is an example of a scalar field. As an example consider the scalar field $g(\mathbf{x}) = \|\mathbf{x}\|^2 = x_1^2 + x_2^2 + x_3^2$, in which the function $g(\mathbf{x})$ gives the square of the length of the position vector $\mathbf{x}$. In two dimensions, a scalar field can be represented by either a graph or a contour map like those shown in Fig. 18.

As with any function that varies from point to point in a domain, we can ask the question: At what rate does the field change as we move from one point to another? It is fairly obvious from the contour map that if one moves from one point to another along a contour then the change in the value of the function is zero (and therefore the rate of change is zero). If one crosses contours then the function value changes. Clearly, the question of rate of change depends upon direction of the line connecting the two points in question.

Consider a scalar field g in three dimensional space evaluated at two points a and b, as shown in Fig. 19. Point a is located at position $\mathbf{x}$ and point b is located at position $\mathbf{x} + \Delta s\,\mathbf{n}$, where $\mathbf{n}$ is a unit vector that points in the direction from a to b and Δs is the distance between them. The *directional derivative* of the function g in the direction $\mathbf{n}$, denoted $Dg \cdot \mathbf{n}$, is the ratio of the difference in the function values at a and b to the distance between the points, as the point b is taken closer and closer to a

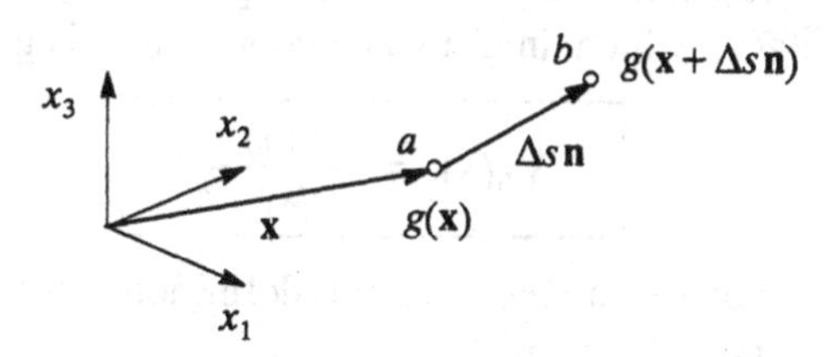

Figure 19 Interpretation of the gradient of a scalar field

$$Dg(\mathbf{x}) \cdot \mathbf{n} \equiv \lim_{\Delta s \to 0} \frac{g(\mathbf{x}+\Delta s\mathbf{n}) - g(\mathbf{x})}{\Delta s} \tag{68}$$

The directional derivative of g can be computed, using the chain rule of differentiation, from the formula

$$Dg(\mathbf{x}) \cdot \mathbf{n} = \frac{d}{d\varepsilon}\big(g(\mathbf{x}+\varepsilon\mathbf{n})\big)_{\varepsilon=0} = \frac{\partial g}{\partial x_i} n_i \tag{69}$$

In essence, the directional derivative determines the one-dimensional rate of change (i.e., $d/d\varepsilon$) of the function at the point $\mathbf{x}$ and just starting to move in the fixed direction $\mathbf{n}$. Because $\mathbf{x}$ and $\mathbf{n}$ are fixed, the derivative is an ordinary one.

Example 6. *Directional Derivative.* Consider the scalar function given by the expression $g(\mathbf{x}) \equiv \mathbf{x} \cdot \mathbf{x} = x_k x_k$. We can compute the directional derivative in the direction $\mathbf{n}$ by Eqn. (69). Noting that the augmented function can be written as $g(\mathbf{x}+\varepsilon\mathbf{n}) = (x_k+\varepsilon n_k)(x_k+\varepsilon n_k)$, we compute the directional derivative as

$$\begin{aligned} Dg(\mathbf{x}) \cdot \mathbf{n} &= \frac{d}{d\varepsilon}\big\{(x_k+\varepsilon n_k)(x_k+\varepsilon n_k)\big\}_{\varepsilon=0} \\ &= \frac{d}{d\varepsilon}\big\{x_k x_k + 2\varepsilon x_k n_k + \varepsilon^2 n_k n_k\big\}_{\varepsilon=0} \\ &= \big\{2x_k n_k + 2\varepsilon n_k n_k\big\}_{\varepsilon=0} = 2x_k n_k \end{aligned}$$

It is also useful to note that $\partial g/\partial x_i = \delta_{ki} x_k + x_k \delta_{ki} = 2x_i$. Then, according to Eqn. (69) again, we have

$$Dg \cdot \mathbf{n} = \frac{\partial g}{\partial x_i} n_i = 2x_i n_i$$

which is identical to the previous result.

From Eqn. (69) it is evident that the partial derivatives of the function g play a key role in determining the rate of change in a particular direction. In fact, the partial derivatives $\partial g/\partial x_i$ give the rate of change of g in the direction of the coordinate axis x_i. These three quantities can be viewed as the components of a vector called the *gradient* of the field. The gradient of a scalar field $g(\mathbf{x})$ is a vector field $\nabla g(\mathbf{x})$, which, in Cartesian coordinates, is given by

$$\boxed{\nabla g(\mathbf{x}) = \frac{\partial g(\mathbf{x})}{\partial x_j}\mathbf{e}_j} \tag{70}$$

where summation on j is implied. With this definition of the gradient, the directional derivative takes the form

$$\begin{aligned} Dg \cdot \mathbf{n} &= \nabla g \cdot \mathbf{n} \\ &= \frac{\partial g}{\partial x_i}\mathbf{e}_i \cdot n_j \mathbf{e}_j = \frac{\partial g}{\partial x_i} n_j \mathbf{e}_i \cdot \mathbf{e}_j = \frac{\partial g}{\partial x_i} n_j \delta_{ij} = \frac{\partial g}{\partial x_i} n_i \end{aligned} \tag{71}$$

We know that the directional derivative of g is zero if $\mathbf{n}$ is tangent to a contour line. Therefore, the vector ∇g must be perpendicular to the contour lines, as shown in Fig. 18b, because $\nabla g \cdot \mathbf{n} = 0$ in that direction. For the direction $\mathbf{n} = \nabla g / \| \nabla g \|$ it is evident from Eqn. (71) that $Dg \cdot \mathbf{n} = \| \nabla g(\mathbf{x}) \|$. Hence, $\| \nabla g(\mathbf{x}) \|$ is the maximum rate of change of the scalar field g.

We can define the *gradient* of a scalar field independent of any coordinate system. Consider an arbitrary region $\mathfrak{B}$ with surface Ω and outward unit normal vector field $\mathbf{n}$, shown in Fig. 17. The gradient is the ratio of the flux $g\mathbf{n}$ over the surface to the volume $\mathcal{V}(\mathfrak{B})$, in the limit as the volume of the region shrinks to zero. To wit

$$\nabla g \equiv \lim_{\mathcal{V}(\mathfrak{B})\to 0} \frac{1}{\mathcal{V}(\mathfrak{B})} \int_\Omega g\,\mathbf{n}\, dA \tag{72}$$

where $\mathcal{V}(\mathfrak{B})$ is the volume of the enclosed surface.

Equation (72) does not depend upon a specific coordinate system. Equation (70) is a formula for the gradient in rectangular Cartesian coordinates. The derivation of Eqn. (70) from Eqn. (72) is very instructive. To compute with Eqn. (72) we need to select a specific region $\mathfrak{B}$ so that we can compute the flux and the volume and take the limit as the volume shrinks to zero. The simplest possible choice is the cuboid with sides parallel to the coordinate planes shown in Fig. 20. The volume of this region is $\mathcal{V}(\mathfrak{B}) = \Delta x_1 \Delta x_2 \Delta x_3$. The surface Ω consists of six rectangles each with constant normal $\mathbf{n}$ pointing in the direction of one of the base vectors. Furthermore, the six faces occur in pairs with normals $\mathbf{n} = \pm\, \mathbf{e}_i$ on which x_i is constant over the entire face (with a value of x_i for the face with normal $-\mathbf{e}_i$ and $x_i + \Delta x_i$ for the face with normal $\mathbf{e}_i$). Hence, we can compute the flux as

$$\int_\Omega g\mathbf{n}\, dA = \sum_{i=1}^{3} \int_{\Omega_i} \left[g(\mathbf{x} + \Delta x_i \mathbf{e}_i)\mathbf{e}_i + g(\mathbf{x})(-\mathbf{e}_i) \right] dA_i \tag{73}$$

where Ω_i is the rectangular region with area A_i over which x_i is constant. Note that $A_1 = \Delta x_2 \Delta x_3$, $A_2 = \Delta x_3 \Delta x_1$, and $A_3 = \Delta x_1 \Delta x_2$ are the areas of the

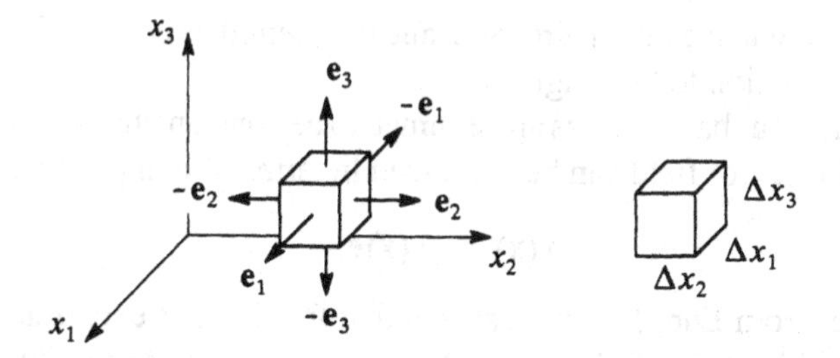

Figure 20 A particular region for the computation of flux and volume needed to compute derivatives in multivariate calculus

faces. Next, we can recognize that the volume is $\mathcal{V}(\mathcal{B}) = A_i \Delta x_i$ (no sum) for any i=1, 2, 3. Finally, we can recognize that

$$\frac{1}{A_i}\int_{\Omega_i} (\cdot)\, dA_i \tag{74}$$

is simply the average of $(\cdot)$ over the integration region Ω_i. In the limit, as the volume and the face areas shrink to zero, the average values will approach the values at $\mathbf{x}$. Therefore, Eqn. (72) can be written as

$$\begin{aligned} \nabla g &= \sum_{i=1}^{3} \lim_{A_i \to 0} \frac{1}{A_i}\int_{\Omega_i} \lim_{\Delta x_i \to 0} \left(\frac{g(\mathbf{x}+\Delta x_i \mathbf{e}_i) - g(\mathbf{x})}{\Delta x_i}\right) dA_i\, \mathbf{e}_i \\ &= \sum_{i=1}^{3} \frac{\partial g(\mathbf{x})}{\partial x_i}\, \mathbf{e}_i \end{aligned} \tag{75}$$

The limiting process for Δx_i can be moved inside the integral over Ω_i because x_i is constant for that integral. This limit is, of course, the partial derivative of g with respect to x_i. That partial derivative is a function of the other two variables which are not constant over that face. However, we then take the limit of the average over the region of integration to give the final result.

As we shall see, this approach will work in essentially identical fashion for developing coordinate expressions for all of the derivatives in this chapter.

Vector fields. A *vector field* is a function $\mathbf{v}(\mathbf{x})$ that assigns a vector to each point $\mathbf{x}$ in a particular domain. The displacement of a body is a vector field. Each point of the body moves by some amount in some direction. The force induced by gravitational attraction is a vector field.

Figure 21 shows two examples of vector fields. The pictures show the vectors at only enough points to get the idea of how the vectors are oriented and sized. The second vector field shown in the figure can be expressed in functional form as

$$\mathbf{v}(\mathbf{x}) = x_1\mathbf{e}_1 + x_2\mathbf{e}_2 \tag{76}$$

The vectors point in the radial direction, and their length is equal to the distance of the point of action to the origin.

In general, if our base vectors are assumed to be constant throughout our domain, then the vector field can be expressed in terms of component functions

$$\mathbf{v}(\mathbf{x}) = v_i(\mathbf{x})\mathbf{e}_i \tag{77}$$

For example, from Eqn. (76) we can see that the explicit expression for the components of the vector field are $v_1(\mathbf{x}) = x_1$, $v_2(\mathbf{x}) = x_2$, and $v_3(\mathbf{x}) = 0$. For curvilinear coordinates, the base vectors are also functions of the coordinates.

There are as many ways to differentiate a vector field as there are ways of multiplying vectors. The analogy between vector multiplication and vector differentiation is given in the following table

Multiplication		Differentiation	
$\mathbf{u} \cdot \mathbf{v}$	dot	div $\mathbf{v}$	divergence
$\mathbf{u} \times \mathbf{v}$	cross	curl $\mathbf{v}$	curl
$\mathbf{u} \otimes \mathbf{v}$	tensor	$\nabla\mathbf{v}$	gradient

As was the case for vector multiplication, each different way to differentiate a vector field yields a result with different character. For example, the divergence of a vector field is a scalar field, while the gradient of a vector field is a tensor field. Each of these derivatives, however, represents the rate of change of the vector field in some sense. Each one can be viewed as the "first derivative" of the vector field. In the sequel, we shall give a definition for each of these derivatives and give an idea of what they physically represent.

The divergence of a vector field. One way to measure of the rate of change of a vector field is the *divergence*. Consider again a domain $\mathcal{B}$ with enclosed volume $\mathcal{V}(\mathcal{B})$ and boundary Ω with unit normal vector $\mathbf{n}$, as shown in Fig. 17. Let us assume that the body lives in a vector field $\mathbf{v}(\mathbf{x})$. Thus, at each point $\mathbf{x}$ in $\mathcal{B}$ there exists a vector $\mathbf{v}(\mathbf{x})$. Let the flux be $\mathbf{v} \cdot \mathbf{n}$ on the boundary Ω. The

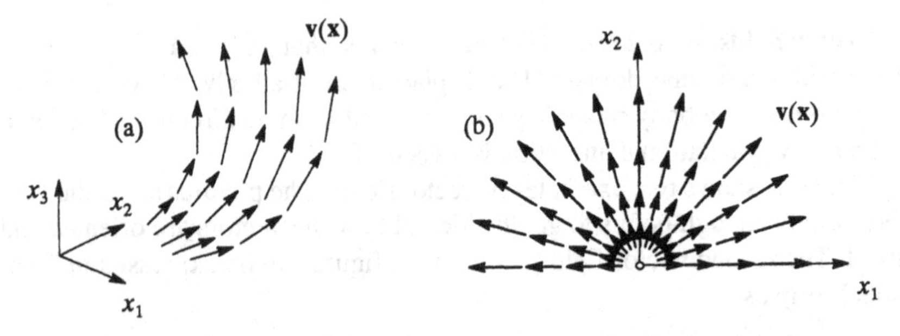

Figure 21 A vector field assigns a vector to each point in a domain

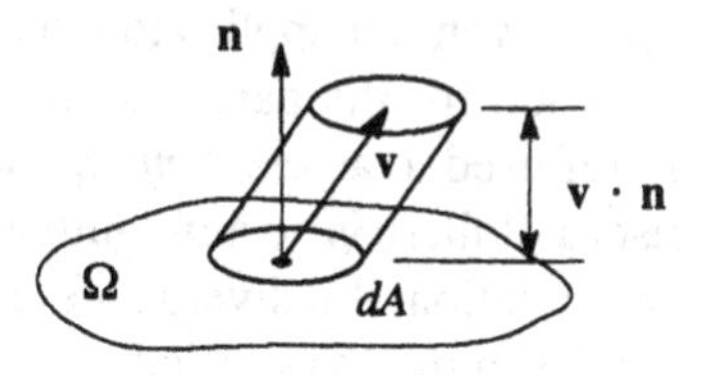

Figure 22 The flux through the area dA in unit time

divergence of the vector field is defined as the limit of the ratio of flux to volume, in the limit as the volume shrinks to zero. To wit

$$\boxed{\operatorname{div}(\mathbf{v}) \equiv \lim_{\mathcal{V}(\mathcal{B})\to 0} \frac{1}{\mathcal{V}(\mathcal{B})} \int_{\Omega} \mathbf{v} \cdot \mathbf{n} \, dA} \tag{78}$$

where dA is the infinitesimal element of area defined on the surface.

We can better understand why the integrand $\mathbf{v} \cdot \mathbf{n}$ is called the *flux* if we think of the vector field $\mathbf{v}$ as the particle velocity in a fluid flow, wherein the vectors would be tangent to particle streamlines. The product $\mathbf{v} \cdot \mathbf{n}$ would then represent the total amount of fluid that escapes through the area dA on the boundary per unit of time, as shown in Fig. 22. The physical significance of the product $\mathbf{v} \cdot \mathbf{n}$ is that the volume of fluid that passes through the area dA in unit time is equal to the base area of the cylinder dA times the height of the cylinder $\mathbf{v} \cdot \mathbf{n}$. Note that streamlines that are tangent to the boundary (i.e., $\mathbf{v} \cdot \mathbf{n} = 0$) do not let any fluid out, while streamlines normal to the boundary let it out most efficiently.

Let us compute an expression for the divergence of a vector field in Cartesian coordinates, again using the simple cuboid shown in Fig. 20. Following the same conventions we can compute the flux as

$$\int_{\Omega} \mathbf{v} \cdot \mathbf{n} \, dA = \sum_{i=1}^{3} \int_{\Omega_i} \left[\mathbf{v}(\mathbf{x} + \Delta x_i \mathbf{e}_i) \cdot \mathbf{e}_i + \mathbf{v}(\mathbf{x}) \cdot (-\mathbf{e}_i) \right] dA_i \tag{79}$$

where, again, Ω_i is the rectangular region with area A_i over which x_i is constant. Substituting $A_i = \mathcal{V}(\mathcal{B})/\Delta x_i$ we get

$$\operatorname{div}(\mathbf{v}) = \sum_{i=1}^{3} \lim_{A_i \to 0} \frac{1}{A_i} \int_{\Omega_i} \lim_{\Delta x_i \to 0} \left(\frac{\mathbf{v}(\mathbf{x} + \Delta x_i \mathbf{e}_i) - \mathbf{v}(\mathbf{x})}{\Delta x_i} \right) dA_i \cdot \mathbf{e}_i \tag{80}$$

Taking the limit of the average of the limit, as before, we arrive at the expression for the divergence in Cartesian coordinates:

$$\boxed{\operatorname{div}(\mathbf{v}) = \frac{\partial \mathbf{v}(\mathbf{x})}{\partial x_i} \cdot \mathbf{e}_i = \frac{\partial v_i(\mathbf{x})}{\partial x_i}} \tag{81}$$

Note that the summation convention applies to indices that are repeated in a quotient. A common notation for the partial derivative is $(\cdot)_{,i} \equiv \partial(\cdot)/\partial x_i$. This notation is usually referred to as the *comma notation* for partial derivatives. This notation is useful if there is no ambiguity the variable of differentiation. In this abbreviated notation, the divergence has the more compact expression $\operatorname{div}(\mathbf{v}) = v_{i,i}$ with summation implied across the comma. It should be evident that the comma notation is convenient for index manipulation.

The gradient of a vector field. Consider again the domain $\mathfrak{B}$ with boundary Ω shown in Fig. 17. The gradient of a vector field $\mathbf{v}(\mathbf{x})$ is a second-order tensor defined as the limit of the ratio of the flux $\mathbf{v} \otimes \mathbf{n}$ over the surface to the volume, as the volume shrinks to zero. To wit

$$\nabla \mathbf{v} \equiv \lim_{\mathcal{V}(\mathfrak{B}) \to 0} \frac{1}{\mathcal{V}(\mathfrak{B})} \int_{\Omega} \mathbf{v} \otimes \mathbf{n} \, dA \tag{82}$$

Again, $\mathcal{V}(\mathfrak{B})$ is the volume of the region $\mathfrak{B}$, Ω is the surface of the region, and $\mathbf{n}$ is the unit normal vector field to the surface. With a construction similar to the one used for the divergence, we can compute a coordinate expression for the gradient. The component expression for $\nabla \mathbf{v}$ in Cartesian coordinates is

$$\nabla \mathbf{v} = \frac{\partial v_i(\mathbf{x})}{\partial x_j} [\mathbf{e}_i \otimes \mathbf{e}_j] \tag{83}$$

where summation is implied for both i and j. Thus, the components of $\nabla \mathbf{v}$ are simply the various partial derivatives of the component functions with respect to the coordinates, that is, the component $[\nabla \mathbf{v}]_{ij}$ gives the rate of change of the ith component of $\mathbf{v}$ with respect to the jth coordinate axis.

We can interpret the gradient of a vector field geometrically by considering the construction shown in Fig. 23. Consider two points a and b that are near to each other (i.e., Δs is very small). The unit vector $\mathbf{n}$ points in the direction from a to b. The value of the vector field at a is $\mathbf{v}(\mathbf{x})$ and the value of the vector field

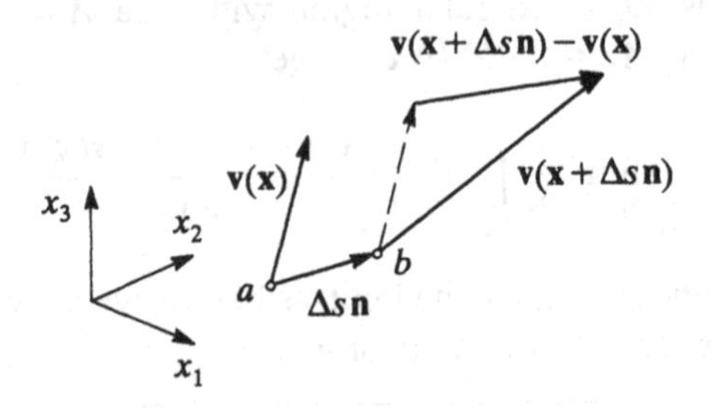

Figure 23 Interpretation of the gradient of a vector field with the directional derivative

at b is $\mathbf{v}(\mathbf{x}+\Delta s\,\mathbf{n})$. Since the vector field changes with position in the domain, these two vectors are different in both length and orientation. If we transport a copy of $\mathbf{v}(\mathbf{x})$ and position it at b (shown dotted), then we can compare the difference between the two vectors. The vector that connects the head of $\mathbf{v}(\mathbf{x})$ to the head of $\mathbf{v}(\mathbf{x}+\Delta s\,\mathbf{n})$ is $\mathbf{v}(\mathbf{x}+\Delta s\,\mathbf{n})-\mathbf{v}(\mathbf{x})$. This vector represents the difference in the vector between points a and b. If we divide this difference by Δs, then we get the rate of change as we move in the specified direction. Finally, taking the limit as Δs goes to zero, we get the directional derivative

$$D\mathbf{v}(\mathbf{x}) \cdot \mathbf{n} \equiv \lim_{\Delta s \to 0} \frac{\mathbf{v}(\mathbf{x}+\Delta s\,\mathbf{n}) - \mathbf{v}(\mathbf{x})}{\Delta s}$$

Like the analogous formula for scalar fields, the quantity $D\mathbf{v}(\mathbf{x}) \cdot \mathbf{n}$ is called the *directional derivative* because it gives the rate of change of the vector field in the direction $\mathbf{n}$. The limiting process above suggests that we can compute the directional derivative as

$$D\mathbf{v}(\mathbf{x}) \cdot \mathbf{n} = \frac{d}{d\varepsilon}\Big[\mathbf{v}(\mathbf{x}+\varepsilon\,\mathbf{n})\Big]_{\varepsilon=0} \tag{84}$$

A straightforward application of the chain rule for differentiation gives

$$D\mathbf{v}(\mathbf{x}) \cdot \mathbf{n} = \big[\nabla\mathbf{v}\big]\mathbf{n} \tag{85}$$

The directional derivative provides the answer to the question: What is the rate of change of the vector field? But Eqn. (85) makes it clear that the tensor $\nabla\mathbf{v}$ contains all of the information needed to assess rate of change in any direction.

Example 7. Consider a vector field given by the following explicit expression $\mathbf{v}(\mathbf{x}) = x_1x_2x_3(x_1\mathbf{e}_1 + x_2\mathbf{e}_2 + x_3\mathbf{e}_3)$. The components of the vector field are given by the following expressions

$$v_1 = x_1^2x_2x_3, \qquad v_2 = x_1x_2^2x_3, \qquad v_3 = x_1x_2x_3^2$$

The gradient of this vector field can be computed from Eqn. (83). The result is the following tensor field

$$\begin{aligned}\nabla\mathbf{v}(\mathbf{x}) &= 2x_1x_2x_3[\mathbf{e}_1 \otimes \mathbf{e}_1] + x_1^2x_3[\mathbf{e}_1 \otimes \mathbf{e}_2] + x_1^2x_2[\mathbf{e}_1 \otimes \mathbf{e}_3] \\ &+ x_2^2x_3[\mathbf{e}_2 \otimes \mathbf{e}_1] + 2x_1x_2x_3[\mathbf{e}_2 \otimes \mathbf{e}_2] + x_1x_2^2[\mathbf{e}_2 \otimes \mathbf{e}_3] \\ &+ x_2x_3^2[\mathbf{e}_3 \otimes \mathbf{e}_1] + x_1x_3^2[\mathbf{e}_3 \otimes \mathbf{e}_2] + 2x_1x_2x_3[\mathbf{e}_3 \otimes \mathbf{e}_3]\end{aligned}$$

The components of the tensor $\nabla\mathbf{v}$ can be put in matrix form as follows

$$\nabla\mathbf{v} \sim \begin{bmatrix} 2x_1x_2x_3 & x_1^2x_3 & x_1^2x_2 \\ x_2^2x_3 & 2x_1x_2x_3 & x_1x_2^2 \\ x_2x_3^2 & x_1x_3^2 & 2x_1x_2x_3 \end{bmatrix}$$

The divergence of a vector field can be computed from Eqn. (81). It is worth noting that the divergence is simply the trace of the gradient

$$\text{div}(\mathbf{v}) = \text{tr}(\nabla \mathbf{v})$$

where the trace is the sum of the diagonal components of the tensor. Therefore, for the present example, $\text{div}(\mathbf{v}) = 6x_1x_2x_3$.

One can define the curl of a vector field in a completely analogous way by considering the flux $\mathbf{v} \times \mathbf{n}$ (see Problem 45). The details are left to the reader.

A comment on notation for derivatives. There are many notations used to characterize operation in vector calculus. In this book we stick to "div" and ∇ (some authors use "grad"). Occasionally it is useful to use a shorthand notation for gradients of scalar and vectors fields

$$\nabla g = \frac{\partial g}{\partial \mathbf{x}}, \qquad \nabla \mathbf{v} = \frac{\partial \mathbf{v}}{\partial \mathbf{x}} \tag{86}$$

While this notation is a bit sloppy it is convenient. For many problems in mechanics we use more than one coordinate system. When we take derivatives we must specify the variable of differentiation (if it is ambiguous). For the divergence we will often use "div" and "DIV" to distinguish between two choices. For the gradient we will often use the notation $\nabla_{\mathbf{x}}(\cdot)$ or $\nabla_{\mathbf{z}}(\cdot)$ to indicate the variable of differentiation.

Divergence of a tensor field. A tensor field is a function that assigns a tensor $\mathbf{T}(\mathbf{x})$ to each point $\mathbf{x}$ in the domain. Consider a tensor field $\mathbf{T}(\mathbf{x})$ on a region $\mathfrak{B}$ with surface Ω having unit normal vector field $\mathbf{n}$. There are many ways to differentiate a tensor field. In solid mechanics we are primarily interested in one way. By analogy with vector differentiation, we define the divergence of a tensor field

$$\boxed{\text{div}\mathbf{T} \equiv \lim_{\mathcal{V}(\mathfrak{B})\to 0} \frac{1}{\mathcal{V}(\mathfrak{B})} \int_{\Omega} \mathbf{T}\mathbf{n}\, dA} \tag{87}$$

where, as before, $\mathcal{V}(\mathfrak{B})$ is the volume of the region $\mathfrak{B}$, Ω is the surface of the region, and $\mathbf{n}$ is the unit normal vector field to the surface. Since the integrand $\mathbf{T}\mathbf{n}$ is a vector, $\text{div}\mathbf{T}$ is a vector.

One can use the definition of the divergence to compute a component expression and to prove the divergence theorem for tensor fields, by following the same arguments we have used for vector fields. Let us compute an expression for the divergence of a tensor field in Cartesian coordinates, again using the simple cuboid shown in Fig. 20. Following the same conventions we can compute the flux as

$$\int_{\Omega} \mathbf{Tn}\, dA = \sum_{i=1}^{3} \int_{\Omega_i} \left[\mathbf{T}(\mathbf{x}+\Delta x_i \mathbf{e}_i)\mathbf{e}_i + \mathbf{T}(\mathbf{x})(-\mathbf{e}_i) \right] dA_i \tag{88}$$

where, again, Ω_i is the rectangular region with area A_i over which x_i is constant. Substituting $A_i = \mathcal{V}(\mathfrak{B})/\Delta x_i$ we get

$$\operatorname{div}(\mathbf{T}) = \sum_{i=1}^{3} \lim_{A_i \to 0} \frac{1}{A_i} \int_{\Omega_i} \lim_{\Delta x_i \to 0} \left(\frac{\mathbf{T}(\mathbf{x}+\Delta x_i \mathbf{e}_i) - \mathbf{T}(\mathbf{x})}{\Delta x_i} \right) dA_i\, \mathbf{e}_i \tag{89}$$

Taking the limit of the average of the limit, as before, we arrive at the expression for the divergence in Cartesian coordinates:

$$\boxed{\operatorname{div}(\mathbf{T}) = \frac{\partial \mathbf{T}}{\partial x_i}\mathbf{e}_i = \frac{\partial}{\partial x_i}(\mathbf{T}\mathbf{e}_i) = \frac{\partial T_{ij}(\mathbf{x})}{\partial x_j}\mathbf{e}_i} \tag{90}$$

It should be evident that all of the forms of the divergence of a tensor field given in Eqn. (90) are equivalent. The convenience of one form over another depends upon the application.

Integral Theorems

The divergence theorem. There is an integration theorem worth mentioning here because it comes up repeatedly in solid mechanics. We call it the *divergence theorem* because it involves the divergence of a vector field. Consider again a region $\mathfrak{B}$ of arbitrary size and shape, with boundary Ω described by its normal vectors $\mathbf{n}$. The divergence theorem can be stated as follows

$$\boxed{\int_{\mathfrak{B}} \operatorname{div} \mathbf{v}\, dV = \int_{\Omega} \mathbf{v} \cdot \mathbf{n}\, dA} \tag{91}$$

This remarkable theorem, also known as *Green's theorem* or *Gauss's theorem*, relates an integral over the volume of a region to an integral over the boundary of that same region. It applies to any sufficiently well-behaved vector field $\mathbf{v}(\mathbf{x})$, and, thus, is very powerful. The proof of the divergence theorem can be carried out along many lines. The one in Schey (1973) is particularly descriptive. Schey's argument goes something as follows.

Partition the region $\mathfrak{B}$ into N small subregions $\mathfrak{B}_i$ each having volume $\mathcal{V}(\mathfrak{B}_i)$, surfaces Ω_i, and unit outward normal vector field $\mathbf{n}_i$, as illustrated in Fig. 24. The surface of a certain subregion is the union of interior surfaces shared with adjacent subregions and (possibly) part of the original exterior surface Ω. The normal vectors along a shared surface between two adjacent subregions point in opposite directions, as shown in the figure. Consequently, if we sum the fluxes over all of the subregions we get

$$\sum_{i=1}^{N} \int_{\Omega_i} \mathbf{v} \cdot \mathbf{n}_i \, dA_i = \int_{\Omega} \mathbf{v} \cdot \mathbf{n} \, dA \tag{92}$$

In other words, the contributions of fluxes across the interior surfaces cancel each other out because there is only one $\mathbf{v}$ at a given point on the surface (provided that $\mathbf{v}$ is a continuous field) while one normal is the negative of the other.

Let us define the "almost divergence" of the vector field to be the finite ratio of flux to volume of subregion $\mathfrak{B}_i$

$$\mathfrak{D}_i[\mathbf{v}] \equiv \frac{1}{\mathcal{V}(\mathfrak{B}_i)} \int_{\Omega_i} \mathbf{v} \cdot \mathbf{n}_i \, dA_i \tag{93}$$

and observe that $\mathfrak{D}_i[\mathbf{v}] \rightarrow \operatorname{div}(\mathbf{v})$ in the limit as $\mathcal{V}(\mathfrak{B}_i) \rightarrow 0$. Multiplying Eqn. (93) through by $\mathcal{V}(\mathfrak{B}_i)$ and summing over all N subregions, we can see from Eqn. (92) that

$$\int_{\Omega} \mathbf{v} \cdot \mathbf{n} \, dA = \sum_{i=1}^{N} \mathfrak{D}_i[\mathbf{v}] \mathcal{V}(\mathfrak{B}_i) \tag{94}$$

This equation holds no matter how many subregions there are in the partition. As the number of partitions is taken larger and larger the size of the subregions shrinks. In the limit as $N \rightarrow \infty$ the discrete elements pass to their infinitesimal limits, that is, $\mathfrak{D}_i[\mathbf{v}] \rightarrow \operatorname{div}(\mathbf{v})$ and $\mathcal{V}(\mathfrak{B}_i) \rightarrow dV$. The limit of the sum is the integral over the volume

$$\lim_{N \rightarrow \infty} \sum_{i=1}^{N} \mathfrak{D}_i[\mathbf{v}] \mathcal{V}(\mathfrak{B}_i) = \int_{\mathfrak{B}} \operatorname{div} \mathbf{v} \, dV \tag{95}$$

thereby completing the proof.

The utility of defining the divergence with the intrinsic formula, Eqn. (78), should be evident from the proof of the divergence theorem. This proof might

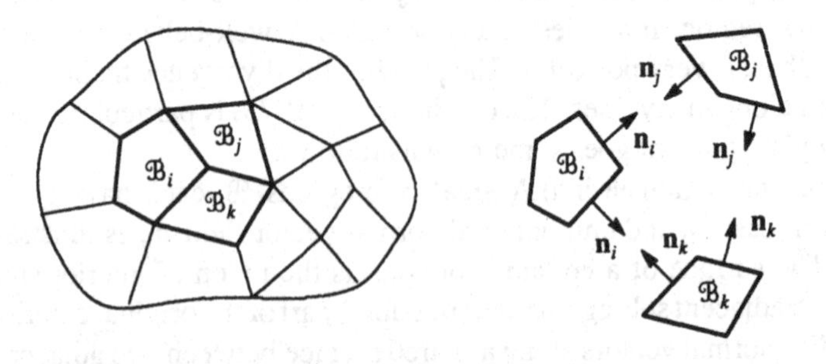

Figure 24 A region in three-dimensional space partitioned into subregions $\mathfrak{B}_i$, each with volume $\mathcal{V}(\mathfrak{B}_i)$, surface Ω_i, and unit outward normal vector field $\mathbf{n}_i$.

not have the level of rigor that a mathematician would like (the limiting process and crossover to infinitesimals being the sloppiest point), but the geometric basis lends it a clarity that is more than adequate for our purposes here.

The divergence theorem holds for any vector field $\mathbf{v}(\mathbf{x})$ that is well behaved. A simple way to think about "well-behavedness" is to consider some of the bad things that might happen on the way to the limit. In particular, any of the objects, like $\mathfrak{D}_i[\mathbf{v}]$, must exist for all possible subdivisions. If the vector field has a singular point ($\mathbf{v} \to \infty$), then eventually the subdivision process will encounter it, and, for the subdomains on whose boundaries the singularity lies, $\mathfrak{D}_i[\mathbf{v}]$ is not defined. Similarly, if the field has a bounded jump along some surface (where $\mathbf{v}^- \neq \mathbf{v}^+$ on opposites sides of the jump), then for those subdomains that have a boundary on the jump surface, the fluxes will not cancel out. Many of these pathologies can be treated by enhancing the integral theorems with features that account for them. We do not have to worry about the pathologies if our vector field $\mathbf{v}$ and its divergence are continuous over the domain $\mathfrak{B}$ and on the surface Ω.

Example 8. The divergence theorem for the gradient of a scalar field is

$$\int_{\mathfrak{B}} \nabla g \, dV = \int_{\Omega} g \, \mathbf{n} \, dA$$

where $\mathfrak{B}$ is a region with surface Ω having unit outward normal vector field $\mathbf{n}$. Verify the relationship by applying it to the function $g = x_1^2 + x_2^2 + x_3^2$ defined on a cylinder of unit radius and unit height, centered at the origin.

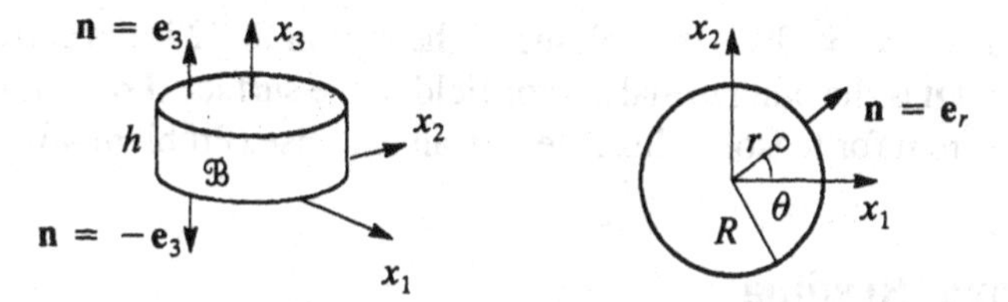

Figure 25 Circular cylinder definition for Example 8.

The integral of the gradient over the volume is best done in cylindrical coordinates. Let $x_1 = r\cos\theta$, $x_2 = r\sin\theta$, and $x_3 = z$. The gradient of g can be computed as $\nabla g = 2\mathbf{x} = 2(r\mathbf{e}_r + z\mathbf{e}_3)$, where $\mathbf{e}_r(\theta) \equiv \cos\theta\,\mathbf{e}_1 + \sin\theta\,\mathbf{e}_2$. The volume integral can be carried out as follows:

$$\int_{\mathfrak{B}} \nabla g \, dV = \int_0^h \int_0^{2\pi} \int_0^R 2\big(r\mathbf{e}_r(\theta) + z\mathbf{e}_3\big)\, r \, dr \, d\theta \, dz = \pi R^2 h^2 \mathbf{e}_3$$

(Observe that the integral of $\mathbf{e}_r(\theta)$ with respect to θ from 0 to 2π is zero). The surface has the following characteristics:

Bottom Surface:	$z = 0$	$\mathbf{n} = -\mathbf{e}_3$	$g = r^2$
Top Surface:	$z = h$	$\mathbf{n} = +\mathbf{e}_3$	$g = r^2 + h^2$
Lateral Surface:	$r = R$	$\mathbf{n} = \mathbf{e}_r$	$g = R^2 + z^2$

The surface integral can be carried out as follows, noting that the integrand for the top and bottom surfaces reduces to $(r^2 + h^2)\mathbf{e}_3 - r^2\mathbf{e}_3 = h^2\mathbf{e}_3$,

$$\int_{\Omega} g\mathbf{n}\, dA = \int_0^{2\pi}\int_0^R h^2\mathbf{e}_3\, r\, dr\, d\theta + \int_0^h\int_0^{2\pi}\left(R^2 + z^2\right)\mathbf{e}_r(\theta)\, R\, d\theta\, dz$$

$$= \pi R^2 h^2 \mathbf{e}_3$$

Clearly, the volume and surface integrals have the same value, as the divergence theorem promises.

There are integral theorems for the gradient of a scalar field, the gradient of a vector field and a tensor field (see next section) that are analogous to the divergence theorem. The statements and proofs of these theorem are left as an exercise (Problem 46).

Divergence theorem for tensor fields. Any tensor field satisfies the following integral theorem (divergence theorem)

$$\boxed{\int_{\mathcal{B}} \operatorname{div}\mathbf{T}\, dV = \int_{\Omega} \mathbf{T}\mathbf{n}\, dA} \tag{96}$$

where, as before, $\mathcal{V}(\mathcal{B})$ is the volume of the region $\mathcal{B}$, Ω is the surface of the region, and $\mathbf{n}$ is the unit normal vector field to the surface. Proof of the divergence theorem for tensor fields is left as an exercise (Problem 46).

Additional Reading

M. E. Gurtin, *An introduction to continuum mechanics*, Academic Press, 1981.

H. M. Schey, *Div, grad, curl, and all that*, Norton, New York, 1973.

J. G. Simmonds, *A brief on tensor analysis*, Springer-Verlag, New York, 1982.

J. H. Wilkinson, *The algebraic eigenvalue problem*, Oxford University Press, Oxford, 1965.

Problems

1. Compute the values of the following expressions

(a) δ_{ii}

(b) $\delta_{ij}\delta_{ij}$

(c) $C_{ij}\delta_{ik}\delta_{jk}$

(d) $\delta_{ab}\delta_{bc}\delta_{cd}\cdots\delta_{xy}\delta_{yz}$ (enough terms to exhaust the whole alphabet)

2. Let two vectors, **u** and **v**, have components relative to some basis as $\mathbf{u} = (5, -2, 1)$ and $\mathbf{v} = (1, 1, 1)$. Compute the lengths of the vectors and the angle between them. Find the area of the parallelogram defined by **u** and **v**.

3. The vertices of a triangle are given by the position vectors **a**, **b**, and **c**. The components of these vectors in a particular basis are $\mathbf{a} = (0,0,0)$, $\mathbf{b} = (1,4,3)$, and $\mathbf{c} = (2,3,1)$.Using a vector approach, compute the area of the triangle. Find the area of the triangle projected onto the plane with normal $\mathbf{n} = (0, 0, 1)$. Find the unit normal vector to the triangle.

4. Let the coordinates of four points a, b, c and d be given by the following position vectors a=(1, 1, 1), b=(2, 1, 1), c=(1, 2, 2), and d=(1, 1, 3) in the coordinate system shown. Find vectors normal to planes abc and bcd. Find the angle between those vectors. Find the area of the triangle abc. Find the volume of the tetrahedron $abcd$.

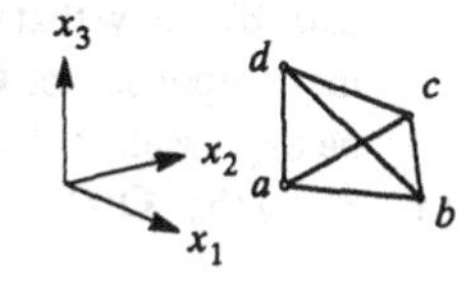

5. Demonstrate that $(\mathbf{u} \times \mathbf{v}) \cdot \mathbf{w} = u_i v_j w_k \varepsilon_{ijk}$ from basic operations on the base vectors.

6. Show that the triple scalar product is skew-symmetric with respect to changing the order in which the vectors appear in the product. For example, show that

$$(\mathbf{u} \times \mathbf{v}) \cdot \mathbf{w} = -(\mathbf{v} \times \mathbf{u}) \cdot \mathbf{w}$$

To generalize this notion, any cyclic permutation (e.g., **u**, **v**, **w** → **w**, **u**, **v**) of the order of the vectors leaves the algebraic sign of the product unchanged, while any acyclic permutation (e.g., **u**, **v**, **w** → **v**, **u**, **w**) of the order of the vectors changes the sign. How does this observation relate to swapping rows of a matrix in the computation of the determinant of that matrix?

7. Use the observation that $\| \mathbf{u}-\mathbf{v} \|^2 = (\mathbf{u}-\mathbf{v}) \cdot (\mathbf{u}-\mathbf{v})$ along with the distributive law for the dot product to show that

$$\mathbf{u} \cdot \mathbf{v} \equiv \tfrac{1}{2}\left(\| \mathbf{u} \|^2 + \| \mathbf{v} \|^2 - \| \mathbf{v}-\mathbf{u} \|^2\right)$$

8. Prove the *Schwarz inequality*, $|\mathbf{u} \cdot \mathbf{v}| \le \| \mathbf{u} \| \, \| \mathbf{v} \|$. Try to prove this inequality without using the formula $\mathbf{u} \cdot \mathbf{v} = \| \mathbf{u} \| \| \mathbf{v} \| \cos\theta(\mathbf{u}, \mathbf{v})$.

9. Show that $[\mathbf{u} \otimes \mathbf{v}]^T = \mathbf{v} \otimes \mathbf{u}$ using the definition of the transpose of a tensor and by demonstrating that the two tensors give the same result when acting on arbitrary vectors **a** and **b**.

10. Show that the components of a tensor **T** and its transpose $\mathbf{T}^T$ satisfy $[\mathbf{T}^T]_{ij} = [\mathbf{T}]_{ji}$.

11. Show that the tensor $\mathbf{T}^T\mathbf{T}$ is symmetric.

12. Consider any two tensors $\mathbf{S}$ and $\mathbf{T}$. Prove the following:
(a) $\det(\mathbf{T}^T) = \det(\mathbf{T})$
(b) $\det(\mathbf{ST}) = \det(\mathbf{S})\det(\mathbf{T})$
(c) $[\mathbf{ST}]^T = \mathbf{T}^T\mathbf{S}^T$
(d) $[\mathbf{ST}]^{-1} = \mathbf{T}^{-1}\mathbf{S}^{-1}$

13. Consider two Cartesian coordinate systems, one with basis $\{\mathbf{e}_1, \mathbf{e}_2, \mathbf{e}_3\}$ and the other with basis $\{\mathbf{g}_1, \mathbf{g}_2, \mathbf{g}_3\}$. Let $Q_{ij} \equiv \mathbf{g}_i \cdot \mathbf{e}_j$ be the cosine of the angle between $\mathbf{g}_i$ and $\mathbf{e}_j$.
(a) Show that $\mathbf{g}_i = Q_{ij}\mathbf{e}_j$ and $\mathbf{e}_j = Q_{ij}\mathbf{g}_i$ relate the two sets of base vectors.
(b) We can define a rotation tensor $\mathbf{Q}$ such that $\mathbf{e}_i = \mathbf{Q}\mathbf{g}_i$. Show that this tensor can be expressed as $\mathbf{Q} \equiv Q_{ij}[\mathbf{g}_i \otimes \mathbf{g}_j]$, that is, Q_{ij} are the components of $\mathbf{Q}$ with respect to the basis $[\mathbf{g}_i \otimes \mathbf{g}_j]$. Show that the tensor can also be expressed in the form $\mathbf{Q} = [\mathbf{e}_i \otimes \mathbf{g}_i]$.
(c) We can define a rotation tensor $\mathbf{Q}^T$, such that $\mathbf{g}_i = \mathbf{Q}^T\mathbf{e}_i$ (the reverse rotation from part (b). Show that this tensor can be expressed as $\mathbf{Q}^T \equiv Q_{ij}[\mathbf{e}_j \otimes \mathbf{e}_i]$, that is, Q_{ij} are the components of $\mathbf{Q}^T$ with respect to the basis $[\mathbf{e}_j \otimes \mathbf{e}_i]$. Show that the tensor can also be expressed in the form $\mathbf{Q}^T = [\mathbf{g}_i \otimes \mathbf{e}_i]$.
(d) Show that $\mathbf{Q}^T\mathbf{Q} = \mathbf{I}$, which implies that the tensor $\mathbf{Q}$ is orthogonal.

14. The components of tensors $\mathbf{T}$ and $\mathbf{S}$ and the components of vectors $\mathbf{u}$ and $\mathbf{v}$ are

$$\mathbf{T} \sim \begin{bmatrix} 1 & 2 & 0 \\ 2 & 0 & 1 \\ 0 & 1 & 2 \end{bmatrix} \quad \mathbf{S} \sim \begin{bmatrix} 0 & -2 & 1 \\ 2 & 0 & -1 \\ -1 & 1 & 0 \end{bmatrix} \quad \mathbf{v} \sim \begin{bmatrix} 1 \\ 1 \\ 1 \end{bmatrix} \quad \mathbf{u} \sim \begin{bmatrix} 1 \\ 1 \\ 2 \end{bmatrix}$$

Compute the components of the vector $\mathbf{Su}$. Find the cosine of the angle between $\mathbf{u}$ and $\mathbf{Su}$. Compute the determinants of $\mathbf{T}$, $\mathbf{S}$, and $\mathbf{TS}$. Compute $T_{ij}T_{ij}$ and $u_iT_{ik}S_{kj}v_j$.

15. Verify that, for the particular case given here, the components of the tensor $\mathbf{T}$ and the components of its inverse tensor $\mathbf{T}^{-1}$ are

$$\mathbf{T} \sim \begin{bmatrix} 2 & -1 & 0 \\ -1 & 2 & -1 \\ 0 & -1 & 2 \end{bmatrix} \qquad \mathbf{T}^{-1} \sim \frac{1}{4}\begin{bmatrix} 3 & 2 & 1 \\ 2 & 4 & 2 \\ 1 & 2 & 3 \end{bmatrix}$$

16. Consider two bases: $\{\mathbf{e}_1, \mathbf{e}_2, \mathbf{e}_3\}$ and $\{\mathbf{g}_1, \mathbf{g}_2, \mathbf{g}_3\}$. The basis $\{\mathbf{g}_1, \mathbf{g}_2, \mathbf{g}_3\}$ is given in terms of the base vectors $\{\mathbf{e}_1, \mathbf{e}_2, \mathbf{e}_3\}$ as

$$\mathbf{g}_1 = \tfrac{1}{\sqrt{3}}(\mathbf{e}_1+\mathbf{e}_2+\mathbf{e}_3), \qquad \mathbf{g}_2 = \tfrac{1}{\sqrt{6}}(2\mathbf{e}_1-\mathbf{e}_2-\mathbf{e}_3), \qquad \mathbf{g}_3 = \tfrac{1}{\sqrt{2}}(\mathbf{e}_2-\mathbf{e}_3)$$

The components of the tensor $\mathbf{T}$ and vector $\mathbf{v}$, relative to the basis $\{\mathbf{e}_1, \mathbf{e}_2, \mathbf{e}_3\}$ are

$$\mathbf{T} \sim \begin{bmatrix} 0 & -1 & 1 \\ 1 & 0 & -1 \\ -1 & 1 & 0 \end{bmatrix} \qquad \mathbf{v} \sim \begin{bmatrix} 1 \\ 2 \\ 3 \end{bmatrix}$$

Compute the components of the vector $\mathbf{Tv}$ in both bases. Compute the nine values of $T_{ij}T_{jk}T_{kl}$ (i.e., for $i, l = 1, 2, 3$). Find the components of the tensor $[\mathbf{T}+\mathbf{T}^T]$. Compute T_{ii}.

17. Consider two bases: $\{\mathbf{e}_1, \mathbf{e}_2, \mathbf{e}_3\}$ and $\{\mathbf{g}_1, \mathbf{g}_2, \mathbf{g}_3\}$, where

$$\mathbf{g}_1 = \mathbf{e}_1 + \mathbf{e}_2 + \mathbf{e}_3, \qquad \mathbf{g}_2 = \mathbf{e}_2 + \mathbf{e}_3, \qquad \mathbf{g}_3 = \mathbf{e}_2 - \mathbf{e}_3$$

Compute Q_{ij} for the given bases. Compute the value of $Q_{ik}Q_{kj}$. Explain why the identity $Q_{ik}Q_{kj} = \delta_{ij}$ does not hold in this case.

Now consider a vector $\mathbf{v} = \mathbf{e}_1 + 2\mathbf{e}_2 + 3\mathbf{e}_3$ and a tensor $\mathbf{T}$ given as

$$\mathbf{T} = [\mathbf{e}_2 \otimes \mathbf{e}_1 - \mathbf{e}_1 \otimes \mathbf{e}_2] + [\mathbf{e}_3 \otimes \mathbf{e}_1 - \mathbf{e}_1 \otimes \mathbf{e}_3] + [\mathbf{e}_3 \otimes \mathbf{e}_2 - \mathbf{e}_2 \otimes \mathbf{e}_3]$$

Compute the components of the vector $\mathbf{Tv}$ in both bases, i.e., find v_i and $\hat{v}_i$ so that the following relationship holds $\mathbf{Tv} = v_i \mathbf{e}_i = \hat{v}_i \mathbf{g}_i$. Find the cosine of the angle between the vector $\mathbf{v}$ and the vector $\mathbf{Tv}$. Find the length of the vector $\mathbf{Tv}$.

18. A general nth-order tensor invariant can be defined as follows

$$f_n(\mathbf{T}) \equiv T_{i_1 i_2} T_{i_2 i_3} \cdots T_{i_n i_1}$$

where $\{i_1, i_2, \ldots, i_n\}$ are the n indices. For example, when $n = 2$ we can use $\{i, j\}$ to give $f_2(\mathbf{T}) = T_{ij}T_{ji}$; when $n = 3$ we can use $\{i, j, k\}$ to give $f_3(\mathbf{T}) = T_{ij}T_{jk}T_{ki}$. Prove that $f_n(\mathbf{T})$ is invariant with respect to coordinate transformation.

19. Use the Cayley-Hamilton theorem to prove that for $n \geq 4$ all of the invariants $f_n(\mathbf{T})$, defined in Problem 18, can be computed from $f_1(\mathbf{T})$, $f_2(\mathbf{T})$, and $f_3(\mathbf{T})$.

20. From any tensor $\mathbf{T}$ one can compute an associated *deviator* tensor $\mathbf{T}_{dev}$ which has the property that the deviator tensor has no trace, i.e., $\text{tr}(\mathbf{T}_{dev}) = 0$. Such a tensor can be obtained from the original tensor $\mathbf{T}$ simply by subtracting $\alpha \equiv \frac{1}{3}\text{tr}(\mathbf{T})$ times the identity from the original tensor, i.e., $\mathbf{T}_{dev} = \mathbf{T} - \alpha\mathbf{I}$. Show that $\text{tr}(\mathbf{T}_{dev}) = 0$. Show that the principal directions of $\mathbf{T}_{dev}$ and $\mathbf{T}$ are identical, but that the principal values of $\mathbf{T}_{dev}$ are reduced by an amount α from those of the tensor $\mathbf{T}$.

21. Consider a tensor $\mathbf{T}$ that has all repeated eigenvalues $\mu_1 = \mu_2 = \mu_3 \equiv \mu$. Show that the tensor $\mathbf{T}$ must have the form $\mathbf{T} = \mu\mathbf{I}$.

22. Prove that the product of a tensor with itself n times can be represented as

$$\mathbf{T}^n = \sum_{i=1}^{3} (\mu_i)^n \, \mathbf{n}_i \otimes \mathbf{n}_i$$

Hint: Observe that $[\mathbf{n}_i \otimes \mathbf{n}_i][\mathbf{n}_j \otimes \mathbf{n}_j] = \delta_{ij}[\mathbf{n}_i \otimes \mathbf{n}_j]$ (no summation implied).

23. Show that the determinant of the tensor $\mathbf{T}$ can be expressed as follows

$$\det(\mathbf{T}) = \tfrac{1}{3}\text{tr}(\mathbf{T}^3) - \tfrac{1}{2}I_T\,\text{tr}(\mathbf{T}^2) + \tfrac{1}{6}(I_T)^3$$

where $I_T = \text{tr}(\mathbf{T}) = T_{ii}$ is the first invariant of $\mathbf{T}$. Use the Cayley-Hamilton theorem.

24. A certain state of deformation at a point in a body is described by the tensor $\mathbf{T}$, having the components relative to a certain basis of

$$\mathbf{T} \sim \begin{bmatrix} 3 & -1 & 0 \\ -1 & 5 & 1 \\ 0 & 1 & 2 \end{bmatrix}$$

Find the eigenvalues and eigenvectors of **T**. Show that the invariants of the tensor **T** are the same in the given basis and in the basis defined by the eigenvectors for the present case.

25. Find the tensor **T** that has eigenvalues $\mu_1 = 1$, $\mu_2 = 2$, and $\mu_3 = 3$ with two of the associated eigenvectors given by

$$\mathbf{n}_1 = \frac{1}{\sqrt{2}}(\mathbf{e}_1 + \mathbf{e}_2), \quad \mathbf{n}_2 = \frac{1}{3}(-2\mathbf{e}_1 + 2\mathbf{e}_2 + \mathbf{e}_3)$$

Is the tensor unique (i.e., is there another one with these same eigenproperties)?

26. Find the tensor **T** that has eigenvalues $\mu_1 = 1$, $\mu_2 = 3$, and $\mu_3 = 3$, with two of the associated eigenvectors given by

$$\mathbf{n}_1 = \frac{1}{\sqrt{3}}(\mathbf{e}_1 + \mathbf{e}_2 + \mathbf{e}_3), \quad \mathbf{n}_2 = \frac{1}{\sqrt{2}}(-\mathbf{e}_2 + \mathbf{e}_3)$$

Are the eigenvectors unique?

27. A certain state of deformation at a point in a body is described by the tensor **T**, having the components relative to a certain basis of

$$\mathbf{T} \sim 10^{-2}\begin{bmatrix} 14 & 2 & 14 \\ 2 & -1 & -16 \\ 14 & -16 & 5 \end{bmatrix}$$

Let the principal values and principal directions be designated as μ and **n**. Show that $\mathbf{n}_1 = (-1, 2, 2)$ is a principal direction and find μ_1. The second principal value is $\mu_2 = 9 \times 10^{-2}$, find $\mathbf{n}_2$. Find μ_3 and $\mathbf{n}_3$ with as little computation as possible.

28. The equation for balance of angular momentum can be expressed in terms of a tensor **T** and the base vectors $\mathbf{e}_i$ as $\mathbf{e}_i \times (\mathbf{T}\mathbf{e}_i) = \mathbf{0}$ (sum on repeated index implied). What specific conditions must the components of the tensor **T** satisfy in order for this equation to be satisfied?

29. The tensor **R** that operates on vectors and reflects them (as in a mirror) with unit normal **n** is given by

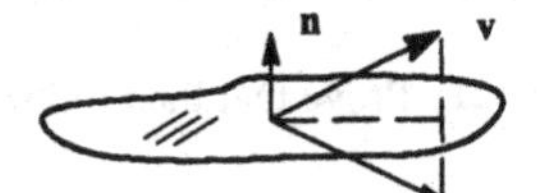

$$\mathbf{R} \equiv \mathbf{I} - 2\,\mathbf{n} \otimes \mathbf{n}$$

Compute the vector that results from [**RR**]**v**. Compute the length of the vector **Rv** in terms of the length of **v**. What is the inverse of the tensor **R**? Compute the eigenvalues and eigenvectors of **R**.

30. Let **v**(**x**) and **u**(**x**) be two vector fields, and **T**(**x**) be a tensor field. Compute the following expressions in terms of the components (v_i, u_i, and T_{ij}) of these fields relative to the basis $\{\mathbf{e}_1, \mathbf{e}_2, \mathbf{e}_3\}$: $\text{div}(\mathbf{T}\mathbf{v})$, $\nabla(\mathbf{u} \cdot \mathbf{T}\mathbf{v})$, $\nabla(\mathbf{T}\mathbf{v})$, and $\mathbf{u} \otimes \mathbf{T}\mathbf{v}$.

31. Evaluate the following expressions:

(a) $\text{div}(\text{div}[\mathbf{x} \otimes \mathbf{x}])$ (b) $\text{div}(\mathbf{x}\,\text{div}(\mathbf{x}\,\text{div}\,\mathbf{x}))$ (c) $\nabla[\|(\nabla\|\mathbf{x}\|^2)\|^2]$

(d) $\text{div}(\mathbf{x} \otimes \text{div}[\mathbf{x} \otimes \mathbf{x}])$ (e) $\nabla(\mathbf{x}\,\text{div}\,\mathbf{x})$ (f) $\nabla[\mathbf{x} \cdot \nabla(\mathbf{x} \cdot \mathbf{x})]$

where $\mathbf{x} = x_1\mathbf{e}_1 + x_2\mathbf{e}_2 + x_3\mathbf{e}_3$ is the position vector in space and all derivatives are with respect to the coordinates x_i.

32. Let $\mathbf{v}(\mathbf{x}) = (x_2 - x_3)\mathbf{e}_1 + (x_3 - x_1)\mathbf{e}_2 + (x_1 - x_2)\mathbf{e}_3$. Evaluate the following expressions: $\nabla\mathbf{v}$, $\nabla(\mathbf{x} \cdot \mathbf{v})$, $\operatorname{div}[\mathbf{x} \otimes \mathbf{v}]$, and $\nabla(\mathbf{x} \times \mathbf{v})$, where $\mathbf{x} = x_i\mathbf{e}_i$ is the position vector. Evaluate the expressions at the point $\mathbf{x} = \mathbf{e}_1 + 2\mathbf{e}_2 + \mathbf{e}_3$.

33. Let $\mathbf{v}(\mathbf{x})$ be given by the following explicit function

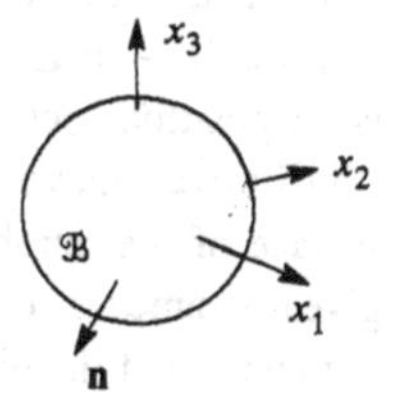

$$\mathbf{v}(\mathbf{x}) = (x_1^2 + x_2x_3)\mathbf{e}_1 + (x_2^2 + x_1x_3)\mathbf{e}_2 + (x_3^2 + x_1x_2)\mathbf{e}_3$$

where $\mathbf{x}$ is the position vector of any point and has components $\{x_1, x_2, x_3\}$ relative to the Cartesian coordinate system shown. The vector field is defined on the spherical region $\mathfrak{B}$ of unit radius as shown in the sketch. Give an explicit expression for the unit normal vector field $\mathbf{n}(\mathbf{x})$ to the surface of the sphere. Compute the gradient of the vector field $\mathbf{v}(\mathbf{x})$. Compute the product $[\nabla\mathbf{v}]\mathbf{n}$, i.e., the gradient of the vector field acting on the normal vector. Compute the divergence of the vector field $\mathbf{v}(\mathbf{x})$. Compute the integral of $\operatorname{div}\mathbf{v}$ over the volume of the sphere. Compute the integral of $\mathbf{v} \cdot \mathbf{n}$ over the surface of the sphere.

34. Let $\mathbf{v}(\mathbf{x})$ be a vector field given by the following explicit function

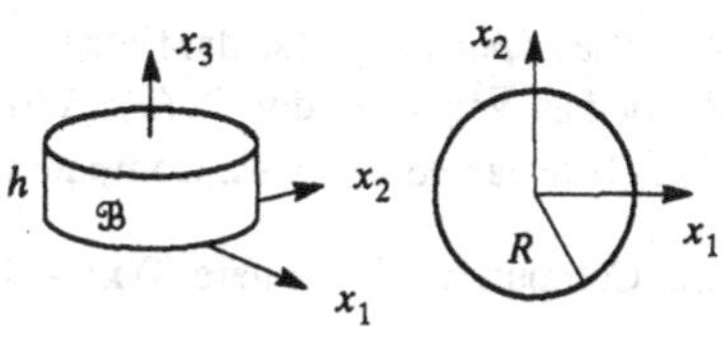

$$\mathbf{v}(\mathbf{x}) = (x_1\mathbf{e}_1 + x_2\mathbf{e}_2)\ln(x_1^2 + x_2^2)$$

where $\ln(\cdot)$ indicates the natural logarithm of $(\cdot)$. The vector field is defined on the cylindrical region $\mathfrak{B}$ of height h and radius R as shown in the sketch. Give an expression for the unit normal vector field $\mathbf{n}(\mathbf{x})$ to the for the cylinder (including the ends). Compute the divergence of the vector field $\mathbf{v}(\mathbf{x})$ and the integral of $\operatorname{div}\mathbf{v}$ over the volume of the cylinder.

35. Consider the scalar field $g(\mathbf{x}) = (\mathbf{x} \cdot \mathbf{x})^2$. Compute $\operatorname{div}\big[\nabla(\operatorname{div}[\nabla g(\mathbf{x})])\big]$.

36. Let $\mathbf{v}(\mathbf{x})$ be given by the following explicit function

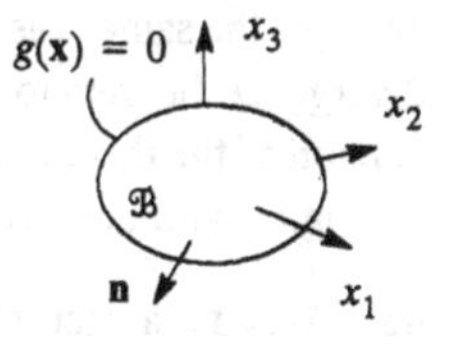

$$\mathbf{v}(\mathbf{x}) = (x_2 + x_3)\mathbf{e}_1 + (x_1 + x_3)\mathbf{e}_2 + (x_1 + x_2)\mathbf{e}_3$$

where $\mathbf{x}$ is the position vector of any point and has components $\{x_1, x_2, x_3\}$ relative to the Cartesian coordinate system as shown. The vector field is defined on the ellipsoidal region $\mathfrak{B}$ whose surface is described by the equation $g(\mathbf{x}) = 2x_1^2 + x_2^2 + 2x_3^2 - 4 = 0$. Give an expression for the unit normal vector field $\mathbf{n}(\mathbf{x})$ to the ellipsoid. Compute the gradient of the vector field $\mathbf{v}(\mathbf{x})$. Compute the product $[\nabla\mathbf{v}]\mathbf{n}$, i.e., the gradient of the vector field acting on the normal vector. Compute the divergence of the vector field $\mathbf{v}(\mathbf{x})$.

37. Evaluate the expression $\operatorname{div}\big[\nabla(\mathbf{x} \cdot \mathbf{A}\mathbf{x})\big]$, where $\mathbf{A}$ is a constant tensor (i.e., it does not depend upon $\mathbf{x}$), and the vector $\mathbf{x}$ has components $\mathbf{x} = x_i\mathbf{e}_i$. The derivatives are to be taken with respect to the independent variables x_i. Express the results in terms of the components of $\mathbf{A}$ and $\mathbf{x}$.

38. Let $g(\mathbf{x}) = e^{-\|\mathbf{x}\|^2}$ be a scalar field in three-dimensional space, where $\|\mathbf{x}\|$ is the distance from the origin to the point $\mathbf{x}$. Qualitatively describe the behavior of the function (a one- or two-dimensional analogy might be helpful). Compute the gradient ∇g of the field. Where does the gradient of the function go to zero?

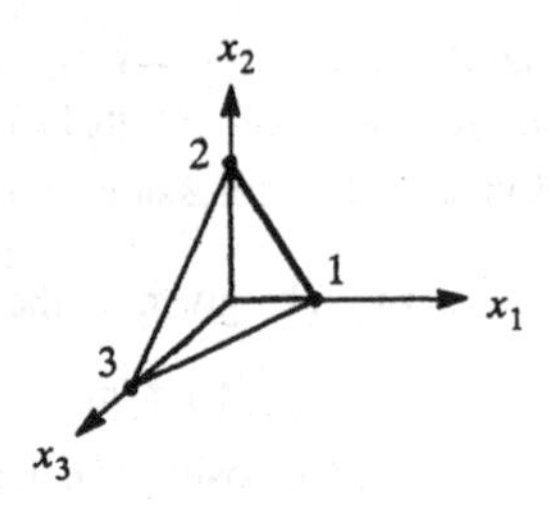

39. Consider a tensor field $\mathbf{T}$ defined on a tetrahedral region bounded by the coordinate planes $x_1 = 0$, $x_2 = 0$, $x_3 = 0$, and the oblique plane $6x_1 + 3x_2 + 2x_3 = 6$, as shown in the sketch. The tensor field has the particular expression $\mathbf{T} = \mathbf{b} \otimes \mathbf{x}$, where $\mathbf{b}$ is a constant vector and $\mathbf{x}$ is the position vector $\mathbf{x} = x_i\,\mathbf{e}_i$. Compute the integral of $\text{div}(\mathbf{T})$ over the volume and the integral of $\mathbf{Tn}$ over the surface of the tetrahedron (and thereby show that they give the same result, as promised by the divergence theorem). Note that the volume of the tetrahedron of the given dimensions is one.

40. Let $\mathbf{v}(\mathbf{x}) = \mathbf{x}$ on a spherical region of radius R, centered at the origin. Compute the integral of $\text{div}(\mathbf{v})$ over the volume of the sphere and compute the integral of the flux $\mathbf{v} \cdot \mathbf{n}$, where $\mathbf{n}$ is the unit normal to the sphere, over the surface of the sphere. Give the result in terms of the radius R. What does this calculation tell you about the ratio of surface area to volume of a sphere?

41. The *Laplacian* of a scalar field is a scalar measure of the second derivative of the field, defined as $\nabla^2 g(\mathbf{x}) \equiv \text{div}(\nabla g(\mathbf{x}))$. Write the component (index) form of the Laplacian of g in Cartesian coordinates. Compute the Laplacian of the scalar field of Problem 38.

42. Compute $\text{div}(\mathbf{T})$, where $\mathbf{T}(\mathbf{x}) = (\mathbf{x} \cdot \mathbf{x})\mathbf{I} - 2\mathbf{x} \otimes \mathbf{x}$ is a tensor field.

43. Let $\mathbf{u}(\mathbf{x})$, $\mathbf{v}(\mathbf{x})$, and $\mathbf{w}(\mathbf{x})$ be vector fields and let $\mathbf{T}(\mathbf{x})$ be a tensor field. Compute the component forms of the following derivatives of products of vectors

(a) $\nabla(\mathbf{u} \cdot \mathbf{v})$
(b) $\text{div}(\mathbf{u} \times \mathbf{v})$
(c) $\nabla(\mathbf{u} \times \mathbf{v})$
(d) $\text{div}(\mathbf{Tv})$
(e) $\nabla(\mathbf{u} \cdot \mathbf{Tv})$
(f) $\nabla(\mathbf{Tv})$
(g) $\text{div}(\mathbf{u} \otimes \mathbf{v})$
(h) $\text{div}([\mathbf{u} \otimes \mathbf{v}]\mathbf{w})$
(i) $\nabla[(\mathbf{u} \times \mathbf{v}) \cdot \mathbf{w}]$

44. Use the same reasoning that was used to derive the three-dimensional version of the divergence theorem to develop (a) a one-dimensional version, and (b) a two-dimensional version of the theorem. Use sketches to illustrate your definitions and draw any possible analogies with the three-dimensional case.

45. Consider a vector field $\mathbf{v}(\mathbf{x})$ on a region $\mathfrak{B}$ with surface Ω having unit normal field $\mathbf{n}$. The "curl" of the vector field can be defined as

$$\text{curl}(\mathbf{v}) \equiv \lim_{\mathcal{V}(\mathfrak{B})\to 0} \frac{1}{\mathcal{V}(\mathfrak{B})} \int_{\Omega} \mathbf{v} \times \mathbf{n}\, dA$$

Show (using the cuboid for $\mathfrak{B}$, as in the text) that the expression for $\text{curl}(\mathbf{v})$ is

$$\text{curl}(\mathbf{v}) = \frac{\partial \mathbf{v}}{x_i} \times \mathbf{e}_i = \left(\frac{\partial v_2}{\partial x_3} - \frac{\partial v_3}{\partial x_2}\right)\mathbf{e}_1 + \left(\frac{\partial v_3}{\partial x_1} - \frac{\partial v_1}{\partial x_3}\right)\mathbf{e}_2 + \left(\frac{\partial v_1}{\partial x_2} - \frac{\partial v_2}{\partial x_1}\right)\mathbf{e}_3$$

Note that many authors define the curl to be the negative of the definition given here, which is easily achieved by using the flux $\mathbf{n} \times \mathbf{v}$ instead. The form presented here seems to be more consistent with our other definitions of derivatives of vector fields.

46. Consider variously a scalar field $g(\mathbf{x})$, a vector field $\mathbf{v}(\mathbf{x})$, and a tensor field $\mathbf{T}(\mathbf{x})$ on a region $\mathfrak{B}$ with surface Ω with unit normal vector field $\mathbf{n}$. Prove the following theorems

$$\int_{\mathfrak{B}} \nabla g \, dV = \int_{\Omega} g \mathbf{n} \, dA, \qquad \int_{\mathfrak{B}} \nabla \mathbf{v} \, dV = \int_{\Omega} \mathbf{v} \otimes \mathbf{n} \, dA, \qquad \int_{\mathfrak{B}} \operatorname{div} \mathbf{T} \, dV = \int_{\Omega} \mathbf{T}\mathbf{n} \, dA$$

47. Use the divergence theorem for a vector field to show the following identities

(a) *Green's first identity* for scalar functions $u(\mathbf{x})$ and $v(\mathbf{x})$, (Hint: Let $\mathbf{v}(\mathbf{x}) = u \nabla v$)

$$\int_{\mathfrak{B}} \left(u\nabla^2 v + \nabla u \cdot \nabla v\right) dV = \int_{\Omega} \mathbf{n} \cdot \left(u\nabla v\right) dA$$

(b) *Green's second identity* for scalar functions $u(\mathbf{x})$ and $v(\mathbf{x})$, (Hint: Let $\mathbf{v}(\mathbf{x}) = u \nabla v - v \nabla u$)

$$\int_{\mathfrak{B}} \left(u\nabla^2 v - v\nabla^2 u\right) dV = \int_{\Omega} \mathbf{n} \cdot \left(u\nabla v - v\nabla u\right) dA$$

48. Many problems are more conveniently formulated and solved in cylindrical coordinates (r, θ, z). In cylindrical coordinates, the components of a vector $\mathbf{v}$ can be expressed as

$$\mathbf{v}(r, \theta, z) = v_r \mathbf{e}_r + v_\theta \mathbf{e}_\theta + v_z \mathbf{e}_z$$

where the components v_r, v_θ, and v_z are each functions of the coordinates (r, θ, z). However, now the base vectors $\mathbf{e}_r(\theta)$ and $\mathbf{e}_\theta(\theta)$ depend upon the coordinate θ. We must account for this dependence of the base vectors on the coordinates when computing derivatives of the vector field.

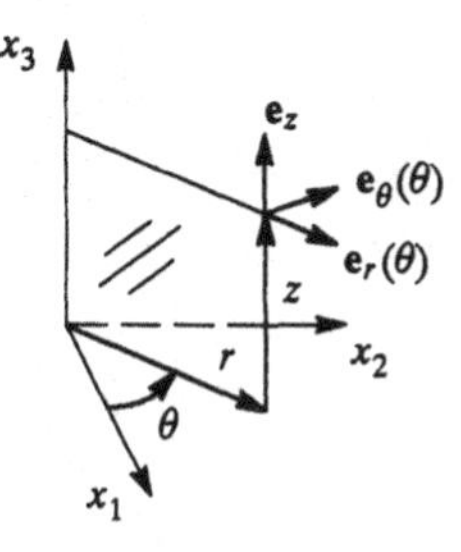

Using the coordinate-free definition of the divergence of a vector field, Eqn. (78), show that the divergence of $\mathbf{v}$ in cylindrical coordinates is given by

$$\operatorname{div} \mathbf{v}(r, \theta, z) = \frac{1}{r}\frac{\partial}{\partial r}\left(r v_r\right) + \frac{1}{r}\frac{\partial v_\theta}{\partial \theta} + \frac{\partial v_z}{\partial z}$$

(Hint: Observe from the figure that $\mathbf{n}_1 = \mathbf{e}_\theta(\theta + \Delta\theta)$ and $\mathbf{n}_2 = -\mathbf{e}_\theta(\theta)$ and are constant over the faces 1 and 2, respectively. The normal vectors $\mathbf{n}_3 = \mathbf{e}_r(\xi)$ and $\mathbf{n}_4 = -\mathbf{e}_r(\xi)$, with $\xi \in [\theta, \theta + \Delta\theta]$, vary over faces 3 and 4. Finally, note that $\mathbf{n}_5 = \mathbf{e}_z$ and $\mathbf{n}_6 = -\mathbf{e}_z$ are constant over faces 5 and 6.)

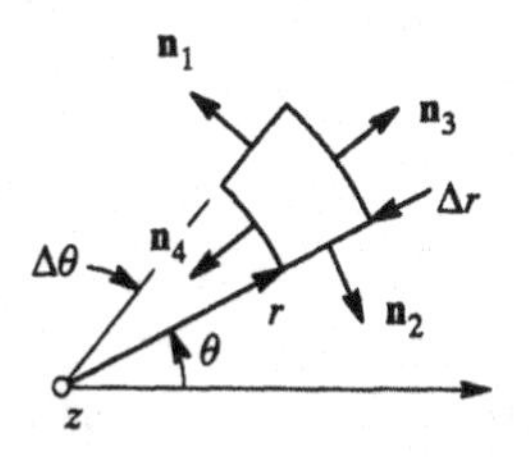

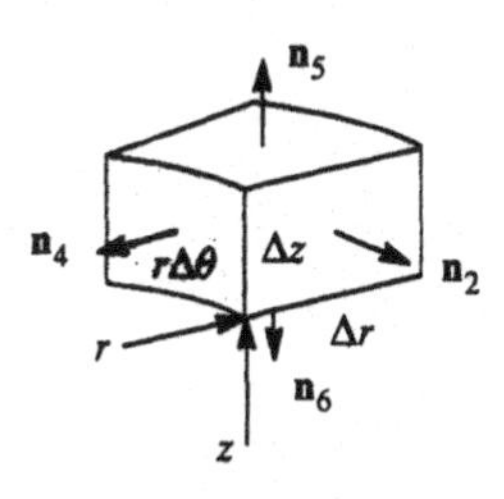

Note that the volume of the wedge is $\mathcal{V}(\mathfrak{B}) = r\Delta\theta\,\Delta r\,\Delta z$ plus terms of higher order that vanish more quickly in the limit as $\mathcal{V}(\mathfrak{B}) \rightarrow 0$.

2 The Geometry of Deformation

Mechanics is the study of bodies in motion. A solid body can be put into motion by any of a variety of actions. Among the most common causes of motion are the application of force or heat to the body. In general, a body in motion undergoes some combination of rigid motion (the distance between two particles does not change) and stretching (the distance between two particles changes). The motion can be either fast or slow. If the motion is slow enough then the resistance to motion caused by the inertia of the body (the so-called D'Alembert forces) can be neglected in the accounting of force equilibrium. We generally refer to this class of motions as *quasi-static*†. If the motion is fast enough then the forces associated with inertial resistance are not negligible and must be included in the accounting of force equilibrium. We refer to this class of problems as *dynamic*. In this book we confine our attention mostly to *quasi-static* motions.

The fundamental reason for studying motion is that the motion accommodates and influences the development of force in the body. One of the fundamental hypotheses of the mechanics of deformable bodies is that materials resist stretching in the sense that the distance between two points in a body can change, but it takes force to get the job done (i.e., to stretch the molecular bonds). In the study of motion we are interested in characterizing the part of

† It is not really possible to have a *static* motion because static implies no motion. Yet we think of static analysis, from the perspective of equilibrium, as a problem for which inertial forces are negligible or zero. Hence, it makes more sense, from the perspective of kinematics, to refer to the motion as *quasi-static*. It will sometimes be convenient to suppress the time dependence of the motion when we are considering the state at a moment in time.

the motion that gives rise to internal force (or stresses, which will be covered in Chapter 3). The relationship between force and deformation will depend upon the constitution of the material (which will be covered in Chapter 4).

Kinematics is the study of the motion of a body independent of the cause of that motion. In this chapter we shall focus on kinematics. One way to characterize motion is to describe the current position of each point in a body relative to the position that point occupied in some known reference configuration (often called the Lagrangian description of motion). We shall call the mathematical description of the motion a ***map*** or, since the motion will almost always include deformation, a ***deformation map.*** The goal of this chapter is to characterize the map and to analyze the deformation implied by the map. This analysis will lead us to the definition of ***strain,*** which will serve as one of the basic descriptors of deformation and will prove useful in the development of constitutive equations for materials.

We motivate our discussion by starting with the simple case of uniaxial motion of a (one-dimensional) rod†. The simple case will help to fix ideas and to connect with concepts from elementary strength of materials. Then, with the help of vector calculus, we generalize the concepts to three-dimensional solid bodies.

Uniaxial Stretch and Strain

Let us begin our discussion of the concept of strain by examining the deformation associated with the elongation of the thin rod shown in Fig. 26. It may be useful to think of the state of the rod at two different instants in time without worrying too much about what happened between those times (or how fast it happened). At the first instant, the rod has length ℓ_o. At the second instant, the rod has length ℓ. The difference in length is simply $\Delta\ell \equiv \ell - \ell_o$.

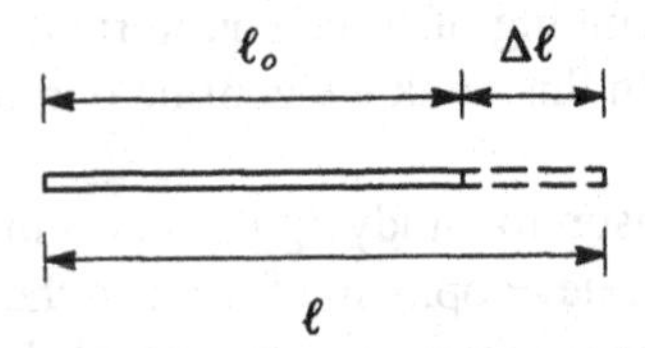

Figure 26 Elongation of a thin rod

There are only two things that can happen to the rod: (1) rigid-body motion (translation and rotation in space), and (2) change in length. Intuitively, the *deformation* or *strain* should be independent of any rigid-body motion of the rod because the rigid motion does not give rise to internal forces. Let us define the *stretch* of the rod as the ratio of deformed length to original length

† We assume that the rod is only capable of elongating and shortening, not bending.

$$\lambda \equiv \frac{\ell}{\ell_o} \tag{97}$$

The stretch is dimensionless and is equal to unity when the rod is the same length before and after the motion. Since change in length is the only part of the motion associated with deformation, and since the stretch completely captures the change in length, we should expect the stretch to completely characterize the deformation part of the motion.

In fact, any function of the stretch will be an acceptable measure of deformation or strain. Unfortunately, this observation implies that there is no unique measure of strain. Indeed, many definitions of strain are currently in use in engineering. For convenience, let us place two requirements on our definitions of strain: (1) The strain must be zero when the rod is unstretched, i.e., when $\lambda = 1$. (2) All strain measures should yield the same values in the neighborhood of $\lambda = 1$. In the following paragraphs we mention a few of the common definitions of strain used in solid and structural mechanics.

The *engineering strain* for the rod is measured as the change in length divided by the original length of the member

$$\varepsilon_{eng} \equiv \frac{\ell - \ell_o}{\ell_o} = \lambda - 1 \tag{98}$$

The *natural* or *"true" strain* is measured as the change in length divided by the deformed (or current) length of the member

$$\varepsilon_{true} \equiv \frac{\ell - \ell_o}{\ell} = 1 - \frac{1}{\lambda} \tag{99}$$

These two strain measures are familiar from elementary strength of materials. Part of the appeal of these strain measures is their linearity with respect to the stretch or its inverse.

The ***Lagrangian*** or ***Green strain*** is measured as half the difference in the squares of the deformed and undeformed lengths divided by the square of the undeformed length

$$E \equiv \frac{1}{2}\left(\frac{\ell^2 - \ell_o^2}{\ell_o^2}\right) = \frac{1}{2}(\lambda^2 - 1) \tag{100}$$

The Lagrangian way of formulating the strain has the same spirit as engineering strain in the sense that change in length is reckoned with respect to original length. Furthermore, we can see that when the change in length is small, i.e., $\lambda - 1 \ll 1$, these two measures of strain are equivalent. The desire to have this equivalence explains why we put the mysterious factor of 2 in the definition of Lagrangian strain. This observation is more evident when we write the Lagrangian strain as $\frac{1}{2}(\lambda - 1)(\lambda + 1)$. Now, when the stretch is very near to unity, $(\lambda + 1) \approx 2$ and $E \approx (\lambda - 1)$. As we shall see in this chapter, there is some theoretical advantage to taking the Lagrangian definition as the basic measure of strain as we generalize the concept to three dimensions and large deformations.

The *Eulerian* or *Almansi strain* is measured as half the difference in the squares of the undeformed and deformed lengths divided by the square of the deformed length

$$e = \frac{1}{2}\left(\frac{\ell^2 - \ell_o^2}{\ell^2}\right) = \frac{1}{2}\left(1 - \frac{1}{\lambda^2}\right) \tag{101}$$

Just as the Lagrangian strain is reminiscent of the engineering strain, the Eulerian strain is reminiscent of the natural strain. Again, the Eulerian definition has certain theoretical advantages over the true strain definition in generalizing to three dimensions and large deformations.

Finally, the *logarithmic strain* can be defined as

$$\varepsilon_{ln} = \ln(\lambda) \tag{102}$$

One interpretation of this measure of strain is to think of a continuous deformation process in which each step i has a change in length $\Delta\ell_i$ that can be divided by the current length ℓ_i to give an incremental true strain $\eta_i = \Delta\ell_i/\ell_i$. If these incremental strains are summed and if we take the limit as the size of the step becomes infinitesimally small we get

$$\varepsilon_{ln} = \lim_{\substack{N\to\infty \\ \Delta\ell_i\to 0}} \sum_{i=1}^{N} \frac{\Delta\ell_i}{\ell_i} = \int_{\ell_o}^{\ell} \frac{d\ell}{\ell} = \ln(\ell) - \ln(\ell_o) = \ln\left(\frac{\ell}{\ell_o}\right) \tag{103}$$

The various measures of uniaxial strain are summarized in Table 1. The choice of which strain measure to use is dictated by how we choose to describe the *constitutive law* governing the relationship between stress and strain in the material. All suitable measures of strain (including the four mentioned here, and many more) are basically equivalent in that they all attempt to characterize the same state of deformation. The difference in the measures of strain starts to show up when you use them to characterize induced stresses from constitutive equations.

Table 1 Different measures of uniaxial strain

Strain measure	Common designation
$\varepsilon_{eng} \equiv \lambda - 1$	Engineering strain
$E \equiv \frac{1}{2}(\lambda^2 - 1)$	Lagrangian or Green strain
$\varepsilon_{true} \equiv 1 - 1/\lambda$	Natural or "true" strain
$e \equiv \frac{1}{2}(1 - 1/\lambda^2)$	Eulerian or Almansi strain
$\varepsilon_{ln} \equiv \ln(\lambda)$	Logarithmic strain

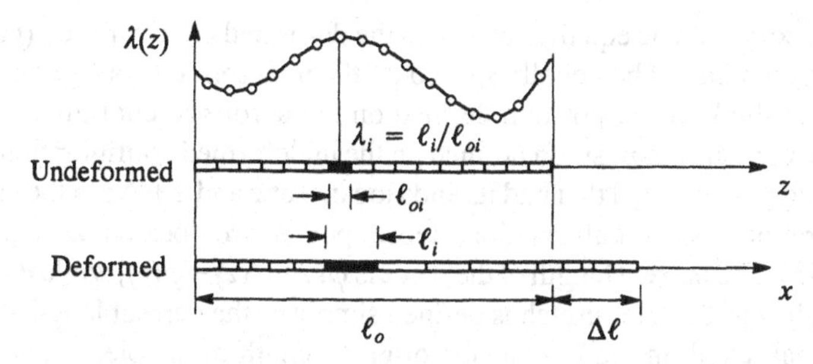

Figure 27 Nonuniform stretch of a thin rod

Nonuniform stretching. We have a couple of hurdles to clear in order to suitably generalize our one-dimensional characterizations of strain. The first hurdle regards the homogeneity of the strain state. This issue is present even for our one-dimensional rod, as shown in Fig. 27.

Let the undeformed rod be marked off in uniform subdivisions. The lack of uniformity of the stretch of the deformed rod is evident because the deformed pieces all have different lengths. The stretch of the *i*th piece can be computed as the ratio of the final length of the piece to the original length of the piece, $\lambda_i = \ell_i/\ell_{oi}$. These points are plotted in the figure. If we were to take the initial subdivision finer and finer, we would get more and more points describing the variation of the stretch along the length of the rod. In the limit, as the number of subdivisions goes to infinity, the description of the stretch approaches the continuous function $\lambda(z)$, where z is the measure of distance in the undeformed configuration.

The deformed length of the rod can be computed as the integral of the stretch over the original length

$$\ell = \int_0^{\ell_o} \lambda(z)\, dz \tag{104}$$

If the deformation is homogeneous, and, hence, $\lambda(z) = \bar{\lambda}$, then Eqn. (104) gives $\bar{\lambda} = \ell/\ell_o$, as we expect. The limits of integration of 0 and ℓ_o make sense because the stretch $\lambda(z)$ is defined as a function of position measured in the undeformed configuration, as shown in Fig. 27.

The independent variable z measures distance linearly in the undeformed configuration, i.e., if you made marks at equal increments of z those marks would be equally spaced on the undeformed configuration. We actually did make those marks in Fig. 27. Notice that the marks are not equally spaced on the deformed configuration because of the nonuniformity of stretching. We could also define an independent variable x that measures distance linearly in the deformed configuration, i.e., if you made a mark at equal increments of x, those marks would be equally spaced on the deformed configuration (but then the marks would not be equally spaced on the undeformed configuration).

We can think of the position defined on the deformed configuration as a function $x = \phi(z)$ of position defined on the undeformed configuration. Consider a segment of a rod that had its ends located at z and $z+\Delta z$ in the original configuration. After deformation, those points are located at $\phi(z)$ and $\phi(z+\Delta z)$. The current length of the piece is $\phi(z+\Delta z)-\phi(z)$, while the original length was Δz. The stretch is defined simply as the current length divided by original length in the limit as the original length of the piece approaches zero. To wit,

$$\lambda(z) \equiv \lim_{\Delta z \to 0} \frac{\phi(z+\Delta z)-\phi(z)}{\Delta z} = \frac{d\phi}{dz} \tag{105}$$

that is, the stretch is the derivative of the function $\phi(z)$ that maps the original coordinate z to the deformed coordinate x. We can view the computation of the current length of the finite bar through the rule for change in variables for integration

$$\ell = \int_0^{\ell} dx = \int_0^{\ell_o} \frac{d\phi}{dz}\, dz = \int_0^{\ell_o} \lambda\, dz \tag{106}$$

which is the same result as Eqn. (104). In this setting, we can think of λ as the Jacobian of the change in variable $dx = \lambda\, dz$.

The introduction of the stretching function $\lambda(z)$ provides a suitable generalization of the definition in Eqn. (97) from homogeneous deformation to nonhomogeneous deformation. In the same sense, we can generalize the concept of strain to nonhomogeneous deformations simply by substituting $\lambda(z)$ into each of the definitions of strain. Thus, we generally think of stretch and strain as properties associated with a point in the bar, and not as properties of the whole bar. This perspective, called *localization*, is central to our study of the geometry of deformation.

The second issue we must face in characterizing deformation is the extension of the concept of strain to three-dimensional solid bodies. This issue is the primary focus of this chapter. We shall see that our definitions of one-dimensional stretch and strain play an important role in three dimensions.

The Deformation Map

The description of the geometry of deformation must begin with a description of the body in question. For our purposes it is sufficient to imagine a continuous, solid body located in three-dimensional space. We must be able to completely characterize the geometry of the body in some configuration in order to make any headway in describing the geometry of deformation. We will call

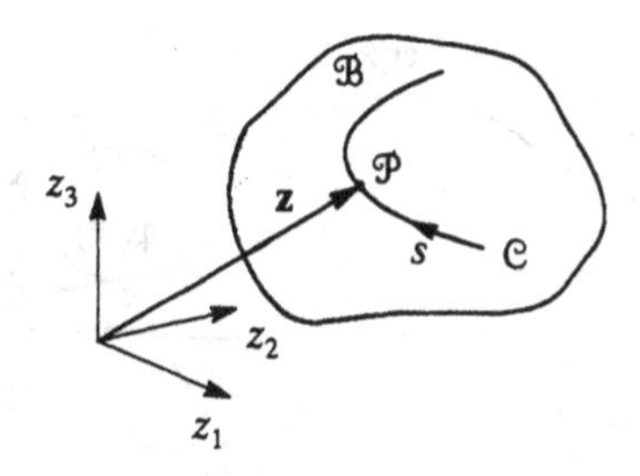

Figure 28 A solid, continuous body in three-dimensional space

the known geometry the *reference configuration*. The reference configuration is often taken to be the unstressed and unstrained configuration of the body, although such a restriction is not necessary. Our primary assumption about the initial geometry is that in this configuration we know the position of every point in the body. A second, equally crucial, assumption is that the body is continuous (as opposed to, say, a collection of discrete particles). The assumption of continuity will allow us to use the tools of differential geometry.

Our prototypical body $\mathfrak{B}$ is illustrated in Fig. 28. The initial geometry has two basic features: the *domain*, which is everything inside the body, and the *boundary*, which is the surface of the body. In the reference configuration $\mathfrak{B}$ we can locate the position of a point, say point $\mathfrak{P}$, by giving its coordinates $\{z_1, z_2, z_3\}$ relative to the origin of the coordinate system. The vector pointing from the origin of coordinates to the point $\mathfrak{P}$ is called the *position vector* $\mathbf{z}$.

Imagine a curve $\mathfrak{C}$ running through the body. For the sake of discussion, let us imagine that we have marked the material along this curve (for example, with a radioactive marker that allows us to see its position with a device like an X-ray machine) during its formation. The curve can be parameterized by a measure of its arc length s. (Imagine that you are an ant walking along the curve. The parameter s is the value that you read on your pedometer as you travel along.) The curve is an important geometric construct because it will provide a connection with our one-dimensional ideas of stretch and strain. There are infinitely many curves passing through the point $\mathfrak{P}$, each one distinguished by its direction at $\mathfrak{P}$. The direction that a curve is heading at any instant is the direction tangent to the curve. As we shall see, these tangent directions will play a key role in the description of strain.

Let us assume that we can characterize the deformation of a body $\mathfrak{B}$ with a *deformation map* $\boldsymbol{\phi}(\mathbf{z})$ as shown in Fig. 29. The deformation map takes the position vector $\mathbf{z}$ and locates the position of that same point in the *deformed configuration* $\boldsymbol{\phi}(\mathfrak{B})$ as

$$\boxed{\mathbf{x} = \boldsymbol{\phi}(\mathbf{z})} \qquad (107)$$

relative to the coordinate system $\{x_1, x_2, x_3\}$. Note that our point $\mathfrak{P}$ is designated as $\boldsymbol{\phi}(\mathfrak{P})$ in the deformed configuration and represents the same material

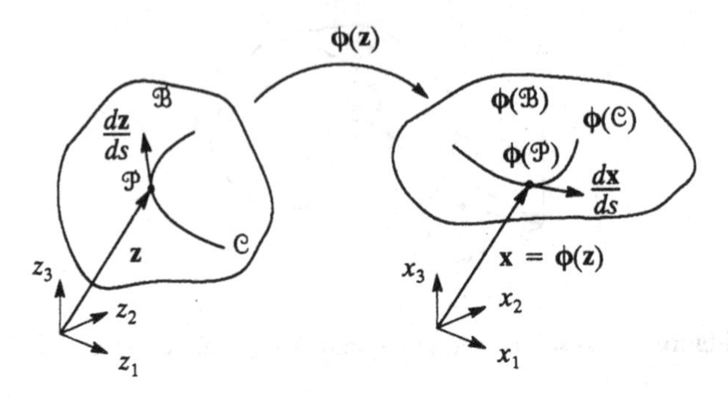

Figure 29 The deformation map

point as $\mathcal{P}$ in the sense that, if we could make a mark at point $\mathcal{P}$ before deformation, we would find that mark at point $\phi(\mathcal{P})$ after deformation. We might also wish to see the effect of the deformation map on an imaginary curve $\mathcal{C}$. A curve is a parameterized sequence of points $\mathbf{z}(s)$, where the scalar variable s measures distance along the curve. We designate the position of the curve in the deformed configuration as $\phi(\mathcal{C})$. The deformed curve is also described by a parameterized sequence of points $\mathbf{x}(s) = \phi(\mathbf{z}(s))$. Note that, like our one-dimensional example, the distance measure s always refers to distance in the undeformed configuration.

The concept of the map is a familiar one. Anyone who has traveled has probably used a map. The cartographer's map is a functional representation of position. Each position on a flat map of the world represents a particular location on the surface of the Earth. A road or a river on a map is analogous to our curve $\mathcal{C}$. The main function of a cartographer's map is to scale down areas so that a region can fit within the confines of a piece of paper. The ideal map would only scale; however, as everyone knows, maps tend to distort areas and distances. Have you ever wondered why Greenland often appears to be as large as South America on some flat world maps? The distortion on a cartographer's map is caused by the function used to map points from the surface of the Earth to the piece of paper or globe. Some mapping functions preserve areas; some preserve straight lines; some preserve none of the above. The amount of distortion is implicit in the mapping function.

The deformation map $\phi(\mathbf{z})$ is very much like a cartographer's map in the sense that it unambiguously locates the position of points on the deformed configuration of the body. It is unlike the cartographer's map in the sense that the mapping function is dictated by the physical processes driving deformation. In mechanics, our aim is often to determine the map from data like applied forces and laws of nature. Whether a cartographer's map or a deformation map, the concept of mapping gives us a way of organizing the process of relating the location of points in two configurations.

The Stretch of a Curve

Our imaginary curve is a good starting point for the definition of strain in a three-dimensional body because we can examine the change in length of this line under the action of the deformation map. From elementary considerations we already know what strain means for the stretching of a line. The arbitrariness of the choice of our curve will allow us to generalize our concept of strain to three dimensions. Let us examine the change in length of the curve $\mathcal{C}$ between two points.

Consider two points on the curve $\mathcal{C}$, one described by the position vector $\mathbf{z}(s)$ and the other by the position vector $\mathbf{z}(s+\Delta s)$, as shown in Fig. 30. The vector connecting the first point to the second is $\Delta\mathbf{z} \equiv \mathbf{z}(s+\Delta s)-\mathbf{z}(s)$, and the length of this vector measures the straight-line distance between the two points. The two points are mapped to the positions $\mathbf{x}(s)$ and $\mathbf{x}(s+\Delta s)$, respectively, in the deformed configuration. The vector connecting the two points in the deformed configuration is $\Delta\mathbf{x} \equiv \mathbf{x}(s+\Delta s)-\mathbf{x}(s)$, and the length of this vector measures the straight-line distance between the two points. In the limit as $\Delta s \rightarrow 0$, the straight-line distance between two points and the distance measured along the arc become equal. Hence, in the limit, the lengths of the vectors $\Delta\mathbf{z}$ and $\Delta\mathbf{x}$ are appropriate measures of the lengths of the respective curves.

In the limit as $\Delta s \rightarrow 0$, the length of the vector $\mathbf{z}(s+\Delta s)-\mathbf{z}(s)$ approaches zero, but the ratio of the length of the vector to the length of the arc approaches unity. Taking the limit of this ratio as $\Delta s \rightarrow 0$, we obtain the expression for the *tangent vector* to the curve

$$\lim_{\Delta s\rightarrow 0} \frac{\mathbf{z}(s+\Delta s)-\mathbf{z}(s)}{\Delta s} = \frac{d\mathbf{z}}{ds} \tag{108}$$

Thus, the derivative of a position vector along a curve is always tangent to the curve. If it is normalized with respect to the measure of distance along the curve, then it is always a unit vector because as $\Delta s \rightarrow 0$, the secant line length approaches the arc length, i.e., $\Delta s \rightarrow \| \mathbf{z}(s+\Delta s)-\mathbf{z}(s) \|$.

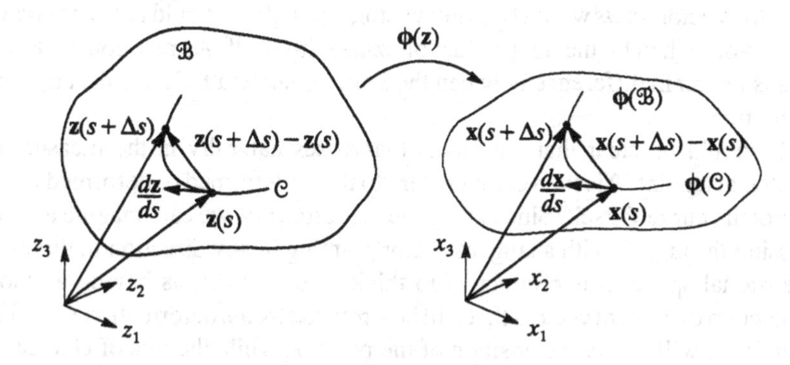

Figure 30 Measuring the distance between two points on a curve

The vector $\mathbf{x}(s+\Delta s)-\mathbf{x}(s)$ can also be normalized with respect to the length of its arc. However, it will be more useful to compute the limit of the ratio of this vector to the length of the arc in the undeformed configuration. This is possible because there is a one-to-one mapping between corresponding points on the original and deformed curves. Thus, we know where to put the head and tail of the vector pointing along the secant of the deformed curve, and we know the corresponding length of arc for the undeformed curve. The result is the tangent vector to the deformed curve

$$\lim_{\Delta s \to 0} \frac{\mathbf{x}(s+\Delta s)-\mathbf{x}(s)}{\Delta s} = \frac{d\mathbf{x}}{ds} \tag{109}$$

The vector $d\mathbf{x}/ds$ is tangent to the deformed curve because it is the limit of secant lines of points taken closer and closer together. The vector $d\mathbf{x}/ds$ does not have unit length because it is reckoned with respect to the arc length between the two points in the undeformed configuration. Since $\| \mathbf{x}(s+\Delta s)-\mathbf{x}(s) \|$ is the length of the secant line between two points on the curve, it approaches the length of the actual deformed curve in the limit as $\Delta s \to 0$. The ratio of $\| \mathbf{x}(s+\Delta s)-\mathbf{x}(s) \|$ to Δs is, therefore, the ratio of the length of the deformed curve to the length of the original curve. Consequently

$$\lim_{\Delta s \to 0} \frac{\| \mathbf{x}(s+\Delta s)-\mathbf{x}(s) \|}{\Delta s} = \lambda(s) \tag{110}$$

Since the square of the length of a vector $\mathbf{v}$ is given by $\| \mathbf{v} \|^2 = \mathbf{v} \cdot \mathbf{v}$, Eqns. (109) and (110) suggest that the square of the stretch of the curve $\mathbf{z}(s)$ is given by the dot product of the tangent vectors in the deformed configuration

$$\lambda^2(s) = \left(\frac{d\mathbf{x}}{ds}\right) \cdot \left(\frac{d\mathbf{x}}{ds}\right) \tag{111}$$

Remark. We could just as easily parameterize the curve with a measure of distance along the curve in the deformed configuration. If this distance is used, then the vector $d\mathbf{x}/ds$ would be a unit vector, while $d\mathbf{z}/ds$ would not. The stretch $1/\lambda^2$ would then be the dot product of $d\mathbf{z}/ds$ with itself. As mentioned earlier, this is the main difference between the Lagrangian and Eulerian descriptions of motion.

It should be clear that the object that arises naturally in the measure of stretch at a point $\mathcal{P}$ is the vector tangent to the undeformed or deformed curve $\mathcal{C}$, not the curve $\mathcal{C}$ itself. Since the curve $\mathcal{C}$ is arbitrary, we can imagine a curve passing through $\mathcal{P}$ with a tangent vector pointing in any direction in three-dimensional space. It is productive to think of each point as having a whole collection of tangent vectors that will be stretched by the deformation map. The map itself will locate the position of the point $\mathcal{P}$, while the rate of change, or gradient, of the map will tell us how the tangent vectors stretch.

The Deformation Gradient

Noting that the parameterized curve in the deformed configuration is determined by the deformation map as $\mathbf{x}(s) = \boldsymbol{\phi}(\mathbf{z}(s))$, one can apply the chain rule for differentiation to relate the vectors tangent to the curves in the deformed and undeformed configurations. In components, noting that the map can be written $x_i(s) = \phi_i\big(z_1(s), z_2(s), z_3(s)\big)$, we can compute the derivative by the chain rule as

$$\frac{dx_i(s)}{ds} = \frac{\partial \phi_i(\mathbf{z})}{\partial z_j} \frac{dz_j(s)}{ds} \tag{112}$$

Note that the partial derivatives $\partial\phi_i / \partial z_j$ are simply the components of the tensor $\nabla\boldsymbol{\phi}$. This tensor plays such an important role in the subsequent developments that we shall give it a special name and symbol. We call

$$\boxed{\mathbf{F}(\mathbf{z}) \equiv \nabla\boldsymbol{\phi}(\mathbf{z})} \tag{113}$$

the *deformation gradient* because it characterizes the rate of change of deformation with respect to the material coordinates $\mathbf{z}$. With this notation, Eqn. (112) can be written in direct notation as

$$\boxed{\frac{d\mathbf{x}}{ds} = \mathbf{F}\frac{d\mathbf{z}}{ds}} \tag{114}$$

The deformation gradient carries the information about the stretching of the domain in the infinitesimal neighborhood of the point $\mathbf{z}$. It also carries information about the rotation of the vector $d\mathbf{z}/ds$. We will often dispense with the notion of the curve and its parameterization and simply refer to the tangent vector as $\mathbf{n}$ or $\mathbf{t}$ or some such notation. Many authors like to use the notation $d\mathbf{z}$ to refer to tangent vectors in the undeformed configuration and $d\mathbf{x}$ to refer to tangent vectors in the deformed configuration. With this notation, the deformation gradient operates as $d\mathbf{x} = \mathbf{F}\,d\mathbf{z}$. This notation has the advantage of reminding us that the tangent vector represents the rate of change of a position vector, but it also hides the role of the arbitrary curve.

The deformation gradient $\mathbf{F}$ is a tensor with the coordinate representation

$$\mathbf{F} \equiv \frac{\partial \phi_i(\mathbf{z})}{\partial z_j}\big[\mathbf{e}_i \otimes \mathbf{g}_j\big] \tag{115}$$

where $\{\mathbf{e}_i\}$ are the base vectors in the deformed configuration and $\{\mathbf{g}_i\}$ are the base vectors in the undeformed configuration. The deformation gradient is often called a two-point tensor because the basis $\mathbf{e}_i \otimes \mathbf{g}_j$ has one leg in the undeformed configuration and one in the deformed configuration. The need for this distinction is clear when we recall that a tensor is an object that operates on a vector to produce another vector. According to Eqn. (114), $\mathbf{F}$ operates on unit tangent vectors in the undeformed configuration to produce tangent vectors in the deformed configuration. Let the kth component of $d\mathbf{z}/ds$ be defined as $n_k \equiv dz_k/ds$, and consider the following component computation

$$\begin{aligned}\mathbf{F}\frac{d\mathbf{z}}{ds} &= F_{ij}\big[\mathbf{e}_i \otimes \mathbf{g}_j\big] n_k \mathbf{g}_k \\ &= F_{ij} n_k \big[\mathbf{e}_i \otimes \mathbf{g}_j\big]\mathbf{g}_k \\ &= F_{ij} n_j \mathbf{e}_i\end{aligned}$$

since $[\mathbf{e}_i \otimes \mathbf{g}_j]\mathbf{g}_k = \delta_{jk}\mathbf{e}_i$ by the definition of the tensor product of vectors. The most natural basis for the vector $d\mathbf{z}/ds$ is $\{\mathbf{g}_i\}$ because the undeformed configuration is defined in that coordinate system. The most natural basis for the vector $d\mathbf{x}/ds$ is $\{\mathbf{e}_i\}$ because the position $\mathbf{x}$ is defined on the deformed configuration. We can see from the previous construction that $F_{ij}n_j\mathbf{e}_i$ is a vector defined on the deformed configuration, as it should be, so the components of the tangent to the deformed curve are $dx_i/ds = F_{ij}n_j$.

It should be clear that the deformation gradient $\mathbf{F}(\mathbf{z})$ is a function of position in the body, since the mapping function will generally not be uniform. To economize the notation we often will not show the explicit functional dependence, and will simply refer to $\mathbf{F}$ with the understanding that it depends on $\mathbf{z}$.

Strain in Three-dimensional Bodies

The stretch of a curve at a point is the ratio of the deformed length of the curve to the original length of that curve, in the neighborhood of the point in question. Let us consider an infinitesimal length of curve in the neighborhood of the point $\mathcal{P}$. The length of the curve is proportional to the length of the tangent vector at that point. Since the length of the tangent vector $\mathbf{n}$ in the undeformed configuration is unity, the stretch is simply the length of the tangent vector $\mathbf{Fn}$ in the deformed configuration. The stretch can be expressed in terms of the deformation gradient by substituting Eqn. (114) into Eqn. (111) as

$$\lambda^2(\mathbf{n}) = \mathbf{Fn}\cdot\mathbf{Fn} \tag{116}$$

for any unit vector $\mathbf{n}$ in the undeformed configuration. From the definition of the transpose of a tensor, we have $\mathbf{Fn}\cdot\mathbf{Fn} = \mathbf{n}\cdot\mathbf{F}^T\mathbf{Fn}$. Let us introduce the *Green deformation tensor* $\mathbf{C}$, defined to be the composition of the transpose of $\mathbf{F}$ operating on $\mathbf{F}$ as follows

$$\boxed{\mathbf{C} \equiv \mathbf{F}^T\mathbf{F}} \tag{117}$$

The stretch of a line oriented in the direction $\mathbf{n}$ in the undeformed configuration can then be computed as

$$\boxed{\lambda^2(\mathbf{n}) = \mathbf{n}\cdot\mathbf{Cn}} \tag{118}$$

Equation (118) holds for any curve with $d\mathbf{z}/ds = \mathbf{n}$, and, hence, enables us to compute the stretch in any direction at a given point.

Recall that our definition of Lagrangian strain is the difference between the square of the deformed length and the square of the original length divided by twice the square of the original length. We can use the same definition for strain in the direction $\mathbf{n}$ as follows

$$E(\mathbf{n}) = \tfrac{1}{2}\left[\lambda^2(\mathbf{n}) - 1\right] \equiv \mathbf{n} \cdot \mathbf{E}\mathbf{n} \tag{119}$$

where the *Lagrangian strain tensor* $\mathbf{E}$ is defined to be half the difference between the Green deformation tensor and the identity tensor $\mathbf{I}$ as follows

$$\boxed{\mathbf{E} \equiv \tfrac{1}{2}\left[\mathbf{C} - \mathbf{I}\right]} \tag{120}$$

A straightforward computation will demonstrate the validity of Eqns. (119) and (120).

Examples

As a bit of relief from all of the preceding abstraction, let us consider some specific cases of deformation maps. Four simple cases of deformation, their deformation maps, and the corresponding deformation gradients are given below. As an exercise, compute the Green deformation tensor $\mathbf{C}$ and the Lagrangian strain tensor $\mathbf{E}$ for each case. These maps are all two dimensional in the sense that there is no action in the third coordinate direction. Assume that each geometric figure has unit thickness. Throughout these examples we will take the base vectors in the deformed configuration to be the same as the base vectors in the undeformed configuration, i.e., $\{\mathbf{g}_i\} = \{\mathbf{e}_i\}$.

Example 9. *Simple extension.* The deformation map for simple, homogenous extension in the z_1 direction is shown in Fig. 31. The explicit mathematical expression for the map is

$$\boldsymbol{\phi}(\mathbf{z}) = (1+\beta)z_1\mathbf{e}_1 + z_2\mathbf{e}_2 + z_3\mathbf{e}_3$$

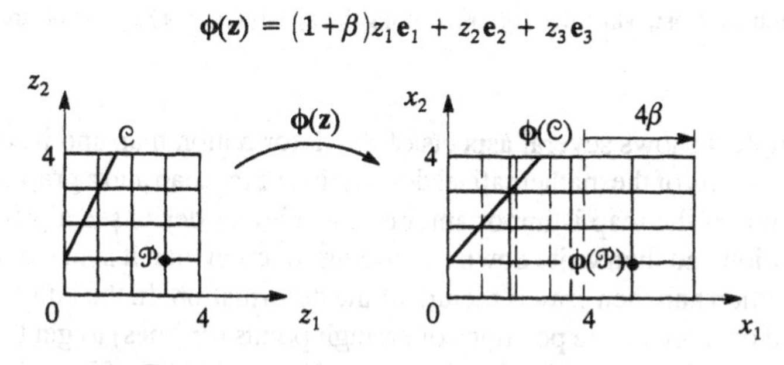

Figure 31 The map for simple extension

Let us examine the action of the map. Each point in the undeformed configuration moves to a point in the deformed configuration with coordinates

$$x_1 = (1+\beta)z_1, \quad x_2 = z_2, \quad x_3 = z_3$$

For example, the point $\mathcal{P}$ is initially located at $\mathbf{z} = (3, 1, 0)$. After deformation it is located at position $\mathbf{x} = (3+3\beta, 1, 0)$.

The curve $\mathcal{C}$ in the undeformed configuration, shown in the figure, has the equation $z_2 = 1+2z_1$. To find the equation of the curve in the deformed configuration we must invert the map, that is, solve for $\mathbf{z}$ in terms of $\mathbf{x}$. For this map this operation is straightforward, yielding

$$z_1 = \frac{x_1}{1+\beta}, \quad z_2 = x_2, \quad z_3 = x_3$$

Substituting these expressions into the equation of the original line, we get the equation for the line in the deformed configuration $\phi(\mathcal{C})$, $x_2 = 1+2x_1/(1+\beta)$. Since the equation of the curve in the deformed configuration is linear, we conclude that the map deforms straight lines into straight lines (we proved it for one line, at least).

The deformation gradient can be computed from the map as

$$\mathbf{F} = \mathbf{I} + \beta[\mathbf{e}_1 \otimes \mathbf{e}_1]$$

The stretch in the direction of the coordinate axes can be computed by noting

$$\mathbf{F}\mathbf{e}_1 = (1+\beta)\mathbf{e}_1, \quad \mathbf{F}\mathbf{e}_2 = (1)\mathbf{e}_2, \quad \mathbf{F}\mathbf{e}_3 = (1)\mathbf{e}_3$$

(note that the base vectors are eigenvectors of $\mathbf{F}$). The stretches are

$$\lambda^2(\mathbf{e}_1) = \mathbf{F}\mathbf{e}_1 \cdot \mathbf{F}\mathbf{e}_1 = (1+\beta)\mathbf{e}_1 \cdot (1+\beta)\mathbf{e}_1 = (1+\beta)^2$$

Therefore, $\lambda(\mathbf{e}_1) = 1+\beta$. The stretches in the other two directions can be computed similarly to show that $\lambda(\mathbf{e}_2) = 1$ and $\lambda(\mathbf{e}_3) = 1$. A line oriented at an angle θ from the z_1 axis points in the direction $\mathbf{n} = \cos\theta\,\mathbf{e}_1 + \sin\theta\,\mathbf{e}_2$. Since we have $\mathbf{F}\mathbf{n} = (1+\beta)\cos\theta\,\mathbf{e}_1 + \sin\theta\,\mathbf{e}_2$, the square of the stretch is

$$\lambda^2(\mathbf{n}) = \mathbf{F}\mathbf{n} \cdot \mathbf{F}\mathbf{n} = (1+\beta)^2\cos^2\theta + \sin^2\theta$$

Hence, we can observe that not all lines stretch by the same amount. The stretch depends upon the orientation of the line. Lines oriented along the z_1 axis ($\theta = 0$) stretch the most, while lines oriented along the z_2 axis ($\theta = \pi/2$) do not stretch.

Example 9 shows several aspects of the deformation map and its analysis. The correlation of the mathematical description of the map and a graphical representation of the map is important, but it is a lot harder to see a picture of a deformation and then write down a mapping function than it is to have a mapping function and then draw a picture of the deformation. In the latter instance one need only locate the positions of enough points (or lines) to get the gist of the mapping. This example also illustrates the simple idea of how the stretch at a given point (in this case all points experience the same deformation, as evidenced by a constant **F** tensor) varies with direction. Note that we did not explicitly compute the components of the tensor **C** to carry out our computations of the stretch.

The next example is also a homogeneous deformation, but has the feature that it couples the motion in the two directions.

Example 10. *Simple shear.* The deformation map for simple, homogenous shearing in the z_1 direction is shown in Fig. 32.

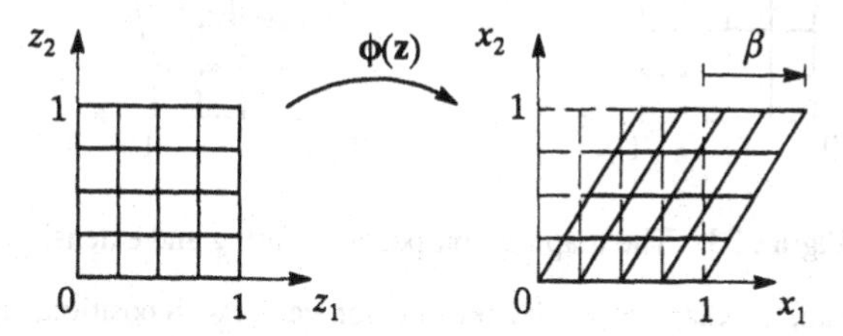

Figure 32 The map for simple shear

The explicit expression for the map is

$$\phi(\mathbf{z}) = (z_1 + \beta z_2)\mathbf{e}_1 + z_2\mathbf{e}_2 + z_3\mathbf{e}_3$$

The action of this map is to shear the block. Lines parallel to the z_1 axis do not rotate but they do translate relative to their original positions. Lines parallel to the z_2 axis rotate.

The deformation gradient is $\mathbf{F} = \mathbf{I} + \beta[\mathbf{e}_1 \otimes \mathbf{e}_2]$. The stretch in the direction of the coordinate axes can be computed by noting

$$\mathbf{F}\mathbf{e}_1 = (1)\mathbf{e}_1, \quad \mathbf{F}\mathbf{e}_2 = \beta\mathbf{e}_1 + \mathbf{e}_2, \quad \mathbf{F}\mathbf{e}_3 = (1)\mathbf{e}_3$$

(Note that $\mathbf{e}_2$ is *not* an eigenvector of **F**). The stretch of $\mathbf{e}_2$ is

$$\lambda^2(\mathbf{e}_2) = \mathbf{F}\mathbf{e}_2 \cdot \mathbf{F}\mathbf{e}_2 = (\beta\mathbf{e}_1 + \mathbf{e}_2) \cdot (\beta\mathbf{e}_1 + \mathbf{e}_2) = 1 + \beta^2$$

$$\Rightarrow \quad \lambda(\mathbf{e}_2) = \sqrt{1+\beta^2}$$

Stretches in other directions can be computed in a similar fashion. For example, the direction $\mathbf{n} = \cos\theta\mathbf{e}_1 + \sin\theta\mathbf{e}_2$, with θ measured from the z_1 axis, gives

$$\lambda^2(\mathbf{n}) = \mathbf{F}\mathbf{n} \cdot \mathbf{F}\mathbf{n} = (\cos\theta + \beta\sin\theta)^2 + \sin^2\theta$$

Both of the previous examples are *linear maps*. A linear map is one that has a constant deformation gradient **F**. Such a motion is called a homogeneous deformation because the state of strain is the same for each point in the body. It is always simple to invert a linear map to give **z** as a function of **x**. It is usually not possible to find an inverse map in closed form for a nonlinear map, although if $\det \mathbf{F} > 0$ everywhere the implicit function theorem guarantees that an inverse mapping exists. The next two examples are not linear maps.

Example 11. *Compound shearing and extension.* A more complicated deformation map, the map for compound shearing and extension, is shown in Fig. 33.

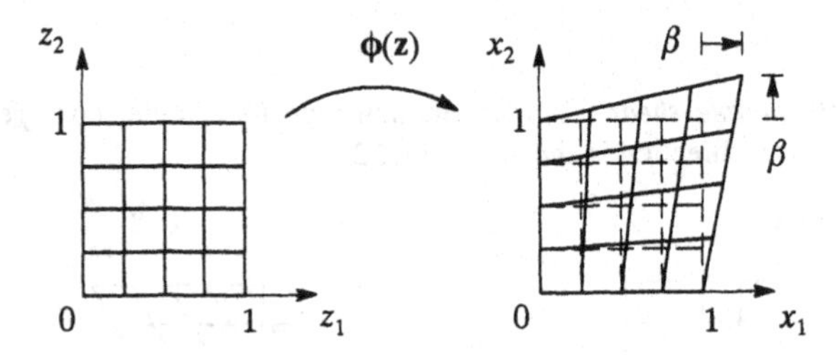

Figure 33 The map for compound shearing and extension

For this map, the character of the deformation varies with position, and the deformation gradient is a function of position. The expression for the map is

$$\boldsymbol{\phi}(\mathbf{z}) = (z_1 + \beta z_1 z_2)\mathbf{e}_1 + (z_2 + \beta z_1 z_2)\mathbf{e}_2 + z_3 \mathbf{e}_3$$

The components of the deformation gradient are

$$\mathbf{F} \sim \begin{bmatrix} 1+\beta z_2 & \beta z_1 & 0 \\ \beta z_2 & 1+\beta z_1 & 0 \\ 0 & 0 & 1 \end{bmatrix}$$

A few attempts to invert this map, i.e., put the map in the form $\mathbf{z} = \boldsymbol{\psi}(\mathbf{x})$, should convince the reader that the business of explicitly inverting a deformation map gets difficult even for some rather simple maps. It is less difficult to imagine that this inverse exists because it is quite clear from the picture that the mapping of every point from the undeformed to deformed configuration is unique. Hence, one should be able to reverse the map, or, in other words, find the place where a point on the deformed configuration came from.

Example 12. *Pure bending.* An even more complicated deformation map, a map for pure bending, is shown in Fig. 34.
The beam is bent until the cross section at the right end reaches around to just touch the cross section at the left end. The explicit expression for the map is

$$\boldsymbol{\phi}(\mathbf{z}) = \big((1-z_2)\sin z_1\big)\,\mathbf{e}_1 + \big(1-(1-z_2)\cos z_1\big)\,\mathbf{e}_2 + z_3\mathbf{e}_3$$

and the components of the deformation gradient are

$$\mathbf{F} \sim \begin{bmatrix} (1-z_2)\cos z_1 & -\sin z_1 & 0 \\ (1-z_2)\sin z_1 & \cos z_1 & 0 \\ 0 & 0 & 1 \end{bmatrix}$$

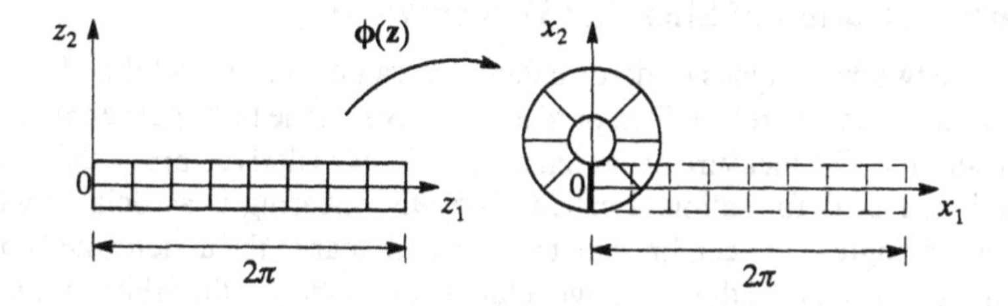

Figure 34 The map for pure bending of a strip into a circle

The action of this map is to take lines parallel to the z_1 axis and deform them into perfect circles. Of course, the only way such a deformation can be accomplished is if those lines change length. We can see how lines stretch by examining the components of the tensor $\mathbf{C} = \mathbf{F}^T\mathbf{F}$ given by

$$\mathbf{C} \sim \begin{bmatrix} (1-z_2)^2 & 0 & 0 \\ 0 & 1 & 0 \\ 0 & 0 & 1 \end{bmatrix}$$

The lines $z_2 = c$ (i.e., fibers oriented along the axis of the beam) get mapped to circular curves with radius $1-c$ centered at (0, 1) in the $x_1 - x_2$ plane, i.e.

$$x_1^2 + (x_2-1)^2 = (1-c)^2$$

To see that this result comes from the given map, substitute $z_2 = c$ into the map to find $x_1 = (1-c)\sin z_1$ and $x_2 = 1 - (1-c)\cos z_1$. Square x_1 and x_2-1 above and add them together to get the equation of the circle.

Transverse lines $z_1 = b$ get mapped to straight lines in the deformed configuration with equation

$$x_2 = -(\cot b)x_1 + 1$$

These deformed lines all pass through the point (0, 1), the center of the circles, and look like radial spokes of a wheel. The deformation and strain tensors show that lines that were initially transverse to the axis of the beam (i.e., along the z_2 axis) do not change in length, and that axial fibers above the axis ($z_2 > 0$) are shortened and axial fibers below the axis ($z_2 < 0$) are lengthened by the deformation. Note that the axial stretch is always positive, while the axial strain can be positive or negative (tensile or compressive).

The deformation tensor also shows us a limit to the deformation map. The stretch of the top fiber of the beam is $\lambda = 1-z_2$. Hence, a beam with a depth greater than 2 cannot be mapped to this position because to do so would require fibers with $z_2 > 1$ to shrink to zero length (or beyond, whatever that means). As a practical hypothesis, we will reject any deformation map that implies the annihilation of material.

The above example is a very special case of beam bending. We will consider more general beam-bending maps later.

Characterization of Shearing Deformation

There are two basic types of deformation that can occur in a solid body. The first is extension wherein a fiber, or material curve, in the body gets either longer or shorter. The measure of stretch is $\lambda(\mathbf{n})$ and was derived previously. The second type of deformation is called *shearing*. Shearing is associated with changes in angles between lines that are not collinear in the undeformed configuration. To examine this issue, we return to our notion of the arbitrary curve in our body — only now we shall consider the deformation to two different curves that are initially orthogonal at the point in question.

Figure 35 shows our body $\mathcal{B}$ subjected to the deformation map $\boldsymbol{\phi}(\mathbf{z})$. Again we examine the deformation of the body in the neighborhood of the point $\mathcal{P}$, which gets mapped to the point $\boldsymbol{\phi}(\mathcal{P})$ in the deformed configuration. Let us consider two curves in the undeformed configuration, designated as $\mathcal{C}_1$ and $\mathcal{C}_2$, that pass through the point $\mathcal{P}$. These curves are orthogonal and are mapped to the curves $\boldsymbol{\phi}(\mathcal{C}_1)$ and $\boldsymbol{\phi}(\mathcal{C}_2)$ in the deformed configuration. The curves are not necessarily orthogonal in the deformed configuration owing to shear.

We saw previously that the unit vector $\mathbf{n} \equiv d\mathbf{z}/ds$, tangent to a curve in the undeformed configuration, is mapped to a vector $d\mathbf{x}/ds = \mathbf{Fn}$, tangent to the deformed curve in the deformed configuration. Thus, the unit vectors $\mathbf{n}_1$ and $\mathbf{n}_2$, tangent to our two curves at the common point $\mathcal{P}$ in the undeformed configuration, get mapped to vectors $\mathbf{Fn}_1$ and $\mathbf{Fn}_2$, tangent to our two curves at the common point $\boldsymbol{\phi}(\mathcal{P})$ in the deformed configuration. We shall consider the change in the angle between these vectors.

We can compute the angle between the deformed tangent vectors $\mathbf{Fn}_1$ and $\mathbf{Fn}_2$ shown in Fig. 35 as

$$\cos\theta(\mathbf{Fn}_1, \mathbf{Fn}_2) = \frac{\mathbf{Fn}_1 \cdot \mathbf{Fn}_2}{\| \mathbf{Fn}_1 \| \| \mathbf{Fn}_2 \|} = \frac{\mathbf{n}_1 \cdot \mathbf{Cn}_2}{\lambda(\mathbf{n}_1)\lambda(\mathbf{n}_2)} \tag{121}$$

Again, the deformation tensor $\mathbf{C}$ plays the key role in assessing the angle between two deformed vectors. If the original vectors are orthogonal, then the change in angle, often referred to as the *shearing angle*, is $\gamma \equiv \pi/2 - \theta$.

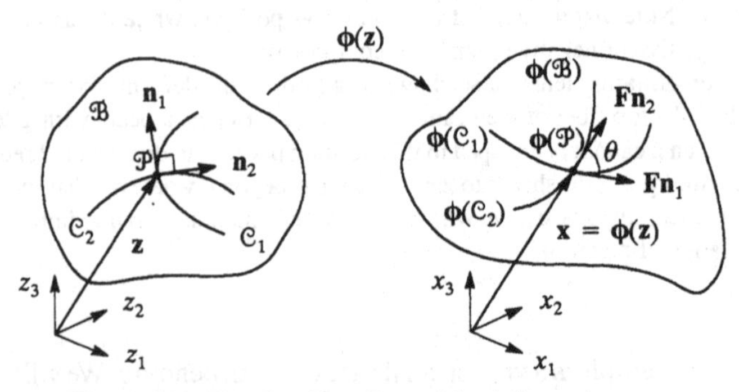

Figure 35 The shearing effect of the deformation map

It would appear that shearing is a deformation process completely distinct from elongation. We shall see later that there is a basic equivalence between shearing and stretching for a three-dimensional solid.

Example 13. *Previous examples revisited.* The shearing of the unit base vectors $\mathbf{e}_1$ and $\mathbf{e}_2$ are shown for three of the four deformation maps given previously. The deformation map for simple, homogenous shearing in the z_1 direction is shown in Fig. 36(a). For simple shear, the base vector $\mathbf{e}_1$ is not stretched by the deformation, but $\mathbf{e}_2$ is; the two vectors are sheared by an amount

$$\cos\theta(\mathbf{Fe}_1, \mathbf{Fe}_2) = \frac{\beta}{\sqrt{1+\beta^2}}$$

Note that if β is small, then the amount of shearing is roughly equal to β.

The deformation map for compound shearing and extension is shown in Fig. 36(b). For compound shearing and extension, the base vectors stretch and shear, and the amount of stretching and shearing depends upon z_1 and z_2. There is no shearing at the origin (0,0,0); shearing and extension increase with distance from the origin. In this case, the orthogonal vectors shear by the amount

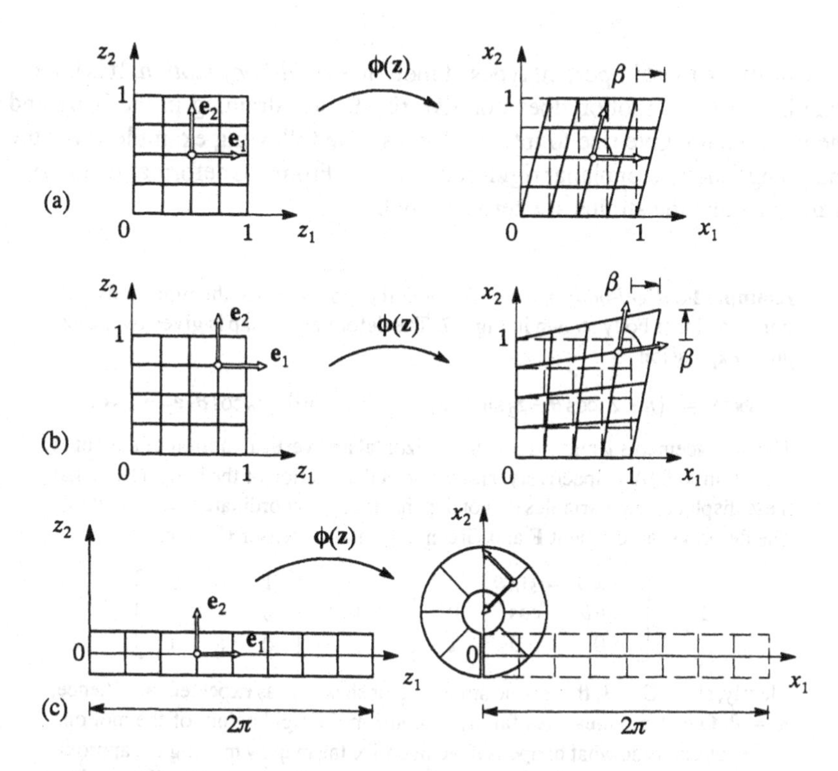

Figure 36 Shearing of base vectors for the example deformation maps (a) simple shear, (b) compound shearing and extension, and (c) pure bending

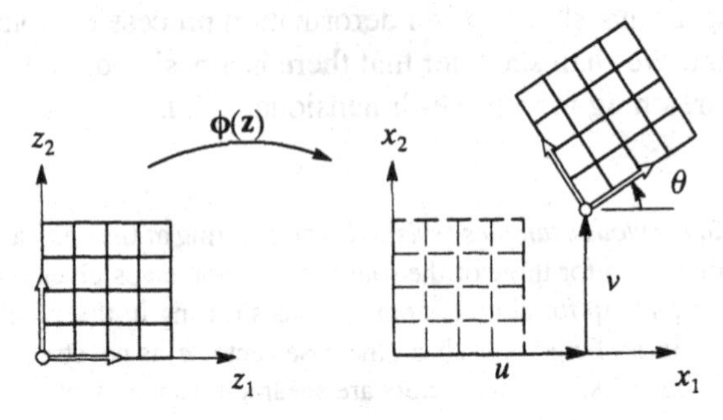

Figure 37 The map for planar rigid-body motion

$$\cos\theta(\mathbf{Fe}_1, \mathbf{Fe}_2) = \frac{\beta(z_1+z_2) + 2\beta^2 z_1 z_2}{\sqrt{1+2\beta z_1+2\beta^2 z_1^2}\ \sqrt{1+2\beta z_2+2\beta^2 z_2^2}}$$

At the origin the base vectors do not shear at all, and the shearing angle increases as we move into the positive quadrant.

The deformation map for pure bending is shown in Fig. 36(c). For pure bending, the orthogonal vectors remain orthogonal to each other after deformation.

One of the most important types of motion is *rigid-body motion*. It is important because this motion does not give rise to any straining in the body and hence does not give rise to internal forces. The following example gives the mapping function for planar rigid-body motion. From this deformation map we can show that the strains are zero everywhere.

Example 14. *Rigid-body motion.* Consider the description of the rigid-body motion of a planar body shown in Fig. 37. The deformation map is given by the explicit expression

$$\boldsymbol{\phi}(\mathbf{z}) = \left(u+z_1\cos\theta - z_2\sin\theta\right)\mathbf{e}_1 + \left(v+z_1\sin\theta+z_2\cos\theta\right)\mathbf{e}_2 + z_3\mathbf{e}_3$$

The displacements u and v track the horizontal and vertical motion of a point at the origin (0,0,0), respectively, and θ tracks the rotation of the body. (Note that these displacement variables do not depend upon the coordinates z_1, z_2, and z_3.) The deformation gradient $\mathbf{F}$ and Green deformation tensor $\mathbf{C}$ are given by

$$\mathbf{F} \sim \begin{bmatrix} \cos\theta & -\sin\theta & 0 \\ \sin\theta & \cos\theta & 0 \\ 0 & 0 & 1 \end{bmatrix} \qquad \mathbf{C} \sim \begin{bmatrix} 1 & 0 & 0 \\ 0 & 1 & 0 \\ 0 & 0 & 1 \end{bmatrix}$$

Clearly, since $\mathbf{C} = \mathbf{I}$, there is no stretching or shearing, as expected, and, hence, $\mathbf{E} = \mathbf{0}$. Our strain measures faithfully confirm the rigid nature of the motion.

Let us examine what happens if we linearize the map by making the approximations $\cos\theta \approx 1$ and $\sin\theta \approx \theta$. Computing the deformation gradient and resulting Green deformation tensor from the linearized map, we obtain

$$\mathbf{F} \sim \begin{bmatrix} 1 & -\theta & 0 \\ \theta & 1 & 0 \\ 0 & 0 & 1 \end{bmatrix}, \quad \mathbf{C} \sim \begin{bmatrix} 1+\theta^2 & 0 & 0 \\ 0 & 1+\theta^2 & 0 \\ 0 & 0 & 1 \end{bmatrix}$$

The consequences of linearizing the deformation map is that rigid-body motion is no longer rigid, i.e., $\mathbf{E} \neq \mathbf{0}$. According to the Green deformation tensor, the stretch in the z_1 and z_2 directions is $\lambda^2 = 1+\theta^2$, not 1 as required by the definition of rigid-body motion. As the block rotates, it becomes larger and larger owing to the artificial stretch induced by the linearization.

If the angle is truly small, that is, $\theta \ll 1$, then the error made in linearization will probably be acceptable. It is on this ground that all of our engineering assumptions of geometrically linear behavior stand. Whenever we make use of the assumption of linearity for a theory that is not really linear, we must define the limit of applicability of our results. When does the linearized theory stop giving us useful results and start giving us garbage? This limit is one of the most difficult pieces of information to come by in engineering. Hence, we issue the warning: *Watch out for rigid bodies that stretch!*

The Physical Significance of the Components of C

We can get a general idea of the physical significance of the components of the deformation tensor $\mathbf{C}$ (and therefore $\mathbf{E}$) by considering the stretching and shearing of the unit base vectors $\{\mathbf{g}_1, \mathbf{g}_2, \mathbf{g}_3\}$. Recall that the ijth component of the tensor $\mathbf{C}$ can be extracted from the tensor as $C_{ij} = \mathbf{g}_i \cdot \mathbf{C}\mathbf{g}_j$. The square of the stretch in the direction of the unit base vector $\mathbf{g}_i$ is given by the expression $\lambda^2(\mathbf{g}_i) = \mathbf{g}_i \cdot \mathbf{C}\mathbf{g}_i$. Thus (no sum on i)

$$\boxed{C_{ii} = \lambda^2(\mathbf{g}_i)} \tag{122}$$

In other words, the diagonal terms of the tensor $\mathbf{C}$ represent the squares of the stretches in the directions of the coordinate axes. The angle of shearing between two base vectors $\mathbf{g}_i$ and $\mathbf{g}_j$ (i not equal to j), deformed by the map, is given by (no sum on repeated indices)

$$\cos\theta(\mathbf{F}\mathbf{g}_i, \mathbf{F}\mathbf{g}_j) = \frac{\mathbf{g}_i \cdot \mathbf{C}\mathbf{g}_j}{\lambda(\mathbf{g}_i)\lambda(\mathbf{g}_j)} = \frac{C_{ij}}{\sqrt{C_{ii}}\sqrt{C_{jj}}} \tag{123}$$

Therefore, the off-diagonal components of the tensor $\mathbf{C}$ are related to the shearing of the three pairs of orthogonal base vectors (no sum on i or j)

$$\boxed{C_{ij} = \lambda(\mathbf{g}_i)\lambda(\mathbf{g}_j)\cos\theta(\mathbf{F}\mathbf{g}_i, \mathbf{F}\mathbf{g}_j)} \tag{124}$$

Notice the role of the stretching of the base vectors in Eqn. (124). The off-diagonal components of $\mathbf{C}$ do not measure purely shearing of the base vectors.

However, for deformations in which the elongations are relatively small, the interpretation of the off-diagonal terms as shearing is quite acceptable.

Strain in Terms of Displacement

For many problems, it is convenient to describe the deformation map in terms of *displacement* from the undeformed configuration. As shown in Fig. 38, if we take the coordinates $\{x_1, x_2, x_3\}$ describing the deformed configuration to be identical to the coordinates $\{z_1, z_2, z_3\}$ describing the undeformed configuration, then the position vectors can be added. Let $\mathbf{u}(\mathbf{z})$ be the displacement vector of a point $\mathcal{P}$ originally at $\mathbf{z}$ and moved to $\mathbf{x}$ under the deformation map. Then the deformation map can be written in the following form

$$\boldsymbol{\phi}(\mathbf{z}) = \mathbf{z} + \mathbf{u}(\mathbf{z}) \tag{125}$$

With this description of the deformation map, we can proceed to compute all of the strain measures that we have computed before. The deformation gradient is given by the expression

$$\boxed{\mathbf{F} = \mathbf{I} + \nabla\mathbf{u}} \tag{126}$$

where the components of the tensor $\nabla\mathbf{u}$ are given by $[\nabla\mathbf{u}]_{ij} = \partial u_i/\partial z_j$. Accordingly, the deformation gradient has components $F_{ij} = \delta_{ij} + u_{i,j}$ (recall that a comma followed by an index j means differentiation with respect to z_j).

The Green deformation tensor is computed from $\mathbf{F}$ as $\mathbf{C} = \mathbf{F}^T\mathbf{F}$ to give the following explicit expression in terms of the displacement $\mathbf{u}$

$$\boxed{\mathbf{C} = \mathbf{I} + \nabla\mathbf{u} + \nabla\mathbf{u}^T + \nabla\mathbf{u}^T\nabla\mathbf{u}} \tag{127}$$

where the tensor $\nabla\mathbf{u}^T \equiv [\nabla\mathbf{u}]^T$ is the transpose of the gradient of the displacement vector $\mathbf{u}$. The tensor $\mathbf{C}$ has components $C_{ij} = \delta_{ij} + u_{i,j} + u_{j,i} + u_{k,i}\, u_{k,j}$.

Finally, the Lagrangian strain tensor can be computed from the Green deformation tensor, and has the expression

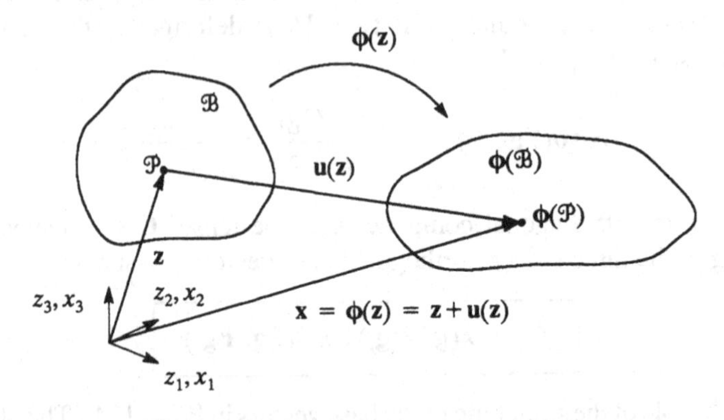

Figure 38 The deformation map in terms of displacement

$$\boxed{\mathbf{E} = \tfrac{1}{2}\left[\nabla\mathbf{u}+\nabla\mathbf{u}^T+\nabla\mathbf{u}^T\nabla\mathbf{u}\right]} \tag{128}$$

with component form $E_{ij} = \frac{1}{2}\left[u_{i,j} + u_{j,i} + u_{k,i}u_{k,j}\right]$. The expression for the Lagrangian strain tensor in terms of displacements allows one to clearly distinguish the linear part of the strain from the nonlinear part. The first two terms in Eqn. (128) constitute the linear part of the strain tensor, while the third term is the nonlinear (quadratic) part. In many problems, the assumption of linearity is useful. We shall call the linear part of **E** the *linearized strain tensor*

$$\boxed{\mathbf{E}_{\text{linear}} \equiv \tfrac{1}{2}\left[\nabla\mathbf{u}+\nabla\mathbf{u}^T\right]} \tag{129}$$

Just as **C** and **E** are, $\mathbf{E}_{\text{linear}}$ is a symmetric, second-order tensor and is a function of the position **z**. The physical interpretation of the components of $\mathbf{E}_{\text{linear}}$ is basically the same as **E**. When strains are small, the differences between **E** and $\mathbf{E}_{\text{linear}}$ are negligible.

Principal Stretches of the Deformation

It seems reasonable to ask whether there are certain directions at a point in our body $\mathcal{B}$ that give extreme values (maximum or minimum) of the stretch (or, more conveniently, the square of the stretch). Recall that the square of the stretch in the direction **n** is given by $\lambda^2(\mathbf{n}) = \mathbf{n}\cdot\mathbf{Cn}$ for unit vectors **n**. As such, the stretch is a quadratic function of the unit direction vector **n** at a given position in space. We can express the question of finding directions of extreme values of stretch as a constrained optimization problem as follows

$$\underset{\mathbf{n}}{\text{extremize}}\ \mathbf{n}\cdot\mathbf{Cn}, \quad \text{subject to } \mathbf{n}\cdot\mathbf{n} = 1 \tag{130}$$

The constraint is needed to make certain that the search is over unit vectors. Otherwise, the solution to the maximization would be infinitely long vectors **n** in basically any direction and the solution to the minimization problem would be the zero vector. The tensor **C** varies with position in space and, therefore, so do the vectors **n** that we seek. However, because we are considering the state at a fixed point we shall suppress the dependence of **C** and **n** on **z** for the purposes of this discussion.

Lagrange had a great idea for setting up a constrained optimization problem. Let us introduce a new independent variable μ and create a function

$$\mathcal{L}(\mathbf{n},\mu) \equiv \mathbf{n}\cdot\mathbf{Cn} - \mu\left(\mathbf{n}\cdot\mathbf{n}-1\right) \tag{131}$$

We call this function the *Lagrangian* (not to be confused with the Lagrangian strain tensor) and we call μ the *Lagrange multiplier.* Lagrange observed that if the vector **n** satisfied the constraint of unit length (i.e., $\mathbf{n}\cdot\mathbf{n} = 1$) then the value of the Lagrangian would be the same as the original function $\lambda^2(\mathbf{n})$ regardless of the value of μ. Hence, extremizing $\mathcal{L}(\mathbf{n},\mu)$ would yield the same vector $\mathbf{n}$ as extremizing $\lambda^2(\mathbf{n})$. In addition, extremizing $\mathcal{L}(\mathbf{n},\mu)$ with respect to μ gives back the equation of constraint. The advantage of the Lagrangian is that we can do an unconstrained optimization to find our unknowns $\mathbf{n}$ and μ.

The necessary condition for an extremum is that the derivative of the Lagrangian, with respect to its arguments, be equal to zero. Clearly, setting the derivative of the Lagrangian, with respect to μ, equal to zero simply gives back the constraint condition that the vector $\mathbf{n}$ be a unit vector (i.e., $\mathbf{n} \cdot \mathbf{n} = 1$). The derivative of the Lagrangian with respect to $\mathbf{n}$ is best done in components. Note that the component expression for the Lagrangian is (summation on repeated indices is implied)

$$\mathcal{L}(\mathbf{n},\mu) = n_i C_{ij} n_j - \mu(n_i n_i - 1) \tag{132}$$

Thus, the derivative of the Lagrangian with respect to n_k can be computed as

$$\begin{aligned} \frac{\partial \mathcal{L}}{\partial n_k} &= \frac{\partial}{\partial n_k}\left(n_i C_{ij} n_j - \mu(n_i n_i - 1)\right) \\ &= \delta_{ik} C_{ij} n_j + n_i C_{ij} \delta_{jk} - \mu(\delta_{ik} n_i + n_i \delta_{ik}) \\ &= C_{kj} n_j + n_i C_{ik} - \mu(n_k + n_k) \\ &= 2(C_{ki} n_i - \mu n_k) \end{aligned}$$

The last step is possible because $C_{ij} = C_{ji}$. Setting the derivatives of the Lagrangian, with respect to $\mathbf{n}$ and μ, equal to zero gives the equations

$$\boxed{\mathbf{C}\mathbf{n} = \mu\mathbf{n}, \qquad \mathbf{n} \cdot \mathbf{n} = 1} \tag{133}$$

from which we can determine $\mathbf{n}$ and μ. Equation (133) is nothing more (and nothing less) that the eigenvalue problem for the tensor $\mathbf{C}$. The solution of the eigenvalue problem is outlined in Chapter 1. Observe that Eqn. (133) represents four equations in four unknowns: the three components of the vector $\mathbf{n}$ and the Lagrange multiplier μ.

As pointed out in Chapter 1, the result of solving the eigenvalue problem is three eigenvalues and their associated eigenvectors (μ_i, $\mathbf{n}_i$, $i = 1,2,3$). The eigenvalues of $\mathbf{C}$ have the three basic possible cases: (1) all eigenvalues distinct, in which case the eigenvectors are all orthogonal, (2) two of the eigenvalues repeated with the third eigenvalue distinct, and (3) all three eigenvalues repeated. This last case occurs for uniform *dilatation*, that is, equal stretch in all three directions with no shearing.

Recall that in Chapter 1 we simply made the assumption that the eigenvectors would be of unit length (a convenience that did not cost us any generality at the time). However, in Chapter 1 we could not ascribe any physical significance to the eigenvalue itself. In the present context, unit length is part of the formulation (the stretch is not equal to $\mathbf{n} \cdot \mathbf{Cn}$ unless $\mathbf{n}$ is a unit vector). The eigenvalue showed up as a result of enforcing the constraint as we search for the directions of extreme stretch.

The physical significance of the eigenvalues and eigenvectors of C. We can use the definitions of shearing and stretching to see the physical significance of the eigenvalues and eigenvectors. Let us compute the square of the stretch of one of the eigenvectors $\mathbf{n}_j$ from the definition of stretch

$$\lambda^2(\mathbf{n}_j) = \mathbf{n}_j \cdot \mathbf{C}\mathbf{n}_j = \mathbf{n}_j \cdot (\mu_j \mathbf{n}_j) = \mu_j \tag{134}$$

since $\mathbf{n}_j \cdot \mathbf{n}_j = 1$. Thus, the eigenvalue μ_j is the square of the stretch in the direction of the eigenvector $\mathbf{n}_j$. One of the important ramifications of this observation is that the eigenvalues of $\mathbf{C}$ cannot be less than or equal to zero. It is not possible to have a zero or negative length ℓ of a line that originally had finite length ℓ_o.

We can also observe that the eigenvectors are not sheared by the deformation. Let us compute the cosine of the angle between two distinct eigenvectors $\mathbf{n}_j$ and $\mathbf{n}_k$ deformed by the map

$$\cos\theta(\mathbf{Fn}_j, \mathbf{Fn}_k) = \frac{\mathbf{n}_j \cdot \mathbf{C}\mathbf{n}_k}{\sqrt{\mu_j}\sqrt{\mu_k}} = \frac{\mathbf{n}_j \cdot (\mu_k \mathbf{n}_k)}{\sqrt{\mu_j}\sqrt{\mu_k}} = \frac{\sqrt{\mu_k}}{\sqrt{\mu_j}}(\mathbf{n}_j \cdot \mathbf{n}_k) = 0 \tag{135}$$

since $\mathbf{n}_j$ is orthogonal to $\mathbf{n}_k$. Since the cosine of the angle between the deformed vectors is zero they must be orthogonal. We can also consider two eigenvectors $\mathbf{n}_j$ and $\mathbf{n}_k$ that are associated with a repeated eigenvalue $\mu_j = \mu_k$. In this case the eigenvectors are not necessarily orthogonal, i.e., in general $\mathbf{n}_j \cdot \mathbf{n}_k \neq 0$. But, Eqn. (135) still informs this case. Take any two eigenvectors that satisfy the eigenvalue problem for the repeated root and observe that $\mathbf{n}_j \cdot \mathbf{n}_k = \cos\alpha$, where α is the angle between the vectors. Now, from Eqn. (135) we have $\cos\theta(\mathbf{Fn}_j, \mathbf{Fn}_k) = \cos\alpha$ because $\sqrt{\mu_k}/\sqrt{\mu_j} = 1$. In other words, the original angle between the vectors remains unchanged by the deformation.

We conclude that any deformation state can always be represented by pure stretching in the principal directions.

The eigenproperties of the Lagrangian strain tensor. The Lagrangian strain tensor is related to the Green deformation tensor as $\mathbf{E} = \frac{1}{2}[\mathbf{C}-\mathbf{I}]$. As a consequence, it has the same eigenvectors, and its eigenvalues are related to the eigenvalues of $\mathbf{C}$. To see that this assertion is true, let us simply multiply the tensor $\mathbf{E}$ by an eigenvector of the tensor $\mathbf{C}$ and see what happens. Let $\mathbf{n}_i$ be an eigenvector of $\mathbf{C}$ with eigenvalue μ_i. Then, by the definition of $\mathbf{E}$, we have

$$\mathbf{E}\mathbf{n}_i = \tfrac{1}{2}[\mathbf{C}-\mathbf{I}]\mathbf{n}_i = \tfrac{1}{2}(\mathbf{C}\mathbf{n}_i - \mathbf{n}_i) = \tfrac{1}{2}(\mu_i - 1)\mathbf{n}_i$$

Recall that an eigenvector is simply a vector that does not change direction when operated on by a tensor. The eigenvalue is the amount by which the vector stretches when operated on by a tensor. Thus, we can see that $\mathbf{n}_i$ is indeed an eigenvector of $\mathbf{E}$ and that the corresponding eigenvalue is given by

$$\gamma_i = \tfrac{1}{2}(\mu_i - 1) \tag{136}$$

such that $\mathbf{E}\mathbf{n}_i = \gamma_i \mathbf{n}_i$. Since μ_i is the square of the stretch in the direction $\mathbf{n}_i$, we see that γ_i is the (scalar) Lagrangian strain in that same direction. Hence, we have $\gamma_i = \frac{1}{2}\left(\lambda^2(\mathbf{n}_i) - 1\right) = E(\mathbf{n}_i)$. Thus, the directions of extreme stretch are also the directions of extreme strain, as we would expect.

The polar decomposition. We can show that the deformation gradient $\mathbf{F}$ (actually, any tensor) can be decomposed into a product of two tensors as

$$\mathbf{F} = \mathbf{R}\mathbf{U} \tag{137}$$

where the action of $\mathbf{U}$ on a vector is to change the length of the vector without changing its direction, and the action of $\mathbf{R}$ is to change the direction of the vector without changing its length. As such, $\mathbf{R}$ is an orthogonal tensor (i.e., it has the properties $\mathbf{R}^T\mathbf{R} = \mathbf{I}$ and $\det(\mathbf{R}) = 1$). Because all of the stretching is accomplished by $\mathbf{U}$, it is a suitable measure of strain or deformation. Since $\mathbf{R}$ is orthogonal we can show that

$$\mathbf{C} = \mathbf{F}^T\mathbf{F} = \left[\mathbf{R}\mathbf{U}\right]^T\left[\mathbf{R}\mathbf{U}\right] = \mathbf{U}^T\left[\mathbf{R}^T\mathbf{R}\right]\mathbf{U} = \mathbf{U}^T\mathbf{U} \tag{138}$$

which shows the relationship between $\mathbf{C}$ and $\mathbf{U}$. The tensor $\mathbf{U}$ is not necessarily symmetric.

The deformational part $\mathbf{U}$ of the polar decomposition can be readily computed from the spectral decomposition of $\mathbf{C}$ if we further specify $\mathbf{U}$ to be a symmetric tensor. First, observe that, for a symmetric $\mathbf{U}$ we have $\mathbf{C} = \mathbf{U}\mathbf{U} = \mathbf{U}^2$. Now let us assume that we know the spectral representation of $\mathbf{C}$ to be

$$\mathbf{C} = \sum_{i=1}^{3} \mu_i\, \mathbf{n}_i \otimes \mathbf{n}_i \tag{139}$$

where μ_i are the eigenvalues of $\mathbf{C}$ and $\mathbf{n}_i$ are the (unit orthogonal) eigenvectors of $\mathbf{C}$. The eigenvalues and eigenvectors can be computed by the methods outlined in Chapter 1. As was previously noted, the eigenvalues of $\mathbf{C}$ have the physical interpretation as the square of the stretch in the direction of the eigenvectors. Let us call $\lambda_i^2 \equiv \mu_i$. Now, it is easy to show that the tensor $\mathbf{U}$ has the spectral representation

$$\boxed{\mathbf{U} = \sum_{i=1}^{3} \lambda_i\, \mathbf{n}_i \otimes \mathbf{n}_i} \tag{140}$$

In fact, to show that this is true, we can simply compute $\mathbf{U}\mathbf{U}$

$$\begin{aligned}\mathbf{UU} &= \sum_{i=1}^{3}\lambda_i\,\mathbf{n}_i\otimes\mathbf{n}_i\sum_{j=1}^{3}\lambda_j\,\mathbf{n}_j\otimes\mathbf{n}_j \\ &= \sum_{i=1}^{3}\sum_{j=1}^{3}\lambda_i\lambda_j\left[\mathbf{n}_i\otimes\mathbf{n}_i\right]\left[\mathbf{n}_j\otimes\mathbf{n}_j\right] \\ &= \sum_{i=1}^{3}\sum_{j=1}^{3}\lambda_i\lambda_j\delta_{ij}\left[\mathbf{n}_i\otimes\mathbf{n}_j\right] = \sum_{i=1}^{3}\lambda_i^2\,\mathbf{n}_i\otimes\mathbf{n}_i\end{aligned}$$

Therefore, to compute **U** we need only compute the eigenvalues and eigenvectors of **C** and then build **U** from its spectral representation. Once **U** is known, we can compute **R** from $\mathbf{F} = \mathbf{RU}$. To wit,

$$\mathbf{R} = \mathbf{FU}^{-1} = \sum_{i=1}^{3}\frac{1}{\lambda_i}\,\mathbf{Fn}_i\otimes\mathbf{n}_i \tag{141}$$

It is also straightforward to show that we can also write $\mathbf{F} = \mathbf{VR}$, where, again, **R** is an orthogonal tensor and **V** represents the deformation. The left Cauchy-Green deformation tensor is

$$\mathbf{b} \equiv \mathbf{FF}^T = \left[\mathbf{VR}\right]\left[\mathbf{VR}\right]^T = \mathbf{VRR}^T\mathbf{V}^T = \mathbf{VV}^T$$

Since $\mathbf{F} = \mathbf{VR} = \mathbf{RU}$, the tensors **V** and **U** are related as $\mathbf{V} = \mathbf{RUR}^T$.

Example 15. Consider a state of deformation at a point characterized by the following deformation gradient (and the corresponding right Cauchy-Green deformation tensor)

$$\mathbf{F} \sim \frac{1}{\sqrt{7}}\begin{bmatrix} 4 & 1 & 2 \\ 2 & 4 & 1 \\ 1 & 2 & 4 \end{bmatrix} \qquad \mathbf{C} \sim \begin{bmatrix} 3 & 2 & 2 \\ 2 & 3 & 2 \\ 2 & 2 & 3 \end{bmatrix}$$

Observe that the vector $\mathbf{n} \equiv (\mathbf{g}_1+\mathbf{g}_2+\mathbf{g}_3)/\sqrt{3}$ is an eigenvector of **C** corresponding to the eigenvalue $\mu = 7$. The tensor **C** also has a repeated eigenvalue $\mu = 1$. Thus, we can write $\mathbf{C} = 7\mathbf{n}\otimes\mathbf{n}+1\left[\mathbf{I}-\mathbf{n}\otimes\mathbf{n}\right]$, the spectral form of **C**, and then the tensor **U** can be written as $\mathbf{U} = \sqrt{7}\,\mathbf{n}\otimes\mathbf{n}+1\left[\mathbf{I}-\mathbf{n}\otimes\mathbf{n}\right]$. We can, therefore express $\mathbf{U}^{-1}$ in the form

$$\mathbf{U}^{-1} = \frac{1}{\sqrt{7}}\mathbf{n}\otimes\mathbf{n} + 1\left[\mathbf{I}-\mathbf{n}\otimes\mathbf{n}\right]$$

which can be expressed in components as

$$\begin{aligned}\mathbf{U}^{-1} &\sim \frac{1}{3\sqrt{7}}\begin{bmatrix} 1 & 1 & 1 \\ 1 & 1 & 1 \\ 1 & 1 & 1 \end{bmatrix} + \frac{1}{3}\begin{bmatrix} 2 & -1 & -1 \\ -1 & 2 & -1 \\ -1 & -1 & 2 \end{bmatrix} \\ &= \frac{1}{3\sqrt{7}}\begin{bmatrix} 1+2\sqrt{7} & 1-\sqrt{7} & 1-\sqrt{7} \\ 1-\sqrt{7} & 1+2\sqrt{7} & 1-\sqrt{7} \\ 1-\sqrt{7} & 1-\sqrt{7} & 1+2\sqrt{7} \end{bmatrix}\end{aligned}$$

Finally, the tensor $\mathbf{R}$ can be computed as $\mathbf{R} = \mathbf{F}\mathbf{U}^{-1}$, to have components

$$\mathbf{R} \sim \frac{1}{21}\begin{bmatrix} 7+5\sqrt{7} & 7-4\sqrt{7} & 7-\sqrt{7} \\ 7-\sqrt{7} & 7+5\sqrt{7} & 7-4\sqrt{7} \\ 7-4\sqrt{7} & 7-\sqrt{7} & 7+5\sqrt{7} \end{bmatrix}$$

It is easy to verify that this tensor is orthogonal, i.e., $\mathbf{R}^T\mathbf{R} = \mathbf{I}$.

It is also interesting to consider the spectral decomposition of the deformation gradient $\mathbf{F}$. Because $\mathbf{F}$ is not symmetric—it is a two-point tensor whose components are described in a mixed basis, as shown in Eqn. (115)—the developments associated with eigenvalue problems of symmetric tensors do not necessarily apply. However, in the present case we can observe that there exists a vector $\mathbf{m}_i$ such that

$$\mathbf{F}\mathbf{n}_i = \lambda_i \mathbf{m}_i \tag{142}$$

where $\mathbf{n}_i$ is a unit eigenvector of $\mathbf{U}$ (and $\mathbf{C}$) and λ_i is an eigenvalue of $\mathbf{U}$ (and the square-root of an eigenvalue of $\mathbf{C}$). Indeed, taking the dot product of each side with respect to itself we can show that (no summation implied)

$$\begin{aligned} (\mathbf{F}\mathbf{n}_i) \cdot (\mathbf{F}\mathbf{n}_i) &= \mathbf{n}_i \cdot \mathbf{C}\mathbf{n}_i = \lambda_i^2 \\ (\lambda_i \mathbf{m}_i) \cdot (\lambda_i \mathbf{m}_i) &= \lambda_i^2 (\mathbf{m}_i \cdot \mathbf{m}_i) \end{aligned} \tag{143}$$

These two results must be identical and, hence, $\mathbf{m}_i$ must be a unit vector. Therefore, operation by $\mathbf{F}$ on the unit vector $\mathbf{n}_i$ results in a vector pointing in the direction $\mathbf{m}_i$ having magnitude λ_i. Therefore, we can write

$$\boxed{\mathbf{F} = \sum_{i=1}^{3} \lambda_i \left[\mathbf{m}_i \otimes \mathbf{n}_i\right]} \tag{144}$$

To prove this result simply operate on the vector $\mathbf{n}_j$

$$\mathbf{F}\mathbf{n}_j = \sum_{i=1}^{3} \lambda_i \left[\mathbf{m}_i \otimes \mathbf{n}_i\right]\mathbf{n}_j = \sum_{i=1}^{3} \lambda_i \delta_{ij} \mathbf{m}_j = \lambda_j \mathbf{m}_j \tag{145}$$

Once again, the two-point nature of the tensor $\mathbf{F}$ is evident. We can substitute Eqn. (145) into Eqn. (141) to give

$$\boxed{\mathbf{R} = \sum_{i=1}^{3} \mathbf{m}_i \otimes \mathbf{n}_i} \tag{146}$$

Change of Volume and Area

We have seen how the deformation map affects the lengths of lines. In fact, we used the notion of change of length of a line under the deformation to define strain, and it gave rise to the tensors **C** and **E** as natural measures of strain in a three-dimensional body. We now examine how surface areas and volumes are affected by the deformation. The main motivation for looking at these topics is that we often need to compute integrals over areas and volumes to obtain global statements of equilibrium.

Preliminary considerations. From Chapter 1 we have formulas for areas and volumes described by pairs and triads of vectors, respectively. Consider a triad of vectors **u**, **v**, and **w** emanating from the same point as shown in Fig. 39. The area of the parallelogram defined by the vectors **u** and **v** is given by $A(\mathbf{u},\mathbf{v}) = \|\mathbf{u}\times\mathbf{v}\|$. The volume of the parallelepiped defined by the vectors is given by $V(\mathbf{u},\mathbf{v},\mathbf{w}) = (\mathbf{u}\times\mathbf{v})\cdot\mathbf{w}$.

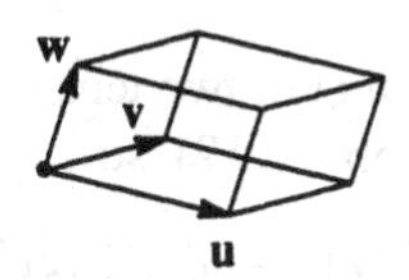

Figure 39 Area and volume are defined by a triad of vectors

We will need two results from tensor analysis in our study of area and volume change. These results are given in the following theorem.

Theorem. Let $\{\mathbf{u},\mathbf{v},\mathbf{w}\}$ be any triad of noncollinear vectors in three-dimensional space. Let **T** be any second-order tensor that operates on a vector **u** to produce a new vector **Tu**. The following identities hold

$$\begin{aligned} \big[(\mathbf{Tu})\times(\mathbf{Tv})\big]\cdot\mathbf{Tw} &= (\det\mathbf{T})\big((\mathbf{u}\times\mathbf{v})\cdot\mathbf{w}\big) && \text{(a)} \\ \mathbf{T}^T\big[(\mathbf{Tu})\times(\mathbf{Tv})\big] &= (\det\mathbf{T})\big[\mathbf{u}\times\mathbf{v}\big] && \text{(b)} \end{aligned} \quad (147)$$

Proof. Let **V** be a tensor defined by three vectors $\mathbf{v}_1$, $\mathbf{v}_2$, and $\mathbf{v}_3$, as

$$\mathbf{V} \equiv \mathbf{v}_j \otimes \mathbf{e}_j$$

where $\mathbf{e}_j$ is the jth base vector. The determinant of the tensor **V** is given by the triple scalar product of the column vectors that define it, i.e.,

$$\det\mathbf{V} = (\mathbf{v}_1\times\mathbf{v}_2)\cdot\mathbf{v}_3 \quad (148)$$

If a tensor **T** acts on each of the columns of **V**, the result can be summarized as $\mathbf{TV} = [\mathbf{Tv}_j]\otimes\mathbf{e}_j$. The determinant of **TV** is therefore

$$\det[\mathbf{TV}] = (\mathbf{Tv}_1\times\mathbf{Tv}_2)\cdot\mathbf{Tv}_3$$

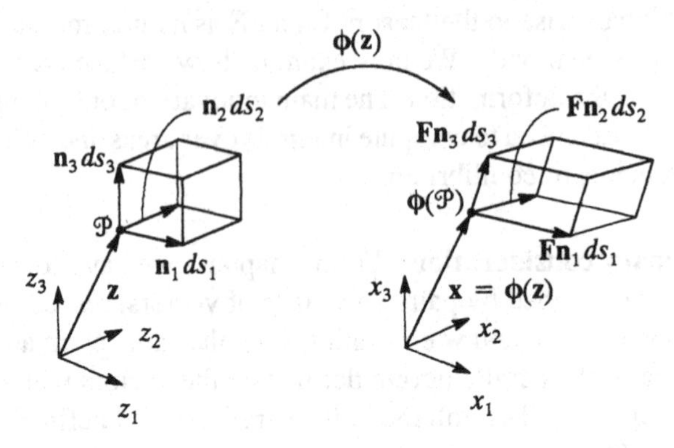

Figure 40 Transformation of volume under the deformation map

The determinant of the product of two tensors is equal to the product of the determinants (see Problem 12), so

$$\det[\mathbf{TV}] = \det\mathbf{T}\det\mathbf{V} \tag{149}$$

Combining Eqns. (148) and (149) proves identity $(147)_a$.

Now let us prove the second identity. Consider the vector $\mathbf{w}$ to be arbitrary. From Eqn. $(147)_a$ we have

$$(\mathbf{Tw})\cdot\big[(\mathbf{Tu})\times(\mathbf{Tv})\big] = (\det\mathbf{T})\big[\mathbf{w}\cdot(\mathbf{u}\times\mathbf{v})\big]$$

Using the definition of the transpose of a tensor we obtain

$$\mathbf{w}\cdot\mathbf{T}^T\big[(\mathbf{Tu})\times(\mathbf{Tv})\big] = (\det\mathbf{T})\big[\mathbf{w}\cdot(\mathbf{u}\times\mathbf{v})\big] \tag{150}$$

Finally, scalar multiplication commutes with the dot product so

$$\mathbf{w}\cdot\mathbf{T}^T\big[(\mathbf{Tu})\times(\mathbf{Tv})\big] = \mathbf{w}\cdot(\det\mathbf{T})\big[(\mathbf{u}\times\mathbf{v})\big] \tag{151}$$

Since Eqn. (151) must hold for any vector $\mathbf{w}$, $(147)_b$ is proved.

Volume change. Figure 40 shows our body $\mathcal{B}$ subjected to the deformation map $\phi(\mathbf{z})$. Again we examine the deformation of the body in the neighborhood of the point $\mathcal{P}$, which gets mapped to the point $\phi(\mathcal{P})$ in the deformed configuration. Let us consider a volume of material described by the triad of unit vectors $\{\mathbf{n}_1, \mathbf{n}_2, \mathbf{n}_3\}$ and the infinitesimal lengths ds_1, ds_2, and ds_3. The vectors can again be thought of as the tangent vectors of three curves in the undeformed configuration, designated as $\mathcal{C}_1$, $\mathcal{C}_2$, and $\mathcal{C}_3$, that pass through the point $\mathcal{P}$. The curves are not actually shown in the figure, but they can be imagined in the same way as previous sketches of the vectors tangent to curves. The variables s_1, s_2, and s_3 measure arc length along those curves. The curves are mapped to the curves $\phi(\mathcal{C}_1)$, $\phi(\mathcal{C}_2)$, and $\phi(\mathcal{C}_3)$ in the deformed configuration. The deformed curves have tangent vectors $\{\mathbf{F}\mathbf{n}_1, \mathbf{F}\mathbf{n}_2, \mathbf{F}\mathbf{n}_3\}$ that are not necessarily orthogonal owing to shearing and are not necessarily of unit length owing to stretching. Let us find the volume of the parallelepiped defined by the triad of vectors $\{\mathbf{Fn}_1, \mathbf{Fn}_2, \mathbf{Fn}_3\}$.

The volume of material in the undeformed configuration is given by

$$dV \equiv (\mathbf{n}_1 \times \mathbf{n}_2) \cdot \mathbf{n}_3 \, ds_1 ds_2 ds_3$$

The volume of the deformed parallelepiped is given by the triple scalar product of the vectors $\{\mathbf{Fn}_1, \mathbf{Fn}_2, \mathbf{Fn}_3\}$. Thus, the volume of the deformed cuboid is

$$dv \equiv (\mathbf{Fn}_1 \times \mathbf{Fn}_2) \cdot \mathbf{Fn}_3 \, ds_1 ds_2 ds_3$$

Using Eqn. $(147)_a$ we can relate the original and deformed volumes as

$$\boxed{dv = (\det \mathbf{F}) \, dV} \tag{152}$$

The determinant of $\mathbf{F}$ is often designated as $\det \mathbf{F} = J$ in the literature. Since it is not physically possible to deform finite volumes into zero or negative volumes, a restriction on any deformation map is that at every point in the domain

$$\det \mathbf{F} > 0 \tag{153}$$

It is also worth noting that, since $\mathbf{C} = \mathbf{F}^T\mathbf{F}$, we have

$$\det(\mathbf{C}) = \det(\mathbf{F}^T\mathbf{F}) = \det(\mathbf{F}^T)\det(\mathbf{F}) = (\det \mathbf{F})^2$$

since the determinant of the product of two tensors is the product of the determinants of the two tensors and since the determinant of the transpose of a tensor is equal to the determinant of the tensor itself.

Example 16. Consider a sphere of unit radius, centered at the origin of coordinates, as shown in Fig. 41(a). The sphere is subject to the deformation map

$$\boldsymbol{\phi}(\mathbf{z}) = (z_1 + \epsilon z_2 z_3)\mathbf{e}_1 + (z_2 + \epsilon z_3 z_1)\mathbf{e}_2 + (z_3 + \epsilon z_1 z_2)\mathbf{e}_3$$

where ϵ is a constant parameter of the motion. Compute the volume of the sphere after deformation in terms of ϵ.

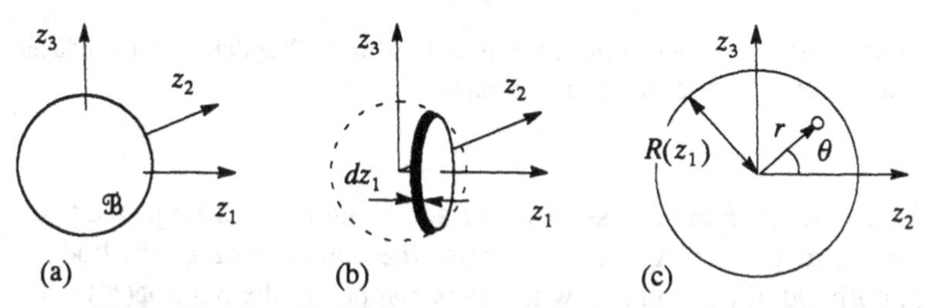

Figure 41 Description of the sphere for the example problem

In accord with Eqn. (152) we can compute the volume v of the deformed sphere as

$$v = \int_{\phi(\mathcal{B})} dv = \int_{\mathcal{B}} \det \mathbf{F} \, dV$$

The deformation gradient of the motion and its determinant can be computed as

$$\mathbf{F} \sim \begin{bmatrix} 1 & \epsilon z_3 & \epsilon z_2 \\ \epsilon z_3 & 1 & \epsilon z_1 \\ \epsilon z_2 & \epsilon z_1 & 1 \end{bmatrix} \qquad \det \mathbf{F} = 1 - \epsilon^2\left(z_1^2 + z_2^2 + z_3^2\right) + 2\epsilon^3 z_1 z_2 z_3$$

To carry out the integral over the original volume $\mathcal{B}$ consider the disk of thickness dz_1 located at z_1 shown in Fig. 41(b). Because $z_1^2 + z_2^2 + z_3^2 = 1$ we can observe that the square of the radius of the disk is $R^2(z_1) = 1 - z_1^2$. Now make a change of variables to $z_1 \equiv z$, $z_2 \equiv r\cos\theta$, $z_3 \equiv r\sin\theta$, as defined in Fig. 41(c). Now the integrand is $\det \mathbf{F} = 1 - \epsilon^2\left(z^2 + r^2\right) + 2\epsilon^3 z\, r^2 \cos\theta \sin\theta$. The volume integral can now be expressed as

$$\int_{\mathcal{B}} \det \mathbf{F} \, dV = \int_{-1}^{1} \int_{0}^{R(z)} \int_{0}^{2\pi} \left(1 - \epsilon^2\left(z^2 + r^2\right) + 2\epsilon^3 z\, r^2 \cos\theta \sin\theta\right) d\theta \, r \, dr \, dz$$

Thus, the integral can be evaluated as

$$\begin{aligned} \int_{\mathcal{B}} \det \mathbf{F} \, dV &= 2\pi \int_{-1}^{1} \int_{0}^{R(z)} \left(\left(1 - \epsilon^2 z^2\right) r - \epsilon^2 r^3\right) dr \, dz \\ &= \frac{\pi}{2} \int_{-1}^{1} \left(2\left(1 - \epsilon^2 z^2\right) R^2(z) - \epsilon^2 R^4(z)\right) dz \\ &= \frac{\pi}{2} \int_{-1}^{1} \left(2\left(1 - \epsilon^2 z^2\right)\left(1 - z^2\right) - \epsilon^2\left(1 - z^2\right)^2\right) dz \\ &= \pi \int_{0}^{1} \left(2\left(1 - z^2\right) - \epsilon^2\left(1 - z^4\right)\right) dz \end{aligned}$$

Therefore, the deformed volume is

$$v = \tfrac{4}{3}\pi\left(1 - \tfrac{3}{5}\epsilon^2\right)$$

The volume of the undeformed sphere is $V = 4\pi/3$. The deformation reduces the volume of the sphere in proportion to ϵ^2.

Area change. Figure 42 shows a surface of our body $\mathcal{B}$ subjected to the deformation map $\boldsymbol{\phi}(\mathbf{z})$. Again we examine the deformation of the body in the neighborhood of the point $\mathcal{P}$, which gets mapped to the point $\boldsymbol{\phi}(\mathcal{P})$ in the deformed configuration. Let us consider a square of material described by the pair of unit vectors $\{\mathbf{n}_1, \mathbf{n}_2\}$ and the infinitesimal lengths ds_1 and ds_2. The vectors can again be thought of as the tangent vectors of two curves in the undeformed configuration, designated as $\mathcal{C}_1$ and $\mathcal{C}_2$, that lie in the surface $\mathcal{S}$ and pass through the point $\mathcal{P}$. The variables s_1 and s_2 measure arc length along those curves. The curves are mapped to the curves $\boldsymbol{\phi}(\mathcal{C}_1)$ and $\boldsymbol{\phi}(\mathcal{C}_2)$, which lie in the deformed surface $\boldsymbol{\phi}(\mathcal{S})$ in the deformed configuration. The deformed curves have tangent vectors $\{\mathbf{Fn}_1, \mathbf{Fn}_2\}$ that are not necessarily orthogonal owing to shearing and are not necessarily of unit length owing to stretching. Let us find the area of the parallelogram defined by the vector pair $\{\mathbf{Fn}_1, \mathbf{Fn}_2\}$.

Let $\mathbf{m}$ be a unit vector normal to the undeformed surface and let $\mathbf{n}$ be a unit vector normal to the deformed surface. These normal vectors can be computed from $\mathbf{n}_1$ and $\mathbf{n}_2$ as follows

$$\mathbf{m} = \frac{\mathbf{n}_1 \times \mathbf{n}_2}{\| \mathbf{n}_1 \times \mathbf{n}_2 \|}, \qquad \mathbf{n} = \frac{\mathbf{Fn}_1 \times \mathbf{Fn}_2}{\| \mathbf{Fn}_1 \times \mathbf{Fn}_2 \|} \neq \mathbf{Fm}$$

where we have specifically noted that $\mathbf{n}$ is not the result of passing $\mathbf{m}$ through the map. The area of the original parallelogram described by the vectors $\mathbf{n}_1$ and $\mathbf{n}_2$ is $dA \equiv \| \mathbf{n}_1 \times \mathbf{n}_2 \| \, ds_1 ds_2 = ds_1 ds_2$. (Note that the scalars ds_1 and ds_2 can be pulled out of the norm operation). The area of the deformed area described by the vectors $\mathbf{Fn}_1$ and $\mathbf{Fn}_2$ is

$$da \equiv \| \mathbf{Fn}_1 \times \mathbf{Fn}_2 \| \, ds_1 ds_2 \tag{154}$$

An *oriented area* in the undeformed configuration can be expressed as

$$\mathbf{m}\,dA = \frac{\mathbf{n}_1 \times \mathbf{n}_2}{\| \mathbf{n}_1 \times \mathbf{n}_2 \|} \| \mathbf{n}_1 \times \mathbf{n}_2 \| \, dA = \left(\mathbf{n}_1 \times \mathbf{n}_2\right) dA$$

and an oriented area in the deformed configuration can be computed as

$$\mathbf{n}\,da = \frac{\mathbf{Fn}_1 \times \mathbf{Fn}_2}{\| \mathbf{Fn}_1 \times \mathbf{Fn}_2 \|} \| \mathbf{Fn}_1 \times \mathbf{Fn}_2 \| \, ds_1 ds_2 = \left(\mathbf{Fn}_1 \times \mathbf{Fn}_2\right) ds_1 ds_2$$

To find the relationship between the two areas in terms of the deformation gradient, let us compute the quantity $\mathbf{F}^T\mathbf{n}\,da$

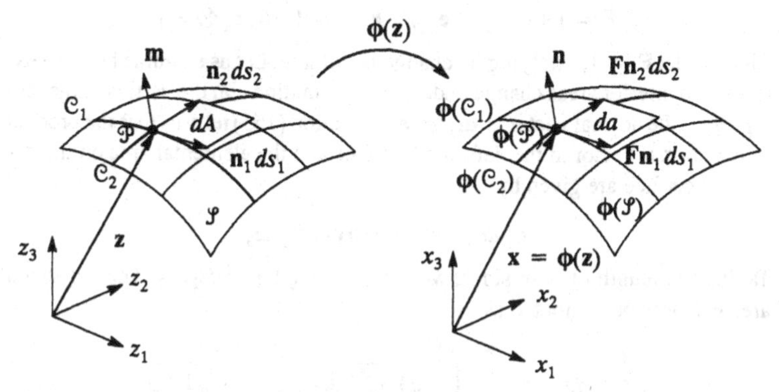

Figure 42 Transformation of surface areas under the deformation map

$$\mathbf{F}^T\mathbf{n}\, da = \mathbf{F}^T(\mathbf{F}\mathbf{n}_1 \times \mathbf{F}\mathbf{n}_2)\, ds_1 ds_2$$

From Eqn. $(147)_b$ we obtain $\mathbf{F}^T\mathbf{n}\, da = (\det \mathbf{F})\mathbf{m}\, dA$ or

$$\boxed{\mathbf{n}\, da = (\det \mathbf{F})\mathbf{F}^{-T}\mathbf{m}\, dA} \tag{155}$$

where $\mathbf{F}^{-T} = [\mathbf{F}^T]^{-1}$ is the inverse of the transpose of the tensor $\mathbf{F}$. The transformation of areas given in Eqn. (155) is often associated with the name *Nanson's formula* or the *Piola transformation*, and it plays a key role in the definition of stress as force per unit area. It is evident from Eqn. (155) that the ratio of deformed area to undeformed area is

$$\frac{da}{dA} = (\det \mathbf{F})\, \| \mathbf{F}^{-T}\mathbf{m} \| \tag{156}$$

Example 17. A four by four square piece of material of unit thickness with a circular hole of unit radius experiences a simple shear deformation as shown in Fig. 43. Take the coordinate axes at the center of the hole and let $\{\mathbf{g}_i\} = \{\mathbf{e}_i\}$. Compute the change in area along along the right edge and on the circle.

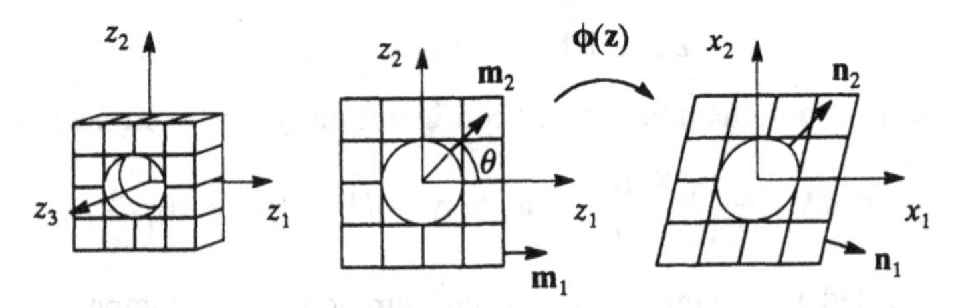

Figure 43 The deformation map for the example problem

The deformation map is given by

$$\phi(\mathbf{z}) = (z_1 + \beta z_2)\mathbf{e}_1 + z_2\mathbf{e}_2 + z_3\mathbf{e}_3$$

The deformation gradient $\mathbf{F}$ and its inverse transpose $\mathbf{F}^{-T}$ are

$$\mathbf{F} = \mathbf{I} + \beta[\mathbf{e}_1 \otimes \mathbf{e}_2], \quad \mathbf{F}^{-T} = \mathbf{I} - \beta[\mathbf{e}_2 \otimes \mathbf{e}_1]$$

Clearly, $\det \mathbf{F} = 1$, implying no change in volume. Let us examine how the exterior and interior areas change under the deformation. On the vertical edge facing right, the normal vector is $\mathbf{m}_1 = \mathbf{e}_1$. Equation (155) tells us that the product of the normal vector to the deformed surface and the elemental area on the deformed surface are given by

$$\mathbf{n}_1\, da_1 = (\mathbf{e}_1 - \beta\mathbf{e}_2)\, dz_2 dz_3$$

Taking the length of both sides, we get $da_1 = \sqrt{1+\beta^2}\, dz_2 dz_3$. The deformed area can now be computed as

$$\int_{\phi(\mathcal{S})} da_1 = \int_0^1 \int_{-2}^2 \sqrt{1+\beta^2}\, dz_2 dz_3 = 4\sqrt{1+\beta^2}$$

We can perform the same operation on the area associated with the cylinder defined by the circle. In this case, the vector normal to the undeformed circle points in the radial direction, i.e, $\mathbf{m}_2 = \cos\theta\,\mathbf{e}_1 + \sin\theta\,\mathbf{e}_2$. The deformed normal vector and deformed element of area are given by

$$\mathbf{n}_2\,da_2 = \left[\cos\theta\,\mathbf{e}_1 + (\sin\theta - \beta\cos\theta)\,\mathbf{e}_2\right] d\theta\,dz_3$$

Proceeding in the same way as before, we find the deformed area of the cylinder

$$\int_{\phi(\mathcal{S})} da_2 = \int_0^1 \int_0^{2\pi} \sqrt{\cos^2\theta + (\sin\theta - \beta\cos\theta)^2}\; d\theta\,dz_3$$

This integral can, of course, be evaluated numerically for specific values of β. For example, when $\beta = 0$, there is no deformation and the area is 2π. When $\beta = 1$ the area is 2.35π, and when $\beta = 2$, the area is 3.19π.

Time-dependent motion

The motion of a solid body is generally a continuous process that evolves with time. In a quasi-static description of a problem, time does not play a central role. In fact, one can think of the "time" t as an orderly means of indexing snapshots of deformed configurations. The rate of deformation in such a case is completely determined from the rate of loading. In dynamic problems or problems with rate-dependent constitutive properties time must be explicitly included in the description of the motion. In these cases velocity and acceleration play an important part in the characterization of the motion.

Velocity and acceleration. Consider a time dependent mapping shown in Fig. 44. The position at time t is given by the mapping $\mathbf{x} = \phi(\mathbf{z}, t)$. At a fixed time t the body is in a configuration that is amenable to the analyses developed earlier in the chapter. A fixed particle in the body, indexed as a point at location $\mathbf{z}$ in the reference configuration, follows a trajectory described by the position vector $\mathbf{x} = \phi(\mathbf{z}, t)$. The velocity of that point is the time rate of change of the position vector

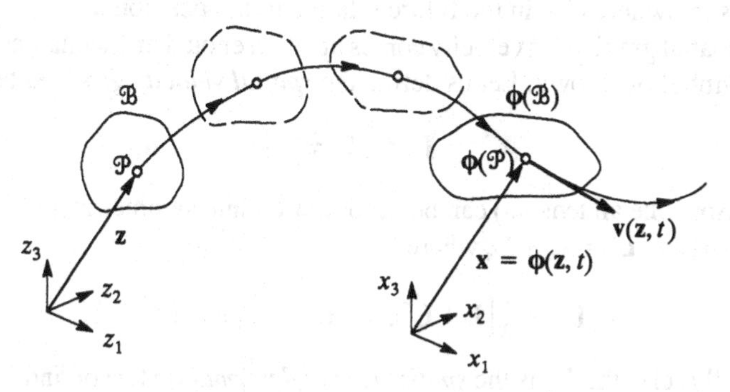

Figure 44 The motion of a body is a time-dependent process

$$\mathbf{v}(\mathbf{z},t) \equiv \frac{\partial \mathbf{x}}{\partial t} = \frac{\partial}{\partial t}\big(\boldsymbol{\phi}(\mathbf{z},t)\big) \tag{157}$$

The velocity vector is tangent to the particle path at the current location of the point. The acceleration of the point is the time rate of change of the velocity

$$\mathbf{a}(\mathbf{z},t) \equiv \frac{\partial \mathbf{v}}{\partial t} = \frac{\partial^2}{\partial t^2}\big(\boldsymbol{\phi}(\mathbf{z},t)\big) \tag{158}$$

These quantities are often called the *material velocity* and *material acceleration* because they record the velocity and acceleration of a material point in the body.

An alternative to the *Lagrangian* description of motion, which tracks material points, is the *Eulerian* description of motion. The Eulerian description considers the motion of points in the neighborhood of a fixed point $\mathbf{x}$ in the deformed configuration. The inverse mapping function $\mathbf{z} = \boldsymbol{\phi}^{-1}(\mathbf{x}(t),t)$ tells us where the particle currently located at position $\mathbf{x}$ in the deformed configuration was located in the reference configuration.

Now the velocity $\mathbf{v}$ can be thought of as being a function of current position $\mathbf{x}$ (which is a function of time) and time. Let us write the velocity of a material point $\mathbf{z}$ as $\mathbf{v}(t) = \hat{\mathbf{v}}(\mathbf{x}(t),t)$. By change of variables the expression for the acceleration takes the form

$$\begin{aligned} \hat{\mathbf{a}} &= \frac{\partial}{\partial t}\big(\hat{\mathbf{v}}(\mathbf{x}(t),t)\big) = \frac{\partial \hat{\mathbf{v}}}{\partial \mathbf{x}}\frac{\partial \mathbf{x}}{\partial t} + \frac{\partial \hat{\mathbf{v}}}{\partial t} \\ &= [\nabla_{\mathbf{x}}\hat{\mathbf{v}}]\,\hat{\mathbf{v}} + \frac{\partial \hat{\mathbf{v}}}{\partial t} \end{aligned} \tag{159}$$

where $\nabla_{\mathbf{x}}(\cdot)$ is the spatial gradient (i.e., derivative with respect to the spatial coordinates $\mathbf{x}$) of $(\cdot)$. Notice that in the Eulerian description we do not keep track of material particles. Although the acceleration $\hat{\mathbf{a}} = \mathbf{a}$ (because there is only one physical acceleration at a certain point at a certain time), the description in terms of current position requires a convective term $[\nabla_{\mathbf{x}}\hat{\mathbf{v}}]\,\hat{\mathbf{v}}$ to account for the fact that the point in question had been somewhere else recently and is headed somewhere else in the future relative to the position $\mathbf{x}$.

The spatial gradient of velocity comes up often enough in mechanics to warrant a symbol of its own. Let us define the *spatial velocity gradient* tensor

$$\mathbf{L} \equiv \big[\nabla_{\mathbf{x}}\hat{\mathbf{v}}\big] \tag{160}$$

This tensor (like all tensors) can be decomposed into symmetric and antisymmetric parts as $\mathbf{L} = \mathbf{L}_s + \mathbf{L}_a$ where

$$\mathbf{L}_s \equiv \tfrac{1}{2}\big[\mathbf{L}+\mathbf{L}^T\big], \qquad \mathbf{L}_a \equiv \tfrac{1}{2}\big[\mathbf{L}-\mathbf{L}^T\big] \tag{161}$$

We usually refer the $\mathbf{L}_s$ as the *spatial rate of deformation* tensor and $\mathbf{L}_a$ as the *spatial spin* tensor.

Strain rates. The strain measures the spatial gradients of the motion, apart from rigid motion, at a snapshot in time. For a deformation that is a function of time we would expect the strains also to be functions of time and, therefore, to change with time. Hence straining must have a rate of change associated with it for a time-dependent motion.

The time rate of change of the deformation gradient can be computed as

$$\dot{\mathbf{F}} = \frac{\partial}{\partial t}\left(\frac{\partial \boldsymbol{\phi}(\mathbf{z},t)}{\partial \mathbf{z}}\right) = \frac{\partial}{\partial \mathbf{z}}\left(\frac{\partial \boldsymbol{\phi}(\mathbf{z},t)}{\partial t}\right) = \nabla_{\mathbf{z}}\mathbf{v} \tag{162}$$

In other words, the rate of change of the deformation gradient is the spatial gradient of the velocity with respect to the reference coordinates $\mathbf{z}$. Recognizing that $\mathbf{v}(\mathbf{z},t) = \hat{\mathbf{v}}(\mathbf{x}(t),t)$ we can make the following observation

$$\frac{\partial}{\partial \mathbf{z}}\big(\mathbf{v}(\mathbf{z},t)\big) = \frac{\partial}{\partial \mathbf{x}}\big(\hat{\mathbf{v}}(\mathbf{x}(t),t)\big)\frac{\partial \mathbf{x}}{\partial \mathbf{z}} = \big[\nabla_{\mathbf{x}}\hat{\mathbf{v}}\big]\mathbf{F} = \mathbf{L}\mathbf{F} \tag{163}$$

Now the rate of change of the Green deformation tensor can be computed by the product rule for differentiation as

$$\begin{aligned}\dot{\mathbf{C}} = \frac{d}{dt}\big(\mathbf{F}^T\mathbf{F}\big) &= \dot{\mathbf{F}}^T\mathbf{F} + \mathbf{F}^T\dot{\mathbf{F}} \\ &= \mathbf{F}^T\mathbf{L}^T\mathbf{F} + \mathbf{F}^T\mathbf{L}\mathbf{F} \\ &= \mathbf{F}^T\big[\mathbf{L}^T + \mathbf{L}\big]\mathbf{F} \\ &= 2\mathbf{F}^T\big[\mathbf{L}_s\big]\mathbf{F}\end{aligned} \tag{164}$$

By the same reasoning the rate of change of the Lagrangian strain can be computed as $\dot{\mathbf{E}} = \mathbf{F}^T\mathbf{L}_s\mathbf{F}$.

Additional Reading

Y. C. Fung, *Foundations of solid mechanics*, Prentice Hall, Englewood Cliffs, N.J., 1965.

M. E. Gurtin, "The linear theory of elasticity," *Mechanics of solids* Vol. II (C. Truesdell, ed.), Springer-Verlag, N.Y., 1972.

L. E. Malvern, *Introduction to the mechanics of a continuous medium*, Prentice Hall, Englewood Cliffs, N.J., 1969.

J. Bonet and R. D. Wood, *Nonlinear continuum mechanics for finite element analysis*, Cambridge University Press, Cambridge, UK, 1997.

Problems

Note: Unless otherwise indicated we shall assume that the base vectors in the deformed and undeformed configurations coincide, i.e., that $\{\mathbf{e}_i\} = \{\mathbf{g}_i\}$.

49. Consider a unit cube in the positive octant with a vertex positioned at the origin of coordinates subjected to the following deformation map

$$\boldsymbol{\phi}(\mathbf{z}) = (z_1 + \epsilon z_2 z_3)\mathbf{e}_1 + (z_2 + \epsilon z_1 z_3)\mathbf{e}_2 + (z_3 + \epsilon z_1 z_2)\mathbf{e}_3$$

where ϵ is a constant. Compute the deformation gradient **F**, the Green deformation tensor **C**, and the Lagrangian strain tensor **E** for the given deformation. Using graph paper, plot the deformed position of a square in the $x_1 - x_2$ plane by locating the positions of a grid of points. (Select a value of ϵ to execute the plot.)

50. The deformation gradient that results from deforming the body shown through a deformation map $\boldsymbol{\phi}(\mathbf{z})$ has the following components relative to the standard basis at the point $\mathcal{P}$

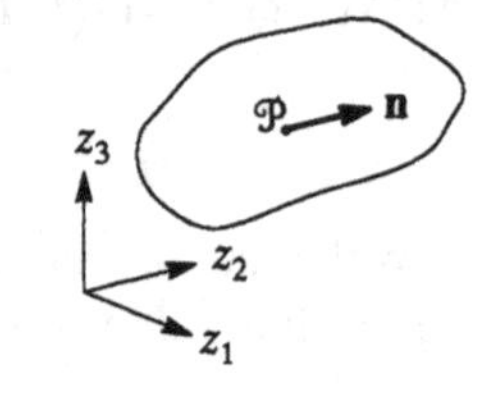

$$\mathbf{F}(\mathcal{P}) \sim \begin{bmatrix} 1.1 & 0.3 & 0.1 \\ 0.1 & 1.2 & 0.2 \\ 0.2 & 0.3 & 1.3 \end{bmatrix}$$

Find the stretch of a line oriented in the direction of the vector $\mathbf{n} = (1,1,0)$ at the point $\mathcal{P}$. What is the value of the Lagrangian strain of that same line at that same point? Calculate the tensors **C** and **E**.

51. Consider a square piece of material of unit thickness with a round hole in it of radius 1. The material is subjected to a deformation described by the map shown in the diagram. The deformation map shown has the following explicit expression

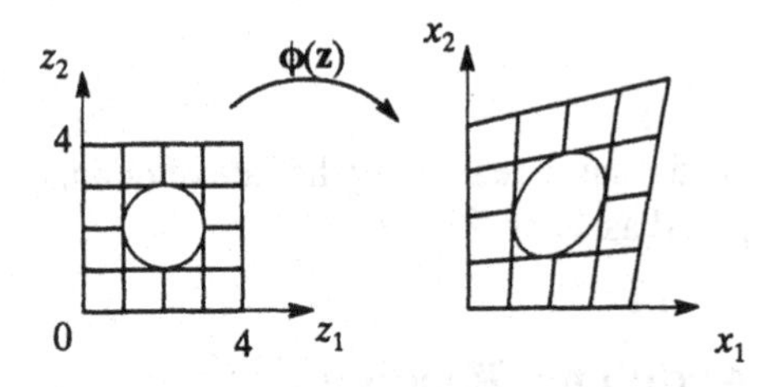

$$\boldsymbol{\phi}(\mathbf{z}) = z_1(1+\beta z_2)\mathbf{e}_1 + z_2(1+3\beta z_1)\mathbf{e}_2 + z_3\mathbf{e}_3$$

Compute the the volume of the hole in the undeformed and deformed configurations. Compute the perimeter area of the square in the undeformed and deformed configurations. Compute the perimeter area of the circle in the undeformed and deformed configurations.

52. Prove that (unit) eigenvectors $\mathbf{n}_1$ and $\mathbf{n}_2$, of the tensor **C**, associated with distinct eigenvalues μ_1 and μ_2, respectively, point in the direction of extreme stretch by computing the stretch for a unit vector $\mathbf{m} = \sin\theta\,\mathbf{n}_1 + \cos\theta\,\mathbf{n}_2$, where θ is a parameter. Plot the stretch in the direction **m** as a function of θ.

53. Consider a square piece of material of unit thickness. The material is subjected to a deformation described by the map shown in the diagram. The deformation map shown has the following explicit expression

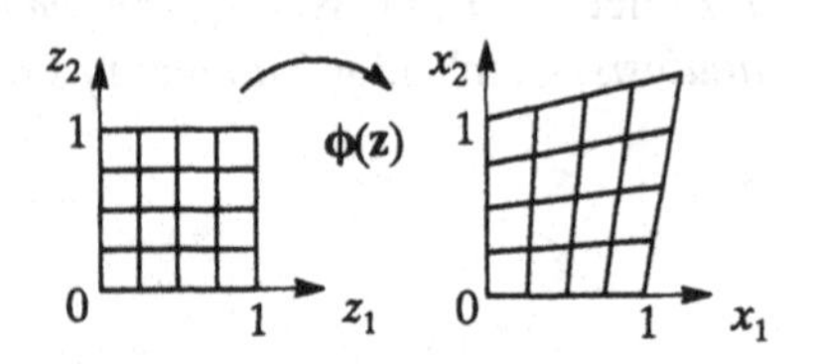

$$\phi(\mathbf{z}) = z_1(1+z_2)\mathbf{e}_1 + z_2(1+3z_1)\mathbf{e}_2 + z_3\mathbf{e}_3$$

Compute the components (with respect to the standard basis $\mathbf{e}_1$, $\mathbf{e}_2$, and $\mathbf{e}_3$) of the Green deformation tensor $\mathbf{C}$ and the Lagrangian strain tensor $\mathbf{E}$ at the point $\mathbf{z} = (1,1,0)$. Find the principal stretches and principal directions of $\mathbf{C}$ at $\mathbf{z} = (1,1,0)$. Find the eigenvalues and eigenvectors of $\mathbf{E}$ at $\mathbf{z} = (1,1,0)$.

54. Prove that it is impossible to deform the vertex of a solid cube into a flat face (e.g., the deformation map shown in the sketch deforms the cube into a tetrahedron with the vertex at a deformed onto the flat plane). Hint: You do not need to find an explicit expression for the map to do this problem. Consider a neighborhood of the point a.

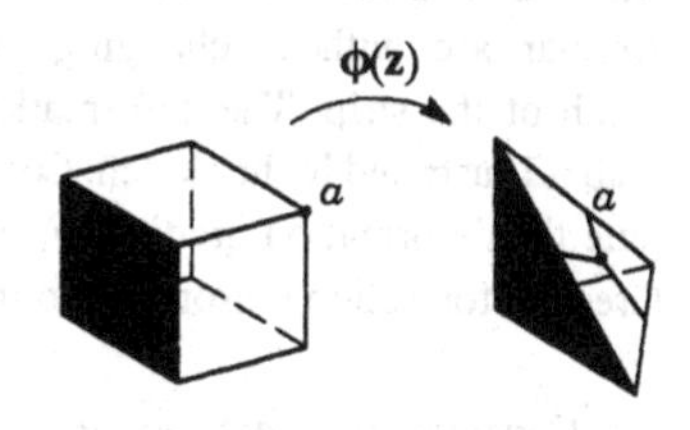

55. A semi-infinite half-space (i.e., the body occupies every point in space that satisfies $z_3 > 0$) has a deformation map given by the following explicit expression

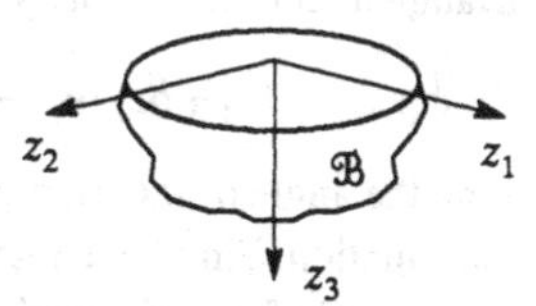

$$\phi(\mathbf{z}) = \left(1+\beta e^{-R}\right)\mathbf{z} + \left(\gamma e^{-R}\right)\mathbf{e}_3$$

where β and γ are constants and R is the distance from the origin to any point with position vector $\mathbf{z}$, that is, $R^2 \equiv \mathbf{z} \cdot \mathbf{z}$. Plot the variation of displacement along the coordinate axes. Compute the displacement of the point that was originally located at $\mathbf{z} = (0, 0, \ln 2)$. Compute the deformation gradient $\mathbf{F}(\mathbf{z})$ of the motion in general and evaluate it at $\mathbf{z} = (0, 0, \ln 2)$. Compute the Green deformation tensor $\mathbf{C}(\mathbf{z})$. Find the value of the stretch of a line in the neighborhood of $\mathbf{z}=(0, 0, \ln 2)$ and initially oriented in the direction $\mathbf{e}_1$.

56. A beam theory is characterized by a specific deformation map that is parameterized by a set of deformation variables that depend only on the axial coordinate z_1. The dependence of the map on z_2 and z_3 is explicit. Let $u(z_1)$ represent the displacement of the centroid of the beam in the z_1 direction, $w(z_1)$ the displacement of the centroid of the beam in the z_2 direction, and $\theta(z_1)$ the rotation of a vector normal to the deformed cross section relative to the horizontal. The deformation map for finite planar motion of the beam then takes the form shown in the diagram. The deformation map has the following mathematical expression

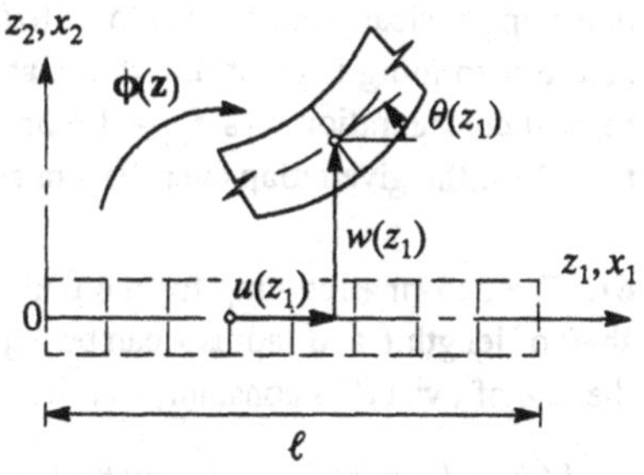

$$\phi(\mathbf{z}) = \left(z_1 + u(z_1) - z_2 \sin\theta(z_1)\right)\mathbf{e}_1 + \left(w(z_1) + z_2\cos\theta(z_1)\right)\mathbf{e}_2 + z_3\mathbf{e}_3$$

Compute the deformation gradient $\mathbf{F}$ of the given deformation map. Compute the Green deformation tensor $\mathbf{C}$, and the Lagrangian strain $\mathbf{E}$. Linearize the deformation map by assuming that $\cos\theta \approx 1$ and $\sin\theta \approx \theta$, and compute $\mathbf{F}$, $\mathbf{C}$, and $\mathbf{E}$ for the linearized kinematic description. Is the strain linear in the displacement variables $u(z_1)$, $w(z_1)$, and $\theta(z_1)$? Linearize $\mathbf{E}$ by neglecting all squares and products of the generalized variables u, w, and θ. What are the consequences of neglecting the higher-order terms?

57. Does the linearized strain tensor ever have the same eigenvectors and eigenvalues as the Lagrangian strain tensor? If so, provide an explicit example.

58. Find the mathematical expression for the map that takes a strip of length 2π and deforms it into a semicircular arc without changing the depth of the strip. The deformation map is illustrated in the sketch. Compute the deformation gradient $\mathbf{F}$, the Green deformation tensor $\mathbf{C}$, and the Lagrangian strain tensor $\mathbf{E}$ for the map.

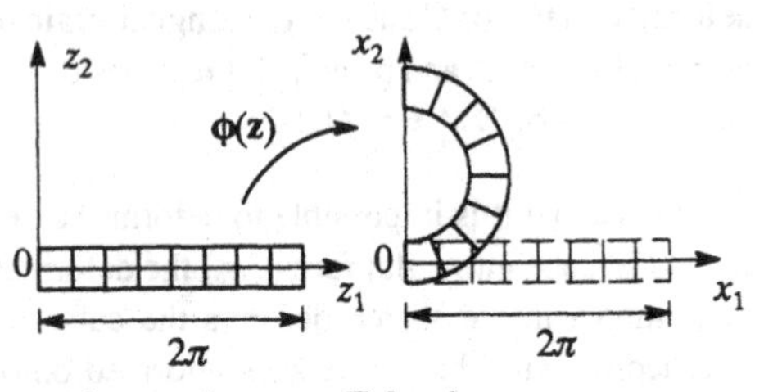

59. Consider the rectangular piece of material with the triangular cutout. The body is subjected to the deformation map

$$\phi(\mathbf{z}) = (z_1 + \beta z_2)\mathbf{e}_1 + z_2\mathbf{e}_2 + z_3\mathbf{e}_3$$

Find the angle of the triangle at the vertex at a before and after deformation. Find the equation describing the inclined line a-b before and after deformation. Find the area of the triangle abc before and after deformation.

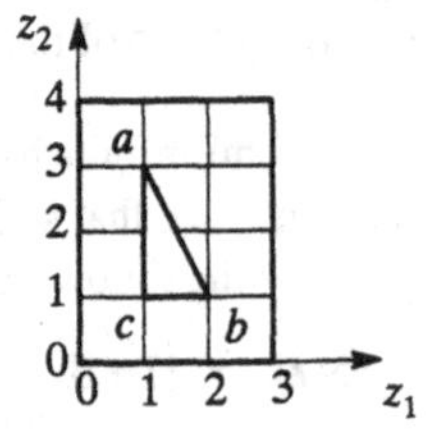

60. Consider a square piece of material of unit thickness. The material is subjected to a deformation described by the following explicit expression

$$\phi(\mathbf{z}) = (\alpha z_1 + \beta z_2)\mathbf{e}_1 + (\gamma z_1 + \delta z_2)\mathbf{e}_2 + z_3\mathbf{e}_3$$

where α, β, γ, and δ are constants. For what values of the constants is the given deformation map physically impossible to realize? Assume that we have scribed a line on the body before deforming it according to the above map. The equation of that line in the undeformed configuration was $z_2 = 1 - 3z_1$. What is the equation of the line after deformation? Will the given map ever deform straight lines into curved lines? Why or why not?

61. The deformation map for the pure twist of a circular shaft of length ℓ and radius r can be expressed in terms of the rate of twist β (a constant) as follows

$$\begin{aligned}\phi(\mathbf{z}) &= \big(z_1\cos(\beta z_3) - z_2\sin(\beta z_3)\big)\,\mathbf{e}_1 \\ &\quad + \big(z_1\sin(\beta z_3) + z_2\cos(\beta z_3)\big)\,\mathbf{e}_2 + z_3\,\mathbf{e}_3\end{aligned}$$

Compute the deformation gradient $\mathbf{F}(\mathbf{z})$. Find the displacement of the point initially located at the position $\mathbf{z} = (r, 0, \ell)$ in the undeformed configuration. Find the volume of the deformed shaft in terms of the angle of twist β. A horizontal line is etched on the surface of the undeformed shaft, parallel to the z_3 axis as shown. Find the length of the line in the deformed configuration.

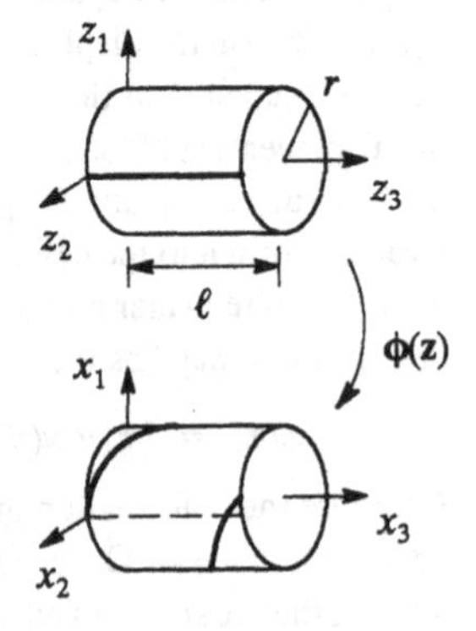

62. Consider the unit cube shown. Let the cube be subjected to the deformation map given by

$$\boldsymbol{\phi}(\mathbf{z}) = (z_1 + z_1 z_2)\mathbf{e}_1 + (z_2 + z_1 z_2)\mathbf{e}_2 + z_3 \mathbf{e}_3$$

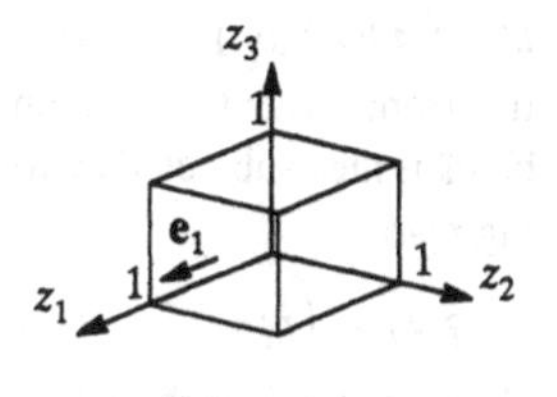

Compute the volume of the cube in the deformed configuration. Find the area in the deformed configuration of the face with normal $\mathbf{e}_1$ in the undeformed configuration.

63. A thin flexible wire of initial length ℓ, originally oriented along the z_3 axis, is wrapped around a hub (with negligible friction between the wire and the hub). The deformation map that accomplishes the motion is given by

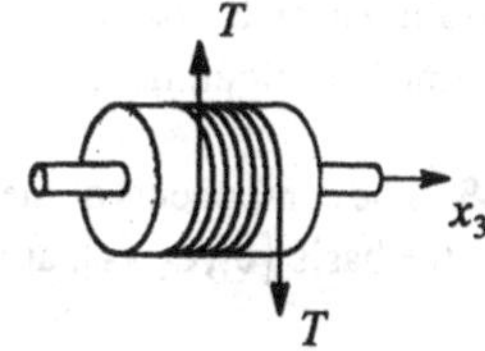

$$\boldsymbol{\phi}(\mathbf{z}) = \sin(\alpha z_3)\mathbf{e}_1 + \cos(\alpha z_3)\mathbf{e}_2 + \beta z_3 \mathbf{e}_3$$

where α and β are known constants. What is the radius of the hub? How many times does the wire wrap around it? What is the spacing between adjacent passes of the wire? What is the length of the wire after it is wrapped?

64. The displacement map of a certain solid body can be expressed as follows:

$$\mathbf{u}(\mathbf{z}) = \alpha z_2 z_3 \mathbf{e}_1 + \alpha z_1 z_3 \mathbf{e}_2 + \alpha z_1 z_2 \mathbf{e}_3$$

where α is a constant. Compute the deformation gradient of the motion. Find the principal stretches at the point originally located at $\mathbf{z} = \{0, 0, 1\}$, in terms of α. Is $\mathbf{n} \sim (1, 1, 0)$ a principal direction for the specified motion? Find the principal (Lagrangian) strains at $\mathbf{z} = \{0, 0, 1\}$ in terms of α.

65. The expansion of a hollow sphere can be described by the deformation map

$$\boldsymbol{\phi}(\mathbf{z}) = \lambda(r)\,\mathbf{z}$$

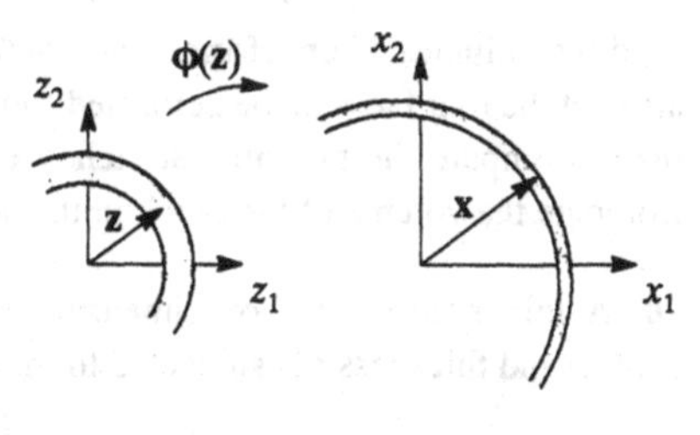

where $\mathbf{z}$ is the position vector of a point in the undeformed configuration and $\lambda(r)$ is a given function of the radial distance $r(\mathbf{z}) \equiv \sqrt{\mathbf{z} \cdot \mathbf{z}}$. Compute the deformation gradient $\mathbf{F}$ for the map. Compute the stretch through the thickness of the sphere in terms of λ, r, and $d\lambda/dr$.

66. The Green deformation tensor that results from deforming the body shown through a deformation map $\boldsymbol{\phi}(\mathbf{z})$ has the following components relative to the standard basis at the point $\mathcal{P}$:

$$\mathbf{C}(\mathcal{P}) \sim \begin{bmatrix} 1.0 & 0.2 & 0.5 \\ 0.2 & 3.0 & 0.2 \\ 0.5 & 0.2 & 2.0 \end{bmatrix}$$

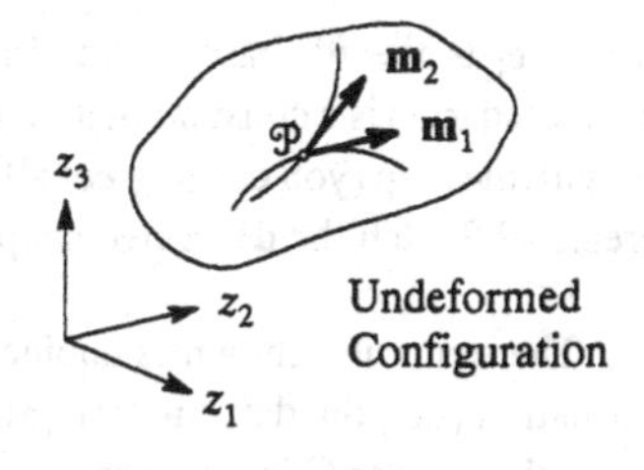

Find the stretch of a line oriented in the direction of the vector $\mathbf{m}_1 = (1,1,1)$ at the point $\mathcal{P}$. Find the angle, after deformation, between two lines with tangent vectors $\mathbf{m}_1 = (1,1,1)$ and $\mathbf{m}_2 = (0,1,1)$ in the undeformed configuration at the point $\mathcal{P}$. Is the vector $\mathbf{m}_1 = (1,1,1)$ an eigenvector of the tensor $\mathbf{C}$ at the point $\mathcal{P}$?

67. A 4 by 3 by 1 in. block of material is scribed with a straight line from corner to corner on one of its broad faces as shown. The block is then subjected to a deformation described by the following map:

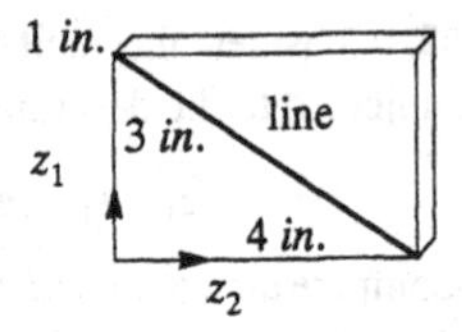

$$\phi(\mathbf{z}) = (z_1+0.2z_2)\mathbf{e}_1 + (z_2+0.3z_3)\mathbf{e}_2 + (z_3+0.1z_1)\mathbf{e}_3$$

Compute the length of the line in the deformed configuration. Compute the Lagrangian strain of the line due to the motion. Compute the Lagrangian strain tensor **E** associated with the motion. Compute the volume of the block in the deformed configuration.

68. The components of the deformation tensor **C** at a certain point in a solid body, relative to the basis $\{\mathbf{e}_1, \mathbf{e}_2, \mathbf{e}_3\}$, are given as

$$\mathbf{C} \sim \frac{1}{10}\begin{bmatrix} 11 & -1 & 0 \\ -1 & 11 & 0 \\ 0 & 0 & 10 \end{bmatrix}$$

Compute the eigenvalues and eigenvectors of **C**. What is the direction in which the stretch of the body is greatest at the given point? What is the magnitude of that stretch? What is the ratio of deformed volume to undeformed volume in the neighborhood of the point?

69. A right tetrahedral block of material, with edges of length 1, 2, and 3 along the coordinate axes, is subjected to a deformation described by the following map:

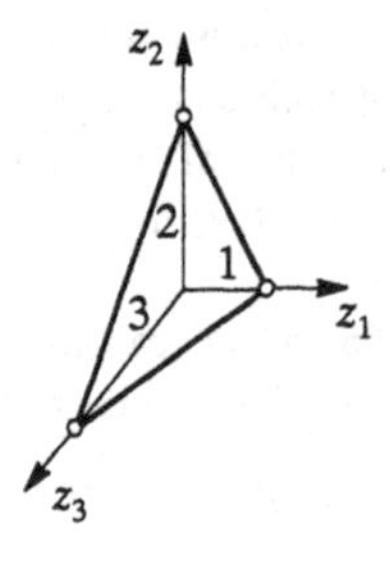

$$\phi(\mathbf{z}) = 6z_1\mathbf{e}_1 + 3z_2\mathbf{e}_2 + 2z_3\mathbf{e}_3$$

Find the volumes of undeformed and deformed bodies. Find the areas of the four faces in the deformed and undeformed configurations. Compute the principal stretches and principal directions. Compute the volume of the block in the deformed configuration.

70. A thin square plate of dimension π (the number 3.14...) and thickness t is subjected to the deformation

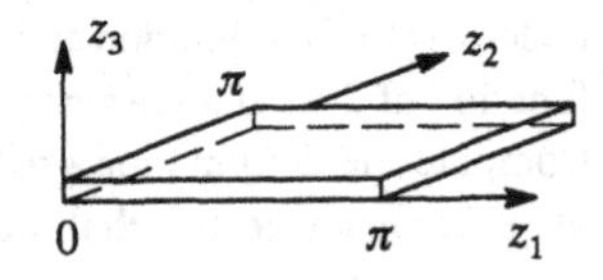

$$\begin{aligned}\phi(\mathbf{z}) &= (z_1-\beta\, z_3 \cos z_1)\mathbf{e}_1 \\ &\quad + z_2\mathbf{e}_2 + (z_3+\beta \sin z_1)\mathbf{e}_3\end{aligned}$$

where $\mathbf{e}_i$ is the ith base vector in the deformed configuration and $\beta \ll 1$ (very small compared to 1) is a constant that describes the motion. Compute the strain tensor associated with the map (you can neglect all terms of order β^2 and higher). Where is the strain the greatest? Sketch the deformed shape of the plate.

71. The unit cube shown is subjected to a homogeneous deformation (i.e., the deformation gradient is constant). The deformation tensor **C** is given by

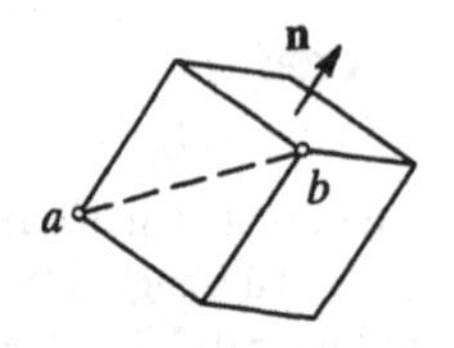

$$\mathbf{C} \equiv \gamma\mathbf{I} - \mathbf{n}\otimes\mathbf{n}$$

where γ is a constant that characterizes the deformation and **n** is a unit vector normal to one of the faces of the cube, as shown on the sketch. Find the principal stretches associated with this state of deformation. Find the stretch λ and the (scalar) Lagrangian strain E of the line ***ab***. What is the smallest possible value of the constant γ for which the deformation is physically reasonable? Explain why smaller values are not possible. If $\det \mathbf{C} = 1$ then the volume of the deformed cube is the same as the volume of the undeformed cube. For what value of γ is the volume unchanged?

72. A 2 by 2 by 2 unit solid cube is subjected to the deformation described by the map (the center of the block is at the origin of coordinates):

$$\boldsymbol{\phi}(\mathbf{z}) = z_1(1+\alpha z_1)\mathbf{e}_1 + z_2(1+\alpha z_1)\mathbf{e}_2 + \beta z_3 \mathbf{e}_3$$

Compute the values of the constants α and β that are consistent with the observation that the *total* volume of the block is unchanged by the deformation. Compute the length of the line ***ab*** in the deformed configuration. Compute the Lagrangian strain tensor $\mathbf{E}$ associated with the motion. Compute the deformed area of the side with original normal $\mathbf{e}_1$.

73. A circular cylinder with initial inside radius of 1 and outside radius of 2 is subjected to a deformation with *displacement* map

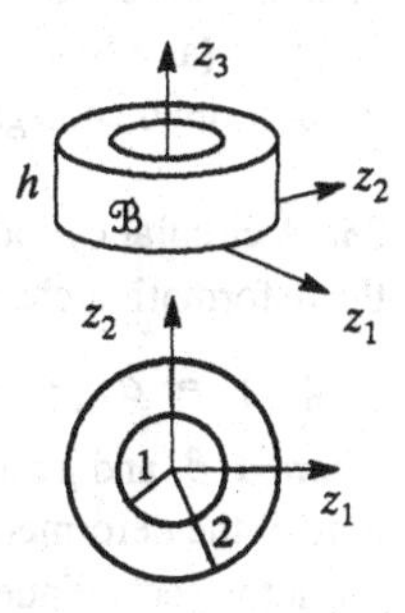

$$\mathbf{u}(\mathbf{z}) = (z_1\mathbf{e}_1 + z_2\mathbf{e}_2)\ln(z_1^2+z_2^2)$$

where $\ln(\cdot)$ indicates the natural logarithm of $(\cdot)$. Find the deformation gradient $\mathbf{F}$ for the given motion. Compute the stretch of the cylinder in the radial direction. Compute the Lagrangian strain of a line in the radial direction. What are the height, inside radius, and outside radius of the cylinder after the deformation?

74. Consider a deformation map $\boldsymbol{\phi}(\mathbf{z})$ given by the explicit expression

$$\boldsymbol{\phi}(\mathbf{z}) = (1+\varepsilon\,\mathbf{z}\cdot\mathbf{z})\,\mathbf{z}$$

Compute the deformation gradient $\mathbf{F}$ of the given motion. Compute the stretch in the radial direction (i.e., in the direction $\mathbf{z}$). Compute the Lagrangian strain tensor $\mathbf{E}$ for the given motion. Is the direction $\mathbf{z}$ an eigenvector of $\mathbf{E}$ or not?

75. A spherical shell in the undeformed configuration has an inside radius of R and an outside radius of $2R$. The shell is subjected to a deformation described by the following map:

$$\boldsymbol{\phi}(\mathbf{z}) = \left(1+\alpha\left(4R^2-\mathbf{z}\cdot\mathbf{z}\right)\right)\mathbf{z}$$

where α is a given constant of the motion and $\mathbf{z}$ is the position vector of a point in the undeformed configuration. Find the displacement of the point originally located at $\mathbf{z}=(0, 0, R)$? Compute the deformation gradient $\mathbf{F}$ of the motion. What is the change in thickness of the shell? How much does the inside surface of the shell stretch? (Note: the stretch is the same in all directions because of the spherical symmetry).

76. A sphere (exploded view shown in sketch) with initial inside radius of 1 and outside radius of 2 is subjected to a deformation with a radially symmetric *displacement* map given by

$$\mathbf{u}(\mathbf{z}) = \beta \mathbf{z} \ln(\mathbf{z} \cdot \mathbf{z})$$

where $\mathbf{z}$ is the position vector and $\ln(\cdot)$ indicates the natural logarithm of $(\cdot)$. Find the deformation gradient $\mathbf{F}$ for the given motion. Compute the stretch of the sphere in the radial direction. What is the inside radius and the outside radius of the sphere after the deformation? Compute the stretch of the sphere in any direction perpendicular to the radial direction and evaluate that stretch at the surface.

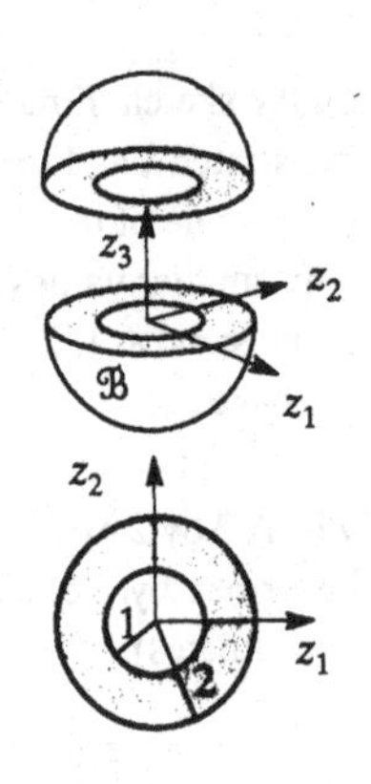

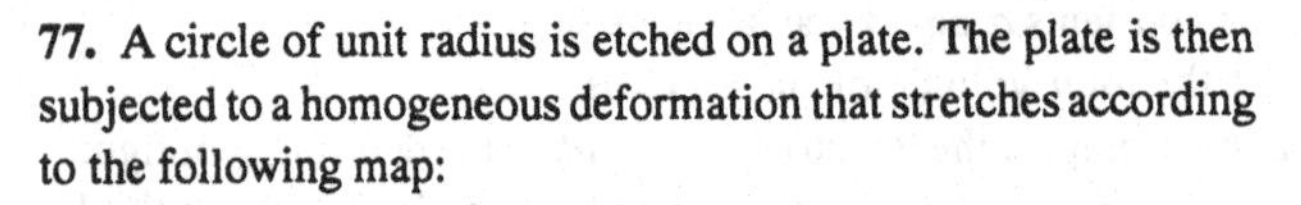

77. A circle of unit radius is etched on a plate. The plate is then subjected to a homogeneous deformation that stretches according to the following map:

$$\boldsymbol{\phi}(\mathbf{z}) = 2z_1\mathbf{e}_1 + z_2\mathbf{e}_2 + z_3\mathbf{e}_3$$

Find the expression for the stretch of the line under the deformation map (as a function of θ). Find the length of the etched line in the deformed configuration.

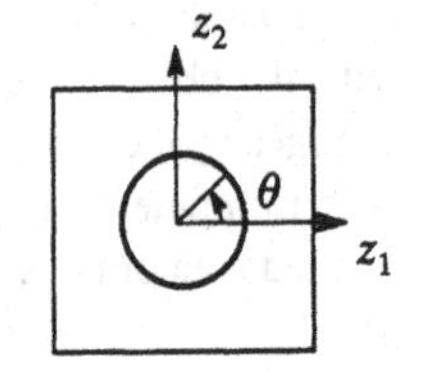

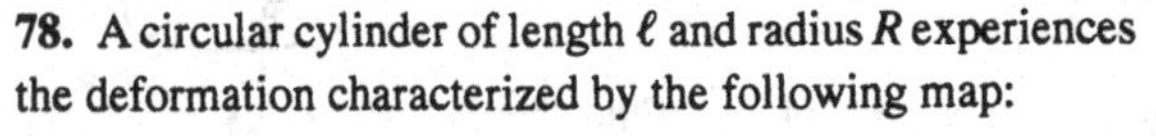

78. A circular cylinder of length ℓ and radius R experiences the deformation characterized by the following map:

$$\boldsymbol{\phi}(\mathbf{z}) = \alpha z_1\mathbf{e}_1 + \beta z_2\mathbf{e}_2 + \gamma z_3\mathbf{e}_3$$

where α, β, and γ are constants of the motion. Find the volume of the deformed cylinder. Find the total surface area of the deformed cylinder. Find the principal stretches of the motion. What are the limits on the constants α, β, and γ?

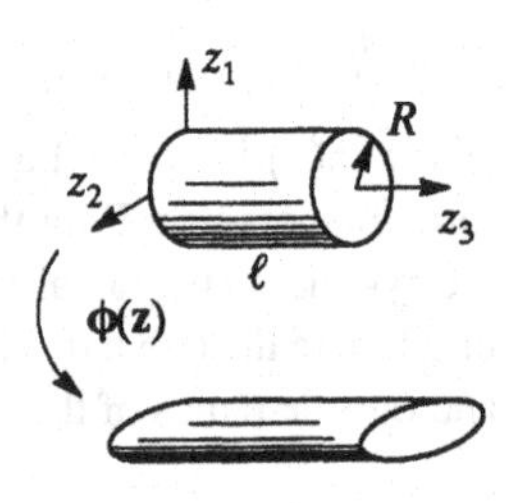

79. Consider a thin (i.e., it has essentially no thickness in the z_3 direction) circular membrane of radius R initially lying in the z_1-z_2 plane as shown in the sketch. Under pressure the membrane deforms into a bubble according to the following deformation map

$$\boldsymbol{\phi}(\mathbf{z}) = z_1\mathbf{e}_1 + z_2\mathbf{e}_2 + \beta\cos\left(\pi\sqrt{z_1^2+z_2^2}/2R\right)\mathbf{e}_3$$

where β is a known constant and R is the radius of the circle. Compute the deformation gradient of the given map. Compute the stretch in the initial radial direction (i.e., the direction of the vector $\mathbf{r} = z_1\mathbf{e}_1 + z_2\mathbf{e}_2$). Also compute the stretch in the direction that is in the initial plane of the membrane but is orthogonal to $\mathbf{r}$ (i.e., tangent to a circle centered at the origin). Are these two directions principal directions? Why or why not? What is the deformed length of the line that was the radial line from the origin to the edge of the circle along

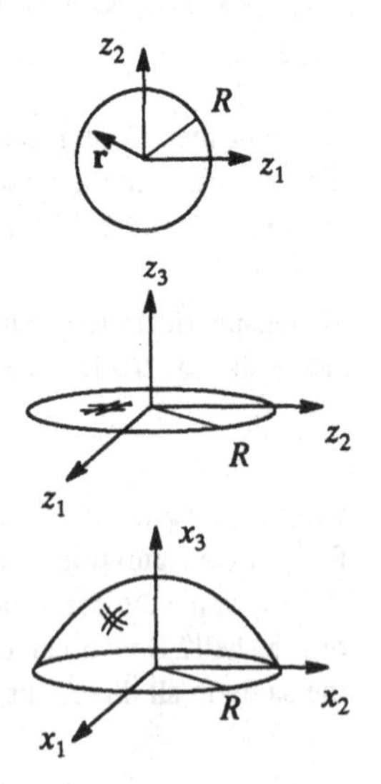

the z_1 direction in the undeformed configuration? What is the slope of the membrane at the edge after deformation?

Note: The stretch through the thickness of the membrane is zero, but that is acceptable because we are assuming that the thickness is very small compared to the diameter of the membrane.

80. Consider the deformation map defined on a sphere of unit radius

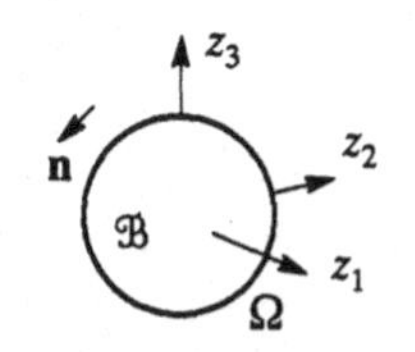

$$\boldsymbol{\phi}(\mathbf{z}) = (1+\varepsilon)\,\mathbf{z} - \beta\,(\mathbf{n}\cdot\mathbf{z})\,\mathbf{n}$$

where ε and β are known (small) constants of the motion and $\mathbf{n}$ is a known constant direction. Compute the ratio of the volume of the sphere after deformation to the volume of the sphere before deformation. Compute the surface area of the sphere after deformation. What makes this calculation complicated? Is the deformed area larger or smaller than the original area? What is the stretch of the sphere in the radial direction? What is the radius of the sphere after the deformation?

3 The Transmission of Force

The transmission of force in a body is basically governed by Newton's laws of conservation of linear and angular momentum. In the static context, with which we are concerned here, these laws amount to the familiar notions of equilibrium of forces and moments. Whereas Newton was primarily concerned with systems of particles, we are concerned with deformable continuous bodies. Consequently, we must introduce an auxiliary concept to model force transmission through the body. The notion of *stress*, as defined by Cauchy, is fundamental to the mechanics of a continuous body, and provides a natural complement to the concept of force. Whereas *tractions* are forces that act on the surface of a body, stress is the measure of the state of force transmission in the interior of the body.

We assume that every piece of a body must be in equilibrium, and, thereby, embrace the concept of the *freebody diagram*, wherein any piece of the body can be isolated from the surrounding material. The effects of the surrounding material on the isolated piece are represented by the traction forces that the surrounding material must exert in order to maintain equilibrium. When isolating a freebody, part of the interior of the body may be exposed as a surface. On that surface, the state of stress must be represented as an equivalent traction. Therein lies the connection between traction forces and stresses. We shall formalize these concepts in the sequel.

The Traction Vector and the Stress Tensor

There are two basic kinds of force that arise in the mechanics of continuous bodies: *body force* (force per unit of volume) and *surface traction* (force per unit of area). To clarify the difference between these two types of force, it is

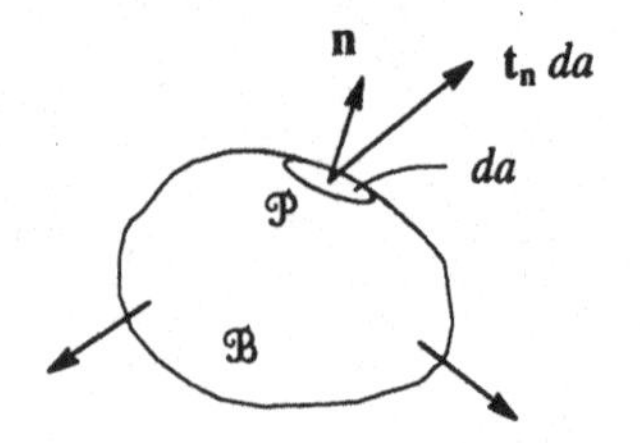

Figure 45 Traction force on a surface

instructive to consider specific examples of these forces. The weight of a body that arises because it possesses a mass density and exists in a gravitational field is an example of a body force; the intensity of the force depends on the mass density of the material. Forces caused by electromagnetic interaction are body forces; the intensity of those forces depends upon the strength of the electromagnetic field. Hence, body forces are caused by action at a distance. The force caused by contact between two bodies is an example of a surface traction. The intensity of the force is related to the area of contact. The analysis of the relationship between stress and surface tractions is our primary concern here.

Consider the body $\mathcal{B}$ shown in Fig. 45. Let $\mathbf{t}_n$ be the *traction vector* field, in a small neighborhood of a point $\mathcal{P}$, acting on the exposed surface with (infinitesimal) area *da* and unit normal vector **n**, shown as a white circle in the figure. As a matter of notational convention, the traction vector will always carry a subscript indicating the normal vector for the plane on which that traction force is acting. The traction is a vector field, and, therefore, has units of force per unit of area. The total force acting on the exposed surface is $\mathbf{t}_n da$ (force per unit area times area), and does not necessarily point in the direction of the normal.

Establishing the relationship between the traction vector and the state of stress at a point depends upon two simple constructions due to Cauchy. The first of these is the "pillbox" construction that helps us formalize the concept of action and reaction. Consider the wafer with face Ω, contour Γ, and thickness ϵ, shown in Fig. 46. We shall consider that the "diameter" of the wafer is $h \gg \epsilon$. Accordingly, the area is $c_\Omega h^2$ and the area of the perimeter is $c_\Gamma \epsilon h$, where c_Ω and c_Γ are fixed constants that depend only on the shape of the wafer. The top face has unit normal **n** and traction field $\mathbf{t}_n$, while the bottom face has

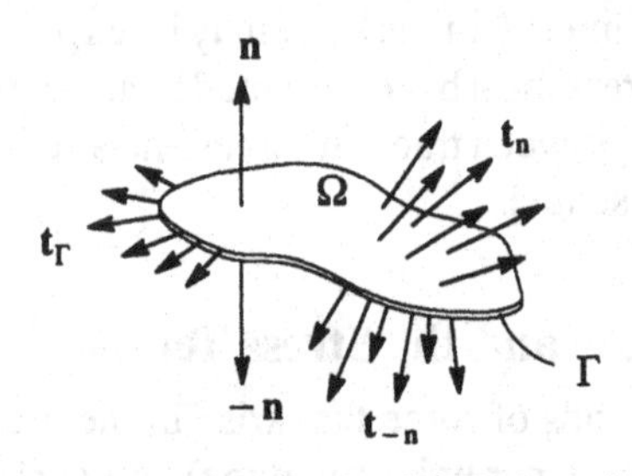

Figure 46 Freebody construction for Cauchy's reciprocal theorem

unit normal $-\mathbf{n}$ and traction field $\mathbf{t}_{-\mathbf{n}}$. The lateral contour has tractions $\epsilon\mathbf{t}_\Gamma$ per unit length along the contour, i.e., $\mathbf{t}_\Gamma$ is the average traction through the thickness of the wafer. In addition, a body force $\mathbf{b}$ is acting on the volume. Again, since the wafer is thin, we can average the body force over the thickness and say that the body force $\epsilon\mathbf{b}$ per unit area is acting on the wafer. Static equilibrium of the wafer requires that the integrals of the tractions of the surface of the body and the integral of the body force over the volume of the body vanish as

$$\int_\Gamma \epsilon\mathbf{t}_\Gamma\, ds + \int_\Omega \mathbf{t}_\mathbf{n}\, dA + \int_\Omega \mathbf{t}_{-\mathbf{n}}\, dA + \int_\Omega \epsilon\mathbf{b}\, dA = \mathbf{0} \tag{165}$$

Taking the limit of Eqn. (165) as $\epsilon \to 0$, the forces $\epsilon\mathbf{t}_\Gamma$ and $\epsilon\mathbf{b}$ become infinitesimally small compared to the forces $\mathbf{t}_\mathbf{n}$ and $\mathbf{t}_{-\mathbf{n}}$. Hence, the first and fourth integrals in Eqn. (165) vanish, and we arrive at the limiting expression

$$\int_\Omega \left(\mathbf{t}_\mathbf{n} + \mathbf{t}_{-\mathbf{n}}\right) dA = \mathbf{0} \tag{166}$$

Since the region Ω of the wafer can be chosen arbitrarily, the integrand must be identically equal to zero, and we must have the *Cauchy reciprocal theorem*

$$\boxed{\mathbf{t}_\mathbf{n} = -\mathbf{t}_{-\mathbf{n}}} \tag{167}$$

expressing that the traction on the area with normal $\mathbf{n}$ is the negative of the traction on the area with normal $-\mathbf{n}$. This theorem should be obvious to anyone schooled in one-dimensional mechanics that has established equilibrium of a segment of a truss bar. It has the Newtonian flavor of "equal and opposite" actions. The theorem will be useful in deriving our next result.

Consider now the *Cauchy tetrahedron* shown in Fig. 47. Let us first examine the geometry of the tetrahedron. The vertices of the tetrahedron are the origin and the points *a*, *b*, and *c*. The lengths of the sides along the coordinate axes are ϵ_1, ϵ_2, and ϵ_3. Let the area of the face having normal vector $-\mathbf{e}_i$ be called a_i. Each of these areas is simply the area of a right triangle. Consequently we have

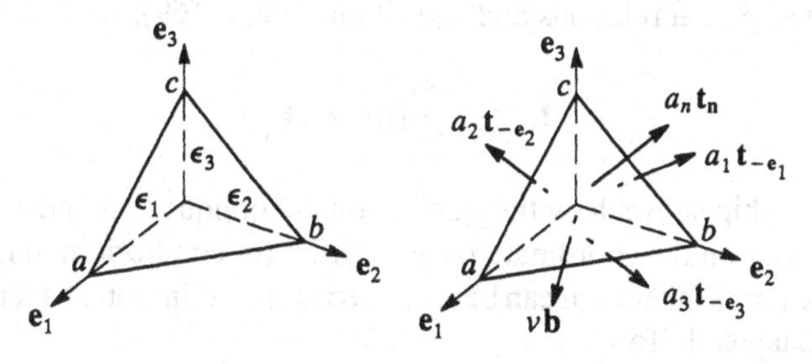

Figure 47 The Cauchy tetrahedron construction

$$a_1 = \tfrac{1}{2}\epsilon_2\epsilon_3, \quad a_2 = \tfrac{1}{2}\epsilon_1\epsilon_3, \text{ and } \quad a_3 = \tfrac{1}{2}\epsilon_1\epsilon_2$$

The volume of the tetrahedron is $v = \frac{1}{6}\epsilon_1\epsilon_2\epsilon_3$. To determine the area of and normal to the oblique face, we need only consider the cross product of the vector pointing from point a to point b with the vector pointing from point a to point c. Let the vector pointing from a to b be $\mathbf{v}_{ab} \equiv \epsilon_2\mathbf{e}_2 - \epsilon_1\mathbf{e}_1$, and the vector pointing from a to c be $\mathbf{v}_{ac} \equiv \epsilon_3\mathbf{e}_3 - \epsilon_1\mathbf{e}_1$. The cross product of these two vectors gives a vector normal to the oblique plane whose length is twice the area of the oblique face of the tetrahedron. That cross product is

$$a_n\mathbf{n} = \tfrac{1}{2}\left[\mathbf{v}_{ab} \times \mathbf{v}_{ac}\right] = \tfrac{1}{2}\left[\left(\epsilon_2\mathbf{e}_2 - \epsilon_1\mathbf{e}_1\right) \times \left(\epsilon_3\mathbf{e}_3 - \epsilon_1\mathbf{e}_1\right)\right]$$

Using the distributive law of multiplication and noting the identities among cross products of the orthogonal unit base vectors, we find the purely geometric relationship between the areas of the sides of the tetrahedron, the unit base vectors, and the unit vector to the oblique side as

$$a_n\mathbf{n} = a_1\mathbf{e}_1 + a_2\mathbf{e}_2 + a_3\mathbf{e}_3 \tag{168}$$

Taking the dot product of Eqn. (168) with the ith base vector, we get the area of the sides of the tetrahedron in terms of the area of the oblique side as

$$a_i = \left(\mathbf{n} \cdot \mathbf{e}_i\right) a_n \tag{169}$$

Consider now the equilibrium of the tetrahedron acted upon by the traction vector fields on the four sides and the body force vector $\mathbf{b}$ throughout the volume. Static equilibrium insists that

$$a_n\mathbf{t}_\mathbf{n} + a_1\mathbf{t}_{-\mathbf{e}_1} + a_2\mathbf{t}_{-\mathbf{e}_2} + a_3\mathbf{t}_{-\mathbf{e}_3} + v\mathbf{b} = \mathbf{0} \tag{170}$$

where the traction and body forces are the average of the field of forces acting over the appropriate domain of action. Dividing Eqn. (170) by a_n and taking the limit as ϵ_1, ϵ_2, and $\epsilon_3 \rightarrow 0$ (holding their ratios constant so as not to distort the geometric shape of the tetrahedron), we find that the ratio of volume to the area of the oblique side is of order ϵ in comparison to the coefficients of the traction vectors, and, hence, the body force term vanishes in the limit. Noting Cauchy's reciprocal relations and substituting Eqn. (169), we find that

$$\mathbf{t}_\mathbf{n} = \sum_{i=1}^{3}(\mathbf{n} \cdot \mathbf{e}_i)\,\mathbf{t}_{\mathbf{e}_i} \tag{171}$$

This relationship shows how the traction on the oblique face must be related to the tractions on the coordinate faces in order for equilibrium to hold. Each of the three terms in the sum can be recast using the definition of a tensor product from Chapter 1. To wit

$$\left(\mathbf{n} \cdot \mathbf{e}_i\right)\mathbf{t}_{\mathbf{e}_i} = \left[\mathbf{t}_{\mathbf{e}_i} \otimes \mathbf{e}_i\right]\mathbf{n} \tag{172}$$

We shall define the *stress tensor* **S** in the following manner

$$\mathbf{S} \equiv \sum_{i=1}^{3} \mathbf{t}_{\mathbf{e}_i} \otimes \mathbf{e}_i \tag{173}$$

Finally, we can summarize the Cauchy stress formula of Eqn. (171) simply as

$$\boxed{\mathbf{t}_{\mathbf{n}} = \mathbf{S}\mathbf{n}} \tag{174}$$

This simple formula embodies our concept of stress. Stress represents the state of force transmission in the interior of a solid body. By cutting a freebody diagram, and thereby exposing a surface with normal **n**, we can see the effect of the stress through the exposed traction vector. Clearly, one can cut a body through a single point in an infinite number of possible ways, but there is only one state of stress at each point. Each way of taking a cut through the point in question is characterized by a plane with a different normal vector **n**.

The tetrahedron construction showed us that, in general, a traction $\mathbf{t}_{\mathbf{n}}$ can be uniquely expressed in terms of three base tractions $\mathbf{t}_{\mathbf{e}_1}$, $\mathbf{t}_{\mathbf{e}_2}$, and $\mathbf{t}_{\mathbf{e}_3}$. Consequently, the description of the transmission of tractions throughout a body is ideally suited to the concept of the second-order tensor. Observe that we have done nothing more than to establish equilibrium of the tetrahedron as Newton would have done for a system of particles. The tensor **S**, defined in the manner above, allows us to speak of the state of stress independent of the orientation of the plane used to cut the freebody diagram.

The physical significance of the components of S. The physical significance of the components of the stress tensor can be seen through a simple computation. The components of **S** are given by

$$S_{ij} \equiv \mathbf{e}_i \cdot \mathbf{S}\mathbf{e}_j = \mathbf{e}_i \cdot \sum_{k=1}^{3} \left[\mathbf{t}_{\mathbf{e}_k} \otimes \mathbf{e}_k \right] \mathbf{e}_j = \mathbf{e}_i \cdot \mathbf{t}_{\mathbf{e}_j} \tag{175}$$

In other words, the *ij*th component of **S** is the *i*th component of the traction vector acting on the face with normal vector $\mathbf{e}_j$, as shown in Fig. 48. The reader should be warned that there are two possible conventions for indexing the stress tensor. Many authors reverse the order of the subscripts on the components of the stress tensor so that the first subscript refers to direction of the normal vector to the plane of action, while the second subscript refers to the component of the traction vector on that plane. Our convention is just the opposite. We will soon see that equilibrium requirements will insist that the stress tensor be symmetric, making the distinction between these two conventions irrelevant.

Some simple states of stress. There are some important special cases of the stress state that deserve to be mentioned. Some of these are analogous with the special homogeneous states of deformation presented in the previous chapter.

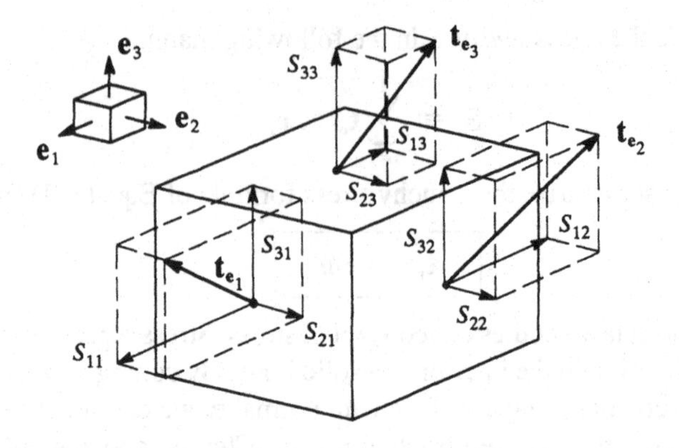

Figure 48 Physical significance of the components of the stress tensor

$\mathbf{S} = p\mathbf{I}$	Hydrostatic pressure
$\mathbf{S} = \sigma\left[\mathbf{e}_1 \otimes \mathbf{e}_1\right]$	Pure tension along $\mathbf{e}_1$
$\mathbf{S} = \tau\left[\mathbf{e}_2 \otimes \mathbf{e}_1 + \mathbf{e}_1 \otimes \mathbf{e}_2\right]$	Pure shear along $\mathbf{e}_1$ and $\mathbf{e}_2$

One way to understand a stress state is to examine the traction vectors on certain planes of a freebody diagram. This approach is a good one because we can draw a vector; we cannot draw a tensor. The three homogeneous stress states are illustrated in Fig. 49. The freebody shown is a 10-sided solid with sides normal to the coordinate axes and with sides whose normals split $\mathbf{e}_1$ and

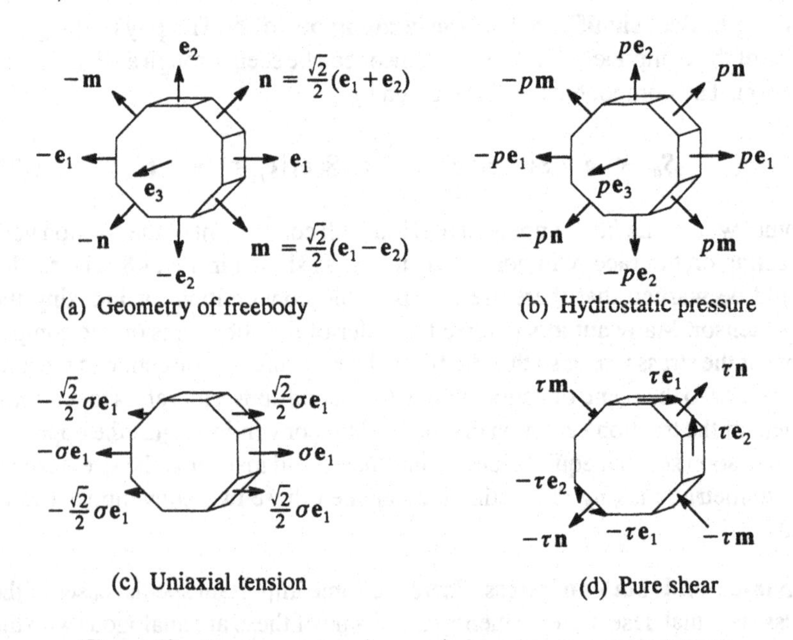

Figure 49 Traction vectors for certain homogeneous stress tensors

$\mathbf{e}_2$. Figure 49 shows: (a) the geometry of the freebody diagram, (b) the hydrostatic pressure stress state, (c) the uniaxial tension stress state, and (d) the pure shear stress state. The action of these stress states can be seen by examining the traction vector that acts on a plane with unit normal vector $\mathbf{v}$, one of the 10 possible normal vectors in each figure. The traction vector for hydrostatic pressure is $\mathbf{t}_v = p\mathbf{v}$, that is, a vector with magnitude p pointing in the direction $\mathbf{v}$. We can see that the traction vectors are all of equal size and point normal to the face in question. The traction vector for pure tension is $\mathbf{t}_v = \sigma(\mathbf{v} \cdot \mathbf{e}_1)\mathbf{e}_1$. The front and back face are free of traction, as are the top and bottom faces. All of the vectors point along the $\mathbf{e}_1$ direction, but the magnitude depends upon the orientation of the face. The traction vector for pure shear can be expressed as $\mathbf{t}_v = \tau[(\mathbf{v} \cdot \mathbf{e}_1)\mathbf{e}_2 + (\mathbf{v} \cdot \mathbf{e}_2)\mathbf{e}_1]$. Again, the front and back faces are free of traction. On the coordinate faces, the traction vector is orthogonal to the normal vector, while on the oblique faces, the traction vector is parallel to the normal vector. This picture suggests that, in some way, pure shearing in one orientation is equivalent to pure tension and compression in another.

Normal and Shearing Components of the Traction

It is sometimes useful to break the traction vector acting on a plane with normal $\mathbf{n}$ into a component normal to the plane and a component in the plane as

$$\mathbf{t}_n = \sigma\mathbf{n} + \tau\mathbf{m} \tag{176}$$

where σ is the magnitude of the normal traction component, τ is the magnitude of the shearing traction component, and $\mathbf{m}$ is a unit vector in the plane normal to $\mathbf{n}$ along which shearing takes place, as shown in Fig. 50. The magnitude of the normal traction can be easily found by taking the dot product of the traction vector with the normal vector

$$\sigma = \mathbf{n} \cdot \mathbf{t}_n = \mathbf{n} \cdot \mathbf{S}\mathbf{n} \tag{177}$$

Note the similarity of this formula and the formula for stretching based on the Green deformation tensor $\mathbf{C}$. The shearing vector can be determined by taking the difference between the traction $\mathbf{t}_n$ and the vector $\sigma\mathbf{n}$

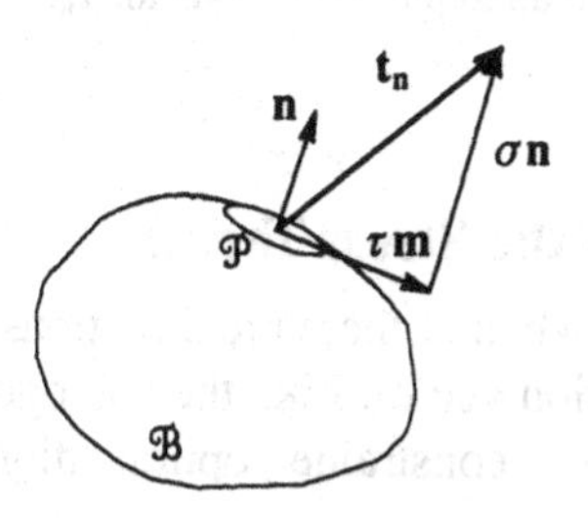

Figure 50 Normal and shearing components of the traction vector

$$\tau\mathbf{m} = \mathbf{Sn} - \sigma\mathbf{n} = [\mathbf{I} - \mathbf{n} \otimes \mathbf{n}]\mathbf{t_n} \tag{178}$$

The magnitude τ can be found by taking the norm of the above vector

$$\tau = \| \mathbf{Sn} - \sigma\mathbf{n} \| = \sqrt{\| \mathbf{t_n} \|^2 - \sigma^2} \tag{179}$$

and the direction of the vector can be found from Eqn. (178) as

$$\mathbf{m} = \frac{1}{\tau}(\mathbf{Sn} - \sigma\mathbf{n}) \tag{180}$$

The computation of normal and shearing components of a traction vector is straightforward if we know the values of the components of the stress tensor and the components of the normal vector. Often these computations can be simply executed using matrix algebra.

Example 18. Consider a plane passing through a point in the body described by the equation $x_1 + 2x_2 = 0$. The state of stress at a point in the body is characterized by the stress tensor $\mathbf{S} = 2[\mathbf{e}_1 \otimes \mathbf{e}_1] - 5[\mathbf{e}_1 \otimes \mathbf{e}_2 + \mathbf{e}_2 \otimes \mathbf{e}_1] + 3[\mathbf{e}_2 \otimes \mathbf{e}_2]$. Let us find the traction vector and its normal and shear components.

First, the normal vector to the plane can be computed as $\mathbf{n} = (\mathbf{e}_1 + 2\mathbf{e}_2)/\sqrt{5}$ (do you know why?). The traction vector is $\mathbf{t_n} = \mathbf{Sn} = (-8\mathbf{e}_1 + \mathbf{e}_2)/\sqrt{5}$. Note that the square of the length of the traction vector is $\| \mathbf{t_n} \|^2 = 13$. The normal component of the traction can be computed from Eqn. (177) as

$$\sigma = \mathbf{n} \cdot \mathbf{t_n} = \frac{1}{\sqrt{5}}(\mathbf{e}_1 + 2\mathbf{e}_2) \cdot \frac{1}{\sqrt{5}}(-8\mathbf{e}_1 + \mathbf{e}_2) = -\frac{6}{5}$$

Knowing the normal components of the traction we can compute the shearing component from Eqn. (179) as

$$\tau = \sqrt{\| \mathbf{t_n} \|^2 - \sigma^2} = \frac{17}{5}$$

Finally, we can compute the direction in which the shearing component acts as

$$\mathbf{m} = \frac{5}{17}\left(\frac{1}{\sqrt{5}}(-8\mathbf{e}_1 + \mathbf{e}_2) + \frac{6}{5}\frac{1}{\sqrt{5}}(\mathbf{e}_1 + 2\mathbf{e}_2)\right) = \frac{1}{\sqrt{5}}(-2\mathbf{e}_1 + \mathbf{e}_2)$$

It is easy to verify that **m** and **n** are orthogonal and that the two components of the traction vector add up to the original vector $\mathbf{t_n}$.

Principal Values of the Stress Tensor

It seems natural to ask whether there are directions that extremize the normal component of the traction vector. Like the analogous question for strain, we can state the problem as a constrained optimization of $\mathbf{n} \cdot \mathbf{Sn}$. Specifically,

$$\underset{\mathbf{n}}{\text{extremize}} = \mathbf{n} \cdot \mathbf{Sn} \quad \text{subject to } \mathbf{n} \cdot \mathbf{n} = 1 \tag{181}$$

The constraint is needed to make certain that the search is over unit vectors. The Lagrangian of the constrained optimization problem is given by

$$\mathcal{L}(\mathbf{n},\mu) = \mathbf{n} \cdot \mathbf{S}\mathbf{n} - \mu(\mathbf{n} \cdot \mathbf{n} - 1) \tag{182}$$

where μ is the Lagrange multiplier. The necessary condition for an extremum is that the derivative of the Lagrangian with respect to its arguments be equal to zero. Clearly, setting the derivative of the Lagrangian with respect to μ equal to zero simply gives back the constraint condition that the vector $\mathbf{n}$ be a unit vector. Setting the derivative of the Lagrangian with respect to $\mathbf{n}$ and μ equal to zero gives the conditions

$$\boxed{\mathbf{S}\mathbf{n} = \mu\mathbf{n}, \qquad \mathbf{n} \cdot \mathbf{n} = 1} \tag{183}$$

It should not be surprising that the search for the direction of extreme normal stress is an eigenvalue problem. It also should be no surprise that all of the techniques for finding the eigenvalues and vectors for $\mathbf{T}$ from Chapter 1 apply equally to $\mathbf{S}$.

There are several interpretations of the principal values μ_i and the principal directions $\mathbf{n}_i$ of the stress tensor. First, we can see that the eigenvalue μ_i is an extreme value of the normal component of the traction vector (no implied sum)

$$\mu_i = \mathbf{n}_i \cdot \mathbf{S}\mathbf{n}_i \equiv \sigma_i \tag{184}$$

From Eqn. (183), we can see that the principal planes are exactly those planes that have no shearing component to the traction vector since (no implied sum)

$$\tau_i = \| \mathbf{S}\mathbf{n}_i - \sigma_i \mathbf{n}_i \| = 0 \tag{185}$$

This observation gives us other ways of stating the question leading to the eigenvalue problem for the stress tensor: (a) Are there coordinate planes, passing through the point in question, on which the traction vector is purely normal to the plane, that is, $\mathbf{t}_\mathbf{n} = \sigma\mathbf{n}$? (b) are there coordinate planes, passing through the point in question, on which the component of shearing traction is identically zero: $[\mathbf{I} - \mathbf{n} \otimes \mathbf{n}]\mathbf{t}_\mathbf{n} = \mathbf{0}$? Each of these statements leads to the same eigenvalue problem, Eqn. (183).

We can determine the principal values and directions for the simple homogeneous states of stress described in Fig. 49 by noting that we are searching for planes having a traction vector pointing in the same direction as the normal vector. For hydrostatic pressure, every plane satisfies this requirement; hence, every direction is a principal direction. Recall that a tensor for which all directions are principal directions has all three eigenvalues equal; in this case they are equal to p. For uniaxial tension we have planes with no traction at all. These are principal planes with corresponding eigenvalue equal to zero. We have two zero eigenvalues for uniaxial tension corresponding to eigenvectors $\mathbf{e}_2$ and $\mathbf{e}_3$. The $\mathbf{e}_1$ direction is also a principal direction with principal value of σ. For pure

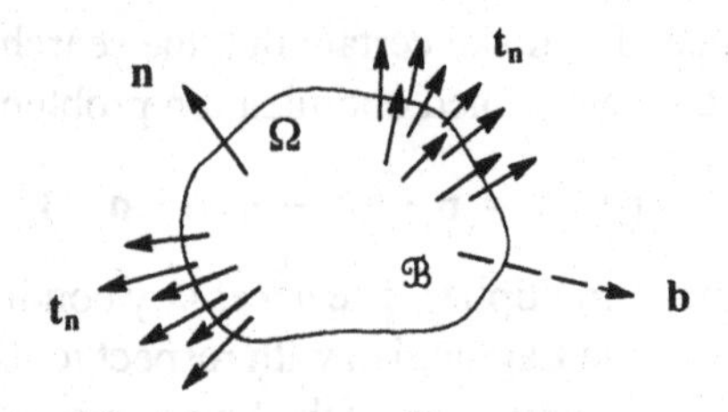

Figure 51 A solid region subjected to surface tractions and body forces

shear, only the $\mathbf{e}_3$ direction corresponds to a zero eigenvalue. The plane with normal $\mathbf{n}$ (and its negative) is a principal plane with eigenvalue equal to τ, while the plane with normal $\mathbf{m}$ (and its negative) is a principal plane with eigenvalue equal to $-\tau$. Clearly, the principal directions are orthogonal for pure shear, as all eigenvalues are distinct.

Differential Equations of Equilibrium

To deduce the general requirements of equilibrium, let us examine the equilibrium of a body $\mathfrak{B}$ with boundary Ω, having normal vector field $\mathbf{n}(\mathbf{x})$ at each point as shown in Fig. 51. The body is subjected to a surface traction field $\mathbf{t}_n(\mathbf{x})$ and a body force field $\mathbf{b}(\mathbf{x})$. Equilibrium of the body $\mathfrak{B}$ requires that

$$\int_{\Omega} \mathbf{t}_n \, dA + \int_{\mathfrak{B}} \mathbf{b} \, dV = \mathbf{0} \tag{186}$$

In other words, the sum of all of the forces acting on the body must be equal to zero for static equilibrium. From Cauchy's formula we have $\mathbf{t}_n = \mathbf{Sn}$. The divergence theorem for a tensor field from Chapter 1 gives the relation

$$\int_{\Omega} \mathbf{Sn} \, dA = \int_{\mathfrak{B}} \operatorname{div} \mathbf{S} \, dV \tag{187}$$

Therefore, we can express Eqn. (186) in the equivalent form

$$\int_{\mathfrak{B}} \left(\operatorname{div} \mathbf{S} + \mathbf{b} \right) dV = \mathbf{0} \tag{188}$$

This argument must hold true for any volume $\mathfrak{B}$ taken as a freebody diagram. Otherwise, it would be possible to find a freebody diagram that does not satisfy equilibrium, in opposition to our definition of the freebody diagram. In order for equilibrium to be satisfied for any body, the integrand must vanish identically, giving the *local form* of equilibrium

$$\boxed{\operatorname{div} \mathbf{S} + \mathbf{b} = \mathbf{0}} \tag{189}$$

The equations of equilibrium of the body, Eqn. (189), are a set of three first-order partial differential equations. If we have a tensor field $\mathbf{S}(\mathbf{x})$, it must satisfy these equations at every point in the body $\mathfrak{B}$ in order to be a solution to our problem. Furthermore, the state of stress must be such that the traction vectors on the surface of our body are equal to the applied tractions where they are prescribed. One can also view the equations of equilibrium as being a set of partial differential equations for the components of the stress tensor. The ith equation is $S_{ij,j} + b_i = 0$, where $(\cdot)_{,j} = \partial(\cdot)/\partial x_j$ is the index notation for partial derivative with respect to the coordinate x_j.

Balance of angular momentum and the symmetry of the stress tensor. For particles, the vanishing of the sum of the moments of the forces is a corollary of the vanishing of the sum of the forces. For solid mechanics, we must make an independent hypothesis that the moment of the forces sum to zero.

Let $\mathbf{r}(\mathbf{x})$ be the position vector from the point O to the point with position vector $\mathbf{x}$. Vanishing of the moment of the surface tractions and the body forces acting on a body $\mathfrak{B}$ with surface boundary Ω can be expressed as

$$\int_{\Omega} \mathbf{r} \times \mathbf{t_n}\, dA + \int_{\mathfrak{B}} \mathbf{r} \times \mathbf{b}\, dV = \mathbf{0} \tag{190}$$

We can transform the surface integral to a volume integral with the divergence theorem. Let us note that $\mathbf{t_n} = \mathbf{Sn}$, and take the dot product of the first term in Eqn. (190) with an arbitrary, constant vector field $\mathbf{h}$ as follows

$$\int_{\Omega} (\mathbf{r} \times \mathbf{Sn}) \cdot \mathbf{h}\, dA = \int_{\Omega} (\mathbf{h} \times \mathbf{r}) \cdot \mathbf{Sn}\, dA = \int_{\Omega} \mathbf{S}^T(\mathbf{h} \times \mathbf{r}) \cdot \mathbf{n}\, dA$$

The first equality is due to the cyclic nature of the triple scalar product, and the second equality comes from the definition of the transpose of a tensor. The divergence theorem allows us to convert the last expression to a volume integral

$$\int_{\Omega} \mathbf{S}^T(\mathbf{h} \times \mathbf{r}) \cdot \mathbf{n}\, dA = \int_{\mathfrak{B}} \operatorname{div}\left[\mathbf{S}^T(\mathbf{h} \times \mathbf{r})\right] dV \tag{191}$$

In order to make further headway, we need to expand the expression for the divergence. Note that for any tensor field $\mathbf{T}$ and any vector field $\mathbf{v}$, the following equality holds (prove this for yourself)

$$\operatorname{div}(\mathbf{T}^T\mathbf{v}) = \mathbf{v} \cdot \operatorname{div}\mathbf{T} + \mathbf{T} \cdot \nabla\mathbf{v}$$

where the scalar or dot product of tensors $\mathbf{A}$ and $\mathbf{B}$ is defined as $\mathbf{A} \cdot \mathbf{B} \equiv A_{ij}B_{ij}$ and is a scalar invariant of the tensor $\mathbf{A}^T\mathbf{B}$; in fact, $\mathbf{A} \cdot \mathbf{B} \equiv \operatorname{tr}(\mathbf{A}^T\mathbf{B})$.

Let us identify the vector $\mathbf{v}$ above with $\mathbf{h} \times \mathbf{r}$ in Eqn. (191). Substituting these expressions into the balance of angular momentum, Eqn. (190), we get

$$\int_{\mathfrak{B}} (\mathbf{h} \times \mathbf{r}) \cdot (\operatorname{div}\mathbf{S} + \mathbf{b})\, dV + \int_{\mathfrak{B}} \mathbf{S} \cdot \nabla(\mathbf{h} \times \mathbf{r})\, dV = \mathbf{0} \qquad (192)$$

The first integral vanishes because $\operatorname{div}\mathbf{S}+\mathbf{b} = \mathbf{0}$ from balance of linear momentum. Noting that $\nabla\mathbf{r} = \mathbf{I}$ (because $\mathbf{r} = \mathbf{x}+\mathbf{c}$, where $\mathbf{c}$ is a constant vector), the tensor $\nabla(\mathbf{h} \times \mathbf{r})$ is the skew-symmetric tensor

$$\nabla(\mathbf{h} \times \mathbf{r}) \equiv \mathbf{H} \qquad (193)$$

where $\mathbf{H}\mathbf{v} = \mathbf{h} \times \mathbf{v}$ for any vector $\mathbf{v}$. Since $\mathbf{S} \cdot \mathbf{H} = \operatorname{tr}(\mathbf{S}^T\mathbf{H}) = \mathbf{e}_j \cdot \mathbf{S}^T\mathbf{H}\mathbf{e}_j$, the expression for balance of angular momentum reduces to

$$\int_{\mathfrak{B}} \mathbf{h} \cdot [\mathbf{e}_j \times \mathbf{S}\mathbf{e}_j]\, dV = 0 \qquad (194)$$

with an implied sum on j. The details of the proof of the equivalence between Eqn. (192) and Eqn. (194) are left as an exercise (see Problem 92). Since $\mathbf{h}$ is arbitrary and since the choice of the region $\mathfrak{B}$ is arbitrary, balance of angular momentum implies

$$\boxed{\mathbf{e}_j \times \mathbf{S}\mathbf{e}_j = \mathbf{0}} \qquad (195)$$

To see what this expression implies, let us compute it in terms of its components

$$\mathbf{e}_j \times \mathbf{S}\mathbf{e}_j = \mathbf{e}_j \times \left(S_{mn}[\mathbf{e}_m \otimes \mathbf{e}_n]\mathbf{e}_j\right) = S_{mn}\delta_{nj}\mathbf{e}_j \times \mathbf{e}_m = S_{mj}\epsilon_{ijm}\mathbf{e}_i$$

Writing out these expressions, we get the explicit relations

$$[S_{23}-S_{32}]\mathbf{e}_1 + [S_{31}-S_{13}]\mathbf{e}_2 + [S_{12}-S_{21}]\mathbf{e}_3 = \mathbf{0}$$

Since the base vectors are independent and nonzero, the only way balance of angular momentum can hold is if the terms in brackets independently vanish, that is, if the components of the stress tensor satisfy $S_{12} = S_{21}$, $S_{31} = S_{13}$, and $S_{23} = S_{32}$. Therefore, balance of angular momentum implies that the stress tensor must be symmetric

$$\boxed{\mathbf{S}^T = \mathbf{S}} \qquad (196)$$

The ramification of the symmetry of the stress tensor is that it really only takes six independent quantities to fully describe the state of stress at a point, rather than nine.

Summary. We have found that the stress tensor $\mathbf{S}$ plays the key role in the description of the transmission of forces through a solid body. Application of the notion of equilibrium of forces and the moments of those forces about an arbitrary point lead to three equations governing the spatial variation of the stress tensor. These important formulas are summarized below both in direct and component notation

$$\mathbf{t_n} = \mathbf{S}\mathbf{n} \qquad \mathbf{t_n} = S_{ij} n_j \mathbf{e}_i$$
$$\operatorname{div}\mathbf{S} + \mathbf{b} = \mathbf{0} \qquad (S_{ij,j} + b_i)\,\mathbf{e}_i = \mathbf{0}$$
$$\mathbf{S} = \mathbf{S}^T \qquad S_{ij} = S_{ji}$$

The first equation allows us to relate applied tractions acting on the surface of the body to the stress field inside the body. It also gives us a vehicle to form freebody diagrams by exposing an interior surface in the body and replacing the tractions exerted by the removed portion of the body. The second equation governs the rate of change of the stress tensor. The third guarantees that balance of moments holds for any piece of the body. The second equation is a first-order partial differential equation for the stress field. Any stress state that satisfies these equations is an equilibrium stress state.

Examples

To gain an appreciation for the requirements of equilibrium and the relationship between the stress tensor and the traction vector we shall examine some simple examples of equilibrium stress states.

Example 19. *Rigid block under its own weight.* Consider a block of uniform density ϱ, height h, and base area $A = \ell^2$, subjected only to the force of gravity and fixed at the base (i.e., at $x_3 = 0$), as shown in Fig. 52.

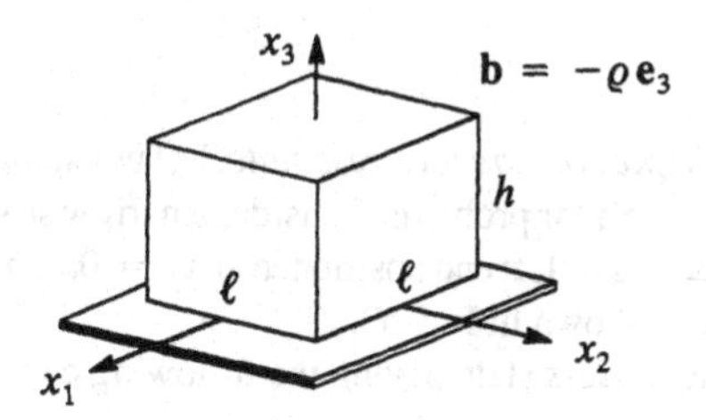

Figure 52 Block stressed under its own weight

The body force $\mathbf{b} = -\varrho\,\mathbf{e}_3$ is constant. The stress tensor is given as follows

$$\mathbf{S}(\mathbf{x}) = \varrho(x_3 - h)[\mathbf{e}_3 \otimes \mathbf{e}_3]$$

The divergence of the stress tensor is $\operatorname{div}\mathbf{S} = S_{ij,j}\,\mathbf{e}_i = S_{33,3}\,\mathbf{e}_3 = \varrho\,\mathbf{e}_3$. Substituting into the equations of equilibrium, we see that the equilibrium equations are satisfied, i.e., $\operatorname{div}\mathbf{S} + \mathbf{b} = \varrho\,\mathbf{e}_3 - \varrho\,\mathbf{e}_3 = \mathbf{0}$. Thus, the stress tensor field satisfies the equations of equilibrium for all points in the body.

We must also verify that the sides and top of the block are free of traction (there are no applied forces). The requirement that a surface be traction-free is quite different than having the stress tensor equal to zero for values of $\mathbf{x}$ on the surface of the body. (Of course, we know that it only makes sense to talk about traction vectors on surfaces.) On the lateral sides, the normal vectors are $\mathbf{e}_1$, $\mathbf{e}_2$, $-\mathbf{e}_1$, and $-\mathbf{e}_2$, all of which give zero tractions when multiplied by the stress tensor. On the top face we have a normal of $\mathbf{e}_3$, but $x_3 = h$, so the stress tensor is zero there. Again, we find no traction on the top face. On the bottom face we have $-\mathbf{e}_3$ so that $\mathbf{S}(-\mathbf{e}_3) = \varrho h \mathbf{e}_3$, a force with magnitude ϱh pointing upward. The traction times the area is $\varrho h \ell^2 \mathbf{e}_3$, the resistance to the total weight of the block.

The problem of finding a stress state that satisfies the differential equations of equilibrium throughout the body and gives the applied tractions at the surface is more difficult than we might imagine, particularly if we are hoping to express these solutions in terms of simple functional forms such as polynomials. To simplify the situation, early researchers often posed the question in reverse: Given a particular functional form, does it solve an interesting problem in mechanics? Taking a function, and verifying that it satisfies all of the governing equations, is generally a simple task. If a function satisfies the equations then, in a certain sense, that function is what you were looking for. As the use of numerical methods in mechanics has grown, the drive to find closed-form solutions to problems has all but disappeared. There is great value in having a closed-form solution to a problem, but there currently exist more effective means to get answers to engineering problems. The reader interested in closed-form solutions to elasticity problems might wish to consult the text by Timoshenko and Goodier (1970).

Example 20. *A simple polynomial stress state.* Let us examine a particular solution to an essentially planar problem. Consider a narrow strip of length ℓ, width b, and depth $2h$, having its left end positioned at $x_1 = 0$, and its middle line positioned at $x_2 = 0$, as shown in Fig. 53.

Let us examine a stress state having the following components

$$\mathbf{S} \sim \frac{3q_o}{4bh^3}\begin{bmatrix} \left(x_1^2 x_2 - \frac{2}{3}x_2^3\right) & x_1\left(h^2 - x_2^2\right) & 0 \\ x_1\left(h^2 - x_2^2\right) & \frac{1}{3}\left(x_2^3 - 3x_2 h^2 - 2h^3\right) & 0 \\ 0 & 0 & 0 \end{bmatrix}$$

We must first verify that the stress components satisfy equilibrium. Let us assume that the body forces are $\mathbf{b} = \mathbf{0}$ for the present problem. Hence, we must have $S_{ij,j} = 0$. Writing out these equations (dividing each one through by the common factor $3q_o/4bh^3$), we have

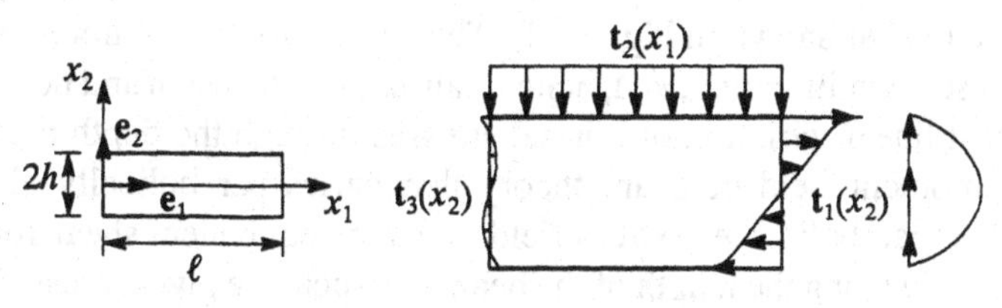

(a) Geometry of strip (c) Tractions on certain boundaries

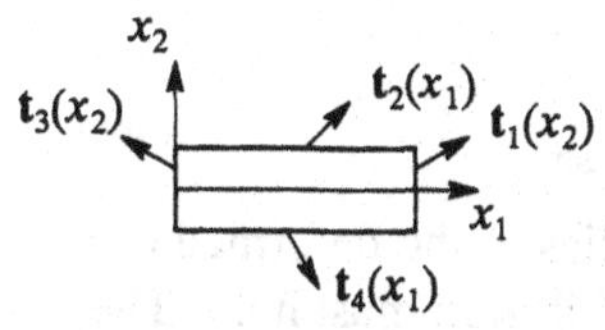

(b) Convention on tractions

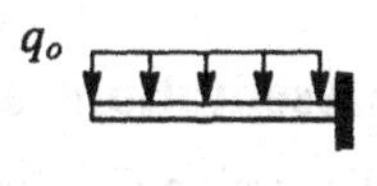

(d) Closest problem from engineering beam theory

Figure 53 Example 20 stress field problem

$$S_{11,1} + S_{12,2} + S_{13,3} = \left(2x_1x_2\right) + \left(-2x_1x_2\right) + 0 = 0$$
$$S_{21,1} + S_{22,2} + S_{23,3} = \left(h^2-x_2^2\right) + \left(x_2^2-h^2\right) + 0 = 0$$
$$S_{31,1} + S_{32,2} + S_{33,3} = 0 + 0 + 0 = 0$$

Thus, equilibrium is satisfied in the interior of the body. It remains to be seen what the surface tractions are. Along the right side we have $x_1 = \ell$ and $\mathbf{n} = \mathbf{e}_1$; therefore, the traction on that face is

$$\mathbf{t}_1(x_2) = \frac{3q_o}{4bh^3}\left(x_2\ell^2 - \tfrac{2}{3}x_2^3\right)\mathbf{e}_1 + \frac{3q_o\ell}{4bh^3}\left(h^2-x_2^2\right)\mathbf{e}_2$$

The distribution of normal tractions (the $\mathbf{e}_1$ component) is predominantly a linear variation with a cubic part superimposed. If the length is large relative to the depth, that is, $\ell \gg h$, then the cubic part becomes relatively small because $|x_2| \leq h$. The shearing tractions (the $\mathbf{e}_2$ component) are parabolic and vanish at the top and bottom fibers of the strip. Along the top we have $x_2 = h$ and $\mathbf{n} = \mathbf{e}_2$; therefore, the traction on that face is

$$\mathbf{t}_2(x_1) = -\frac{q_o}{b}\mathbf{e}_2$$

giving a constant normal traction along the length of the strip. Along the left side we have $x_1 = 0$ and $\mathbf{n} = -\mathbf{e}_1$; therefore, the traction on that face is

$$\mathbf{t}_3(x_2) = \frac{q_o x_2^3}{2bh^3}\mathbf{e}_1$$

The bottom side has $x_2 = -h$ and $\mathbf{n} = -\mathbf{e}_2$; therefore, the traction on that face is $\mathbf{t}_4(x_1) = \mathbf{0}$.

The astute student of beam theory will recognize this problem as being similar to a beam, free at the left end and fixed at the right end, subject to a uniform transverse load as shown in Fig. 53(d). The cubic variation of tractions at the left end is self-equilibrating, and, hence, causes no net tension and no moment. The almost linear variation of normal stresses through the depth creates the bending moment field of beam theory that varies parabolically along the length. The parabolic shear stress field gives the equivalent shear force that varies linearly along the length of the beam. As such, the given stress field can be viewed as a nearly exact solution to a beam-bending problem.

Alternative Representations of Stress

All the preceding discussion of stress applies to the deformed configuration $\boldsymbol{\phi}(\mathbf{z})$. That is the configuration where equilibrium must hold. The deformed configuration is the natural configuration in which to characterize stress. Since Cauchy had so much to do with the definition of stress, we call the stress $\mathbf{S}$ the *Cauchy stress tensor*. As we noted in Chapter 2, there can be computational advantages in referring all quantities back to the undeformed configuration of the body because often that configuration has geometric features and symmetries that are lost going through the deformation. A volleyball, for example, initially has a nice spherical shape that is lost under the force of a hand spiking it. An automobile generally has a nicer geometry before a crash than after. We will often want to analyze initially straight beams that become curved under loading, or initially flat plates that become curved surfaces under loading. Referring back to the undeformed configuration is really nothing more than a change of variable, with the deformation map describing that change.

If we know the map from the undeformed configuration to the deformed configuration then we can relate geometric quantities in the two configurations. Specifically, we know how areas are mapped by the deformation. Since traction vectors are nothing more than force per unit of area, we might expect that the Piola transformation plays a role in defining other stress tensors. Like strain, stress can be defined in many ways. We examine two alternatives in this section.

The first Piola-Kirchhoff stress tensor. Let us suppose that we have a traction vector $\mathbf{t}_{\mathbf{n}}\, da$ on a plane with normal $\mathbf{n}$ in the deformed configuration, as shown in Fig. 54. We can trace back through the deformation map $\boldsymbol{\phi}(\mathbf{z})$ what the corresponding plane was in the undeformed configuration. If the plane has area da and normal $\mathbf{n}$ in the deformed configuration, then it had area dA and normal $\mathbf{m}$ in the undeformed configuration. The relationship between the two areas and normals is given by Nanson's formula (see Chapter 2). Let us define a traction vector $\mathbf{t}^{o}_{\mathbf{m}}$ in the undeformed geometry that results in the same total force as the traction in the deformed configuration. To wit

$$\mathbf{t}^{o}_{\mathbf{m}}\, dA \equiv \mathbf{t}_{\mathbf{n}}\, da \tag{197}$$

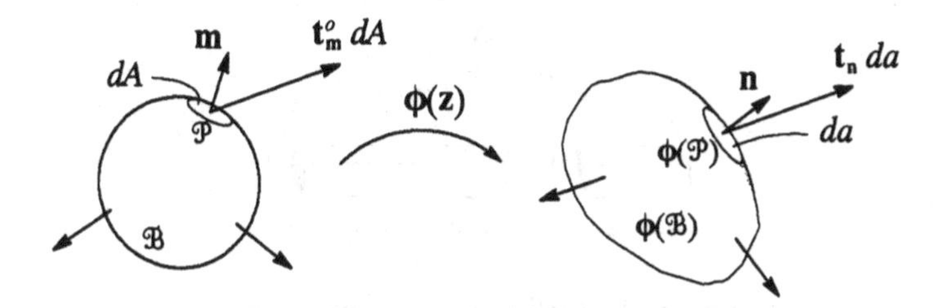

Figure 54 Definition of the first Piola-Kirchhoff stress tensor

Clearly, the two traction vectors are the same except for magnitude, the difference in magnitude resulting from the different areas to which they are reckoned. The traction vector $\mathbf{t}^o_\mathbf{m}$ is the total force on the surface per unit of undeformed area, while $\mathbf{t}_\mathbf{n}$ is the total force per unit of deformed area. From Cauchy's formula we know that $\mathbf{t}_\mathbf{n} = \mathbf{Sn}$, where $\mathbf{S}$ is the Cauchy stress tensor. We also know that the transformation of areas is given by $\mathbf{n}\,da = J\mathbf{F}^{-T}\mathbf{m}\,dA$, where $J \equiv \det \mathbf{F}$. We can get an analogous relationship and an alternative definition of stress by substituting this expression into Eqn. (197)

$$\mathbf{t}^o_\mathbf{m}\,dA = \mathbf{t}_\mathbf{n}\,da = \mathbf{Sn}\,da = J\mathbf{SF}^{-T}\mathbf{m}\,dA \tag{198}$$

Let us define the first Piola-Kirchhoff stress tensor $\mathbf{P}$ in terms of the Cauchy stress tensor and the deformation gradient $\mathbf{F}$ as

$$\mathbf{P} \equiv J\mathbf{SF}^{-T} \tag{199}$$

where, again, $J = \det \mathbf{F}$. The components of the first Piola-Kirchhoff stress tensor are defined relative to a tensor basis as follows

$$\mathbf{P} = P_{ij}\big[\mathbf{e}_i \otimes \mathbf{g}_j\big] \tag{200}$$

(You can explain why by noting the component expressions of $\mathbf{S}$ and $\mathbf{F}$). Why did we define the first Piola-Kirchhoff stress tensor in such a strange way? We did it so that it would satisfy a Cauchy-like relationship analogous to $\mathbf{t}_\mathbf{n} = \mathbf{Sn}$. Notice that, according to Eqn. (198), we have the relationship

$$\mathbf{t}^o_\mathbf{m} = \mathbf{Pm} \tag{201}$$

where $\mathbf{m}$ is a unit vector defined in the undeformed configuration. Now it is a simple matter to recognize that we can establish equilibrium for a region $\phi(\mathcal{B})$ by the formula for change of variables for integration as follows

$$\int_{\phi(\Omega)} \mathbf{t}_\mathbf{n}(\mathbf{x})\,da + \int_{\phi(\mathcal{B})} \mathbf{b}(\mathbf{x})\,dv = \int_{\Omega} \mathbf{t}^o_\mathbf{m}(\mathbf{z})\,dA + \int_{\mathcal{B}} \mathbf{b}^o(\mathbf{z})\,dV \tag{202}$$

where $\mathbf{b}^o(\mathbf{z}) = J\mathbf{b}(\phi(\mathbf{z})) = J\mathbf{b}(\mathbf{x})$ is the body force defined with respect to the undeformed configuration, since $dv = J\,dV$. Consider, for example, the rec-

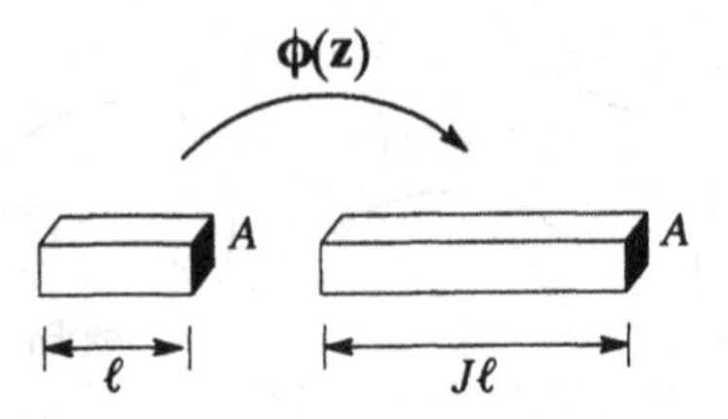

Figure 55 Illustration of conservation of mass

tangular parallelepiped shown in Fig. 55, subjected to a deformation map that preserves the area A but increases the length from ℓ to $J\ell$ (hence, the ratio of deformed volume to original volume is J). Let the body force be the action of mass in a gravitational field, that is, $\mathbf{b}^o = \varrho_o \mathbf{g}$, where ϱ_o is the initial density of the material and $\mathbf{g}$ is a constant vector representing the gravitational force per unit of mass. The body force in the deformed configuration is $\mathbf{b} = \varrho_o \mathbf{g}/J$. The interpretation, of course, is that the density of the material is less in the deformed configuration (for $J > 1$). The current density is $\varrho = \varrho_o/J$. In general, conservation of mass implies that

$$\int_{\phi(\mathcal{B})} \varrho(\mathbf{x})\, dv = \int_{\mathcal{B}} \varrho(\phi(\mathbf{z}))\, J\, dV \equiv \int_{\mathcal{B}} \varrho_o(\mathbf{z})\, dV \tag{203}$$

Hence, by definition of ϱ relative to ϱ_o, we again have $\varrho = \varrho_o/J$ simply as a consequence of the formula for change of variables in integration.

The divergence theorem can be applied to the area integrals just as it was before to get local equations of equilibrium. These equations also can be expressed in terms of the undeformed geometry and are summarized in the following box

$\mathbf{t}^o_{\mathbf{m}} = \mathbf{P}\mathbf{m}$	$\mathbf{t}^o_{\mathbf{m}} = P_{ij} m_j \mathbf{e}_i$
$\text{DIV}\,\mathbf{P} + \mathbf{b}^o = \mathbf{0}$	$\left(P_{ij,j} + b^o_i\right)\mathbf{e}_i = \mathbf{0}$
$\mathbf{P}\mathbf{F}^T = \mathbf{F}\mathbf{P}^T$	$P_{ij} F_{kj} = F_{ij} P_{kj}$

The divergence operator, in the present case, involves derivatives with respect to the coordinates $\mathbf{z}$ of the undeformed configuration because it follows from the analogy with earlier derivation that

$$\text{DIV}\,\mathbf{P} \equiv \lim_{\mathcal{V}(\mathcal{B})\to 0} \frac{1}{\mathcal{V}(\mathcal{B})} \int_{\Omega} \mathbf{P}\mathbf{m}\, dA \tag{204}$$

where Ω is the boundary of the region $\mathcal{B}$ and has unit normal vector field $\mathbf{m}$. It is straightforward to demonstrate that the expression for the divergence of $\mathbf{P}$ in Cartesian coordinates is given by the formula

$$\text{DIV}\,\mathbf{P} = \sum_{i=1}^{3} \frac{\partial \mathbf{P}(\mathbf{z})}{\partial z_i}\,\mathbf{g}_i \tag{205}$$

where $\{\mathbf{g}_1, \mathbf{g}_2, \mathbf{g}_3\}$ are the base vectors for the $\mathbf{z}$ coordinate system. To remind us that we are differentiating with respect to $\mathbf{z}$ rather than $\mathbf{x}$, we will denote the divergence as DIV (as opposed to div used previously when differentiating with respect to $\mathbf{x}$).

The symmetry condition on the first Piola-Kirchhoff stress tensor is expressed as $\mathbf{P}\mathbf{F}^T = \mathbf{F}\mathbf{P}^T$, and arises naturally from balance of angular momentum, which, when expressed in the undeformed configuration, reduces to

$$\sum_{j=1}^{3} \mathbf{F}\mathbf{g}_j \times \mathbf{P}\mathbf{g}_j = \mathbf{0} \tag{206}$$

The derivation of Eqn. (206) is nearly identical to that of Eqn. (195), with most of the differences accruing from the substitution $\mathbf{t}_n\,da = \mathbf{t}^o_m\,dA$. The other change is that the position vectors $\mathbf{r}(\mathbf{x})$ must be referred back to the undeformed configuration as $\mathbf{r}(\boldsymbol{\phi}(\mathbf{z}))$. When it comes time to take the derivative of $\mathbf{r}$ with respect to z_j, we must use the chain rule to get

$$\frac{\partial \mathbf{r}(\boldsymbol{\phi}(\mathbf{z}))}{\partial z_j} = \frac{\partial \mathbf{r}(\boldsymbol{\phi})}{\partial \boldsymbol{\phi}} \frac{\partial \boldsymbol{\phi}(\mathbf{z})}{\partial \mathbf{z}} \frac{\partial \mathbf{z}}{\partial z_j} = \mathbf{F}\mathbf{g}_j \tag{207}$$

Since $\mathbf{r}(\mathbf{z}) = \boldsymbol{\phi}(\mathbf{z}) - \mathbf{c}$ is the position vector in the deformed configuration, shifted from the origin by a constant vector $\mathbf{c}$ to get to the point about which moments are summed, the derivative $\partial \mathbf{r}/\partial \boldsymbol{\phi} = \mathbf{I}$.

We must, of course, satisfy equilibrium in the deformed configuration. However, we usually know more about the geometry of the undeformed configuration than we do about the deformed configuration because the undeformed geometry is usually given as part of the problem data (e.g., find the stresses and deformations of an initially unstressed and undeformed body whose geometry is completely described in the undeformed configuration). Hence, you might prefer the first Piola-Kirchhoff stress tensor to the Cauchy stress tensor, even though they measure exactly the same state of stress.

Both expressions of the equilibrium equations establish equilibrium in the deformed configuration. Often we are given a set of loads and are asked to find the deformation map, an inherently nonlinear problem. If the deformations are small, a linear approximation to our problem is appropriate. As the deformations get small, the deformation map approaches the identity $\mathbf{I}$, and, hence, the difference between the two stress tensors $\mathbf{P}$ and $\mathbf{S}$ vanishes. Accordingly, for a linearized problem, we speak only of *the* stress tensor. There is only one.

The second Piola-Kirchhoff stress tensor. One of the unsettling observations about the first Piola-Kirchhoff stress tensor is that the symmetry condition that arises from the balance of angular momentum involves the deformation gradient **F**, and, hence, cannot be trivially satisfied. We can define a new stress tensor defined on the reference configuration that does have the same kind of symmetry that the Cauchy stress tensor does. This tensor is called the *second Piola-Kirchhoff stress tensor* $\mathbf{\Sigma}$, and is defined as

$$\mathbf{\Sigma} \equiv \mathbf{F}^{-1}\mathbf{P} = J\mathbf{F}^{-1}\mathbf{S}\mathbf{F}^{-T} \tag{208}$$

where $J \equiv \det \mathbf{F}$. The physical significance of the tensor $\mathbf{\Sigma}$ is not as clear as **S** and **P**, but it has some other advantages in computation. The second Piola-Kirchhoff stress tensor has the component form

$$\mathbf{\Sigma} = \Sigma_{ij}\left[\mathbf{g}_i \otimes \mathbf{g}_j\right] \tag{209}$$

Again, the basis is inherited from the tensor description.

It should be clear that we could go on defining measures of stress endlessly, just as we could with strain. The two additional stresses we have defined are useful when the time comes to relate stress to strain through constitutive equations. The first Piola-Kirchhoff stress **P** is most naturally related to the deformation gradient **F**, and the second Piola-Kirchhoff stress $\mathbf{\Sigma}$ is most naturally related to the Lagrangian strain **E**. The relationships between the different stress measures are summarized as follows

$$\begin{aligned} \mathbf{\Sigma} &= \mathbf{F}^{-1}\mathbf{P} = J\mathbf{F}^{-1}\mathbf{S}\mathbf{F}^{-T} \\ \mathbf{F}\mathbf{\Sigma} &= \mathbf{P} = J\mathbf{S}\mathbf{F}^{-T} \\ \frac{1}{J}\mathbf{F}\mathbf{\Sigma}\mathbf{F}^{T} &= \frac{1}{J}\mathbf{P}\mathbf{F}^{T} = \mathbf{S} \end{aligned}$$

Example 21. *Comparison of the three stress tensors.* Consider a bar of unit area and length ℓ with its longitudinal axis along $\mathbf{g}_1$. The bar is subject to uniaxial tension and a deformation that stretches the bar by an amount λ and rotates it by an amount θ, as shown in Fig. 56.

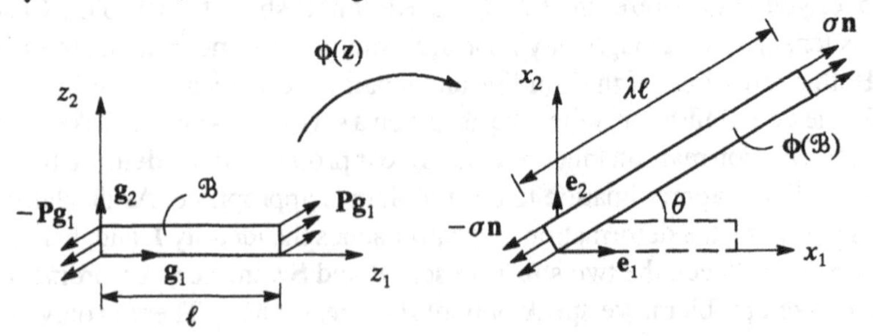

Figure 56 Example of the first Piola-Kirchhoff stress tensor

The bar does not bend. Let us assume that there are no body forces. The Cauchy stress tensor is given by $\mathbf{S} = \sigma[\mathbf{n} \otimes \mathbf{n}]$, where σ is the constant intensity of the stress. The unit normal vector to the deformed cross-section of the bar can be written as $\mathbf{n} = \cos\theta\mathbf{e}_1 + \sin\theta\mathbf{e}_2$. The traction vector acting on the face with normal $\mathbf{n}$ is then $\mathbf{Sn} = \sigma\mathbf{n}$, as shown in the figure. The deformation map is

$$\phi(\mathbf{z}) = (\lambda z_1\cos\theta - a z_2\sin\theta)\mathbf{e}_1 + (\lambda z_1\sin\theta + a z_2\cos\theta)\mathbf{e}_2 + z_3\mathbf{e}_3$$

where λ is the constant proportion of stretch of the bar. The components of the deformation gradient can be computed as

$$\mathbf{F} \sim \begin{bmatrix} \lambda\cos\theta & -a\sin\theta & 0 \\ \lambda\sin\theta & a\cos\theta & 0 \\ 0 & 0 & 1 \end{bmatrix}$$

and the determinant of the deformation gradient can be evaluated as $\det\mathbf{F} = a\lambda$, indicating that the ratio of the deformed volume to the original volume is $a\lambda$. Since the length of the bar changed by λ we can conclude that a gives the ratio of deformed to undeformed cross-sectional area. We can compute the components of the first Piola-Kirchhoff stress tensor as follows

$$\mathbf{P} = J\mathbf{S}\mathbf{F}^{-T} \sim \sigma\begin{bmatrix} \cos^2\theta & \cos\theta\sin\theta & 0 \\ \cos\theta\sin\theta & \sin^2\theta & 0 \\ 0 & 0 & 0 \end{bmatrix}\begin{bmatrix} a\cos\theta & -\lambda\sin\theta & 0 \\ a\sin\theta & \lambda\cos\theta & 0 \\ 0 & 0 & a\lambda \end{bmatrix}$$

which gives the result

$$\mathbf{P} = a\sigma\cos\theta[\mathbf{e}_1 \otimes \mathbf{g}_1] + a\sigma\sin\theta[\mathbf{e}_2 \otimes \mathbf{g}_1]$$

The second Piola-Kirchhoff stress tensor can be computed as $\boldsymbol{\Sigma} = \mathbf{F}^{-1}\mathbf{P}$

$$\boldsymbol{\Sigma} = \frac{a\sigma}{\lambda}[\mathbf{g}_1 \otimes \mathbf{g}_1]$$

We can clearly see the lack of symmetry of $\mathbf{P}$ in the preceding example (actually, because of the mixed basis of $\mathbf{P}$, it makes no sense to talk about symmetry in the first place). Tangent vectors to the undeformed longitudinal axis are $\mathbf{g}_1$. These vectors map to vectors $\mathbf{n} = \mathbf{F}\mathbf{g}_1$ in the deformed configuration. The vector $\mathbf{P}\mathbf{g}_1 = a\sigma(\cos\theta\mathbf{e}_1 + \sin\theta\mathbf{e}_2)$ is the traction vector acting on the cross section, but referred to the undeformed configuration, as shown in the figure. Notice that it points in the same direction as $\mathbf{Sn}$. It differs in magnitude from $\mathbf{Sn}$ because the area changed by the factor a. Clearly, the traction vectors $\mathbf{P}\mathbf{g}_1$ at the right end of the bar equilibrate the traction vectors $-\mathbf{P}\mathbf{g}_1$ at the left end in the sense of balance of linear momentum. However, it appears that these tractions do not satisfy vanishing of the moment of the forces. The resolution of the apparent paradox comes from recognizing how balance of angular momentum is implemented for the first Piola-Kirchhoff stress. Specifically, balance of angular momentum is assured by the symmetry condition $\mathbf{P}\mathbf{F}^T = \mathbf{F}\mathbf{P}^T$, which is certainly true in the present case. Therefore, the net moment of the forces does indeed vanish in the deformed configuration. It makes no sense to ask that those same forces vanish in a Newtonian sense when referred back to the undeformed configuration. We have derived the sense in which those tractions satisfy equilibrium.

The preceding example illustrates the differences between the three stress tensors. It is important to notice where each is defined, either on the deformed configuration or on the undeformed configuration, and what governing equations they satisfy. Like the different strain measures of Chapter 2, the stress measures all describe exactly the same state of stress (nature does not know what coordinate system you will choose to describe the body). The preference of one stress tensor over another will be dictated by the choice of constitutive model and the computational strategy. Theoretically, all formulations are equivalent.

Additional Reading

Y. C. Fung, *Foundations of solid mechanics*, Prentice Hall, Englewood Cliffs, N.J., 1965.

M. E. Gurtin, "The linear theory of elasticity," *Mechanics of solids*, Vol. II (C. Truesdell, ed.), Springer-Verlag, New York, 1972.

L. E. Malvern, *Introduction to the mechanics of a continuous medium*, Prentice Hall, Englewood Cliffs, N.J., 1969.

I. S. Sokolnikoff, *Mathematical theory of elasticity*, 2nd ed., McGraw-Hill, New York, 1956.

S. P. Timoshenko and J. N. Goodier, *Theory of elasticity*, McGraw-Hill, New York, 1970.

Problems

81. The stress tensor $\mathbf{S}$ at a certain point in a body has components with respect to a set of coordinate axes $\{x_1, x_2, x_3\}$ of

$$\mathbf{S} \sim \begin{bmatrix} 5 & 3 & -8 \\ 3 & 0 & -3 \\ -8 & -3 & 11 \end{bmatrix}$$

On a plane whose normal $\mathbf{n}$ makes equal acute angles with the coordinate axes, find the traction vector $\mathbf{t}_n$, the component of the traction vector that is normal to the plane, and the shearing component of the traction vector.

82. Resolve Problem 81 with S_{22} changed to $10\sqrt{3}$.

83. Find the principal values and principal directions of the two stress tensors having components with respect to the standard basis of

$$\mathbf{S} \sim \begin{bmatrix} 3 & 1 & 2 \\ 1 & -6 & 0 \\ 2 & 0 & 15 \end{bmatrix} \qquad \mathbf{S} \sim \begin{bmatrix} 20 & -5 & 0 \\ -5 & -10 & 0 \\ 0 & 0 & 0 \end{bmatrix}$$

84. The condition called *plane stress* is characterized by the stress state $S_{33} = S_{23} = S_{13} = 0$. Show that if the remaining stress components are given by

$$S_{11} = \frac{\partial^2 \psi(x_1, x_2)}{\partial x_2^2}, \quad S_{22} = \frac{\partial^2 \psi(x_1, x_2)}{\partial x_1^2}, \quad S_{12} = -\frac{\partial^2 \psi(x_1, x_2)}{\partial x_1 \partial x_2}$$

and the body force $\mathbf{b} = \mathbf{0}$, then the equations of equilibrium are satisfied for any sufficiently smooth function $\psi(x_1, x_2)$. How smooth must the function be?

85. The state of stress at a point is characterized by the stress tensor $\mathbf{S}$, given below

$$\mathbf{S} \sim \begin{bmatrix} 4 & -4 & 0 \\ -4 & 4 & 0 \\ 0 & 0 & 8 \end{bmatrix}$$

Consider the vectors $\mathbf{n}$ and $\mathbf{m}$ given by

$$\mathbf{n} = \frac{1}{\sqrt{3}}(\mathbf{e}_1 - \mathbf{e}_2 - \mathbf{e}_3), \quad \mathbf{m} = \frac{1}{\sqrt{2}}(\mathbf{e}_1 + \mathbf{e}_2)$$

Are the two given vectors $\mathbf{n}$ and $\mathbf{m}$ eigenvectors of $\mathbf{S}$? Find the principal stresses for the given stress tensor $\mathbf{S}$.

86. Consider the tetrahedron shown in the figure, with edges along the coordinate axes of length 4, 2, and 1, respectively. The state of stress in the tetrahedron is given by the expression

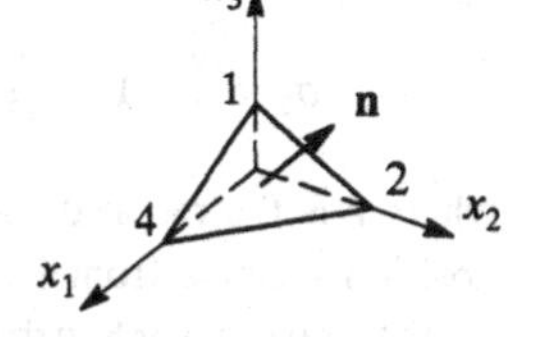

$$\mathbf{S}(\mathbf{x}) = S_o[\mathbf{x} \otimes \mathbf{x}]$$

where S_o is a constant and $\mathbf{x}$ is the position vector. The equation of the oblique plane is $x_1 + 2x_2 + 4x_3 = 4$. Compute the body force $\mathbf{b}$ required for the tetrahedron to be in equilibrium. Compute the tractions on the four faces of the tetrahedron required for equilibrium.

87. Find an expression for the following derivatives of the principal invariants with respect to tensor components

$$\frac{\partial I_S}{\partial S_{mn}}, \qquad \frac{\partial II_S}{\partial S_{mn}}, \quad \text{and} \quad \frac{\partial III_S}{\partial S_{mn}}$$

88. Consider the sphere of radius R shown in the figure. The state of stress in the sphere is given by the stress field

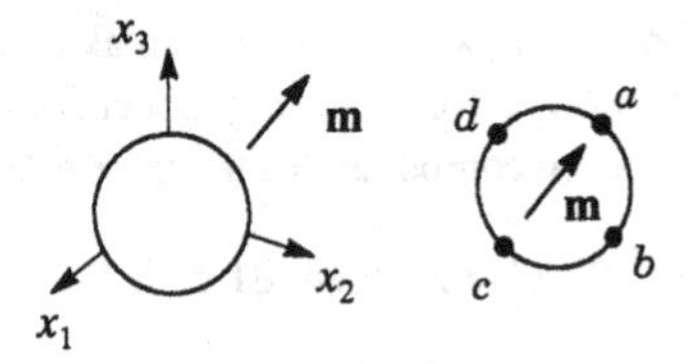

$$\mathbf{S}(\mathbf{x}) = S_o\big[\mathbf{m} \otimes \mathbf{x} + \mathbf{x} \otimes \mathbf{m}\big]$$

where S_o is a constant, $\mathbf{x}$ is the position vector of the point in question, and $\mathbf{m}$ is a constant unit vector field. What is the body force $\mathbf{b}(\mathbf{x})$ required for equilibrium? Compute the tractions acting on the surface of sphere. Sketch the traction vectors at points a, b, c, and d shown on the figure (line segment ca points in the direction of $\mathbf{m}$).

89. Consider a state of stress $\mathbf{S}$ that has principal values $\{\sigma_1, \sigma_2, \sigma_3\}$ with corresponding (orthogonal) principal directions $\{\mathbf{n}_1, \mathbf{n}_2, \mathbf{n}_3\}$. Let us consider one of the eight (octahedral) planes whose normal vector $\mathbf{m}_i$ makes equal angles with the principal directions (one of the eight vectors is shown in the sketch). Show that the normal component of the traction on any of the eight octahedral planes is given by $\sigma = (\sigma_1+\sigma_2+\sigma_3)/3$. Show that the shearing component of the traction on any of the eight octahedral planes is

$\mathbf{n}_3$, $\mathbf{m}_1$, $\mathbf{n}_2$, $\mathbf{n}_1$

$$\mathbf{m}_1 = \frac{1}{\sqrt{3}}(\mathbf{n}_1+\mathbf{n}_2+\mathbf{n}_3)$$

$$\tau^2 = \tfrac{1}{9}\big[(\sigma_1-\sigma_2)^2 + (\sigma_2-\sigma_3)^2 + (\sigma_3-\sigma_1)^2\big] \equiv \tau_{oct}^2$$

Express τ_{oct} in terms of the principal invariants of the stress tensor I_S and II_S.

90. A thick-walled sphere of inside radius 1 and outside radius 2 is subjected to an internal pressure of magnitude p. The principal directions of stress are the radial and tangential directions. The principal values of stress are given by the expressions

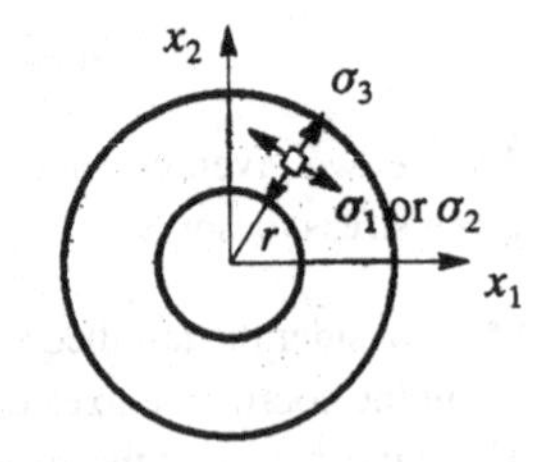

$$\sigma_1 = \sigma_2 = \frac{p}{7}\left(1 + \frac{4}{r^3}\right), \quad \sigma_3 = \frac{p}{7}\left(1 - \frac{8}{r^3}\right)$$

where r is the radial distance to an arbitrary point (with position vector $\mathbf{x}$) from the center of the sphere, i.e., $r^2 = x_1^2+x_2^2+x_3^2$. Find the expression for the stress tensor $\mathbf{S}$ in the cartesian coordinate system $\{x_1, x_2, x_3\}$. Prove that the outside surface of the sphere is traction free. Find the body force $\mathbf{b}$ that must be present to maintain equilibrium.

91. Consider a state of stress $\mathbf{S}$ that has principal values $\{\sigma_1, \sigma_2, \sigma_3\}$ with corresponding (orthogonal) principal directions $\{\mathbf{n}_1, \mathbf{n}_2, \mathbf{n}_3\}$. Let us consider a plane parallel to $\mathbf{n}_3$ described by the normal vector

$$\mathbf{m} \equiv \cos\theta\,\mathbf{n}_1 + \sin\theta\,\mathbf{n}_2$$

parameterized by the angle θ as shown in the figure. Show, using the spectral decomposition theorem, that the traction vector and its normal component, acting on this plane, are given by

$$\mathbf{t}_{\mathbf{m}}(\theta) = \sigma_1 \cos\theta\,\mathbf{n}_1 + \sigma_2 \sin\theta\,\mathbf{n}_2, \qquad \sigma(\theta) = \sigma_1 \cos^2\theta + \sigma_2 \sin^2\theta$$

Show that the shear and normal components of the traction vector satisfy the relationship

$$\tau^2 + \sigma^2 = \sigma_1^2 \cos^2\theta + \sigma_2^2 \sin^2\theta$$

Now let

$$\bar{\sigma} \equiv \tfrac{1}{2}(\sigma_1 + \sigma_2), \quad \text{and} \quad \varrho \equiv \tfrac{1}{2}(\sigma_1 - \sigma_2)$$

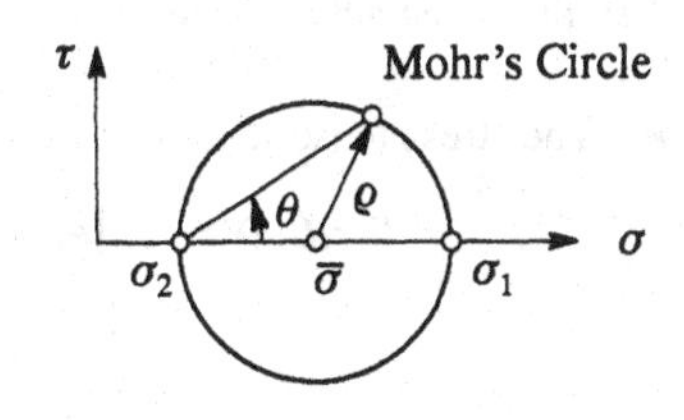

Show that the shear and normal components of the traction vector satisfy the relationship

$$\tau^2 + (\sigma - \bar{\sigma})^2 = \varrho^2$$

Note that this problem proves that the shear and normal components of the traction vector on a plane with any value of θ lies on a circle of radius ϱ, centered at $(\bar{\sigma}, 0)$ in the $\tau - \sigma$ plane. This result, discovered by Otto Mohr, is usually called Mohr's circle and is shown in the figure above. Clearly, the same results hold for all three pairings of the principal directions.

92. Prove the identity $\mathbf{S} \cdot \nabla(\mathbf{h} \times \mathbf{x}) = \mathbf{h} \cdot [\mathbf{e}_j \times \mathbf{S}\mathbf{e}_j]$, where $\mathbf{h}$ is constant.

93. A block of material is subjected to a homogeneous state of stress described by the constant stress tensor with $\mathbf{S} = 10[\mathbf{e}_1 \otimes \mathbf{e}_1] - 2[\mathbf{e}_1 \otimes \mathbf{e}_2 + \mathbf{e}_2 \otimes \mathbf{e}_1] + 5[\mathbf{e}_2 \otimes \mathbf{e}_2]$. The triangular wedge shown is cut out of the block as a freebody. Compute the tractions that must act on each side of the freebody diagram. Demonstrate that the freebody is in overall equilibrium. Assume that the block has unit width.

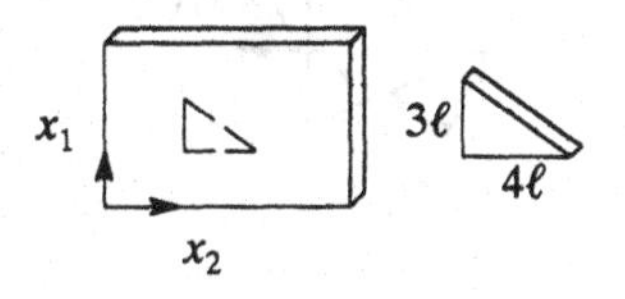

94. A triangular prism of material (with base b, height h, and unit thickness) has an internal stress given by the stress field

$$\mathbf{S}(\mathbf{x}) = \frac{\varrho}{bh}(x_1 - h)(bx_2 + hx_1 - bh)[\mathbf{e}_2 \otimes \mathbf{e}_2]$$

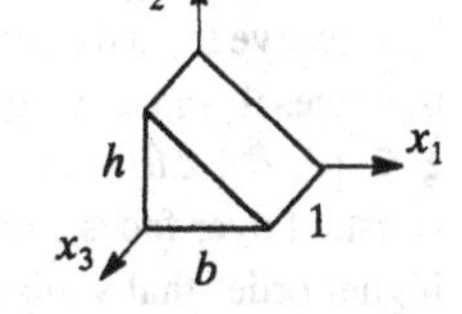

where ϱ is the (constant) unit weight of the material and $\mathbf{e}_i$ is the unit base vector in the direction of the coordinate axis x_i. Find the body force $\mathbf{b}$ required for equilibrium. Find the tractions of all of the faces of the prism. Sketch the normal (σ) and tangential (τ) components of traction on the three faces whose normals are orthogonal to the $\mathbf{e}_3$ direction.

95. A spherical shell has an inside radius of R and an outside radius of $2R$. In the center of the sphere there is a magnetic core that sets up a stress field in the shell. The state of stress in the shell is

$$\mathbf{S}(\mathbf{x}) = \frac{\varrho}{3r}\left(1 - \frac{8R^3}{r^3}\right)[\mathbf{x} \otimes \mathbf{x}]$$

where ϱ is a magnetic constant of the material, $\mathbf{x}$ is the position vector, and r is the radial distance to the point $\mathbf{x}$ defined as

$$r \equiv \sqrt{\mathbf{x} \cdot \mathbf{x}} \quad \text{(Note that } \partial r/\partial x_j = x_j/r\text{)}$$

Find the body force vector field $\mathbf{b}$ in the shell. What is the pressure at the inside surface of the shell? Take a freebody of the shell by slicing it along the plane $x_3 = 0$. What are the tractions $\mathbf{t}$ on the shell that must act at the slice?

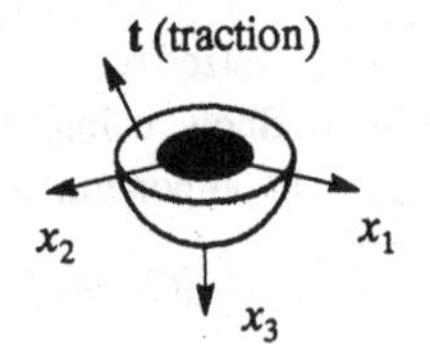

96. The stress tensor $\mathbf{S}$ can be expressed in cylindrical coordinates (r, θ, z) as

$$\begin{aligned} \mathbf{S}(r,\theta,z) = \; & S_{rr}[\mathbf{e}_r \otimes \mathbf{e}_r] + S_{r\theta}[\mathbf{e}_r \otimes \mathbf{e}_\theta] + S_{rz}[\mathbf{e}_r \otimes \mathbf{e}_z] \\ & + S_{\theta r}[\mathbf{e}_\theta \otimes \mathbf{e}_r] + S_{\theta\theta}[\mathbf{e}_\theta \otimes \mathbf{e}_\theta] + S_{\theta z}[\mathbf{e}_\theta \otimes \mathbf{e}_z] \\ & + S_{zr}[\mathbf{e}_z \otimes \mathbf{e}_r] + S_{z\theta}[\mathbf{e}_z \otimes \mathbf{e}_\theta] + S_{zz}[\mathbf{e}_z \otimes \mathbf{e}_z] \end{aligned}$$

where the components (e.g., S_{rz}) are each functions of the coordinates (r, θ, z). However, now the base vectors $\mathbf{e}_r(\theta)$ and $\mathbf{e}_\theta(\theta)$ depend upon the coordinate θ.

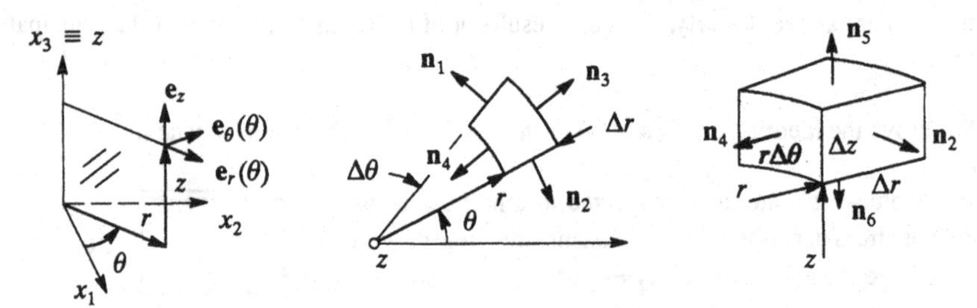

Using the coordinate-free definition of the divergence of a tensor field, Eqn. (87), show that the divergence of $\mathbf{S}$ in cylindrical coordinates is given by

$$\operatorname{div}\mathbf{S}(r,\theta,z) = \frac{1}{r}\frac{\partial}{\partial r}\left(r\mathbf{S}\mathbf{e}_r\right) + \frac{1}{r}\frac{\partial}{\partial \theta}\left(\mathbf{S}\mathbf{e}_\theta\right) + \frac{\partial}{\partial z}\left(\mathbf{S}\mathbf{e}_z\right)$$

Observe from the figure that $\mathbf{n}_1 = \mathbf{e}_\theta(\theta + \Delta\theta)$ and $\mathbf{n}_2 = -\mathbf{e}_\theta(\theta)$, and are constant over the faces 1 and 2, respectively. The normal vectors $\mathbf{n}_3 = \mathbf{e}_r(\xi)$ and $\mathbf{n}_4 = -\mathbf{e}_r(\xi)$ with $\xi \in [\theta,\ \theta + \Delta\theta]$ varying over faces 3 and 4. Finally, note that $\mathbf{n}_5 = \mathbf{e}_z$ and $\mathbf{n}_6 = -\mathbf{e}_z$ are constant over faces 5 and 6. The volume of the wedge is $\Delta V = r\Delta\theta\Delta r\Delta z$ plus terms of higher order that vanish more quickly in the limit as $\Delta V \to 0$.

To compute the component expression for the divergence of the stress tensor, we must expand the vectors $\mathbf{S}\mathbf{e}_r$, $\mathbf{S}\mathbf{e}_\theta$, and $\mathbf{S}\mathbf{e}_z$. Show that

$$\begin{aligned}
\mathbf{S}\mathbf{e}_r &= S_{rr}\mathbf{e}_r + S_{\theta r}\mathbf{e}_\theta + S_{zr}\mathbf{e}_z \\
\mathbf{S}\mathbf{e}_\theta &= S_{r\theta}\mathbf{e}_r + S_{\theta\theta}\mathbf{e}_\theta + S_{z\theta}\mathbf{e}_z \\
\mathbf{S}\mathbf{e}_z &= S_{rz}\mathbf{e}_r + S_{\theta z}\mathbf{e}_\theta + S_{zz}\mathbf{e}_z
\end{aligned}$$

Before we take derivatives, we must observe that in terms of the standard constant basis $\{\mathbf{e}_1, \mathbf{e}_2, \mathbf{e}_3\}$, the radial and angular base vectors have the form

$$\begin{aligned}
\mathbf{e}_r(\theta) &= \cos\theta\,\mathbf{e}_1 + \sin\theta\,\mathbf{e}_2 \\
\mathbf{e}_\theta(\theta) &= -\sin\theta\,\mathbf{e}_1 + \cos\theta\,\mathbf{e}_2
\end{aligned}$$

and, therefore, $\partial\mathbf{e}_r/\partial\theta = \mathbf{e}_\theta$ and $\partial\mathbf{e}_\theta/\partial\theta = -\mathbf{e}_r$. Show that the component expression of the divergence of $\mathbf{S}$ is

$$\begin{aligned}
\operatorname{div}\mathbf{S} &= \left(\frac{\partial S_{rr}}{\partial r} + \frac{1}{r}\frac{\partial S_{r\theta}}{\partial \theta} + \frac{\partial S_{rz}}{\partial z} + \frac{1}{r}\left[S_{rr} - S_{\theta\theta}\right]\right)\mathbf{e}_r \\
&+ \left(\frac{\partial S_{\theta r}}{\partial r} + \frac{1}{r}\frac{\partial S_{\theta\theta}}{\partial \theta} + \frac{\partial S_{\theta z}}{\partial z} + \frac{1}{r}\left[S_{r\theta} + S_{\theta r}\right]\right)\mathbf{e}_\theta \\
&+ \left(\frac{\partial S_{zr}}{\partial r} + \frac{1}{r}\frac{\partial S_{z\theta}}{\partial \theta} + \frac{\partial S_{zz}}{\partial z} + \frac{1}{r}\left[S_{zr}\right]\right)\mathbf{e}_z
\end{aligned}$$

$$S e_1 = S_{11} e_1 + S_{21} e_2 + S_{31} e_3$$

$$S e_2 = S_{12} e_1 + S_{22} e_2 + S_{32} e_3$$

$$S e_3 = S_{13} e_1 + S_{23} e_2 + S_{33} e_3$$

[illegible] we take derivatives, we must observe that in terms of the [illegible] the radial and angular base vectors have the form

$$e_r(\theta) = \cos\theta e_1 + \sin\theta e_2$$

$$e_\theta(\theta) = -\sin\theta e_1 + \cos\theta e_2$$

and [illegible] $\partial e_r/\partial\theta = e_\theta$ [illegible] that the component [illegible] of the [illegible]

[illegible]

4 Elastic Constitutive Theory

Within the confines of continuum mechanics, a purely geometric argument leads to the definition of strain and the concept of balance of momentum leads to the definition of stress. The relationship between strain and the motion does not depend upon stress. The relationship between stress and the applied force does not depend upon strain. As such, the equations of kinematics and equilibrium do not completely characterize the mechanical response of a solid body. We must introduce another relationship to complete the theory. An equation that relates stress and strain is called a *constitutive hypothesis* or *constitutive model*.

A continuum constitutive model is simply a mathematical relationship among certain of the fields that appear in our theory (e.g., strain and stress). The mathematical relationship generally depends upon a set of parameters (material constants) that must be established empirically. In other words, if we wish to establish the values of the material parameters, we must go to the laboratory, perform tests, and fit the model to the data.† There are, however, certain theoretical restrictions to which a constitutive model must adhere, and there are certain ways of stating our assumptions about material behavior that are more productive than others. We shall examine a few of these features of constitutive theory in this chapter.

The idea that force and deformation are related is intuitive. When you pull on a rubber band it stretches; the harder you pull it, the more it stretches. This cause and effect is the feature of mechanical response that the equations of kinematics and equilibrium alone do not address. The simple motivation behind

---

† Contrast this situation with kinematics and equilibrium in which there is no room for empiricism.

the mechanical response of materials is that all materials are made up of elementary particles (atoms and molecules) and these particles are held together by atomic and molecular bonds. When subjected to force, these bonds stretch and allow the particles to move relative to one another. The aggregate effect of the relative motion of the particles is observed as macroscopic deformation. Continuum mechanics homogenizes the discrete nature of materials with the intent of capturing the essential macroscopic features of the response that results from the interaction of the microscopic particles.

Constitutive theory generally means finding a mathematical framework (or parameterization) that covers an entire class of qualitative material response and renders the distinction among materials to be simply different values of the material constants. Among those classes we have models of elasticity, plasticity, viscoelasticity, viscoplasticity, and many others. An elastic material will return to its initial configuration upon unloading; a plastic material generally will not. A viscoelastic material will eventually return to its initial configuration upon unloading, but it takes some time to relax back to that condition. A viscoplastic material generally will not return to its unstressed configuration and will take some time to get to whatever configuration it returns to upon unloading. The configuration adopted by a stressed elastic material does not depend upon the history of loading; the configuration of a plastic material does. Here we shall consider only elastic materials, and, further, primarily those with linear behavior. This class of materials, however small it may be, is quite important to the field of mechanics.

One of the fundamental hypotheses underlying the modeling of constitutive behavior is that cause and effect between force and deformation occurs only at the local level. We call this the ***axiom of locality*** which simply posits that

stress (at a point) depends upon strain (at a point).

This simple hypothesis is not provable (hence the designation *axiom*), and has been the subject of great debate by those concerned with the behavior of materials.† It is, however, the result of centuries of observation. Where it leads to useful results, it has been embraced by the engineering community.

You can imagine the progress in thought that led up to the axiom of locality by considering a uniaxial tension test, shown schematically in Fig. 57. A bar of initial length ℓ and cross-sectional area A is pulled with a force P resulting

† There are some well-known situations where this hypothesis does not seem adequate. For example, strain localization, wherein deformations are highly concentrated (essentially over zero volume), is possible in the mathematical theory when an increase in strain is associated with a decrease in stress (often associated with the term *strain-softening*). Although strain concentration can occur in nature (e.g., necking in a tension bar) it is always associated with a finite volume of material. Extending the axiom of locality from "at a point" to "in the neighborhood of a point" is one way of resolving this problem. This issue is beyond the scope of the topics covered in this chapter.

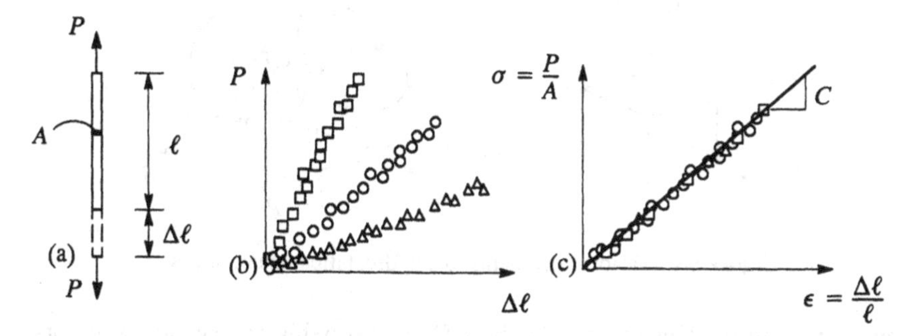

Figure 57 The relationship between force and extension

in an elongation of amount $\Delta\ell$. In 1678, Robert Hooke recognized the linearity between force P and change in length $\Delta\ell$, and recorded it in his law *ut tensio sic vis* (the power is in proportion to the extension). He ascribed this linear behavior to all materials, although the bulk of his experiments were on springs. One can imagine carrying out this experiment with bars made of the same material but having different areas and different lengths. One might still observe the linearity that Hooke observed but with different slopes (as illustrated by the squares, circles, and triangles in the figure).

The next leap of insight is to normalize the force P by the cross-sectional area A (i.e., to give what we now call stress) and the change in length $\Delta\ell$ by the length ℓ (i.e., to give what we now call strain) and observe the tidy organization of the data. All of the dots line up and it appears reasonable to characterize the relationship between σ and ϵ as a straight line with slope C. To wit

$$\sigma = C\epsilon \tag{210}$$

This equation is a mathematical model that represents the observed data. The parameter C is the empirical constant of the model.

The differences in the responses observed in the plot of P versus $\Delta\ell$ must, therefore, be due to the geometry of the test piece and not the constitution of the material. The slope C on the other hand must be a property of the material. One could repeat the test with a different material to confirm that the constant C is different for different materials.

It wasn't until 1807 that Thomas Young recognized the universal modulus that bears his name, and even then his concept of the modulus was quite different from how we define it today. In 1826, Navier presented the definition of the modulus, which we call C here, as we use it today.

Linear elasticity in one dimension. Our task here is to generalize the one-dimensional observation of linear elasticity, Eqn. (210), to three-dimensional solid bodies. One of the key observations on elastic bodies is that the state of stress does not depend upon the strain history. The final state does not depend

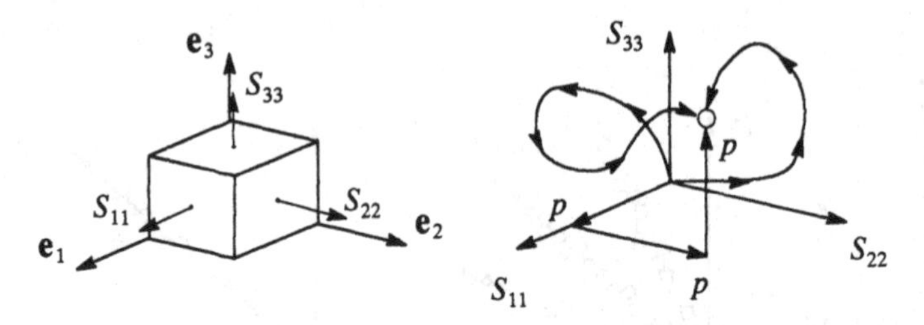

Figure 58 Various load paths to a hydrostatic state of stress

upon any intermediate stage of loading. One reason this observation is important is that if the state of stress were dependent upon the strain path, it would be difficult to ensure that the body would return to the unstressed configuration upon unloading. This issue is not a problem for the uniaxial case, since there is only one path — the straight line through the origin. However, in a three-dimensional body, we can imagine various paths to the same state of stress. For example, consider a hydrostatic pressure $\mathbf{S} = p\mathbf{I}$. The most obvious way to get to this state of stress is to apply uniform pressure to all sides simultaneously, increasing from zero to p. We can also imagine getting to this state of stress by applying pure tension p in the x_1 direction, then superposing pure tension p in the x_2 direction while keeping the already achieved stress constant, then superposing pure tension p in the x_3 direction while keeping the already achieved stress constant, as shown in Fig. 58. In general, we might think of the components of stress as being functions of a time-like parameter τ such that

$$\mathbf{S}(\tau) = S_{ij}(\tau)\left[\mathbf{e}_i \otimes \mathbf{e}_j\right] \tag{211}$$

There are many paths that lead to a hydrostatic state of stress of magnitude p. All of these paths wind up at the same place in stress space, but the paths are considerably different. For example, we might have

$$S_{11}(\tau) = p\tau, \quad S_{22} = p\sin(\pi\tau/2), \quad S_{33} = p\tau^2 \tag{212}$$

with all other components equal to zero. At time $\tau = 1$ we arrive at the hydrostatic state of stress.

Path independence of the stress state can be guaranteed very simply by assuming the existence of a *strain energy function* $W(\epsilon)$ from which we can compute the stress by differentiation with respect to strain as

$$\sigma = \frac{dW(\epsilon)}{d\epsilon} \tag{213}$$

To observe the path independence let us compute the work done (the area under the stress strain curve, as shown in Fig. 59) in going from the strain state ϵ_1 to the strain state ϵ_2. The work done is the integral of the stress $\sigma(\epsilon)$ from ϵ_1 to ϵ_2, which can be carried out as

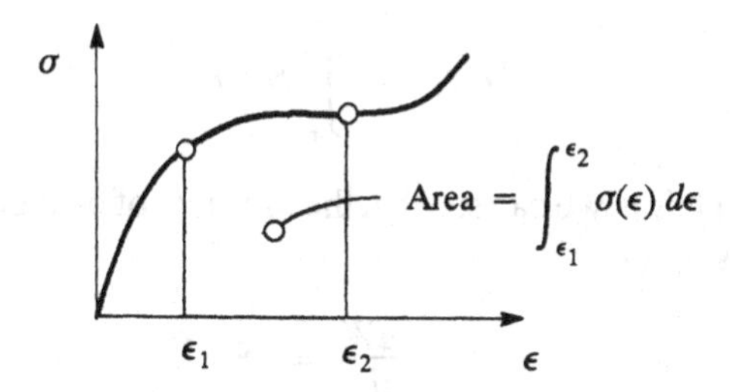

Figure 59 Work done is the area under the stress-strain curve

$$\int_{\epsilon_1}^{\epsilon_2} \sigma(\epsilon)\, d\epsilon = \int_{\epsilon_1}^{\epsilon_2} \frac{dW(\epsilon)}{d\epsilon}\, d\epsilon = \int_{\epsilon_1}^{\epsilon_2} dW(\epsilon) = W(\epsilon_2) - W(\epsilon_1) \tag{214}$$

Through this construction, the work done is a function only of the difference in strain energy at the two end states; it is not a function of the path between them. It should be obvious that if the terminal state of strain ϵ_2 is the same as the initial state ϵ_1 then the total work done over the path is exactly zero. This feature assures that the material can return to its original state upon removal of excitation.

A material defined in this manner is called *hyperelastic*. The uniaxial, linearly elastic material with modulus C has a quadratic strain energy function $W(\epsilon) = \frac{1}{2}C\epsilon^2$. From Eqn. (213) one can easily confirm that $\sigma = C\epsilon$ for this case. Notice that the slope of the stress strain curve is given by

$$C = \frac{\partial \sigma}{\partial \epsilon} = \frac{\partial^2 W(\epsilon)}{\partial \epsilon^2} \tag{215}$$

Although there are other models of elasticity, hyperelasticity is clearly the most important.

The extension to three dimensions. The concept of the strain energy function is easy to generalize to three dimensions. Let us assume that we are working with the small deformations so that the nonlinear measures of strain cannot be distinguished from the linearized strain tensor. Further, for small deformations, the Cauchy stress and the first and second Piola-Kirchhoff stress tensors are the same. Let us refer simply to the strain as $\mathbf{E}$ and to the stress as $\mathbf{S}$.

To create a model that has the quality of path independence observe that the work done in going from a state at time τ_1 to a state at time τ_2 is the integral of the stress power† $\mathbf{S} \cdot \dot{\mathbf{E}}$, where $\dot{\mathbf{E}} \equiv d\mathbf{E}/d\tau$ indicates derivative with respect to time and the dot product of tensors is $\mathbf{A} \cdot \mathbf{B} \equiv A_{ij}B_{ij}$. Hence, we can write the strain energy as

---

† The *stress power* is the rate of change of internal mechanical work. It is independent of the constitutive hypothesis and appears in the principles of conservation of energy (along with kinetic energy, dissipation, and external energy).

$$W(\tau) \equiv \int_{\tau_1}^{\tau} \mathbf{S} \cdot \dot{\mathbf{E}}\, dt \tag{216}$$

Now, by construction, we can see that the time rate of change of the strain energy function is simply

$$\dot{W} \equiv \frac{dW}{d\tau} \equiv \mathbf{S} \cdot \dot{\mathbf{E}} \tag{217}$$

Finally, we can observe that the integral of the rate of change of the strain energy from the state at time τ_1 to a state at time τ_2 is simply

$$\int_{\tau_1}^{\tau_2} \mathbf{S} \cdot \dot{\mathbf{E}}\, d\tau = \int_{\tau_1}^{\tau_2} \dot{W}\, d\tau = W(\tau_2) - W(\tau_1) \tag{218}$$

thereby assuring path independence.

Now let us think of the strain energy as a function of strain and make the formal identification of the dependence on time as

$$W(\tau) \equiv W\big(\mathbf{E}(\tau)\big) \tag{219}$$

We can compute the time rate of change of W by the chain rule for differentiation as (sum on repeated indices implied)

$$\dot{W} = \frac{\partial W}{\partial E_{ij}} \dot{E}_{ij} \tag{220}$$

Comparing this result with Eqn. (217) we see that the components of the stress tensor are simply the derivatives of the strain energy function with respect to the components of strain. To wit,

$$\boxed{S_{ij} = \frac{\partial W(\mathbf{E})}{\partial E_{ij}}} \tag{221}$$

This result is the appropriate generalization of the one-dimensional concept of the strain energy function. Again, the strain energy function is created in a way that assures path independence of the state, in accord with our definition of hyperelasticity.

Elasticities. We can compute the *elasticities* (the generalization of the slope of the stress-strain curve) as

$$C_{ijkl} \equiv \frac{\partial S_{ij}}{\partial E_{kl}} = \frac{\partial^2 W(\mathbf{E})}{\partial E_{ij}\, \partial E_{kl}} \tag{222}$$

The derivative of a second-order tensor component with respect to another second-order tensor component is an object with four indices. We can think of this object with four indices as being the components of a fourth-order tensor. To be a proper tensor it must be expressible in terms of a tensor basis and the tensor itself must transform according to the rules for change of basis. Let us define the *elasticity tensor* to be

$$\mathbf{C} \equiv C_{ijkl}\left[\mathbf{e}_i \otimes \mathbf{e}_j \otimes \mathbf{e}_k \otimes \mathbf{e}_l\right] \tag{223}$$

where $\left[\mathbf{e}_i \otimes \mathbf{e}_j \otimes \mathbf{e}_k \otimes \mathbf{e}_l\right]$ is the *ijkl*th fourth-order base tensor. There is an obvious complexity of fourth-order tensors over second-order tensors, but there are some similarities also. Whereas a second-order tensor was introduced for the purposes of providing an object that operates on a vector to produce another vector, a fourth-order tensor is an object that operates on a second-order tensor (strain, in this case) to produce another second-order tensor (stress, in this case).†

In order to see how a fourth-order tensor works, we must define a new tensor product of vectors. Let **a**, **b**, **s**, **t**, **u**, and **v** be vectors. Let these vectors define a second-order tensor $\mathbf{a} \otimes \mathbf{b}$ and a fourth-order tensor $\left[\mathbf{s} \otimes \mathbf{t} \otimes \mathbf{u} \otimes \mathbf{v}\right]$. The fourth-order tensor, as defined, inherits a meaning (other than four letters separated by tensor product symbols) only through a definition of how it operates. The result of the fourth-order tensor operating on the second-order tensor is defined as follows

$$\left[\mathbf{s} \otimes \mathbf{t} \otimes \mathbf{u} \otimes \mathbf{v}\right]\left[\mathbf{a} \otimes \mathbf{b}\right] \equiv (\mathbf{u} \cdot \mathbf{a})(\mathbf{v} \cdot \mathbf{b})\left[\mathbf{s} \otimes \mathbf{t}\right] \tag{224}$$

Clearly, the result is a second-order tensor. A particular manifestation of this tensor product, and the one of primary interest to us as we do component computations, is the following relationship between base vectors

$$\begin{aligned}\left[\mathbf{e}_i \otimes \mathbf{e}_j \otimes \mathbf{e}_k \otimes \mathbf{e}_l\right][\mathbf{e}_m \otimes \mathbf{e}_n] &= (\mathbf{e}_k \cdot \mathbf{e}_m)(\mathbf{e}_l \cdot \mathbf{e}_n)[\mathbf{e}_i \otimes \mathbf{e}_j] \\ &= \delta_{km}\delta_{ln}[\mathbf{e}_i \otimes \mathbf{e}_j]\end{aligned} \tag{225}$$

With this relationship, we can compute the effect of the elasticity tensor operating on the strain tensor in components. For linear elasticity we get

$$\begin{aligned}\mathbf{S} &= \mathbf{CE} \\ &= C_{ijkl}[\mathbf{e}_i \otimes \mathbf{e}_j \otimes \mathbf{e}_k \otimes \mathbf{e}_l]\, E_{mn}[\mathbf{e}_m \otimes \mathbf{e}_n] \\ &= C_{ijkl}E_{kl}[\mathbf{e}_i \otimes \mathbf{e}_j]\end{aligned} \tag{226}$$

† A few observations about the general notion of tensors are worth noting. First, it is possible to define a tensor of any order. Second, the operation of a tensor can be more general than what we have described here. In particular, a tensor of order n can be defined as an object that operates on tensors of order m (necessarily less than or equal to n) to produce tensors of order $n - m$. For example, a fourth-order tensor could be defined as an object that operates on vectors (first-order tensor) to produce third-order tensors. A second-order tensor can be viewed as an object that operates on a second-order tensor to produce a scalar (i.e., zeroth-order tensor).

Therefore, the components of the stress tensor can be computed from the components of the elasticity tensor and the strain tensor as $S_{ij} = C_{ijkl}E_{kl}$.

For a nonlinear stress–strain relationship, we can observe that the rate of change of stress can be computed, by the chain rule for differentiation, as

$$\dot{S}_{ij} = \frac{d}{dt}\left(\frac{\partial W}{\partial E_{ij}}\right) = \frac{\partial^2 W}{\partial E_{ij}\partial E_{kl}}\dot{E}_{kl} = C_{ijkl}\dot{E}_{kl} \tag{227}$$

Hence, $\dot{\mathbf{S}} = \mathbf{C}\dot{\mathbf{E}}$. The elasticity tensor operates on the strain rate and produces the stress rate. The elasticity tensor is, in general, a function of the stress or strain in a nonlinear model. A constitutive model that relates the strain rate to the stress rate directly is often called ***hypoelastic***. Not all hypoelastic constitutive relationships are hyperelastic but, as Eqn. (227) demonstrates, all hyperelastic constitutive relationships can be put into rate form (and hence are hypoelastic).

The elasticity tensor has $3 \times 3 \times 3 \times 3 = 81$ components C_{ijkl}. However, not all of these are independent. Since the stress and strain tensors are symmetric, there are only six independent components of each. Thus, instead of $9 \times 9 = 81$ components, the elasticity tensor has only $6 \times 6 = 36$ independent components. Furthermore, since the order of differentiation of the strain energy function with respect to the components of the strain tensor is immaterial, the elasticity tensor is symmetric with respect to *ij* and *kl*. A symmetric six by six matrix has only 21 independent terms (the diagonals and those terms above the diagonal). Thus, the last symmetry means that there are really only 21 independent components in the elasticity tensor. With some assumptions on preferential directions in the material, or *isotropies*, we can further reduce the number of independent parameters in our model.

Our elastic material is linear if the strain energy function is quadratic. In components (summation convention implied), we have

$$\boxed{W(\mathbf{E}) = \tfrac{1}{2}E_{ij}C_{ijkl}E_{kl}} \tag{228}$$

Isotropy

A material is said to be isotropic if its properties do not depend upon certain preferential directions. Another way to say this is to insist that the elasticity tensor be invariant with respect to coordinate transformation. There is a rather straightforward way to assure that the elasticity tensor is isotropic. If the strain energy function depends only on the invariants of the strain tensor, then the resulting constitutive model will also be invariant. Hence, we must have

$$W(\mathbf{E}) = W\big(f_1(\mathbf{E}), f_2(\mathbf{E}), f_3(\mathbf{E})\big) \tag{229}$$

where the invariants of the strain tensor are

$$f_1(\mathbf{E}) = \operatorname{tr}(\mathbf{E}) = E_{ii}$$
$$f_2(\mathbf{E}) = \operatorname{tr}(\mathbf{E}^2) = E_{ij}E_{ji}$$
$$f_3(\mathbf{E}) = \operatorname{tr}(\mathbf{E}^3) = E_{ij}E_{jk}E_{ki}$$

The stress can be computed as the derivative of the strain energy as follows

$$\mathbf{S} = \frac{\partial W}{\partial \mathbf{E}} = \frac{\partial W}{\partial f_1}\frac{\partial f_1}{\partial \mathbf{E}} + \frac{\partial W}{\partial f_2}\frac{\partial f_2}{\partial \mathbf{E}} + \frac{\partial W}{\partial f_3}\frac{\partial f_3}{\partial \mathbf{E}} \tag{230}$$

where the derivative of a scalar with respect to a tensor is a tensor with components $[\partial f/\partial \mathbf{E}]_{ij} = \partial f/\partial E_{ij}$. To complete the derivation we need the derivatives of the invariants with respect to the strain. These derivatives are straightforward to compute in components. To wit,

$$\frac{\partial f_1(\mathbf{E})}{\partial E_{mn}} = \frac{\partial E_{ii}}{\partial E_{mn}} = \delta_{im}\delta_{in} = \delta_{mn}$$

$$\frac{\partial f_2(\mathbf{E})}{\partial E_{mn}} = \frac{\partial E_{ij}}{\partial E_{mn}}E_{ji} + E_{ij}\frac{\partial E_{ji}}{\partial E_{mn}} = 2E_{mn} \tag{231}$$

$$\frac{\partial f_3(\mathbf{E})}{\partial E_{mn}} = \frac{\partial E_{ij}}{\partial E_{mn}}E_{jk}E_{ki} + E_{ij}\frac{\partial E_{jk}}{\partial E_{mn}}E_{ki} + E_{ij}E_{jk}\frac{\partial E_{ki}}{\partial E_{mn}} = 3E_{nk}E_{km}$$

These results can be summarized in direct notation as

$$\frac{\partial f_1(\mathbf{E})}{\partial \mathbf{E}} = \mathbf{I}, \quad \frac{\partial f_2(\mathbf{E})}{\partial \mathbf{E}} = 2\mathbf{E}, \quad \frac{\partial f_3(\mathbf{E})}{\partial \mathbf{E}} = 3\mathbf{E}^2 \tag{232}$$

Using these results in Eqn. (230) we arrive at the most general isotropic elastic constitutive model

$$\boxed{\mathbf{S} = \frac{\partial W}{\partial f_1}\mathbf{I} + 2\frac{\partial W}{\partial f_2}\mathbf{E} + 3\frac{\partial W}{\partial f_3}\mathbf{E}^2} \tag{233}$$

The constitutive model given by Eqn. (233) can be put in a slightly different form by noting that $\mathbf{E}^2$ can be expressed in terms of $\mathbf{E}$ and $\mathbf{E}^{-1}$ through the Cayley-Hamilton theorem as the following example shows.

Example 22. *Alternative form for hyperelastic constitutive equation.* The Cayley-Hamilton theorem states that (see Chapter 1)

$$\mathbf{E}^3 - I_E\mathbf{E}^2 + II_E\mathbf{E} - III_E\mathbf{I} = \mathbf{0}$$

where $I_E = f_1$, $2II_E = f_1^2 - f_2$, and $6III_E = f_1^3 - 3f_1f_2 + 2f_3$ are the invariants that show up in the eigenvalue problem for principal strains. We can rewrite the equation as

$$\mathbf{E}\left[\mathbf{E}^2 - I_E\mathbf{E} + II_E\mathbf{I} - III_E\mathbf{E}^{-1}\right] = \mathbf{0}$$

from which we can deduce that

$$\mathbf{E}^2 = I_E\mathbf{E} - II_E\mathbf{I} + III_E\mathbf{E}^{-1}$$

Now $\mathbf{E}^2$ can be replaced with this expression in the stress-strain relationship.

Linear, isotropic elasticity. If we want a *linear* constitutive law, the strain energy must be a purely quadratic function of strain. Consequently, it can depend only upon f_1^2 and f_2 (f_1 is only linear in the components of $\mathbf{E}$, while f_3 is cubic in the components of $\mathbf{E}$). Thus, our strain energy function must have the form

$$W(\mathbf{E}) = a_1 f_1^2(\mathbf{E}) + a_2 f_2(\mathbf{E}) \tag{234}$$

where a_1 and a_2 are material parameters. Now

$$\frac{\partial}{\partial f_1}\left(f_1^2\right) = 2f_1 = 2\,\mathrm{tr}\left(\mathbf{E}\right), \quad \frac{\partial}{\partial f_2}\left(f_2\right) = 1 \tag{235}$$

Renaming the parameters $\lambda \equiv 2a_1$ and $\mu \equiv a_2$, from Eqn. (233) we get the final form of the linearly elastic constitutive equations, in direct notation

$$\boxed{\mathbf{S} = \lambda\left(\mathrm{tr}\,\mathbf{E}\right)\mathbf{I} + 2\mu\,\mathbf{E}} \tag{236}$$

This constitutive model is often referred to as *Hooke's law* even though Robert Hooke undoubtedly never saw anything like it. These equations embody the assumptions of hyperelasticity and linearity and represent, without question, the most widely used constitutive model ever conceived. The two material constants λ and μ are called the *Lamé parameters* for their discoverer G. Lamé, although Cauchy might have been the first to express the equations of elastic constitution with two constants.

Example 23. *Elasticity tensor for linear elasticity.* From the developments above it is straightforward to compute the components of the elasticity tensor for a linear isotropic elastic material. The components of the stress tensor are given by $S_{ij} = \lambda E_{aa}\delta_{ij} + 2\mu E_{ij}$. The elasticity tensor can be computed by differentiation to give

$$\begin{aligned} C_{ijkl} &= \frac{\partial}{\partial E_{kl}}\left(\lambda E_{aa}\delta_{ij} + 2\mu E_{ij}\right) \\ &= \lambda\delta_{ak}\delta_{al}\delta_{ij} + 2\mu\delta_{ik}\delta_{jl} \\ &= \lambda\delta_{kl}\delta_{ij} + 2\mu\delta_{ik}\delta_{jl} \end{aligned}$$

Observe that the moduli are constant, as expected for a linear model. Because the strain tensor is symmetric, the elasticity tensor is often written as

$$C_{ijkl} = \lambda\delta_{ij}\delta_{kl} + \mu\left[\delta_{ik}\delta_{jl}+\delta_{il}\delta_{jk}\right]$$

The term in brackets is expressed as shown in order to assure symmetry with respect to the indices ij and kl. Any fourth-order tensor with components in this form is invariant with respect to coordinate transformation, and hence is often called an *isotropic fourth-order tensor.*

The constitutive equations given by Eqn. (236) can be easily inverted to give strain in terms of stress. First, compute the trace of both sides of the equation to get the result $\text{tr}(\mathbf{S}) = (3\lambda+2\mu)\text{tr}(\mathbf{E})$. Now we can substitute this result for $\text{tr}(\mathbf{E})$ in the equation to get

$$\mathbf{E} = -\frac{\lambda}{2\mu(3\lambda+2\mu)}\text{tr}(\mathbf{S})\mathbf{I} + \frac{1}{2\mu}\mathbf{S} \tag{237}$$

This form of Hooke's law is convenient when stresses are prescribed and the task is to compute the associated strains. Some of the problems at the back of the chapter generalize the concept to situations where some of the components of stress and some of the components of strain are prescribed and the task is to compute the remaining, unknown, components.

Definitions of Elastic Moduli

The interesting observation about the linear elastic constitutive equations we have just derived is that there is no mention of the famous modulus of Thomas Young. We can, however, derive such a result from our basic equations expressed in terms of the Lamé constants. There is a lesson in doing so. We shall see that the constitutive equations of linear elasticity can be expressed a number of different ways, all valid and equivalent, each with its own definition of the moduli. The key difference among them is the experiment we would be inclined to do to find the constants. We will consider two important cases here.

Young's modulus and Poisson's ratio. The first experiment that we will imagine is the uniaxial tension test, which will provide us with a means of directly measuring *Young's modulus* and *Poisson's ratio*. Let the axis of applied tension σ be along x_1. We thus induce a state of stress $\mathbf{S} = \sigma[\mathbf{e}_1 \otimes \mathbf{e}_1]$. The stress tensor has components $S_{11} = \sigma$, $S_{22} = S_{33} = S_{12} = S_{13} = S_{23} = 0$. According to our constitutive equations the components of stress are

$$
\begin{aligned}
S_{11} &= (\lambda+2\mu)E_{11} + \lambda(E_{22}+E_{33}) = \sigma \\
S_{22} &= (\lambda+2\mu)E_{22} + \lambda(E_{11}+E_{33}) = 0 \\
S_{33} &= (\lambda+2\mu)E_{33} + \lambda(E_{11}+E_{22}) = 0
\end{aligned}
\tag{238}
$$

$$
S_{12} = 2\mu E_{12} = 0, \quad S_{13} = 2\mu E_{13} = 0, \quad S_{23} = 2\mu E_{23} = 0
$$

The last three equations yield $E_{12} = E_{13} = E_{23} = 0$. The second and third equations can be used to express the strains E_{22} and E_{33} in terms of E_{11}. Solving these equations, we get

$$
E_{22} = E_{33} = -\frac{\lambda}{2(\lambda+\mu)} E_{11} \tag{239}
$$

Let us assume that we have measured the axial strain $E_{11} = \epsilon$. Eqn. (239) can be substituted back into the first of Eqns. (238) to give a relationship between the applied axial stress and the measured axial strain as

$$
\sigma = \frac{\mu(3\lambda+2\mu)}{\lambda+\mu}\epsilon \equiv C\epsilon \tag{240}
$$

giving $C \equiv \mu(3\lambda+2\mu)/(\lambda+\mu)$. We call this constant Young's modulus and observe that it can be directly measured as $C = \sigma/\epsilon$ in a uniaxial tension test. Young's modulus C has units of stress f/l^2.

Let us also define Poisson's ratio ν as the negative of the ratio between the lateral strain and the axial strain in a uniaxial tension test. To wit

$$
\nu \equiv -\frac{E_{22}}{E_{11}} = -\frac{E_{33}}{E_{11}} = \frac{\lambda}{2(\lambda+\mu)} \tag{241}
$$

To determine Poisson's ratio from a uniaxial tension test one of the lateral strains E_{22} or E_{33} must also be measured.

Young's modulus and Poisson's ratio provide two suitable independent material constants for our linear elastic constitutive equations. One can find λ and μ in terms of C and ν from their definitions as

$$
\lambda = \frac{C\nu}{(1+\nu)(1-2\nu)}, \quad \mu = \frac{C}{2(1+\nu)} \tag{242}
$$

The linear elastic constitutive equations can be expressed in terms of C and ν as follows

$$
\boxed{\mathbf{S} = \frac{C\nu}{(1+\nu)(1-2\nu)}\mathrm{tr}(\mathbf{E})\mathbf{I} + \frac{C}{1+\nu}\mathbf{E}} \tag{243}
$$

Another useful form of the preceding equations is to invert them and express strain in terms of stress. Since the equations are linear, this inversion is straightforward. The end result is

$$\boxed{\mathbf{E} = -\frac{\nu}{C}\operatorname{tr}(\mathbf{S})\mathbf{I} + \frac{1+\nu}{C}\mathbf{S}} \tag{244}$$

The bulk and shear moduli. We can imagine an experiment wherein the material is subjected to pure pressure, and the change in volume is measured. Such a test is important, for example, in measuring the properties of geotechnical materials (it is impossible to perform a tension test on a granular material). Let us re-examine the change in volume for small strains. Let us assume that we have a volume of material V subjected to a homogeneous deformation, i.e., $\mathbf{F}$ is constant. The assumption of homogeneity of deformation is reasonable because, in an experiment, we generally try to induce the simplest state possible in order to measure the quantity of interest in the most direct way possible. The deformed volume of the body is $v = (\det \mathbf{F})V$, in accord with the results of Chapter 2. Let us denote the change in volume as $\Delta V = v - V$. We have the following relationships among the deformation gradient $\mathbf{F}$, the Green deformation tensor $\mathbf{C}$, and the Lagrangian strain tensor $\mathbf{E}$

$$\det \mathbf{F} = \sqrt{\det \mathbf{C}} = \sqrt{\det[\mathbf{I}+2\mathbf{E}]} \tag{245}$$

Therefore, the ratio of the deformed volume to the original volume is

$$\frac{V+\Delta V}{V} = \sqrt{\det[\mathbf{I}+2\mathbf{E}]} \tag{246}$$

One can expand the determinant of the tensor $\mathbf{I}+2\mathbf{E}$ to find (see Problem 106)

$$\det[\mathbf{I}+2\mathbf{E}] = 1+2I_E+4II_E+8III_E \tag{247}$$

If strains are small, then $I_E \gg II_E \gg III_E$, since the first is linear in $\mathbf{E}$, the second quadratic, and the third cubic. Therefore, to a first approximation, we have $\det[\mathbf{I}+2\mathbf{E}] \approx 1+2I_E$. To finish our derivation, we need to deal with the square root in Eqn. (246). We can use a Taylor series expansion to show that $\sqrt{1+2x} \approx 1+x$ when $x \ll 1$ (prove this for yourself!). Using this result in Eqn. (246), we find that, for small strains, the ratio e of change in volume to original volume is measured by the trace of the strain tensor

$$\boxed{\frac{\Delta V}{V} \equiv e \approx \operatorname{tr}(\mathbf{E})} \tag{248}$$

We generally refer to $e = \Delta V/V$ as the *dilatation*. The dilatation is a quantity that is readily measurable in an experiment. It is also easy to apply a constant pressure and measure its value.

Let us compute the trace of the stress tensor from the constitutive equations

$$\operatorname{tr}(\mathbf{S}) = \lambda\operatorname{tr}(\mathbf{E})\operatorname{tr}(\mathbf{I})+2\mu\operatorname{tr}(\mathbf{E}) = (3\lambda+2\mu)\operatorname{tr}(\mathbf{E}) \equiv 3K\operatorname{tr}(\mathbf{E})$$

We call the constant $K \equiv \lambda + \frac{2}{3}\mu$ the *bulk modulus*. If we do an experiment in which a hydrostatic pressure p is applied, then the stress is $\mathbf{S} = p\mathbf{I}$, and the trace of the stress is $\text{tr}(\mathbf{S}) = 3p$. If we measure the change in volume (and the original volume, of course), then e is known. Hence, we can directly measure the bulk modulus as $K = p/e$ in an isotropic pressure test. The bulk modulus has units of pressure f/l^2.

To provide a complement to the volumetric part of the constitutive equations, let us subtract the trace of the stress from the stress tensor. Let the *deviatoric stress* be defined as $\mathbf{S}' \equiv \mathbf{S} - \frac{1}{3}\text{tr}(\mathbf{S})\mathbf{I}$ so that $\text{tr}(\mathbf{S}') = 0$. Let us compute the deviatoric part of the stress from our constitutive equations

$$\begin{aligned} \mathbf{S}' &= \mathbf{S} - \tfrac{1}{3}\text{tr}(\mathbf{S})\mathbf{I} \\ &= \lambda\,\text{tr}(\mathbf{E})\mathbf{I} + 2\mu\mathbf{E} - \left(\lambda + \tfrac{2}{3}\mu\right)\text{tr}(\mathbf{E})\mathbf{I} \\ &= 2\mu\left(\mathbf{E} - \tfrac{1}{3}\text{tr}(\mathbf{E})\mathbf{I}\right) = 2\mu\mathbf{E}' \end{aligned}$$

where $\mathbf{E}' \equiv \mathbf{E} - \frac{1}{3}\text{tr}(\mathbf{E})\mathbf{I}$ is the *deviatoric strain*, with $\text{tr}(\mathbf{E}') = 0$. We can write the constitutive equations as the sum of bulk and shear parts as

$$\boxed{\mathbf{S} = Ke\mathbf{I} + 2\mu\mathbf{E}'} \tag{249}$$

These equations are exactly equivalent to the original equations expressed in terms of the Lamé parameters. The constants K and μ are a suitable alternative pair of elastic material parameters. For any given state of strain $\mathbf{E}$ we can compute the stress from Eqn. (249) by first computing e and $\mathbf{E}'$, and then substituting into the equation for $\mathbf{S}$. Note that any state of strain is amenable to this decomposition. One can invert this relationship to get

$$\boxed{\mathbf{E} = \frac{p}{3K}\mathbf{I} + \frac{1}{2\mu}\mathbf{S}'} \tag{250}$$

where $p \equiv \frac{1}{3}\text{tr}(\mathbf{S})$ is the pressure and $\mathbf{S}' = \mathbf{S} - p\mathbf{I}$ is the stress deviator. We can compute the state of strain by first computing p and $\mathbf{S}'$ and then substituting into the equation for $\mathbf{E}$.

Example 24. *Triaxial Test.* A common test to determine the elastic constants K and μ is the triaxial test configuration, shown in Fig. 60. In this test, a confining pressure σ_2 (usually compressive) is applied around the sides of the cylinder, and an axial pressure σ_1 (usually compressive) is applied on the ends of the cylinder. The stress and strain tensors have the form

$$\begin{aligned} \mathbf{S} &= \sigma_1[\mathbf{e}_1 \otimes \mathbf{e}_1] + \sigma_2[\mathbf{e}_2 \otimes \mathbf{e}_2 + \mathbf{e}_3 \otimes \mathbf{e}_3] \\ \mathbf{E} &= \epsilon_1[\mathbf{e}_1 \otimes \mathbf{e}_1] + \epsilon_2[\mathbf{e}_2 \otimes \mathbf{e}_2 + \mathbf{e}_3 \otimes \mathbf{e}_3] \end{aligned}$$

The pressure is then $p = \frac{1}{3}(\sigma_1 + 2\sigma_2)$ and the deviator stress is

$$\mathbf{S}' = \tfrac{1}{3}(\sigma_1 - \sigma_2)[3\mathbf{e}_1 \otimes \mathbf{e}_1 - \mathbf{I}]$$

Similarly, the volume change is $e = (\epsilon_1 + 2\epsilon_2)$ and the deviator strain is

$$\mathbf{E}' = \tfrac{1}{3}(\epsilon_1 - \epsilon_2)[3\mathbf{e}_1 \otimes \mathbf{e}_1 - \mathbf{I}]$$

where ϵ_1 and ϵ_2 are the axial and lateral straining of the sample (change of dimension over original dimension). This test gives adequate information to compute the bulk and shear moduli as $K = p/e$ and $2\mu = (\sigma_1 - \sigma_2)/(\epsilon_1 - \epsilon_2)$.

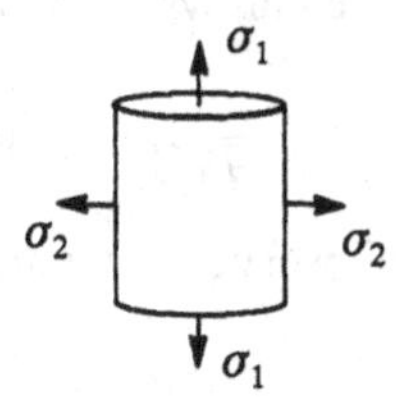

Figure 60 The triaxial test configuration

The purpose of the preceding discussion is twofold. First, we have seen that while there is only one constitutive model for linear elasticity, there are many equivalent forms of it. Second, the different forms are dictated by the experiment we use to define the constants of the model. All of the constants are related (Problem 98).

Elastic Constitutive Equations for Large Strains

In all of the preceding discussion we assumed that the deformations were small with the convenience that the Lagrangian strain and the linearized strain were essentially the same, and the Cauchy stress was indistinguishable from the first and second Piola-Kirchhoff stress tensors. When deformations are not small, the distinction among the various strain and stress tensors is important. The problem of finding suitable constitutive equations is more complicated because there are many more choices for a nonlinear model.

The mathematical model of hyperelasticity can again be built by defining a strain energy function that is the integral of the stress power between times τ_1 and $\tau > \tau_1$. Let us define the stress power to be the product of the first Piola-Kirchhoff stress with the time rate of change of the deformation gradient, i.e., $\mathbf{P} \cdot \dot{\mathbf{F}}$. Now let the strain energy function be defined as

$$\Psi(\tau) \equiv \int_{\tau_1}^{\tau} \mathbf{P} \cdot \dot{\mathbf{F}}\, dt \tag{251}$$

implying that $\dot{\Psi} = \mathbf{P} \cdot \dot{\mathbf{F}}$. Following exactly the same argument as we did for small strains, observing that $\Psi(\tau) = \Psi(\mathbf{F}(\tau))$ and using the chain rule for differentiation, we find that

$$\mathbf{P} = \frac{\partial \Psi(\mathbf{F})}{\partial \mathbf{F}} \tag{252}$$

with component expression $P_{ij} = \partial \Psi / \partial F_{ij}$.

As was pointed out in Chapter 1, the deformation gradient **F** carries information about the stretching of the material, but it also carries information about the rigid body motion. Using the argument that the constitutive equations should be invariant under superimposed rigid-body motions, we can show that the strain energy can depend on the deformation gradient only insofar as it appears in the form $\mathbf{F}^T\mathbf{F}$. Thus, we can write (at a given point the values of the functions are identical)

$$\Psi(\mathbf{F}) = \overline{\Psi}(\mathbf{C}) = \overline{\overline{\Psi}}(\mathbf{E})$$

where $2\mathbf{E}+\mathbf{I} = \mathbf{C} = \mathbf{F}^T\mathbf{F}$. Observe that since $\mathbf{A}\cdot\mathbf{B} = \mathrm{tr}\left(\mathbf{A}^T\mathbf{B}\right)$ we can write the stress power as

$$\begin{aligned}
\mathbf{P}\cdot\dot{\mathbf{F}} &= \mathrm{tr}\left(\mathbf{P}^T\dot{\mathbf{F}}\right) \\
&= \mathrm{tr}\left(\mathbf{P}^T\mathbf{F}^{-T}\mathbf{F}^T\dot{\mathbf{F}}\right) \\
&= \mathrm{tr}\left(\left(\mathbf{F}^{-1}\mathbf{P}\right)^T\left(\mathbf{F}^T\dot{\mathbf{F}}\right)\right) \\
&= \mathrm{tr}\left(\boldsymbol{\Sigma}^T\left(\mathbf{F}^T\dot{\mathbf{F}}\right)\right) = \boldsymbol{\Sigma}\cdot\left(\mathbf{F}^T\dot{\mathbf{F}}\right)
\end{aligned}$$

Now we can observe that because $\dot{\mathbf{C}} = \dot{\mathbf{F}}^T\mathbf{F}+\mathbf{F}^T\dot{\mathbf{F}}$ and because $\boldsymbol{\Sigma}$ is a symmetric tensor we have the following equivalence for the stress power

$$\mathbf{P}\cdot\dot{\mathbf{F}} = \boldsymbol{\Sigma}\cdot\left(\mathbf{F}^T\dot{\mathbf{F}}\right) = \tfrac{1}{2}\boldsymbol{\Sigma}\cdot\dot{\mathbf{C}} = \boldsymbol{\Sigma}\cdot\dot{\mathbf{E}}$$

A relatively straightforward computation shows that the second Piola-Kirchhoff stress can be computed from the energy as

$$\boxed{\boldsymbol{\Sigma} = 2\frac{\partial \overline{\Psi}(\mathbf{C})}{\partial \mathbf{C}} = \frac{\partial \overline{\overline{\Psi}}(\mathbf{E})}{\partial \mathbf{E}}} \tag{253}$$

Thus, we can express the functional form of the constitutive equations for the first and second Piola-Kirchhoff stress tensors in terms of the Green deformation tensor **C**, or equivalently through the Lagrangian strain tensor **E**.

As before, we can show that for an isotropic material, the constitutive equation of the second Piola-Kirchhoff stress has the form

$$\boldsymbol{\Sigma} = 2\psi_1\mathbf{I}+4\psi_2\mathbf{C}+6\psi_3\mathbf{C}^2 \tag{254}$$

where $\psi_i = \partial\overline{\Psi}/\partial f_i$ are the derivatives of $\overline{\Psi}\left(f_1(\mathbf{C}), f_2(\mathbf{C}), f_3(\mathbf{C})\right)$ with respect to the invariants. Hence, they are each functions of $f_1(\mathbf{C})$, $f_2(\mathbf{C})$, and $f_3(\mathbf{C})$.

An example of a particular constitutive equation for large strains is the so-called *Mooney-Rivlin* material that has a stored energy function of

$$\overline{\Psi}(\mathbf{C}) = a(I_C - 3) + b(II_C - 3) \tag{255}$$

where I_C and II_C are the first and second principal invariants of the tensor **C**, and a and b are material constants. The reason for subtracting 3 from each of the invariants is simply to have zero energy when there are no strains (recall that no strain means $\mathbf{C} = \mathbf{I}$). The similarity between the form of this function and the one we used for linear elasticity should be noted. Because the energy depends only on the invariants of the deformation tensor, the material is isotropic. While this stored energy function leads to a linear constitutive relationship, the behavior of a body with this constitutive model would not be linear because neither **C** nor **Σ** will necessarily be linearly related to the applied forcing function. The Mooney-Rivlin constitutive model has been successfully used to model the behavior of rubber and rubberlike materials.

There is a middle ground that deserves mention. Some problems exhibit large motions, but with small strains. In such cases, it is often appropriate to model the constitutive equations after those of the linear theory. You must take care in doing so, however, as the following example illustrates.

Example 25. *Saint-Venant-Kirchhoff constitutive model.* Consider the following strain energy function (in terms of the Lagrangian strain)

$$\overline{\Psi}(\mathbf{E}) \equiv \tfrac{1}{2}\lambda(\mathrm{tr}\,\mathbf{E})^2 + \mu\,\mathrm{tr}(\mathbf{E}^2)$$

The similarity with the strain energy function of the linear theory should be evident. The constitutive equation takes the form

$$\mathbf{\Sigma} = \lambda\,\mathrm{tr}(\mathbf{E})\,\mathbf{I} + 2\mu\,\mathbf{E}$$

which appears to be a generalization of Hooke's law to finite deformations.

To understand the limitations of this model consider a homogenous uniaxial deformation with deformation gradient $\mathbf{F} = \mathbf{I} + (\gamma - 1)\mathbf{n} \otimes \mathbf{n}$. This deformation has a stretch of γ in the direction **n** and a stretch of 1 in the directions perpendicular to **n**. Note that $\det \mathbf{F} = \gamma$. The Lagrangian strain can be computed as

$$\mathbf{E} = \tfrac{1}{2}(\gamma^2 - 1)\,\mathbf{n} \otimes \mathbf{n}$$

Noting that $\mathrm{tr}(\mathbf{n} \otimes \mathbf{n}) = 1$ we find that

$$\mathbf{\Sigma} = \tfrac{1}{2}\lambda(\gamma^2 - 1)\mathbf{I} + \mu(\gamma^2 - 1)\,\mathbf{n} \otimes \mathbf{n}$$

Let us compute the component p of the Cauchy stress in the direction **n**. Noting the relationship between the Cauchy stress and second Piola-Kirchhoff stress (from Chapter 3) we have

$$\begin{aligned} p \equiv \mathbf{n} \cdot \mathbf{S}\mathbf{n} &= \frac{1}{J}\,\mathbf{n} \cdot \mathbf{F}\mathbf{\Sigma}\mathbf{F}^T\mathbf{n} \\ &= \frac{1}{J}(\mathbf{F}^T\mathbf{n}) \cdot \mathbf{\Sigma}(\mathbf{F}^T\mathbf{n}) \\ &= \frac{1}{\gamma}(\gamma\mathbf{n}) \cdot \mathbf{\Sigma}\,(\gamma\mathbf{n}) = \gamma\,\mathbf{n} \cdot \mathbf{\Sigma}\mathbf{n} \end{aligned}$$

where $J = \det \mathbf{F} = \gamma$ and $\mathbf{F}^T\mathbf{n} = \gamma\mathbf{n}$. Finally, substituting into the equation for the second Piola-Kirchhoff stress we find that

$$p(\gamma) = \left(\tfrac{1}{2}\lambda + \mu\right)\left(\gamma^3 - \gamma\right)$$

Observe that the stiffness $dp/d\gamma$ goes to zero at a stretch of $\gamma_{crit} = \sqrt{1/3}$ and that the stress p goes to zero as the bar shrinks to zero length. Both of these phenomena are physically unreasonable. Both are artifacts of the Saint-Venant-Kirchhoff model.

The main reason that the Saint-Venant-Kirchhoff model fails in the previous example is that it does not treat change in volume appropriately. For small deformations, change in volume is proportional to $\mathrm{tr}(\mathbf{E})$. For large deformations the ratio of deformed volume to original volume is $J = \det \mathbf{F}$. The Saint-Venant-Kirchhoff model does not respond appropriately in the limit as the volume shrinks.

One minor modification to the Saint-Venant-Kirchhoff model greatly improves its performance. Let the strain energy density be

$$\overline{\overline{\Psi}}(\mathbf{E}) \equiv \tfrac{1}{2}\lambda\left(\ln J\right)^2 + \mu\,\mathrm{tr}\left(\mathbf{E}^2\right) \tag{256}$$

where $J = \det \mathbf{F}$ and $\ln(\cdot)$ represents the natural logarithm of $(\cdot)$. Now the second Piola-Kirchhoff stress has the form

$$\mathbf{\Sigma} = \lambda J \ln J\left[2\mathbf{E} + \mathbf{I}\right]^{-1} + 2\mu\,\mathbf{E} \tag{257}$$

When deformations are small (i.e., $J \approx 1$) this model reverts to Hooke's law (as does the Saint-Venant-Kirchhoff model). There are numerous finite elasticity models that revert to Hooke's law in the limit of small deformations. Holzapfel (2000) provides an excellent discussion of these models.

Limits to Elasticity

Few materials exhibit elastic response indefinitely. At some level of stress or strain, materials start to exhibit irrecoverable strains. There are many constitutive models aimed at capturing yielding, cracking, evolution of porosity, and other microscopic phenomena that manifest at the macroscopic level (and show up as observable features of the mechanical response, e.g., in the stress-strain curve).

One of the most important continuum nonlinear material models is inelasticity. First conceived for metals, inelasticity has been applied to a wide range of materials from concrete to granular solids. While the development of inelastic constitutive models is beyond the scope of this book, it is useful to make some observations on the limits to elastic behavior.

Most models of inelasticity posit that the material responds elastically over a certain range of stresses and strains and that the accrual of inelastic (non-re-

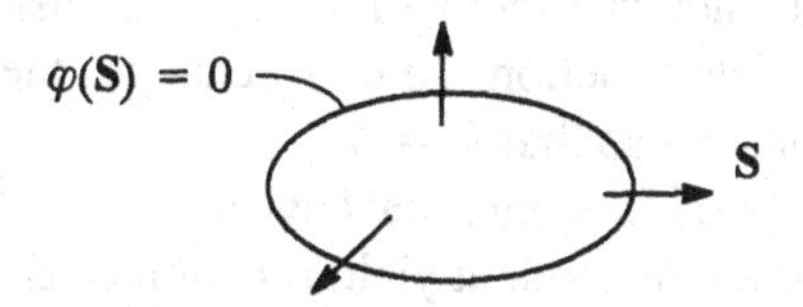

Figure 61 The yield surface describes the limit of elastic behavior

coverable) strains begins only upon reaching a certain state of stress or strain. The critical state is often referred to as the *yield surface*, which is illustrated in Fig. 61. The yield surface is a surface in, say, stress space that satisfies the scalar equation $\varphi(\mathbf{S}) = 0$. The *yield function* $\varphi(\mathbf{S})$ has the property that the interior of the elastic domain satisfies $\varphi(\mathbf{S}) < 0$ and the exterior of the elastic domain satisfies $\varphi(\mathbf{S}) > 0$. In many models the direction of inelastic straining is taken to be in the direction of the normal to the yield surface (i.e., the so-called *normality rule* of plastic deformation).

One of the most popular yield functions is the one due to von Mises that is based on the concept that yielding is independent of hydrostatic pressure. Specifically, the von Mises yield function is

$$\varphi(\mathbf{S}) \equiv \sqrt{\mathbf{S}' \cdot \mathbf{S}'} - k \tag{258}$$

where $\mathbf{S}' \equiv \mathbf{S} - \frac{1}{3}\operatorname{tr}(\mathbf{S})\mathbf{I}$ is the deviator stress, k is a material constant, and the dot product of tensors is $\mathbf{A} \cdot \mathbf{B} = A_{ij}B_{ij}$.

Example 26. *Uniaxial yield test.* To get an idea of the meaning of the constant k in Eqn. (258) one can imagine a uniaxial test with the load oriented in the direction $\mathbf{n}$. The stress tensor is $\mathbf{S} = \sigma[\mathbf{n} \otimes \mathbf{n}]$. The deviator stress is

$$\mathbf{S}' = \tfrac{2}{3}\sigma[\mathbf{n} \otimes \mathbf{n}] - \tfrac{1}{3}\sigma[\mathbf{I} - \mathbf{n} \otimes \mathbf{n}]$$

The yield function can be computed in this particular case to be

$$\varphi(\mathbf{S}) = \sqrt{\mathbf{S}' \cdot \mathbf{S}'} - k = \sqrt{\tfrac{2}{3}}\,\sigma - k = 0$$

If we call σ_y the yield stress in axial tension, then the parameter k has the interpretation

$$k = \sqrt{\tfrac{2}{3}}\,\sigma_y$$

One can repeat this thought experiment for a pure shear loading with stress tensor $\mathbf{S} = \tau[\mathbf{n} \otimes \mathbf{m} + \mathbf{m} \otimes \mathbf{n}]$ for perpendicular directions $\mathbf{n}$ and $\mathbf{m}$. If we call τ_y the yield stress in pure shear, then we get

$$k = \sqrt{2}\,\tau_y$$

From these two results we can observe the well-known result $\sigma_y = \sqrt{3}\,\tau_y$.

Some authors prefer to include a factor of three halves inside the square-root in the definition of the yield function. To do so changes the interpretation of k by a multiplicative constant so that $k = \sigma_y$.

There are many other yield functions that have proven useful in engineering computations. The pressure-dependent yield function of Drucker and Prager has found application in granular materials. The anisotropic generalization of the von Mises yield function, due to Hill has found application in composite materials. For a more complete account of the issues associated with inelasticity, particularly from a computational point of view, the reader should consult Simo and Hughes (1998).

Additional Reading

Y. C. Fung, *Foundations of solid mechanics,* Prentice Hall, Englewood Cliffs, N.J., 1965.

M. E. Gurtin, "The linear theory of elasticity," *Mechanics of solids,* Vol. II (C. Truesdell, ed.), Springer-Verlag, New York, 1972.

G. A. Holzapfel, *Nonlinear solid mechanics: a continuum approach for engineering,* John Wiley & Sons, New York, 2000.

L. E. Malvern, *Introduction to the mechanics of a continuous medium,* Prentice Hall, Englewood Cliffs, N.J., 1969.

J. E. Marsden and T. J. R. Hughes, *Mathematical foundations of elasticity,* Prentice Hall, Englewood Cliffs, N.J., (1983). (Now available in a Dover edition.)

Simo, J. C. and T. J. R. Hughes, *Computational Inelasticity,* Springer-Verlag, New York, 1998.

I. S. Sokolnikoff, *Mathematical theory of elasticity,* 2nd ed., McGraw-Hill, New York, 1956.

S. P. Timoshenko, *History of strength of materials,* McGraw-Hill, New York, 1953. (Reprinted by Dover, 1983.)

Problems

97. The constitutive equations for a three-dimensional isotropic, linearly elastic material can be expressed in the form $S_{ij} = \lambda E_{kk}\delta_{ij} + 2\mu E_{ij}$ where the subscripts i, j, and k range over the values 1, 2, and 3. Find equivalent expressions for the constitutive equations that already reflect the plane stress condition $S_{33} = S_{23} = S_{13} = 0$, that is, find new material constants λ^* and μ^* such that the two-dimensional relationship can be written as

$$S_{\alpha\beta} = \lambda^* E_{\gamma\gamma}\delta_{\alpha\beta} + 2\mu^* E_{\alpha\beta}$$

where the Greek indices range only over the values 1 and 2. Express the new constants (λ^*, μ^*) in terms of the constants (λ, μ) of the three-dimensional theory.

98. Demonstrate that the following relationships between the elastic constants λ, μ, C, K, and ν hold for an isotropic, linearly elastic material

$$\lambda = \frac{2\mu\nu}{1-2\nu} = \frac{\mu(C-2\mu)}{3\mu-C} = \frac{C\nu}{(1+\nu)(1-2\nu)} = \frac{3K\nu}{1+\nu}$$

$$K = \lambda + \tfrac{2}{3}\mu = \frac{\mu C}{3(3\mu-C)} = \frac{\lambda(1+\nu)}{3\nu} = \frac{C}{3(1-2\nu)}$$

$$C = 2\mu(1+\nu) = \frac{\mu(3\lambda+2\mu)}{\lambda+\mu} = \frac{\lambda(1+\nu)(1-2\nu)}{\nu} = \frac{9K\mu}{3K+\mu}$$

$$\mu = \frac{C}{2(1+\nu)} = \tfrac{3}{2}(K-\lambda) = \frac{3K(1-2\nu)}{2(1+\nu)} = \frac{\lambda(1-2\nu)}{2\nu}$$

$$\nu = \frac{\lambda}{2(\lambda+\mu)} = \frac{C}{2\mu} - 1 = \frac{3K-2\mu}{2(3K-\mu)} = \frac{3K-C}{6K}$$

We can observe that, in each case, one of the elastic constants is expressed in terms of two of the others from the set of five constants. There are some natural limits to the values that the constitutive parameters can take. Assume that under compressive hydrostatic pressure it is impossible for the volume to increase, and that in uniaxial tension it is impossible for a bar to get shorter. What do these hypotheses imply about the other moduli?

99. Show that the isotropic elasticity tensor with components

$$C_{ijkl} = \lambda\delta_{ij}\delta_{kl} + \mu\left[\delta_{ik}\delta_{jl} + \delta_{il}\delta_{jk}\right]$$

is invariant with respect to coordinate transformation since the components of the tensor in the two coordinate systems are related by $\mathsf{C}_{abcd} = C_{ijkl}Q_{ai}Q_{bj}Q_{ck}Q_{dl}$, where, as usual, $Q_{ij} = \mathbf{g}_i \cdot \mathbf{e}_j$ are the components of the orthogonal change-of-basis tensor. (Hint: another way to view change of basis is $\mathbf{g}_i = Q_{ij}\mathbf{e}_j$.)

100. Consider a linearly elastic, isotropic material with Lamé parameters λ and μ, subjected to the following displacement map $\mathbf{u}(\mathbf{x}) = \beta(x_1^2\mathbf{e}_1 + x_2^2\mathbf{e}_2)$. Assume that the linearized strain tensor is adequate to characterize the strain field, and compute the body forces required to satisfy equilibrium.

101. Arrange the six independent stress and strain components in column matrices as follows: $\mathbf{S} = (S_{11}, S_{22}, S_{33}, S_{12}, S_{23}, S_{13})^T$ and $\mathbf{E} = (E_{11}, E_{22}, E_{33}, 2E_{12}, 2E_{23}, 2E_{13})^T$. Assume that the constitutive equations of linear elasticity hold. Show that the constitutive equations can be expressed in matrix form as $\mathbf{S} = \mathbf{DE}$, where $\mathbf{D}$ is a six by six matrix.

102. Consider the thin rectangular sheet with Young's modulus $C = 1000$ ksi and Poisson's ratio $\nu = 0.45$. The sheet is subjected to a uniform state of stress through the tractions given in the sketch. The thickness of the sheet before the tractions were applied was 0.1 in. What is the thickness of the sheet after the tractions are applied?

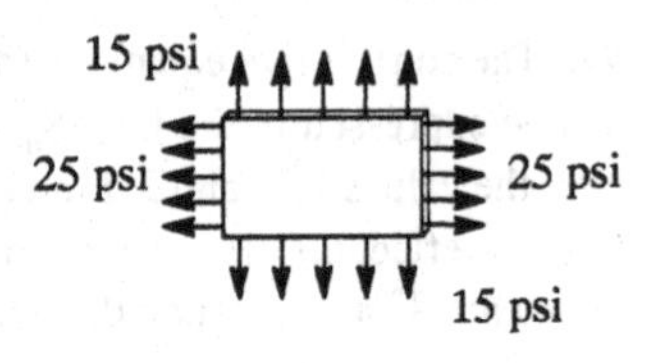

103. Consider the thin rectangular sheet with Young's modulus $C=1000$ ksi and Poisson's ratio $\nu=0.2$. The sheet is fixed between two immovable frictionless plates and is subjected to a uniform state of stress through the tractions around the edges as shown in the sketch. The thickness of the sheet before the tractions were applied was 0.3 in. What is the state of stress in the sheet after the tractions are applied? What are the reacting tractions provided by the plates? Find the ratio of the change in volume to the original volume of the sheet.

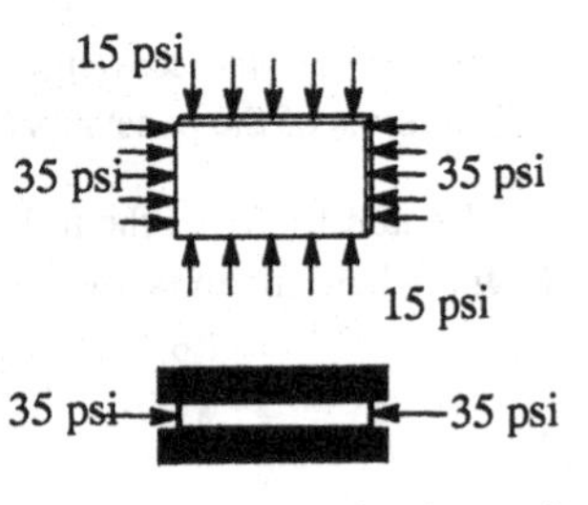

104. A disk made of isotropic, linearly elastic material is subjected to a known uniform pressure p around its perimeter. The faces of the disk are clamped between immovable, frictionless plates so that the strain through the thickness is zero. Assume that the stress state is homogeneous throughout the disk and that the Lamé constants are known. Find the tractions t acting on the faces.

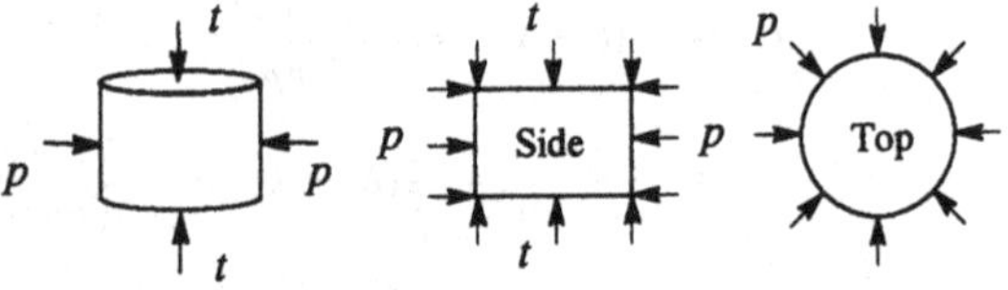

105. In a triaxial test, a cylindrical specimen is subjected to a uniform pressure σ_1 on the ends of the cylinder and a uniform pressure σ_2 on the sides. The change in height Δh and the change in diameter Δd are measured. Let $\varepsilon_1 \equiv \Delta h/h$ and $\varepsilon_2 \equiv \Delta d/d$, where h is the original height and d the original diameter. The values measured in a test are given in the diagram. Assume that the material is linear, isotropic, and elastic. What is the volume of the deformed cylinder? Compute the value of the bulk and shear moduli (K and μ) for this sample.

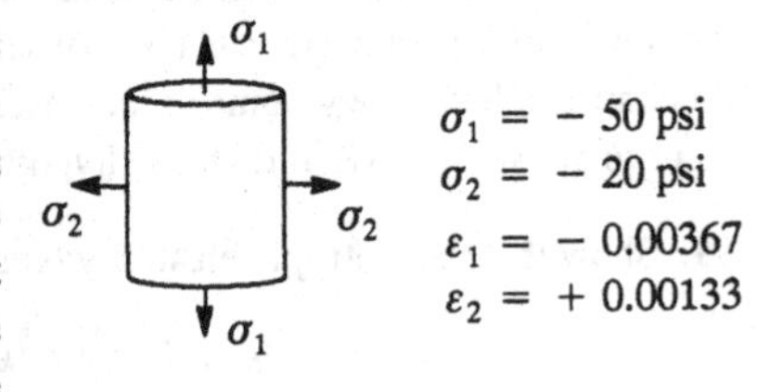

106. Prove the identity $\det[\mathbf{I}+2\mathbf{E}] = 1+2I_E+4II_E+8III_E$. (Hint: Use the component expression for the determinant of a tensor. The $\epsilon-\delta$ identity from Chapter 1 may also be useful.)

107. Consider a beam of length ℓ with its axis oriented along the z_3 direction. The cross section of the beam lies in the $z_1 - z_2$ plane, and its second moment of the area is equal to I. The beam is subjected to equal and opposite end moments of magnitude M, bending it about the axis with second moment of the area I. The beam is made of elastic material with moduli C and ν. The displacement field in the beam is given by the expression

$$\mathbf{u(z)} = \frac{M}{CI}\Big[\tfrac{1}{2}\big(z_3^2+\nu z_1^2-\nu z_2^2\big)\mathbf{e}_1 + \nu z_1 z_2 \mathbf{e}_2 - z_1 z_3 \mathbf{e}_3\Big]$$

Assume that the applied moment is small enough relative to CI that the displacements are quite small. Compute the components of the strain tensor. Compute the components of the stress tensor from the strain tensor and the linear elastic constitutive equations. Verify that the stress field satisfies the equations of equilibrium.

108. Consider the displacement map $\mathbf{u}(\mathbf{z})$ for a sphere of unit radius, given by the explicit expression

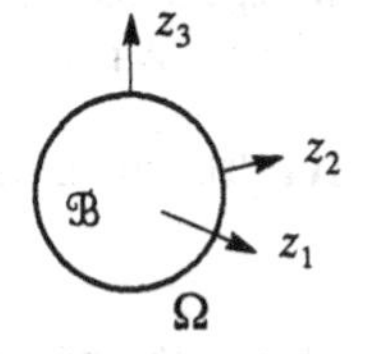

$$\mathbf{u}(\mathbf{z}) = \varepsilon(\mathbf{z}\cdot\mathbf{z})\mathbf{z}$$

where ε is a (very small) constant of the motion. Assume that the material is isotropic and linearly elastic with material constants λ and μ (i.e., the Lamé parameters). Compute the body force $\mathbf{b}$ required to maintain equilibrium. Compute the traction forces $\mathbf{t}$ that must be acting on the surface of the sphere. Determine the principal stress field associated with the given motion.

109. Let the elasticity tensor be given by $C_{ijkl} = \lambda\delta_{ij}\delta_{kl} + \mu\left[\delta_{ik}\delta_{jl}+\delta_{il}\delta_{jk}\right]$. Show that the expression $S_{ij} = C_{ijkl}E_{kl}$ reduces to Eqn. (236).

110. A cube of isotropic elastic material, having Lamé constants $\lambda = 1000$ psi and $\mu = 1000$ psi is in a homogeneous (i.e., does not vary with position) state of stress given by a stress tensor with components

$$\mathbf{S} \sim \begin{bmatrix} 10 & 2 & 1 \\ 2 & 5 & 1 \\ 1 & 1 & S_{33} \end{bmatrix}$$

Find the stress component S_{33} that is consistent with the observation that the cube decreases in volume (from the stress-free state) by 5%. Now compute the components of the deviatoric stress tensor and the strain components E_{13} and E_{33}.

111. A block of elastic material, having Lamé constants $\lambda = 1000$ psi and $\mu = 1000$ psi is subjected to a lateral compressive pressure of $\sigma = 80$ psi and clamped between two frictionless rigid plates that reduce the height of the block to 99% of its original height. Compute the total force required on the plates to accomplish the motion. Compute the volume of the block after deformation. Compute the change in the area of the block on the faces in contact with the plates.

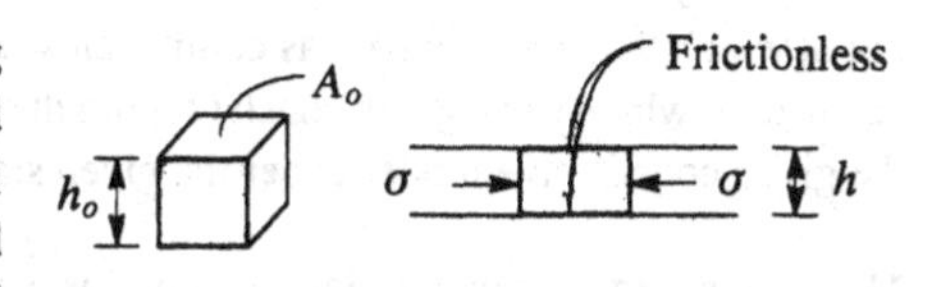

112. Consider a body $\mathfrak{B}$ subjected to the following displacement map:

$$\mathbf{u}(\mathbf{z}) = \beta\left(z_1^2-2z_2z_3\right)\mathbf{e}_1 + \beta\left(z_2^2+2z_1z_3\right)\mathbf{e}_2 + \beta\left(z_3^2-2z_1z_2\right)\mathbf{e}_3$$

where β is a (very small) constant. Find the stress tensor associated with this motion, assuming that the material is linear, isotropic, and elastic with moduli λ and μ and that the stress is zero when the displacement is zero. Find the body force field required to maintain equilibrium for the given motion. Find the principal values of the (linearized) strain tensor.

113. The state of the deformation at a certain point in a solid body is such that it has the following principal strains $\epsilon_1 = \varepsilon$ and $\epsilon_2 = \epsilon_3 = 3\varepsilon$, where ε is a known value. The principal directions associated with these principal strains are known to be $\mathbf{n}_1$, $\mathbf{n}_2$, and $\mathbf{n}_3$. Assuming linear, isotropic response, find the principal values and principal directions of the stress tensor $\mathbf{S}$. The material constants are $\lambda = 1000$ psi and $\mu = 1000$ psi. What is the average pressure p at the point in question? What is the change in volume in the neighborhood of the point in question?

114. The strain energy function of a nonlinear hyperelastic material is given by

$$W(\mathbf{E}) = \tfrac{1}{2}aE_{ii}E_{jj} + \tfrac{1}{2}bE_{ij}E_{ij} + \tfrac{1}{3}cE_{ij}E_{jk}E_{ki}$$

where a, b, and c are material constants and $\mathbf{E} = E_{ij}[\mathbf{e}_i \otimes \mathbf{e}_j]$ is the strain tensor. Find the stress tensor $\mathbf{S}$ as a function of the strain $\mathbf{E}$ implied by the strain energy function.

115. The strain energy function of a nonlinear hyperelastic material is given by

$$W(\mathbf{E}) = aE_{ii}\ln\left(1+E_{jj}\right) + \tfrac{3}{2}bE_{ij}E_{ij}$$

where ln(.) indicates the natural logarithm of (.), a and b are known material constants, and $\mathbf{E} = E_{ij}[\mathbf{e}_i \otimes \mathbf{e}_j]$ is the strain tensor. Find the stress tensor $\mathbf{S}$ as a function of the strain $\mathbf{E}$ implied by the strain energy function. Consider a hydrostatic state of stress with pressure p in which the stress tensor is given by $\mathbf{S} = p\mathbf{I}$. Set up a relationship between the change in volume and the pressure p. What is the pressure required to decrease the volume to 95% of the original volume (assume that the linearized strain tensor is adequate)?

116. The strain energy function of a nonlinear hyperelastic material is given by the (component) expression $W(\mathbf{E}) = a_0E_{ii}E_{jj}E_{kk} + a_1E_{ij}E_{jk}E_{ki}$, where a_0 and a_1 are known material constants, and E_{ij} is the ijth component of the strain tensor $\mathbf{E}$. Find the stress tensor $\mathbf{S}$ as a function of the strain $\mathbf{E}$ implied by the strain energy function. Is the material isotropic? Explain your answer. Is the material linear? Explain. Consider a uniform state of shearing in which the strain tensor has components $E_{12} = E_{21} = \gamma$ and all other components equal zero, where γ is a given constant. Find the principal values of the stress tensor $\mathbf{S}$ for the given constitutive model under the given state of strain.

117. The strain energy function of a nonlinear hyperelastic material is given by

$$W(\mathbf{E}) = \ln\left(1+\alpha E_{ij}E_{ij}\right) + \beta\left(e^{E_{ii}} - E_{ii}\right)$$

where ln(.) indicates the natural logarithm of (.) and $e^{(.)}$ indicates the exponential of (.), α and β are known material constants, and E_{ij} is the ijth component of the strain tensor $\mathbf{E}$. Find the stress tensor $\mathbf{S}$ as a function of the strain $\mathbf{E}$ implied by the strain energy function. How do the constants α and β relate to the Lamé parameters of linear isotropic elasticity? Consider a uniform state of dilation in which the strain tensor is given by $\mathbf{E} = \varepsilon\mathbf{I}$, where ε is a constant. Find the principal values of the stress tensor as a function of ε.

118. The strain energy function of a nonlinear hyperelastic material is given by the expression $W(e,\gamma) = a_0e^2 + a_1\gamma + a_2e\gamma$, where a_0, a_1, and a_2 are known material constants, and the scalar invariant strain measures e and γ, which are functions of the strain tensor $\mathbf{E}$, are defined as $e \equiv \mathrm{tr}(\mathbf{E})$ and $\gamma \equiv \mathrm{tr}(\mathbf{E}'\mathbf{E}')$, where $\mathbf{E}' \equiv \mathbf{E} - e\mathbf{I}/3$ is the deviator strain. Observe that $\partial e/\partial\mathbf{E} = \mathbf{I}$ and $\partial\gamma/\partial\mathbf{E} = 2\mathbf{E}'$. Find the stress tensor $\mathbf{S}$ as a function of the strain $\mathbf{E}$ implied by the strain energy function. Consider a state of hydrostatic pressure $\mathbf{S} = p\mathbf{I}$, where p is a given pressure. Find the relationship among p, e and γ. Next consider a sample of the material subjected to a state of pure shear strain described by $\mathbf{E} = g[\mathbf{n} \otimes \mathbf{m} + \mathbf{m} \otimes \mathbf{n}]$, where g is a given constant describing the motion and $\mathbf{m}$ and $\mathbf{n}$ are given orthogonal unit vectors. Will there be a change in volume of the sample? Do you expect that you would need a confining pressure to execute this motion? Why?

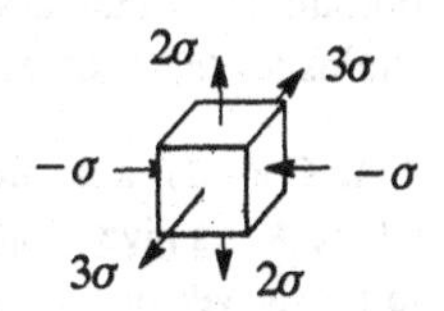

119. A cube of elastic material, having Lamé constants $\lambda = 1000$ psi and $\mu = 1000$ psi is subjected to purely normal tractions on its faces as shown in the sketch. Compute the value of σ required to change the volume of the block by 2% of its original volume.

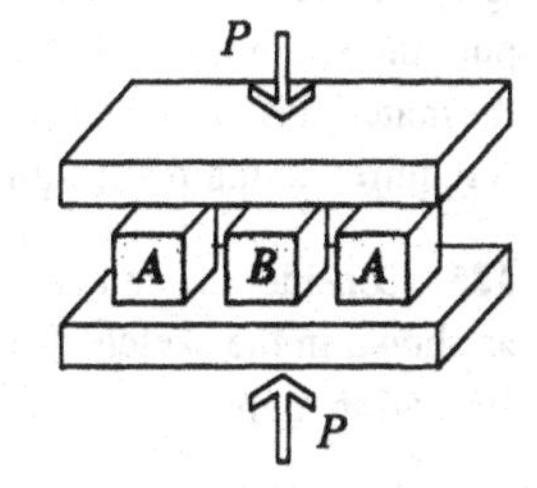

120. Three unit cubes $(1 \times 1 \times 1)$ are uniformly compressed between two rigid plates with an aggregate force of P. The change in height is the same for all three cubes. The two outer cubes are made of material A, while the inner cube is made of material B. Both of the materials are linearly elastic with Lamé constants

$$\lambda_A = 1000 \text{ psi}, \quad \mu_A = 1000 \text{ psi}$$

$$\lambda_B = 500 \text{ psi}, \quad \mu_B = 2000 \text{ psi}$$

Compute the force P required to change the volume of the middle block by 3% of its original volume. What is the final area of the compressed face of the outer cubes?

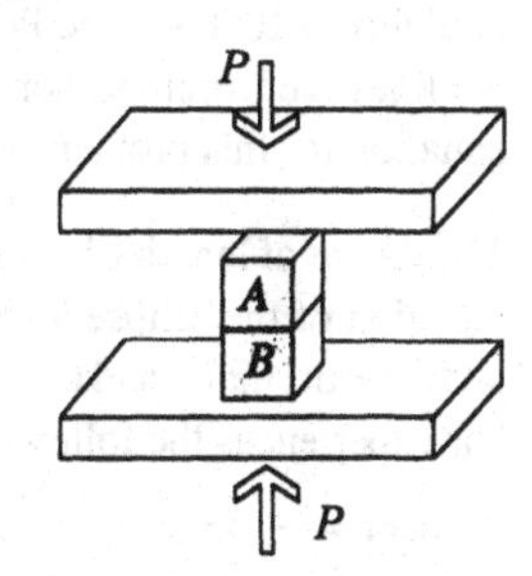

121. Two cubes with dimensions $2 \times 2 \times 2$ are uniformly compressed between two rigid plates with an aggregate force of P. Assume that there is no friction between any of the contacting surfaces. The top cube is made of material A, while the bottom cube is made of material B. Both of the materials are linearly elastic with Lamé constants

$$\lambda_A = 1000 \text{ psi}, \quad \mu_A = 1000 \text{ psi}$$

$$\lambda_B = 500 \text{ psi}, \quad \mu_B = 2000 \text{ psi}$$

Compute the force P required to change the total volume of the two cubes by 5% of the original volume. What are the final dimensions of the two cubes?

122. The strain at a point in a body is given by

$$\mathbf{E} \sim 10^{-3} \begin{bmatrix} 2 & 3 & 4 \\ 3 & 5 & 1 \\ 4 & 1 & 1 \end{bmatrix}$$

Find the components of the stress tensor assuming linear, isotropic, elastic material behavior, with $\lambda = 16{,}000$ ksi and $\mu = 11{,}000$ ksi.

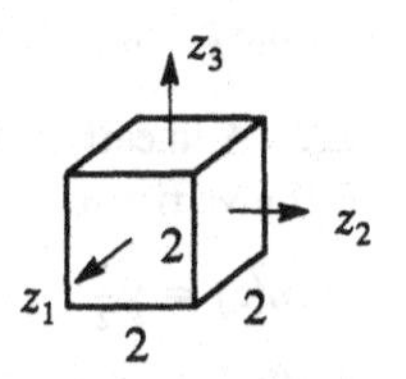

123. A 2 by 2 by 2 unit solid cube, centered at the origin of coordinates, is subjected to the deformation described by the map:

$$\boldsymbol{\phi}(\mathbf{z}) = \left(z_1 + \tfrac{1}{3}az_1^3\right)\mathbf{e}_1 + \left(z_2 + \tfrac{1}{3}az_2^3\right)\mathbf{e}_2 + bz_3\mathbf{e}_3$$

Compute the values of the constants a and b that are consistent with the observations that the total volume of the block is unchanged by the deformation and the total area of the side with original normal $\mathbf{e}_1$ decreases by 5% due to the deformation. Assuming that the cube is made of a linear, elastic, isotropic material with Lamé parameters λ and μ, find the body forces and surface tractions required for equilibrium. (You may assume that the linearized strain tensor is adequate to describe the strains for this problem).

124. Consider a displacement map $\mathbf{u}(\mathbf{z})$ given by the explicit expression $\mathbf{u}(\mathbf{z}) = \varepsilon \mathbf{A}\mathbf{z}$, where $\mathbf{A}$ is a given constant tensor and ε is a given scalar (which is very small compared to 1). The vector $\mathbf{z}$ is the position vector of a point in the undeformed configuration. Compute the strain tensor $\mathbf{E}$ of the given motion. Compute the stress tensor $\mathbf{S}$ assuming that the material is linear and elastic and has Lamé parameters λ and μ. Compute the body force $\mathbf{b}$ required to maintain equilibrium with the stress.

125. Consider the unit cube with vertex at the origin of coordinates as shown in the sketch. The cube is subjected to the following deformation map:

$$\boldsymbol{\phi}(\mathbf{z}) = \left(z_1 + z_2 \sin\gamma\right)\mathbf{e}_1 + z_2 \cos\gamma\, \mathbf{e}_2 + z_3 \mathbf{e}_3$$

Note that γ is a constant. Compute the tractions and body forces required to achieve the given deformation for the specific shearing angle of γ=0.2 rad assuming that the material is linear, elastic, and isotropic with Young's modulus of 1000 psi and Poisson's ratio of 0.499. Does it make any difference if you use the linearized strain tensor as opposed to the Lagrangian strain tensor in the constitutive equation for this problem? Explain.

126. A bar of length ℓ has an elliptical cross section. The equation of the ellipse is $b^2 z_1^2 + a^2 z_2^2 = a^2 b^2$, where a and b are the major and minor semi-axis dimensions. The bar experiences the following displacement map:

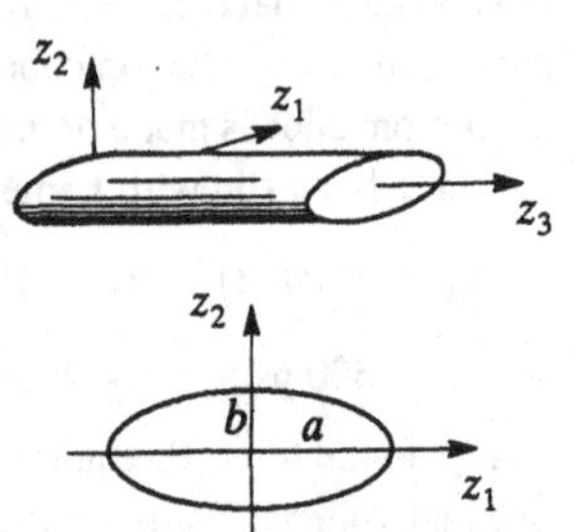

$$\mathbf{u}(\mathbf{z}) = -\beta z_2 z_3 \mathbf{e}_1 + \beta z_1 z_3 \mathbf{e}_2 - \beta c z_1 z_2 \mathbf{e}_3$$

where β and c are constants. Find the stress tensor associated with this motion, assuming that the material is linear, isotropic, and elastic with moduli λ and μ. Find the body force required for equilibrium. What value must the constant c have in order that the lateral surface of the bar be traction-free?

127. Consider a displacement map $\mathbf{u}(\mathbf{z}) = \left[\mathbf{z} \otimes \mathbf{z}\right]\mathbf{a}$, where $\mathbf{a}$ is a given constant vector (which has a magnitude very small compared to 1). The vector $\mathbf{z}$ is the position vector of a point in the undeformed configuration. Compute the *linearized* strain tensor $\mathbf{E}$ of the given motion. Compute the stress tensor $\mathbf{S}$ assuming that the material is linear and elastic and has Lamé parameters λ and μ. Compute the body force $\mathbf{b}$ required to maintain equilibrium for the given motion.

128. A linearly elastic solid body is subjected to forces that give rise to the following displacement map:

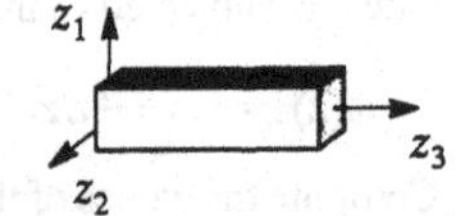

$$\mathbf{u}(\mathbf{z}) = \gamma \tfrac{1}{2}\left(z_3^2 + \beta z_1^2 - \beta z_2^2\right)\mathbf{e}_1 + \gamma\beta z_1 z_2 \mathbf{e}_2 - \gamma z_1 z_3 \mathbf{e}_3$$

where $\gamma \ll 1$ (i.e., very small) and β are constants describing the motion. Assume that the elastic response is adequately characterized by Hooke's law with known material constants λ and μ (the Láme parameters). Find β in terms of the constants λ and μ such that $S_{11} = 0$ (**S** is the stress tensor). Find S_{22} and S_{33} for the conditions given previously. Find the traction on the surface with normal $-\mathbf{e}_3$ at $z_3 = 0$.

129. The state of stress **S** as a function of position **x** in a certain solid body is given by the expression $\mathbf{S}(\mathbf{x}) = \mathbf{x} \otimes \mathbf{B}\mathbf{x}$, where **B** is a given constant tensor. Find the body force (as a function of position **x**) required to maintain equilibrium of the body. Express the result in both index and direct (vector) notation. What are the restrictions, if any, on the constant tensor **B** in order for the stress field **S** to be an admissible stress state? (Please describe any restrictions explicitly in terms of the components of **B**, not in terms of **x** and **S**.)

5

Boundary Value Problems in Elasticity

All problems in solid mechanics require three basic components: (a) equations of geometry of deformation relating the displacements (i.e., the map) to strains; (b) equations of equilibrium relating the applied tractions and body forces to the stresses; and (c) equations of constitution relating stresses to strains. All of these equations are necessary to the statement of mechanics problems like the torsion of a bar or the bending of a beam, but they are not sufficient to solve such problems. In addition to these equations, which describe what is happening inside the body, we must also describe what is happening on the surface, or boundary, of the body. These boundary conditions and generally comprise given data about the displacements and applied tractions on the surface of the body. The combination of domain equations and boundary conditions is called a *boundary value problem*.

There are two important facts one must know about the specification of boundary conditions. First, we can specify as given data *either* the displacement of a certain point *or* the traction applied at that point; we can never specify both the displacement and the traction at a certain point. If the displacement is known, as it is for a fixed point, for example, then the traction at that point is unknown. We usually call such unknown tractions "reaction forces" because they develop in accord with whatever the equations of equilibrium require to react to the applied forces. At a point where traction is applied, the displacement cannot be known a priori.

The second important fact is that not all specifications of boundary conditions are acceptable. For example, we cannot apply tractions to the entire surface of the body willy-nilly. Unless those tractions are very specially prescribed, it may not be possible to satisfy equilibrium. Furthermore, any two displacement fields that differed by only a rigid-body motion would satisfy the governing equations. Hence, the position of the body in space would not be uniquely determined. Properly specified boundary conditions give rise to what we call a *well-posed* boundary value problem, while improperly specified boundary conditions give rise to an *ill-posed* boundary value problem. In simple terms, a well-posed boundary value problem is one that we can solve (at least theoretically; the practical aspects of carrying out the mathematical manipulations may be well beyond our capability for many "solvable" problems).

We have developed the necessary governing equations for the domain of the body, so we shall proceed to state the boundary value problem of three-dimensional elasticity. The resulting system of partial differential equations are difficult to solve in a classical sense (i.e., find fields that exactly satisfy all of the differential equations at every point in the body) for all but a few special cases. We shall recast the equations into a very different format called the *principle of virtual work*. This principle will lead directly to some powerful approximate methods of solution, among which we find the finite element method.

We will first state the general boundary value problem for three-dimensional elasticity. For the purpose of illustration, we specialize these equations to a one-dimensional version, which we call the *little boundary value problem*. We use the one-dimensional problem to contrast the classical and variational approaches to stating a boundary value problem and to warm up to the principle of virtual work. Finally, we recast the three-dimensional equations of equilibrium as a principle of virtual work, showing that the steps are identical to the one-dimensional case.

Throughout this chapter we shall be concerned primarily with the linear theory of elasticity. The key issue will be the understanding of the relationship between classical and variational formulations of the equations that govern the response of structural systems. We extend the ideas to finite deformation at the end of the chapter.

Boundary Value Problems of Linear Elasticity

Consider the body $\mathfrak{B}$ shown in Fig. 62. It has displacements prescribed over part of its boundary, and tractions (forces) prescribed over the remaining part of its boundary (remember, it is not possible to prescribe both the displacement and the traction at the same place). It is subject to body forces $\mathbf{b}(\mathbf{x})$. Let us call

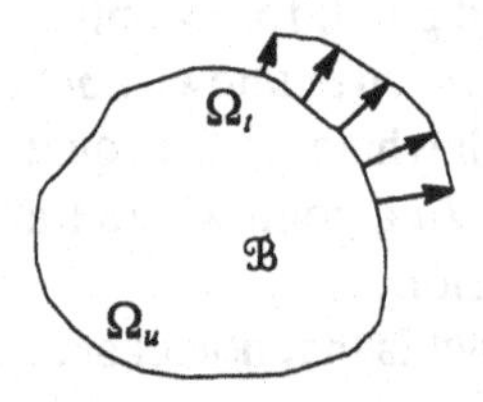

Figure 62 A body subjected to prescribed displacements and tractions

the part of the boundary where displacements $\hat{\mathbf{u}}$ are prescribed Ω_u, and the part of the boundary where tractions $\hat{\boldsymbol{\tau}}$ are prescribed Ω_t. (Note: The free surfaces of a body are places where the traction is prescribed to be zero.) The entire boundary comprises the two parts $\Omega = \Omega_u \cup \Omega_t$. The boundary value problem of linear elasticity is specified by the following equations

$$
\begin{array}{rcll}
\operatorname{div}\mathbf{S}+\mathbf{b} &=& \mathbf{0} & \text{in } \mathfrak{B} \\
\mathbf{E} &=& \frac{1}{2}\left[\nabla\mathbf{u}+\nabla\mathbf{u}^T\right] & \text{in } \mathfrak{B} \\
\mathbf{S} &=& \lambda\operatorname{tr}(\mathbf{E})\mathbf{I}+2\mu\mathbf{E} & \text{in } \mathfrak{B} \\
\mathbf{Sn} &=& \hat{\boldsymbol{\tau}} & \text{on } \Omega_t \\
\mathbf{u} &=& \hat{\mathbf{u}} & \text{on } \Omega_u
\end{array}
\tag{259}
$$

This set of equations has, as given data, the prescribed body forces $\mathbf{b}$, boundary tractions $\hat{\boldsymbol{\tau}}$, and boundary displacements $\hat{\mathbf{u}}$, as well as the material properties λ and μ. The unknowns of the problem are the displacement field $\mathbf{u}(\mathbf{x})$, the stress field $\mathbf{S}(\mathbf{x})$ and the strain field $\mathbf{E}(\mathbf{x})$. In terms of components, we have 15 unknown scalar fields (three displacement components, six stress components, and six strain components. The domain equations of Eqn. (259) provide 15 component equations to find the 15 unknowns. The equations are differential equations so there are constants of integration. It is useful to note that there are exactly three independent components to a displacement $\mathbf{u}_o$ for which $\nabla\mathbf{u}_o$ is skew symmetric (and hence $\mathbf{E}$ is zero). There are exactly three independent components to a stress tensor $\mathbf{S}_o$ for which $\operatorname{div}\mathbf{S}_o = \mathbf{0}$ (self-equilibrating). Hence we need six boundary condition components to solve the boundary value problem. The boundary conditions provide these six conditions.

The equations are a set of partial differential equations relating the unknown variables. Some of these variables can be eliminated in favor of the others by differentiation and substitution. In particular, we can find a set of equations involving only the unknown displacement field $\mathbf{u}(\mathbf{x})$. These equations are usually referred to as the *Navier equations* (actually, Navier did not get them quite right, and Cauchy came to his rescue). These equations, expressed in component form, are

$$
\begin{array}{rcll}
(\lambda+\mu)\dfrac{\partial^2 u_k}{\partial x_k \partial x_i} + \mu\dfrac{\partial^2 u_i}{\partial x_k \partial x_k} + b_i &=& 0 & \text{in } \mathfrak{B} \\
\lambda\dfrac{\partial u_k}{\partial x_k} n_i + \mu\left(\dfrac{\partial u_i}{\partial x_j} + \dfrac{\partial u_j}{\partial x_i}\right) n_j &=& \hat{\tau}_i & \text{on } \Omega_t \\
u_i &=& \hat{u}_i & \text{on } \Omega_u
\end{array}
\tag{260}
$$

We often refer to such a system of equations as a *three-field* theory because there are three unknown functions of $\mathbf{x}$ that we are trying to find, the three components of $\mathbf{u}(\mathbf{x})$. These equations have been expressed in index notation. The summation convention is applied to repeated indices.

Integration of the domain equation $(260)_a$ leads to six constants of integration (it is a system of three second-order partial differential equations). There are essentially three traction conditions and three displacement conditions with which to establish those constants. This count of unknowns is somewhat artificial. For example, a case where displacements are prescribed on the entire surface appears not to have enough conditions to establish a unique solution. However, in such a case, the change in volume is prescribed. Establishing the conditions for the solution to a partial differential equation is extremely important and sometimes tricky. Some of these issues will be clearer in the context of specific problems and specialized theories.

Since all bodies in the physical world are three-dimensional, you would think that it would be sufficient to simply learn to solve Eqns. (260). From an engineering point of view, such an approach is not practical. These equations are difficult to solve analytically (that is, to find closed-form expressions for the field $\mathbf{u}(\mathbf{x})$ in terms of defined functions like cosines, sines, exponentials, and the like). We shall see that we can actually solve these equations with the finite element method, but we still will not want to view all problems as three-dimensional because the resulting systems of equations will still be too large to solve on today's computers (perhaps some day this will no longer be an issue). Hence, we are led to making assumptions about the behavior of our bodies to simplify the above equations.

Most of these simplifications constitute a reduction in the dimensionality of the problem. Such a reduction is accomplished either by making assumptions about the stress or strain fields, by making assumptions about the displacement field (i.e., the map), or from known symmetries of the problem. ***Plane stress*** and ***plane strain*** are two-dimensional theories, the first of which makes the assumptions about the stress field of the form $S_{33} = 0$, $S_{23} = S_{32} = 0$, and $S_{13} = S_{31} = 0$ (as far as stress is concerned, there is no action taking place in the x_3 direction), while the second makes the assumptions about the strain field of the form $E_{33} = 0$, $E_{23} = E_{32} = 0$, and $E_{13} = E_{31} = 0$ (as far as strain is concerned, there is no action taking place in the x_3 direction). ***Axisymmetric problems*** (i.e., axisymmetric bodies with axisymmetric loads) can be reduced to two dimensions if cylindrical coordinates $\{r, z, \theta\}$ are used because the solution does not depend upon θ. ***Beam theory*** makes an assumption like "cross sections of the beam that were plane before deformation remain plane after deformation" which leads to one-dimensional differential equations where the generalized displacement variables are functions of only the axial coordinate. ***Plate theory*** is essentially a two-dimensional beam theory. ***Shell theory*** is essentially a plate theory in which the original geometry is not flat. These specialized theories allow us to make enormous strides in understanding the mechanics of structural systems, and it is worthwhile to spend a fair amount of effort in understanding exactly how they relate to the general three-dimensional theory.

Let us first examine a problem for which the exact three-dimensional equations of elasticity are satisfied in order to see how all of the individual components play out in the problem specification. One of the few problems that has a simple representation and also exactly satisfies the complete three-dimensional equations of linear elasticity is the torsion of a circular shaft. Here we shall simply give the expression for the map (i.e., the displacement field) and demonstrate that it satisfies all of the equations of linear elasticity.

Example 27. *Torsion of a circular shaft.* Consider a prismatic circular shaft with the x_3 coordinate axis along the central axis of the shaft as shown in Fig. 63. The shaft has length ℓ and radius R, has no body force $\mathbf{b}$, is fixed at one end, and has a stress distribution equipollent to a pure torque $\mathbf{T}$ applied to one end.

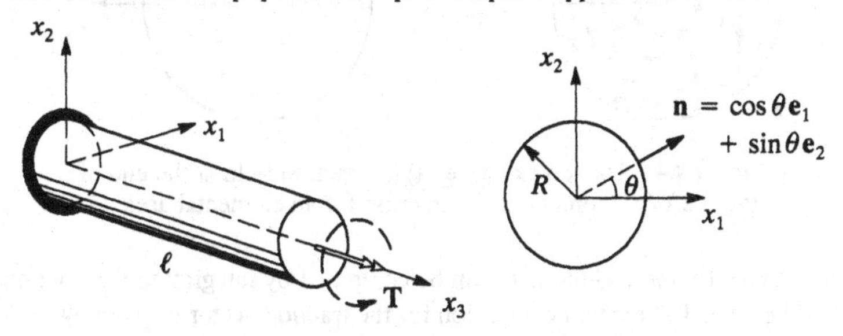

Figure 63 Pure torsion of a circular shaft

The displacement field for the torsion problem is given in terms of the constant β, which measures the angle of rotation of the cross section per unit of length, that is, the rate of twist of the shaft. The displacement map is given by

$$\mathbf{u}(\mathbf{x}) = -\beta x_2 x_3 \mathbf{e}_1 + \beta x_1 x_3 \mathbf{e}_2$$

From the displacement field we can compute the strain tensor as

$$\begin{aligned}\mathbf{E} &= \tfrac{1}{2}\left[\nabla\mathbf{u} + \nabla\mathbf{u}^T\right] \\ &= \tfrac{1}{2}\beta\left(-x_2\left[\mathbf{e}_1 \otimes \mathbf{e}_3 + \mathbf{e}_3 \otimes \mathbf{e}_1\right] + x_1\left[\mathbf{e}_2 \otimes \mathbf{e}_3 + \mathbf{e}_3 \otimes \mathbf{e}_2\right]\right)\end{aligned}$$

From the constitutive equations $\mathbf{S} = \lambda\,\mathrm{tr}(\mathbf{E})\mathbf{I} + 2\mu\mathbf{E}$ we get the stress field

$$\mathbf{S} = \mu\beta\left(-x_2\left[\mathbf{e}_1 \otimes \mathbf{e}_3 + \mathbf{e}_3 \otimes \mathbf{e}_1\right] + x_1\left[\mathbf{e}_2 \otimes \mathbf{e}_3 + \mathbf{e}_3 \otimes \mathbf{e}_2\right]\right)$$

We can compute the divergence of the stress tensor and substitute it into the equilibrium equations to verify that $\mathrm{div}\,\mathbf{S} = \partial(\mathbf{S}\mathbf{e}_i)/\partial x_i = \mathbf{0}$. Therefore, we have shown that the body satisfies the equilibrium, kinematic, and constitutive relationships of linear elasticity at every point in the domain.

It remains only to be shown that the boundary conditions are satisfied. Obviously, at $x_3 = 0$, all of the displacement components are equal to zero, as required by the fixed end. The lateral sides of the shaft are free of traction. It will be convenient to express some of the following results in polar coordinates using the transformation $x_1 = r\cos\theta$ and $x_2 = r\sin\theta$. Points on the lateral surface are defined by the condition $r = R$; the vector normal to the lateral surface is given by $\mathbf{n} = \cos\theta\,\mathbf{e}_1 + \sin\theta\,\mathbf{e}_2$. Thus, the traction on the lateral surface is

$$\mathbf{Sn} = \mu\beta(-R\sin\theta\cos\theta + R\cos\theta\sin\theta)\,\mathbf{e}_3 = \mathbf{0}$$

At the cross section with $x_3 = \ell$ we can compute the resultant of the traction field to show that it has no net force on the section, and we can compute the resultant of the moment of the tractions to show that it is equivalent to a torque acting in the axial direction as assumed. The traction field at $x_3 = \ell$ is shown in Fig. 64.

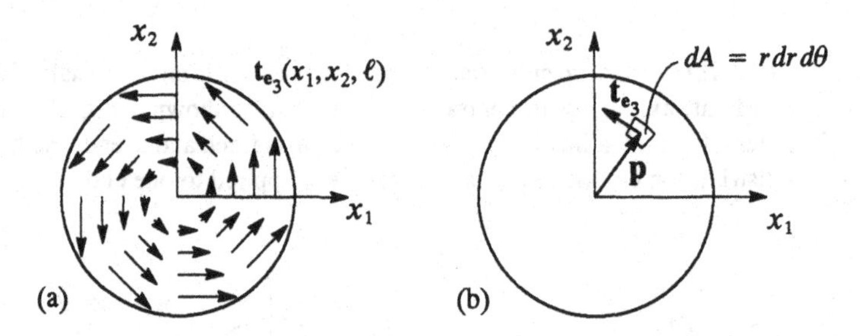

Figure 64 Torsion example (a) he traction field at the end of the shaft (b) moment of the traction for an elemental area dA

The resultant force and moment can be computed by integrating the tractions over the area. The explicit expression for the traction vector is given by

$$\mathbf{t}_{\mathbf{e}_3} = \mathbf{Se}_3 = \mu\beta(-x_2\mathbf{e}_1 + x_1\mathbf{e}_2)$$

acting at position $\mathbf{p} = x_1\mathbf{e}_1 + x_2\mathbf{e}_2$. The resultant force $\mathbf{R}$ can be computed as the resultant of the tractions over the cross-sectional area, in polar coordinates,

$$\mathbf{R} = \int_A \mathbf{t}_{\mathbf{e}_3}\,dA = \int_0^{2\pi}\int_0^R \mu\beta(-r\sin\theta\,\mathbf{e}_1 + r\cos\theta\,\mathbf{e}_2)\,r\,dr\,d\theta = \mathbf{0}$$

Noting that $\mathbf{p} \times \mathbf{t}_{\mathbf{e}_3} = \mu\beta(x_1^2 + x_2^2)\mathbf{e}_3 = \mu\beta r^2\mathbf{e}_3$, the resultant torque $\mathbf{T}$ is

$$\mathbf{T} = \int_A \mathbf{p} \times \mathbf{t}_{\mathbf{e}_3}\,dA = \mathbf{e}_3\int_0^{2\pi}\int_0^R \mu\beta r^3\,dr\,d\theta = \tfrac{1}{2}\mu\beta\pi R^4\,\mathbf{e}_3$$

These results are in exact accord with the formulas from elementary strength of materials. We solved this problem in a rather backwards fashion, having been given the map. In most situations, we will be given the surface tractions, body forces, and the known surface displacements, with the goal of finding the map. Such a problem is considerably more difficult to solve. However, the present example serves to show how all of the various parts of the boundary value problem of elasticity must be satisfied.

A Little Boundary Value Problem

We can often gather a great deal of insight from examining the simplest possible case of a general theory, especially if the essence of the general case is basically preserved. In the general boundary value problem of linear elasticity, we are faced with mathematical difficulties at every turn. Some of these difficulties are of great importance, while others are simply a nuisance. Reducing the issue to its simplest case helps to separate the important ideas from the merely tedious. Let us examine the specific one-dimensional version of our boundary value problem of elasticity illustrated in Fig. 65. For simplicity, consider that the bar has length ℓ with $x\,(=x_3)$ measured from the fixed end, and that it has *unit area*. The bar is subjected to body forces $b(x)$ and a traction t_ℓ at the end $x = \ell$. The traction t_o acting at $x = 0$ is, as yet, an unknown reaction force. The movement at $x = 0$ is known and is equal to u_o (along the axis of the bar, of course).

The bar is in uniaxial tension or compression. Thus, the only nonzero stress component is the axial component $S_{33} \equiv \sigma(x)$. The primary strain of interest is the axial strain $E_{33} \equiv \epsilon(x)$. There is no shear stress; therefore, there is no shear strain. The constitutive equations indicate that there will be lateral straining in the amount $E_{11} = E_{22} = -\nu E_{33}$, but these strains will play a secondary role in the present problem. The constitutive equation can be expressed as $\sigma(x) = C\epsilon(x)$, where C is Young's modulus. We can state the boundary value problem as

$$\left.\begin{aligned} \sigma'(x) + b(x) &= 0 \\ \epsilon(x) - u'(x) &= 0 \\ \sigma(x) - C\epsilon(x) &= 0 \end{aligned}\right\} \quad \text{for } x \in [0,\ell]$$
$$\begin{aligned} u(0) &= u_o \qquad \text{at } x = 0 \ (\Omega_u) \\ \sigma(\ell) &= t_\ell \qquad \text{at } x = \ell \ (\Omega_t) \end{aligned} \tag{261}$$

where a prime denotes ordinary differentiation with respect to x, that is $(\cdot)' = d(\cdot)/dx$.

This boundary value problem constitutes a system of first-order ordinary differential equations in the unknown functions $\sigma(x)$, $\epsilon(x)$, and $u(x)$. Through standard reduction techniques, these equations can be recast as a second-order differential equation in the unknown displacement $u(x)$ alone, just as we did

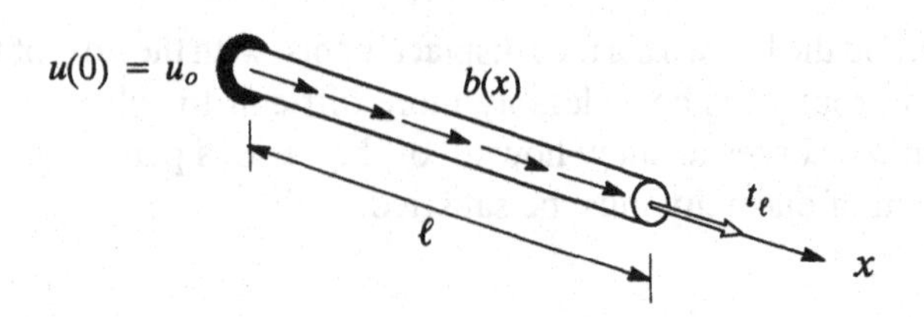

Figure 65 A one-dimensional boundary value problem

to get the Navier equations of three-dimensional elasticity. Substituting the strain-displacement equation into the constitutive equation, and that result into the equilibrium equation, the boundary value problem takes the following form

$$\boxed{\begin{gathered} (Cu')' + b = 0 \quad \text{for } x \in [0, \ell] \\ u(0) = u_o \\ Cu'(\ell) = t_\ell \end{gathered}} \tag{262}$$

Example 28. *Classical solution to the little boundary value problem.* To see what is at stake in solving a boundary value problem, let us take the particular case of a constant body force $b(x) = b_o$, traction $t_\ell = \hat{\tau}$, and fixed boundary $u_o = 0$. We solve the domain equation by successive integration to obtain

$$u(x) = -\frac{b_o}{2C}x^2 + a_0 + a_1 x$$

Integrating twice has left us with two arbitrary constants of integration, a_0 and a_1. The boundary conditions are precisely the conditions needed to single out a particular solution from an entire family of solutions that satisfy the differential equations in the domain (i.e., the body $\mathfrak{B}$). Using the given boundary conditions, we find that $u(0) = 0 = a_0$ and $Cu'(\ell) = \hat{\tau} = -b_o\ell + Ca_1$, which gives

$$a_0 = 0 \quad \text{and} \quad a_1 = \frac{1}{C}\left(b_o\ell + \hat{\tau}\right)$$

We have found the *map* to our problem: the displacement field $u(x)$. It is

$$u(x) = \frac{1}{2C}\left[2\hat{\tau}x + b_o\left(2x\ell - x^2\right)\right]$$

It is quite simple to verify that this map satisfies the governing boundary value problem. All we need to do is to differentiate it and substitute back into the governing equations. As a matter of fact, one should never go to all the trouble of finding the solution to a differential equation and then not take the extra few minutes to verify that it is indeed the solution by differentiating and substituting. Here we get

$$u'(x) = \frac{1}{C}\left[\hat{\tau} + b_o(\ell - x)\right], \quad u''(x) = -\frac{b_o}{C}$$

We can see by inspection that these quantities satisfy the differential equation and two boundary conditions

$$Cu''(x) + b_o = 0, \quad u(0) = 0, \quad Cu'(\ell) = \hat{\tau}$$

The stress field can now be found as $\sigma(x) = Cu'(x) = \hat{\tau} + b_o(\ell - x)$. The reaction at the left end is $t_o = -\sigma(0) = -\hat{\tau} - b_o\ell$, as computed through the Cauchy relation. The displacement field is, in this case, a quadratic function of x. The stress is, then, a linear function because it is proportional to the derivative of the displacement. The reaction is, of course, the one that satisfies overall equilibrium of the applied forces.

We say that this solution of the boundary value problem in the preceding example satisfies the problem in the *strong*, or *classical*, sense; it satisfies every equation at every point x in the domain and on the boundary. A strong, or classical, solution is what most people think of when they think of solving a differential equation. In the next section, we will recast our little boundary value problem into another form that will lead us to a different definition of the solution of the differential equation called the *weak*, or *variational*, sense.

Work and Virtual Work

The concept of virtual work is a specialization of the physical concept of real work, which is the product of force and the distance that force moves in the direction of its action. Consider a force $\mathbf{f}(s)$ as it moves from point a to point b along a curved path parameterized by s, as shown in Fig. 66. The curve can be viewed as a path in three-dimensional space and the parameter s can be taken as the time or distance. The direction and magnitude of the force $\mathbf{f}$ may vary as it moves along the path. The total work done by the force in moving from point a to point b is given by the line integral

$$W \equiv \int_a^b \mathbf{f}(s) \cdot \mathbf{t}(s)\, ds \tag{263}$$

where $\mathbf{t}(s) = d\mathbf{x}/ds$ is the unit vector tangent to the curve at s. There are two things we need to notice about work. First, work is a scalar quantity, not a vector quantity. Therefore, we never need to worry about invariance with respect

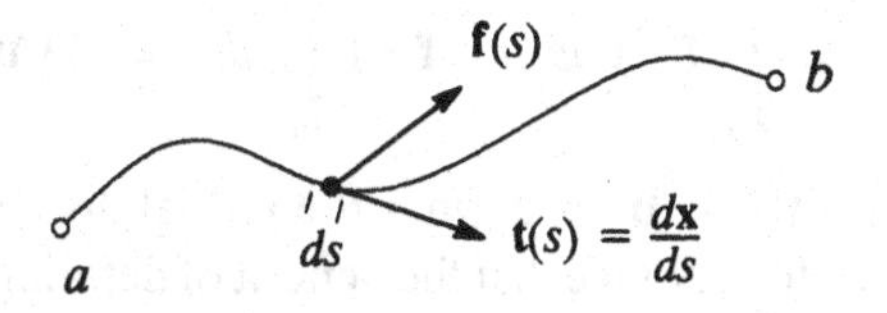

Figure 66 A force traversing a curved path

to the coordinate system. Second, the only quantities involved in the definition of work are forces and displacements (not stresses and strains). The units of work are always force times distance. If you consider the arc-length parameter s as a measure of time, then, according to how we defined the displacement of the force, you will see that we are really integrating force times velocity over the time it takes to go from point a to point b.

The principle of virtual work will be a thought experiment that we will perform on a mechanical system. The designation virtual refers to the imaginary nature of our experiment. Insofar as our experiment is a virtual one, we can construct certain aspects of the experiment to suit our mathematical needs (just as in a real experiment we contrive a testing system that most directly measures the quantity of interest). To wit, we make two assumptions in defining what we shall call the virtual displacement.

(a) The force $\mathbf{f}$ is held constant throughout the virtual displacement, and is equal to the actual value of the force at the moment the thought experiment begins.

(b) The virtual displacement is restricted to be a motion along a straight path with a constant velocity (i.e., the virtual velocity is constant). We shall designate the virtual displacement by $\bar{\mathbf{u}}$ as the vector from the starting point of the experiment a to the ending point b, as shown in Fig. 67.

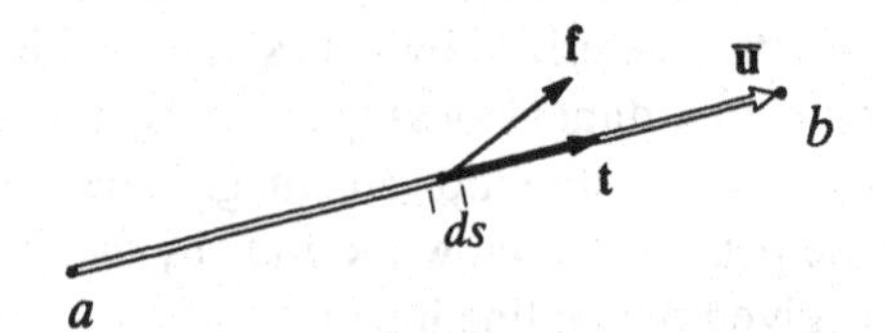

Figure 67 The definition of virtual displacement

If we take the motion of the force to be in the direction of the unit vector $\mathbf{t}$ pointing along the line connecting point a to point b, then the virtual work $\overline{W}$ (we designate all virtual quantities with overbars) done by the constant force can be computed from Eqn. (263) as

$$\overline{W} \equiv \int_a^b \mathbf{f} \cdot \mathbf{t}\, ds = \mathbf{f} \cdot \mathbf{t} \int_a^b ds \equiv \mathbf{f} \cdot \bar{\mathbf{u}} \tag{264}$$

where $\bar{\mathbf{u}} \equiv [s(b) - s(a)]\mathbf{t} = \mathbf{b} - \mathbf{a}$ defines the virtual displacement to be the vector pointing from $\mathbf{a}$ to $\mathbf{b}$. Notice that the benefit of defining virtual displacement as we have done is that we will not have to compute line integrals in order to use the concept of work.

The point a is known and is associated with the real position of the force before our thought experiment. The arbitrariness of the virtual displacement is then totally manifested in the choice of the point b to which the force moves. Since the virtual displacement must be along a straight line, the virtual displacement can be completely characterized by describing how far to go and in what direction. Thus, the virtual displacement vector is a completely appropriate choice for that description.

In the above definition of virtual work, we were concerned only with the virtual displacement of a single point a, and the force associated with that point. A solid body is made up of many such points, each of which can be subjected to a virtual displacement. As such, virtual displacements will be a field $\bar{\mathbf{u}}(\mathbf{x})$ defined over the domain of our body. While the virtual displacement field can be chosen arbitrarily (by definition), we will generally find it useful to require the field to be continuous, that is, two neighboring points (or regions) cannot be displaced in a manner implying cleavage of the material.

The Principle of Virtual Work for the Little Boundary Value Problem

Now that we have a definition of virtual work, let us create the so-called principle of virtual work. As we shall see, this principle is actually a tautology (in the same sense that $0=0$ is a tautology). The derivation of the principle of virtual work amounts to computing the work done by the external loads when subjected to a virtual displacement field, and manipulating the resulting expression with legal mathematical operations. In the process, we see that the concept of internal virtual work and virtual strain appear naturally. The statement of the principle of virtual work requires a result from the calculus of variations, which we develop here.

Let us reconsider our little boundary value problem described in Fig. 65. The forces that are acting on the bar are the body force $b(x)$ and the tractions at the two ends: the unknown reaction t_o acting at $x = 0$ and the applied traction t_ℓ acting at $x = \ell$. Let us subject this bar to a virtual displacement field $\bar{u}(x)$ and compute the work done by the external forces, or *external virtual work*, as

$$\boxed{\overline{W}_E \equiv t_o\bar{u}(0) + t_\ell\bar{u}(\ell) + \int_0^\ell b\bar{u}\,dx} \tag{265}$$

If the forces t_o, t_ℓ, and $b(x)$ are known, then the external virtual work can be computed for any virtual displacement $\bar{u}(x)$. We can manipulate this expression to put it in an equivalent, but much more useful, form. Let us begin with the definition of external virtual work and add zero to the expression in the particular form $[\sigma(0)\bar{u}(0) - \sigma(0)\bar{u}(0) + \sigma(\ell)\bar{u}(\ell) - \sigma(\ell)\bar{u}(\ell)]$ to give

$$\overline{W}_E = \left[t_o + \sigma(0)\right]\overline{u}(0) + \left[t_\ell - \sigma(\ell)\right]\overline{u}(\ell) + \left[\sigma(\ell)\overline{u}(\ell) - \sigma(0)\overline{u}(0)\right] + \int_0^\ell b\overline{u}\,dx$$

Clearly, the added expression does not change the right side because we have added and subtracted exactly the same terms. Now let us note that the one-dimensional version of the divergence theorem gives

$$\sigma(\ell)\overline{u}(\ell) - \sigma(0)\overline{u}(0) = \int_0^\ell \left(\sigma\overline{u}\right)'\,dx \tag{266}$$

(Recall that a prime indicates differentiation with respect to x.) Next, we note that the derivative of the product can be expressed as $\left(\sigma\overline{u}\right)' = \sigma'\overline{u} + \sigma\overline{u}'$. Using all of these results we can finally write the external virtual work in the form

$$\overline{W}_E = \left[t_o + \sigma(0)\right]\overline{u}(0) + \left[t_\ell - \sigma(\ell)\right]\overline{u}(\ell) + \int_0^\ell \left(\sigma' + b\right)\overline{u}\,dx + \int_0^\ell \sigma\overline{u}'\,dx$$

While it may appear that we have done nothing more than create new terms in an already simple expression for external virtual work, the last line of the derivation shows some interesting things. The first two terms remind us of the Cauchy expression relating traction to stress at the end points of the bar. The third term reminds us of the equation of equilibrium in the domain of the bar. Indeed, the classical equilibrium equations for this problem are

$$\sigma' + b = 0, \quad \sigma(0) + t_o = 0, \quad \sigma(\ell) - t_\ell = 0$$

In the expression for virtual work, each of these expressions is multiplied by the virtual displacement; the domain part is multiplied by the virtual displacement and integrated over the domain.

The last term in the last line of the derivation is the mysterious one. Let us call this term the *internal virtual work*

$$\boxed{\overline{W}_I \equiv \int_0^\ell \sigma\overline{u}'\,dx} \tag{267}$$

Notice that internal virtual work is nothing more than a definition. In particular, it has no relation to the concept of work other than the fact that it was the result of legal mathematical manipulations of the original definition of the work of the external forces. Some authors like to view the internal virtual work in terms of stress times "virtual strain," defining *virtual strain* as

$$\overline{\epsilon}(x) \equiv \overline{u}'(x) \tag{268}$$

The similarity of this expression to the definition of real strain, as the gradient of displacement, justifies calling $\bar{\varepsilon}(x)$ the virtual strain. It is precisely the strain that would be caused by the virtual displacement, if the virtual displacement really occurred. With this identification, the internal virtual work takes the equivalent form

$$\overline{W}_I = \int_0^\ell \sigma \bar{\varepsilon}\, dx \tag{269}$$

We are now almost ready to frame the principle of virtual work. Let us rewrite the last line of the derivation on the external virtual work as follows

$$\begin{aligned} \overline{W}_E - \overline{W}_I = {} & \left[t_o + \sigma(0)\right] \bar{u}(0) \\ & + \left[t_\ell - \sigma(\ell)\right]\bar{u}(\ell) + \int_0^\ell \left(\sigma' + b\right)\bar{u}\, dx \end{aligned} \tag{270}$$

where we understand that $\overline{W}_E$ is to be computed from Eqn. (265) and $\overline{W}_I$ is to be computed from Eqn. (267). Let us make the following key observation.

> **Observation.** If the stresses and tractions in the bar satisfy the equations of equilibrium; that is, if
>
> $$\sigma' + b = 0, \quad \sigma(0) + t_o = 0, \quad \sigma(\ell) - t_\ell = 0$$
>
> then the external virtual work must be equal to the internal virtual work, $\overline{W}_E - \overline{W}_I = 0$, for any (admissible) virtual displacement, $\bar{u}(x)$.

The observation simply states that if the right side of Eqn. (270) is zero, then the left side must be also. We threw in a restriction on $\bar{u}(x)$ called *admissibility* that will haunt us every time we make such a statement. What we are really hedging against here is unquantifiable indeterminate expressions like $\infty \cdot 0$, ∞ / ∞, and $0/0$. We can generally enforce admissibility by choosing virtual displacements that are sufficiently smooth and well behaved.

The above observation itself is not all that useful. If we happened to be lucky enough to know the stress field, then we could select virtual displacements at random and demonstrate that internal virtual work always balanced external virtual work. It is important to realize that for *every* choice of $\bar{u}(x)$ the equation $\overline{W}_E - \overline{W}_I = 0$ provides a valid equilibrium equation. In fact, this observation is can be used to find reaction forces for certain problems (see, for example, Problem 135).

Usually, we are not given the stress field; rather, we are trying to find it. What the principle of virtual work does is to reverse the observation to say if the external virtual work is equal to the internal virtual work for *all* admissible virtual displacements, then the system is in equilibrium. The subtle swapping of the word "any" for the word "all" is not a trivial operation. To do it, we need the fundamental theorem of the calculus of variations.

The fundamental theorem of the calculus of variations. Consider the functional defined in the following manner

$$G(g, a, b, \bar{u}) \equiv \int_0^{\ell} g(x)\bar{u}(x)\, dx + a\bar{u}(0) + b\bar{u}(\ell) \tag{271}$$

where $g(x)$ is some, as yet, unknown function of x, a and b are two, as yet, unknown constants, and $\bar{u}(x)$ is any of a variety of possible functions of x taken from a collection of admissible functions, $\mathfrak{F}(0, \ell)$. The *fundamental theorem of the calculus of variations* is the assertion that

$$\boxed{\begin{array}{l} \text{If } G(g, a, b, \bar{u}) = 0 \quad \forall \bar{u} \in \mathfrak{F}(0, \ell) \\ \text{then } g(x) = 0, \ a = 0, \text{ and } b = 0 \end{array}} \tag{272}$$

In other words, if $G = 0$ for all admissible choices of the function $\bar{u}(x)$ (the notation $\forall$ means "for all"), then $g(x) = 0$ must hold for each point x, and $a = 0$ and $b = 0$ must also hold. This is precisely the kind of statement we need to reverse the order of our observation above.

The proof of the fundamental theorem of the calculus of variations goes as follows. The function $\bar{u}(x)$ is arbitrary, and we must satisfy $G = 0$ for all of them. Let us first consider a subset of those functions, those being all functions $\bar{u}(x)$ that satisfy $\bar{u}(0) = 0$ and $\bar{u}(\ell) = 0$. For these functions, the last two terms of G do not appear. Since the equation must hold for all $\bar{u}(x)$, it must certainly hold for the function $\bar{u}(x) = g(x)$ at every point except at the ends, where it is defined to be zero. For this particular choice, $G = 0$ gives

$$\int_0^{\ell} g^2\, dx = 0 \tag{273}$$

If the integral of the square of a function is zero, then that function must be identically zero because the function $g^2(x)$ lies entirely above the axis. The integral measures the area under the curve between the limits of integration. The only curve that can be entirely above the axis and have zero area is the curve $g(x) = 0$. That proves the first conclusion in Eqn. (272).

Now let us ease up on our restrictions for $\bar{u}(x)$ to include those additional functions that are not zero at $x = 0$ (but still satisfy $\bar{u}(\ell) = 0$). Since we already have proved that $g(x) = 0$, we have $a\bar{u}(0) = 0$ for nonzero values of $\bar{u}(0)$. The only way this can be satisfied is if $a = 0$. That proves the second conclusion in Eqn. (272). Finally, let us remove all restrictions from $\bar{u}(x)$. Since we already have $g(x) = 0$ and $a = 0$, we must now satisfy $b\bar{u}(\ell) = 0$ for nonzero $\bar{u}(\ell)$.

The only way that this condition can be satisfied is if $b = 0$. That proves the third conclusion in Eqn. (272).

The extension of the fundamental theorem of the calculus of variations to vector fields and three dimensions is straightforward. Consider the functional

$$G(\mathbf{v}, \mathbf{w}, \mathbf{u}) \equiv \int_{\mathcal{B}} \mathbf{v} \cdot \mathbf{u} \, dV + \int_{\Omega} \mathbf{w} \cdot \mathbf{u} \, dA \tag{274}$$

defined on a solid region $\mathcal{B}$ with boundary Ω. The vector fields $\mathbf{v}(\mathbf{x})$ and $\mathbf{w}(\mathbf{x})$ are, as yet, unknown, and the vector field $\mathbf{u}(\mathbf{x}) \in \mathcal{F}(\mathcal{B})$ is any arbitrary function taken from our bag of admissible functions $\mathcal{F}(\mathcal{B})$. The fundamental theorem of the calculus of variations suggests that

$$\begin{array}{c} \text{If } G(\mathbf{v}, \mathbf{w}, \mathbf{u}) = 0 \quad \forall \mathbf{u} \in \mathcal{F}(\mathcal{B}) \\ \text{then } \mathbf{v}(\mathbf{x}) = \mathbf{0} \text{ in } \mathcal{B}, \text{ and } \mathbf{w}(\mathbf{x}) = \mathbf{0} \text{ on } \Omega \end{array} \tag{275}$$

The proof is just like the one-dimensional version. First, restrict the collection of functions $\mathcal{F}(\mathcal{B})$ to $\mathcal{F}_e(\mathcal{B})$, a collection of functions taken from $\mathcal{F}(\mathcal{B})$, each of which satisfy the condition $\mathbf{u}(\mathbf{x}) = \mathbf{0}$ on the boundary Ω. With this reduced set of functions, the boundary integral always vanishes. Let us consider a vector field $\mathbf{u}(\mathbf{x}) = \mathbf{v}(\mathbf{x})$ at every point in the domain except on the boundary, where it is zero. Setting $G = 0$ in Eqn. (274) gives

$$\int_{\mathcal{B}} (\mathbf{v} \cdot \mathbf{v}) \, dV = \int_{\mathcal{B}} \| \mathbf{v} \|^2 \, dV = 0 \tag{276}$$

The square of the length of a vector is always positive. Thus, the only possible vector field for which the integral of $\| \mathbf{v} \|^2$ vanishes is the field $\mathbf{v}(\mathbf{x}) = \mathbf{0}$, thereby proving the domain part of the fundamental theorem. Now remove the restriction that $\mathbf{u}(\mathbf{x}) = \mathbf{0}$ on the boundary Ω. Since we already have $\mathbf{v}(\mathbf{x}) = \mathbf{0}$, only the integral over the boundary can be nonnegative. Let us choose a function for which $\mathbf{u}(\mathbf{x}) = \mathbf{w}(\mathbf{x})$ on the boundary. Now setting $G = 0$ in Eqn. (274) suggests

$$\int_{\Omega} \| \mathbf{w} \|^2 \, dA = 0 \tag{277}$$

thereby implying that $\mathbf{w}(\mathbf{x}) = \mathbf{0}$ on the boundary. The functional G can appear in many different forms. It should be clear from the above developments how to prove the fundamental theorem of the calculus of variations.

What is a functional anyway? In the development of the fundamental theorem of the calculus of variations, we introduced an object G that we called a *functional*. Perhaps it would be a good idea to say exactly what a functional is before we go any further. A functional is an operator that takes as its input a function (which itself has an independent variable). As such, a functional is a function of a function. A functional always operates on a function in such a way that it produces a number. The definite integral of an ordinary function is one of the most prevalent examples of a functional. The value of an ordinary function at a certain point is a functional. The maximum of an ordinary function is a functional.

Example 29. *Evaluation of a functional.* Consider an example of a functional $J(u)$, which operates on functions $u(x)$ that are defined on the real segment [0,1]

$$J(u) = u(0) + \int_0^1 u^2(x)\, dx \tag{278}$$

The action of the functional is to take a function $u(x)$ and add the value of that function at the point $x = 0$ to the integral of the square of the function between the limits 0 and 1. The result is a number. For example, consider the particular function $u(x) = 1+3x$, then the value of the functional can be computed as

$$\begin{aligned} J(1+3x) &= (1+3(0)) + \int_0^1 (1+3x)^2\, dx \\ &= 1 + \left[x+3x^2+3x^3\right]_0^1 = 8 \end{aligned}$$

The functional can be evaluated for any other function defined on [0, 1]. For example, consider $u(x) = \sin \pi x$. Now

$$\begin{aligned} J(\sin \pi x) &= \sin \pi(0) + \int_0^1 (\sin \pi x)^2\, dx \\ &= 0 + \frac{1}{2\pi}\left[\pi x - \sin \pi x \cos \pi x\right]_0^1 = \tfrac{1}{2} \end{aligned}$$

It is evident that the result of evaluating a functional is always a number and that number can be computed by simply substituting each incidence of the function $u(x)$ in the functional.

A functional is like an ordinary function in many ways. For example, if you have an ordinary function, you must define the region on which it applies. In the previous example, the function $u(x)$ depends upon a single variable. Its domain is the segment $[0,1] \subset \mathbb{R}$ (this notation means that the interval is a *subset* of the real line). For physically motivated problems, it is generally quite clear what is the domain of the function. We must characterize the domain of a functional, too. This specification amounts simply to stating which functions are allowed as inputs to the functional and which ones are not. You could, for example, allow only functions that are continuous, and exclude all functions that have a jump discontinuity (or worse) somewhere within the domain of the function. For our purposes, the specification is generally quite simple, following the general principle that we will not allow functions that cause our functional to compute infinite values. It is usually quite clear from the form of the functional what we should allow and what we should exclude.

What is an admissible function? In stating the fundamental theorem of the calculus of variations, we appealed to the notion of *admissible* functions, and we gathered all possible admissible functions into what we called a *collection* of admissible functions. We gave the collection of admissible functions the name $\mathfrak{F}(\mathfrak{B})$, which told us two things: (a) which functions are in the collection and which functions are not and (b) the domain over which those functions must be defined for the present purposes. The contents of the collection will vary from application to application depending upon what mathematical objects appear in the functional G. Hence, for each G we must establish what the collection contains. The region over which the functions must be defined is quite important, but is usually obvious from the particular problem specification. For example, in Eqn. (274), the vector field $\mathbf{v}(\mathbf{x})$ must be defined throughout the region $\mathfrak{B}$, while the vector field $\mathbf{w}(\mathbf{x})$ is defined on the boundary Ω.

We often classify functions in terms of their certain special characteristics, e.g., the polynomials, the trigonometric functions, exponentials, and the like. Such classifications are much less appropriate here. The function

$$u(x) = e^x + x^3 - \cos 3x$$

may be a perfectly suitable function in our collection, and it crosses several of the classical lines of categorizing functions. A better way to visualize the admissible functions in a collection is to think of the graph of the function. Four different functions are shown in Fig. 68. Function (a) is a smooth, continuous function that varies according to no particular classical functional form (although we could approximate its variation using classical functions). Function (b) is continuous everywhere except for the jump discontinuity at p. The function varies linearly between points of slope discontinuity. Between those points we would consider the function to be quite smooth, but because of the exis-

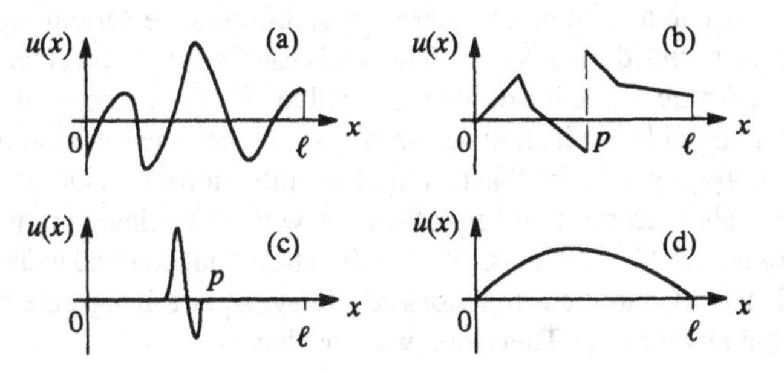

Figure 68 Four functions defined on the interval $[0, \ell]$

tence of the jump discontinuity and the kinks, the function, overall, is not very smooth. Function (c) is another smooth function. It is peculiar because the value of the function is zero over most of the domain. Function (d) is a parabola, a function that we can write as $u(x) = x(\ell - x)$, a classical mathematical form.

How, then, should we classify the functions in our collection? There are basically two classification schemes that are important to the principle of virtual work. The simpler way is to assess its square-integrability. If the integral of the square of the function over the domain is finite, then we call the function *square-integrable*. The criterion for square-integrability is

$$\int_{\mathfrak{B}} u^2(x)\, dx < \infty$$

The square of a function is easy to visualize, as the value at each point is computed by squaring the value of the original function. Clearly, the square of the function is a completely positive function. We are computing the area under this function. If the area is not finite, then the function is not square-integrable. All four of the functions in Fig. 68 are square-integrable. The function $u(x) = 1/x$ is not square-integrable on the interval (0, 1), but it is on the interval (1, 2). Can you explain why?

Another way of classifying functions is in terms of their smoothness. We can assess smoothness by examining the continuity of the derivatives of the function. Differentiation amplifies the roughness of a function. If we differentiate a rough function enough times, we will eventually get a discontinuous function. Function (b) is already a discontinuous function because of the jump at p. If we take the derivative of function (b) we get discontinuities at each of the kinks. The derivative is not defined at p. Function (c) is continuous everywhere, but its first derivative is not continuous at p. A function whose mth derivative is continuous belongs to the collection of functions called $\mathcal{C}^m$ functions. A function that is continuous, but has kinks, belongs to $\mathcal{C}^o$. A function whose first derivative is continuous, but whose first derivative has kinks, belongs to $\mathcal{C}^1$. If a function belongs to $\mathcal{C}^1$, then it also belongs to $\mathcal{C}^o$ because it satisfies all the requirements for $\mathcal{C}^o$.

We can extend the notion of square-integrability to the derivatives of functions, too. If the mth derivative of a function is square-integrable, then the function belongs to the collection of functions called $\mathcal{H}^m$. For example, if a function is square-integrable but its first derivative is not, then the function belongs to $\mathcal{H}^o$. Certainly, any function that belongs to a collection with more stringent requirements also belongs to those collections with less stringent requirements. For example, $\mathcal{H}^1 \subset \mathcal{H}^o$, that is, all of the functions that you find in $\mathcal{H}^1$ will also be found in $\mathcal{H}^o$. Continuous functions are always square-integrable, but the reverse is not always true. Therefore, we must have

$$\mathcal{C}^m \subset \mathcal{H}^m$$

This last observation is important because it is almost always easier to determine whether a function and its derivatives are continuous than it is to determine whether they are square-integrable. When we get into the business of approximation, we will generally select functions that meet the requirement of continuity and hence guarantee square-integrability. In what follows, we will always refer to the generic collection $\mathcal{F}(\mathcal{B})$ and use our common sense about the problem to decide which functions are in the collection.

Example 30. *Dirac delta function.* The function shown in Fig. 69 is zero to the left of $x = -\varepsilon$, ramps up linearly to the peak at the origin, and then ramps back down linearly to zero at $x = \varepsilon$.

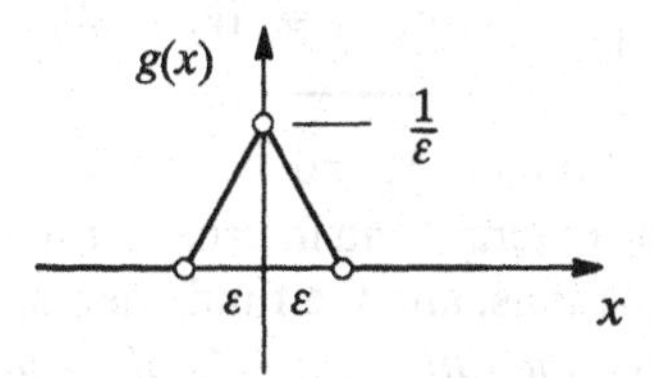

Figure 69 A Dirac delta "function" (which is not square integrable)

In the limit (as ε goes to zero) this function becomes the the Dirac delta function. The explicit functional form for this function is (for the parts that are not zero)

$$g(x) \equiv \begin{cases} \dfrac{1}{\varepsilon^2}(x+\varepsilon) & -\varepsilon \leq x \leq 0 \\ \dfrac{1}{\varepsilon^2}(\varepsilon - x) & 0 \leq x \leq \varepsilon \end{cases}$$

This function is integrable. It should be obvious that the integral of this function is 1 (one half the base times the height). This result is independent of the value of ε. The integral can be computed explicitly (noting symmetry) as

$$\int_{-\infty}^{\infty} g(x)\,dx = \frac{2}{\varepsilon^2}\int_0^{\varepsilon}(\varepsilon - x)\,dx = \frac{2}{\varepsilon^2}\left[x\varepsilon - \frac{x^2}{2}\right]_0^{\varepsilon} = 1$$

Certainly, this integral exists in the limit as $\varepsilon \to 0$. Let us also compute the integral of the square of this function (again, noting the symmetry of the function)

$$\int_{-\infty}^{\infty} g^2(x)\,dx = \frac{2}{\varepsilon^4}\int_0^{\varepsilon}(\varepsilon - x)^2\,dx = \frac{2}{\varepsilon^4}\left[x\varepsilon^2 - x^2\varepsilon + \frac{x^3}{3}\right]_0^{\varepsilon} = \frac{2}{3\varepsilon}$$

In the limit as $\varepsilon \to 0$, the integral of the square of the function approaches infinity. Thus, the Dirac delta function is not square integrable.

One of the merits of the Dirac delta function is that it has the property that

$$\int_0^{\ell} f(x)\,\delta(x - x_o)\,dx = f(x_o) \tag{279}$$

for $0 \le x_o \le \ell$. The Dirac delta function is used frequently to model point loads in a setting where the loading is thought of as a distributed load (i.e., a function of x). The Dirac delta function is also important because it is, in essence, what results from taking the second derivative of a function with a kink (the first derivative has a jump, often modeled with the *Heaviside step function*). The importance of this observation will be evident when we discuss the finite element method.

The little principle of virtual work. Let us define the functional G to be the difference between the internal and external virtual work

$$\boxed{G(\sigma, \bar{u}) \equiv \overline{W}_I - \overline{W}_E} \tag{280}$$

where $\overline{W}_E$ is given by Eqn. (265) and $\overline{W}_I$ is given by Eqn. (267). Clearly, the difference between the external and internal virtual work is also given by Eqn. (270). From these equations, and the fundamental theorem of the calculus of variations, we can state *the little principle of virtual work* as follows

$$\boxed{\begin{array}{c} \text{If } G(\sigma, \bar{u}) = 0 \quad \forall \bar{u}(x) \in \mathcal{F}(0, \ell) \\ \text{then } \sigma' + b = 0, \quad -\sigma(0) = t_o, \text{ and } \sigma(\ell) = t_\ell \end{array}} \tag{281}$$

where $\mathcal{F}(0, \ell)$ is a collection of functions admissible for use as virtual displacements. The requirements for membership in this collection are not all that stringent. Obviously, the definition of internal virtual work involves the first derivative of the virtual displacement. We want to make sure that the integral that has this term in it exists. The requirement for admissibility is

$$\int_0^\ell (\bar{u}')^2 \, dx < \infty \tag{282}$$

Hence, the functions must belong to $\mathcal{H}^1(0, \ell)$. We will see this kind of requirement every time we state a principle of virtual work. The idea is always the same: Look for the highest derivative on a function in G. Let us say that the function has been differentiated n times. The space of admissible functions will include all functions whose nth (and lower) derivatives are square integrable. As we shall soon see, it is sometimes advantageous to further restrict the functions, but any further restriction is a convenience rather than a necessity.

Here is what the principle of virtual work says: If we satisfy the virtual-work equation

$$\int_0^\ell \left(\sigma \bar{u}' - b \bar{u}\right) dx - t_o \bar{u}(0) - t_\ell \bar{u}(\ell) = 0 \tag{283}$$

for all functions $\bar{u} \in \mathcal{F}(0, \ell)$, then the equilibrium equations $\sigma' + b = 0$, and the Cauchy relations $t_o = -\sigma(0)$ and $t_\ell = \sigma(\ell)$ are automatically satisfied. In other words, we have managed to swap a differential equation for an integral equation. We call the integral equation the *weak form* of the differential equation. Here is the catch. The strong form and the weak form are only identically equivalent if the weak form is really satisfied for all choices of the virtual displacement field $\bar{u}(x)$. Since $\bar{u}(x)$ is a field, there are an infinite number of possible variations of this function. Making sure that Eqn. (283) is satisfied for all possible choices of $\bar{u}(x)$ would seem an impossible task, and indeed there aren't very many problems for which we can accomplish this task.

There is a distinct advantage to the weak form. Put in simple terms, integration is a very forgiving process, whereas differentiation is not. Integration acts to smooth rough things out, while differentiation always makes rough things rougher. An approximation can be viewed as a rough thing. Thus, if we approximate the solution to our problem, then the weak form will forgive us but the strong form will not. The advantage of the weak form of the differential equation is in its power of approximation. Indeed, this is the basis of the finite element method.

Weighted Residuals. It seemed as though we did a lot of mysterious defining of terms and manipulation of equations to come up with the expression for $G(\sigma, \bar{u})$. There is another way of getting it. Simply take the equilibrium equation in the domain, multiply it by the virtual displacement and integrate it over the domain, and add the boundary terms multiplied by the virtual displacement evaluated at the appropriate location. The negative of the result is G. To wit

$$G(\sigma, \bar{u}) \equiv -\int_0^\ell \left(\sigma' + b\right) \bar{u}\, dx - \left[t_o + \sigma(0)\right] \bar{u}(0) - \left[t_\ell - \sigma(\ell)\right] \bar{u}(\ell) \tag{284}$$

This functional has an interesting physical interpretation. If $\sigma(x)$ is not the solution to the problem at hand, then $\sigma' + b$ represents the *residual force* at each point x. It is the amount by which the equilibrium equation is not satisfied. Likewise, $t_o + \sigma(0)$ is the amount by which the stress field fails to meet the traction boundary condition at the left end and $t_\ell - \sigma(\ell)$ is the amount by which the stress field fails to meet the traction boundary condition at the right end. Thus, each term in square brackets in Eqn. (284) represents an equilibrium residual. The virtual displacement field $\bar{u}(x)$ can be viewed as a weighting function and the functional $G(\sigma, \bar{u})$ a *weighted residual* representing, in some sense, the aggregate failure of the stress field to satisfy equilibrium. Because of this interpretation, methods based upon this functional are often referred to as the *method of weighted residuals.*

We can show the correspondence with the form of the functional derived earlier by taking any term in the domain part that has a derivative on the real field variable and integrating by parts to transfer the derivative to the virtual displacement. In the present case, the term $\sigma'(x)\,\overline{u}(x)$ must be integrated by parts. The result is

$$G(\sigma,\overline{u}) \equiv \int_0^{\ell} (\sigma\overline{u}' - b\overline{u})\,dx - t_o\overline{u}(0) - t_{\ell}\overline{u}(\ell) \tag{285}$$

Notice that the boundary terms that come from the integration by parts cancel some of the existing boundary terms. Equation (285) is exactly the difference between the internal and external virtual work. This approach is often the most effective way to get the functional G required to state the principle of virtual work (particularly for differential equations where the notion of mechanical work does not apply).

Example 31. *Getting the classical differential equation from the virtual-work functional.* Consider the (virtual-work) functional for a one-dimensional boundary value problem, defined on the range $[0, \ell]$, given by

$$G(u,\overline{u}) = \int_0^{\ell} (Au''\overline{u}'' - Bu'\overline{u}' + Cu\overline{u} - b\overline{u})\,dx$$

where A, B, C, and b are known constants, $u(x)$ is the unknown field, and $\overline{u}(x)$ is its virtual counterpart. What is the classical differential equation governing the response of the system? What form must the boundary conditions have?

In order to apply the fundamental theorem of the calculus of variations we must integrate the first two terms by parts to put the functional in the form

$$G(u,\overline{u}) = \int_0^{\ell} \mathcal{L}(u)\,\overline{u}\,dx + \big[\mathcal{P}(u)\,\overline{u}\big]_0^{\ell} + \big[\mathcal{Q}(u)\,\overline{u}'\big]_0^{\ell}$$

where $\mathcal{L}(u)$, $\mathcal{P}(u)$, and $\mathcal{Q}(u)$ are differential operators. Only in this form can we deduce that the governing equation in the domain is $\mathcal{L}(u) = 0$. For the present case note that

$$\begin{aligned}(u'''\overline{u})' &= u''''\overline{u} + u'''\overline{u}' = u''''\overline{u} + (u''\overline{u}')' - u''\overline{u}'' \\ (u'\overline{u})' &= u''\overline{u} + u'\overline{u}'\end{aligned}$$

Thus, we can write

$$\begin{aligned}u''\overline{u}'' &= u''''\overline{u} + (u''\overline{u}')' - (u'''\overline{u})' \\ -u'\overline{u}' &= u''\overline{u} - (u'\overline{u})'\end{aligned}$$

Finally, substituting these expressions into the original functional and carrying out the integrals of the exact differentials we obtain

$$\begin{aligned}G(u,\overline{u}) = \int_0^{\ell} &(Au'''' + Bu'' + Cu - b)\,\overline{u}\,dx \\ &+ \big[Au''\overline{u}'\big]_0^{\ell} - \big[(Au''' + Bu')\overline{u}\big]_0^{\ell}\end{aligned}$$

Using the fundamental theorem of the calculus of variations we can conclude that the variational statement $G(u,\bar{u}) = 0$ for all $\bar{u}$ is equivalent to the classical (strong) form of the differential equation

$$Au'''' + Bu'' + Cu - b = 0, \quad 0 \le x \le \ell$$

We also learn something about the classical boundary conditions from the boundary terms. In fact,

$$\left[Au''\bar{u}'\right]_0^\ell = 0 \rightarrow \text{ either } Au'' = 0 \text{ or } \bar{u}' = 0 \text{ at } 0 \text{ and } \ell$$

$$\left[(Au'''+Bu')\bar{u}\right]_0^\ell = 0 \rightarrow \text{ either } Au'''+Bu' = 0 \text{ or } \bar{u} = 0 \text{ at } 0 \text{ and } \ell$$

The significance of the conclusions $\bar{u} = 0$ and $\bar{u}' = 0$ will be more evident from our discussion of essential and natural boundary conditions in the next section. Suffice it to say at this point, that the boundary terms that result from integration by parts always provide information about the classical boundary conditions of the problem.

Essential and Natural Boundary Conditions

The principle of virtual work holds for any constitutive model since constitutive equations did not enter the derivation. A completely displacement-based expression for the virtual-work functional can be found by directly implementing the constitutive equation $\sigma = Cu'$ into the original virtual-work functional. In doing so, we change the argument of the functional from σ to u. The virtual-work functional then has the form

$$G(u,\bar{u}) \equiv \int_0^\ell \left(Cu'\bar{u}' - b\bar{u}\right) dx - t_o\bar{u}(0) - t_\ell\bar{u}(\ell) \qquad (286)$$

Now the principle of virtual work states that equilibrium will be satisfied if

$$G(u,\bar{u}) = 0 \quad \forall \bar{u} \in \mathcal{F}(0,\ell) \qquad (287)$$

The only unknowns in this equation are the displacement field $u(x)$ and the reaction force t_o. We can get rid of the reaction by choosing only functions that satisfy the condition $\bar{u}(0) = 0$ (note that the boundary condition on the real displacement is $u(0) = u_o$). This assumption does, of course, weaken the principle of virtual work. If you go back to the proof of the fundamental theorem of the calculus of variations, you can see that we have sacrificed the conclusion that $t_o = -\sigma(0)$. Is this a serious sacrifice or not? Since equilibrium is still guaranteed at all other points in the domain, we have sacrificed only the equation that actually computes the reaction force. Equilibrium must still hold at this point. If we simply compute the reaction with the equation $t_o = -\sigma(0)$ instead of letting the equation be satisfied automatically, then we have lost nothing. Thus, let us modify our functional to be

$$G(u,\bar{u}) \equiv \int_0^{\ell} \left(Cu'\bar{u}' - b\bar{u} \right) dx - t_{\ell}\bar{u}(\ell) \tag{288}$$

and our statement of the equilibrium condition to be

$$G(u,\bar{u}) = 0 \quad \forall \bar{u} \in \mathcal{F}_e(0,\ell) \tag{289}$$

where $\mathcal{F}_e(0,\ell)$ is a subset of functions from $\mathcal{F}(0,\ell)$ containing only functions that satisfy the homogeneous *essential* boundary condition $\bar{u}(0) = 0$. As you might guess, the subscript *e* stands for "essential boundary conditions satisfied." Essential boundary conditions are also often called *displacement* boundary conditions. They are the boundary conditions on the lower-order derivatives of the displacements, since we use the substitution $\sigma = Cu'$. The boundary condition $t_{\ell} = \sigma(\ell)$ is called a *natural* boundary condition (sometimes called a *force* boundary condition) because the principle of virtual work guarantees that it will be satisfied naturally. Our variational approach to problems (i.e., virtual work) will always distinguish between these two kinds of boundary conditions.

Interestingly, we cannot specify $\bar{u}(x)$ at any other point. For example, we cannot set $\bar{u}(\ell) = 0$ because, if we did, the principle of virtual work could not distinguish between problems with different values of the applied end load t_{ℓ}. In particular, it could not distinguish a problem with a zero end load from one with a nonzero end load. Clearly, this is too great a sacrifice. Similarly, if the value of $\bar{u}(x)$ were specified at any other point along the length of the bar, we would have a point where a concentrated body force of any value could be placed without changing the value of the virtual work. Again, the principle of virtual work would be unable to distinguish between fundamentally different problems.

The Principle of Virtual Work for 3D Linear Solids

The development of the general principle of virtual work for a three-dimensional solid body is quite similar to the development for the little boundary value problem. All of the essential steps are the same, but some of the mathematical manipulations are a little more involved. In following the derivation, you would do well to refer back to the one-dimensional case to keep perspective.

Consider a body $\mathfrak{B}$ having boundary Ω with unit normal vector field $\mathbf{n}(\mathbf{x})$. The forces that are acting on the solid are the body force $\mathbf{b}(\mathbf{x})$ and the tractions $\mathbf{t}(\mathbf{x})$ acting on the surface of the body. Here we will not distinguish between the prescribed tractions and the reacting tractions. In order to derive the principle of virtual work, we must compute the work done by the body forces and surface tractions as they move through a virtual displacement field. Let us subject the body to a virtual displacement field $\bar{\mathbf{u}}(\mathbf{x})$ and compute the work done by the external forces

$$\overline{W}_E \equiv \int_{\Omega} \mathbf{t} \cdot \overline{\mathbf{u}}\, dA + \int_{\mathfrak{B}} \mathbf{b} \cdot \overline{\mathbf{u}}\, dV \tag{290}$$

where the integrals are carried out over the surface and throughout the volume of the body, as indicated. In the first term, the virtual displacements $\overline{\mathbf{u}}$ are the values of the virtual displacement field on the surface of the body. As in the little boundary value problem, we can derive several equivalent forms of the expression for the external work. Let us start with Eqn. (290) and add zero to the right side in the form of an integral of $(\mathbf{Sn}-\mathbf{Sn}) \cdot \overline{\mathbf{u}}$ over the boundary, where $\mathbf{n}$ is the vector field normal to the boundary of the region Ω. We get

$$\overline{W}_E = \int_{\Omega} (\mathbf{t}-\mathbf{Sn}) \cdot \overline{\mathbf{u}}\, dA + \int_{\Omega} \mathbf{Sn} \cdot \overline{\mathbf{u}}\, dA + \int_{\mathfrak{B}} \mathbf{b} \cdot \overline{\mathbf{u}}\, dV \tag{291}$$

In the second step let us apply the divergence theorem to the second term, noting that $\mathbf{Sn} \cdot \overline{\mathbf{u}} = \mathbf{S}^T\overline{\mathbf{u}} \cdot \mathbf{n}$ and that $\mathbf{S}^T = \mathbf{S}$, to get

$$\int_{\Omega} \mathbf{Sn} \cdot \overline{\mathbf{u}}\, dA = \int_{\mathfrak{B}} \operatorname{div}(\mathbf{S}\overline{\mathbf{u}})\, dV$$

Substituting these results into Eqn. (291) gives

$$\overline{W}_E = \int_{\Omega} (\mathbf{t}-\mathbf{Sn}) \cdot \overline{\mathbf{u}}\, dA + \int_{\mathfrak{B}} \left(\operatorname{div}(\mathbf{S}\overline{\mathbf{u}}) + \mathbf{b} \cdot \overline{\mathbf{u}}\right) dV \tag{292}$$

We can expand the divergence of the product of a tensor times a vector to get

$$\operatorname{div}(\mathbf{S}\overline{\mathbf{u}}) = (\operatorname{div}\mathbf{S}) \cdot \overline{\mathbf{u}} + \mathbf{S} \cdot \nabla\overline{\mathbf{u}}$$

In components, this expression is $[S_{ij}\overline{u}_i]_{,j} = S_{ij,j}\,\overline{u}_i + S_{ij}\overline{u}_{i,j}$. Substituting this result into Eqn. (292) gives

$$\begin{aligned}\overline{W}_E = &\int_{\Omega} (\mathbf{t}-\mathbf{Sn}) \cdot \overline{\mathbf{u}}\, dA \\ &+ \int_{\mathfrak{B}} (\operatorname{div}\mathbf{S} + \mathbf{b}) \cdot \overline{\mathbf{u}}\, dV + \int_{\mathfrak{B}} \mathbf{S} \cdot \nabla\overline{\mathbf{u}}\, dV\end{aligned} \tag{293}$$

Like its one-dimensional counterpart, Eqn. (293) contains some interesting terms. The first two terms are related to boundary equilibrium and domain equilibrium. The third term we shall define to be the internal virtual work. We can clearly see the analogy with Eqn. (267) in the following definition of internal virtual work

$$\overline{W}_I \equiv \int_{\mathfrak{B}} \mathbf{S} \cdot \nabla\overline{\mathbf{u}}\, dV \tag{294}$$

The internal virtual work can be expressed in the form of stress times virtual strain if we define the virtual strain to be the strain that would occur if the virtual displacement were to take place. Let us define the virtual strain to be

$$\bar{\mathbf{E}} \equiv \tfrac{1}{2}\left[\nabla\bar{\mathbf{u}} + \nabla\bar{\mathbf{u}}^T\right]$$

With this definition, Eqn. (294) can be equivalently expressed as

$$\overline{W}_I \equiv \int_{\mathfrak{B}} \mathbf{S}\cdot\bar{\mathbf{E}}\,dV$$

since the stress tensor is symmetric, and, hence, $\mathbf{S}\cdot\nabla\bar{\mathbf{u}} = \mathbf{S}\cdot\frac{1}{2}\left[\nabla\bar{\mathbf{u}}+\nabla\bar{\mathbf{u}}^T\right]$. Again, it is not necessary to introduce the virtual strain, but it provides a convenient way to view the internal virtual work. The concept of internal virtual work is a natural consequence of the mathematical developments and does not necessarily need any physical motivation.

We are now ready to state the principle of virtual work. Let us define the functional G to be the difference between the internal virtual work and the external virtual work

$$G(\mathbf{S},\bar{\mathbf{u}}) \equiv \overline{W}_I - \overline{W}_E$$

where $\overline{W}_I$ is given by Eqn. (294) and $\overline{W}_E$ is given by Eqn. (290). Therefore, the virtual-work functional has the explicit form

$$\boxed{G(\mathbf{S},\bar{\mathbf{u}}) \equiv \int_{\mathfrak{B}} \left(\mathbf{S}\cdot\nabla\bar{\mathbf{u}} - \mathbf{b}\cdot\bar{\mathbf{u}}\right)dV - \int_{\Omega} \mathbf{t}\cdot\bar{\mathbf{u}}\,dA} \tag{295}$$

Clearly, from Eqn. (293), the difference between the internal and external virtual work is also given by

$$\overline{W}_E - \overline{W}_I = \int_{\Omega} \left(\mathbf{t}-\mathbf{S}\mathbf{n}\right)\cdot\bar{\mathbf{u}}\,dA + \int_{\mathfrak{B}} \left(\operatorname{div}\mathbf{S}+\mathbf{b}\right)\cdot\bar{\mathbf{u}}\,dV$$

From the fundamental theorem of the calculus of variations, we obtain the principle of virtual work for a three-dimensional solid

$$\boxed{\begin{array}{c}\text{If } G(\mathbf{S},\bar{\mathbf{u}}) = 0 \quad \forall\bar{\mathbf{u}} \in \mathfrak{F}(\mathfrak{B}) \\ \text{then } \operatorname{div}\mathbf{S}+\mathbf{b} = \mathbf{0} \text{ in } \mathfrak{B} \text{ and } \mathbf{S}\mathbf{n} = \mathbf{t} \text{ on } \Omega\end{array}} \tag{296}$$

where $\mathfrak{F}(\mathfrak{B})$ is the collection of admissible functions defined over the domain of the body. Since the definition of the internal virtual work involves the gradient of the virtual displacements, we will require that the gradient of our virtual displacement functions be square-integrable over the domain. Thus, the collection $\mathfrak{F}(\mathfrak{B})$ contains all functions that satisfy

$$\int_{\mathcal{B}} \| \nabla\bar{\mathbf{u}} \|^2 \, dV < \infty$$

Here is what the principle of virtual work says: If we satisfy the virtual-work equation $G(\mathbf{S}, \bar{\mathbf{u}}) = 0$ for all arbitrary admissible virtual displacement fields, then the domain and boundary (Cauchy) equilibrium relations are automatically satisfied. Again, we have traded a differential equation for an integral equation with all of the advantages that accrue.

Linear elasticity. In solving particular problems, it will again be advantageous to restrict our virtual displacement functions to be zero on that portion of the boundary where displacements are prescribed. This proscription will annihilate the unknown tractions that act over that part of the boundary. Also, the principle of virtual work does not involve the constitutive equations of the material, but those relationships can be implemented in a classical sense into the virtual-work equation. This substitution allows us to express the virtual-work functional in terms of only the displacement fields. The equations of linear elasticity lead to the following virtual-work functional

$$G(\mathbf{u}, \bar{\mathbf{u}}) \equiv \int_{\mathcal{B}} \left(\nabla\bar{\mathbf{u}} \cdot \mathbf{C}\nabla\mathbf{u} - \mathbf{b} \cdot \bar{\mathbf{u}} \right) dV - \int_{\Omega_t} \hat{\mathbf{t}} \cdot \bar{\mathbf{u}} \, dA \tag{297}$$

and the statement of the principle of virtual work is: If $G(\mathbf{u}, \bar{\mathbf{u}}) = 0$ for all virtual displacements $\bar{\mathbf{u}} \in \mathcal{F}_e(\mathcal{B})$, then $\mathbf{u}$ is an equilibrium configuration. The admissible functions contained in $\mathcal{F}_e(\mathcal{B})$ are simply those in $\mathcal{F}(\mathcal{B})$ restricted to have $\bar{\mathbf{u}} = \mathbf{0}$ on Ω_u. Note that the stress can be written in terms of displacement as $\mathbf{S} = \mathbf{C}\nabla\mathbf{u}$ because the elasticity tensor is symmetric in the tensor components $C_{ijkl} = C_{ijlk}$. Therefore, we have

$$S_{ij} = C_{ijkl} E_{kl} = \tfrac{1}{2} C_{ijkl} \left(u_{k,l} + u_{l,k} \right) = C_{ijkl} u_{k,l}$$

The particular form of the internal virtual work for an isotropic linearly elastic material can be obtained by recognizing that isotropic elastic constitutive relations can be written as $\mathbf{S} = \lambda(\operatorname{div}\mathbf{u})\mathbf{I} + \mu(\nabla\mathbf{u} + \nabla\mathbf{u}^T)$ and that (accounting for symmetry of $\mathbf{S}$) $\mathbf{S} \cdot \nabla\bar{\mathbf{u}} = \operatorname{tr}(\mathbf{S}\nabla\bar{\mathbf{u}})$. Thus,

$$\mathbf{S} \cdot \nabla\bar{\mathbf{u}} = \lambda(\operatorname{div}\mathbf{u})(\operatorname{div}\bar{\mathbf{u}}) + \mu\left[\nabla\mathbf{u} + \nabla\mathbf{u}^T\right] \cdot \nabla\bar{\mathbf{u}}$$

and the virtual-work functional takes the form

$$G(\mathbf{u}, \bar{\mathbf{u}}) = \int_{\mathcal{B}} \left(\lambda(\operatorname{div}\mathbf{u})(\operatorname{div}\bar{\mathbf{u}}) + \mu\left[\nabla\mathbf{u} + \nabla\mathbf{u}^T\right] \cdot \nabla\bar{\mathbf{u}} \right) dV - \int_{\mathcal{B}} \mathbf{b} \cdot \bar{\mathbf{u}} \, dV - \int_{\Omega_t} \hat{\mathbf{t}} \cdot \bar{\mathbf{u}} \, dA$$

These functionals will be useful in developing numerical methods for solving problems, as we shall see in the next Chapter.

Finite Deformation Version of the Principle of Virtual Work—Reference Configuration

The arguments that lead up to the principle of virtual work for small deformations carry over to the case of finite deformation. In fact, if we think of $\mathbf{S}$ as the Cauchy stress and the region $\mathfrak{B}$ as the current configuration, then all of the preceding developments are appropriate to the finite deformation setting (with possible exception of the dubious use of Hooke's law in that setting, as pointed out in Chapter 4). As we observed in Chapter 3, one can cast the equations of equilibrium in either the current or reference configuration. The classical differential equations in the two configurations are summarized in Table 2.

Table 2 Equilibrium in reference and deformed configurations

	Reference Configuration	Current Configuration
Linear momentum	$\mathrm{DIV}\,\mathbf{P}+\mathbf{b}^o = \mathbf{0}$	$\mathrm{div}\,\mathbf{S}+\mathbf{b} = \mathbf{0}$
Cauchy tractions	$\mathbf{Pm} = \mathbf{t}^o$	$\mathbf{Sn} = \mathbf{t}$
Angular momentum	$\mathbf{PF}^T = \mathbf{FP}^T$	$\mathbf{S} = \mathbf{S}^T$

In these equations $\mathbf{P}$ and $\mathbf{S}$ are the first Piola-Kirchhoff and Cauchy stress tensors, respectively, $\mathbf{F}$ is the deformation gradient, $\mathbf{b}$ is the body force in the current configuration, $\mathbf{b}^o \equiv J\mathbf{b}$ is the body force in the reference configuration ($J = \det \mathbf{F}$), $\mathbf{t}$ is the (applied or reacting) traction on the surface in the current configuration having unit normal vector $\mathbf{n}$, and $\mathbf{t}^o$ is the (applied or reacting) traction on a surface in the reference configuration having unit normal vector $\mathbf{m}$. Note that the surface tractions satisfy, by definition, $\mathbf{t}^o\, dA = \mathbf{t}\, da$, with da being the elemental area in the current configuration and dA being the elemental area in the reference configuration. As was pointed out in Chapter 2 the elemental volumes are related as $dv = J\, dV$. Recall that

$$\mathrm{DIV}\,\mathbf{P} = \frac{\partial}{\partial z_k}\big(\mathbf{P}\mathbf{g}_k\big), \qquad \mathrm{div}\,\mathbf{S} = \frac{\partial}{\partial x_k}\big(\mathbf{S}\mathbf{e}_k\big) \tag{298}$$

where $\{x_k\}$ and $\{\mathbf{e}_k\}$ are the coordinates and base vectors in the current configuration and $\{z_k\}$ and $\{\mathbf{g}_k\}$ are the coordinates and base vectors in the reference configuration.

With this background we are ready to state the principle of virtual work in the finite deformation setting. Observe that the external virtual work can be expressed in the current configuration, in accord with Eqn. (290), as

$$\overline{W}_E \equiv \int_{\phi(\Omega)} \mathbf{t} \cdot \overline{\mathbf{u}}\, da + \int_{\phi(\mathfrak{B})} \mathbf{b} \cdot \overline{\mathbf{u}}\, dv \tag{299}$$

Note that we have explicitly indicated that the work is computed as integrals over the current configuration (which is where equilibrium must hold). Using the relationships $\mathbf{b}^o = J\mathbf{b}$ and $\mathbf{t}^o\, dA = \mathbf{t}\, da$, as a simple change of variable gives the equivalent expression

$$\overline{W}_E \equiv \int_{\Omega} \mathbf{t}^o \cdot \overline{\mathbf{u}}\, dA + \int_{\mathfrak{B}} \mathbf{b}^o \cdot \overline{\mathbf{u}}\, dV \tag{300}$$

The divergence theorem gives

$$\begin{aligned} \int_{\Omega} \mathbf{Pm} \cdot \overline{\mathbf{u}}\, dA &= \int_{\mathfrak{B}} \mathrm{DIV}\left(\mathbf{P}^T\overline{\mathbf{u}}\right) dV \\ &= \int_{\mathfrak{B}} \left(\mathrm{DIV}\,\mathbf{P} \cdot \overline{\mathbf{u}} + \mathbf{P} \cdot \nabla\overline{\mathbf{u}}\right) dV \end{aligned} \tag{301}$$

where $\mathbf{P} \cdot \nabla\overline{\mathbf{u}} = P_{ij}\,\partial\overline{u}_i/\partial z_j$. It will be useful, for reasons identical to those of the earlier derivation, to define the *internal virtual work* as

$$\overline{W}_I \equiv \int_{\mathfrak{B}} \mathbf{P} \cdot \nabla\overline{\mathbf{u}}\, dV = \int_{\mathfrak{B}} \mathbf{\Sigma} \cdot \mathbf{F}^T\nabla\overline{\mathbf{u}}\, dV \tag{302}$$

where $\mathbf{\Sigma} = \mathbf{F}^{-1}\mathbf{P}$ is the second Piola-Kirchhoff stress tensor. We can use the various stress tensors interchangeably, with the main convenience accruing in the expression of the constitutive equations (which is not part of the principle of virtual work).

Bringing all of these results together we can compute the difference between external and internal work, which we will take as the very definition of our virtual-work functional. A straightforward application of Eqns. (300), (301), and (302) gives

$$\overline{W}_E - \overline{W}_I = \int_{\mathfrak{B}} \left(\mathrm{DIV}\,\mathbf{P} + \mathbf{b}^o\right) \cdot \overline{\mathbf{u}}\, dV + \int_{\Omega} \left(\mathbf{t}^o - \mathbf{Pm}\right) \cdot \overline{\mathbf{u}}\, dA \tag{303}$$

If we define our virtual-work functional as $G(\mathbf{P}, \overline{\mathbf{u}}) \equiv \overline{W}_I - \overline{W}_E$ then we are in position to state a finite deformation version of the principle of virtual work. Explicitly, let us define

$$G(\mathbf{P}, \overline{\mathbf{u}}) \equiv \int_{\mathfrak{B}} \left(\mathbf{P} \cdot \nabla\overline{\mathbf{u}} - \mathbf{b}^o \cdot \overline{\mathbf{u}}\right) dV - \int_{\Omega} \mathbf{t}^o \cdot \overline{\mathbf{u}}\, dA \tag{304}$$

If $G(\mathbf{P}, \bar{\mathbf{u}}) \equiv 0$ for all admissible virtual displacements $\bar{\mathbf{u}}$, then, by virtue of Eqn. (303) and the fundamental theorem of the calculus of variations, it must be true that $\text{DIV}\,\mathbf{P} + \mathbf{b}^o = 0$ everywhere in the domain and $\mathbf{Pm} = \mathbf{t}^o$ everywhere on the surface.

Virtual Strains. In the linear theory of virtual work we identified the gradient of the virtual displacement that appears in the internal virtual work as the "virtual strain." What is the situation relative to the finite deformation case?

Examining Eqn. (302) we see that the integrand of the internal virtual work takes either of two equivalent forms: $\mathbf{P} \cdot \nabla\bar{\mathbf{u}}$ for the first Piola-Kirchhoff stress or $\boldsymbol{\Sigma} \cdot \mathbf{F}^T\nabla\bar{\mathbf{u}}$ for the second Piola Kirchhoff stress. Since $\mathbf{P}$ and $\boldsymbol{\Sigma}$ are stresses, that leaves $\nabla\bar{\mathbf{u}}$ or $\mathbf{F}^T\nabla\bar{\mathbf{u}}$ as candidates for virtual strain. One way to think of virtual strain is that it is the strain associated with the virtual displacement. In finite deformation that displacement takes place on top of an existing displacement. Keeping with the "virtual velocities" idea mentioned at the beginning of the chapter we might think of the virtual displacement as being the directional derivative of the strain in the direction of the virtual motion. The following example pursues this idea.

Example 32. *Finite deformation version of virtual strain.* In Chapter 4 we learned that $\mathbf{F}$ was the deformation measure conjugate, in the sense of energy, to the stress $\mathbf{P}$ and that $\mathbf{E}$ (the Lagrangian strain) was the deformation measure conjugate to the stress $\boldsymbol{\Sigma}$. Let us compute the directional derivatives of these strains in the direction of the virtual displacement. For the deformation gradient we get the "virtual strain"

$$\bar{\mathbf{F}} = \frac{d}{d\varepsilon}\left[\nabla(\boldsymbol{\phi}(\mathbf{z}) + \varepsilon\bar{\mathbf{u}}(\mathbf{z}))\right]_{\varepsilon=0} = \nabla\bar{\mathbf{u}} \equiv \mathbf{H}$$

where $\mathbf{H} \equiv \nabla\bar{\mathbf{u}}$ is defined for notational convenience in the next calculation. For the Lagrangian strain we can compute the virtual strain

$$\begin{aligned}
\bar{\mathbf{E}} &= \frac{d}{d\varepsilon}\left[\frac{1}{2}\left[(\mathbf{F}+\varepsilon\mathbf{H})^T(\mathbf{F}+\varepsilon\mathbf{H}) - \mathbf{I}\right]\right]_{\varepsilon=0} \\
&= \frac{1}{2}\left[\mathbf{H}^T(\mathbf{F}+\varepsilon\mathbf{H}) + (\mathbf{F}+\varepsilon\mathbf{H})^T\mathbf{H}\right]_{\varepsilon=0} \\
&= \frac{1}{2}\left[\mathbf{H}^T\mathbf{F} + \mathbf{F}^T\mathbf{H}\right]
\end{aligned}$$

Observing that, since $\boldsymbol{\Sigma}$ is symmetric we have $\boldsymbol{\Sigma} \cdot \bar{\mathbf{E}} = \boldsymbol{\Sigma} \cdot \mathbf{F}^T\mathbf{H}$, thereby showing that $\bar{\mathbf{E}}$ is the appropriate virtual strain.

Closure

Throughout the remainder of the book, we shall develop specialized versions of the principle of virtual work. In particular, we shall examine the linear theories of beams and plates, as well as some nonlinear theories of beams. For each theory there will be an appropriate expression of the virtual-work functional. All of these theories will be consistent with the general three-dimensional expression of the virtual-work functional considered here. Thus, we end this chapter, having dispensed the simplest and most complicated versions of the linear boundary value problems of elasticity, with the promise to visit these issues again for each of the theories of structural mechanics that we consider.

Additional Reading

F. Hartmann, *The mathematical foundation of structural mechanics*, Springer-Verlag, New York, 1985.

L. E. Malvern, *Introduction to the mechanics of a continuous medium*, Prentice Hall, Englewood Cliffs, N.J., 1969.

J. T. Oden, *Mechanics of elastic structures*, McGraw-Hill, New York, 1967.

I. S. Sokolnikoff, *Mathematical theory of elasticity*, 2nd ed., McGraw-Hill, New York, 1956.

I. Stakgold, *Green's functions and boundary value problems*, Wiley, New York, 1979.

Problems

130. Consider the uniaxial rod shown below, fixed at $x = 0$, free at $x = \ell$, and subjected to the linearly varying body force indicated. The rod is made from a composite material with a variable elastic modulus $C(x) = C_o(2 - x/\ell)$, making it twice as stiff at $x = 0$ as it is at $x = \ell$. The governing differential equation for a rod with variable modulus is

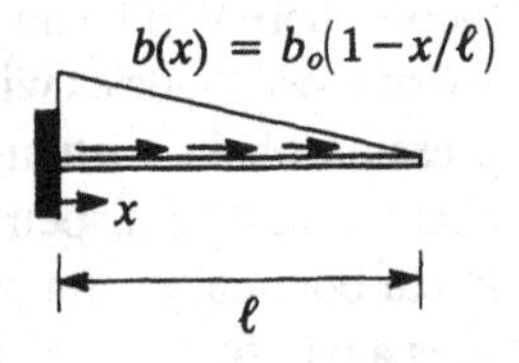

$$\big(C(x)u'\big)' + b(x) = 0$$

where a prime indicates differentiation with respect to x. Find the exact (classical) solution to the problem by directly integrating the governing equations.

131. Consider the rod of unit length and modulus $C(x)$ that varies as shown in the sketch. The rod is fixed at the left end, is free at the right end, and is subjected to a linearly varying body force $b(x)$ as shown. Consider the following displacement map: $u(x) = a\left(x^3 + 2x^2 - 3x\right)$ where a is some constant. Is the displacement map a solution to the given problem? Why or why not?

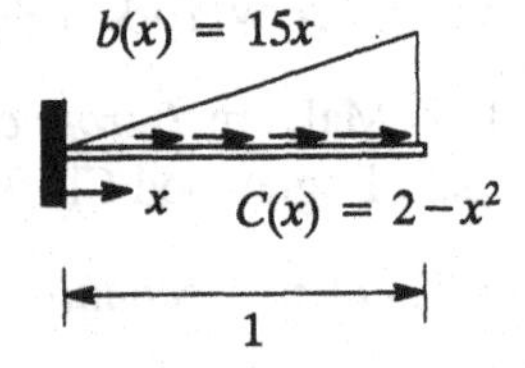

132. Prove that $\mathbf{S} \cdot \overline{\mathbf{E}} = \mathbf{S} \cdot \nabla\overline{\mathbf{u}}$ when the virtual strain is defined to be the strain that would occur if the virtual displacement actually took place, i.e., $\overline{\mathbf{E}} = \frac{1}{2}\left[\nabla\overline{\mathbf{u}} + \nabla\overline{\mathbf{u}}^T\right]$. Upon what property of the stress tensor $\mathbf{S}$ does this identity rely?

133. Show that $\nabla\overline{\mathbf{u}} \cdot \mathbf{C}\nabla\mathbf{u} = \lambda(\operatorname{div}\mathbf{u})(\operatorname{div}\overline{\mathbf{u}}) + \mu\left[\nabla\mathbf{u} + \nabla\mathbf{u}^T\right] \cdot \nabla\overline{\mathbf{u}}$ for an isotropic, linear, elastic material. Express this equation in component form.

134. Carry out the derivation of the principle of virtual work for the case in which the real displacements are known and a system of virtual forces are applied to the body, and thereby deduce the *principle of virtual forces*. Specifically, apply virtual body forces $\overline{\mathbf{b}}$ and virtual surface tractions $\overline{\mathbf{t}}$, and define the complementary external virtual work as

$$\hat{W}_E \equiv \int_{\mathfrak{B}} \mathbf{u} \cdot \overline{\mathbf{b}}\, dV + \int_{\Omega} \mathbf{u} \cdot \overline{\mathbf{t}}\, dA$$

where $\mathbf{u}$ is the real displacement of the body. Perform a derivation similar to the one for the principle of virtual displacements to demonstrate that an appropriate definition of complementary internal virtual work is

$$\hat{W}_I \equiv \int_{\mathfrak{B}} \overline{\mathbf{S}} \cdot \mathbf{E}\, dV$$

where $\overline{\mathbf{S}}$ is the virtual stress associated with the applied virtual force system and $\mathbf{E}$ is the strain tensor associated with the real displacements. Prove the *principle of virtual forces*, which states that if $\hat{W}_E = \hat{W}_I$ for all virtual stresses $\overline{\mathbf{S}}$ in equilibrium with the applied virtual forces $\overline{\mathbf{b}}$ and $\overline{\mathbf{t}}$, then $\mathbf{E} = \frac{1}{2}\left[\nabla\mathbf{u} + \nabla\mathbf{u}^T\right]$. State precisely the conditions that must hold in order for the principle to be valid.

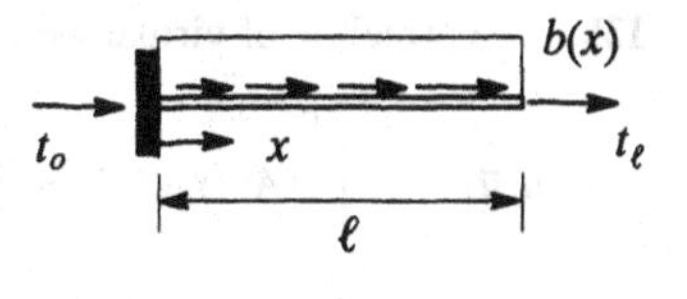

135. The virtual-work functional for the little boundary value problem is given by

$$G(\sigma, \bar{u}) = \int_0^\ell (\sigma\bar{u}' - b\bar{u})\, dx - t_o\bar{u}(0) - t_\ell\bar{u}(\ell)$$

The body force $b(x)$ and the traction at the free end t_ℓ are known while the stress $\sigma(x)$ and the reaction t_o are unknown. Using the principle of virtual work, select a virtual displacement field that allows you to compute the reaction force in terms of only the known applied forces. Give the expression for t_o in terms of $b(x)$ and t_ℓ.

136. In the little boundary value problem, we saw that the virtual-work functional could be stated as a weighted residual functional. A weighted residual functional for a three-dimensional solid body $\mathfrak{B}$ with boundary Ω (with normal vector field **n**) can be defined as

$$G(\mathbf{S}, \mathbf{w}) \equiv -\int_{\mathfrak{B}} (\text{div}\,\mathbf{S} + \mathbf{b}) \cdot \mathbf{w}\, dV - \int_{\Omega} (\mathbf{t} - \mathbf{S}\mathbf{n}) \cdot \mathbf{w}\, dA$$

where $\text{div}\,\mathbf{S} + \mathbf{b}$ is the equilibrium residual in the domain $\mathfrak{B}$, $\mathbf{t} - \mathbf{S}\mathbf{n}$ is the equilibrium residual on the boundary Ω, and $\mathbf{w}$ is an arbitrary weighting function. Show that the weighted residual functional is identical to the virtual-work functional given in Eqn. (295), and, therefore, that the arbitrary weighting function is identical to the virtual displacement, i.e., $\mathbf{w} = \bar{\mathbf{u}}$.

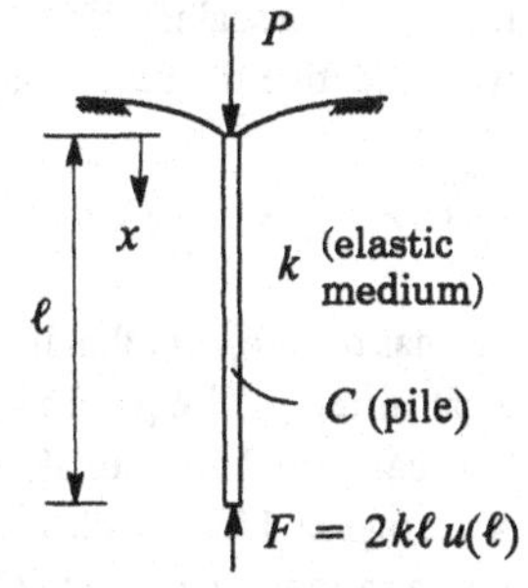

137. Consider the pile of length ℓ, constant modulus C (w/ unit area), embedded in an elastic medium with modulus k (force per unit displacement per unit length), and subjected to a load P at $x = 0$. The pile is elastically restrained at the end $x = \ell$ giving an end force of the amount $F = 2k\ell u(\ell)$ as shown. The governing differential equation for the system is $Cu''(x) - ku(x) = 0$. What must be the value of the constant a for the solution to have the form $u(x) = Ae^{ax} + Be^{-ax}$?

What are the values of the constants A and B that satisfy the problem shown in the figure? Does this function $u(x)$ represent a classical solution to the given problem? Why or why not? Are there any other solutions to this specific problem?

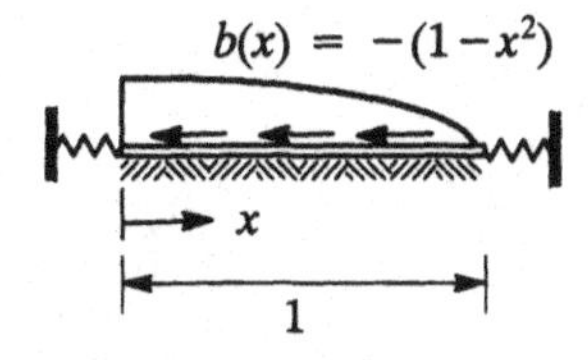

138. Consider the rod of length $\ell = 1$ and constant modulus $C = 1$. The rod is restrained by an elastic spring of modulus $k = 1$ at each end and rests on an elastic foundation, also with modulus $k = 1$. The rod is subjected to a quadratically varying body force as shown. The displacement $u(x)$, positive in the x direction, is governed by the following differential equation $u'' - u = 1 - x^2$. What are the boundary conditions for this problem? Is the following displacement function a classical solution to this problem?

$$u(x) = 1 + x^2 - 2e^x - \tfrac{1}{2}e^{-x}$$

Why or why not? If it is not then modify it so that it is.

139. The principle of virtual work for a certain boundary value problem can be stated as

$$G(u,\bar{u}) \equiv \int_0^\ell \left[A(x)\,u''\bar{u}'' + B(x)\,u\bar{u} - b(x)\,\bar{u}\right]dx = 0 \qquad \text{for all } \bar{u}(x) \in \mathfrak{F}(0,\ell)$$

where A, B, and b are known functions of x, $u(x)$ is the unknown field, and a prime denotes derivative with respect to x. What is the classical differential equation that is equivalent to this variational statement?

140. The classical (4th order) differential equation and boundary conditions for a certain boundary value problem are

$$Au'''' + Bu'' + Cu = b \qquad \text{for all } x \in [0,\ell]$$

$$u(0) = 0, \quad u(\ell) = 0, \quad Au''(0) = 0, \quad Au''(\ell) = 0$$

where A, B, C, and b are known constants, $u(x)$ is the unknown field, and a prime denotes derivative with respect to x. Find an expression for the virtual-work functional associated with the classical differential equation. In other words, find the functional G that has the property that the statement "$G(u,\bar{u}) = 0$ for all $\bar{u} \in \mathfrak{F}_e$" is equivalent to the classical differential equation and the highest derivative that appears in G is second order. Describe any restrictions that must be placed on $\mathfrak{F}_e$.

141. Consider the solid spherical region $\mathfrak{B}$ with surface Ω having a unit normal vector field $\mathbf{n}$, as shown in the sketch. Assume that there exists a scalar field $w(\mathbf{x})$, of the position vector $\mathbf{x}$, for which we can define the functional

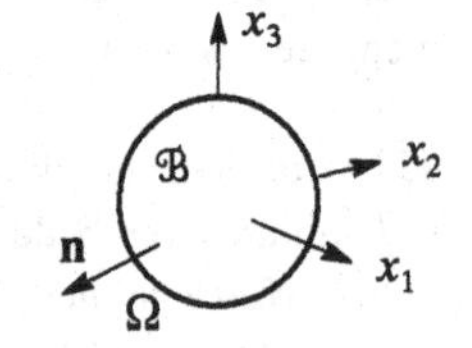

$$G(w,v) \equiv \int_{\mathfrak{B}} \left(\nabla v \cdot \nabla w - v\right)dV - \int_{\Omega} t\,v\,dA$$

that has the property that if $G(u,v) = 0$ for all (virtual) scalar functions $v(\mathbf{x})$ then the classical differential equations governing the real field $w(\mathbf{x})$ are satisfied (i.e., $G(w,v)$ is a "virtual-work" functional). Note that the scalar field $t(\mathbf{x})$ is defined on the surface of the solid region. Find the classical governing differential equation for $w(\mathbf{x})$ that is implied by the variational statement "$G(u,v) = 0$ for all v". Determine what must be the relationship between $t(\mathbf{x})$ and $w(\mathbf{x})$ on the surface of the sphere.

6
The Ritz Method of Approximation

The principle of virtual work is a beautiful alternative to the Newton-Cauchy view of mechanics. This beauty notwithstanding, the principle of virtual work, in its basic form, is not very useful. The simple truth is that it is impossibly difficult to implement the part of the principle that says, "for all $\bar{\mathbf{u}} \in \mathcal{F}(\mathcal{B})$." Furthermore, the displacement $\mathbf{u}(\mathbf{x})$ that solves the problem may not be one of the named and tabulated functions of classical mathematics (e.g., polynomial, trigonometric, and exponential). For a continuous system, the "for-all" statement implies proving that the functional is zero for an infinite number of virtual displacement functions. This aspect of the continuous system stands in stark contrast to a discrete system of N degrees of freedom where the for-all statement means to prove it for N linearly independent vectors, a decidedly finite operation. It is the nonfinite aspect of the principle of virtual work that causes problems for practical computations.

In 1908, Walter Ritz offered an idea that would put some real power into the principle of virtual work. His idea was simple. Let the unknown function be approximately represented as a linear combination of known functions. For a scalar function of a scalar variable we can write

$$u(x) \approx \sum_{n=1}^{N} a_n h_n(x) \tag{305}$$

where a_n is a constant and $h_n(x)$ is a known function of x. Since the functions are known, any variation in $u(x)$ must come from varying the values of the coefficients. Thus, the Ritz approximation reduces the continuous problem to a discrete problem with N degrees of freedom. Clearly, if we approximate the virtual displacement field with a Ritz approximation, the "for-all" statement of the principle of virtual work will be finite and, hence, manageable for practical computations.

The beauty of the Ritz idea is that it can be applied to any theory for which we can write a virtual-work functional. All we need to do in each particular case is to make an appropriate selection of the functions $h_n(x)$. Clearly, these functions must have the same character and domain of definition as the real function $u(x)$. As such, we can often use intuitive knowledge of the physical system to great advantage in constructing a suitable set of base functions for simple problems. For more complex problems, our intuition often forsakes us, and we are left in need of a systematic way of constructing a suitable basis and assessing the accuracy of the approximate solution that it produces. In this chapter, we shall briefly examine the issue of selecting base functions and illustrate the details of implementing the Ritz method.

To get the basic idea behind the Ritz method, we will continue to explore the little boundary value problem introduced in Chapter 5. Using simple polynomial base functions, we will illustrate the workings of the method for the little boundary value problem with a sinusoidally varying load. This simple example will serve to demonstrate important concepts like the nature of the approximation and the concept of convergence to the true solution. We discuss the basic problem with the polynomial base functions and offer two alternatives. The first alternative lies in the concept of orthogonal functions. The second alternative is the use of lightly coupled local functions. The second alternative is the basis of the popular *finite element method*. In this book, we resist the temptation to explore the finite element method in great detail because there are many good books on the subject and because it would take us too far afield in our study. It is important, however, to see the connection between the Ritz method and the finite element method.

The Ritz Approximation for the Little Boundary Value Problem

An approximate solution to our little boundary value problem can be found if we approximate our unknown function $u(x)$ as a linear combination of a finite set of *known* base functions $\mathcal{H}_N = \{h_1(x), \ldots, h_N(x)\}$ as

$$u(x) = \sum_{n=1}^{N} a_n h_n(x) = \mathbf{a} \cdot \mathbf{h}(x) \tag{306}$$

where N is the number of terms in the expansion, $\mathbf{a} \equiv [a_1, \ldots, a_N]^T$ is an array of the unknown constant coefficients, and $\mathbf{h} \equiv [h_1, \ldots, h_N]^T$ is an array of the known base functions. Note that the dot product defined here is a generalization to N dimensions of the dot product in three-dimensional space. In matrix notation we can also write the dot product of arrays in standard matrix form as $\mathbf{a} \cdot \mathbf{h} = \mathbf{a}^T\mathbf{h} = \mathbf{h}^T\mathbf{a}$, where $(\cdot)^T$ is the matrix transpose of $(\cdot)$.

The approximate function must satisfy all essential boundary conditions. Let us again, for the sake of discussion, consider the problem that has a prescribed motion at the left end of $u(0) = u_o$ (essential boundary condition) and a prescribed traction at the right end $\sigma(\ell) = t_\ell$ (natural boundary condition). From Eqn. (306) we express the essential boundary condition in terms of the approximation as

$$u(0) = \sum_{n=1}^{N} a_n h_n(0) = \mathbf{h}(0) \cdot \mathbf{a} = u_o \qquad (307)$$

We will use this equation as part of the solution process. Again, there is no need to implement the natural boundary conditions because the principle of virtual work will try to satisfy those (they are equilibrium equations).

The basis $\mathcal{H}_N$ can be composed of any known functions, but we intend to get more than simply an ad hoc numerical approximation from this approach. If the base functions are carefully selected then we can develop a strategy that will yield a sequence of numerical approximations of ever increasing accuracy. A good numerical method always comes equipped with a means of deciding when the approximation is accurate enough and a systematic approach for improving the accuracy if it is not. A uniform approximation can be achieved if the base functions form a *complete approximating subspace* of functions with square integrable first derivatives on the domain $x \in [0, \ell]$. One such basis is the polynomials $\{1, \xi, \xi^2, \xi^3, \ldots\}$, where $\xi \equiv x/\ell$.† There is a theorem, due to Weierstrass, that essentially says that any function can be approximated as a linear combination of polynomials. Fourier showed that any function can be approximated by a linear combination of trigonometric functions (usually called *Fourier series*). Hence, another suitable basis is given by the trigonometric functions $\{1, \sin n\pi\xi, \cos n\pi\xi, \text{for } n = 1, 2, 3, \ldots\}$. Another important approximating subspace is the so-called finite element functions, which will be described later in this chapter.

Implicit in the ideas of Weierstrass and Fourier (and in finite elements) is the notion of a *complete approximating subspace.* Practically, what that means is that you cannot leave any of the terms out without risking the ruination of the approximation. As we shall see later in this chapter, there is an analogy between function spaces and vector spaces (which are usually easier to visualize because there are more geometric hooks to hang your understanding on). Missing a function (say, for example, we construct a polynomial approximation and we elect to leave out the term ξ^2 in the series expansion) is like leaving out a base vector in a vector space. The remaining vectors do not span the space and

---

† We can express the basis as polynomials in x, but x has units of length. Each base function in the basis $\{1, x, x^2, \ldots\}$ has different units and, consequently, each coefficient a_n in the Ritz expansion will also have different units because each term in approximation of u must have units of length. If we express the basis in terms of the dimensionless variable $\xi \equiv x/\ell$ then all of the coefficients in the expansion will have the same units.

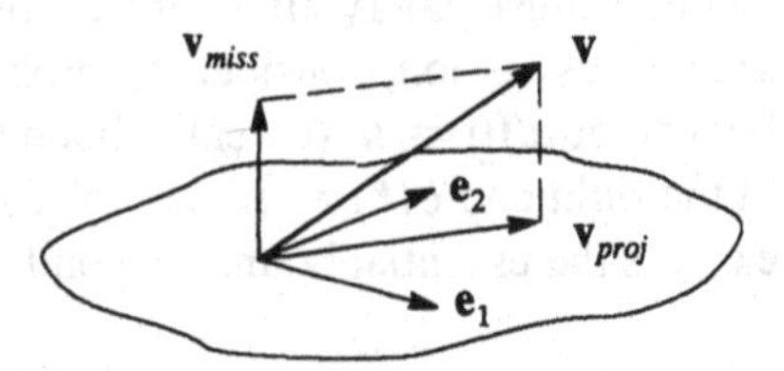

Figure 70 Representation of a vector in a basis that does not span three-dimensional space

so it is not possible to represent all other vectors with that basis. Any vector expressed in components relative to an incomplete basis will be missing, as illustrated in Fig. 70.

This figure shows a basis $\{\mathbf{e}_1, \mathbf{e}_2\}$ that does not span three-dimensional space (i.e., it is missing the base vector $\mathbf{e}_3$). We can, as always, write a component form of the vector as $\mathbf{v} = v_\alpha \mathbf{e}_\alpha$ (where now the summation only extends to $\alpha = 1, 2$ because there are only two base vectors). Of course, if we do this component representation we only get $\mathbf{v}_{proj}$ (the projection of $\mathbf{v}$ onto the plane) and we annihilate $\mathbf{v}_{miss}$ because we have no base vector to represent it. In essence, the basis simply projects out any component that it cannot represent.

The situation is very similar when we represent functions as a linear combination of base functions. If we omit a term, like the ξ^2 mentioned previously, then the approximation will not be able to represent the "quadratic features" of the function in question (u in the present case). If the function has an essential quadratic behavior and the basis does not contain the quadratic function then the approximation will never succeed at representing the function, no matter how many other (non-quadratic) functions we include in the expansion. Fortunately, it is rather obvious in the case of the polynomials what constitutes completeness. It is not as obvious for other bases (even the trigonometric basis leaves some open questions like: What about fractional values of n?).

Technically, we can only assure the approximation of functions, in the sense of Weierstrass and Fourier, if we use an infinite series. In our numerical approximations it is never practical to include an infinite number of terms and so we truncate the series at N terms. We will find that, for most problems we face, we can obtain excellent results with a finite-dimensional space of base functions. To understand when we might fail, consider the task of representing the function $g(x) = \sin 100\pi\xi$ for $\xi \in [0, 1]$ with a truncated Fourier series that includes the functions $\{\sin n\pi\xi,\ n = 1, \ldots, 10\}$. The function $g(x)$ simply oscillates with too great a frequency for any of the base functions to capture it. In general, low-order base functions will do best at representing smooth and slowly varying functions. We will often know enough about the nature of our forcing function, the body force $b(x)$ in the case of the little boundary value problem, to make reasonable assessments of the adequacy of our basis.

Discretization of the principle of virtual work. We can effect a solution if we also approximate our virtual displacements. One choice is to approximate the virtual displacement functions with the same basis as the real displacement function. This approach is called the *Galerkin approximation*. There are many other possible choices. As we shall see, the Galerkin approximation leads to a system of equations that has a symmetric coefficient matrix.

Let the virtual displacement be expressed in the form

$$\boxed{\bar{u}(x) = \sum_{n=1}^{N} \bar{a}_n h_n(x) = \bar{\mathbf{a}} \cdot \mathbf{h}(x)} \tag{308}$$

where $\bar{\mathbf{a}} \equiv [\bar{a}_1, \ldots, \bar{a}_N]^T$ is an array of arbitrary (virtual) constants. With base functions known, it is simple to find the first derivatives of the displacements and virtual displacements. These are

$$u'(x) = \mathbf{a} \cdot \mathbf{h}'(x), \qquad \bar{u}'(x) = \bar{\mathbf{a}} \cdot \mathbf{h}'(x)$$

Herein lies the beauty of the Ritz method. It is generally easy to differentiate known functions. The variety of the real and virtual displacements comes from the coefficients of the series expansion. Since these coefficients are constants, they play a very simple role in the processes of differentiation and integration.

Recall from Eqn. (286) that the virtual-work functional for the little boundary value problem has the expression

$$G(u, t_o, \bar{u}) \equiv \int_0^{\ell} \left(Cu'\bar{u}' - b\bar{u}\right) dx - t_o\bar{u}(0) - t_\ell\bar{u}(\ell)$$

with the associated statement that if $G(u, \bar{u}) = 0$ for all $\bar{u}$, then $u(x)$ represents an equilibrium configuration. Substituting the Ritz approximations given in Eqns. (306) and (308), we obtain a discrete version of the functional (a discrete functional is simply an ordinary function). The functional G reduces to the following expression

$$G(\mathbf{a}, t_o, \bar{\mathbf{a}}) = \bar{\mathbf{a}}^T\left(\mathbf{K}\mathbf{a} - t_o\mathbf{b} - \mathbf{f}\right) \tag{309}$$

where $\mathbf{K}$ is an N by N matrix and $\mathbf{f}$ and $\mathbf{b}$ are N by 1 matrices defined as

$$\mathbf{K} \equiv \int_0^{\ell} C[\mathbf{h}'][\mathbf{h}']^T dx, \quad \mathbf{f} \equiv t_\ell\mathbf{h}(\ell) + \int_0^{\ell} b\,\mathbf{h}\,dx, \quad \mathbf{b} \equiv \mathbf{h}(0) \tag{310}$$

As with the real displacement field, all of the variety in our virtual displacement field comes from the coefficients $\bar{\mathbf{a}}$; the base functions are known and fixed. Therefore, the principle of virtual work reduces to the solution of a discrete system of linear equations to determine the real displacements $\mathbf{a}$ as the following lemma describes.

Lemma. Let $G(\mathbf{w}, \bar{\mathbf{v}}) \equiv \bar{\mathbf{v}}^T \mathbf{w}$ be a function of the real vector $\mathbf{w} \in \mathbb{R}^N$ and the virtual vector $\bar{\mathbf{v}} \in \mathbb{R}^N$. The condition

$$G(\mathbf{w}, \bar{\mathbf{v}}) = 0 \quad \forall \bar{\mathbf{v}} \tag{311}$$

is satisfied if and only if $\mathbf{w} = \mathbf{0}$.

Proof. The proof is by counterexample. Assume that $\mathbf{w} \neq \mathbf{0}$. Since G must be zero for all $\bar{\mathbf{v}}$ then it must certainly be zero for the specific choice $\bar{\mathbf{v}} = \mathbf{w}$. However, in this case $\bar{\mathbf{v}}^T \mathbf{w} = \mathbf{w}^T \mathbf{w} = \| \mathbf{w} \|^2 \neq 0$ in violation of the original assumption. Therefore, $\mathbf{w}$ must be zero. ❑

Corollary. *Discrete principle of virtual work.* Let $G(\mathbf{a}, t_o, \bar{\mathbf{a}})$ be the discrete virtual-work functional given in Eqn. (309), which was obtained by applying the Ritz method to the continuous functional $G(u, t_o, \bar{u})$. Then

$$G(\mathbf{a}, t_o, \bar{\mathbf{a}}) = 0 \quad \forall \bar{\mathbf{a}} \quad \Rightarrow \quad \mathbf{K}\mathbf{a} - t_o \mathbf{b} = \mathbf{f} \tag{312}$$

One ramification of the discrete principle of virtual work is that we do not need to know (or solve for) the virtual displacements $\bar{\mathbf{a}}$. In fact, all virtual-work functionals are linear in the virtual displacement and the action described by Eqn. (312) will always be a feature of the discretization process.

Solving the discrete equations. Equation $\mathbf{K}\mathbf{a} - t_o \mathbf{b} = \mathbf{f}$ represents N component equations in $N+1$ unknowns (N for the components of $\mathbf{a}$ and one for the reaction force t_o). We need another equation in order to solve this problem. The additional equation comes from the essential boundary condition given by Eqn. (307), which, in view of the definitions in Eqn. (310) we can write as a scalar equation $\mathbf{b}^T \mathbf{a} = u_o$. The final set of equations has the structure (often called a *bordered system*)

$$\begin{bmatrix} \mathbf{K} & -\mathbf{b} \\ \mathbf{b}^T & 0 \end{bmatrix} \begin{bmatrix} \mathbf{a} \\ t_o \end{bmatrix} = \begin{bmatrix} \mathbf{f} \\ u_o \end{bmatrix} \tag{313}$$

One option available for solving this system of equations is to simply treat it as a system of $N+1$ equations with $N+1$ unknowns and apply any of a number of techniques (e.g., Gaussian elimination) to carry out the solution. However, it is important to note that $\mathbf{a}$ and t_o have different units and that can lead to some ill-conditioning of the system matrix.

We can solve this system of equations by assuming that $\mathbf{a} \equiv \hat{\mathbf{a}}_1 + t_o \hat{\mathbf{a}}_2$, where $\hat{\mathbf{a}}_1 = \mathbf{K}^{-1}\mathbf{f}$ and $\hat{\mathbf{a}}_2 = \mathbf{K}^{-1}\mathbf{b}$. With these definitions we can see that

$$\begin{aligned} \mathbf{K}\mathbf{a} - t_o \mathbf{b} &= \mathbf{K}\left(\hat{\mathbf{a}}_1 + t_o \hat{\mathbf{a}}_2\right) - t_o \mathbf{b} \\ &= \mathbf{K}\left(\mathbf{K}^{-1}\mathbf{f} + t_o \mathbf{K}^{-1}\mathbf{b}\right) - t_o \mathbf{b} = \mathbf{f} \end{aligned} \tag{314}$$

thereby verifying that $\mathbf{Ka} - t_o\mathbf{b} = \mathbf{f}$ is satisfied for *any value* of t_o. To determine t_o we use the second equation $\mathbf{b}^T\mathbf{a} = u_o$ as follows

$$\mathbf{b}^T\left(\hat{\mathbf{a}}_1 + t_o\hat{\mathbf{a}}_2\right) = u_o \tag{315}$$

From this equation we can determine the reaction force t_o to be

$$t_o = \frac{u_o - \mathbf{b}^T\hat{\mathbf{a}}_1}{\mathbf{b}^T\hat{\mathbf{a}}_2} \tag{316}$$

With the reaction force determined, the complete expression for the displacement can be computed from $\mathbf{a} = \hat{\mathbf{a}}_1 + t_o\hat{\mathbf{a}}_2$.

Another approach to essential boundary conditions. We can simplify our implementation of the essential boundary conditions by writing

$$u(x) = \hat{u}(x) + \mathbf{h}(x) \cdot \mathbf{a}, \qquad \bar{u}(x) = \mathbf{h}(x) \cdot \bar{\mathbf{a}} \tag{317}$$

where $\hat{u}(x)$ is some known function that satisfies the essential boundary conditions and the base functions $\mathbf{h}(x)$ are selected to satisfy the *homogenous essential boundary condition* $\mathbf{h}(0) = \mathbf{0}$ (that is, each component function satisfies the equation $h_n(0) = 0$). If that is the case, then the virtual displacement satisfies the homogeneous boundary condition $\bar{u}(0) = 0$ and the term in the virtual-work functional $t_o\bar{u}(0)$ vanishes. The function $\hat{u}(x)$ does not need to satisfy any of the governing equations in the domain or the natural boundary conditions. There are no restrictions on $\hat{u}(x)$ but one should generally select the simplest possible function that satisfies the essential boundary conditions.

For a case in which the motion at the left end is prescribed to be $u(0) = u_o$ but the right end has an applied traction then an appropriate choice would be the function $\hat{u}(x) = u_o$. For a case in which the motion is prescribed at both ends, i.e., $u(0) = u_o$ and $u(\ell) = u_\ell$, then an appropriate choice would be the linear function $\hat{u}(x) = u_o\left(1 - x/\ell\right) + u_\ell x/\ell$.

An easy way to determine a boundary function $\hat{u}(x)$ and base functions $\mathbf{h}(x)$ that satisfy the homogeneous essential boundary conditions is to start with an expansion in terms of the complete basis $\mathcal{H}_N = \{1, x, x^2, \ldots, x^N\}$. With this approximation, we simply substitute the essential boundary conditions into the Ritz expansion, eliminate one displacement parameter for each essential boundary condition by substitution, regroup terms, and make the appropriate identifications, as the following example shows.

Example 33. *Nonzero boundary displacements.* Consider a rod free at $x = 0$ with a prescribed displacement $u(\ell) = 2$. Let us find the functions $h_n(x)$ and $\hat{u}(x)$ using a quadratic approximation. First, let $\xi \equiv x/\ell$ and take

$$u(\xi) = a_0 + a_1\xi + a_2\xi^2$$

Now apply the essential boundary condition $u(1) = 2$.

$$u(1) = a_0 + a_1 + a_2 = 2, \quad \Rightarrow \quad a_0 = 2 - a_1 - a_2$$

Substitute the expression for a_0 into the original approximation and regroup

$$\begin{aligned} u(\xi) &= (2 - a_1 - a_2) + a_1 \xi + a_2 \xi^2 \\ &= 2 + a_1(\xi - 1) + a_2(\xi^2 - 1) \end{aligned}$$

From this expression we can clearly see that $\hat{u}(\xi) = 2$, and the appropriate base functions are $h_1(\xi) \equiv \xi - 1$ and $h_2(\xi) \equiv \xi^2 - 1$. Observe that both of the base functions satisfy the *homogeneous* essential boundary condition $h_n(1) = 0$.

Notice that the approximation in Example 33 is quadratic, which generally requires three terms, but only two base functions are needed because the essential boundary condition is enforced up front. If we start with an N-term approximation and there are M essential boundary conditions, then we can expect to have $N-M$ base functions. With the above approach, we always start with the complete set of base functions. After implementation of the essential boundary conditions, any term that does not multiply an unknown coefficient must be part of $\hat{u}(x)$, and $h_n(x)$ is everything that multiplies a_n.

We will generally use this simplification in our computations, but it is worth emphasizing that it is a convenience and not a limitation imposed by the principle of virtual work (some authors give the impression that the base function *must* satisfy the homogenous essential boundary conditions in order to be "admissible"). We will discover that for two- and three-dimensional problems the convenience is a bit more attractive because in those problems the reacting tractions are fields and not simply constants as they are in the one-dimensional case. If we can eliminate the reaction forces by restricting the virtual displacement to satisfy the homogeneous essential boundary conditions then we can avoid interpolating the reaction forces. It will always be possible to recover the reaction forces from the stresses and the Cauchy relationship on the boundary of the domain, as pointed out in Chapter 5.

Convergence of the Ritz method. The Ritz method provides a systematic method for discretizing a continuous problem in mechanics. It also provides a means of improving the solution. Our strategy will be roughly as follows. Pick a set of base functions and a degree of approximation (i.e., the number of terms N). Compute the coefficients $\mathbf{a}$ and from those the displacement field and stress field. Assess the quality of the solution by computing the equilibrium domain residual $r_1(x) \equiv \sigma' + b$ and a boundary residual $r_2 \equiv \sigma(\ell) - t_\ell$. If equilibrium is satisfied then the residuals should vanish. If not, then the residuals provide a measure of failure to satisfy the equations of equilibrium. We can set up a criterion for solution adequacy as

$$err \equiv \beta_1 \int_0^{\ell} r_1^2(x)\, dx + \beta_2 r_2^2 \tag{318}$$

where β_1 and β_2 are weights that establish the importance of satisfying equilibrium in the domain as opposed to satisfying equilibrium at the points on the boundary where tractions are prescribed. If $err < tol$ (where tol is some predefined tolerance) then the solution is adequate. If not, then it needs to be improved by adding more terms to the approximation. The following example shows some of the features of convergence of the Ritz method for a simple problem. We select a sinusoidally varying force because a polynomial approximation will not give the exact solution with a finite number of terms.

Example 34. *Convergence of the Ritz method.* Consider the one-dimensional rod of length ℓ and modulus C, shown in Fig. 71, to be fixed at the left end (i.e., $u(0) = 0$), free at the right end (i.e., $\sigma(\ell) = Cu'(\ell) = 0$), and subjected to a sinusoidal body force $b(x) = b_o \sin \pi x/\ell$.

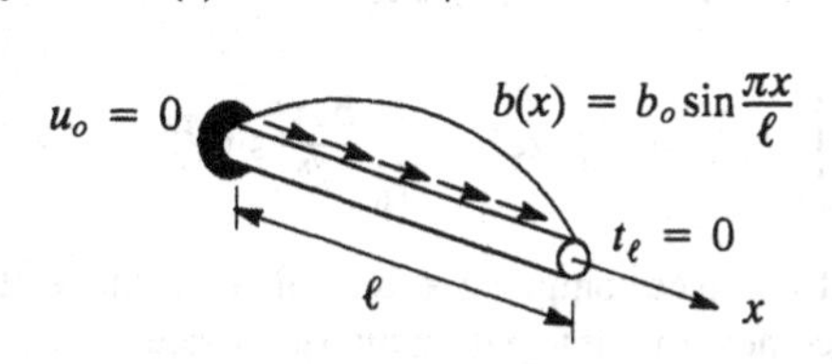

Figure 71 Example problem with sinusoidal body force

The exact solution to this problem can be found by directly integrating the governing equations as we did in Chapter 5. The displacement and stress fields are given by (check these solutions by substituting them into the classical equations)

$$u(x) = \frac{b_o \ell^2}{\pi^2 C}\left(\sin \frac{\pi x}{\ell} + \frac{\pi x}{\ell}\right), \qquad \sigma(x) = \frac{b_o \ell}{\pi}\left(\cos \frac{\pi x}{\ell} + 1\right)$$

Note that the exact solution, in addition to satisfying the above boundary conditions, has the following features at the ends of the rod

$$u(\ell) = \frac{b_o \ell^2}{\pi C}, \qquad \sigma(0) = \frac{2 b_o \ell}{\pi}$$

In what follows, we shall use the end displacement and the reaction as measures of accuracy of our approximate solutions.

Our goal is to construct an approximate solution to the given problem using the principle of virtual work. Let us consider the approximation base functions to be the simple polynomials. To satisfy the essential boundary conditions, we must omit the constant function $h_o(x) = 1$ from the set of base functions. The next four higher-order polynomials can be expressed as

$$h_1(x) = \frac{x}{\ell}, \quad h_2(x) = \frac{x^2}{\ell^2}, \quad h_3(x) = \frac{x^3}{\ell^3}, \quad h_4(x) = \frac{x^4}{\ell^4}$$

Note that the base functions are scaled by the length so that all of the base functions are dimensionless. As a result, the coefficients a_n all have dimensions of length. This scaling is not required, but to do so simplifies the subsequent calculations.

We shall solve the given problem four times, the first time taking only the first term in the approximation, i.e., $u(x) = a_1 h_1(x)$, the second time taking the first two terms, i.e., $u(x) = a_1 h_1(x) + a_2 h_2(x)$, and so forth, up to four terms. We will look at how the solution improves as we take more terms in the approximation series. Let us write our approximation for the real and virtual displacements in accord with Eqns. (306) and (308), respectively. Let N be the number of terms in our approximation, $h_n(x)$ be the nth base function, and the coefficients a_n be the primary unknowns. Since the derivatives of the base functions can be written as $h_n'(x) = nx^{n-1}/\ell^n$, the coefficient matrix $\mathbf{K}$ can be computed in general terms to have components with the following values

$$K_{mn} = \int_0^\ell C h_m' h_n' \, dx = \frac{C}{\ell}\left(\frac{mn}{m+n-1}\right)$$

The right-side matrix $\mathbf{f}$ can be computed to be

$$f_n = \int_0^\ell b_o h_n \sin\frac{\pi x}{\ell} \, dx = \int_0^\ell \frac{b_o x^n}{\ell^n} \sin\frac{\pi x}{\ell} \, dx = \frac{b_o \ell}{\pi} I_n$$

where the value of I_n can be computed recursively from the relationship (which is a formula that comes from integrating by parts twice)

$$I_n = 1 - \left(\frac{n^2 - n}{\pi^2}\right) I_{n-2}$$

for the values $n = 2, 3, 4, \ldots$, knowing the first two terms to be $I_0 = 2$ and $I_1 = 1$ (which are easy to compute explicitly).

The system of equations that results, $\mathbf{Ka} = \mathbf{f}$, is given below for the cases of the one-term expansion, two-term expansion, three-term expansion, and four-term expansion

$$\frac{C}{\ell}\begin{bmatrix} 1 & 1 & 1 & 1 \\ 1 & \frac{4}{3} & \frac{6}{4} & \frac{8}{5} \\ 1 & \frac{6}{4} & \frac{9}{5} & \frac{12}{6} \\ 1 & \frac{8}{5} & \frac{12}{6} & \frac{16}{7} \end{bmatrix} \begin{bmatrix} a_1 \\ a_2 \\ a_3 \\ a_4 \end{bmatrix} = \frac{b_o \ell}{\pi} \begin{bmatrix} 1 \\ 1 - \frac{4}{\pi^2} \\ 1 - \frac{6}{\pi^2} \\ 1 - \frac{12}{\pi^2} + \frac{48}{\pi^4} \end{bmatrix} \tag{319}$$

Each system of equations, corresponding to a different order of approximation N, is shaded and bracketed slightly differently. Clearly, all of the $\mathbf{K}$ and $\mathbf{f}$ components computed for the one-term case are apropos to the two-term case because the base function for the first term of the two-term expansion is the same as the base function for the one-term expansion. Similarly, all of the values for the two-term case are still valid for the three-term case. We need only compute the new quantities that appear from the addition of a new base function.

Table 3 Displacement coefficients for different order approximations

N	a_1	a_2	a_3	a_4	$\frac{\sigma(0)}{b_o\ell/\pi}$	$\frac{\sigma(\ell)}{b_o\ell/\pi}$
1	1.00000				1.000	1.000
2	2.21585	-1.21585			2.216	-0.216
3	2.21585	-1.21585	0.00000		2.216	-0.216
4	1.99096	0.13349	-2.24891	1.12446	1.991	0.009

We can solve each of the four systems of equations to give the value of the coefficients **a** for the various expansions. These coefficients, normalized by the value $b_o\ell^2/\pi C$ and computed to six digits, are given in Table 3.

There are a few things worth mentioning about the approximations. First, the value of the end displacement for all of the approximate solutions is exact

$$u(\ell) = \sum_{n=1}^{N} a_n h_n(\ell) = \sum_{n=1}^{N} a_n = \frac{b_o\ell^2}{\pi C}$$

This conformance to the exact solution at the end point is a peculiar feature of the one-dimensional problem. Do not expect it to happen for every problem. Note that the displacements are not exact at any other point except the fixed end (where we insisted that it be exact). Second, the third-order approximation gave rise to a zero coefficient a_3, meaning that the quadratic approximation is exactly the same as the cubic approximation. In essence, the approximation scheme rejected the extra term because it could not help improve the approximation.

The end tractions are also given in Table 3 for each approximation. Clearly, for the lower-order approximations the traction-free boundary has a nonzero traction on it. In fact, the first-order approximation simply splits the difference, placing half of the reaction to the applied load on each end of the rod. The second-order approximation is significantly better with $\sigma(\ell) = -0.216 b_o\ell/\pi$. As previously mentioned, the third-order approximation is no better than the second. The fourth-order approximation gives nearly exact compliance with the traction-free boundary condition with $\sigma(\ell) = 0.009 b_o\ell/\pi$. The principle of virtual work guarantees the satisfaction of equilibrium in some sense. When we make approximations (i.e., do not enforce the work equation for all possible virtual displacement fields), we compromise the satisfaction of these equilibrium equations. The convergence to the traction-free condition at the right end, as the order of approximation increases, shows how the Ritz method realizes one of the basic promises of the principle of virtual work.

The displacement and stress fields for the example problem are plotted in Fig. 72 for the four approximations. We can see that all of the displacement approximations are equal to the exact displacement at the end $x = \ell$. The quadratic displacement field is almost indistinguishable from the exact displacement field. The stresses converge more slowly than the displacements. We can observe a difference between the exact and approximate stress fields for $N = 2$. The quartic approximation $N = 4$ is very close to the exact stress field. It is evident from the approximate stress distributions that the principle of virtual work attempts to find the best stress field in an average sense. For the linear displacement field,

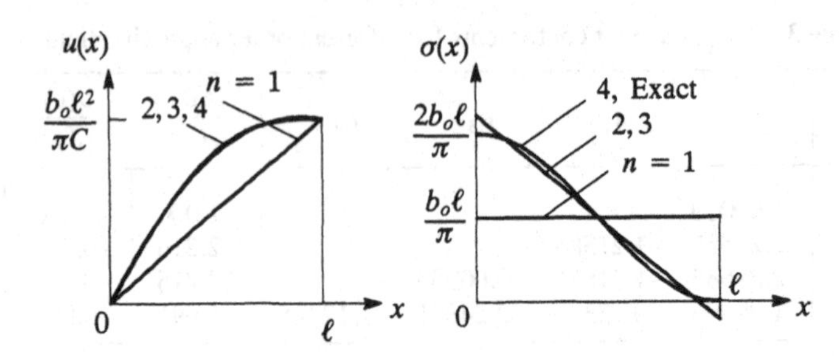

Figure 72 Displacement and stress approximations for example problem

the stress is constant and equal to the average value of the exact stress field over the entire domain. It is also evident from symmetry considerations why the coefficient of the cubic term in the cubic approximation was discarded by the functional.

The preceding example serves to demonstrate how the Ritz approximation method can be used to solve the virtual-work form of the governing differential equations. Inherent in the Ritz method is the choice of base functions and the concept of convergence of the approximate solution to the exact solution. The approximate solution will never get worse with the addition of more base functions. Although exact correspondence with the classical form of the differential equations is guaranteed by the principle of virtual work only in the limit as all virtual displacements are considered, excellent approximations can often be obtained with very few terms in the approximation. The best basis for one problem may not necessarily be the best for another problem, but if the basis is complete, adding base functions should eventually give good results.

The Ritz method is well-suited to implementation in a program that does symbolic or numerical calculations (e.g., MATHEMATICA). Indeed, for problems with more than a few base functions it is not practical to do these calculations by hand. The following example shows how the calculations for the little boundary value problem (with the added feature of an elastic foundation) can be laid out in a MATHEMATICA program. Some of the syntax that are used in the program include

`u = h.a` $\rightarrow \mathbf{u} = \mathbf{h}\cdot\mathbf{a} = \mathbf{h}^T\mathbf{a}$

`D[h, x]` $\rightarrow \frac{d}{dx}\mathbf{h}(x) = \mathbf{h}'(x)$

`Outer[Times, h, h]` $\rightarrow \mathbf{h}\otimes\mathbf{h} = \mathbf{h}\mathbf{h}^T$

`Integrate[g, {x, 0, 1}]` $\rightarrow \int_0^1 g(x)\,dx$

Two symbols next to each other without a symbol between implies scalar multiplication. The command `Clear[x]` simply clears any assigned value to the variable x (so that it can be symbolically manipulated after that point). Some of the other commands, like Inverse[ ] and Plot[ ] should be obvious (but in the case of plotting, there are many more options that one can use to refine the graphic presentation).

The example includes an elastic foundation, which provides resistance to motion in proportion to the amount of displacement at that point (i.e., a distributed spring). The foundation resistance $f(x) = ku(x)$ opposes the motion, thereby contributing to an "equivalent" body force $b_{eff} \equiv b - ku$. The contribution to the virtual work, then, is simply $-b_{eff}\bar{u} \equiv -b\bar{u} + ku\bar{u}$. Consequently, we can see that the elastic foundation actually adds to the stiffness matrix (because it involves both the real displacement u and the virtual displacement $\bar{u}$).

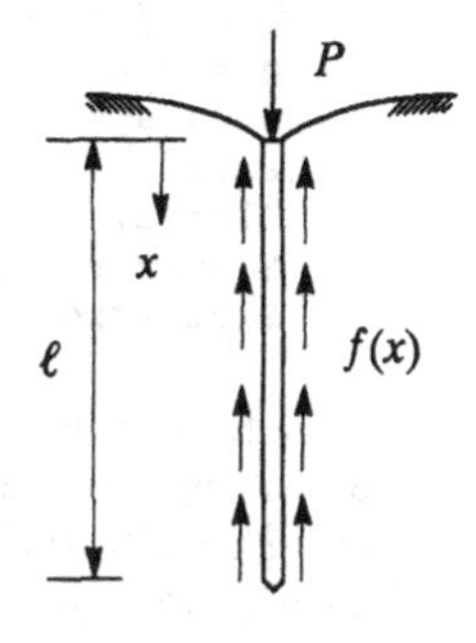

Example 35. *MATHEMATICA program for the Ritz method.* Consider the rod of length ℓ and let $\xi \equiv x/\ell$. The rod has an elastic modulus that varies as $C(\xi) = C_o(2-\xi)$, and it is subjected to a load P at the end $x = 0$ and a body force $b(\xi) = b_o(1-\xi)$. The rod is embedded in an elastic medium such that the force developed along the length is proportional to the displacement i.e., $f(x) = k_o u(x)$. Let the elastic constants be related by $\beta \equiv k_o\ell/C_o$ and let the force constants be related to the modulus and length as $\gamma \equiv P\ell/C_o$ and $\varrho \equiv b_o\ell^2/C_o$, where β, γ, and ϱ are dimensionless problem parameters. The rod is pointed so that the traction at the end $x = \ell$ is zero.

Find an approximation to the displacement field using the Ritz method with a *four-term* approximation with the base functions $\mathbf{h}(\xi) = [1, \xi, \xi^2, \xi^3]$. Note that there are no essential boundary conditions for this problem. The virtual-work functional for this problem is (after dividing through by C_o/ℓ)

$$G(u,\bar{u}) = \int_0^1 \big((2-\xi)u'\bar{u}' + \beta u\bar{u} - \varrho(1-\xi)\bar{u}\big)\,d\xi - \gamma\bar{u}(0)$$

where $(\cdot)' = d(\cdot)/d\xi$. Using $u = \mathbf{h}(\xi)\cdot\mathbf{a}$ and $\bar{u} = \mathbf{h}(\xi)\cdot\bar{\mathbf{a}}$, the virtual work functional $G(u,\bar{u})$ becomes $G(\mathbf{a},\bar{\mathbf{a}}) = \bar{\mathbf{a}}^T[\mathbf{K}\mathbf{a} - \mathbf{f}]$, with system matrices

$$\mathbf{K} \equiv \int_0^1 \big[(2-\xi)\mathbf{h}'(\xi)\otimes\mathbf{h}'(\xi) + \beta\,\mathbf{h}(\xi)\otimes\mathbf{h}(\xi)\big]\,d\xi$$

$$\mathbf{f} \equiv \gamma\mathbf{h}(0) + \varrho\int_0^1 (1-\xi)\mathbf{h}(\xi)\,d\xi$$

The discrete principle of virtual work is satisfied if $\mathbf{K}\mathbf{a} = \mathbf{f}$. The following MATHEMATICA program solves the problem for $\beta = 1$, $\gamma = 1$, and $\varrho = 1$. Note that the comments in italics and brackets are not part of the code.

[Establish the variable parts of b and C and input values of β, γ, and ϱ]

```
load = (1 - x)
modulus = (2 - x)
beta = 1
gamma = 1
rho = 1
```

[Set base functions and compute derivatives]

```
h = {1, x, x^2, x^3}
hp = D[h,x]
```

[Compute stiffness **K** *and load vector* **f***]*

```
K1 = Integrate[ modulus Outer[ Times,hp,hp],{x,0,1}]
K2 = Integrate[ Outer[ Times,h ,h ],{x,0,1}]
K  = K1 + beta K2
f2 = Integrate[ load h ,{x,0,1}]
x = 0
f1 = h
f  = gamma f1 + rho f2
```

[Solve equations for coefficients **a** *and compute displacements, stresses, etc.]*

```
a  = Inverse[K].f
Clear[x]
u = h.a
stress = mod hp.a
error =  D[stress,x]  - beta u + load
Plot[u,      {x,0,1}]
Plot[stress,{x,0,1}]
Plot[error, {x,0,1}]
```

Note that **x** in the MATHEMATICA code is ξ in the above equations. The displacement field is computed as $u(\xi) = \mathbf{h}(\xi) \cdot \mathbf{a}$ and the stress field is computed as $\sigma(\xi) = C(\xi)\mathbf{h}'(\xi) \cdot \mathbf{a}$, once the values of the interpolation parameters **a** are known. The error in the classical differential equation is computed as $error = \sigma' - k_o u + b$. Finally, the code provides for plotting of the results. Note that it is also possible to compute the stress at $x = 0$ or to integrate the square of the error from 0 to 1 to get a better understanding of the approximation error.

The Ritz method can, of course, be programmed in virtually any computer language. The syntax can vary from one language to another, but the basic organization of the calculations is the same.

What is wrong with the basis? There is a problem with the simple polynomial base functions used in the preceding example. The problem is not entirely evident from results shown in Fig. 72 because we stopped at a four-term approximation, but it is there and it does warrant some consideration. The root of the problem is that the higher-order base functions look pretty much the same. The first seven polynomial base functions are plotted in Fig. 73. The functions are increasingly difficult to distinguish from each other with regard to their shape as n gets large.

The problem manifests in the conditioning of the matrix **K** that we must invert in order to find the coefficients **a** from $\mathbf{Ka} = \mathbf{f}$. As we increase the order

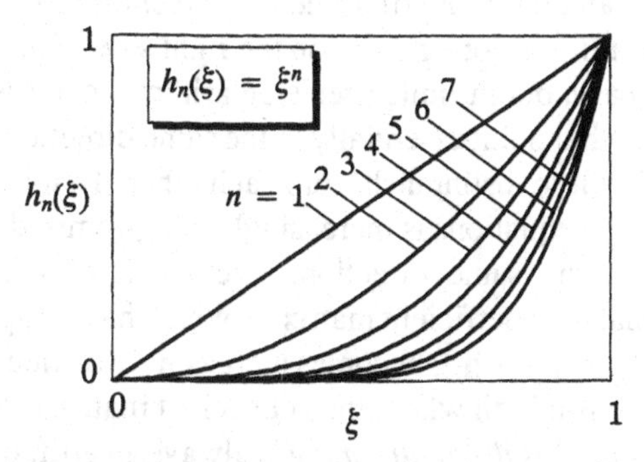

Figure 73 The first seven simple polynomial base functions

of our approximation, the matrix **K** gets harder and harder to invert accurately. For the present problem, it would be virtually impossible to get accurate results for **a** for an approximation with $N \geq 7$ on a finite-precision computer with eight digits of accuracy. In the following section, we discuss this problem in a little more detail and suggest curing it by improving the base functions through a process called orthogonalization. In the subsequent section, we consider an alternative solution to the problem called the finite element approximation.

Orthogonal Ritz Functions

The problem with base functions that are nearly alike is very much akin to the problem of representing the components of a vector with respect to base vectors that point in nearly the same direction. The closer the base vectors are to being collinear, the more difficult it is to accurately compute the components of the vector in question with respect to that basis. Let us assume that we have a set of *non-orthogonal* base vectors $\{\mathbf{h}_1, \mathbf{h}_2, \mathbf{h}_3\}$ spanning $\mathbb{R}^3$, as shown in Fig. 74. Let us compute the components of a vector $\mathbf{v} = v_j \mathbf{h}_j$ (summation implied) with respect to this basis. In other words, we want to find the coefficients v_j.

The equations for making this computation can be found by taking the dot product of the component equation with each of the base vectors. To wit

$$\left(\mathbf{h}_i \cdot \mathbf{h}_j\right) v_j = \mathbf{h}_i \cdot \mathbf{v}$$

These equations are often called the *normal equations*. They are nothing more than a linear system of three equations for the three unknowns $\{v_1, v_2, v_3\}$. The

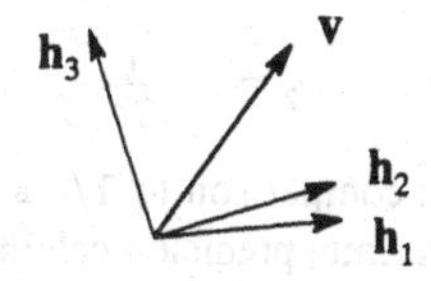

Figure 74 Finding the components of a vector with respect to a non-orthogonal basis

solution can be obtained by inverting the coefficient matrix and multiplying that inverse by the three by one matrix on the right side of the equations. This operation sounds easy, but in finite precision arithmetic, disaster lurks.

For base vectors that point essentially in the same direction, the three equations are very difficult to distinguish from each other. That is, the linear independence of the three equations is increasingly compromised as the base vectors point more in the same direction. The result of this loss of linear independence is that the coefficient matrix **A**, which has components given by the expressions $A_{ij} = (\mathbf{h}_i \cdot \mathbf{h}_j)$, becomes increasingly difficult to invert accurately. In fact, there is a limit where the coefficient matrix is singular in finite precision arithmetic. This *ill-conditioning* is always an artifact of base vectors that are too much alike. The best base vectors are orthogonal.

Example 36. *Normal equations and ill-conditioning.* Consider a certain vector $\mathbf{v} = \mathbf{e}_1 + \mathbf{e}_2 + \mathbf{e}_3$ and a basis described by the base vectors (not orthonormal)

$$\mathbf{h}_1 = \mathbf{e}_1 + \varepsilon\,\mathbf{e}_2, \qquad \mathbf{h}_2 = \mathbf{e}_1 - \varepsilon\,\mathbf{e}_2, \qquad \mathbf{h}_3 = \mathbf{e}_3$$

where ε is a parameter and $\{\mathbf{e}_1, \mathbf{e}_2, \mathbf{e}_3\}$ are the standard (orthonormal) base vectors. Let us compute the components of $\mathbf{v}$ in the basis $\{\mathbf{h}_1, \mathbf{h}_2, \mathbf{h}_3\}$, i.e., let us find the values v_i such that $\mathbf{v} = v_i \mathbf{h}_i$ (sum implied). The normal equations for this basis can be computed as

$$\begin{bmatrix} 1+\varepsilon^2 & 1-\varepsilon^2 & 0 \\ 1-\varepsilon^2 & 1+\varepsilon^2 & 0 \\ 0 & 0 & 1 \end{bmatrix} \begin{bmatrix} v_1 \\ v_2 \\ v_3 \end{bmatrix} = \begin{bmatrix} 1+\varepsilon \\ 1-\varepsilon \\ 1 \end{bmatrix} \tag{320}$$

These equations can be solved to yield

$$v_1 = \tfrac{1}{2}\left(1 + \tfrac{1}{\varepsilon}\right), \qquad v_2 = \tfrac{1}{2}\left(1 - \tfrac{1}{\varepsilon}\right), \qquad v_3 = 1 \tag{321}$$

These values of the coefficients can be substituted back into $\mathbf{v} = v_i \mathbf{h}_i$ to give

$$\mathbf{v} = \tfrac{1}{2}\left(1 + \tfrac{1}{\varepsilon}\right)(\mathbf{e}_1 + \varepsilon\,\mathbf{e}_2) + \tfrac{1}{2}\left(1 - \tfrac{1}{\varepsilon}\right)(\mathbf{e}_1 - \varepsilon\,\mathbf{e}_2) + \mathbf{e}_3 = \mathbf{e}_1 + \mathbf{e}_2 + \mathbf{e}_3$$

as expected. This calculation simply demonstrates that it is possible to compute components of a vector with respect to a non-orthonormal basis.

Let us now consider the case where ε is very small. It should be evident that by adjusting the value ε to be closer to zero we make the base vectors $\mathbf{h}_1$ and $\mathbf{h}_2$ point in the same direction; as $\varepsilon \to 0$ they both point in the direction $\mathbf{e}_1$. In this case, Eqn. (321) would yield the approximate values of the coefficients

$$v_1 \approx \frac{1}{2\varepsilon}, \qquad v_2 \approx -\frac{1}{2\varepsilon}, \qquad v_3 = 1$$

because 1 would be small in comparison to $1/\varepsilon$ and would, consequently, be truncated in the roundoff in a finite precision calculation. Now, from $\mathbf{v} = v_i \mathbf{h}_i$

$$\mathbf{v} = \frac{1}{2\varepsilon}(\mathbf{e}_1 + \varepsilon\,\mathbf{e}_2) - \frac{1}{2\varepsilon}(\mathbf{e}_1 - \varepsilon\,\mathbf{e}_2) + \mathbf{e}_3 = \mathbf{e}_2 + \mathbf{e}_3$$

The roundoff has completely annihilated the $\mathbf{e}_1$ component of the vector $\mathbf{v}$! In reality the outcome of the computation depends upon the algorithm used to solve the normal equations. On a finite precision computer the coefficient matrix of the normal equations is nearly singular and any algorithm used to solve the equations will, at best, give strange results and, at worst, fail to give results at all.

The accuracy of the solution of a system of equations with coefficient matrix **A** deteriorates as **A** gets closer to being singular because of the deleterious effects of roundoff error in a finite precision calculation. A good measure of the invertibility of a matrix or the solvability of a system of linear equations is the *condition number* $\varrho(\mathbf{A})$, defined as the ratio of the largest eigenvalue μ_{max} of **A** divided by the smallest eigenvalue μ_{min} of **A**, that is

$$\varrho(\mathbf{A}) \equiv \frac{\mu_{max}(\mathbf{A})}{\mu_{min}(\mathbf{A})} \tag{322}$$

The closer $\varrho(\mathbf{A})$ is to unity, the better-conditioned is the matrix. The larger $\varrho(\mathbf{A})$ is, the more ill-conditioned is **A**. The best-conditioned matrix is an orthogonal matrix **Q**. Recall that an orthogonal matrix satisfies $\mathbf{Q}^{-1} = \mathbf{Q}^T$. The condition number of an orthogonal matrix is exactly 1. The identity matrix **I** is a particular case of an orthogonal matrix.

The eigenvectors (not normalized) of the coefficient matrix in Example 36, are $\boldsymbol{\phi}_1 \sim (1, -1, 0)$, $\boldsymbol{\phi}_2 \sim (0, 0, 1)$, and $\boldsymbol{\phi}_3 \sim (1, 1, 0)$, associated with the eigenvalues $\mu_1 = 2\varepsilon^2$, $\mu_2 = 1$, and $\mu_3 = 2$. Therefore, the condition number for the coefficient matrix is $\varrho(\mathbf{A}) = 1/\varepsilon^2$ which gets very large as $\varepsilon \to 0$. The condition number is an indicator of the trouble with solving those equations.

The problem with the basis in Example 34 can be seen by examining the condition number of the matrix **K** as the number of terms in the approximation increases. The maximum and minimum eigenvalues of the matrix for each order of approximation are given in Table 4. Clearly, for the case $N = 1$, the matrix is one by one and the maximum and minimum eigenvalues are the same. As the order of the approximation increases, the condition number increases dramatically, an order of magnitude for each increment in the order of approximation.

Table 4 Condition of **K** for the Example 34 problem, simple polynomial basis

N	μ_{max}	μ_{min}	$\varrho(\mathbf{K})$
1	1.00000	1.000000	1.0
2	2.18046	0.152870	14.3
3	3.79646	0.013580	279.6
4	5.88341	0.000845	6959.5

Gram-Schmidt orthogonalization of vectors in $\mathbb{R}^N$. For independent vectors that are not orthogonal, we can always produce an orthogonal set of vectors using the Gram-Schmidt orthogonalization procedure. In this procedure, we put the base vectors in a certain order, with one of them designated as being the first. A vector orthogonal to this first vector is produced by taking the original second vector and projecting out the component of the vector that lies along the first vector. The new orthogonal vector replaces the old second vector in the set. A third orthogonal vector is produced by taking the third original vector and projecting out its components along both of the previous two orthogonal vectors. The procedure continues until an entire orthogonal basis has been produced. Let us assume that our non-orthogonal basis is composed of the following set of vectors spanning $\mathbb{R}^N$: $\{\mathbf{h}_1, \mathbf{h}_2, \ldots, \mathbf{h}_N\}$. We wish to produce a new orthogonal set of vectors $\{\mathbf{g}_1, \mathbf{g}_2, \ldots, \mathbf{g}_N\}$. First set $\mathbf{g}_1 = \mathbf{h}_1$. Then, for $n = 2, 3, \ldots, N$, compute the remaining vectors sequentially from the formula

$$\mathbf{g}_n = \mathbf{h}_n - \sum_{j=1}^{n-1}\left(\frac{\mathbf{g}_j \cdot \mathbf{h}_n}{\mathbf{g}_j \cdot \mathbf{g}_j}\right)\mathbf{g}_j \tag{323}$$

Each new vector is orthogonal to all previous vectors, as can be shown by taking the dot product between any of them ($k < n$)

$$\begin{aligned}\mathbf{g}_n \cdot \mathbf{g}_k &= \mathbf{h}_n \cdot \mathbf{g}_k - \sum_{j=1}^{n-1}\left(\frac{\mathbf{g}_j \cdot \mathbf{h}_n}{\mathbf{g}_j \cdot \mathbf{g}_j}\right)\mathbf{g}_j \cdot \mathbf{g}_k \\ &= \mathbf{h}_n \cdot \mathbf{g}_k - \left(\frac{\mathbf{g}_k \cdot \mathbf{h}_n}{\mathbf{g}_k \cdot \mathbf{g}_k}\right)\mathbf{g}_k \cdot \mathbf{g}_k = 0\end{aligned}$$

The proof depends upon the observation that $\mathbf{g}_j \cdot \mathbf{g}_k = 0$ if $j \neq k$, i.e., that orthogonality holds for all of the vectors already computed. Observe that the case $n = 2$ works because there is only one term in the sum. Now, the orthogonality of the remaining vectors follows by induction.

The new vectors are not necessarily of unit length, but can easily be made so by dividing each vector by its own length. In fact, the best approach to computing an orthonormal basis is a two-stage process

$$\hat{\mathbf{g}}_n = \mathbf{h}_n - \sum_{j=1}^{n-1}(\mathbf{g}_j \cdot \mathbf{h}_n)\mathbf{g}_j, \qquad \mathbf{g}_n = \hat{\mathbf{g}}_n / \| \hat{\mathbf{g}}_n \| \tag{324}$$

Observe that this approach eliminates the need for having the normalizing term $\mathbf{g}_j \cdot \mathbf{g}_j$ in the denominator of each term in the sum because $\mathbf{g}_j \cdot \mathbf{g}_j = 1$.

Orthogonal functions. The jump from vector spaces to function spaces is a big one, but much of what is true of vector spaces carries over by analogy to function spaces. Each one of our base functions is an element in our function space, analogous to a base vector in a finite dimensional space. Certainly, if our base functions tend to line up, their ability to resolve the components of another function will be less well conditioned than if the base functions are all very different. We can see from Fig. 73 that the simple polynomial base functions qualitatively appear to line up. Is there a way to assess this quality of functions? For vectors, we assess similarity in the orientation of vectors with the dot product. We can do basically the same for functions. Let $u(x)$ and $v(x)$ be two scalar-valued functions defined on the real segment $[a, b] \subset \mathbb{R}$. The *inner product* of the two functions is defined to be

$$\boxed{\langle u, v \rangle \equiv \int_a^b u(x)\, v(x)\, dx} \tag{325}$$

If $\langle u, v \rangle = 0$ then we say that the two functions are orthogonal. The length, or *norm*, of a function is given by its inner product with itself $\| u \|^2 \equiv \langle u, u \rangle$. Clearly, a function has zero length only if it is zero at every point in its domain. For two vectors of unit length, their inner product is a direct measure of how much they "line up." With the introduction of the notion of the inner product, we can proceed to talk about the components of a function with respect to a set of base functions, just as we do for vectors. We can also cure the problem of loss of independence of the base functions.

Gram-Schmidt orthogonalization of functions. The idea of orthogonalization can be extended to functions. Let us assume that we have a given set of base functions $\{h_1, \ldots, h_N\}$, e.g., the polynomial basis $\{1, \xi, \xi^2, \ldots, \xi^N\}$ on the domain $\xi \in [0, 1]$, and that we wish to produce a new set of orthogonal base functions $\{g_1, \ldots, g_N\}$. Assume that we have produced the first $n-1$ orthogonal functions $g_1, \ldots, g_{n-1}$ and we now want to compute g_n from h_n. We know that the new function will be a linear combination of the previous orthogonal functions (which span exactly the same space as $\{h_1, \ldots, h_{n-1}\}$) and the next function h_n. Let us write this observation as

$$g_n = a_{nn} h_n - \sum_{j=1}^{n-1} a_{nj}\, g_j \tag{326}$$

where the constants a_{nj} for $j = 1, \ldots, n$ are yet to be determined. The $n-1$ conditions of orthogonality are $\langle g_n, g_i \rangle = 0$ for $i = 1, \ldots, n-1$. The orthogonality of the functions allows the determination of the coefficients a_{nj} for the index values $j = 1, \ldots, n-1$ as

$$a_{nj} = a_{nn} \frac{\langle h_n, g_j \rangle}{\langle g_j, g_j \rangle} \tag{327}$$

The constant a_{nn} is arbitrary and can be set to any convenient value or can be set to meet a convenient criterion, e.g., $g_n(1) = 1$ or $\langle g_n, g_n \rangle = 1$.

The Gram-Schmidt orthogonalization algorithm can be summarized as follows. Set $g_1 = h_1$. Compute the remaining functions, $g_2, \ldots, g_N$, as

$$g_n = h_n - \sum_{j=1}^{n-1} \frac{\langle h_n, g_j \rangle}{\langle g_j, g_j \rangle} g_j \tag{328}$$

We can demonstrate the orthogonality by computing $\langle g_n, g_k \rangle$ for $k < n$ as

$$\begin{aligned} \langle g_n, g_k \rangle &= \langle h_n, g_k \rangle - \sum_{j=1}^{n-1} \frac{\langle g_j, h_n \rangle}{\langle g_j, g_j \rangle} \langle g_j, g_k \rangle \\ &= \langle h_n, g_k \rangle - \frac{\langle g_k, h_n \rangle}{\langle g_k, g_k \rangle} \langle g_k, g_k \rangle = 0 \end{aligned}$$

Example 37. *Gram-Schmidt orthogonalization of functions.* Consider the original basis $\{\xi, \xi^2, \xi^3, \xi^4\}$ defined on the domain $\xi \in [0, 1]$. Let us generate a set of orthogonal functions from Eqn. (328) starting with $g_1 = \xi$. Scale the functions to have a value of unity at the right end, i.e., $g_n(1) = 1$. We obtain the functions given in Fig. 75.

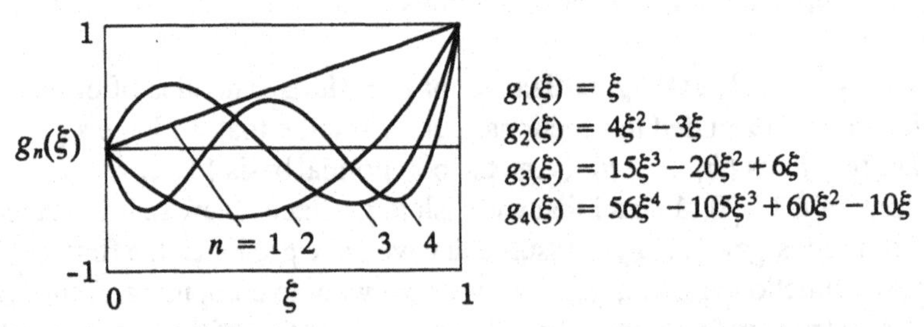

Figure 75 The first four orthogonal polynomials generated from ξ

We can observe that these functions appear quite different from the base functions shown in Fig. 73. You can almost see the orthogonality. These functions, unlike their progenitors $\{\xi, \xi^2, \xi^3, \xi^4\}$, have inflection points, and the higher the index number on the function, the more inflection points the function has. It is important to observe that $g_n(x)$ is still an nth-order polynomial. Orthogonalization does not introduce any higher-order functions.

The orthogonal functions are not necessarily of unit length, but can easily be made so by dividing each functions by its own length. In fact, like the discrete case, one can compute an orthonormal basis as a two-stage process

$$\hat{g}_n = h_n - \sum_{j=1}^{n-1} \langle g_j, h_n \rangle g_j, \quad g_n = \hat{g}_n / \sqrt{\langle \hat{g}_n, \hat{g}_n \rangle} \tag{329}$$

There is one important point to make about the orthogonal basis. The functions themselves are orthogonal because we forced them to be. Their first derivatives are not necessarily orthogonal, and, hence, the coefficient matrix **K** will not be diagonal but it will be reasonably well conditioned, as the following example demonstrates.

Example 38. *Revisit Example* 34 *with an orthogonal basis.* Let us use the base functions given in Fig. 75 in a four-term Ritz approximation of the problem solved in Example 34. The system of equations that results, **Ka** = **f**, with the new base functions, is

$$\frac{C}{\ell}\begin{bmatrix} 1 & 1 & 1 & 1 \\ 1 & \frac{19}{3} & \frac{13}{3} & \frac{27}{5} \\ 1 & \frac{13}{3} & \frac{43}{3} & 9 \\ 1 & \frac{27}{5} & 9 & 25 \end{bmatrix}\begin{bmatrix} a_1 \\ a_2 \\ a_3 \\ a_4 \end{bmatrix} = \frac{b_o\ell}{\pi}\begin{bmatrix} 1 \\ 1-\frac{16}{\pi^2} \\ 1-\frac{10}{\pi^2} \\ 1-\frac{282}{\pi^2}+\frac{2688}{\pi^4} \end{bmatrix}$$

Compare these equations with those given in Eqn. (319). The result of solving these equations is exactly the same as before. Note that the values of the coefficients a_i differ from the values computed with the previous basis, but the final approximate expression for $u(x)$ is identical in each case. The condition number is $\varrho(\mathbf{K}) = 39.36$ rather than the previous value of 6959.5 for the four-term case.

For the little boundary value problem (without elastic foundation) we can observe that the stiffness matrix **K** has components

$$K_{ij} = \int_0^\ell Ch'_i\, h'_j\, dx \tag{330}$$

For a problem with constant modulus C it is evident that we could generate a diagonal stiffness matrix **K** if the first derivatives of the base functions were orthogonal rather than the functions themselves. We could, by a procedure analogous to the one above, produce functions that have orthogonal first derivatives, and, as a consequence, get a diagonal **K** as the following example demonstrates.

Example 39. *Basis with orthogonal first derivatives.* Consider the polynomial basis $\{\xi, \xi^2, \xi^3\}$ defined on the domain $\xi \in [0, 1]$. Let us generate a set of functions whose first derivatives are orthogonal, starting with $g_1 = \xi$. Note that the derivative is $g_1' = 1$ and that $\langle g_1', g_1' \rangle = 1$. Let the second function be

$$g_2 = \xi^2 - a_{21}\xi \qquad g_2' = 2\xi - a_{21}$$

and compute the coefficient a_{21} from $\langle g_1', g_2' \rangle = 0$ (the orthogonality condition). Thus,

$$a_{21} = \int_0^1 2\xi \, d\xi, \quad \Rightarrow \quad a_{21} = 1, \quad \Rightarrow \quad g_2 = \xi^2 - \xi$$

Note that $\langle g_2', g_2' \rangle = 1/3$. Now take the third function in the form

$$g_3 = \xi^3 - a_{31}\xi - a_{32}(\xi^2 - \xi), \quad g_3' = 3\xi^2 - a_{31} - a_{32}(2\xi - 1)$$

and compute the coefficients a_{31} and a_{32} from the orthogonality conditions $\langle g_i', g_3' \rangle = 0$ for $i = 1, 2$. Thus,

$$a_{31} = \int_0^1 3\xi^2 \, d\xi = 1, \quad \frac{1}{3}a_{32} = \int_0^1 3\xi^2(2\xi - 1)\, d\xi = \frac{1}{2}$$

$$\Rightarrow \quad g_3 = \xi^3 - \frac{3}{2}\xi^2 + \frac{1}{2}\xi$$

Note that $\langle g_3', g_3' \rangle = 1/20$. We can normalize these three functions so that the inner products $\langle g_i', g_i' \rangle = 1$ for $i = 1, 2, 3$. The normalized functions are

$$g_1 = \xi, \qquad g_2 = \sqrt{3}\,(\xi^2 - \xi), \qquad g_3 = \sqrt{5}\,(2\xi^3 - 3\xi^2 + \xi)$$

If we apply these functions to the problem in Example 34 then $\mathbf{K} = C\mathbf{I}$ (which means that it is perfectly conditioned with $\varrho(\mathbf{K}) = 1$). The components of the vector $\mathbf{f}$ can be computed as

$$f_i = \int_0^1 g_i b_o \sin \pi\xi \, d\xi i \quad \Rightarrow \quad \mathbf{f} \sim \frac{b_o \ell}{\pi} \begin{bmatrix} 1 \\ -\frac{4\sqrt{3}}{\pi^2} \\ 0 \end{bmatrix}$$

Again, the results are the same as any other cubic polynomial base functions.

An advantage of the basis with orthogonal first derivatives in Example 39 is that the matrix $\mathbf{K}$ was diagonal. The solution of the equations in such a case is trivial, with the coefficients given by $a_i = f_i / K_{ii}$ (no sum on repeated indices). This case is not that important because the orthogonality can be disturbed by a non-constant modulus $C(x)$ or by additional terms like an elastic foundation, as in Example 35.

One thing that is evident from the preceding discussion is that while orthogonal functions have merits (especially with respect to conditioning of the equations) they have drawbacks too. The first drawback is that there is significant computation involved in finding the orthogonal functions. The second drawback is that each function generally has many terms. For example, the *n*th-order orthogonal polynomial has non-zero coefficients for *all* of the lower-order terms, making them rather cumbersome to work with.

The trigonometric functions are orthogonal over certain domains and do not have some of the drawbacks of the orthogonal polynomials. Functions approximated with trigonometric functions are often referred to as *Fourier series* approximations.

Example 40. *Fourier series.* Consider the basis $\{\xi,\ \sin n\pi\xi\ ,\ n = 1, 2, \ldots, N\}$ defined on the domain $\xi \in [0,\ 1]$. Let us solve the little boundary value problem with unit length, constant modulus C, displacement restrained at $\xi = 0$, zero traction at $\xi = 1$, and with body force $b(\xi)$.

$$G(u, \overline{u}) = \int_0^1 \left(Cu'\overline{u}' - b\overline{u}\right) \ell d\xi$$

Let the real and virtual displacements be approximated as infinite series

$$u = a_0\xi + \sum_{n=1}^{\infty} a_n \sin n\pi\xi, \quad \overline{u} = \overline{a}_0\xi + \sum_{m=1}^{\infty} \overline{a}_m \sin m\pi\xi$$

The derivatives of displacement are

$$u' = a_0 + \sum_{n=1}^{\infty} n\pi a_n \cos n\pi\xi, \quad \overline{u}' = \overline{a}_0 + \sum_{m=1}^{\infty} m\pi\overline{a}_m \cos m\pi\xi$$

Substituting these expressions into the virtual-work functional we find that

$$K_{00} = C, \quad K_{nn} = \tfrac{1}{2}Cn^2\pi^2, \quad K_{mn} = 0 \text{ for } m \neq n$$

$$f_0 = \int_0^1 \xi b(\xi)\, d\xi, \quad f_n = \int_0^1 b(\xi) \sin n\pi\xi\, d\xi$$

Because the matrix $\mathbf{K}$ is diagonal, the displacement coefficients are then

$$a_0 = \frac{1}{C}\int_0^1 \xi b(\xi)\, d\xi, \quad a_n = \frac{2}{n^2\pi^2 C}\int_0^1 b(\xi) \sin n\pi\xi\, d\xi$$

These equations hold for any function $b(\xi)$. Take as a specific example the linearly increasing loading function $b(\xi) = b_o\xi$. The coefficients in this case are

$$a_0 = \frac{b_o}{3C}, \quad a_n = -\frac{2b_o}{n^3\pi^3 C}\cos n\pi$$

The displacement takes the final form

$$u = \frac{b_o}{3C}\left(\xi + \frac{6}{\pi^3}\left(\sin\pi\xi - \tfrac{1}{8}\sin 2\pi\xi + \tfrac{1}{27}\sin 3\pi\xi - \tfrac{1}{64}\sin 4\pi\xi + \cdots\right)\right)$$

The beauty of Fourier series approximations of displacement is evident in the rate of convergence of the displacement. In the present case the terms decrease as a reciprocal of n^3. The stress converges only as n^2 because the derivative of sin $n\pi\xi$ throws off another n in each term. The loading function in the example is very smooth and Fourier series converges quickly for smooth functions. In cases where the loading function is discontinuous the convergence is slower. In some cases the stress will converge very slowly (i.e., lots of terms will be required for an accurate solution) or not at all. Again, when the modulus is variable, the diagonality of **K** is lost.

The Finite Element Approximation

A traditional education in mathematics generally leaves one with a bias toward functions that have names, like polynomials or exponentials, whose domain of action is the entire region of interest. The polynomial base functions of the previous section are examples of such functions. Each of the base functions is nonzero over the entire region $[0, \ell]$.

There is a very interesting alternative for the definition of Ritz base functions that has some great advantages for solving problems in structural mechanics. Rather than insisting that a base function be an nth-order polynomial or a trigonometric function, we will allow the piecing together of some of the simplest members of these classes of functions. For example, the function $g(x)$ shown in Fig. 76 is a piecewise linear, continuous function. The function is linear between the points x_{n-1} and x_n (often called *nodes*), but the overall function between the endpoints x_0 and x_N describes a nonlinear variation that characterizes the entire function. The function can be defined as

$$g(x) = g_{n-1}\left(\frac{x_n - x}{x_n - x_{n-1}}\right) + g_n\left(\frac{x - x_{n-1}}{x_n - x_{n-1}}\right), \quad x_{n-1} \leq x \leq x_n \qquad (331)$$

where g_n is the ordinate of the function at the point x_n. The N linear segments meet at their nodes. Thus, the function is continuous. The first derivative of the function is the slope, which is well defined for every point in the domain except at the nodes $\{x_0, \ldots, x_N\}$. At these points the function has kinks, so the first derivative jumps there. The first derivative has the functional form

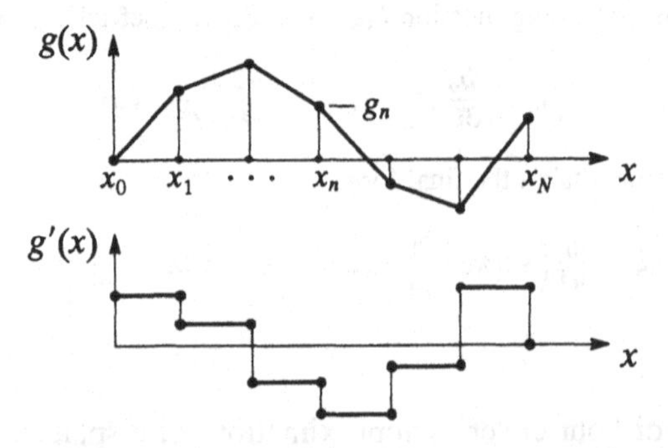

Figure 76 A piecewise linear, continuous function and its derivative

$$g'(x) = \frac{g_n - g_{n-1}}{x_n - x_{n-1}}, \quad x_{n-1} < x < x_n \tag{332}$$

Even though the function $g'(x)$ is not defined at the nodes, it is clearly integrable. The integral of the function is the shaded area under the curve, shown in Fig. 76. It is easy to show that the function is also square-integrable. The second derivative of the function $g''(x)$ is not well defined because of the discontinuities at the nodes in the first derivative. This piecewise linear function is a good example of a function that belongs to the class $\mathcal{C}^0$ (continuous zeroth derivative) but not $\mathcal{C}^1$ (continuous first derivative).

It should be evident that we might piece together functions of any variety. The explicit functional form between the nodes could be a higher-order polynomial, an exponential, or a hyperbolic cosine. No matter how smooth the function is between the nodes, the smoothness of the function overall is limited by the kinks at the nodes.

Finite elements for the little boundary value problem. Let us proceed to demonstrate how we can use piecewise linear functions to construct a basis for a Ritz approximation, and thereby generate an approximate solution to the little boundary value problem. Consider a rod of length ℓ, with $u(0) = 0$, traction free at the right end $t_\ell = 0$, and subjected to the distributed load $b(x) = b_o$. Let us divide the length of the rod into N equal segments (elements) and label the nodes $\{x_0, \ldots, x_N\}$. The nodes are located at the points $x_n = n\ell/N$, starting with $x_0 = 0$ and ending with $x_N = \ell$. We will examine the finite element approximation for different numbers of elements.

We can construct a set of piecewise linear base functions, often called the *roof functions*, as shown in Fig. 77. The function $h_n(x)$ is zero, except in the neighborhood of the node x_n, where it ramps up to a value of one and back down again at the adjacent nodes. The nth base function has the expression

$$h_n(x) = \begin{cases} \dfrac{x - x_{n-1}}{x_n - x_{n-1}}, & x_{n-1} \le x \le x_n \\ \dfrac{x_{n+1} - x}{x_{n+1} - x_n}, & x_n \le x \le x_{n+1} \\ 0, & \text{elsewhere} \end{cases} \tag{333}$$

Clearly, the difference between the nodes is constant, so $x_{n+1} - x_n = \ell/N$. The first derivative of the nth base function has the explicit expression

$$h_n'(x) = \begin{cases} N/\ell, & x_{n-1} < x < x_n \\ -N/\ell, & x_n < x < x_{n+1} \\ 0, & \text{elsewhere} \end{cases} \tag{334}$$

The most interesting feature of these base functions is that each one has the value zero over much of the domain $[0, \ell]$. Furthermore, the part that is zero for

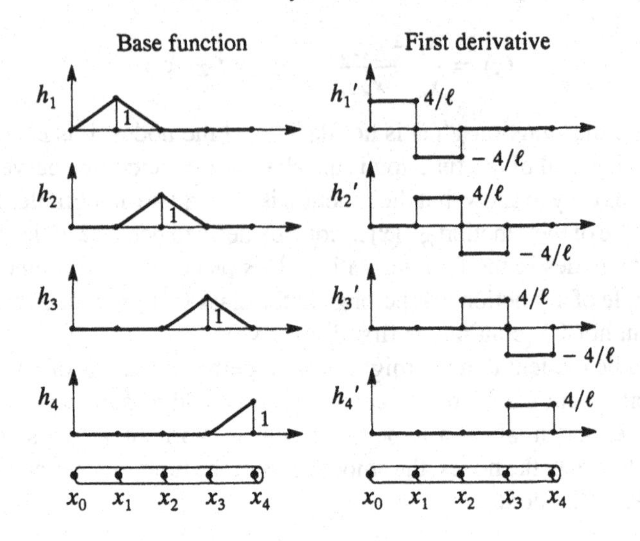

Figure 77 Four piecewise-linear Ritz base functions and their derivatives

one function is different from the part that is zero for another. Thus, the product of two functions is nonzero only over a very small region. The product of some of the functions is zero everywhere! This observation suggests that we will reap some benefits in the computation of **K** and **f** for the Ritz method. Base functions of this variety are often called *local* functions (or functions with compact support), because they are nonzero only in local regions of the domain.

The functions are also called *finite element* base functions. The notion of the "element" comes from the observation that $h_{n-1}(x)$ and $h_n(x)$ are both nonzero only for the region of the bar between the nodes at x_{n-1} and x_n. We shall call that region element n.

As we have done previously, we can use the finite element base functions to construct an approximation of the real and virtual displacement functions. These approximations have the usual form

$$u(x) = \sum_{n=1}^{N} a_n h_n(x), \qquad \bar{u}(x) = \sum_{n=1}^{N} \bar{a}_n h_n(x) \tag{335}$$

The displacement function that results from this approximation is shown schematically in Fig. 78. Since the nth base function has the value of unity at x_n and is zero at all of the other nodes, the coefficient a_n can be interpreted as the value of the displacement at that node, i.e., $a_n = u(x_n)$. Thus, the primary unknowns for the problem are the nodal displacements. Let us compute the coefficient matrices **K** and **f** using the finite element basis. The mnth component of **K** and the mth component of **f** are given by the integrals

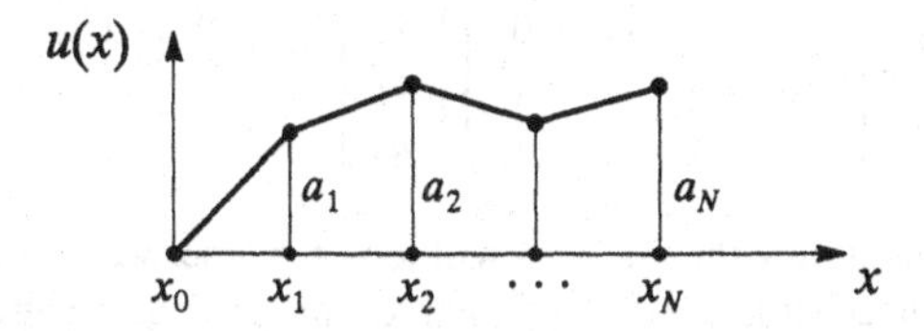

Figure 78 Displacement approximation with finite element base functions

$$K_{mn} = \int_0^\ell C h_m' h_n' \, dx, \qquad f_m = \int_0^\ell b_o h_m \, dx \tag{336}$$

These integrals are simple to compute. The main advantage is evident if we break the integral into a sum of integrals over each element as

$$K_{mn} = \sum_{i=1}^{N} \int_{x_{i-1}}^{x_i} C h_m' h_n' \, dx, \qquad f_m = \sum_{i=1}^{N} \int_{x_{i-1}}^{x_i} b_o h_m \, dx \tag{337}$$

Most of the terms in the sum will be zero. In fact, $K_{mn} = 0$ in element i for all except the four times when m or n equal $i-1$ or i. An example of the computation of $\mathbf{K}$ and $\mathbf{f}$ follows.

Example 41. *Computation of coefficients for the finite element basis.* Consider the finite element basis with N elements of equal length. Let us solve the little boundary value problem with unit length, constant modulus C, displacement restrained at $\xi = 0$, zero traction at $\xi = 1$, and with body force $b(\xi) = b_o$.

$$K_{23} = \int_0^\ell C h_2' h_3' \, dx = \int_{2\ell/N}^{3\ell/N} C\left(-\frac{N}{\ell}\right)\left(\frac{N}{\ell}\right) dx = -\frac{CN}{\ell}$$

For the diagonal elements, $K_{nn} = 2CN/\ell$ if $n \neq N$ and $K_{NN} = CN/\ell$. The off-diagonal terms K_{mn} $(m \neq n)$ are zero if $|m-n| > 1$, since the base functions this far apart have no nonzero region in common. For the terms with $|m-n| = 1$, $K_{mn} = -CN/\ell$. Since the body force $b(x)$ is constant, the integral of f_m is simply proportional to the area under the mth base function. An example computation of a component of $\mathbf{f}$ goes as follows

$$f_2 = \int_0^\ell b_o h_2 \, dx = \int_{\ell/N}^{2\ell/N} b_o\left(\frac{x-\ell/N}{\ell/N}\right) dx + \int_{2\ell/N}^{3\ell/N} b_o\left(\frac{3\ell/N-x}{\ell/N}\right) dx = \frac{b_o\ell}{N}$$

Accordingly, $f_n = b_o\ell/N$ if $n \neq N$ and $f_N = b_o\ell/2N$. You should verify that these values are correct.

The equations of equilibrium, $\mathbf{Ka} = \mathbf{f}$, for the case of $N = 4$, are

$$\frac{4C}{\ell}\begin{bmatrix} 2 & -1 & 0 & 0 \\ -1 & 2 & -1 & 0 \\ 0 & -1 & 2 & -1 \\ 0 & 0 & -1 & 1 \end{bmatrix}\begin{bmatrix} a_1 \\ a_2 \\ a_3 \\ a_4 \end{bmatrix} = \frac{b_o\ell}{2(4)}\begin{bmatrix} 2 \\ 2 \\ 2 \\ 1 \end{bmatrix} \qquad (338)$$

It should be evident how the matrices will look for other values of N. Note in particular the banded structure of the **K** matrix. This particular matrix can be inverted in closed form (not many can, so enjoy this one) as follows

$$\mathbf{K}^{-1} = \frac{\ell}{4C}\begin{bmatrix} 1 & 1 & 1 & 1 \\ 1 & 2 & 2 & 2 \\ 1 & 2 & 3 & 3 \\ 1 & 2 & 3 & 4 \end{bmatrix} \qquad (339)$$

The solutions to the problem for the values $N = 1,\ldots, 5$ are shown in Fig. 79.

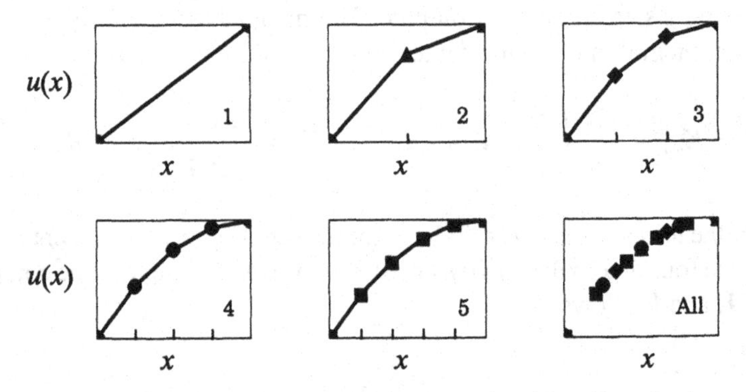

Figure 79 Little boundary value problem with uniform load. Results for five different finite element approximations

There are some interesting observations to be made about the Ritz solution to this problem. The first observation is that the nodal displacements $a_n = u(x_n)$ all turn out to be exactly correct (as shown in the previous chapter, the exact solution to the problem is $u(x) = b_o(2x\ell - x^2)/2C$). We can see this clearly in Fig. 79 in the plot labeled "All," which superimposes the solutions for all five cases. We can compute the stresses at the two ends from the approximate solution as

$$\sigma_N(0) = \left(\frac{2N-1}{N}\right)\frac{b_o\ell}{2}, \qquad \sigma_N(\ell) = \frac{b_o\ell}{2N}$$

where $\sigma_N(x)$ is the approximate stress field for a finite element approximation of order N. In the limit as $N \to \infty$, both of these values converge to their proper limits of $\sigma(0) = b_o\ell$ and $\sigma(\ell) = 0$. It is not really fair to compare the performance of the finite element basis with the polynomial basis $\{x, x^2, \ldots, x^N\}$, with each base function defined on the entire region, because we know that the exact solution is quadratic for the present example. Hence, the stresses would be exact for the polynomial basis for $N \geq 2$. For $N = 5$, the tractions at the ends are still in error by 10% for the finite element approximation.

One of the principal advantages of the finite element basis is that the functions are fairly close to being orthogonal. For example, if $|m-n| > 1$, then the functions and their first derivatives are exactly orthogonal. Adjacent functions, however, are not orthogonal. The consequence of this "near orthogonality" is that we can expect the **K** matrix to be fairly well-conditioned. The condition numbers for the five approximations of lowest order are given in Table 5. Note the slow growth of the condition number with increasing order of approximation.

Table 5 Condition of **K** for the example problem, finite element basis

N	μ_{max}	μ_{min}	$\varrho(\mathbf{K})$
1	1.00000	1.000000	1.00
2	2.61803	0.381966	6.85
3	3.24698	0.198062	16.39
4	3.53209	0.120615	29.28
5	3.68251	0.081014	45.46

Non-homogenous boundary conditions. Another great advantage of the finite element base functions that is not really amplified by the one-dimensional example is the simplicity of satisfying the essential boundary conditions. Because the functions are local, all of the interior functions automatically satisfy homogeneous essential boundary conditions. The only base functions that interact with the boundary are those associated with elements adjacent to the boundary. To satisfy a nonzero essential boundary condition $u(0) = u_o$, we need to include the known function $\hat{u}(x)$ in the Ritz approximation of $u(x)$, in accord with Eqn. (306). The boundary condition can be satisfied with the local function shown in Fig. 80.

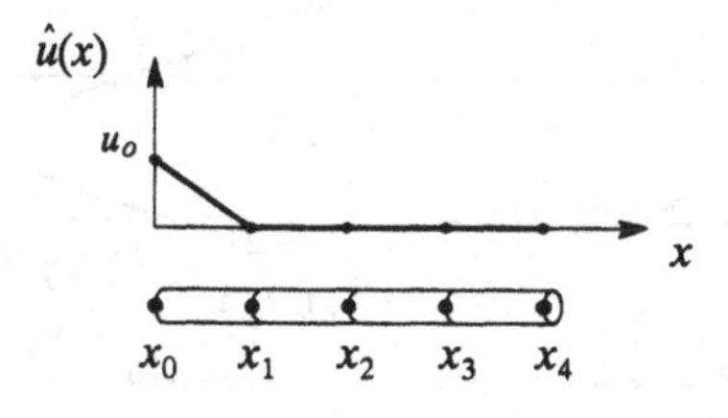

Figure 80 Function for nonzero boundary displacement

Lagrangian finite element base functions. We can produce higher-order finite element base functions by the same reasoning that produced the piecewise linear functions. Consider the problem of passing a quadratic function through the three points x_k, x_m, and x_n that is zero at the points x_m and x_n and has unit value at the point x_k. Lagrange offered the following function (and its derivative)

$$\mathcal{L}_2(x) = \frac{(x-x_m)(x-x_n)}{(x_k-x_m)(x_k-x_n)}, \qquad \mathcal{L}_2'(x) = \frac{(x-x_m)+(x-x_n)}{(x_k-x_m)(x_k-x_n)}$$

Functions of this variety are often called ***quadratic Lagrangian functions***. In this form it is obvious that (1) the function is quadratic, (2) the values are zero at x_m and x_n and (3) the function has unit value at x_k. In fact, it should be evident how to create a general nth order Lagrangian function that passes through zero at $n-1$ specified points x_i and has unit value at one specified point x_k

$$\mathcal{L}_n(x) = \prod_{i=1,\, i\neq k}^{n-1} \frac{x-x_i}{x_k-x_i}$$

We can use the quadratic Lagrangian functions to form a finite element basis. In this case we need three nodes to describe the quadratic variation of each Lagrangian segment. As such, each "element" will have three nodes. We will piece the quadratic functions together to form a Ritz basis in a manner that assures continuity of the base functions but not their derivatives. Hence, the basis is $\mathcal{C}^0$ just like the piecewise linear finite element base functions are. In a global sense, the quadratic base functions are not smoother than the linear ones because they both have kinks at the inter-element boundaries.

Let us demonstrate the idea by producing base functions for a rod of length ℓ divided into four segments that constitute two quadratic elements. The base functions and their derivatives are shown in Fig. 81. Observe that there are three basic curve shapes that make up the base functions (i.e., ones with unit value on the left, one with unit value in the middle, and ones with unit value on the right). To make a continuous base function we must piece together the

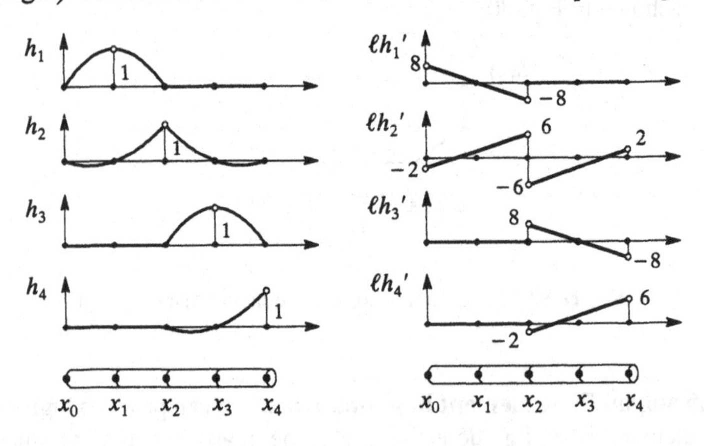

Figure 81 Quadratic Lagrangian finite element base functions and derivatives

functions as shown. Observe that the base functions now overlap for three adjacent functions, which means that the band width of **K** will be five compared with three for the linear finite element basis.

The explicit expressions for the base functions for the case with $\ell = 1$ are

$$h_1 = \begin{cases} -16x^2+8x & 0 \le x \le 1/2 \\ 0 & 1/2 \le x \le 1 \end{cases}$$

$$h_2 = \begin{cases} 8x^2-2x & 0 \le x \le 1/2 \\ 8x^2-14x+6 & 1/2 \le x \le 1 \end{cases}$$

$$h_3 = \begin{cases} 0 & 0 \le x \le 1/2 \\ -16x^2+24x-8 & 1/2 \le x \le 1 \end{cases}$$

$$h_4 = \begin{cases} 0 & 0 \le x \le 1/2 \\ 8x^2-10x+3 & 1/2 \le x \le 1 \end{cases}$$

The derivatives of these functions are straightforward to compute. Application of the quadratic Lagrangian base functions is left as an exercise for the reader (see Problem 152).

Finite element shape functions and automatic assembly of equations. There are some practical merits of the finite element method that are not completely evident from the definition of the base functions in the Ritz context. Let us illustrate the key idea behind implementation of the finite element method using the one-dimensional roof functions described previously. In particular, consider the setup shown in Fig. 82(a), which shows the typical situation involving the region of element "e" (i.e., the domain $x_i \le x \le x_j$ with i and j being two adjacent nodes in the mesh). The only base functions that contribute to element "e" are h_i (which is ramping down in that region) and h_j (which is ramping up in that region). All of the other base functions are zero in the domain of element "e". The parts of the base functions that contribute to element "e" are shown as the darker line in the sketch.

One can observe that all of the roof base functions can be built from the *shape functions* shown in Fig. 82(b). The shape functions have the specific, and very simple, expressions

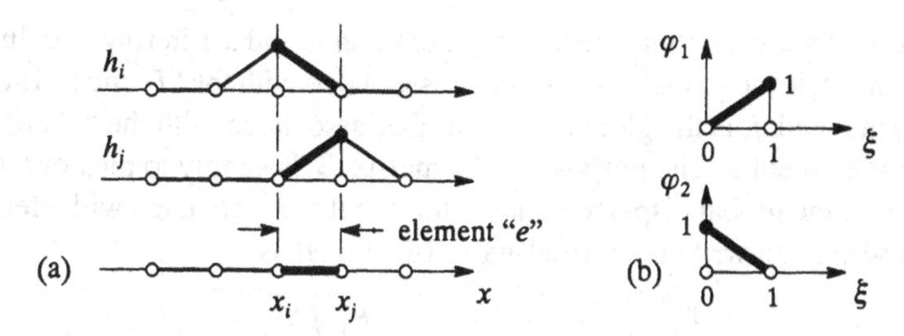

Figure 82 Relationship between the finite element base functions and the finite element shape functions

$$\varphi_1(\xi) = \xi, \qquad \varphi_2(\xi) = 1-\xi \tag{340}$$

We can use φ_1 and φ_2 to build the base functions h_i and h_j by a simple change of variable to shift and stretch ξ to cover x over the region of the element. In particular, we can let

$$\xi = \frac{x-x_i}{x_j-x_i} = \frac{1}{\ell_e}(x-x_i), \qquad dx = \ell_e d\xi \tag{341}$$

where ℓ_e is the length of the element. When $x = x_i$ then $\xi = 0$ (i.e., the left end of the element) and when $x = x_j$ then $\xi = 1$ (i.e., the right end of the element). Now we can think of the base function h_i as

$$h_i = \begin{cases} \varphi_1, & x \in e_{i-1} \\ \varphi_2, & x \in e_i \\ 0, & \text{elsewhere} \end{cases} \tag{342}$$

where e_{i-1} indicates the element to the left of node i and e_i indicates the element to the right of node i. With these definitions we can set up the discretization of the virtual-work functional a little differently. Let us define the matrix

$$\boldsymbol{\varphi} \equiv [\varphi_1, \varphi_2]^T \tag{343}$$

and note that, by the chain rule (and with the convention that a prime means differentiation of a function with respect to its argument) we have

$$\frac{d\boldsymbol{\varphi}}{dx} = \frac{d\boldsymbol{\varphi}}{d\xi}\frac{d\xi}{dx} = \frac{1}{\ell_e}\boldsymbol{\varphi}' \tag{344}$$

In particular, we can note that in the region of element e the real and virtual displacement fields can be written as

$$u_e = \boldsymbol{\varphi}^T \mathbf{B}_e^T \mathbf{a}, \qquad \bar{u}_e = \boldsymbol{\varphi}^T \mathbf{B}_e^T \bar{\mathbf{a}} \tag{345}$$

where $\mathbf{a} \equiv [a_1, \ldots, a_N]^T$ and $\bar{\mathbf{a}} \equiv [\bar{a}_1, \ldots, \bar{a}_N]^T$ are arrays containing the nodal unknowns and their virtual counterparts and

$$\mathbf{B}_e^T \equiv \begin{bmatrix} 0 & \cdots & \overset{i_e}{1} & \overset{j_e}{0} & \cdots & 0 \\ 0 & \cdots & 0 & 1 & \cdots & 0 \end{bmatrix} \tag{346}$$

is a $2 \times N$ matrix with a one in row 1, column i_e and a 1 in row 2, column j_e. Note that i_e is the global node number associated with the "i" end (left end) of element e and j_e is the global node number associated with the "j" end (right end) of element e. The purpose of the matrix $\mathbf{B}_e$ is simply to pick out the two entries in the global displacement vector $\mathbf{a}$ that are associated with element e.

Now we can write the virtual-work functional as

$$G(u,\bar{u}) = \int_0^\ell \left(Cu'\bar{u}' - b\bar{u}\right) dx = \sum_{e=1}^{M} \int_{x_{i_e}}^{x_{j_e}} \left(Cu'_e\bar{u}'_e - b\bar{u}_e\right) dx \tag{347}$$

Using the change of variable for each element and substituting Eqn. (345) we can write the discrete virtual-work functional in the form

$$G(\mathbf{a},\bar{\mathbf{a}}) = \bar{\mathbf{a}}^T \sum_{e=1}^{M} \left[\mathbf{B}_e \int_0^1 \boldsymbol{\varphi}' \frac{C}{\ell_e} \boldsymbol{\varphi}'^T \, d\xi \; \mathbf{B}_e^T \mathbf{a} - \mathbf{B}_e \int_0^1 \boldsymbol{\varphi} b \ell_e \, d\xi \right] \tag{348}$$

Note that in the first term each differentiation threw off a $1/\ell_e$, as indicated in Eqn. (344), and the change of variable gave $dx = \ell_e d\xi$. We can write this result more compactly if we identify the *element stiffness matrix* and the *element force* as

$$\mathbf{k}_e \equiv \int_0^1 \boldsymbol{\varphi}' \frac{C}{\ell_e} \boldsymbol{\varphi}'^T \, d\xi, \quad \mathbf{f}_e \equiv \int_0^1 \boldsymbol{\varphi} b \ell_e \, d\xi \tag{349}$$

Note that in the present application $\mathbf{k}_e$ is 2×2 and $\mathbf{f}_e$ is 2×1. Now the discrete virtual-work functional takes the simple form

$$G(\mathbf{a},\bar{\mathbf{a}}) = \bar{\mathbf{a}}^T \left[\sum_{e=1}^{M} \mathbf{B}_e \mathbf{k}_e \mathbf{B}_e^T \mathbf{a} - \sum_{e=1}^{M} \mathbf{B}_e \mathbf{f}_e \right] \tag{350}$$

Comparison with our earlier results shows that the stiffness matrix and right side vectors are computed as

$$\mathbf{K} = \sum_{e=1}^{M} \mathbf{B}_e \mathbf{k}_e \mathbf{B}_e^T, \quad \mathbf{f} = \sum_{e=1}^{M} \mathbf{B}_e \mathbf{f}_e \tag{351}$$

The summations over the elements are often called the *assembly process*. This calculation is seldom done with an explicit matrix multiplication. In fact, the matrices $\mathbf{B}_e$ are not even explicitly formed. Rather, for each element we record the global node numbers associated with the element. Let the array **ix** have N rows and 2 columns. We put $\mathbf{ix}(e,1) = i_e$ and $\mathbf{ix}(e,2) = j_e$, the global node numbers associated with the left and right end of the element, respectively. let the array **id** have N rows and 1 column. Let $\mathbf{id}(n)$ be the global equation number for node n. The MATLAB code given in Table 6 gives the algorithm for direct assembly of the equations (i.e., the assembly of **K** and **f**). Note that this code assumes there are N unknowns and M elements. It also assumes that there is a routine to call to get the element stiffness matrix and element force.

Example 42. *Computation of element stiffness matrix and force vector.* Consider an element of length ℓ_e with constant modulus C and constant body force b. Let us compute the element matrices according to Eqn. (349). First note that the matrix $\boldsymbol{\varphi}' = [1, -1]$, which is constant. Carrying out the integrals we get

$$\mathbf{k}_e = \frac{C}{\ell_e} \begin{bmatrix} 1 & -1 \\ -1 & 1 \end{bmatrix} \qquad \mathbf{f}_e = \frac{b\ell_e}{2} \begin{bmatrix} 1 \\ 1 \end{bmatrix}$$

Table 6 MATLAB code for assembly process

```
    K = zeros(N,N); f = zeros(N,1);

%.. Loop over all elements to assemble K and f
      for n = 1:M

%.... Find the i-node, j-node
        inode = ix(n,1); jnode = ix(n,2);

%.... Construct the assembly pointer array
        ii(1) = id(inode); ii(2) = id(jnode);

%.... Retrieve element stiffness matrix for element "n"
        [ke,fe] = get stiffness (...)

%.... Assemble element stiffness and force vector
        for i=1:2
          for j=1:2
            K(ii(i),ii(j)) = K(ii(i),ii(j)) + ke(i,j);
          end % loop on j
          f(ii(i)) = f(ii(i)) + fe(i);
        end % loop on i

      end % loop on n
```

The Ritz Method for Two- and Three-dimensional Problems

We can, of course, make the same sort of Ritz approximation of the displacements in a three-dimensional problem that we did for the one-dimensional problem, but now the specification of the base functions is considerably more complicated. The main complicating factor is the specification of appropriate base functions for irregularly shaped domains. Hence, even though the spirit of approximation is the same, we will seldom try to compute in this fashion. The finite element form of the base functions will turn out to be much better suited for performing these approximate calculations for three-dimensional solid bodies. The above warning notwithstanding, let us see how similar the Ritz method appears in three dimensions. Let the displacement field $\mathbf{u}(\mathbf{x})$ and the virtual displacement field $\bar{\mathbf{u}}(\mathbf{x})$ be approximated by three-dimensional vector base functions $\{\mathbf{h}_1(\mathbf{x}), \ldots, \mathbf{h}_N(\mathbf{x})\}$ as

$$\mathbf{u}(\mathbf{x}) = \hat{\mathbf{u}}(\mathbf{x}) + \sum_{n=1}^{N} a_n \mathbf{h}_n(\mathbf{x}), \qquad \bar{\mathbf{u}}(\mathbf{x}) = \sum_{n=1}^{N} \bar{a}_n \mathbf{h}_n(\mathbf{x}) \tag{352}$$

where, as before, the scalar constants $\mathbf{a} = [a_1, \ldots, a_N]$ are unknown and the scalar constants $\bar{\mathbf{a}} = [\bar{a}_1 \ldots, \bar{a}_N]$ are arbitrary. The vector functions $\mathbf{h}_n(\mathbf{x})$ have the same vector character as the displacement fields, and they are chosen to satisfy the homogeneous essential boundary conditions $\mathbf{h}_n(\mathbf{x}) = \mathbf{0}$ on Ω_u. Computing the divergence and gradient of $\mathbf{h}_n(\mathbf{x})$ is straightforward.

Two-dimensional membrane problem. An example of a two-dimensional problem is the stretched membrane under lateral load. The virtual-work functional for this problem has the expression

$$G(u,\bar{u}) = \int_{\Omega} \left(T\,\nabla u \cdot \nabla \bar{u} - p\bar{u}\right) dA \tag{353}$$

where T is the tension in the membrane, $u(x_1, x_2)$ is the transverse deflection, and p is the transverse load. Let the real and virtual displacements be

$$u(\mathbf{x}) = \sum_{n=1}^{N} a_n h_n(\mathbf{x}), \qquad \bar{u}(\mathbf{x}) = \sum_{n=1}^{N} \bar{a}_n h_n(\mathbf{x}) \tag{354}$$

Then the stiffness matrix $\mathbf{K}$ and force vector $\mathbf{f}$ have components

$$K_{ij} = \int_{\Omega} T\,\nabla h_i \cdot \nabla h_j \, dA, \qquad f_i = \int_{\Omega} p h_i \, dA \tag{355}$$

The Ritz approximation leads to the usual discrete version of the functional G given by

$$G(\mathbf{a},\bar{\mathbf{a}}) = \bar{\mathbf{a}}^T\left(\mathbf{K}\mathbf{a} - \mathbf{f}\right) \tag{356}$$

just as it did for the one-dimensional boundary value problem. The discrete version of the principle of virtual work suggests that $\mathbf{a}$ represents an equilibrium configuration if and only if $G(\mathbf{a},\bar{\mathbf{a}}) = 0$ for all $\bar{\mathbf{a}}$, just as it did for the one-dimensional problem. Again, there is no restriction on $\bar{\mathbf{a}}$. Therefore, the principle of virtual work implies $\mathbf{K}\mathbf{a} = \mathbf{f}$. In the jargon of structural analysis, the matrix $\mathbf{K}$ is often referred to as the *stiffness matrix*, and the vector $\mathbf{f}$ is often referred to as the *load vector.*

Example 43. *Membrane problem.* Consider a square stretched membrane of unit length on each side with tension T and lateral load p. Approximate the deflection with the expression $u(x_1, x_2) = a \sin \pi x_1 \sin \pi x_2$ with a similar approximation for the virtual deflection. Estimate the deflection.

There is only one term in the expansion so there is only one base function. Hence, $h_1(x_1, x_2) = \sin \pi x_1 \sin \pi x_2$ and we can compute the gradient as

$$\nabla h_1 = \begin{bmatrix} \pi \cos \pi x_1 \sin \pi x_2 \\ \pi \sin \pi x_1 \cos \pi x_2 \end{bmatrix}$$

Letting $\theta = \pi x_1$ and $\psi = \pi x_2$, the stiffness can be computed as

$$K = \int_0^1 \int_0^1 T\pi^2 \left(\cos^2 \pi x_1 \sin^2 \pi x_2 + \cos^2 \pi x_2 \sin^2 \pi x_1\right) dx_1 dx_2$$

$$= T \int_0^\pi \int_0^\pi \left(\cos^2 \theta \sin^2 \psi + \cos^2 \psi \sin^2 \theta\right) d\theta \, d\psi$$

$$= 2T \int_0^\pi \cos^2 \theta \, d\theta \int_0^\pi \sin^2 \psi \, d\psi = \frac{\pi^2 T}{2}$$

Similarly, the force can be computed as

$$f = \int_0^1 \int_0^1 p \sin \pi x_1 \sin \pi x_2 \, dx_1 dx_2$$

$$= p \int_0^\pi \sin \theta \, d\theta \int_0^\pi \sin \psi \, d\psi = 4p$$

Therefore, the coefficient a must be $a = f/K = 4p/\left(\pi^2 T/2\right) = 8p/\pi^2 T$. Thus, the approximate deflection is

$$u(x_1, x_2) = \frac{8p}{\pi^2 T} \sin \pi x_1 \sin \pi x_2$$

Finite element interpolation in two dimensional problems. The concept of the finite element basis can be extended to two- and three-dimensional domains. In fact, the finite element method really comes into its own for these problems because of the difficulty of establishing a Ritz basis with non-compact functions, particularly for irregular domains. The membrane problem provides a nice illustration of the generalization of the roof functions to two dimensions because the transverse displacement of the membrane is a scalar unknown.

Figure 83 shows a square region divided into 9 elements and 16 nodes (with the numbering convention shown in Fig. 83(d). A typical finite element base function (the generalization of the roof function to two dimensions) is shown in Fig. 83(a). The function has unit value at node 7 and is zero at all of the other nodes. The variation is bilinear in the four elements associated with node 7. The finite element base functions can be built from four finite elements shape functions, one of which is shown in Fig. 83(b). The shape functions are defined on a unit square element shown in Fig. 83(c). The four element shape functions for two dimensional problems are†

$$\begin{aligned} \varphi_1 &= (1-\xi)(1-\eta) \qquad & \varphi_2 &= \xi(1-\eta) \\ \varphi_3 &= (1-\xi)\eta \qquad & \varphi_4 &= \xi\eta \end{aligned} \tag{357}$$

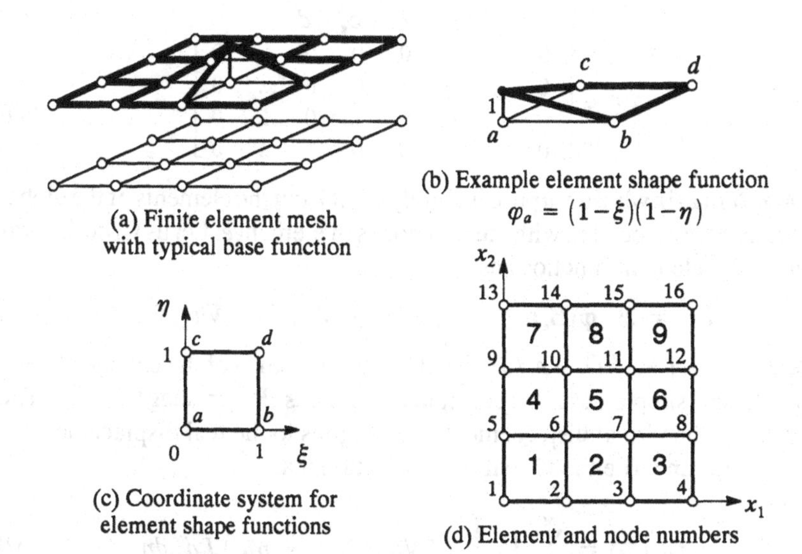

Figure 83 Illustration of finite element basis and finite element shape functions for membrane problem

The element domain can be mapped to the unit square through a bilinear function much like we did for the one-dimensional case. For the rectangular mesh shown, the change of variable can be expressed as

$$\xi = \frac{x_1 - x_1^a}{x_1^b - x_1^a}, \quad \eta = \frac{x_2 - x_2^a}{x_2^c - x_2^a}, \quad dx_1 dx_2 = \ell_1 \ell_2 d\xi \, d\eta \tag{358}$$

where ℓ_1 and ℓ_2 are the actual element dimensions. A more general element coordinate (*isoparametric*) mapping can be accomplished with the element shape functions (see any book on finite elements).

With the element shape functions we can describe the element displacement like we did for the one-dimensional problem. Let $\boldsymbol{\varphi} \equiv [\varphi_1, \varphi_2, \varphi_3, \varphi_4]^T$. In the region of element e the real and virtual displacement fields can be written as

$$u_e = \boldsymbol{\varphi}^T \mathbf{B}_e^T \mathbf{a}, \qquad \overline{u}_e = \boldsymbol{\varphi}^T \mathbf{B}_e^T \overline{\mathbf{a}} \tag{359}$$

where $\mathbf{a} \equiv [a_1, \ldots, a_N]^T$ and $\overline{\mathbf{a}} \equiv [\overline{a}_1, \ldots, \overline{a}_N]^T$ are arrays containing the nodal unknowns and their virtual counterparts and

† Note that we have associated the element node labels *a, b, c, d* with the numbers 1, 2, 3, 4 so that we can index them numerically. The figure seems clearer with the alphabetic labels. In a computational setting the numerical indexing is better.

$$\mathbf{B}_e^T \equiv \begin{array}{c} \\ \left[\begin{array}{cccccccc} & & a_e & b_e & c_e & d_e & & \\ 0 & \dots & 1 & 0 & 0 & 0 & \dots & 0 \\ 0 & \dots & 0 & 1 & 0 & 0 & \dots & 0 \\ 0 & \dots & 0 & 0 & 1 & 0 & \dots & 0 \\ 0 & \dots & 0 & 0 & 0 & 1 & \dots & 0 \end{array} \right] \end{array} \tag{360}$$

is a $4 \times N$ matrix whose purpose is simply to pick out the elements of the global vector that are associated with the four nodes of element e. Let us write the gradient of the element functions as

$$\nabla u_e = \mathfrak{B}^T(\boldsymbol{\varphi})\mathbf{B}_e^T\mathbf{a}, \qquad \mathfrak{B}^T(\boldsymbol{\varphi}) = \mathbf{J}\left[\nabla\varphi_1, \nabla\varphi_2, \nabla\varphi_3, \nabla\varphi_4\right] \tag{361}$$

where $\nabla\varphi_i = [\partial\varphi_i/\partial\xi,\ \partial\varphi_i/\partial\eta]^T$ and $\mathbf{J}$ is the Jacobian of the change of variables. In the simple case of rectangular elements $\mathbf{J}^{-1} = \operatorname{diag}[\ell_1,\ \ell_2]$. The gradient of the virtual displacement is analogous to the real displacement.

Now we can write the virtual-work functional as

$$G(u, \bar{u}) = \sum_{e=1}^{M} \int_0^1 \int_0^1 \left(T\,\nabla u_e \cdot \nabla \bar{u}_e - p\,\bar{u}_e\right) J\, d\xi\, d\eta \tag{362}$$

where $J \equiv \ell_1 \ell_2$. Substituting Eqn. (361) into the virtual-work functional gives a discrete functional identical to Eqn. (350) if we define the element stiffness matrix and force vector as

$$\mathbf{k}_e \equiv \int_0^1 \int_0^1 T\,\mathfrak{B}(\boldsymbol{\varphi})\mathfrak{B}^T(\boldsymbol{\varphi})\, J\, d\xi\, d\eta, \quad \mathbf{f}_e \equiv \int_0^1 \int_0^1 p\,\boldsymbol{\varphi}\, J\, d\xi d\eta \tag{363}$$

The element stiffness matrix is 4×4; the force vector is 4×1. The assembly of the element matrices into the global $\mathbf{K}$ and $\mathbf{f}$ matrices is identical to the one dimensional problem, except that now the **ix** must have 4 columns to record the global node number associated with the four element nodes a, b, c, d.

Three-dimensional elasticity. The virtual-work functional for three-dimensional elasticity is

$$\begin{aligned} G(\mathbf{u}, \bar{\mathbf{u}}) = & \int_{\mathfrak{B}} \left(\lambda\left(\operatorname{div}\mathbf{u}\right)\left(\operatorname{div}\bar{\mathbf{u}}\right) + \mu\left[\nabla\mathbf{u} + \nabla\mathbf{u}^T\right] \cdot \nabla\bar{\mathbf{u}}\right) dV \\ & - \int_{\mathfrak{B}} \mathbf{b} \cdot \bar{\mathbf{u}}\, dV - \int_{\Omega_t} \hat{\mathbf{t}} \cdot \bar{\mathbf{u}}\, dA \end{aligned} \tag{364}$$

We can substitute the approximate displacements from Eqn. (352) to establish the discrete principle of virtual work for this theory. Let us define the N by N matrix $\mathbf{K}$ to have components

$$K_{ij} = \int_{\mathfrak{B}} \left[\lambda\left(\operatorname{div}\mathbf{h}_i\right)\left(\operatorname{div}\mathbf{h}_j\right) + \mu\left(\nabla\mathbf{h}_i + \nabla\mathbf{h}_i^T\right) \cdot \nabla\mathbf{h}_j\right] dV \tag{365}$$

and the N by 1 matrix $\mathbf{f}$ to have components (assuming that $\hat{\mathbf{u}}(\mathbf{x}) = \mathbf{0}$)

$$f_j = \int_{\Omega_t} \mathbf{t} \cdot \mathbf{h}_j \, dA + \int_{\mathfrak{B}} \mathbf{b} \cdot \mathbf{h}_j \, dV \tag{366}$$

The equations of structural analysis are generally formulated in a manner that gives the unknowns $\mathbf{a}$ the character of nodal displacements. We can see that this interpretation is not always appropriate here because the unknowns $\mathbf{a}$ are simply the coefficients of the base function expansion for the displacement field. There are base functions $\mathbf{h}_n(\mathbf{x})$ that give the unknowns the character of nodal displacements, for example, the finite element base functions. The coefficients $\mathbf{a}$ are often called *generalized displacements* because the displacement at any point is a linear combination of these constants.

Example 44. Consider the $2\ell \times 2\ell \times h$ block, fixed at the base, i.e., with $\mathbf{u}(x_1, x_2, 0) = \mathbf{0}$, and subjected to the uniform traction $\mathbf{t} = -q_0 \mathbf{e}_3$ along its top surface shown in Fig. 84.

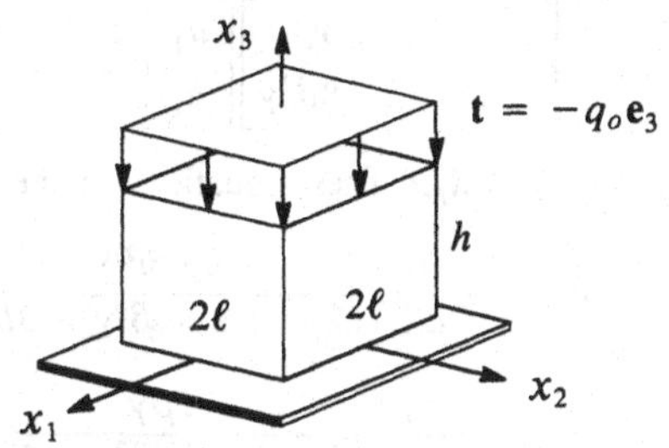

Figure 84 Example 44: block subjected to surface tractions

We shall assume that the material has elastic constants λ and μ. Let us find an approximation to the deformation map induced by the loading with a Ritz approximation to the elasticity equations. Let us assume a displacement field

$$\mathbf{u}(\mathbf{x}) = a_1\left(x_3 \ell \, \mathbf{e}_3\right) + a_2\left(x_1 x_3 \mathbf{e}_1 + x_2 x_3 \mathbf{e}_2\right)$$

This two-term approximation is clearly a crude one, but should suffice to illustrate the computations involved in the three-dimensional Ritz approach. The approximate displacement field satisfies the essential boundary condition. The first term allows linearly varying motion in the x_3 direction. The second term allows motions in the $x_1 - x_2$ plane that increase linearly with x_3. At the very least, this map contains the feature of restraining lateral motion at the base, as the fixity requires. Let us compute the response.

The base functions are $\mathbf{h}_1(\mathbf{x}) = x_3 \ell \, \mathbf{e}_3$ and $\mathbf{h}_2(\mathbf{x}) = x_1 x_3 \mathbf{e}_1 + x_2 x_3 \mathbf{e}_2$. For our computations we need the divergence and gradient of these vector functions

$$\begin{aligned} \operatorname{div} \mathbf{h}_1(\mathbf{x}) &= \ell & \qquad \nabla \mathbf{h}_1(\mathbf{x}) &= \ell \, \mathbf{e}_3 \otimes \mathbf{e}_3 \\ \operatorname{div} \mathbf{h}_2(\mathbf{x}) &= 2x_3 & \qquad \nabla \mathbf{h}_2(\mathbf{x}) &= x_3 \mathbf{e}_1 \otimes \mathbf{e}_1 + x_3 \mathbf{e}_2 \otimes \mathbf{e}_2 \\ &&&\quad + x_1 \mathbf{e}_1 \otimes \mathbf{e}_3 + x_2 \mathbf{e}_2 \otimes \mathbf{e}_3 \end{aligned}$$

Now let us compute the terms $\left[\nabla \mathbf{h}_i + \nabla \mathbf{h}_i^T\right] \cdot \nabla \mathbf{h}_j$ required for the formation of $\mathbf{K}$ in accord with Eqn. (365)

$$\left[\nabla \mathbf{h}_1 + \nabla \mathbf{h}_1^T\right] \cdot \nabla \mathbf{h}_1 = 2\ell^2$$

$$\left[\nabla \mathbf{h}_1 + \nabla \mathbf{h}_1^T\right] \cdot \nabla \mathbf{h}_2 = \left[\nabla \mathbf{h}_2 + \nabla \mathbf{h}_2^T\right] \cdot \nabla \mathbf{h}_1 = 0$$

$$\left[\nabla \mathbf{h}_2 + \nabla \mathbf{h}_2^T\right] \cdot \nabla \mathbf{h}_2 = x_1^2 + x_2^2 + 4x_3^2$$

The components of the matrix $\mathbf{K}$ are given by

$$K_{ij} = \int_0^h \int_{-\ell}^{\ell} \int_{-\ell}^{\ell} \left(\lambda \operatorname{div} \mathbf{h}_i \operatorname{div} \mathbf{h}_j + \mu \left[\nabla \mathbf{h}_i + \nabla \mathbf{h}_i^T\right] \cdot \nabla \mathbf{h}_j\right) dx_1\, dx_2\, dx_3$$

and the components of $\mathbf{f}$ are given by

$$f_i = \int_{-\ell}^{\ell} \int_{-\ell}^{\ell} -q_o \mathbf{e}_3 \cdot \mathbf{h}_i(x_1, x_2, h)\, dx_1\, dx_2$$

Carrying out the indicated integrations gives the following system of equations for the unknown coefficients a_1 and a_2

$$\tfrac{4}{3}\mu h \ell^4 \begin{bmatrix} 3(2+\gamma) & 3\beta\gamma \\ 3\beta\gamma & 2+8\beta^2\gamma \end{bmatrix} \begin{bmatrix} a_1 \\ a_2 \end{bmatrix} = -4 q_o h \ell^3 \begin{bmatrix} 1 \\ 0 \end{bmatrix}$$

where $\beta \equiv h/\ell$ and $\gamma \equiv \lambda/\mu$. These equations can be solved to give

$$a_1 = -\frac{q_o}{\mu \ell}\left(\frac{2+8\beta^2\gamma}{(2+\gamma)(2+8\beta^2\gamma)-3\beta^2\gamma}\right)$$

$$a_2 = +\frac{q_o}{\mu \ell}\left(\frac{3\beta\gamma}{(2+\gamma)(2+8\beta^2\gamma)-3\beta^2\gamma}\right)$$

It is interesting to consider the axial deformation of the block for the special case $\gamma = 0$ (which is the same as $\nu = 0$). Under these circumstances, we get $a_1 = -q_o/2\mu\ell = -q_o/C\ell$ and $a_2 = 0$, where C is Young's modulus. The displacement at the top is then $u_3(x_1, x_2, h) = -q_o h/C$ which is exactly the value we would compute from elementary strength of materials.

We can also investigate the case $\beta \gg 1$ (a long, slender column). In this case, we have $a_1 \to -[8q_o(1-2\nu)(1+\nu)]/[(8-11\nu)C\ell]$ and $a_2 = 0$. For $\nu = 0$ the result is the same as before. But the Poisson effect manifests for nonzero values of Poisson's ratio. It is interesting to note that for $\nu = 1/2$ (the incompressible limit) the axial displacement goes to zero. Lateral strain is prevented at the base. If the volume cannot change then deformation (within the confines of this simple approximation) is not possible.

Clearly, we can investigate this approximate solution in greater detail to reveal that there are some tractions on the sides of the block that should be traction-free (just compute the strain from the displacement field, substitute into the constitutive equations, and apply the Cauchy relationship to the surfaces in question), but the assumed approximation appears to capture some features of the problem at hand.

We can see from the above developments that the principle of virtual work gives us a vehicle for finding a stiffness matrix and an equivalent load vector for problems other than beam and truss elements. For linear theories of beams and trusses, the stiffness matrix is often formulated by exactly solving the governing boundary value problem and using these results to find a relationship between the end forces and the end (i.e., nodal) displacements (and rotations). Since there are so few exact solutions to two- and three-dimensional problems, this approach does not work very well in higher dimensions, even for specialized theories like plates and shells. Because the virtual-work approach does not rely on the exact solution to the governing boundary value problem, it represents a powerful approach to some rather difficult problems.

There is little doubt that the finite element method is the preferred approach to implementing the Ritz approximation method. The finite element basis can be easily adapted to unusual boundary geometries, and the requisite computations can be easily organized into a general-purpose computer program. The sole purpose of the simple examples presented here is to introduce the finite element concept as a bona fide Ritz approximation. Our main concern in this book is mechanics. The ready availability of a general approach to computation is important to the study of mechanics because it helps to keep the focus on the relevance of the theories that we encounter along the way.

The student of mechanics should not approach every new problem or theory wondering whether one of the virtuosos of mechanics has managed to find a solution for a particular problem. The Ritz approximation provides this positive context, and we shall exploit it throughout the remainder of the text. For this purpose, it is not productive to quibble over what is the best set of base functions to use. Rather, keep in mind that there are many alternatives available. For the simple problems that we solve in this book, the choice of the basis is usually secondary, and we will often use the simple polynomial basis.

Additional Reading

I. Fried, *Numerical solution of differential equations*, Academic Press. New York, 1979.

T. J. R. Hughes, *The finite element method: Linear static and dynamic finite element analysis*, Prentice Hall, Englewood Cliffs, N.J., 1987.

H. L. Langhaar, *Energy methods in applied mechanics*, Wiley, New York, 1962.

A. R. Mitchell and R. Wait, *The finite element method in partial differential equations*. Wiley, New York, 1977.

G. Strang and G. J. Fix,*An analysis of the finite element method*, Prentice Hall, Englewood Cliffs, N.J., 1973.

Problems

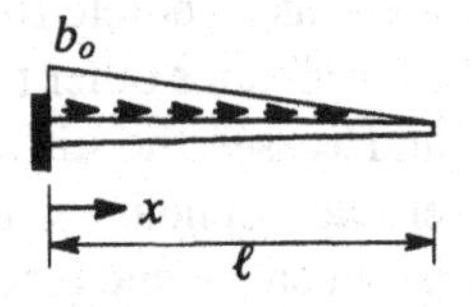

142. Consider the uniaxial rod shown in the sketch, fixed at $x = 0$, free at $x = \ell$, and subjected to the linearly varying body force indicated. The rod has a variable elastic modulus $C(x) = C_o(2-x/\ell)$, making it twice as stiff at $x = 0$ as it is at $x = \ell$. Using the principle of virtual work, find the expression for the displacement $u(x)$ and stress $\sigma(x)$ for the given body force and variable modulus, approximating the real and virtual displacements with polynomials.

143. Reconsider the nonprismatic rod of problem 142 subjected to the linearly varying body force. However, consider the condition in which the rod is fixed at both ends with prescribed end displacements of $u(0) = u_o$ and $u(\ell) = u_1$. Solve the problem with the Ritz method using polynomial base functions.

144. Consider using a basis for the virtual displacement different from the basis used for the real displacement. What would be the ramifications of using a *different number of terms* in the expansions for real and virtual displacements? That is

$$u(x) = \sum_{n=1}^{N} a_n h_n(x), \quad \bar{u}(x) = \sum_{n=1}^{M} \bar{a}_n h_n(x)$$

where $N \neq M$. What happens if $N > M$? What happens if $N < M$? Perform some computations on the little boundary value problem to investigate this issue.

145. Consider using a basis for the virtual displacement different from the basis used for the real displacement. What would be the ramifications of using *different base functions* for the real and virtual displacements? That is

$$u(x) = \sum_{n=1}^{N} a_n h_n(x), \quad \bar{u}(x) = \sum_{n=1}^{N} \bar{a}_n g_n(x)$$

where $g_n(x) \neq h_n(x)$. Perform some computations on the little boundary value problem to investigate this issue using, for example, polynomials for the real displacements and trigonometric functions for the virtual displacements.

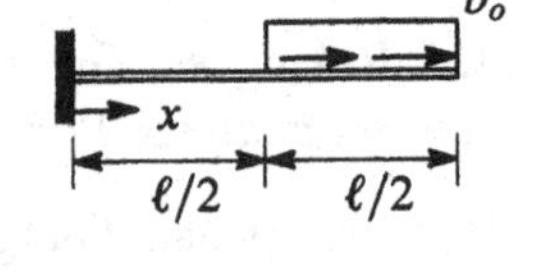

146. The uniaxial rod shown has unit area, length ℓ, and elastic modulus C. The body force is characterized by

$$b(x) = \begin{cases} 0 & 0 < x \leq \ell/2 \\ b_o & \ell/2 < x \leq \ell \end{cases}$$

Assume that the real and virtual displacements can be approximated by the expressions

$$u(\xi) = a_0 + a_1\xi + a_2\xi^2 \qquad \bar{u}(x) = \bar{a}_0 + \bar{a}_1\xi + \bar{a}_2\xi^2$$

where $\xi \equiv x/\ell$. Using the principle of virtual work, compute the displacement field $u(x)$ and the tractions at the two ends $\sigma(\ell)$ and $\sigma(0)$.

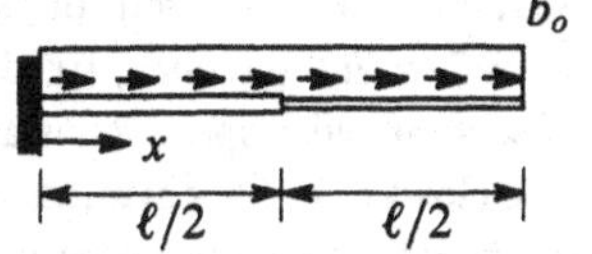

147. The uniaxial rod shown has unit area and length ℓ. It is fixed at the left end, is free at the right end, and is subjected to a constant body force field $b(x) = b_o$ along its length. The elastic modulus $C(x)$ is characterized by

$$C(x) = \begin{cases} 2C & 0 < x \leq \ell/2 \\ C & \ell/2 < x \leq \ell \end{cases}$$

Find the classical solution to the governing differential equation. Using the principle of virtual work, compute a stress field $\sigma(x)$ assuming a Ritz approximation as follows

$$u(\xi) = a_0 + a_1\xi + a_2\xi^2 \qquad \bar{u}(x) = \bar{a}_0 + \bar{a}_1\xi + \bar{a}_2\xi^2$$

where $\xi \equiv x/\ell$. Because the modulus changes abruptly at $x = \ell/2$, the stresses and strains are discontinuous at that point. Why is this discontinuity a problem for the polynomial base functions suggested? What happens if you increase the order of the approximation?

148. Using a piecewise linear finite element basis, resolve Problem 147. Does the finite element basis suffer from the same problem as the polynomial basis? Why? What general conclusions can we make about the smoothness of the approximation?

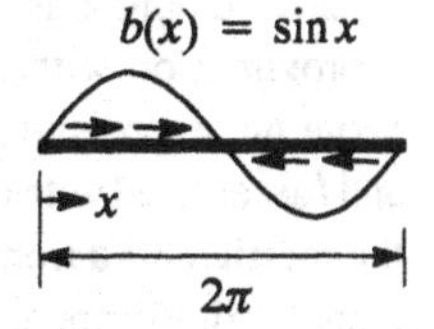

149. Consider the rod of length 2π and constant modulus $C = 1$, free at both ends and subjected to the sinusoidal body force, as shown. The general classical solution for the given loading is $u(x) = a_0 + a_1 x + \sin x$. Show that the given solution satisfies the governing differential equation for the bar, and state the essential and natural boundary conditions. Use the boundary conditions to find the integration constants a_0 and a_1. Explain any peculiar features of the solution to this problem. Use a polynomial Ritz basis to find a two-term approximate solution for the displacement field, using the principle of virtual work. Explain any peculiar features of the Ritz approximate solution.

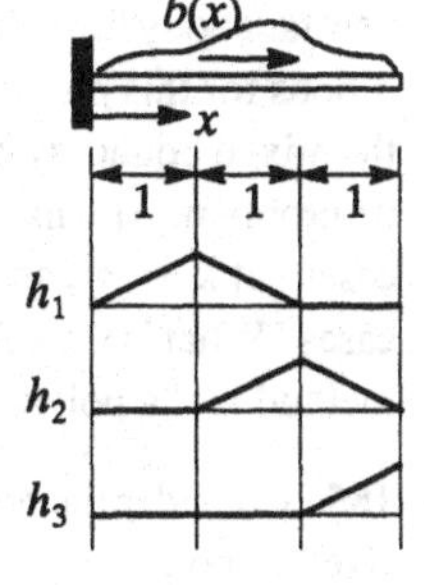

150. Consider the rod of length 3, constant unit modulus $C = 1$ (and unit area), fixed at $x = 0$. The rod is subjected to a certain (unspecified) distribution of body force b, as shown. Three piecewise linear finite element basis functions are shown in the sketch. The functional expressions for the basis function $h_i(x)$ is

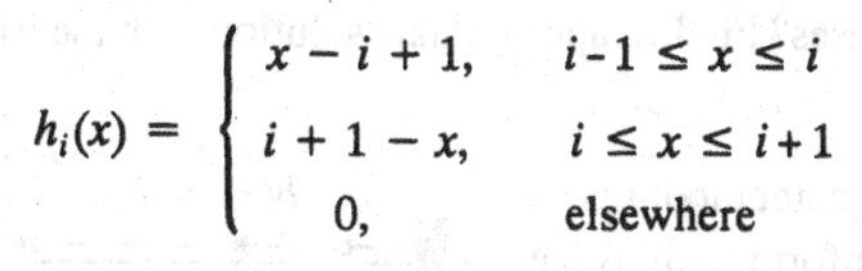

$$h_i(x) = \begin{cases} x - i + 1, & i-1 \leq x \leq i \\ i + 1 - x, & i \leq x \leq i+1 \\ 0, & \text{elsewhere} \end{cases}$$

An approximate displacement field can be constructed from the base functions as $u(x) = a_1 h_1(x) + a_2 h_2(x) + a_3 h_3(x)$. Find the stiffness matrix $\mathbf{K}$ consistent with this approximation and the principle of virtual work. Assume that the values of the coefficients are $a_1 = 1$, $a_2 = 2$, and $a_3 = 4$. Plot the stress field associated with the approximation. Find the equivalent force vector $\mathbf{f}$.

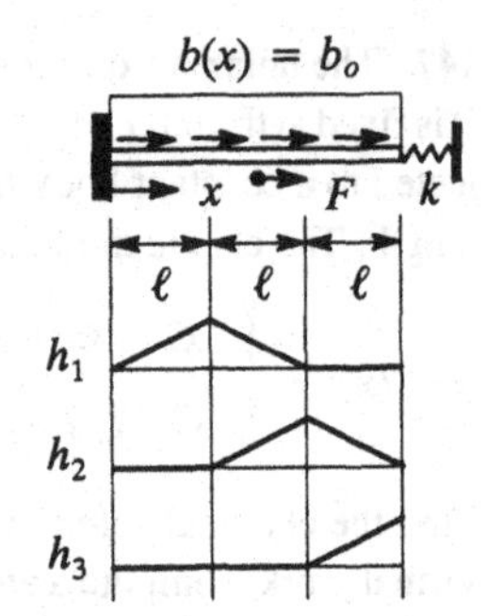

151. Consider the rod of length 3ℓ, constant modulus $C = 2k\ell$, and unit area, fixed at $x = 0$ and spring supported at $x = 3\ell$, with spring constant k. The rod is subjected to a point load $F = 3b_o\ell$ at midspan and a uniform body force b_o, as shown. Three piecewise linear finite element basis functions are shown in the sketch. The functional expressions for the basis function $h_i(x)$ is

$$h_i(x) = \begin{cases} x/\ell - i + 1, & i-1 \le x/\ell \le i \\ i + 1 - x/\ell, & i \le x/\ell \le i+1 \end{cases}$$

Set up the equilibrium equations implied by the principle of virtual work using the Ritz method (*i.e.*, find $\mathbf{K}$ and $\mathbf{f}$). Express your answer in terms of k, b_o, and ℓ (not F and C).

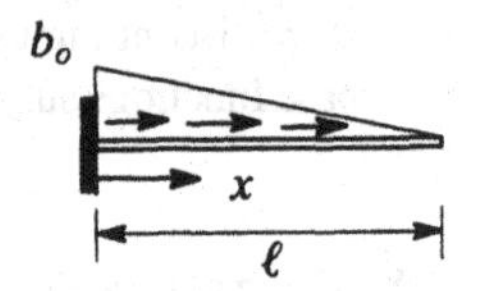

152. Solve the problem of the rod subjected to a triangular load shown in the sketch using the quadratic Lagrangian finite element base functions. Use at least two quadratic elements (i.e., five nodes with four segments of length $\ell/4$) to carry out the solution.

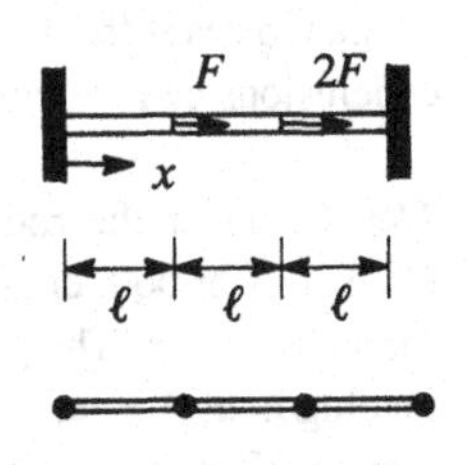

153. Consider the rod of length 3ℓ and constant modulus C, fixed at both ends, and subjected to point loads of magnitude F and $2F$ at the third points, as shown. Use a piecewise linear finite element approximation with nodes at the ends and at the third points. Write the expressions for the base functions $h_i(x)$. Compute $\mathbf{K}$ and $\mathbf{f}$ associated with the discrete virtual-work function. Compute the coefficients $\mathbf{a}$ from $\mathbf{Ka} = \mathbf{f}$. Sketch the approximate displacement field. Compute the approximate stress field.

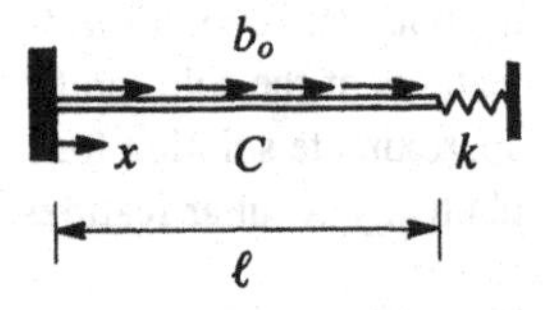

154. Consider the rod of length ℓ and constant modulus C. The rod is fixed at the left end and restrained by an elastic spring of modulus k at the right end. The spring accrues force equal to the product of spring constant and stretch of the spring, i.e., $f_s = ku(\ell)$. The rod is subjected to a constant body force $b(x) = b_o$, as shown. What are the essential and natural boundary conditions for this problem? (Hint: Take a freebody diagram of the right end of the rod to get the mixed boundary condition at that end). Find the classical solution to the boundary value problem. At what point is the strain in the rod the greatest? Consider the two limiting cases (1) $k \to \infty$, and (2) $k \to 0$. What are the boundary conditions in these two limiting cases? What is the solution in these two cases? Find an approximate solution with the Ritz method and a polynomial approximation.

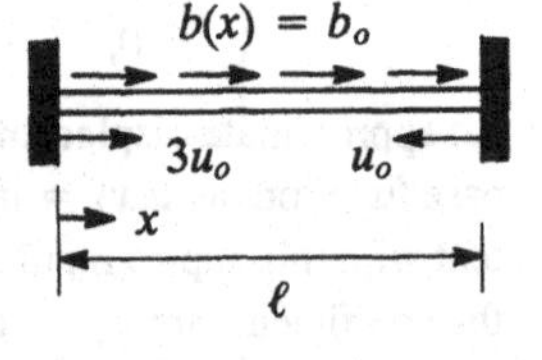

155. Consider the rod of length ℓ and constant modulus C, fixed at both ends, and subjected to a uniform body force b_o as shown. The left end moves to the right by an amount $3u_o$ and the right end moves to the left by an amount u_o. What are the essential and natural boundary conditions. Compute the displacement field $u(x)$ using the principle of virtual work and the Ritz method with a quadratic approximation. Sketch the approximate displacement field. Compute the approximate stress field.

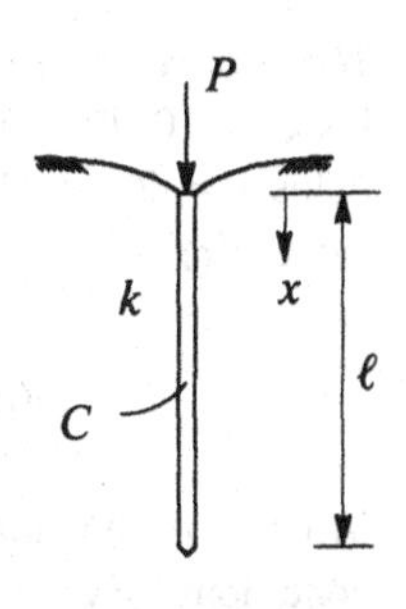

156. Consider the rod of unit length $\ell = 1$, constant unit modulus $C = 1$ (w/ unit area), embedded in an elastic medium that gets stiffer with depth. The elastic medium has variable modulus given by $k(x) = 12(1+x)$, and the resistance to motion is linearly proportional to the displacement. The rod is subjected to a unit load at $x = 0$, i.e., $P = 1$, and is traction free at the end $x = 1$. The classical governing differential equation for the displacement field $u(x)$ of the rod is $u'' - 12(1+x)u = 0$. Calculate an approximate value of the displacement at the point of load using a linear approximation of the displacement field. Use the Ritz method to carry out the calculations. Is the approximate solution a good one? Why or why not? Does the accuracy of the approximation depend upon the relative flexibilities of the rod and the elastic medium? How?

157. Consider the rod of Problem 156, now with length ℓ, constant modulus C (w/ unit area), subjected to a load P at $x = 0$. The rod is embedded in an elastic medium such that the force developed is linearly proportional to the displacement at each point. The modulus k is constant. The elastic constants are related by $k\ell^2/C = 1$. The rod is pointed so that the traction at the end $x = \ell$ is zero. Find the virtual-work functional for the given problem. What are the essential and natural boundary conditions? Find an approximation to the displacement field using the Ritz method with a *two-term* approximation with the following base functions $h_1(x) = e^{ax}$, $h_2(x) = e^{-ax}$, where $a \equiv \sqrt{k/C}$.

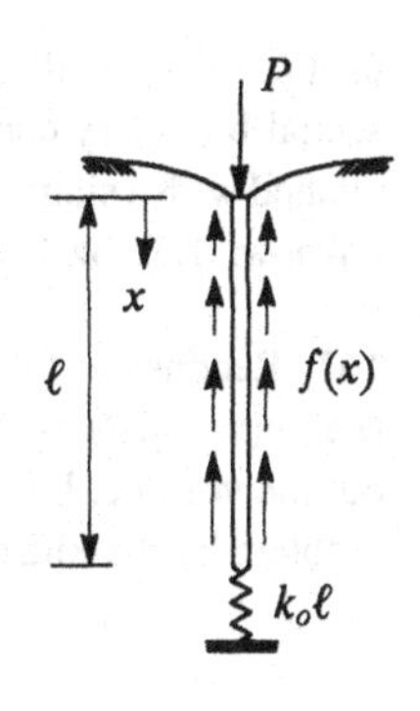

158. Consider the rod of length ℓ, constant modulus C (w/ unit area), subjected to a load P at $x = 0$. The rod is embedded in an elastic medium that provides a resisting force proportional to the displacement at each point with a modulus $k(x)$ that increases linearly with depth, so that the force is $f(x) = k_o x u(x)/\ell$. The end resistance can be modeled as a spring of modulus $k_o\ell$. The elastic constants are related by $k_o\ell^2/C = 1$. Find the virtual-work functional for the given problem. What are the essential and natural boundary conditions? Find the displacement field using the Ritz method with a *two-term* polynomial approximation.

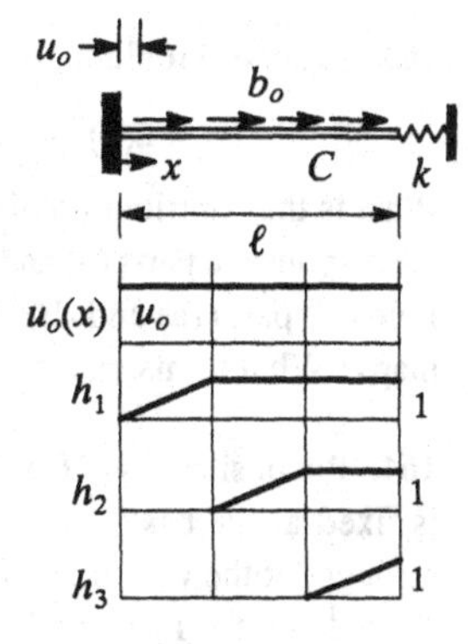

159. Consider the rod of length ℓ and constant modulus C. The rod is fixed at the left end and restrained by a linear elastic spring of modulus k at the right end. The elastic constants are related by $k\ell/C = 2$. The rod is subjected to a constant body force $b(x) = b_o$ and a prescribed displacement at the left end of u_o, as shown. Set up, the discrete equations $\mathbf{Ka} = \mathbf{f}$ that result from applying the Ritz method to the principle of virtual work using the base functions shown in the sketch. Express the answer in terms of k, ℓ, b_o, and u_o. Do you expect the solution using the Ritz method to be the exact classical solution to the boundary value problem? What base functions would you need to add to make the Ritz approximation exact?

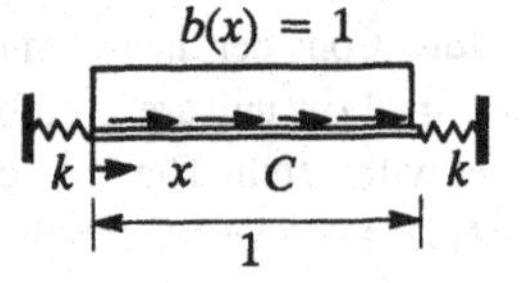

160. Consider the rod of length $\ell = 1$ and a constant modulus $C = 10$. The rod is restrained by an elastic spring of modulus $k = 2$ at each end and is subjected to a constant body force $b(x) = 1$, as shown. The virtual-work functional is

$$G(u, \bar{u}) = \int_0^\ell (Cu'\bar{u}' - b\bar{u})\,dx + ku(0)\bar{u}(0) + ku(\ell)\bar{u}(\ell)$$

What are the boundary conditions for this problem? Are they essential or natural boundary conditions? Explain. Set up, the system of equations that result from using a quadratic approximation of the displacement field with the Ritz method. Consider solving this problem using the trigonometric approximation

$$u(x) = a_1 \sin \pi x + a_2 \sin 2\pi x + a_3 \sin 3\pi x$$

Is this approximation likely to give a good solution to the problem or not?

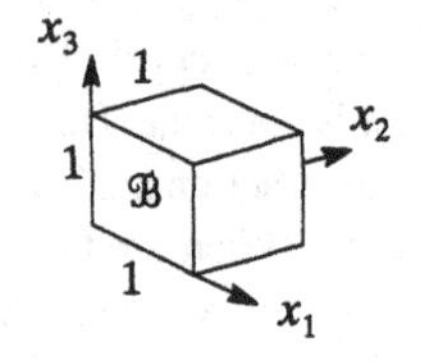

161. Consider the solid cubical region $\mathfrak{B}$ shown in the sketch having unit dimensions. Let the scalar field $w(\mathbf{x})$ characterize the response of the system. The field w is a function of the position vector $\mathbf{x}$. If we define the functional

$$G(w, v) \equiv \int_{\mathfrak{B}} (\nabla w \cdot \nabla v - 3v)\,dV$$

then $G = 0$ (for all v) is a "virtual-work" statement of the equations governing w. The essential boundary conditions are such that $w = 0$ on the coordinate faces. Use the Ritz method with a single term approximation of the form $w(\mathbf{x}) = a_0\, x_1\, x_2\, x_3$ to determine the unknown field w. Describe how you would improve the approximation.

162. Reconsider the $2\ell \times 2\ell \times h$ block shown in Fig. 84. The block is fixed at the base, (i.e., $\mathbf{u}(x_1, x_2, 0) = \mathbf{0}$) and subjected to a body force (self-weight) of $\mathbf{b} = -\varrho_o \mathbf{e}_3$ throughout the volume. Let the material have independent elastic constants λ and μ. Solve the problem by the Ritz method using the following assumption about the displacement field

$$\mathbf{u}(\mathbf{x}) = a_1 x_3 \ell\, \mathbf{e}_3 + a_2 (x_1 x_3 \mathbf{e}_1 + x_2 x_3 \mathbf{e}_2)$$

163. Resolve Problem 162 with the following assumed displacement field

$$\mathbf{u}(\mathbf{x}) = a_1 x_3 \ell\, \mathbf{e}_3 + a_2 x_3^2 \mathbf{e}_3 + a_3 (x_1 x_3 \mathbf{e}_1 + x_2 x_3 \mathbf{e}_2)$$

What is the contribution of the term $(x_1 x_3 \mathbf{e}_1 + x_2 x_3 \mathbf{e}_2)$ (i.e., the lateral displacement) to the response in parts (a) and (b)? Compute the stress tensor $\mathbf{S}$ implied by the displacement fields of parts (a) and (b). What body forces and surface tractions are implied by these maps? What terms would you add to the solution to improve the Ritz approximation?

164. Reconsider the $2\ell \times 2\ell \times h$ block shown in Fig. 84. As in Problem 162, the block is fixed at the base, (i.e., $\mathbf{u}(x_1, x_2, 0) = \mathbf{0}$) and subjected to a body force $\mathbf{b} = -\varrho_o \mathbf{e}_3$ throughout the volume. Let the material have independent elastic constants λ and μ. Let $\xi_1 = \frac{1}{2}(x_1/\ell + 1)$, $\xi_2 = \frac{1}{2}(x_2/\ell + 1)$ and $\xi_3 = x_3/h$ be a change of variables that maps the block onto the unit cube with one vertex at the origin of the coordinate system (ξ_1, ξ_2, ξ_3). Define the following functions

$$\phi_1(\boldsymbol{\xi}) = \xi_1\xi_2\xi_3 \qquad \phi_5(\boldsymbol{\xi}) = \xi_1\xi_2(1-\xi_3)$$

$$\phi_2(\boldsymbol{\xi}) = (1-\xi_1)\xi_2\xi_3 \qquad \phi_6(\boldsymbol{\xi}) = (1-\xi_1)\xi_2(1-\xi_3)$$

$$\phi_3(\boldsymbol{\xi}) = \xi_1(1-\xi_2)\xi_3 \qquad \phi_7(\boldsymbol{\xi}) = \xi_1(1-\xi_2)(1-\xi_3)$$

$$\phi_4(\boldsymbol{\xi}) = (1-\xi_1)(1-\xi_2)\xi_3 \qquad \phi_8(\boldsymbol{\xi}) = (1-\xi_1)(1-\xi_2)(1-\xi_3)$$

These functions have the property that, at each of the eight vertices, one of the functions has unit value while the others are zero. They are, in fact, the finite element base functions for a hexahedron element. Let the displacement be approximated as

$$\mathbf{u}(\boldsymbol{\xi}) = \sum_{i=1}^{8} \mathbf{a}_i \phi_i(\boldsymbol{\xi})$$

where $\mathbf{a}_i$ is a vector constant with component expression $\mathbf{a}_i = a_{ij}\mathbf{e}_j$ (no sum on j). What are the base functions $\mathbf{h}_i(\boldsymbol{\xi})$ associated with this expansion? What is the physical significance of the coefficient vector $\mathbf{a}_i$? What does the essential boundary condition $\mathbf{u}(\xi_1, \xi_2, 0) = \mathbf{0}$ imply about the values of the coefficients in the expansion? Solve the block problem using the base functions identified, as restricted by the essential boundary condition.

165. Consider a cube of dimension $2 \times 2 \times 2$ fixed at the base and subjected to a body force $\mathbf{b} = -\varrho_o\mathbf{e}_3$. Describe a method for refining the finite element approximation by establishing a local coordinate system for each element that allows the creation of the finite element base functions from the eight basic element functions $\phi_i(\boldsymbol{\xi})$ described in Problem 164. Notice that each element is associated with eight nodes while the entire block is associated with 27 nodes. Continuity of displacements can be assured by associating the element base functions with element nodal displacements (i.e., finite element functions) and by associating elements nodal displacements with a common global displacement parameter where elements share a common node.

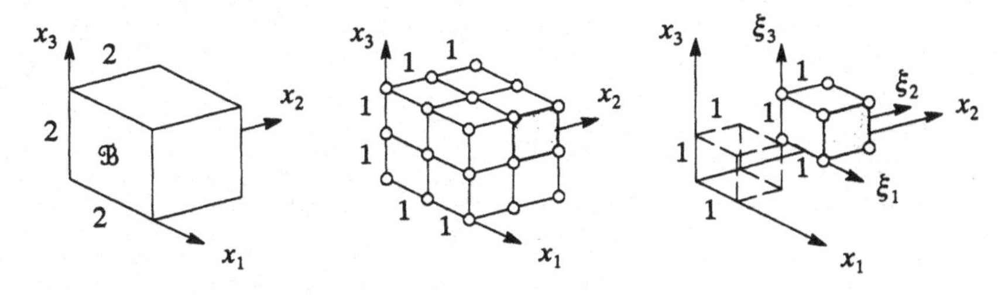

166. What is the appropriate definition of $\mathbf{f}$ in Eqn. (366) when there is a nonzero boundary displacement term $\hat{\mathbf{u}}(\mathbf{x})$ in the Ritz approximation?

7 The Linear Theory of Beams

The equations describing the mechanics of a three-dimensional continuum are formidable to solve even for a simple constitutive model like isotropic hyperelasticity. Even in the age of computers and the finite element method, it is still not feasible to treat every solid body as a three-dimensional continuum. Bodies with certain geometric features are amenable to a reduction from three dimensions to fewer dimensions, from the perspective of the governing differential equations. These bodies are usually called beams (one dimension), plates (two dimensions, flat), and shells (two dimensions, curved). These reduced theories comprise a subset of solid mechanics generally referred to as *structural mechanics*. Among the theories of structural mechanics, beam theory is the simplest.

This chapter really has two, complementary purposes. First, the chapter provides a careful and thorough derivation of the equations of linear beam theory in the context of three-dimensional solid mechanics. Although the equations of beam theory can be obtained without the machinery of solid mechanics (see, for example, any book on elementary strength of materials), to do so is, in itself, a little lesson in solid mechanics; a relevant application of the general theory. The great merit of approaching the derivation this way is that we can clearly see where are the strengths and limitations of beam theory.

Second, the ordinary differential equations of beam theory are much more likely to yield classical solutions than are the partial differential equations of the three-dimensional theory. The ordinary differential equations are a bit more sophisticated than the little boundary value problem and there are some new features (e.g., we need multiple fields, displacement and rotation, to describe the kinematics of motion). Beams are, of course, amenable to the principle of virtual work so we can use beam theory as another opportunity to apply the Ritz method. Because classical and variational approaches are both relatively easy, beam theory provides a fertile opportunity for the study of the relationship between classical and variational methods.

This chapter examines the foundations of linear beam theory. We first introduce the notion of resultants (net force and net moment) of the traction vector field over a cross section and deduce equations of net equilibrium by satisfying the three-dimensional equilibrium equations $\text{div}\,\mathbf{S}+\mathbf{b} = \mathbf{0}$ in an average sense. We then introduce a kinematic hypothesis that describes the motion of the beam in terms of parameters that vary only along the axial coordinate. Finally, we develop elastic constitutive equations for stress and strain resultants by introducing the three-dimensional constitutive equations into the definitions of the resultants. Once the general theory is laid out, we examine two special cases of planar motion (the Timoshenko beam and the Bernoulli-Euler beam). We consider both classical and variational statements of the theory and illustrate the differences between the two with several computational examples.

Notation. A beam is a long, slender cylindrical body.† A planar slice through the undeformed beam, perpendicular to the longitudinal axis, is a two-dimensional surface that we will call a *cross section*. We shall choose to describe our beam in accord with the convention shown in Fig. 85. Note that the x_3 coordinate axis coincides with the axis of the beam. Therefore, any beam cross section will lie in a plane parallel to the $x_1 - x_2$ plane. The cross section, which we shall call Ω, is a closed geometric figure and, hence, possesses geometric properties like area and moments of the area. The cross section has a

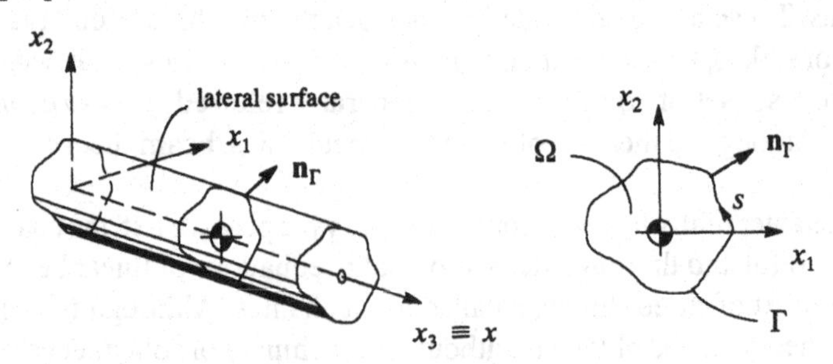

Figure 85 A beam occupying three-dimensional space

† In the derivation of the equations of beam theory it will be evident that the equations apply strictly to prismatic beams (i.e., beams of constant cross section). For prismatic beams the normal to the boundary of the cross section $\mathbf{n}_\Gamma$ has no component in the axial direction, which is important when we use the divergence theorem. We will find later that we can ease up on that restriction to include beams with slowly varying cross section. This is an important approximation that significantly extends the range of problems amenable to beam theory.

boundary, which we shall call Γ, that can be parameterized by its arc length s. The boundary has a normal $\mathbf{n}_\Gamma$ at every point, and this normal is unique (except possibly at a finite number of corners). The surface of the beam consists of its two ends (i.e., the cross sections located at $x_3 = 0$ and $x_3 = \ell$) and the lateral surface. We shall assume that either the motion or the tractions can be prescribed at the beam ends, but that the motion cannot be prescribed on the lateral surface.† On the lateral surface only the tractions $\mathbf{t}_\Gamma$ are prescribed.

For those quantities that are functions of only the axial coordinate, that is, any resultant or generalized displacement, we shall designate the coordinate simply as $x_3 = x$. Derivatives of such quantities are always ordinary derivatives, and we shall often use the notation $(\cdot)' = d(\cdot)/dx$ for the derivative.

Equations of Equilibrium

A beam is subject to the same requirements of equilibrium as every other body, namely $\operatorname{div}\mathbf{S} + \mathbf{b}$ everywhere inside the domain and $\mathbf{Sn} = \mathbf{t_n}$ on the surface of the domain. The concept that distinguishes a beam from a continuum is the *stress resultant*. A stress resultant represents the aggregate effect of all of the traction forces acting on a cross section. We shall find that a single net resultant is not adequate to describe those tractions, so we shall also use the first moment of these tractions about some point in the cross section. We can deduce equations of equilibrium for the resultants from the three-dimensional theory.

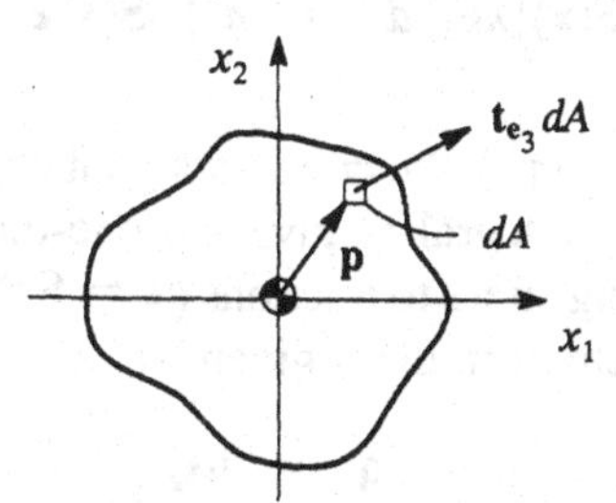

Figure 86 Traction vector acting on a typical section

The resultant force and moment can be computed by integrating the tractions over the cross-sectional area, as shown in Fig. 86. The traction vector acting on a plane with normal $\mathbf{e}_3$ is given by $\mathbf{t}_{\mathbf{e}_3}(\mathbf{x}) = \mathbf{S}(\mathbf{x})\mathbf{e}_3$. The location of this traction vector in the plane can be described by the position vector relative to the x_3 axis $\mathbf{p}(x_1, x_2) \equiv x_1\mathbf{e}_1 + x_2\mathbf{e}_2$, a vector with no component in the axial direction. The *resultant force* $\mathbf{Q}(x)$ and the *resultant moment* $\mathbf{M}(x)$ are computed

† The prescription of motion at certain points along a beam is possible, but it really works out only as a consequence of the kinematic constraint associated with the kinematic hypothesis ("plane sections remain plane"). In fact, a case of interest in structures is a beam with an intermediate support that restricts the displacement of a cross section but not its rotation (often called a "continuous beam"). This subtle point is discussed in detail in the section entitled *Boundary Conditions*, later in the chapter.

as the integral of the tractions and the first moment of the tractions over the cross section as

$$\mathbf{Q}(x) \equiv \int_{\Omega} \mathbf{t}_{\mathbf{e}_3}(\mathbf{x})\, dA, \qquad \mathbf{M}(x) \equiv \int_{\Omega} \mathbf{p}(x_1,x_2) \times \mathbf{t}_{\mathbf{e}_3}(\mathbf{x})\, dA \tag{367}$$

where $\mathbf{t}_{\mathbf{e}_3} = \mathbf{S}\mathbf{e}_3$ and Ω is the area of the cross section. The integration over the cross section eliminates the dependence on the x_1 and x_2 coordinates and leaves $\mathbf{Q}$ and $\mathbf{M}$ as functions of only the axial coordinate x.

Cauchy relations for stress resultants. From their definition, the stress resultants $\mathbf{Q}$ and $\mathbf{M}$ appear to be vectors, and they will behave like vectors in almost every regard. However, these vectors characterize the state of stress in the beam, and, therefore, we must examine how the Cauchy relations implied in $\mathbf{t}_{\mathbf{n}} = \mathbf{S}\mathbf{n}$ manifest for the beam. If we take a thin slice of a beam, we can see that there are two faces, one with normal $\mathbf{e}_3$ and one with normal $-\mathbf{e}_3$, as shown in Fig. 87. There are tractions acting on both faces. Let us compute the resultant traction force $\hat{\mathbf{q}}_n$ acting on the face with normal $\mathbf{n} = n\mathbf{e}_3$, where n is either $+1$ (front face or positive x_3 direction) or -1 (back face or negative x_3 direction)

$$\hat{\mathbf{q}}_n = \int_{\Omega} \mathbf{S}(\mathbf{x})\big(n\mathbf{e}_3\big)\, dA = n\int_{\Omega} \mathbf{S}(\mathbf{x})\mathbf{e}_3\, dA = n\mathbf{Q}(x)$$

Since n does not depend upon the cross-sectional coordinates x_1 and x_2, it can be factored out of the integral to give the one-dimensional version of the Cauchy formula relating stress to tractions ($\mathbf{t} = \mathbf{S}\mathbf{n}$). For resultant forces, the one-dimensional Cauchy formula is given by

$$\hat{\mathbf{q}}_n = n\mathbf{Q} \tag{368}$$

An identical argument produces an equivalent result for the moments. The resultant traction moment $\hat{\mathbf{m}}_n$ acting on the face with normal $\mathbf{n} = n\mathbf{e}_3$ is related to the resultant moment as

$$\hat{\mathbf{m}}_n = n\mathbf{M} \tag{369}$$

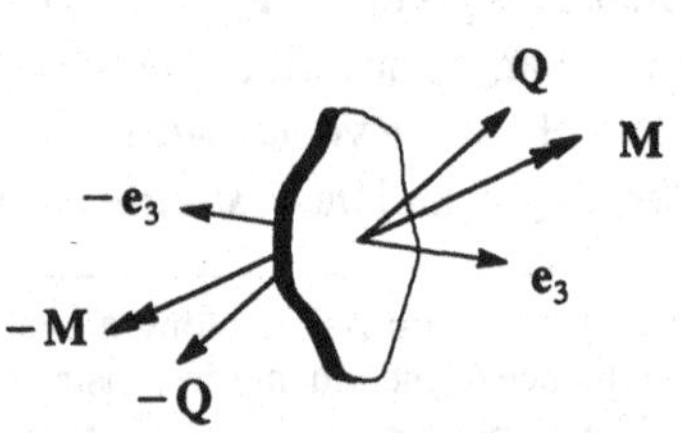

Figure 87 Cauchy relations for stress resultants

As was the case for three-dimensional solids, the Cauchy relationship tells us how to take freebody diagrams in the sense that it tells us what force to place at an exposed section to represent the state of stress resultants there. Since $\mathbf{Q}$ and $\mathbf{M}$ are vectors, they inherently have direction and magnitude, and add with the head-to-tail rule of vector addition. We must have the Cauchy relations in order to draw Fig. 87 correctly. Note that the minus signs on the back face come from n in the Cauchy formula.

Equilibrium of force resultants. The beam is subject to the equations of equilibrium $\operatorname{div}\mathbf{S}+\mathbf{b} = \mathbf{0}$, $\mathbf{S} = \mathbf{S}^T$, and $\mathbf{Sn} = \mathbf{t}_n$, but it remains to be seen how these equations from the three-dimensional theory manifest in beam theory. In particular, what are the governing differential equations for $\mathbf{Q}$ and $\mathbf{M}$ and how do the applied body forces $\mathbf{b}$ and the tractions on the lateral surface $\mathbf{t}_\Gamma$ enter the theory? It turns out that we can derive governing equations for $\mathbf{Q}$ and $\mathbf{M}$ by insisting that $\operatorname{div}\mathbf{S}+\mathbf{b} = \mathbf{0}$ in an average sense over the cross section. To see how this comes about, let us compute the integral of $\operatorname{div}\mathbf{S}+\mathbf{b}$ over the cross section:

$$\begin{aligned}
\int_\Omega \left(\operatorname{div}\mathbf{S}+\mathbf{b}\right) dA &= \int_\Omega \left(\frac{\partial \mathbf{S}\mathbf{e}_3}{\partial x_3} + \frac{\partial \mathbf{S}\mathbf{e}_\alpha}{\partial x_\alpha} + \mathbf{b}\right) dA \\
&= \frac{\partial}{\partial x_3}\int_\Omega \mathbf{S}\mathbf{e}_3\, dA + \int_\Omega \frac{\partial \mathbf{S}\mathbf{e}_\alpha}{\partial x_\alpha}\, dA + \int_\Omega \mathbf{b}\, dA \\
&= \frac{\partial \mathbf{Q}}{\partial x} + \int_\Gamma \mathbf{S}\mathbf{n}_\Gamma\, ds + \int_\Omega \mathbf{b}\, dA
\end{aligned}$$

where the implied summation on a Greek subscript is assumed to be a sum from 1 to 2, as opposed to the implied summation on a Latin subscript, which goes from 1 to 3. Hence, $\operatorname{div}\mathbf{S} = \partial\mathbf{S}\mathbf{e}_i/\partial x_i = \partial\mathbf{S}\mathbf{e}_\alpha/\partial x_\alpha + \partial\mathbf{S}\mathbf{e}_3/\partial x_3$. We pulled the derivative with respect to x_3 out of the first integral over the cross section because that integral only involves the coordinates x_1 and x_2. We then applied the two-dimensional version of the divergence theorem to make the following transformation

$$\int_\Omega \frac{\partial \mathbf{S}\mathbf{e}_\alpha}{\partial x_\alpha}\, dA = \int_\Gamma \mathbf{S}\mathbf{n}_\Gamma\, ds = \int_\Gamma \mathbf{t}_\Gamma\, ds$$

where Γ is the lateral contour of the cross section, parameterized by s, and $\mathbf{n}_\Gamma$ is the unit vector normal to the lateral contour and in the plane of the cross section (i.e., it has no component along $\mathbf{e}_3$). According to the Cauchy relations, the vector $\mathbf{t}_\Gamma = \mathbf{S}\mathbf{n}_\Gamma$ represents the traction on the lateral surface. These tractions are the prescribed loads on the surface of the beam. Since the body forces $\mathbf{b}$ are also known, let us define the applied loading per unit of length as

$$\boxed{\mathbf{q}(x) \equiv \int_{\Omega} \mathbf{b}\, dA + \int_{\Gamma} \mathbf{t}_{\Gamma}\, ds} \tag{370}$$

Noting the above definitions, we have

$$\int_{\Omega} \big(\operatorname{div}\mathbf{S}+\mathbf{b}\big)\, dA = \frac{d\mathbf{Q}}{dx} + \mathbf{q} \tag{371}$$

One can observe from this equation that if $\mathbf{Q}' + \mathbf{q} = \mathbf{0}$, then the three-dimensional equilibrium equations $\operatorname{div}\mathbf{S}+\mathbf{b} = \mathbf{0}$ are satisfied on average over the cross section in question. Clearly, this equivalence holds for all x.

Equilibrium of moments. We can follow the same approach to the equilibrium of moments. Let us integrate $\mathbf{p} \times \big(\operatorname{div}\mathbf{S}+\mathbf{b}\big)$ over the cross section to get

$$\begin{aligned}\int_{\Omega} \mathbf{p} \times \big(\operatorname{div}\mathbf{S}+\mathbf{b}\big)\, dA &= \int_{\Omega} \mathbf{p} \times \left(\frac{\partial \mathbf{S}\mathbf{e}_3}{\partial x_3} + \frac{\partial \mathbf{S}\mathbf{e}_\alpha}{\partial x_\alpha} + \mathbf{b} \right) dA \\ &= \frac{\partial}{\partial x_3} \int_{\Omega} \mathbf{p} \times \mathbf{S}\mathbf{e}_3\, dA + \int_{\Omega} \mathbf{p} \times \frac{\partial \mathbf{S}\mathbf{e}_\alpha}{\partial x_\alpha}\, dA + \int_{\Omega} \mathbf{p} \times \mathbf{b}\, dA\end{aligned}$$

Again, we want to use the two-dimensional divergence theorem to convert the second term on the right side to an integral over the lateral contour. To achieve this result, let us note that, by the rule for differentiation of a product, we have

$$\frac{\partial}{\partial x_\alpha}\big(\mathbf{p} \times \mathbf{S}\mathbf{e}_\alpha\big) = \frac{\partial \mathbf{p}}{\partial x_\alpha} \times \mathbf{S}\mathbf{e}_\alpha + \mathbf{p} \times \frac{\partial \mathbf{S}\mathbf{e}_\alpha}{\partial x_\alpha}$$

again with summation implied on α over the range of 1 to 2. From the definition of $\mathbf{p}$ we know that $\partial \mathbf{p}/\partial x_\alpha = \mathbf{e}_\alpha$. We also know that balance of angular momentum of the continuum implies (see Eqn. (195) in Chapter 3) that

$$\mathbf{e}_i \times \mathbf{S}\mathbf{e}_i = \mathbf{e}_3 \times \mathbf{S}\mathbf{e}_3 + \mathbf{e}_\alpha \times \mathbf{S}\mathbf{e}_\alpha = \mathbf{0}$$

with summation implied over the appropriate ranges in accord with our convention $\{i = 1, 2, 3$ and $\alpha = 1, 2\}$. Using these relationships, we find that

$$\mathbf{p} \times \frac{\partial \mathbf{S}\mathbf{e}_\alpha}{\partial x_\alpha} = \frac{\partial}{\partial x_\alpha}\big(\mathbf{p} \times \mathbf{S}\mathbf{e}_\alpha\big) + \mathbf{e}_3 \times \mathbf{S}\mathbf{e}_3$$

Integrating this equation over the cross-sectional area gives

$$\int_{\Omega} \mathbf{p} \times \frac{\partial \mathbf{S}}{\partial x_\alpha}\mathbf{e}_\alpha\, dA = \int_{\Omega} \frac{\partial}{\partial x_\alpha}\big(\mathbf{p} \times \mathbf{S}\mathbf{e}_\alpha\big)\, dA + \int_{\Omega} \mathbf{e}_3 \times \mathbf{S}\mathbf{e}_3\, dA$$

Applying the two-dimensional divergence theorem to the first term on the right side and recognizing that the second term can be integrated explicitly, we finally arrive at the identity

$$\int_{\Omega} \mathbf{p} \times \frac{\partial \mathbf{S}}{\partial x_a} \mathbf{e}_a \, dA = \int_{\Gamma} \mathbf{p} \times \mathbf{t}_{\Gamma} \, ds + (\mathbf{e}_3 \times \mathbf{Q})$$

Again, since the body force $\mathbf{b}$ and the lateral tractions $\mathbf{t}_{\Gamma} = \mathbf{S}\mathbf{n}_{\Gamma}$ are known as given data, we shall define the applied moment per unit length as

$$\boxed{\mathbf{m}(x) \equiv \int_{\Omega} \mathbf{p} \times \mathbf{b} \, dA + \int_{\Gamma} \mathbf{p} \times \mathbf{t}_{\Gamma} \, ds} \tag{372}$$

Noting the above identities, we have

$$\int_{\Omega} \mathbf{p} \times (\operatorname{div}\mathbf{S} + \mathbf{b}) \, dA = \frac{d\mathbf{M}}{dx} + (\mathbf{e}_3 \times \mathbf{Q}) + \mathbf{m} \tag{373}$$

These results show that if $\mathbf{M}' + (\mathbf{e}_3 \times \mathbf{Q}) + \mathbf{m} = \mathbf{0}$, then the average of the first moment of $\operatorname{div}\mathbf{S} + \mathbf{b}$ over the cross section equals zero. It is also interesting to note that the balance of angular momentum $\mathbf{e}_i \times \mathbf{S}\mathbf{e}_i = \mathbf{0}$ played an important role in this derivation.

The equilibrium equations for the stress resultants are summarized in the following box

$$\boxed{\begin{aligned} \mathbf{Q}' + \mathbf{q} &= \mathbf{0} \\ \mathbf{M}' + (\mathbf{e}_3 \times \mathbf{Q}) + \mathbf{m} &= \mathbf{0} \end{aligned}} \tag{374}$$

where the applied loads $\mathbf{q}$ and $\mathbf{m}$ are given by Eqns. (370) and (372), respectively. These equations constitute a set of first-order ordinary differential equations in the unknown vector fields $\mathbf{Q}$ and $\mathbf{M}$.

It is important to appreciate the limitations of the one-dimensional equilibrium equations. On the average, they assure the same equilibrium requirements as the three-dimensional equations. However, within a cross section these equations overlook some of the details. An analogy from probability and statistics may be useful here: Imagine that we have a data sample, say the scores of *n* students in a class on an examination. To characterize the performance of the class on the exam we usually compute the two lowest-order moments of the statistical distribution—the ***mean*** and ***standard deviation***—to capture the overall nature of the statistical distribution of the data. Clearly, we could compute higher moments of the data to get more information (skewedness, kurtosis, etc.) about its distribution, but often the low-order statistics capture the bulk of what we want to know about the distribution. If the data are normally distributed, then the mean and the standard deviation are enough to exactly characterize the statistical distribution, but inadequate to reconstruct any individual score. If we had the *n* moments of the distribution then we would have enough information to reconstruct the data sample from the moments of the distribution (rather like a Fourier transform and its inverse). If the data are not normally distributed, but are close to it, then the mean and standard deviation capture the main features of the data but miss the higher-order wiggles. If the distribution is considerably different from normal (e.g., a bimodal distribution) then the low-order statistics are inappropriate to characterize the distribution.

In our problem, if the tractions $\mathbf{t}_{\mathbf{e}_3}$ vary linearly over the cross section, then $\mathbf{Q}$ and $\mathbf{M}$ completely characterize the state of stress. If the stress components have higher-order wiggles, then our resultant equations of equilibrium miss them. If the variation of stress is not dominated by the constant and linear terms, then beam theory simply does not provide an adequate model of three-dimensional behavior. Centuries of observation have borne out the validity of beam theory for long slender bodies.

Example 45. *Computation of beam loading.* A solid cylindrical beam of radius R, length ℓ, unit weight ϱ_b is submerged halfway in a fluid of unit weight ϱ_o. Recall that the pressure at any point in a fluid is proportional to the depth h. Compute the resultant applied load $\mathbf{q}(x)$ and the resultant applied moment $\mathbf{m}(x)$ that would be appropriate in order to treat the problem using beam theory.

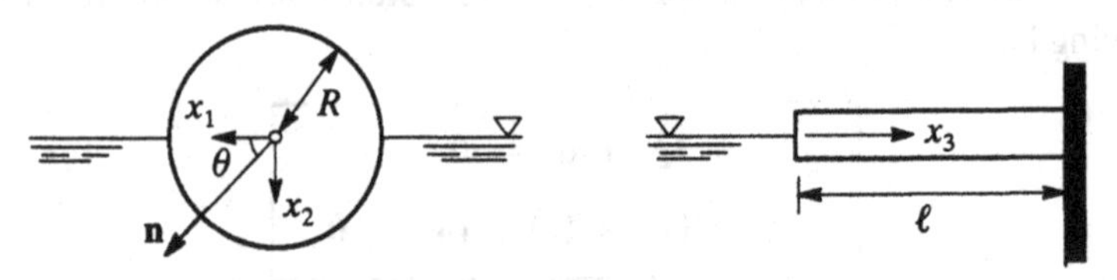

The surface normal vector can be parameterized as $\mathbf{n} = \cos\theta\,\mathbf{e}_1 + \sin\theta\,\mathbf{e}_2$. The pressure acting on the beam surface is $p = \varrho_o R\sin\theta$. The stress tensor is simply $\mathbf{S} = -p\mathbf{I}$. The surface traction and body force vectors are given by,

$$\mathbf{t}_\Gamma = \mathbf{S}\mathbf{n} = -\varrho_o R\sin\theta(\cos\theta\,\mathbf{e}_1 + \sin\theta\,\mathbf{e}_2), \qquad \mathbf{b} = \varrho_b\mathbf{e}_2$$

The applied force per unit length is

$$\mathbf{q}(x) = \int_0^\pi \mathbf{t}_\Gamma\, R\,d\theta + \int_0^{2\pi}\int_0^R \mathbf{b}\, r\,dr\,d\theta = \pi R^2\left(\varrho_b - \tfrac{1}{2}\varrho_o\right)\mathbf{e}_2$$

Noting that $\mathbf{p}\times\mathbf{t}_\Gamma = \mathbf{0}$ and $\mathbf{p}\times\mathbf{b} = \varrho_b r\cos\theta\,\mathbf{e}_3$, the applied moment per unit length is

$$\begin{aligned}\mathbf{m}(x) &= \int_0^\pi (\mathbf{p}\times\mathbf{t}_\Gamma)\,R\,d\theta + \int_0^{2\pi}\int_0^R (\mathbf{p}\times\mathbf{b})\,r\,dr\,d\theta \\ &= \int_0^{2\pi}\int_0^R \varrho_b r^2\cos\theta\,dr\,d\theta\,\mathbf{e}_3 = \mathbf{0}\end{aligned}$$

The Kinematic Hypothesis

The stress resultants provide a vehicle to reduce the equilibrium equations to one dimension. However, the motion of a solid body $\mathbf{u}(\mathbf{x})$ is also a function of spatial position (and therefore, inherently three dimensional). To really define a beam, we need an extra ingredient called the *kinematic hypothesis*.

A kinematic hypothesis is nothing more than a constraint placed on the deformation map. We assume that the body moves in a very specific manner, an assumption that must be verified either by observation of nature or by examining the consequences of imposing the constraints with a theory that does not make those assumptions (i.e., the general three-dimensional theory).

The basic idea behind beam theory is the hypothesis that cross sections that are plane before deformation remain plane after deformation, the so-called *plane-sections hypothesis.* (Although not often stated explicitly, an equally important assumption is that those plane sections do not distort in their own planes, either.) This hypothesis is central to the computation of deflections in beams. Although Galileo (1564-1642) had made the first contributions to beam theory, his results concerned only the static equilibrium of beams. The crucial plane-sections hypothesis did not appear until nearly one hundred years later. It goes back nearly three centuries to Jacob Bernoulli (1654-1705), who did not quite get it right (but came close enough to get partial credit). Two generations later, the great mathematician Leonhard Euler (1707-1783) also made significant contributions to the theory of deflection curves of beams, but made no significant improvements on Bernoulli's kinematic hypothesis. Navier (1785-1836) was the one who finally clarified the issue of the kinematic hypothesis and put beam theory on the solid ground on which it now rests. Beam theory is perhaps the most successful theory in all of structural mechanics, forming the basis of what we call *structural analysis*, the structural engineer's bread-and-butter.

Let us examine the motion of a typical cross section. The plane-sections hypothesis suggests that a cross section will move as a rigid body, neither changing in shape nor deviating from flatness. There are many ways of tracking the motion of a rigid body in three-dimensional space. The method that is most useful here is to select a point, say the point $\mathcal{O}$ (whichi is the origin of coordinates in the $x_1 - x_2$ plane), marked by the target in Fig. 88, and to keep track of the motion of that point. As described in the figure, the point displaces by an amount $\mathbf{w}$. It takes three quantities to keep track of the motion of the point, the three components of the vector $\mathbf{w}$. Keeping track of the motion of a single point is not sufficient to describe the motion of the plane because the body also rotates.

We must also keep track of the vectors that record the orientation of the cross section in space. The cross-sectional plane is completely characterized by its normal vector and two independent vectors that lie in the plane. In the undeformed configuration, these three vectors are the base vectors $\mathbf{e}_1$, $\mathbf{e}_2$, and $\mathbf{e}_3$.

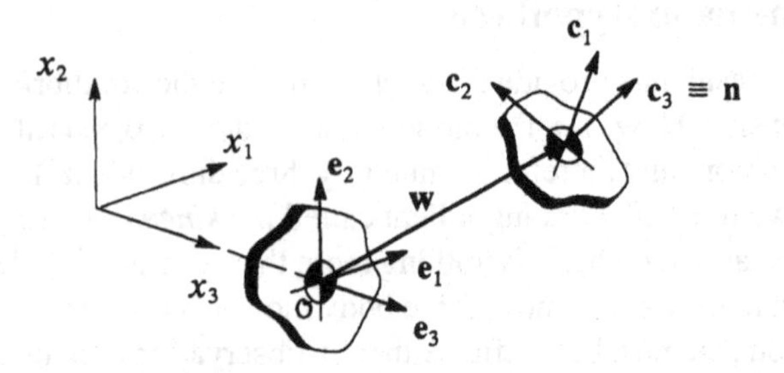

Figure 88 Tracking the motion of a rigid plane in three dimensions

In the deformed configuration these vectors become $\mathbf{c}_1$, $\mathbf{c}_2$, and $\mathbf{c}_3 \equiv \mathbf{n}$, each of which can be determined from the original vectors by a rotation in three-dimensional space. Let $\mathbf{\Lambda}(x_3)$ be a tensor that rotates vectors (without stretching them) in three-dimensional space. In particular, let the rotation be precisely that needed to orient the cross section as $\mathbf{c}_i = \mathbf{\Lambda}(x_3)\mathbf{e}_i$. The tensor $\mathbf{\Lambda}(x_3)$ is an orthogonal tensor and is, therefore, completely characterized by three independent parameters, the so-called Euler angles θ_1, θ_2, and θ_3. Therefore, it takes three parameters to orient the cross section, and, thus, a total of six to uniquely track the motion of the cross section. From those six quantities, we can find the location of any other particle on the body through the map

$$\boldsymbol{\phi}(\mathbf{x}) = x_3\mathbf{e}_3 + \mathbf{w}(x_3) + \mathbf{\Lambda}(x_3)\mathbf{p}(x_1, x_2) \tag{375}$$

The first term takes us from the origin to the appropriate cross section, the second term takes us to the displaced origin of the cross section, and the third term takes us to the position within the cross section that was originally at the position $\mathbf{p}(x_1, x_2) = x_1\mathbf{e}_1 + x_2\mathbf{e}_2$ in the undeformed configuration, i.e., $\mathbf{p}$ locates the position of a point in the cross section relative to the point $\mathcal{O}$.

If the rotation of a cross section is small, the deformation map can be simplified. In particular, for small rotations we have $\mathbf{\Lambda p} \approx \mathbf{p} + \boldsymbol{\theta} \times \mathbf{p}$, where the vector $\boldsymbol{\theta} = \theta_i\mathbf{e}_i$ is called the *rotation vector* (see Problem 174). We can now describe our deformation through a displacement map. Let $\mathbf{u}(\mathbf{x})$ be the displacement of a point originally located at position $\mathbf{x}$ in our undeformed beam. The displacement is given by

$$\boxed{\mathbf{u}(\mathbf{x}) = \mathbf{w}(x_3) + \boldsymbol{\theta}(x_3) \times \mathbf{p}(x_1, x_2)} \tag{376}$$

Observe the explicit dependence of the map on the axial coordinate x_3 and the cross-sectional coordinates x_1 and x_2. The displacement vector $\mathbf{w} = w_i\mathbf{e}_i$, with components $\{w_1, w_2, w_3\}$, and the rotation vector $\boldsymbol{\theta} = \theta_i\mathbf{e}_i$, with components $\{\theta_1, \theta_2, \theta_3\}$, are collectively called the ***generalized displacements*** and are functions only of the axial coordinate x_3. The displacement map can be written out in terms of its components as follows

$$u_1(x_1,x_2,x_3) = w_1(x_3) - x_2\theta_3(x_3)$$
$$u_2(x_1,x_2,x_3) = w_2(x_3) + x_1\theta_3(x_3)$$
$$u_3(x_1,x_2,x_3) = w_3(x_3) - x_1\theta_2(x_3) + x_2\theta_1(x_3)$$

The physical significance of the generalized displacements can be seen by examining the individual terms of the map. Figure 89 shows the displaced beam projected onto the x_1 - x_3 plane. Clearly, the component w_3 measures the displacement along the axis of the beam while the component w_1 measures displacement transverse to the beam axis in the x_1 direction. The component θ_2 measures rotation about the x_2 axis and has a positive sense according to the right-hand rule. Displacements are, of course, positive if they are in the direction of the coordinate axes. Consider the displacement of the point $\mathcal{P}$ a distance x_1 from the axis of the beam. For the purpose of illustration, suppose that the motion is planar, i.e., $w_2 = 0$, $\theta_1 = 0$, and $\theta_3 = 0$. Relative to the point $\mathcal{O}$, the point $\mathcal{P}$ moves in the negative x_3 direction by $x_1 \sin\theta_2 \approx x_1\theta_2$, and in the negative x_1 direction by $x_1(1-\cos\theta_2) \approx 0$. Because the motion is planar there is no motion in the x_2 direction (out of the plane of the page). Clearly, this is the motion that our deformation map captures.

If we have an explicit expression for the deformation map, it is simple to compute the strains implied by that map. Here we shall confine our attention to the linearized strain tensor $\mathbf{E} = \frac{1}{2}[\nabla\mathbf{u} + \nabla\mathbf{u}^T]$. The gradient of $\mathbf{u}$ is the tensor

$$\nabla\mathbf{u}(\mathbf{x}) = \mathbf{u}_{,i} \otimes \mathbf{e}_i$$

Recall that for $\alpha = 1, 2$, the derivative of the position vector $\mathbf{p}$ is given by $\partial\mathbf{p}/\partial x_\alpha = \mathbf{e}_\alpha$. Therefore, we can compute

$$\nabla\mathbf{u}(\mathbf{x}) = \left(\mathbf{w}' + \boldsymbol{\theta}' \times \mathbf{p}\right) \otimes \mathbf{e}_3 - \left(\mathbf{e}_\alpha \times \boldsymbol{\theta}\right) \otimes \mathbf{e}_\alpha \tag{377}$$

with summation implied on the Greek subscript. Before we use $\nabla\mathbf{u}$ to compute the linearized strain tensor, observe that

$$\boldsymbol{\Theta} \equiv (\mathbf{e}_i \times \boldsymbol{\theta}) \otimes \mathbf{e}_i = -\left(\boldsymbol{\theta} \times \mathbf{e}_i\right) \otimes \mathbf{e}_i = -[\boldsymbol{\theta} \times]$$

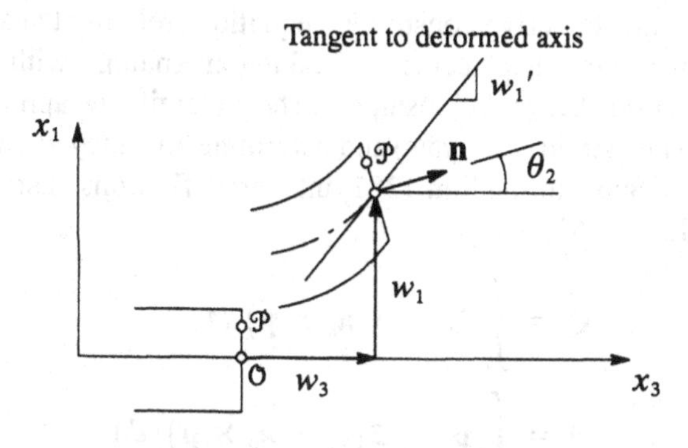

Figure 89 Components of displacement in a plane

(with sum implied on i) is skew-symmetric, that is, $\boldsymbol{\Theta}+\boldsymbol{\Theta}^T = \mathbf{0}$. Thus, we can rewrite Eqn. (377) in the following form

$$\nabla \mathbf{u}(\mathbf{x}) = \left(\mathbf{w}' + \mathbf{e}_3 \times \boldsymbol{\theta} + \boldsymbol{\theta}' \times \mathbf{p}\right) \otimes \mathbf{e}_3 - \boldsymbol{\Theta} \tag{378}$$

We are now in a position to compute the linearized strain tensor as

$$\mathbf{E} = \tfrac{1}{2}\left[\left(\boldsymbol{\epsilon}_o + \boldsymbol{\varkappa}_o \times \mathbf{p}\right) \otimes \mathbf{e}_3 + \mathbf{e}_3 \otimes \left(\boldsymbol{\epsilon}_o + \boldsymbol{\varkappa}_o \times \mathbf{p}\right)\right] \tag{379}$$

where

$$\boxed{\begin{aligned} \boldsymbol{\epsilon}_o(x) &\equiv \mathbf{w}'(x) + \mathbf{e}_3 \times \boldsymbol{\theta}(x) \\ \boldsymbol{\varkappa}_o(x) &\equiv \boldsymbol{\theta}'(x) \end{aligned}} \tag{380}$$

are *strain resultants* for the linear beam. It is evident that there is no strain in the plane of a cross section because $E_{11} = E_{12} = E_{22} = 0$, in accord with the assumption that the cross-sectional plane is rigid.

Constitutive Relations for Stress Resultants

The strains imply stresses through the constitutive relationships. Let us assume that the material is linearly elastic and isotropic so that the stress-strain relationship is $\mathbf{S} = \lambda\,\mathrm{tr}(\mathbf{E})\mathbf{I} + 2\mu\mathbf{E}$. We are interested in the resultant tractions on a cross section with normal $\mathbf{e}_3$. Therefore, it is sufficient to compute

$$\mathbf{t}_{\mathbf{e}_3} = \mathbf{S}\mathbf{e}_3 = \lambda\,\mathrm{tr}(\mathbf{E})\mathbf{e}_3 + 2\mu\,\mathbf{E}\mathbf{e}_3$$

From Eqn. (379) we can compute

$$\begin{aligned} 2\mathbf{E}\mathbf{e}_3 &= \left[\mathbf{I} + \mathbf{e}_3 \otimes \mathbf{e}_3\right]\left(\boldsymbol{\epsilon}_o + \boldsymbol{\varkappa}_o \times \mathbf{p}\right) \\ \mathrm{tr}(\mathbf{E}) &= \mathbf{e}_3 \cdot \left(\boldsymbol{\epsilon}_o + \boldsymbol{\varkappa}_o \times \mathbf{p}\right) \end{aligned}$$

Therefore, the expression for the traction on a cross section with normal $\mathbf{e}_3$ is

$$\mathbf{S}\mathbf{e}_3 = \left[(\lambda+\mu)\mathbf{e}_3 \otimes \mathbf{e}_3 + \mu\mathbf{I}\right]\left(\boldsymbol{\epsilon}_o + \boldsymbol{\varkappa}_o \times \mathbf{p}\right) \tag{381}$$

where $\mathbf{I}$ is the identity tensor. We will use this expression in our definitions for the stress resultants to derive constitutive equations relating the stress resultants to our generalized displacements. Making an analogy with constitutive equations in general, these relationships will help identify the appropriate measures of strain (i.e., strain resultants) and determine how they are related to the stress resultants. Substituting Eqn. (381) into our definitions of stress resultant and moment, Eqn. (367), we get

$$\begin{aligned} \mathbf{Q} &= \int_\Omega \boldsymbol{\Xi}\left(\boldsymbol{\epsilon}_o + \boldsymbol{\varkappa}_o \times \mathbf{p}\right) dA \\ \mathbf{M} &= \int_\Omega \mathbf{p} \times \left(\boldsymbol{\Xi}\left(\boldsymbol{\epsilon}_o + \boldsymbol{\varkappa}_o \times \mathbf{p}\right)\right) dA \end{aligned} \tag{382}$$

where $\boldsymbol{\Xi} \equiv (\lambda + \mu)[\mathbf{e}_3 \otimes \mathbf{e}_3] + \mu \mathbf{I}$. We can explicitly integrate out the $x_1 - x_2$ dependence of these expressions at a typical cross section located at a fixed value of the coordinate x_3. Observing that only the vector $\mathbf{p}$ depends upon x_1 and x_2, we find that these integrations are rather simple. First, some modest rearrangement of the equations is in order. It is useful to recognize that the operation of the cross product of two vectors can be viewed as the operation of a skew-symmetric tensor times a vector, $\mathbf{w} \times \mathbf{v} = \mathbf{W}\mathbf{v}$, where the tensor $\mathbf{W}$ is given by

$$\mathbf{W} \sim \begin{bmatrix} 0 & -w_3 & w_2 \\ w_3 & 0 & -w_1 \\ -w_2 & w_1 & 0 \end{bmatrix}$$

We therefore can think of the quantity $\mathbf{w} \times \equiv \mathbf{W}$ as a second order tensor. Accordingly, we can write

$$[\mathbf{p} \times] \sim \begin{bmatrix} 0 & 0 & x_2 \\ 0 & 0 & -x_1 \\ -x_2 & x_1 & 0 \end{bmatrix}$$

Noting that $[\mathbf{p} \times]^T = -[\mathbf{p} \times]$ we can write the resultants in the form

$$\begin{aligned} \mathbf{Q} &= \int_\Omega \left(\boldsymbol{\Xi}\boldsymbol{\epsilon}_o + \boldsymbol{\Xi}[\mathbf{p} \times]^T \boldsymbol{\varkappa}_o \right) dA \\ \mathbf{M} &= \int_\Omega \left([\mathbf{p} \times]\boldsymbol{\Xi}\boldsymbol{\epsilon}_o + [\mathbf{p} \times]\boldsymbol{\Xi}[\mathbf{p} \times]^T \boldsymbol{\varkappa}_o \right) dA \end{aligned} \tag{383}$$

Carrying out the integrals in Eqn. (382), we obtain the constitutive equations for stress resultants

$$\boxed{\begin{aligned} \mathbf{Q} &= \text{`}\boldsymbol{\epsilon}_o + \mathsf{S}\boldsymbol{\varkappa}_o \\ \mathbf{M} &= \mathsf{S}^T\boldsymbol{\epsilon}_o + \mathsf{I}\boldsymbol{\varkappa}_o \end{aligned}} \tag{384}$$

where the tensors A, S, and I are given by

$$\begin{aligned} \mathsf{A} &\equiv \int_\Omega \boldsymbol{\Xi}\, dA, & \mathsf{S} &\equiv \int_\Omega \boldsymbol{\Xi}[\mathbf{p} \times]^T dA \\ \mathsf{S}^T &\equiv \int_\Omega [\mathbf{p} \times]\boldsymbol{\Xi}\, dA, & \mathsf{I} &\equiv \int_\Omega [\mathbf{p} \times]\boldsymbol{\Xi}[\mathbf{p} \times]^T dA \end{aligned} \tag{385}$$

where $\boldsymbol{\Xi} \equiv (\lambda + \mu)[\mathbf{e}_3 \otimes \mathbf{e}_3] + \mu \mathbf{I}$. The explicit components of the constitutive property tensors A, S, and I depend only on the elastic constants of the material and the geometric properties of the cross section. Let us make the following definitions of cross-sectional geometric properties

$$A \equiv \int_\Omega dA, \quad S_\alpha \equiv \int_\Omega x_\alpha \, dA, \quad I_{\alpha\beta} \equiv \int_\Omega x_\alpha x_\beta \, dA \tag{386}$$

where, as usual, α and β take values from $\{1, 2\}$. We usually refer to A as the area of the cross section, S_α as the first moment of the area, and $I_{\alpha\beta}$ as the second moment of the area. There is only one area, but there are two first moments, and three distinct second moments. Often we use the notation $J \equiv I_{11} + I_{22}$ to designate the *polar moment of the area*. The components of the constitutive tensors have the final expressions (work these out for yourself)

$$\mathbf{A} \sim \begin{bmatrix} \mu A & 0 & 0 \\ 0 & \mu A & 0 \\ 0 & 0 & EA \end{bmatrix} \quad \mathbf{S} \sim \begin{bmatrix} 0 & 0 & -\mu S_2 \\ 0 & 0 & \mu S_1 \\ ES_2 & -ES_1 & 0 \end{bmatrix} \quad \mathbf{I} \sim \begin{bmatrix} EI_{22} & -EI_{12} & 0 \\ -EI_{12} & EI_{11} & 0 \\ 0 & 0 & \mu J \end{bmatrix} \tag{387}$$

where the notation $E \equiv \lambda + 2\mu$ has been introduced for notational simplicity.

The constitutive tensors can be considerably simplified by a judicious choice of coordinate axes. If the origin of coordinates in the cross-sectional plane is taken to be the centroid of the section, then both of the first moments of the area S_α vanish (and, hence, so does the entire tensor $\mathbf{S}$ and its transpose $\mathbf{S}^T$). Further, if the axes are taken to coincide with the principal axes of the cross section, then the product of inertia I_{12} vanishes, rendering the tensor $\mathbf{I}$ diagonal. With such a choice of coordinate axes, the relationships between the stress resultants and the strain resultants simplify considerably. Therefore, in what follows, we shall always make that choice (unless specifically indicated otherwise).

The constitutive equations are of interest not only because they relate the generalized displacements with the stress resultants, but also because they help us identify the concept of strain resultant. The strain resultant is the one-dimensional counterpart of the strain tensor in the three-dimensional theory. As the name indicates, it is the net result of all of the local straining across the cross section—an average, if you will. With the canonical choice of coordinate axes (principal, centroidal axes), the stress resultant $\mathbf{Q}$ is linearly related to the deformation measure $\boldsymbol{\epsilon}_o$. Accordingly, we shall view this quantity as the associated strain resultant. Similarly, the moment $\mathbf{M}$ is linearly related to the measure of deformation $\boldsymbol{\varkappa}_o$. We shall consider $\boldsymbol{\varkappa}_o$ to be the strain resultant associated with $\mathbf{M}$.

These resultants have a clear physical interpretation. Let us write out the components of each of these strain resultants

$$\begin{aligned} \boldsymbol{\epsilon}_o &= (w_1' - \theta_2)\mathbf{e}_1 + (w_2' + \theta_1)\mathbf{e}_2 + w_3'\mathbf{e}_3 \\ \boldsymbol{\varkappa}_o &= \theta_1'\mathbf{e}_1 + \theta_2'\mathbf{e}_2 + \theta_3'\mathbf{e}_3 \end{aligned}$$

Consider again the case of planar deformation in the x_1 - x_3 plane, shown in Fig. 90. For planar motion, we have null displacements and rotations (and their derivatives) for all quantities that give rise to motion out of the plane. Accordingly, $w_2 = 0$, $\theta_1 = 0$, and $\theta_3 = 0$. We can see that w_3' measures the rate of axial stretch, i.e., the net axial strain, of the beam. It is associated with the axial force Q_3. The quantity w_1' is the slope of the deformed axis of the beam. As we can clearly see in the figure, the tangent to the deformed axis does not necessarily coincide with the direction of the normal vector **n**. The angle between these two lines is due to shearing of the beam. The strain resultant $w_1' - \theta_2$ directly measures this component of deformation, and is associated with the shear force Q_1. The rate of change of the rotation of the normal vector **n** is θ_2', the curvature of the beam flexing about the x_2 axis. The curvature is associated with the bending moment M_2 about the x_2 axis. Note that this expression for the curvature is exact (whereas the second derivative of the transverse deflection is an approximation to the curvature). By extension, the meanings of the other terms in the three-dimensional case are evident.

The shear strain resultant $w_2' + \theta_1$ has a sign for the rotation term that is different than the shear strain in the other direction. The reason for this difference is due to right-hand-rule convention for the rotations. The resultant shearing angle is always measured as the angle between the tangent to the deformed axis and the normal to the cross section. The rotation angle is always measured relative to the undeformed axis of the beam. Figure 90 shows positive values for the displacements, displacement gradients, and rotations for two cases: (a) planar deformation in the x_1 - x_3 plane and (b) planar deformation in the x_2 - x_3 plane. The shearing angle is shown shaded. Note that, for the first case, the x_2 axis is directed out of the page, while in the second case the x_1 axis is directed into the page. In both cases, a positive transverse displacement is upward, in the direction of the associated coordinate direction. The rate of change of the transverse displacement, or the slope of the deformed axis, is positive if it points in the direction up and to the right. On the other hand, according to the right-hand rule, the rotation θ_2 is positive if it is anticlockwise, while the rotation θ_1 is positive if it is clockwise. Thus, in the first case, the shear angle is the difference between these two positive quantities, while in the second case, the shear angle is the sum of these two positive quantities.

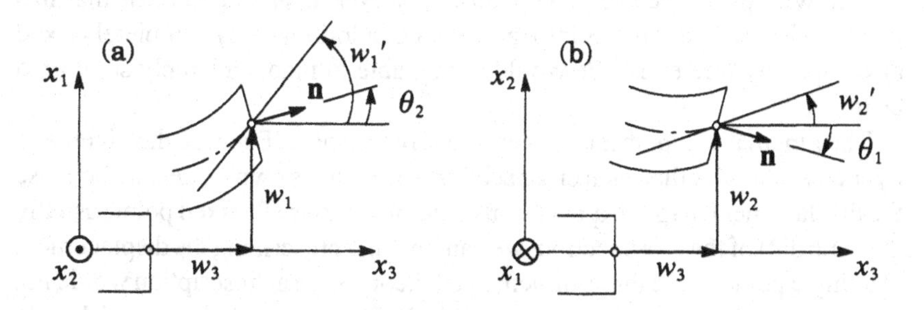

Figure 90 Why is there a sign difference in the two resultant shear strains?

Boundary Conditions

Like any three-dimensional body, a beam has a surface on which either tractions are prescribed or displacements are prescribed. For the three-dimensional theory, we referred to these portions of the boundary of the body $\mathcal{B}$ as Ω_t and Ω_w respectively, and any point on the boundary was a candidate for either prescribed traction or prescribed displacement, but not both. The beam is a special case of a three-dimensional body because of the kinematic hypothesis, and this restriction affects all aspects of the theory, including the prescription of boundary conditions.

We have already addressed the issue of prescribing tractions on the lateral surfaces of the beam. In fact, the definitions of applied force and moment include the resultants of these prescribed tractions. We must also consider the tractions that exist at the ends of the beam. Let $\mathbf{t}_o$ be the applied traction field (and Ω_o the cross section) at $x_3 = 0$, and $\mathbf{t}_\ell$ the applied traction field (and Ω_ℓ the cross section) at $x_3 = \ell$. Computing the net force acting at these cross sections leads to the definition of the end resultant forces as

$$\mathbf{q}_o \equiv \int_{\Omega_o} \mathbf{t}_o \, dA, \quad \mathbf{q}_\ell \equiv \int_{\Omega_\ell} \mathbf{t}_\ell \, dA \tag{388}$$

These forces represent the net resultant of the applied tractions at the two ends of the beam. The resultant moments are similarly defined as

$$\mathbf{m}_o \equiv \int_{\Omega_o} \mathbf{p} \times \mathbf{t}_o \, dA, \quad \mathbf{m}_\ell \equiv \int_{\Omega_\ell} \mathbf{p} \times \mathbf{t}_\ell \, dA \tag{389}$$

These moments are the first moment of the applied tractions about the axis at the two ends of the beam.

Technically, we should consider the ends of the beam as having either applied tractions or prescribed displacements because our theory is expressed in terms of resultants, and, as such, is not equipped to differentiate between a region of an end section with prescribed tractions and a region of that same end section with prescribed displacements. However, it should be clear that this point of view would force us into the corner of admitting only completely fixed or completely free ends. We would not be able to model a simply supported beam!

The kinematic hypothesis comes to our rescue here. Because the kinematic hypothesis implies that each cross section is rigid in its own plane, and because the displacement map is expressed in terms of motions of a fixed point (usually the centroid) of the cross section, we can imagine prescribing the displacement at a single point. In the three-dimensional theory, such a prescription would not be admissible because a point force is a finite force applied over a vanishingly small area and, thus, leads to infinite tractions and stresses. The assumption of rigidity, while not really justifying the concept of a point load, certainly allows the theory to accommodate it. In view of this special feature of beam theory, we can now imagine a cross section where the net displacement $\mathbf{w}$ is prescribed, but the net rotation is not. This is the condition known as the *simple support*, which plays such an important role in structural engineering. For the sake of argument, let us assume that we are talking about the cross section Ω_o. If the displacement $\mathbf{w}(0) = \mathbf{0}$ is known, then some corresponding force must be unknown. We can demonstrate through a virtual-work argument that the unknown force is $\mathbf{q}_o$. Further, since the cross section is free to rotate, there must be some force that is prescribed. Again, we can demonstrate through a virtual-work argument that the prescribed force is $\mathbf{m}_o$.

Such a condition of mixed boundary conditions can only be realized through a condition of constraint. Imagine simply that our beam is attached to a rigid plate at the end. The tractions that the beam feels are those transmitted to it from the rigid plate. Now we can imagine that the rigid plate is attached to a ball-and-socket joint that is free to rotate in any direction, but is not free to translate. This device constitutes our three-dimensional version of the simple support. Since beam theory actually provides the rigid plate, we need not worry about its physical implementation to carry out beam calculations.

Since we are in the business of concocting support devices for our rigid plate, why not imagine a whole collection of such devices. How about one that rotates about the x_2 axis, but not about the x_1 axis? How about one that is free to translate in the x_2 direction, but not in the x_1 direction? We have six generalized displacements (including rotations) at each section. We can imagine a device that independently prescribes the associated force or displacement for each one. Hence, each component of the end resultant vectors can exist as either a prescribed force or a reaction force if the corresponding displacement is prescribed. For the beam in three dimensions, we must prescribe either the force or the displacement at each end point. Thus, we always have exactly 12 boundary conditions. For the planar problem, this number reduces to six. These conditions are always exactly enough to determine the constants of integration that we get when we solve the governing differential equations.

The Limitations of Beam Theory

Unfortunately, beam theory is not completely consistent with the three-dimensional theory. Every time we constrain a system, we pay a price. In the present case, we constrained the deformation map so that cross sections of the beam remain rigid. We pay for this simplification in the constitutive equations and in the satisfaction of equilibrium locally within a cross section.

Poisson's effect. The ramification of the rigid cross section assumption is that the normal and shear strain components in the plane of a cross section vanish; in our coordinate system that means $E_{11} = E_{22} = 0$ and $E_{12} = E_{21} = 0$. This constraint on the strains is not a problem for the shear strain components because the shear stresses and strains are uncoupled in the constitutive equations (for example, $S_{12} = 2\mu E_{12}$). Therefore, vanishing shear strains simply implies vanishing shear stress. The normal strains are another story.

The constitutive equations for isotropic hyperelasticity have an inherent coupling of the normal strains and stresses. The stresses are given by Hooke's law as $\mathbf{S} = \lambda \operatorname{tr}(\mathbf{E})\mathbf{I} + 2\mu\mathbf{E}$. Since $\operatorname{tr}(\mathbf{E}) = E_{33}$ for kinematic hypothesis of the beam, we have $S_{11} = S_{22} = \lambda E_{33}$. Now, from Hooke's law $S_{33} = (\lambda + 2\mu)E_{33}$ so $S_{11} = S_{22} = \nu S_{33}$, where ν is Poisson's ratio. Thus, the constraint induces normal stresses in the plane of the cross section owing to Poisson's effect (tension in one direction causes lateral contraction of the dimensions perpendicular to the direction of tension). Observational evidence on the behavior of beams would indicate that these stresses tend to be rather small. In fact, it is possible that the presence of these stresses will violate the traction boundary conditions on the lateral surface of the beam. For a beam with a traction-free lateral surface, we can argue that the normal stresses S_{11} and S_{22} are very small because the cross-sectional dimensions are small compared with the length of the beam. We would like to make the assumption that $S_{11} = S_{22} = 0$, but that violates the precept of mechanics that we can specify either the motion or the force at a point, but not both.

Let us examine what would have happened if we had made the assumption of vanishing normal stress and not the assumption of vanishing normal strain. We have done it before. We made exactly that assumption for the uniaxial tension test in order to recast the constitutive equations in terms of Young's modulus and Poisson's ratio. When we made the assumption that $S_{11} = S_{22} = 0$, we got the constitutive relationship $S_{33} = CE_{33}$, rather than $S_{33} = (\lambda + 2\mu)E_{33}$, which results for the vanishing strain assumption. This gives us a way to partially recover from our difficulties. In the constitutive relationships for beams, the quantity $E = \lambda + 2\mu$ appears repeatedly. If we simply substitute the value of Young's modulus $E = C$ instead, then the results of beam theory accord well with observation.

Equilibrium inconsistencies. Because we are working with stress resultants, we consider the equilibrium of the stress field over a cross section only in an average sense. On average, the equations of equilibrium are exactly consistent with the three-dimensional theory. However, locally we may fail to satisfy equilibrium. For the beam, the most obvious failure concerns the distribution of shear stresses over the cross section.

The kinematic hypothesis suggests that the shear strains will be constant over the cross section. In reality, owing to the presence of shear stresses, the cross section must warp out of its plane. The restraint of warping gives beam theory slightly more stiffness than the three-dimensional theory or observation would indicate. The local equilibrium equations suggest that the normal stress is related to the gradient of the shear stress over the cross section. If the normal stress is linear, as the kinematic hypothesis suggests, then the shear stress must be quadratic rather than constant.

The inconsistency can be neutralized by modifying the constitutive equations slightly. The modulus for the shear strains that comes from the three-dimensional theory is simply the product μA of the Lamé parameter and the cross-sectional area. If we adjust the area by multiplying by what has become known as the *shear coefficient* to give an effective area A', then the results of beam theory are better when shear deformations are important. For a rectangular cross section, the shear coefficient is approximately 5/6. In general, this coefficient depends upon the cross-sectional geometry. The stress inconsistency remains, but shear stresses can be computed from three-dimensional equilibrium equations from the normal stresses as a post-processing task, if needed. Most elementary texts take an energy approach to determine a modification factor for the shear area. An alternative approach is to modify the kinematic hypothesis to include warping. Such an addition to the kinematic hypothesis leads to a more accurate theory, but for most beam geometries, this refinement is hardly necessary.

Torsion. Along the same lines as the transverse shears discussed in the last section, one of the most significant problems with beam theory is that the plane-sections hypothesis overestimates the torsional stiffness μJ, which comes out to be the shear modulus times the polar moment of the area for any cross section. We can demonstrate that the estimate is exact only for circular cross sections. For other cross-sectional shapes, out-of-plane warping must accompany the displacement and rotation of the cross section in order to satisfy the traction conditions on the lateral surface of the beam.

To illustrate the effects of torsional warping, let us consider a bar fixed at the left end and subjected to a pure torque $T\mathbf{e}_3$ (i.e., about the beam axis) at the right end. This problem is often called the *Saint-Venant torsion problem*. The deformation map for pure twisting of a beam about its axis is

$$\mathbf{u}(\mathbf{x}) = \alpha x_3(\mathbf{e}_3 \times \mathbf{p}_o) + \alpha\psi(x_1,x_2)\mathbf{e}_3 \tag{390}$$

where $\mathbf{p}_o = (x_1 - a_1)\mathbf{e}_1 + (x_2 - a_2)\mathbf{e}_2$, α is the rate of twist, and $\psi(x_1,x_2)$ is, as yet, an unknown function. The point $\mathbf{a}$ is the center of twist of the cross section. According to the map, the angle of twist is zero at the end $x_3 = 0$ and increases at the rate α to a maximum at the right end. The vector $\mathbf{e}_3 \times \mathbf{p}_o$ is perpendicular to both $\mathbf{p}_o$ and $\mathbf{e}_3$. Thus, for small α, the first term of the displacement map represents the displacement of a point initially located at $\mathbf{p}_o$ owing to a pure rotation about the center of twist. This component of displacement takes place entirely in the plane of the cross section. The second term of the map gives displacement out of the plane of the cross section. As such, the function $\psi(x_1,x_2)$ is called the *warping function* and represents the departure from the plane-sections hypothesis. Figure 91 illustrates some features of the torsion map for a rectangular section.

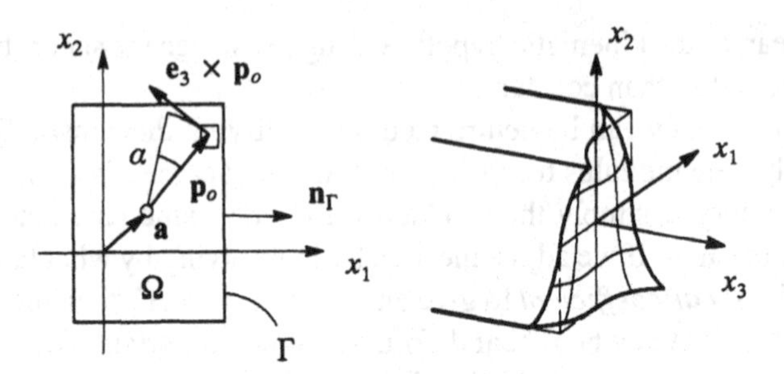

Figure 91 The linearized torsion map with warping

We can compute the linearized strain from the displacement map and substitute the result into the linear elastic constitutive equations to get the stress

$$\mathbf{S} = a\mu\left[\left(\nabla\psi + \mathbf{e}_3 \times \mathbf{p}_o\right) \otimes \mathbf{e}_3 + \mathbf{e}_3 \otimes \left(\nabla\psi + \mathbf{e}_3 \times \mathbf{p}_o\right)\right] \tag{391}$$

The divergence of the stress tensor can be computed from the above expression and substituted into the equilibrium equation $\operatorname{div}\mathbf{S} + \mathbf{b} = \mathbf{0}$, with $\mathbf{b} = \mathbf{0}$, to give the following equation governing the warping function

$$\operatorname{div}\left(\nabla\psi\right) = 0 \quad \text{in } \Omega \tag{392}$$

where $\operatorname{div}\left(\nabla\psi\right) = \nabla^2\psi = \psi_{,aa}$ is the Laplacian of the warping function. The tractions on the lateral contour Γ, with normal vector field $\mathbf{n}_\Gamma = n_1\mathbf{e}_1 + n_2\mathbf{e}_2$, must vanish because no tractions are applied there. Thus, from $\mathbf{S}\mathbf{n}_\Gamma = \mathbf{0}$ we get the additional requirement for the warping function

$$\left(\nabla\psi + \mathbf{e}_3 \times \mathbf{p}_o\right) \cdot \mathbf{n}_\Gamma = 0 \quad \text{on } \Gamma \tag{393}$$

The term $\nabla\psi \cdot \mathbf{n}_\Gamma$ gives the rate of change of ψ in the normal direction $\mathbf{n}_\Gamma$. The second term is $\mathbf{e}_3 \times \mathbf{p}_o \cdot \mathbf{n}_\Gamma = -\left(x_2 - a_2\right)n_1 + \left(x_1 - a_1\right)n_2$, which gives a clue as to why the circular cross section does not warp and noncircular ones do. The normal vector to the circular cross section has $n_1 = x_1/r$ and $n_2 = x_2/r$, where r is the radius of the circle. Clearly, $\mathbf{e}_3 \times \mathbf{p}_o \cdot \mathbf{n}_\Gamma = 0$ for the circular cross section with $\mathbf{a} = \mathbf{0}$. Consequently, the function $\psi(x_1, x_2) = 0$ satisfies Eqns. (392) and (393).

There is no net resultant on a cross section. Hence, the integral of $\mathbf{S}\mathbf{e}_3$ over the cross-sectional area should be zero. A straightforward computation with Eqn. (391) shows that $\mathbf{S}\mathbf{e}_3 = a\mu\left(\nabla\psi + \mathbf{e}_3 \times \mathbf{p}_o\right)$. Therefore

$$\int_\Omega \left(\nabla\psi + \mathbf{e}_3 \times \mathbf{p}_o\right) dA = \mathbf{0} \tag{394}$$

This equation is sufficient to establish the location of the center of twist $\mathbf{a}$. If the cross section is symmetric, then the center of twist lies along the axis of symmetry. The integral of the moments of the tractions about the origin $(\mathbf{p}_o \times \mathbf{S}\mathbf{e}_3)$ must be equal to the total applied torque $T\mathbf{e}_3$, hence

$$T\mathbf{e}_3 = \alpha\mu \int_\Omega \mathbf{p}_o \times \left(\nabla\psi + \mathbf{e}_3 \times \mathbf{p}_o\right) dA \tag{395}$$

This result tells us that the torque T is proportional to the rate of twist α, but that the proportionality constant is not the shear modulus times the polar moment of inertia J. Let us define the torsional stiffness to be $\mu\bar{J}$ where

$$\bar{J} \equiv \mathbf{e}_3 \cdot \int_\Omega \mathbf{p}_o \times \left(\nabla\psi + \mathbf{e}_3 \times \mathbf{p}_o\right) dA \tag{396}$$

It is straightforward to show that

$$\mathbf{e}_3 \cdot \mathbf{p}_o \times \left(\mathbf{e}_3 \times \mathbf{p}_o\right) = -\mathbf{e}_3 \cdot [\mathbf{p}_o \times][\mathbf{p}_o \times]\mathbf{e}_3 = \mathbf{p}_o \cdot \mathbf{p}_o \tag{397}$$

Therefore, the torsion constant can be expressed in the form

$$\bar{J} \equiv \int_\Omega \left(\xi_1 \frac{\partial\psi}{\partial\xi_2} - \xi_2 \frac{\partial\psi}{\partial\xi_1} + \xi_1^2 + \xi_2^2 \right) dA \tag{398}$$

where $\boldsymbol{\xi} = \mathbf{x} - \mathbf{a}$. The consequence of the definition of torsional stiffness will be evident in the following example.

Example 46. ***Torsion of an elliptical cross section.*** Consider an elliptical cross section, shown in Fig. 92, with major and minor semi-axes of a and b,

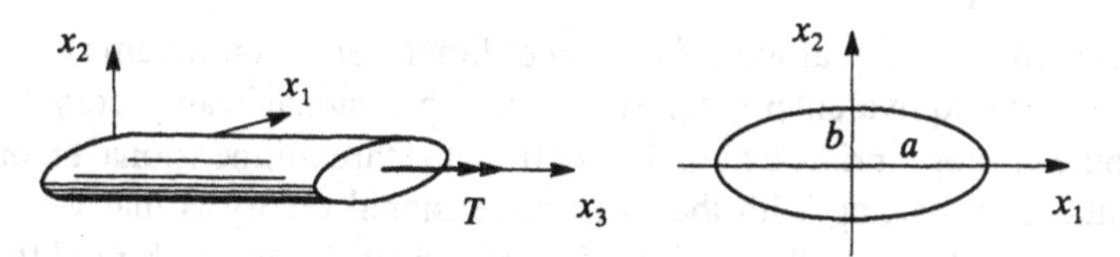

Figure 92 Torsion of an elliptical beam

The boundary of the cross section is described by the equation

$$b^2x_1^2 + a^2x_2^2 - a^2b^2 = 0$$

Therefore, the normal vector $\mathbf{n}_\Gamma$ has components $n_1 = nb^2x_1$ and $n_2 = na^2x_2$, where n is the scaling factor required to make the normal vector a unit vector (which plays no role in this calculation). The warping function for this cross section can be taken as (see, for example Sokolnikoff, 1956)

$$\psi(x_1, x_2) = -cx_1x_2$$

where $c \equiv (a^2-b^2)/(a^2+b^2)$. A straightforward calculation will verify that $\nabla^2\psi = 0$ and $\left(\nabla\psi + \mathbf{e}_3 \times \mathbf{p}_o\right) \cdot \mathbf{n}_\Gamma = 0$ on the boundary of the ellipse. The torsional stiffness $\bar{J}$ can be computed by observing that

$$\mathbf{e}_3 \cdot \left(\mathbf{p}_o \times \nabla\psi + \mathbf{p}_o \times \left(\mathbf{e}_3 \times \mathbf{p}_o\right)\right) = (1-c)x_1^2 + (1+c)x_2^2$$

Carrying out the indicated integral over the area of the cross section in Eqn. (396), we obtain the torsional stiffness

$$\bar{J} = \frac{\pi a^3 b^3}{a^2+b^2}$$

Observe that the polar moment of the area of an ellipse is $J = \pi a b\left(a^2+b^2\right)/4$. Let the ratio of the major to minor dimension of the ellipse be $\gamma \equiv a/b$. Then the ratio of torsional stiffness to the polar moment of the area is

$$\frac{\bar{J}}{J} = \frac{4\gamma^2}{\left(1+\gamma^2\right)^2}$$

This ratio is always less than or equal to one with equality only for a circular cross section ($\gamma = 1$). For $\gamma \gg 1$ the torsional stiffness is $\bar{J} \approx \pi a b^3$.

It should be evident that the first task of solving a torsion problem is the determination of the warping function $\psi(x_1, x_2)$. There are a few alternative approaches to solving this two-dimensional boundary value problem. The reader should consult Sokolnikoff (1956) for a detailed discussion of this classical problem. In the context of beam theory, we might be satisfied to replace μJ with $\mu \bar{J}$ and use beam theory without further modification. Problem 171 explores the issue of the effect of warping restraint.

The Principle of Virtual Work for Beams

The principle of virtual work for a three-dimensional continuum can be used to develop an equivalent principle for a one-dimensional beam theory. We shall compute the appropriate external work from the three-dimensional theory. The advantage of starting with the three-dimensional theory is that we need to know only that work is the product of force and displacement. Straightforward operations will yield the result that work for a beam includes terms computed as the product of moment and rotation. The key to reducing the principle of virtual work to one dimension lies in our kinematic hypothesis and our definition of stress resultants.

The displacement map is constrained by the kinematic hypothesis. We will find it convenient to construct our virtual displacement field in accord with the same hypothesis. Hence, our (three-dimensional) virtual displacement field can be expressed as

$$\bar{\mathbf{u}}(\mathbf{x}) = \bar{\mathbf{w}}(x_3) + \bar{\boldsymbol{\theta}}(x_3) \times \mathbf{p}(x_1, x_2) \tag{399}$$

where $\bar{\mathbf{w}}(x_3)$ and $\bar{\boldsymbol{\theta}}(x_3)$ represent the virtual displacements and rotations of the beam. The external virtual work is simply the product of the applied body forces and tractions with their respective virtual displacements, integrated over the volume and surface of the beam

$$\overline{W}_E \equiv \int_0^\ell \int_\Omega \mathbf{b} \cdot \overline{\mathbf{u}}\, dA\, dx_3 + \int_0^\ell \int_\Gamma \mathbf{t}_\Gamma \cdot \overline{\mathbf{u}}\, ds\, dx_3 + \int_{\Omega_o} \mathbf{t}_o \cdot \overline{\mathbf{u}}\, dA + \int_{\Omega_\ell} \mathbf{t}_\ell \cdot \overline{\mathbf{u}}\, dA \tag{400}$$

where $\mathbf{b}$ is the body force, $\mathbf{t}_\Gamma$ is the applied traction on the lateral contour of the beam, $\mathbf{t}_o$ is the applied traction field acting on the cross section Ω_o at $x_3 = 0$, and $\mathbf{t}_\ell$ is the applied traction field acting on the cross section Ω_ℓ at $x_3 = \ell$. Substituting Eqn. (399) into (400) and carrying out the appropriate integrals over Ω, Γ, Ω_o, and Ω_ℓ, we obtain a one-dimensional expression for the external virtual work. Note that

$$\mathbf{b} \cdot \overline{\mathbf{u}} = \mathbf{b} \cdot \left(\overline{\mathbf{w}} + \overline{\boldsymbol{\theta}} \times \mathbf{p}\right) = \overline{\mathbf{w}} \cdot \mathbf{b} + \overline{\boldsymbol{\theta}} \cdot \left(\mathbf{p} \times \mathbf{b}\right)$$

Each of the integrands in Eqn. (400) can be handled in a similar fashion. The first two terms in Eqn. (400) can be expressed as

$$\int_0^\ell \overline{\mathbf{w}} \cdot \left(\int_\Omega \mathbf{b}\, dA + \int_\Gamma \mathbf{t}_\Gamma\, ds\right) dx_3 + \int_0^\ell \overline{\boldsymbol{\theta}} \cdot \left(\int_\Omega \mathbf{p} \times \mathbf{b}\, dA + \int_\Gamma \mathbf{p} \times \mathbf{t}_\Gamma\, ds\right) dx_3$$

Observe that the terms in parentheses are precisely our definitions of the resultant of the applied loads $\mathbf{q}(x)$ and $\mathbf{m}(x)$, respectively. We can use a similar argument for the third and fourth terms in the definition of external virtual work. These terms can be rearranged to read

$$\overline{\mathbf{w}}(0) \cdot \int_{\Omega_o} \mathbf{t}_o\, dA + \overline{\boldsymbol{\theta}}(0) \cdot \int_{\Omega_o} \mathbf{p} \times \mathbf{t}_o\, dA + \overline{\mathbf{w}}(\ell) \cdot \int_{\Omega_\ell} \mathbf{t}_\ell\, dA + \overline{\boldsymbol{\theta}}(\ell) \cdot \int_{\Omega_\ell} \mathbf{p} \times \mathbf{t}_\ell\, dA$$

Clearly, the four cross-sectional integrals are precisely our definition of the resultants of the tractions on these sections $\mathbf{q}_o$, $\mathbf{m}_o$, $\mathbf{q}_\ell$, and $\mathbf{m}_\ell$. Thus, in the context of the kinematic hypothesis, these four terms exactly account for all of the virtual work done by the traction forces on the ends of the beam. We can also see that $\mathbf{q}_o$, $\mathbf{q}_\ell$, $\mathbf{m}_o$, and $\mathbf{m}_\ell$ are the natural forces conjugate to the virtual displacements $\overline{\mathbf{w}}(0)$, $\overline{\mathbf{w}}(\ell)$, $\overline{\boldsymbol{\theta}}(0)$, and $\overline{\boldsymbol{\theta}}(\ell)$, respectively, in the sense that they completely characterize the work done.

Combining these results, we can see that the external virtual work done by the forces acting on our beam in going through a constrained virtual displacement $\overline{\mathbf{u}} = \overline{\mathbf{w}} + \overline{\boldsymbol{\theta}} \times \mathbf{p}$ is

$$\boxed{\begin{aligned}\overline{W}_E = & \int_0^{\ell} \big(\mathbf{q}(x)\cdot\overline{\mathbf{w}}(x) + \mathbf{m}(x)\cdot\overline{\boldsymbol{\theta}}(x)\big)\,dx \\ & + \mathbf{q}_o\cdot\overline{\mathbf{w}}(0) + \mathbf{q}_\ell\cdot\overline{\mathbf{w}}(\ell) \\ & + \mathbf{m}_o\cdot\overline{\boldsymbol{\theta}}(0) + \mathbf{m}_\ell\cdot\overline{\boldsymbol{\theta}}(\ell)\end{aligned}} \tag{401}$$

Again, $\mathbf{q}(x)$ is the net applied force along the axis of the beam, $\mathbf{m}(x)$ is the net applied moment along the axis of the beam; $\mathbf{q}_o$ is the net applied force and $\mathbf{m}_o$ the net applied moment acting at $x_3 = 0$; and $\mathbf{q}_\ell$ is the net applied force and $\mathbf{m}_\ell$ the net applied moment acting at $x_3 = \ell$. All of these quantities have been defined previously, and are shown in Fig. 93. Again, we see that the kinematic hypothesis allows us to integrate out the cross-sectional dependence, leaving us with quantities depending only on the axial coordinate. The crucial observation is that, within the context of the kinematic hypothesis, this expression for the external virtual work is exactly consistent with the three-dimensional theory.

Observe that moment is the natural dual of rotation, in the sense of virtual work. In addition to work done by forces multiplied by their respective displacements, we must include moments multiplied by their respective rotations in our accounting for the work done by the system. All theories that introduce the concept of moment have this feature, including plates and shells. Of course, both moment and rotation must be reckoned with respect to the same axis (i.e., they must both use the same $\mathbf{p}$ in their definition).

The principle of virtual work is a valuable tool with which to consider the conjugateness of stress and strain resultants. We saw in the derivation of the principle of virtual work that a measure of internal virtual work involving the product of stress and strain appeared naturally. It had the form

$$\overline{W}_I \equiv \int_{\mathfrak{B}} \mathbf{S}\cdot\nabla\overline{\mathbf{u}}\,dV \tag{402}$$

You might expect that if a reduced theory is truly compatible with the three-dimensional theory, then an analogous expression for internal virtual work in terms of the resultant quantities should result. In fact, we can use this equiva-

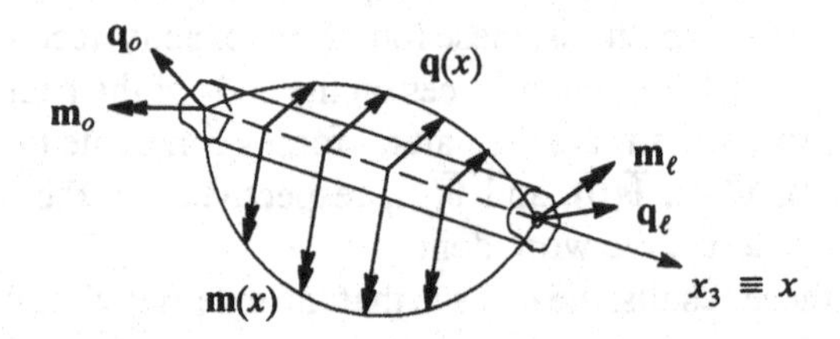

Figure 93 The applied forces on a beam

lence to define which resultant strain measures are appropriately conjugate to the defined stress resultant measures. This equivalence is particularly important since we defined stress resultants without regard to the specific kinematic hypothesis, and the kinematic hypothesis had nothing to do with the definition of stress resultants.

Let us substitute the virtual strain implied by our kinematic map. From Eqn. (378) we have

$$\overline{W}_I = \int_0^\ell \int_\Omega \mathbf{S} \cdot \left[\left(\overline{\mathbf{w}}' + \mathbf{e}_3 \times \overline{\boldsymbol{\theta}} + \overline{\boldsymbol{\theta}}' \times \mathbf{p} \right) \otimes \mathbf{e}_3 - \overline{\boldsymbol{\Theta}} \right] dA\, dx_3$$

$$= \int_0^\ell \int_\Omega \left(\overline{\mathbf{w}}' + \mathbf{e}_3 \times \overline{\boldsymbol{\theta}} + \overline{\boldsymbol{\theta}}' \times \mathbf{p} \right) \cdot \mathbf{S}\mathbf{e}_3 \, dA\, dx_3$$

$$= \int_0^\ell \left(\left(\overline{\mathbf{w}}' + \mathbf{e}_3 \times \overline{\boldsymbol{\theta}} \right) \cdot \int_\Omega \mathbf{S}\mathbf{e}_3 dA + \overline{\boldsymbol{\theta}}' \cdot \int_\Omega \mathbf{p} \times \mathbf{S}\mathbf{e}_3 dA \right) dx_3$$

where we have noted that the product $\mathbf{S} \cdot \overline{\boldsymbol{\Theta}} = S_{ij}\overline{\Theta}_{ij} = 0$ because $\mathbf{S}$ is symmetric and $\boldsymbol{\Theta}$ is skew-symmetric (verify that this is always the case). Recognizing the definitions of resultant force and resultant moment, we find that the expression for internal virtual work takes the following form

$$\boxed{\overline{W}_I = \int_0^\ell \left(\mathbf{Q} \cdot \overline{\boldsymbol{\epsilon}}_o + \mathbf{M} \cdot \overline{\boldsymbol{\kappa}}_o \right) dx} \tag{403}$$

where $\overline{\boldsymbol{\epsilon}}_o \equiv \overline{\mathbf{w}}' + \mathbf{e}_3 \times \overline{\boldsymbol{\theta}}$ and $\overline{\boldsymbol{\kappa}}_o \equiv \overline{\boldsymbol{\theta}}'$ are the virtual strain resultants.

The final form of the internal virtual work is interesting and important. Each term in the expression is analogous to stress times virtual strain. In the present case, this analogy translates to stress resultant times virtual strain resultant. Thus, we can see that the resultant strains are conjugate to the resultant stresses in the sense of virtual work. Notice that the demonstration of conjugateness did not involve the constitutive equations. The principle of virtual work is a very powerful method for finding what the appropriate strain measure should be according to how stress is defined.

The statement of the principle of virtual work is basically the same as for the little boundary value problem: If the external work is equal to the internal work for all virtual displacements satisfying the strain displacement relationships, then the equations of equilibrium are automatically satisfied. As usual, let us define a functional G to be the difference between the internal and external virtual work. This functional has the form $G(\mathbf{s}, \overline{\mathbf{v}}) \equiv \overline{W}_I - \overline{W}_E$, where the force resultants are represented as $\mathbf{s} \equiv [\mathbf{Q}, \mathbf{M}]^T$ and the virtual displacements and rotations are represented as $\overline{\mathbf{v}} \equiv [\overline{\mathbf{w}}, \overline{\boldsymbol{\theta}}]^T$. The principle of virtual work is

$$\text{If } G(\mathbf{s},\bar{\mathbf{v}}) = 0 \quad \forall \bar{\mathbf{v}} \in \mathfrak{B}(0,\ell), \quad \text{then}$$

$$\begin{aligned} \mathbf{Q}' + \mathbf{q} &= \mathbf{0} & \mathbf{Q}(0) &= -\mathbf{q}_o & \mathbf{Q}(\ell) &= \mathbf{q}_\ell \\ \mathbf{M}' + \mathbf{e}_3 \times \mathbf{Q} + \mathbf{m} &= \mathbf{0} & \mathbf{M}(0) &= -\mathbf{m}_o & \mathbf{M}(\ell) &= \mathbf{m}_\ell \end{aligned}$$

where $\mathfrak{B}(0,\ell)$ is our collection of admissible virtual displacements. As usual, the principle of virtual work gives us a vehicle for making approximations. The main difference with the little boundary value problem is that we must specify the basis functions for $\mathbf{w}$ and $\boldsymbol{\theta}$ independently from each other because they are independent fields.

We can express the virtual-work functional for the three-dimensional beam in terms of only displacements by substituting the constitutive equations. To economize notation let us define

$$\mathbf{D} = \begin{bmatrix} \mathbf{A} & \mathbf{S} \\ \mathbf{S}^T & \mathbf{I} \end{bmatrix} \quad \mathbf{e} = \begin{bmatrix} \boldsymbol{\epsilon}_o \\ \boldsymbol{\varkappa}_o \end{bmatrix} \quad \mathbf{f} = \begin{bmatrix} \mathbf{q} \\ \mathbf{m} \end{bmatrix} \quad \mathbf{v} = \begin{bmatrix} \mathbf{w} \\ \boldsymbol{\theta} \end{bmatrix} \tag{404}$$

where $\boldsymbol{\epsilon}_o = \mathbf{w}' + \mathbf{e}_3 \times \boldsymbol{\theta}$ and $\boldsymbol{\varkappa}_o = \boldsymbol{\theta}'$ are the strain resultants associated with the real displacement. The virtual-work functional for a beam takes the form

$$G(\mathbf{v},\bar{\mathbf{v}}) = \int_0^\ell \left(\bar{\mathbf{e}}^T \mathbf{D} \mathbf{e} - \mathbf{f}^T \bar{\mathbf{v}}\right) dx - \mathbf{f}_o^T \bar{\mathbf{v}}(0) - \mathbf{f}_\ell^T \bar{\mathbf{v}}(\ell) \tag{405}$$

where $\mathbf{f}_o = [\mathbf{q}_o, \mathbf{m}_o]^T$ and $\mathbf{f}_\ell = [\mathbf{q}_\ell, \mathbf{m}_\ell]^T$. This expression can be used in conjunction with the Ritz method to generate approximate solutions to the three-dimensional beam problem.

The Planar Beam

A great number of practical problems can be idealized as planar problems. The assumption of planar behavior comes at a fairly high price, the cost of which we can clearly see from the equations for the beam in three dimensions. First, the loading must be such that it does not excite out-of-plane motions. Second, the cross sections must be symmetric with respect to the plane of loading. Clearly, the centroid of the section will lie on the line of symmetry, and this line should be taken as the coordinate axis.

A planar beam could, of course, lie in either of the two planes $x_1 - x_3$ or $x_2 - x_3$. The beam has no way of knowing about the coordinate system we choose to describe it. Thus, the results must be the same either way. Here we shall take the plane of the problem to be the $x_1 - x_3$ plane. Let us make some notational simplifications for discussing the planar problem. Let the axial displacement be called $w_3 \equiv u$, the transverse displacement $w_1 \equiv w$, and the rotation of the cross section $\theta_2 \equiv \theta$. Let the axial force be called $Q_3 \equiv N$, the shear force $Q_1 \equiv Q$, and the bending moment $M_2 \equiv M$. Let us further assume that the axes are centroidal. The equations governing the planar (Timoshenko beam) problem are

$$\boxed{\begin{aligned} N' + n &= 0 & \qquad N &= EAu' \\ Q' + q &= 0 & \qquad Q &= GA(w' - \theta) \\ M' + Q + m &= 0 & \qquad M &= EI\theta' \end{aligned}} \tag{406}$$

where, in addition to the terms already defined, we have let $n(x)$ represent the applied axial load, $q(x)$ the applied transverse load, and $m(x)$ the applied moment. Each of these is a scalar function of x and, hence, their direction is fixed (n along the axis, q perpendicular to the axis, and m out of the plane of the page). We have also used E to stand for Young's modulus and $G = \mu$ to stand for the shear modulus of the material. As already defined, A is the cross-sectional area and I is the second moment of the area about the centroidal axis.

The first thing to notice about the linear planar beam equations is that the axial components of force and displacement are uncoupled from the shear and bending components, but that the shear and bending components are coupled to each other. This feature is one that makes beam theory interesting. The second thing to notice is that the equations constitute a system of six first-order differential equations. The equations can be recast in terms of only displacement variables at the price of raising the order of the differential equations. The shear and bending equations can be rewritten by substituting the constitutive equations (assuming that the moduli EI and GA are constant) into the equilibrium equations as follows

$$\begin{aligned} EI\theta'' + GA(w' - \theta) + m &= 0 \\ GA(w'' - \theta') + q &= 0 \end{aligned} \tag{407}$$

The equations can be recast into a form more favorable for direct integration by differentiating the first equation once and subtracting the second equation. The resulting equation can be used with the first equation to give the equivalent system

$$\begin{aligned} EI\theta''' &= q - m' \\ GAw' &= GA\theta - EI\theta'' - m \end{aligned} \tag{408}$$

The first of these equations can be integrated directly to obtain an expression for the rotation $\theta(x)$. This function will be known, except for the three constants of integration. Substituting the results into the second equation and integrating once gives the expression for $w(x)$, with one additional constant. Because the system is essentially a fourth-order differential equation, it will always involve four integration constants that must be determined from boundary conditions.

There are a variety of possible boundary conditions, some of which are shown in Fig. 94. Notice that a boundary condition can be one of two types:

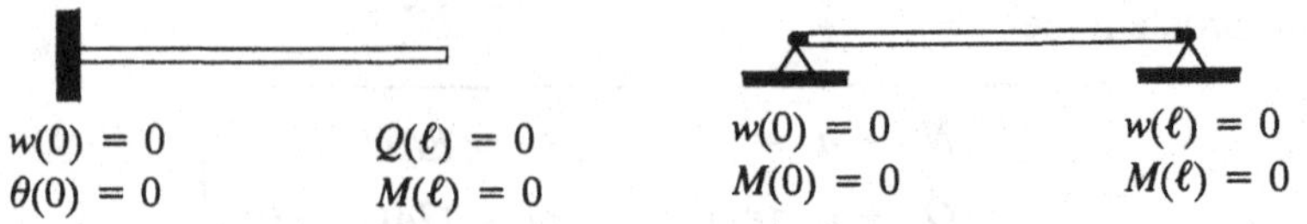

$w(0) = 0$ $\theta(0) = 0$ $Q(\ell) = 0$ $M(\ell) = 0$ $w(0) = 0$ $M(0) = 0$ $w(\ell) = 0$ $M(\ell) = 0$

Figure 94 Boundary conditions for a cantilever and simple beam

conditions on the displacement (or rotation, which we think of as a generalized displacement) or conditions on the force (or moment, which we think of as a generalized force). Boundary conditions, in terms of force, must be translated into statements involving displacements in order to be implemented into our theory. This translation can be accomplished with the constitutive equations. Thus, the condition $Q(\ell) = 0$ can be expressed as $GA\big(w'(\ell) - \theta(\ell)\big) = 0$, and the condition $M(\ell) = 0$ can be written as $EI\theta'(\ell) = 0$.

We must, in general, consider the possibility that end loads will be prescribed. Certainly, at points of fixity (places where the displacement is prescribed), end reaction forces accrue. At the end $x = 0$, let us call the applied (or reacting) axial force n_o, shear force q_o, and moment m_o. At the end $x = \ell$, let us call the applied (or reacting) axial force n_ℓ, shear force q_ℓ, and moment m_ℓ. These forces are related to the internal stress resultants through the Cauchy relations

$$\begin{aligned} -n_o &= N(0), & -q_o &= Q(0), & -m_o &= M(0) \\ n_\ell &= N(\ell), & q_\ell &= Q(\ell), & m_\ell &= M(\ell) \end{aligned} \tag{409}$$

The positive sense of the three internal stress resultants is shown in Fig. 95.

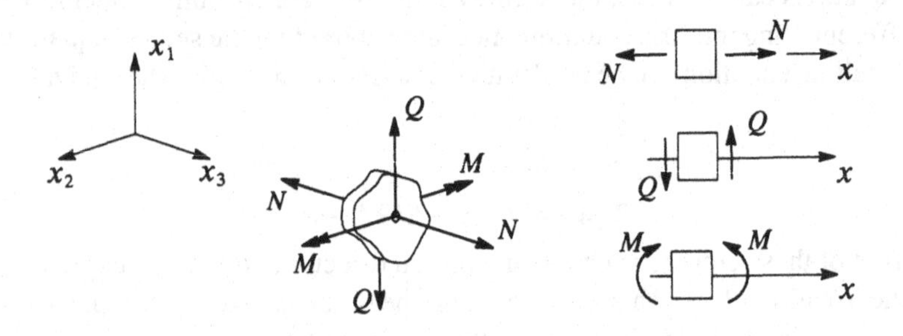

Figure 95 The convention for positive stress resultants for a planar beam

Virtual-work functional for the planar Timoshenko beam. We are now ready to state the principle of virtual work for a beam with shear deformation. For convenience, let $\mathbf{u} \equiv [u, w, \theta]^T$ and $\overline{\mathbf{u}} = [\overline{u}, \overline{w}, \overline{\theta}]^T$. Let

$$\mathbf{D} = \begin{bmatrix} EA & 0 & 0 \\ 0 & GA & 0 \\ 0 & 0 & EI \end{bmatrix} \quad \mathbf{e} = \begin{bmatrix} u' \\ w' - \theta \\ \theta' \end{bmatrix} \quad \mathbf{q} = \begin{bmatrix} n \\ q \\ m \end{bmatrix} \tag{410}$$

With these definitions, the virtual-work functional $G(\mathbf{u}, \bar{\mathbf{u}}) \equiv \overline{W}_I - \overline{W}_E$ for the planar beam can be written as

$$G(\mathbf{u}, \bar{\mathbf{u}}) = \int_0^\ell \left(\bar{\mathbf{e}}^T \mathbf{D} \mathbf{e} - \mathbf{q}^T \bar{\mathbf{u}} \right) dx - \mathbf{q}_o^T \bar{\mathbf{u}}(0) - \mathbf{q}_\ell^T \bar{\mathbf{u}}(\ell) \tag{411}$$

where $\mathbf{q}_o = [n_o, q_o, m_o]^T$ and $\mathbf{q}_\ell = [n_\ell, q_\ell, m_\ell]^T$.

According to the principle of virtual work, the system is in equilibrium if internal work is equal to external work for all choices of the virtual displacement field. Thus, equilibrium can be stated as

$$\boxed{G(\mathbf{u}, \bar{\mathbf{u}}) = 0 \quad \forall \bar{\mathbf{u}} \in \mathcal{T}(0, \ell)} \tag{412}$$

where $\mathcal{T}(0, \ell)$ represents the collection of all of the admissible functions, defined for values of x between 0 and ℓ, from which we can choose our virtual displacements. We have chosen the letter $\mathcal{T}$ to represent the collection to remind us that we are talking about a Timoshenko beam. This collection contains functions that are well enough behaved that the highest derivative that appears in G is square-integrable. In the present case, only first derivatives of each of the three functions appear. Any continuous function will satisfy the requirement of square-integrability (i.e., kinks in the function are allowed), but any function with a jump discontinuity will not.

As was mentioned in the section on the little boundary value problem, this statement of virtual work is most powerful if we can exclude the unknown reaction forces from the expression for external work. This exclusion can be accomplished if we simply insist that the virtual displacement corresponding to an unknown reaction be equal to zero at that point so that the product of the two is zero. Once this is done, the functional G involves only known forces and unknown displacements. We often see this restriction stated formally as a restricted set of functions $\mathcal{T}_e \equiv \{\bar{\mathbf{u}} \mid \bar{\mathbf{u}} = \mathbf{0} \text{ on } \Omega_u\}$, which reads: the collection of all functions $\bar{\mathbf{u}}$ that are zero on that portion of the boundary where displacements are prescribed (and tractions are, therefore, unknown).

The beam boundary conditions are special because of the cross-sectional rigidity constraint, and Ω_u must be interpreted accordingly. For example, if the beam is completely fixed at the end $x = 0$ and is simply supported at the end $x = \ell$, then the space of admissible functions would have the displacements and rotations equal to zero at $x = 0$, but only the displacements equal to zero at $x = \ell$. Choosing a virtual rotation field that vanished at $x = \ell$ might severely impair the ability of the principle of virtual work in distinguishing between different problems having different applied moments at that end (the value of zero is but one of the many choices).

Example 47. *Classical solution of the Timoshenko beam.* Consider the cantilever beam of length ℓ, flexural modulus EI, and shear modulus GA subjected to a uniform transverse load q shown in Fig. 96.

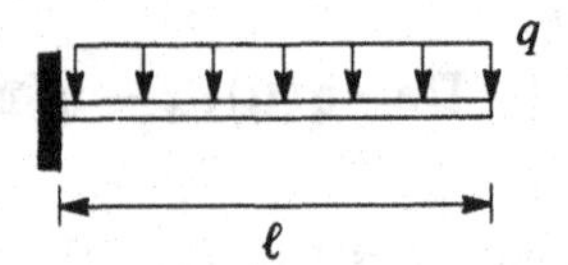

Figure 96 Cantilever beam example

Integrating the rotation equation three times, we obtain the result

$$EI\theta(x) = -\tfrac{1}{6}qx^3 + \tfrac{1}{2}a_1x^2 + a_2x + a_3 \tag{413}$$

This expression can be differentiated to give $EI\theta'' = -qx + a_1$. Substituting into the deflection equation, we have $GAw' = GA\theta - EI\theta''$ so

$$GAw' = \frac{GA}{EI}\Big[-\tfrac{1}{6}qx^3 + \tfrac{1}{2}a_1x^2 + a_2x + a_3\Big] + qx - a_1 \tag{414}$$

Integrating this expression once gives

$$GAw(x) = \frac{GA}{EI}\Big[-\tfrac{1}{24}qx^4 + \tfrac{1}{6}a_1x^3 + \tfrac{1}{2}a_2x^2 + a_3x\Big] + \tfrac{1}{2}qx^2 - a_1x + a_4$$

The boundary conditions are $w(0) = 0$, $\theta(0) = 0$, $w'(\ell) - \theta(\ell) = 0$ (shear force vanishes), $\theta'(\ell) = 0$ (bending moment vanishes). From the first of these we can conclude that $a_4 = 0$, and from the second of these we obtain $a_3 = 0$. From the third boundary condition we find, from Eqn. (414), that $a_1 = q\ell$. Finally, from the fourth boundary condition we find, from the derivative of Eqn. (413), that $a_2 = -q\ell^2/2$. Thus, letting $\xi \equiv x/\ell$, the expressions for deflection and rotation are given by

$$w(\xi) = \frac{q\ell^4}{24EI}\big(-\xi^4 + 4\xi^3 - (6-\beta)\xi^2 - 2\beta\xi\big)$$

$$\theta(\xi) = \frac{q\ell^3}{6EI}\big(-\xi^3 + 3\xi^2 - 3\xi\big)$$

where we have defined $\beta \equiv 12EI/GA\ell^2$ to be the dimensionless ratio of bending to shear stiffness. Note that the condition $\beta \rightarrow 0$ reflects high shear stiffness and, hence, less deflection owing to shear. Substituting these expressions back into the constitutive equations, we can determine the distribution of the shear force and the bending moment

$$Q(\xi) = GA\big(w' - \theta\big) = q\ell(\xi - 1)$$

$$M(\xi) = EI\theta' = \frac{q\ell^2}{2}\big(-\xi^2 + 2\xi - 1\big)$$

Note that the resultants Q and M do not depend upon β since the problem is statically determinate (that is, we could have integrated the equilibrium equations directly and found the two integration constants without appealing to the displacement boundary conditions).

The importance of shearing deformation. How important is shear for a typical case? Consider a rectangular cross section of depth h and width b. For this case, the value of the shear ratio is $\beta \equiv (E/G)(h^2/\ell^2)$. Clearly, there are two aspects that are important to the decision whether or not shear is important. The first aspect is related to the material properties. For an isotropic material, the ratio E/G is approximately 2, although for a material like rubber this ratio is much larger. For a typical beam, the ratio of depth to length is usually not less than $h/\ell = 1/10$. Consequently, we do not expect β to be much larger than 0.02 for typical situations. However, there are situations with either very short beams or rubber-like materials where shear deformation can be important.

A Ritz approximation for the planar beam. Approximate solutions to the planar beam problem can be constructed with the Ritz method. There are two field variables, $w(x)$ and $\theta(x)$, so each must be expressed as a linear sum of basis functions. Each must have a virtual displacement counterpart $\bar{w}(x)$ and $\bar{\theta}(x)$, and those virtual displacements must have a basis function expansion. As usual, the approximation of the real displacement fields must satisfy the essential boundary conditions, and the approximation of the virtual displacement fields must satisfy the homogeneous essential boundary conditions. Let us write the approximations as follows

$$w(x) = w_o(x) + \mathbf{h}(x) \cdot \mathbf{a}, \quad \theta(x) = \theta_o(x) + \mathbf{g}(x) \cdot \mathbf{b}$$

where $w_o(x)$ and $\theta_o(x)$ are known functions that satisfy the nonhomogeneous boundary conditions, $\mathbf{h} = [h_1, \ldots, h_N]^T$ and $\mathbf{g} = [g_1, \ldots, g_M]^T$ are the base functions for $w(x)$ and $\theta(x)$, respectively, and the constants $\mathbf{a} = [a_1, \ldots, a_N]^T$ and $\mathbf{b} = [b_1, \ldots, b_M]^T$ are the unknowns. The virtual displacement fields can be expressed in a similar manner as

$$\bar{w}(x) = \mathbf{h}(x) \cdot \bar{\mathbf{a}}, \quad \bar{\theta}(x) = \mathbf{g}(x) \cdot \bar{\mathbf{b}}$$

It is, of course, possible to select the $h_i(x)$ and $g_i(x)$ from the same class of functions, but there is no need to (in fact, it may be preferable not to) have the same number of terms in the expansions for $w(x)$ and $\theta(x)$. Thus, in general, $N \neq M$. Finally, we should recall that our choice to approximate the virtual displacement fields with the same basis functions as the real displacement fields is the Galerkin approach.

If we substitute these approximations into the virtual-work functional, Eqn. (411), we get the following discrete version of the functional (neglecting the axial contribution)

$$G(\mathbf{a}, \mathbf{b}, \bar{\mathbf{a}}, \bar{\mathbf{b}}) = \bar{\mathbf{a}}^T\left[\mathbf{K}^{aa}\mathbf{a} + \mathbf{K}^{ab}\mathbf{b} - \mathbf{f}^a\right] + \bar{\mathbf{b}}^T\left[\mathbf{K}^{ba}\mathbf{a} + \mathbf{K}^{bb}\mathbf{b} - \mathbf{f}^b\right] \quad (415)$$

where the elements of the coefficient matrices $\mathbf{K}^{aa}$, $\mathbf{K}^{ab}$, $\mathbf{K}^{ba}$, and $\mathbf{K}^{bb}$ can be computed from the basis functions as follows

$$\mathbf{K}^{aa} = \int_0^\ell GA[\mathbf{h}'][\mathbf{h}']^T dx, \qquad \mathbf{K}^{ab} = -\int_0^\ell GA[\mathbf{h}'][\mathbf{g}]^T dx$$

$$\mathbf{K}^{ba} = -\int_0^\ell GA[\mathbf{g}][\mathbf{h}']^T dx, \qquad \mathbf{K}^{bb} = \int_0^\ell \left[EI[\mathbf{g}'][\mathbf{g}']^T + GA[\mathbf{g}][\mathbf{g}]^T\right] dx$$

and the elements of the load matrices $\mathbf{f}^a$ and $\mathbf{f}^b$can be computed as

$$\mathbf{f}^a = \int_0^\ell \left(q\mathbf{h} - GA\left(w_o' - \theta_o\right)\mathbf{h}'\right) dx + q_o\mathbf{h}(0) + q_\ell\mathbf{h}(\ell)$$

$$\mathbf{f}^b = \int_0^\ell \left(m\mathbf{g} + GA\left(w_o' - \theta_o\right)\mathbf{g} - EI\theta_o'\mathbf{g}'\right) dx + m_o\mathbf{g}(0) + m_\ell\mathbf{g}(\ell)$$

If $G(\mathbf{a}, \mathbf{b}, \bar{\mathbf{a}}, \bar{\mathbf{b}}) = 0$ for all choices of $(\bar{\mathbf{a}}, \bar{\mathbf{b}})$ then the principle of virtual work satisfies equilibrium to the degree possible within the context of the approximating basis. From the (discrete) fundamental theorem of the calculus of variations, $G(\mathbf{a}, \mathbf{b}, \bar{\mathbf{a}}, \bar{\mathbf{b}}) = 0$ if and only if

$$\begin{aligned} \mathbf{K}^{aa}\mathbf{a} + \mathbf{K}^{ab}\mathbf{b} &= \mathbf{f}^a \\ \mathbf{K}^{ba}\mathbf{a} + \mathbf{K}^{bb}\mathbf{b} &= \mathbf{f}^b \end{aligned} \tag{416}$$

These equations serve to determine the unknown constants **a** and **b**, as illustrated by the following example.

Example 48. *Ritz method for Timoshenko beam.* Consider again the cantilever beam subjected to uniform load q, shown in Fig. 96. Recall that the beam has length ℓ, bending modulus EI, and shear modulus GA. Let us solve the problem using the Ritz method. Let the transverse displacement and rotation fields, and their virtual counterparts, be approximated as

$$w(x) = a_1 x + a_2 \frac{x^2}{\ell}, \quad \theta(x) = b_1 \frac{x}{\ell}$$

$$\bar{w}(x) = \bar{a}_1 x + \bar{a}_2 \frac{x^2}{\ell}, \quad \theta(x) = \bar{b}_1 \frac{x}{\ell}$$

We can identify the basis functions as $h_1 = x$, $h_2 = x^2/\ell$, and $g_1 = x/\ell$. We have chosen to normalize the functions by ℓ in the manner shown so that all of the constants in the Ritz expansion are dimensionless. Let $\beta \equiv 12EI/GA\ell^2$ be a dimensionless ratio of bending modulus to shear modulus. The discrete equilibrium equations, i.e., Eqn. (416), have the explicit form

$$\frac{GA\ell}{12}\begin{bmatrix} 12 & 12 & -6 \\ 12 & 16 & -8 \\ -6 & -8 & 4+\beta \end{bmatrix}\begin{bmatrix} a_1 \\ a_2 \\ b_1 \end{bmatrix} = -\frac{q_o\ell^2}{6}\begin{bmatrix} 3 \\ 2 \\ 0 \end{bmatrix}$$

Solving these equations, we obtain

$$a_1 = -\frac{q\ell}{GA}, \quad a_2 = \frac{q\ell}{GA}\left(\frac{\beta-2}{2\beta}\right), \quad b_1 = -\frac{q\ell}{GA}\left(\frac{2}{\beta}\right)$$

and, therefore, noting that $\beta GA = 12EI/\ell^2$, the approximate solution

$$w(x) = \frac{q\ell^4}{24EI}\left((\beta-2)\frac{x^2}{\ell^2} - 2\beta\frac{x}{\ell}\right), \quad \theta(x) = -\frac{q\ell^3}{6EI}\left(\frac{x}{\ell}\right)$$

Observe that the rotation field does not involve the shear modulus GA, in accord with the result of the classical solution. Note that the tip deflection, for $\beta = 0$ (i.e., no shear deformation) is $q\ell^4/12EI$, which is 33% less than the classical solution of $q\ell^4/8EI$. Problem 175 examines the Ritz approximation to this problem in more detail.

The Bernoulli-Euler Beam

We can observe from the preceding discussion that shear deformations are often negligible. If we make the assumption that they vanish altogether, we can reduce the number of unknown functions in our theory from two to one, namely the deflection $w(x)$. The assumption that shear deformations are zero can be expressed as $w'(x) - \theta(x) = 0$, from our definition of shear deformation. If we introduce this constraint, then the governing equations of the Bernoulli-Euler beam take the form

$$\begin{aligned} N' + n &= 0 & \quad N &= EAu' \\ Q' + q &= 0 & \quad Q &= \text{reaction} \\ M' + Q + m &= 0 & \quad M &= EIw'' \end{aligned}$$

It would appear that the only thing that happened was that we changed the moment constitutive equation and completely lost our shear constitutive equation. The shear is now a reaction force associated with the constraint, and, thus, computable only from an equilibrium equation. However, now we can make the same substitutions as before and derive a single equation for transverse bending involving only the unknown w

$$\left(EIw''\right)'' = q - m' \tag{417}$$

This equation is somewhat simpler to solve than the shear beam equations, but it still gives rise to four constants of integration. We must still get these constants from the boundary conditions. The constraint changes the boundary conditions somewhat. First, conditions on rotation are now expressed as conditions on the first derivative of w. Conditions on the moment are now conditions on the second derivative of w. The important change is the condition on shear. Since we lost our constitutive equation in implementing the constraint we must find the shear Q from equilibrium. Equilibrium of moments gives the relationship $Q = -M' - m = -EIw''' - m$. Thus, a condition on shear can be translated to a condition on the third derivative of w.

Example 49. *Classical solution for Bernoulli-Euler beam.* Consider again the cantilever beam subjected to a uniform transverse load q shown in Fig. 96. Let us assume that shear deformations are negligible, and, hence, that the Bernoulli-Euler theory is appropriate. Integrating Eqn. (417) four times, we obtain

$$EIw(x) = -\tfrac{1}{24}qx^4 + \tfrac{1}{6}a_1x^3 + \tfrac{1}{2}a_2x^2 + a_3x + a_4$$

The boundary conditions are $w(0) = 0$, $w'(0) = 0$, $-EIw'''(\ell) = 0$ (shear force vanishes), $EIw''(\ell) = 0$ (bending moment vanishes). From the first of these we can conclude that $a_4 = 0$, and from the second of these we obtain $a_3 = 0$. From the third boundary condition we find that $a_1 = q\ell$. Finally, from the fourth boundary condition we find, from the derivative of Eqn. (413), that $a_2 = -q\ell^2/2$. Thus, letting $\xi \equiv x/\ell$, the expressions for deflection and rotation are given by

$$w(\xi) = \frac{q\ell^4}{24EI}\left(-\xi^4 + 4\xi^3 - 6\xi^2\right)$$

Substituting these expressions back into the constitutive equations, we can determine the distribution of the shear force and the bending moment

$$Q(\xi) = -EIw''' = q\ell(\xi - 1)$$
$$M(\xi) = EIw'' = \frac{q\ell^2}{2}\left(-\xi^2 + 2\xi - 1\right)$$

These results correspond exactly with those obtained previously for the Timoshenko beam in the limit as $\beta \to 0$.

Virtual work for the planar Bernoulli-Euler beam. We are now ready to state the principle of virtual work for a beam without shear deformation. For convenience, let $\mathbf{u} \equiv [u, w, w']^T$ and $\bar{\mathbf{u}} \equiv [\bar{u}, \bar{w}, \bar{w}']^T$. Let

$$\mathbf{D} = \begin{bmatrix} EA & 0 \\ 0 & EI \end{bmatrix} \qquad \mathbf{e} = \begin{bmatrix} u' \\ w'' \end{bmatrix} \qquad \mathbf{q} = \begin{bmatrix} n \\ q \\ m \end{bmatrix} \tag{418}$$

With these definitions, the virtual-work functional $G(\mathbf{u}, \bar{\mathbf{u}}) \equiv \overline{W}_I - \overline{W}_E$ for the planar Bernoulli-Euler beam can be written as

$$G(\mathbf{u},\bar{\mathbf{u}}) = \int_0^{\ell} \left(\bar{\mathbf{e}}^T \mathbf{D}\mathbf{e} - \mathbf{q}^T\bar{\mathbf{u}}\right) dx - \mathbf{q}_o^T\bar{\mathbf{u}}(0) - \mathbf{q}_\ell^T\bar{\mathbf{u}}(\ell) \tag{419}$$

where, as before, $\mathbf{q}_o = [n_o, q_o, m_o]^T$ and $\mathbf{q}_\ell = [n_\ell, q_\ell, m_\ell]^T$. Note that everywhere the variable θ appeared in the previous theory, it has been replaced by w' in the current theory. Since there is no shear deformation, the term involving shear vanishes identically. According to the principle of virtual work, the system is in equilibrium if

$$\boxed{G(\mathbf{u},\bar{\mathbf{u}}) = 0 \quad \forall \bar{\mathbf{u}} \in \mathcal{B}(0,\ell)} \tag{420}$$

where $\mathcal{B}(0,\ell)$ is the collection of functions from which the virtual displacement fields $\bar{u}(x)$ and $\bar{w}(x)$ can be chosen. This collection of functions is different from $\mathcal{T}(0,\ell)$ for a variety of reasons. First, $\mathcal{T}(0,\ell)$ contained functions for u, w, and θ, while $\mathcal{B}(0,\ell)$ has only functions for u and w. Thus, obviously, the dimension of the two spaces is different. A more important difference is the restriction implied by the order of derivatives that appear in the Bernoulli-Euler version of the principle of virtual work. There are second derivatives of w and $\bar{w}$ in the expression for G (whereas only first derivatives appeared in Timoshenko beam theory). Thus, any function whose second derivative is square-integrable is admissible. Now our space rejects functions with kinks in them because the first derivative of such a function would be discontinuous at the point of the kink and, consequently, the second derivative would not be square-integrable. The shear beam theory allowed kinks as being a physically reasonable result of shear deformation. The Bernoulli-Euler theory does not allow kinks because the constraint $w' - \theta = 0$ implies more smoothness in the solution.

Again, for practical applications we generally restrict the virtual displacement to be zero on that portion of the boundary where displacements are prescribed. Like the shear beam, the part of the boundary where "displacement" is prescribed for the Bernoulli-Euler beam is any point where the displacement or rotation is known a priori. Therefore, the essential boundary conditions involve both w and w'. Some common boundary conditions for the prismatic Bernoulli-Euler beam include

$$\begin{array}{|rlrll|}\hline w &= 0 & w' &= 0 & \text{fixed end} \\ w &= 0 & w'' &= 0 & \text{simple support} \\ w'' &= 0 & w''' &= 0 & \text{free end} \\ \hline \end{array} \tag{421}$$

Note that you cannot prescribe the displacement *and* the shear at a point; nor can you prescribe the slope *and* the moment at a point.

A Ritz approximation for the Bernoulli-Euler beam. We can apply the Ritz method to the virtual-work functional for the Bernoulli-Euler beam. Let us approximate the real and virtual transverse displacement as

$$w(x) = w_o(x) + \mathbf{h}(x) \cdot \mathbf{a}, \qquad \overline{w}(x) = \mathbf{h}(x) \cdot \overline{\mathbf{a}} \tag{422}$$

where the functions $h_n(x)$ are known base functions, selected from the collection of admissible functions. Substituting these expression into the virtual-work functional, and carrying out the integrals, we obtain the discrete form of the functional (ignoring axial deformation)

$$G(\mathbf{a}, \overline{\mathbf{a}}) = \overline{\mathbf{a}}^T \big[\mathbf{K}\mathbf{a} - \mathbf{f} \big]$$

where the matrices $\mathbf{K}$ and $\mathbf{f}$ are

$$\mathbf{K} = \int_0^\ell EI[\mathbf{h}''][\mathbf{h}'']^T \, dx$$

$$\mathbf{f} = \int_0^\ell \big[q\mathbf{h} + m\mathbf{h}' - EIw_o''\mathbf{h}'' \big] \, dx + q_o\mathbf{h}(0) + q_\ell\mathbf{h}(\ell) + m_o\mathbf{h}'(0) + m_\ell\mathbf{h}'(\ell)$$

Since $G(\mathbf{a}, \overline{\mathbf{a}})$ must vanish for all $\overline{\mathbf{a}}$, the equilibrium equation that results from the principle of virtual work is $\mathbf{K}\mathbf{a} = \mathbf{f}$ as before. Like the little boundary value problem, this equation determines the coefficients of the approximation for the field $w(x)$. Once this field is known, the moments and shears can be computed by differentiation.

Mixed boundary conditions. It is possible to generalize the notion of boundary conditions beyond those that are expressed as pure constraints on motion or force. Boundary conditions that involve combinations of force and displacement are called *mixed boundary conditions*. Mathematically, we wish to include conditions (at either $x = 0$ or $x = \ell$) of the general form

$$c_0 w + c_1 w' + c_2 w'' + c_3 w''' = 0 \tag{423}$$

where $c_0, \ldots, c_3$ are constants associated with the problem description (i.e., they are not unknowns). Although Eqn. (423) applies to Bernoulli-Euler beam theory, this same generalization is possible for any theory. The key observation is that the number of derivatives included must be one less than the order of the differential equation.

The mixed boundary condition is an artifice of modeling that replaces a truncated part of the system by a spring. The most common types of mixed boundary conditions come from linear springs. The spring constant k must be specified as part of the problem description as the following example illustrates.

Example 50. *Mixed boundary conditions.* Consider the beam shown in Fig. 97. The left end of the beam is restrained from translation. The moment developed by the spring is related to the rotation at that point by $M_s = k_\theta \theta_s$, where $\theta_s = w'(0)$ is the rotation experienced by the spring. The right end of the beam is free to translate horizontally and to rotate, but the spring elastically restrains vertical motion. The force developed by the spring is related to the deflection at that point by $F_s = k_w w_s$, where $w_s = w(\ell)$ is the deflection experienced by the spring.

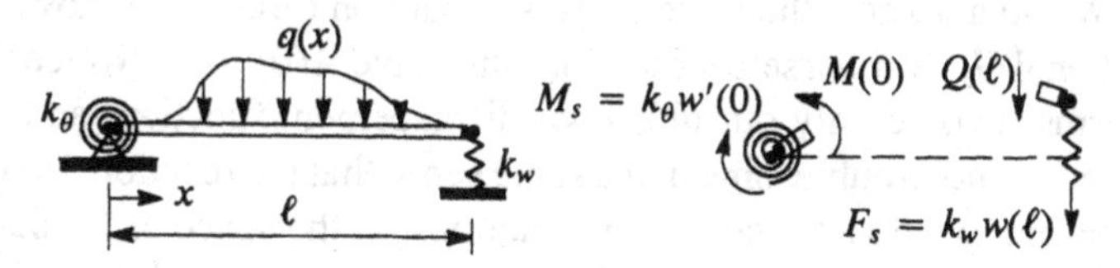

Figure 97 An example of a beam with elastic supports

To find the appropriate boundary conditions, we must take a freebody diagram of the ends of the beam, as shown in Fig. 97. All of the displacement and force quantities are drawn in their positive sense: The moment $M(0)$ is anticlockwise because it acts on the positive x face of the cross section. The shear $Q(\ell)$ is down because it acts on the negative x face of the cross section (see Fig. 95 for the sign convention). The spring forces are shown resisting positive motions. The equations of equilibrium give the appropriate mixed boundary conditions. At the left end of the beam we are considering in our example, we have the condition $M_s - M(0) = 0$, while at the right end we have $F_s + Q(\ell) = 0$. These relationships must be restated in terms of the displacement w as

$$w(0) = 0 \qquad k_\theta w'(0) - EIw''(0) = 0$$
$$w''(\ell) = 0 \qquad k_w w(\ell) - EIw'''(\ell) = 0$$

These four boundary conditions can be used to determine the four constants of integration that appear when we solve the governing differential equation, just as we did for pure boundary conditions.

What if we wanted to include the springs in the principle of virtual work? We must include the virtual work done by the springs in going through their respective motions. To wit, let us take

$$G(w, \bar{w}) = \int_0^\ell \left(EIw''\bar{w}'' - q\bar{w}\right) dx - q_o\bar{w}(0) - m_\ell \bar{w}'(\ell) + k_\theta w'(0)\bar{w}'(0) + k_w w(\ell)\bar{w}(\ell) \tag{424}$$

Two of the boundary terms have been replaced by elastic spring terms. Why do those terms have a positive sign in the work expression? If the displacement is positive, then the force induced in the spring acts in the negative direction, and vice versa. If you think about it, the springs are really elastic elements, like the beam itself. Thus, the work associated with them is more like internal work. These terms have a character more like the internal work than the external work. It does not matter what name you call them, so long as the work is properly accounted for.

Let us now reconsider the question of restricting our space of virtual displacement functions on the part of the boundary where displacements are prescribed. In the present problem, we know the moment $m_\ell = 0$ because it is prescribed. We do not know the force q_o; it is a reaction force. To remove it from our functional G, we must select functions that have $\overline{w}(0) = 0$. Notice that our spring terms involve only our unknown displacement function, and, hence, these terms are not troublesome in the same sense that the reaction force terms are. Therefore, we do not need any restrictions on the space of functions for virtual displacements to take care of these terms. In fact, if we did restrict these terms, we would impair the ability of the principle of virtual work to distinguish among similar problems with different spring constants; clearly, an untenable proposition.

Structural Analysis

One of the most important applications of beam theory is in matrix structural analysis of frames. A frame is an assemblage of beam elements that are connected together at their ends. The main additional feature in structural analysis over the analysis we have done for the single elements is the communication of force from one element to the next. In many ways this problem is very much like the application of finite elements in the previous chapter. The element stiffness matrix and force vector can be computed from element shape functions and then assembled into the global equations, as shown in Chapter 6.

To fix ideas, let us consider the typical framed structure shown in Fig. 98. The structure consists of nine members rigidly connected at eight joints (shown as squares in the figure). Element e is shown separated from the structure in (b) and (c). Each element has an "i" end and a "j" end. The i end is asso-

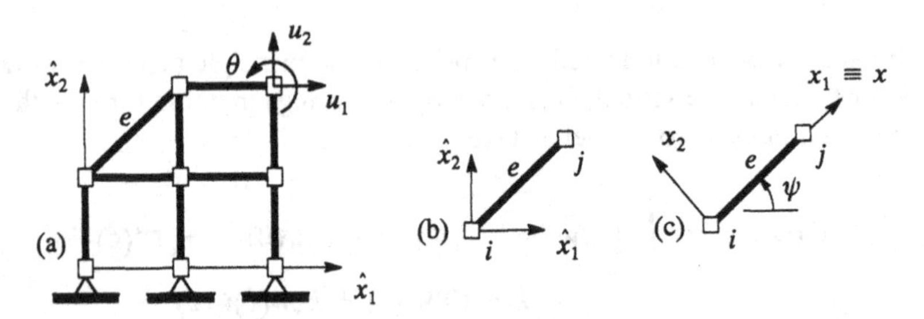

Figure 98 Structural frames: (a) an example of a reticular structure and global coordinates, (b) element e in global coordinates, (c) element e in local coordinates

ciated with global node number $\mathbf{ix}(e,1) = i_e$ and the j end is associated with global node number $\mathbf{ix}(e,2) = j_e$. For the planar frame each node has three degrees of freedom, which are in the order $\hat{\mathbf{u}} = [u_1, u_2, \theta]^T$, where u_1 is the displacement in the direction $\hat{x}_1$ and u_2 is the displacement in the direction $\hat{x}_2$. These degrees-of-freedom can be expressed in the local element coordinates as $\mathbf{u} = [u, w, \theta]^T$, where u is the displacement in the direction x_1 and w is the displacement in the direction x_2. The local displacements can be computed from the global displacements as $\mathbf{u} = \mathbf{T}\hat{\mathbf{u}}$, where

$$\mathbf{T} = \begin{bmatrix} \cos\psi & \sin\psi & 0 \\ -\sin\psi & \cos\psi & 0 \\ 0 & 0 & 1 \end{bmatrix} \tag{425}$$

where ψ is the angle measured from x_1 to $\hat{x}_1$. The global equation numbers are assigned in node order (the global node numbering is arbitrary). The three degrees of freedom of global node n are $\mathbf{id}(n,i) = 3(n-1)+i$ for $i = 1,2,3$ (i.e., the three degrees-of-freedom at that node) and $n = 1,\ldots, N$ (the number of nodes in the structure).

Let $\mathbf{u}_e = [u, w, w']^T$ be the displacement field within the element (in local element coordinates). For element e we can compute the real and virtual displacement fields as

$$\mathbf{u}_e(\xi) = \mathbf{h}^T(\xi)\mathbf{B}_e^T\mathbf{a}, \qquad \bar{\mathbf{u}}_e(\xi) = \mathbf{h}^T(\xi)\mathbf{B}_e^T\bar{\mathbf{a}} \tag{426}$$

where $\mathbf{a} \equiv [\mathbf{a}_1^T,\ldots, \mathbf{a}_N^T]^T$ and $\bar{\mathbf{a}} \equiv [\bar{\mathbf{a}}_1^T,\ldots, \bar{\mathbf{a}}_N^T]^T$ are arrays containing the nodal unknowns and their virtual counterparts. Note that $\mathbf{a}_i$ is a 3×1 matrix containing the unknown nodal displacement $\hat{\mathbf{u}}_i$ for node i so that the matix $\mathbf{a}$ has dimension $3N \times 1$. The matrix

$$\mathbf{B}_e^T \equiv \begin{bmatrix} \mathbf{0} & \cdots & \overset{i_e}{\mathbf{T}} & \overset{j_e}{\mathbf{0}} & \cdots & \mathbf{0} \\ \mathbf{0} & \cdots & \mathbf{0} & \mathbf{T} & \cdots & \mathbf{0} \end{bmatrix} \tag{427}$$

is a $6 \times 3N$ matrix with a 3×3 transformation matrix $\mathbf{T}$ at the block associated with node $\mathbf{id}(i_e, 1{:}3)$ and a 3×3 identity matrix $\mathbf{T}$ at the block associated with node $\mathbf{id}(j_e, 1{:}3)$. The purpose of the matrix $\mathbf{B}_e$ is simply to pick out the displacement degrees-of-freedom from the global vector that are associated with element e and rotate them to the local frame. The interpolation matrix $\mathbf{h}$ is

$$\mathbf{h}^T(\xi) = \begin{bmatrix} h_1 & 0 & 0 & h_4 & 0 & 0 \\ 0 & h_2 & h_3 & 0 & h_5 & h_6 \\ 0 & h_2' & h_3' & 0 & h_5' & h_6' \end{bmatrix} \tag{428}$$

where the beam element shape functions, shown in Fig. 99, are given by

$$\begin{aligned} h_1 &= 1-\xi & h_2 &= 1-3\xi^2+2\xi^3 & h_3 &= \xi(1-2\xi+\xi^2)\ell \\ h_4 &= \xi & h_5 &= \xi^2(3-2\xi) & h_6 &= \xi^2(\xi-1)\ell \end{aligned} \tag{429}$$

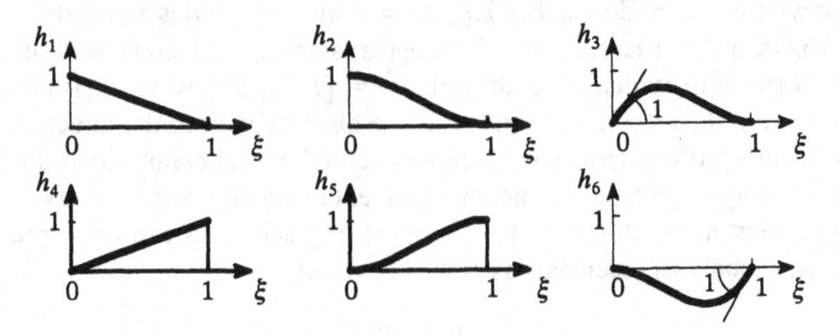

Figure 99 Planar Bernoulli-Euler beam shape functions

and $\xi \equiv x/\ell$. The choice of these shape functions is not arbitrary. In order to establish continuity of the displacement field from one element to the next we must use an interpolation that controls the displacements and rotations at the ends of the elements. The functions in Eqns. (429) are precisely the ones that can accomplish this goal. We must use a cubic interpolation for the field w because the theory computes the rotations as $\theta = w'$ and the rotations must be continuous at the nodes. Hence, we must have continuity of the first derivatives of the interpolation functions for w. The roof functions are adequate to interpolate the axial displacement. Additional shape functions could be used to improve the interpolation, but the function and its first derivative must vanish at the element end so as not to introduce excessive interelement continuity.

Let the element strains be $\mathbf{e} = [u', w'']^T$. The element real and virtual strain resultants can be computed from the interpolation for element e as

$$\mathbf{e}_e = \frac{1}{\ell_e}\mathcal{B}_e^T(\xi)\mathbf{B}_e^T\mathbf{a}, \qquad \bar{\mathbf{e}}_e = \frac{1}{\ell_e}\mathcal{B}_e^T(\xi)\mathbf{B}_e^T\bar{\mathbf{a}} \tag{430}$$

where

$$\mathcal{B}^T(\xi) = \begin{bmatrix} h_1' & 0 & 0 & h_4' & 0 & 0 \\ 0 & h_2'' & h_3'' & 0 & h_5'' & h_6'' \end{bmatrix} \tag{431}$$

Note that the prime in Eqn. (431) indicates differentiation with respect to the argument ξ. The term $1/\ell_e$ comes from the fact that the strains are derivatives with respect to x and $d\xi/dx = 1/\ell$. Using the change of variable for each element and substituting the interpolations from Eqns. (426) and (430) we can write the discrete virtual-work functional in the form

$$G(\mathbf{a}, \bar{\mathbf{a}}) = \bar{\mathbf{a}}^T \sum_{e=1}^{M}\left[\mathbf{B}_e \int_0^1 \frac{1}{\ell_e}\mathcal{B}_e\mathbf{D}_e\mathcal{B}_e^T\, d\xi\, \mathbf{B}_e^T\mathbf{a} - \mathbf{B}_e \int_0^1 \mathbf{h}_e\mathbf{q}_e\ell_e\, d\xi\right] \tag{432}$$

with $\mathbf{q}_e = [n, q, m]^T$ being the distributed loads for element e, and $\mathbf{D}$ the constitutive matrix from Eqn. (418). Note that we used $dx = \ell_e d\xi$. We can write this result more compactly if we identify the *element stiffness matrix* and the *element force* as

$$\mathbf{k}_e \equiv \int_0^1 \frac{1}{\ell_e} \mathfrak{B}_e \mathbf{D}_e \mathfrak{B}_e^T \, d\xi, \quad \mathbf{f}_e \equiv \int_0^1 \mathbf{h}_e \mathbf{q}_e \ell_e \, d\xi \tag{433}$$

Note that in the present application $\mathbf{k}_e$ is 6×6 and $\mathbf{f}_e$ is 6×1. Now the discrete virtual-work functional takes the simple form

$$G(\mathbf{a}, \bar{\mathbf{a}}) = \bar{\mathbf{a}}^T \left[\sum_{e=1}^{M} \mathbf{B}_e \mathbf{k}_e \mathbf{B}_e^T \mathbf{a} - \sum_{e=1}^{M} \mathbf{B}_e \mathbf{f}_e \right] \tag{434}$$

Comparison with our earlier results shows that the stiffness matrix and right side vectors are computed as

$$\mathbf{K} = \sum_{e=1}^{M} \mathbf{B}_e \mathbf{k}_e \mathbf{B}_e^T, \quad \mathbf{f} = \sum_{e=1}^{M} \mathbf{B}_e \mathbf{f}_e \tag{435}$$

Again, the summation over the elements is accomplished with the standard *assembly process*. The main difference from the assembly described in Chapter 6 is that the $\mathbf{B}_e$ matrix contains the local-to-global transformation $\mathbf{T}$. For this case the element stiffness matrix and element force vector can be converted to the global frame as $\hat{\mathbf{k}}_e = \hat{\mathbf{T}}^T \mathbf{k}_e \hat{\mathbf{T}}$ and $\hat{\mathbf{f}}_e = \hat{\mathbf{T}}^T \mathbf{f}_e$ where

$$\hat{\mathbf{T}} = \begin{bmatrix} \mathbf{T} & \mathbf{0} \\ \mathbf{0} & \mathbf{T} \end{bmatrix} \tag{436}$$

The transformed matrices can then be assembled directly as before. The MATLAB code segment introduced in Chapter 6 has been slightly modified for the present case and is shown in Table 7. As before, this code assumes there are N unknowns (three times the number of nodes) and M elements. It also assumes that there is a routine to call to get the element stiffness matrix and element force (and assumes that this routine takes care of rotating these matrices to the global frame).

Some of the nodal displacements are restrained by boundary conditions. These represent known values of some of the coefficients $\mathbf{a}$. Once the equations are assembled we have a linear system of equations $\mathbf{Ka} = \mathbf{f}$. The known values of $\mathbf{a}$ can be multiplied by their associated columns of $\mathbf{K}$ and subtracted from both sides of the equation. The equations associated with the restrained degree-of-freedom have reaction forces on the right side in $\mathbf{f}$. These equations can be used to determine those reaction forces.

The preceding developments capture the essence of matrix structural analysis. Extending these ideas to the three-dimensional beam and the Timoshenko beam is straightforward, but a bit more tedious. Because the virtual-work functional for the Timoshenko beam involves, at most, first derivatives of the fields, it is possible to interpolate those fields with the roof functions. However, this interpolation suffers from a phenomenon called *shear locking* in which the discrete structure is far too stiff because of a deficiency in the numerical approxi-

Table 7 MATLAB code for the assembly process for a planar frame

```
     K = zeros(N,N); f = zeros(N,1);

%.... Loop over all elements to assemble K and f
      for n = 1:M

%...... Find the i-node, j-node
        inode = ix(n,1); jnode = ix(n,2);

%...... Construct the assembly pointer array
        ii(1:3) = id(inode,1:3); ii(4:6) = id(jnode,1:3);

%...... Retrieve element stiffness matrix for element "n"
        [ke,fe] = get stiffness (...)

%...... Assemble element stiffness and force vector
        for i=1:6
          for j=1:6
            K(ii(i),ii(j)) = K(ii(i),ii(j)) + ke(i,j);
          end % loop on j
          f(ii(i)) = f(ii(i)) + fe(i);
        end % loop on i

      end % loop on n
```

mation. Most textbooks on the finite element method have a good description of the locking problem (it also affects low-order 3D elements that are nearly incompressible through a similar phenomenon called *volumetric locking*). One cure for locking is to use higher-order interpolation functions. A cubic $\mathcal{C}^0$ Lagrangian interpolation is sufficient to eliminate shear locking in Timoshenko beam elements.

Additional Reading

H. Goldstein, *Classical mechanics*, 2nd ed. Addison-Wesley, Reading, Mass, 1980.

J. T. Oden, *Mechanics of elastic structures*, McGraw-Hill, New York.(1967).

I. S. Sokolnikoff, *Mathematical theory of elasticity*, 2nd ed., McGraw-Hill, New York, 1956.

Problems

167. The beam shown below has a rectangular cross section of depth 4 and width 2, and has a length of 50 length units. It has a uniform mass density that gives rise to a constant body force of $\mathbf{b}(x) = -2\mathbf{e}_2$ (force units per length units cubed), and is subjected to a surface traction on its top surface that is bilinear with respect to x_2 and $x_3 \equiv x$ reaching a maximum value of 15 (force units per length units squared), as shown.

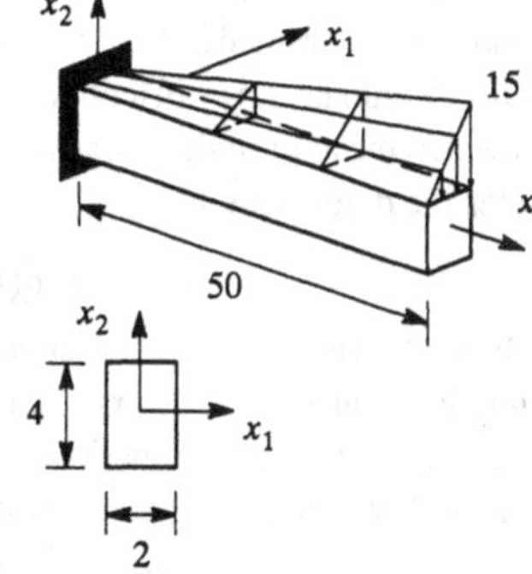

Find an expression for the applied tractions $\mathbf{t}_\Gamma(\mathbf{x})$. Find the resultant applied loads $\mathbf{q}(x)$ and $\mathbf{m}(x)$ equivalent to the surface tractions and body force. Find the distribution of resultant force $\mathbf{Q}(x)$ and resultant moment $\mathbf{M}(x)$ along the beam. Find the displacements $\mathbf{w}(x)$ and the rotations $\boldsymbol{\theta}(x)$ along the beam.

168. Consider the beam with square cross section, of dimension h by h and length ℓ. The beam has Young's modulus C and shear modulus μ. The beam is subjected to horizontal tractions on its top face, as shown. The body forces acting on the beam are negligible. The coordinate axes shown are principal and centroidal.

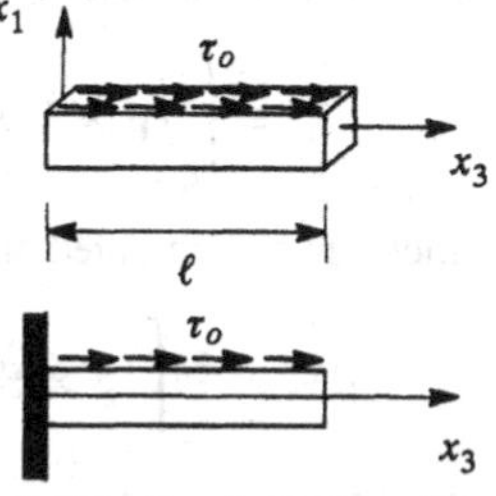

Find expressions for the applied force and moment per unit length of beam, $\mathbf{q}(x)$ and $\mathbf{m}(x)$, where $x = x_3$ is the axial coordinate. Find the displacement and rotation field caused by the loading by integrating the governing beam equations (that is, find the classical solution).

169. Consider a beam of length ℓ with square 2×2 cross section. The beam is subjected to the applied traction field over the cross section at the end of the beam

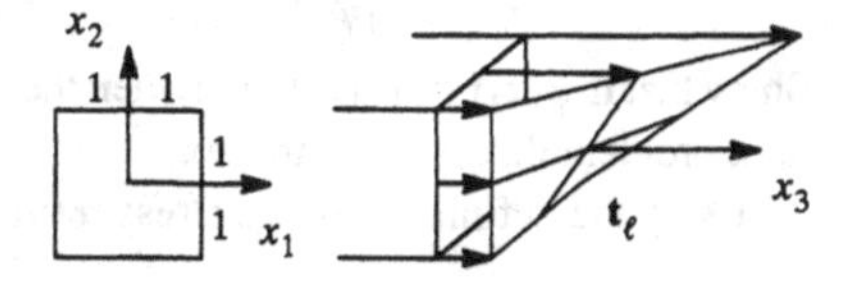

$$\mathbf{t}_\ell = \tau\, x_1 \left(1 + x_2^2\right) \mathbf{e}_3$$

where τ is the known magnitude of the loading. Find the resultant force and moment acting on the end of the beam. Assume that $\mathbf{u}(x_1, x_2, 0)$, i.e., the beam is fixed at $x_3 = 0$. Find the resultant force field $\mathbf{Q}(x)$ and the resultant moment field $\mathbf{M}(x)$ that equilibrate the applied forces. Compute the displacement and rotation fields that result from the applied loads.

170. The hollow box beam shown has a square cross section of dimension b, thickness $t \ll b$, and unit weight ϱ_b. It is submerged in a fluid of unit weight ϱ_o. Recall that the pressure at any point in a static fluid is proportional to the depth h according to the relationship $p = h\varrho_o$. The unit weight of the air inside the beam can be taken as zero. The end is capped so that fluid cannot get inside. Plot the typical traction field $\mathbf{t}_\Gamma$ acting on the lateral surface of the beam. Compute the resultant applied load $\mathbf{q}(x)$ and the resultant applied moment $\mathbf{m}(x)$ that would be appropriate in order to treat the problem using beam theory.

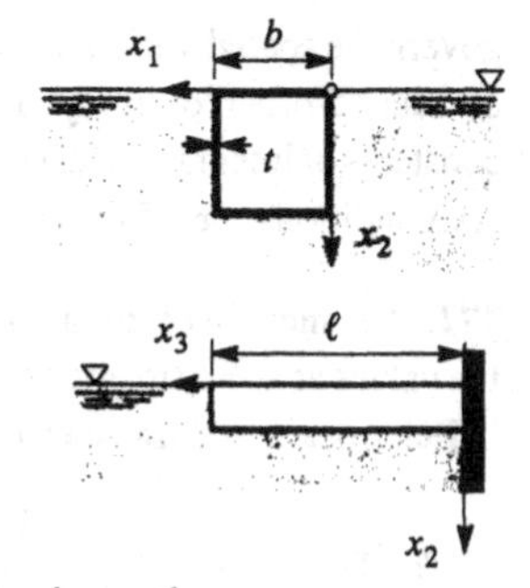

171. The Saint-Venant torsion problem is restricted to problems with constant rate of twist and traction-free lateral surfaces. One feature of this solution is that, at the fixed end, the rotation is restrained but the out-of-plane warping is not. Physically, such a boundary condition would be very difficult to realize. One solution to this problem is to create a model in which the amplitude of warping is independent from the rate of twist of the beam. Consider the deformation map

$$\mathbf{u}(\mathbf{x}) = \theta(x_3)\big(\mathbf{e}_3 \times \mathbf{p}\big) + \varphi(x_3)\psi(x_1,x_2)\mathbf{e}_3$$

where $\theta(x_3)$ is the angle of twist, $\varphi(x_3)$ is the amplitude of warping, $\psi(x_1,x_2)$ is the warping function, and $\mathbf{p} = x_1\mathbf{e}_1 + x_2\mathbf{e}_2$. Further, assume that the warping function $\psi(x_1,x_2)$ is the Saint-Venant warping function derived in the text. Compute the strain tensor and use the linear elastic constitutive equations to show that the tractions on a cross section are

$$\mathbf{S}\mathbf{e}_3 = \mu\big[\varphi\nabla\psi + \theta'\mathbf{e}_3 \times \mathbf{p}\big] + E\varphi'\psi\mathbf{e}_3$$

where $E \equiv \lambda + 2\mu$, and a prime denotes differentiation with respect to x_3.
In addition to the polar moment of inertia J, define the cross-sectional properties

$$J_1 \equiv \int_\Omega \big(\mathbf{p} \times \nabla\psi\big) \cdot \mathbf{e}_3\, dA, \quad J_2 \equiv \int_\Omega \big(\nabla\psi \cdot \nabla\psi\big)\, dA, \quad J_3 \equiv \int_\Omega \psi^2\, dA$$

which can be computed once $\psi(x_1,x_2)$ is known. Define the stress resultants

$$T \equiv \mathbf{e}_3 \cdot \int_\Omega \mathbf{p} \times \mathbf{S}\mathbf{e}_3\, dA, \quad W \equiv \mathbf{e}_3 \cdot \int_\Omega \psi\mathbf{S}\mathbf{e}_3\, dA, \quad B \equiv \int_\Omega \nabla\psi \cdot \mathbf{S}\mathbf{e}_3\, dA$$

where T is the usual torque. The stress resultants W and B are often called the *bi-moment* and the *bi-shear*. Substitute the expression for $\mathbf{S}\mathbf{e}_3$ to show that

$$T = \mu J_1\varphi + \mu J\theta', \quad W = EJ_3\varphi', \quad B = \mu J_2\varphi + \mu J_1\theta'$$

Show that if $\varphi(x_3) = \theta'(x_3) = a$, then the above results are consistent with the Saint-Venant problem discussed in the text.

Using the definitions of the stress resultants, compute T' and W' and show that

$$T' + t = 0, \qquad W' - B + w = 0$$

What are the appropriate definitions of the applied loads $t(x)$ and $w(x)$?

Substitute the resultant constitutive equations into the resultant equilibrium equations, and show that the equations

$$\mu J\theta'' + \mu J_1\varphi' + t = 0, \qquad EJ_3\varphi'' - \mu J_2\varphi - \mu J_1\theta' + w = 0$$

govern the spatial variation of rotation θ and warping φ. These equations constitute a pair of second-order ordinary differential equations, and, therefore, we can expect four constants of integration that must be found from boundary conditions. What are the boundary conditions for a free end and a fixed end?

172. The method of initial parameters integrates the governing equations and substitutes the values at $x = 0$ to give the general form of the displacement function. For the Bernoulli-Euler beam, the transverse deflection can be computed as

$$w_{\mathrm{BE}}(x) = w_o + \theta_o x + \frac{M_o x^2}{2EI} - \frac{Q_o x^3}{6EI} + \frac{1}{EI}\int_0^x \Big[\tfrac{1}{6}(x-\xi)^3 q(\xi) - \tfrac{1}{2}(x-\xi)^2 m(\xi)\Big]\, d\xi$$

where $w_o = w(0)$, $\theta_o = w'(0)$, $M_o = M(0)$, $Q_o = Q(0)$ are the initial parameters. Verify that the expression satisfies the governing differential equations of Bernoulli-Euler beam theory. This equation is particularly useful for those cases where M_o and Q_o can be determined from overall equilibrium. Use the method of initial parameters to solve the problem of the cantilever beam under uniform load given as an example in the text.

173. The method of initial parameters can be applied to the Timoshenko beam to give

$$w_T(x) = w_{BE}(x) + \frac{Q_o x}{GA} - \frac{1}{GA}\int_0^x (x-\xi)q(\xi)\,d\xi$$

where $w_T(x)$ is the deflection according to Timoshenko beam theory and $w_{BE}(x)$ is the deflection according to Bernoulli-Euler beam theory (as given in Problem 172). Verify that the expression satisfies the governing differential equations of Timoshenko beam theory. Use the method of initial parameters to solve the problem of the cantilever beam under uniform load given as an example in the text.

174. The three-dimensional rotation tensor $\mathbf{\Lambda}$ can be expressed in terms of three parameters e_1, e_2, and e_3 as

$$\mathbf{\Lambda}(e_1, e_2, e_3) = \begin{bmatrix} e_0^2 + e_1^2 - e_2^2 - e_3^2 & 2(e_1 e_2 + e_0 e_3) & 2(e_1 e_3 - e_0 e_2) \\ 2(e_1 e_2 - e_0 e_3) & e_0^2 - e_1^2 + e_2^2 - e_3^2 & 2(e_2 e_3 + e_0 e_1) \\ 2(e_1 e_3 + e_0 e_2) & 2(e_2 e_3 - e_0 e_1) & e_0^2 - e_1^2 - e_2^2 + e_3^2 \end{bmatrix}$$

where the parameter e_0 has been introduced for convenience. This fourth parameter does not represent an independent parameter, but rather satisfies the constraint equation $e_0^2 + e_1^2 + e_2^2 + e_3^2 = 1$. These parameters are called the *Euler parameters*. Demonstrate that the tensor $\mathbf{\Lambda}$ is orthogonal by showing that $\mathbf{\Lambda}^{-1} = \mathbf{\Lambda}^T$. Show that for small values of the parameters e_1, e_2, and e_3 the tensor can be expressed in the form $\mathbf{\Lambda} \approx \mathbf{I} + \mathbf{W}$, where $\mathbf{I}$ is the identity and $\mathbf{W}$ is a skew-symmetric tensor. Show, therefore, that when the three parameters are small, they can be viewed as the components of the rotation vector $\boldsymbol{\theta}$, with $\theta_i = 2e_i$, and that $\mathbf{W} = \boldsymbol{\theta}\times$.

175. Reconsider the cantilever beam of length ℓ, fixed at $x = 0$, with bending modulus EI and shear modulus GA solved as an example in the text (Fig. 96). Examine the results of the Ritz method as you increase the number of basis functions taken from the sets

$$h_i(x) \in \left\{x, \frac{x^2}{\ell}, \frac{x^3}{\ell^2}, \ldots, \frac{x^N}{\ell^{N-1}}\right\}, \quad g_i(x) \in \left\{\frac{x}{\ell}, \frac{x^2}{\ell^2}, \frac{x^3}{\ell^3}, \ldots, \frac{x^M}{\ell^M}\right\}$$

In particular, find general expressions for the ijth components of $\mathbf{K}^{aa}$, $\mathbf{K}^{ab}$, $\mathbf{K}^{ba}$, $\mathbf{K}^{bb}$, and the ith components of $\mathbf{f}^a$ and $\mathbf{f}^b$ when $h_i(x) = x^i/\ell^{i-1}$ and $g_i(x) = x^i/\ell^i$. Solve the problem for $(N, M) = (2, 2), (3, 2), (3, 3), (4, 3), (4, 4)$. What do you expect to happen for higher-order approximations? Comment on the differences in the solutions obtained when $M = N-1$ versus those obtained with $M = N$.

176. Prove that $\mathbf{S} \cdot \mathbf{W} = 0$, or $S_{ij}W_{ij} = 0$, when $\mathbf{S}$ is a symmetric tensor and $\mathbf{W}$ is a skew-symmetric tensor. Note that $\mathbf{S} \cdot \mathbf{W} = \mathbf{e}_i \cdot \mathbf{S}\mathbf{W}\mathbf{e}_i$.

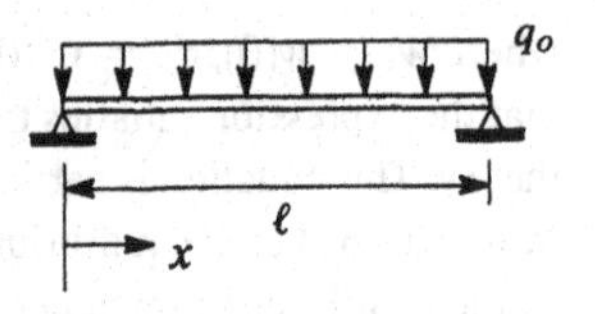

177. The prismatic beam shown has a cross section that is symmetric with respect to the plane of the page. The cross section has axial modulus EA, shear modulus GA, and flexural modulus EI. The beam is subjected to a uniform transverse load $q(x) = -q_o$. Find the displacements and rotations for the beam by directly integrating the governing equations.

178. Resolve Problem 177 after making the Bernoulli-Euler assumption that $w' = \theta$ (i.e., there is no shear deformation). What is the difference between the two solutions?

179. Use a polynomial basis to find an approximate solution to Problem 177 using the principle of virtual work for the beam including shear deformation. Which terms should you include? What order approximation is adequate?

180. Repeat the virtual-work computation in Problem 179 for the Bernoulli-Euler beam.

181. Carry out the integrations in Eqn. (385) to show that the tensors given in Eqn. (387) result.

182. Consider the beam in Problem 177. Find an expression for the transverse displacement $w(x)$ using the principle of virtual work, using a quartic polynomial basis. Note that the problem has two essential boundary conditions and two natural boundary conditions.

183. The principle of virtual work does not require that the assumed displacement functions satisfy the natural boundary conditions a priori. Is there an advantage to satisfying the natural boundary conditions, too? What happens in Problem 182 if we do enforce the natural boundary conditions?

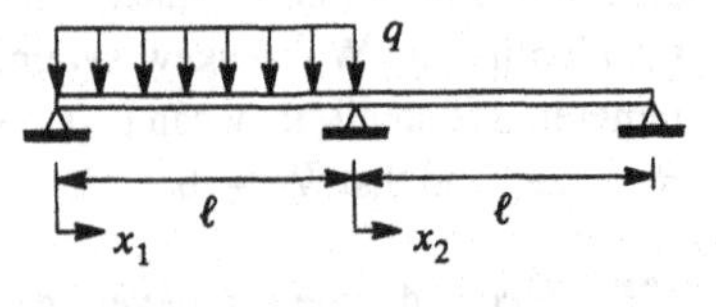

184. A continuous beam is one that has one or more intermediate supports. The extra boundary conditions are in excess of the four end conditions. Describe an approach to solving the following problem that exactly satisfies the differential equations everywhere in the domain, as well as the boundary and intermediate conditions. Find the classical solution to the given problem by integrating the governing differential equations. (Hint: It is useful to describe the solution independently in each segment and to enforce continuity by equating state variables at the place where the two segments join.)

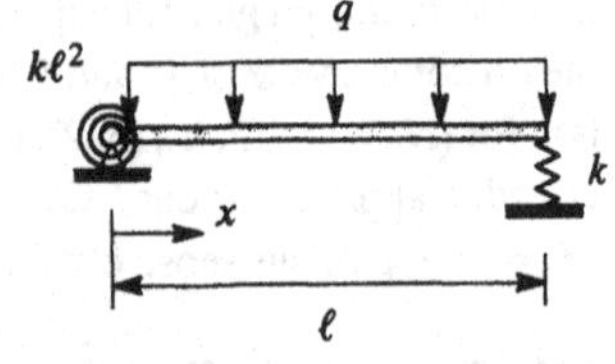

185. The following prismatic beam has a cross section that is symmetric with respect to the plane of the page. The cross section has flexural modulus EI. Axial and shear deformations can be neglected (i.e., use Bernoulli-Euler beam theory). The beam is subjected to a uniform transverse load q acting downward. The beam has deformable spring supports at the ends. At the left end, the support prevents translation in the vertical and horizontal directions and the spring elastically restrains rotations. The moment developed by the spring is related to the rotation at that point by $M_s = k\ell^2\theta_s$, where $\theta_s = \theta(0)$ is the rotation experienced by the spring. The right end of the beam is free to translate horizontally and to rotate, but the spring elastically restrains vertical motion. The force developed by the spring is related to the deflection at that point by $F_s = kw_s$, where $w_s = w(\ell)$ is the deflection experienced by the spring. What are the appropriate boundary conditions for this problem? Solve the problem by integrating the differential equations and using the boundary conditions to find the constants of integration. Revise the principle of virtual work to account for the work done by the springs. Estimate the deflection of the beam using a two-term polynomial expansion for the transverse deflection. That is, assume the real and virtual transverse deflections to be of the form

$$w(x) = a_1 x + a_2 \frac{x^2}{\ell}, \quad \overline{w}(x) = \overline{a}_1 x + \overline{a}_2 \frac{x^2}{\ell}$$

What constraints do the assumed displacement field add to the problem?

186. In the derivations of beam theory, both in a classical sense and in a variational sense, no mention was made of concentrated forces. Describe a way to account for concentrated forces in solving the classical differential equations (for example, for a Bernoulli-Euler beam). Describe how the concentrated forces should be implemented into the principle of virtual work.

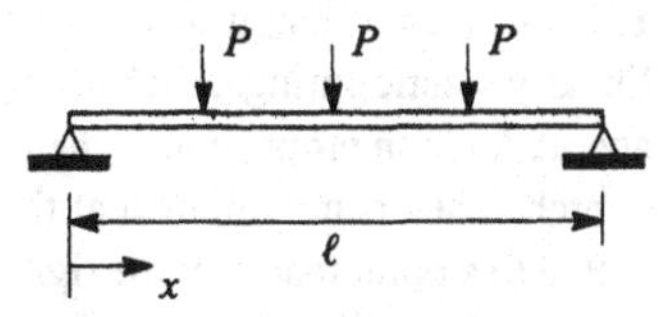

187. A beam on a Winkler elastic foundation accrues force in the foundation in proportion to the deflection of the beam according to $f(x) = kw(x)$, where k is the modulus of the foundation.

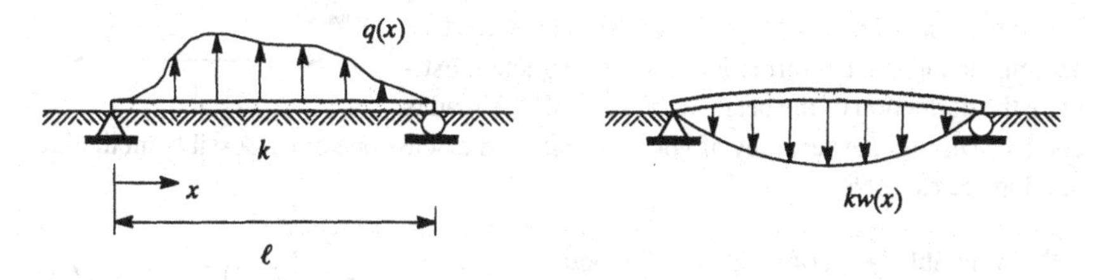

Show that a simply supported Bernoulli-Euler beam on an elastic foundation is governed by the following differential equation and boundary conditions

$$EIw^{iv} + kw = q(x)$$

$$w(0) = 0, \quad w''(0) = 0, \quad w(\ell) = 0, \quad w''(\ell) = 0$$

Verify that the $w(x) = e^{\beta x}(a_1 \cos\beta x + a_2 \sin\beta x) + e^{-\beta x}(a_3 \cos\beta x + a_4 \sin\beta x)$ is the displacement field that satisfies the homogeneous differential equation, if $4\beta^4 \equiv k/EI$.

188. Consider the beam on a Winkler elastic foundation of Problem 187, subjected to a uniform load q. Show that the principle of virtual work, accounting for the work done by the elastic foundation, is

$$G(w, \overline{w}) = \int_0^\ell \left[EIw''\overline{w}'' + kw\overline{w} - q\overline{w}\right] dx = 0$$

Find the displacement map of the system using the Ritz method, assuming that the real and virtual displacements are approximated as

$$w(x) = \sum_{n=1}^{N} a_n \sin\frac{n\pi x}{\ell}, \quad \overline{w}(x) = \sum_{n=1}^{N} \overline{a}_n \sin\frac{n\pi x}{\ell}$$

189. A semi-infinite beam on a Winkler elastic foundation extends to infinity in one direction. Find the classical solution to the problem of a beam of modulus EI on a foundation with modulus k subjected to a concentrated force P at $x = 0$. There is no boundary at the right end of the beam, but you can argue that $a_1 = a_2 = 0$ to have finite displacements. Plot the deflected shape of the beam.

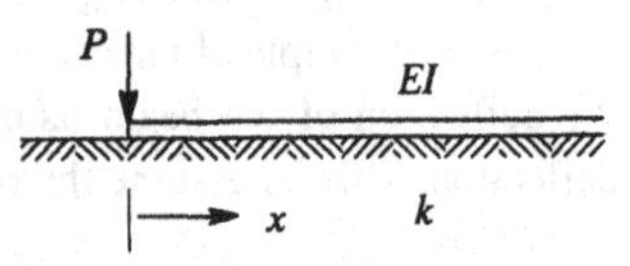

190. A beam of length ℓ and modulus EI rests on two linearly elastic springs, each of modulus k. The springs accrue force in proportion to the amount by which they stretch. The beam is pinned at the left end and is subjected to a point load P at the right end. Axial and shear deformations of the beam can be neglected. What is the virtual-work form of the equilibrium equations? What are the essential and natural boundary conditions? Use the Ritz method to find an approximation of the displacement field using the two-term polynomial $w(x) = a_1 x + a_2 x^2/\ell$.

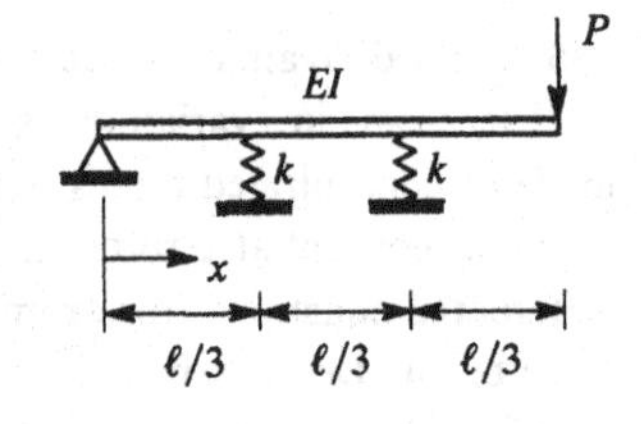

191. Consider the beam of modulus EI, pinned at one end, free at the other, and restrained by a rotational spring as shown. The beam is subjected to a tip load P at the free end. Shear and axial deformations can be neglected. Estimate the deflection of the beam at the point where load is applied. Discuss the accuracy of your estimate and discuss one other possible method for making the estimate.

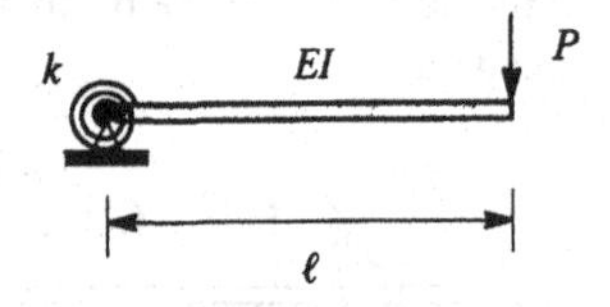

192. A flexible beam of length ℓ and modulus EI is welded to a rigid beam of length ℓ, and rests on an elastic foundation of modulus $k = 60EI/\ell^4$. The beam is simply supported and is subjected to a transverse force q over the rigid part of the span. The elastic foundation accrues a transverse force in proportion to the transverse displacement w. Shear and axial deformations in the beam are negligible. Write the virtual-work functional G for the system. What are the essential and natural boundary conditions for the flexible beam? Find an approximate solution for the displacement $w(x)$ using a two-term polynomial Ritz basis.

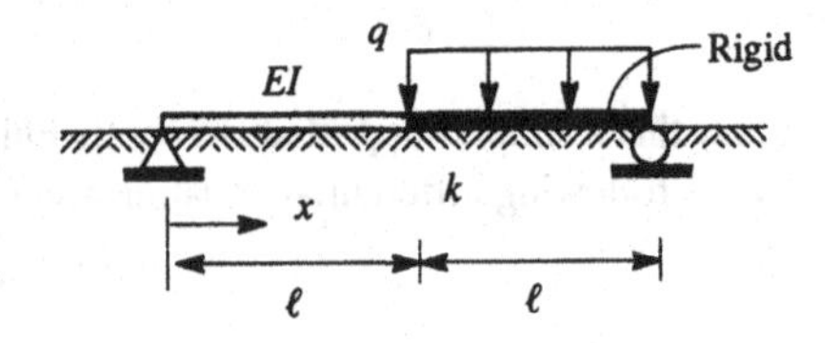

193. Consider the beam of modulus EI, fixed at one end, pinned at the other. The beam is subjected to a prescribed displacement of w_ℓ at the right end. Shear and axial deformations can be neglected. Find the expression for the displacement $w(x)$ that satisfies

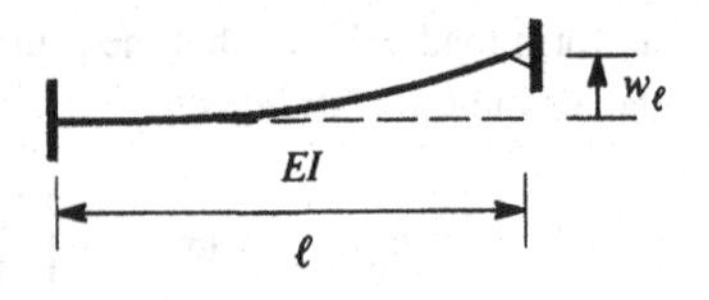

the governing equations exactly. Approximately solve the problem using the principle of virtual work, assuming a cubic polynomial deflection field.

194. A beam of unit length and variable modulus is fixed at the right end and is subjected to a moment M at the left end. The bending and shear stiffnesses of the beam are variable with $EI(x) = EI_0(1+x)$ and $GA(x) = GA_0(1+x)$, where EI_0 and GA_0 are known constants. Recall that the governing equations for the Timoshenko beam are given by Eqns. (406). Find the deflection and rotation at the left end of the beam (i.e., at $x = 0$) by finding the *classical solution* to the governing differential equations.

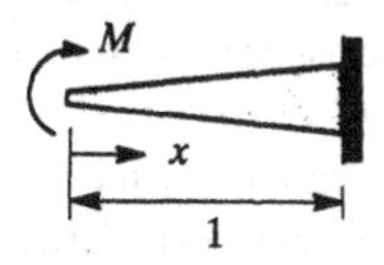

195. A beam of length ℓ and modulus EI rests on two linearly elastic springs, each of modulus k. The springs accrue force in proportion to the amount by which they stretch. The beam is pinned at both ends and is subjected to a concentrated moment M at the right end. Axial and shear deformations of the beam can be neglected. What is the virtual-work form of the equilibrium equations? What are the essential and natural boundary conditions? Use the Ritz method to find a polynomial approximation of the displacement field.

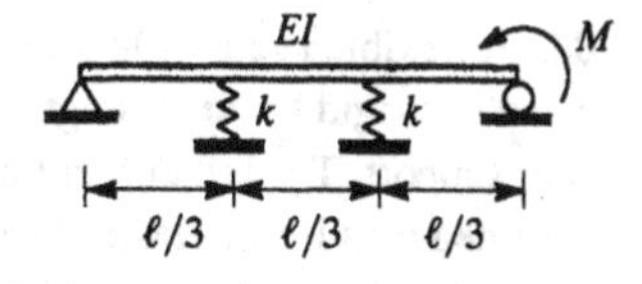

196. Consider a beam of length ℓ, elastic moduli E and G and rectangular cross section of width $2a$ and depth $2b$. Let c be a (very small) constant and let $\xi \equiv x_3/\ell$ be the normalized axial coordinate. The beam has the following displacement and rotation fields

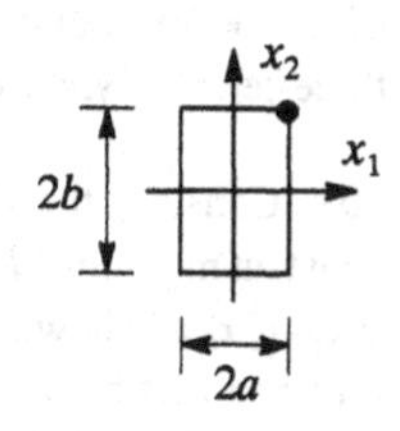

$$\theta_1 = c(6\xi - 6\xi^2), \quad \theta_2 = c(3\xi^2 - 2\xi), \quad \theta_3 = 0$$

$$w_1 = c\ell(\xi^3 - \xi^2), \quad w_2 = c\ell(2\xi^3 - 3\xi^2), \quad w_3 = 0$$

Find the resultant moment and resultant force at $\xi = 1$, i.e. $\mathbf{M}(\ell)$ and $\mathbf{Q}(\ell)$. Find the total displacement $\mathbf{u}$ of the point located at the position $\mathbf{x} = (a, b, \ell)$.

197. A beam of length ℓ is fixed at the left end, free at the right end, and is subjected to a concentrated transverse load P at the right end. The bending and shear stiffnesses of the beam are EI and GA, respectively. What is the virtual-work functional for the system? What are the essential and natural boundary conditions? Let $\xi \equiv x/\ell$. Find the deflection and rotation fields for the given loading by the Ritz method using the following approximation

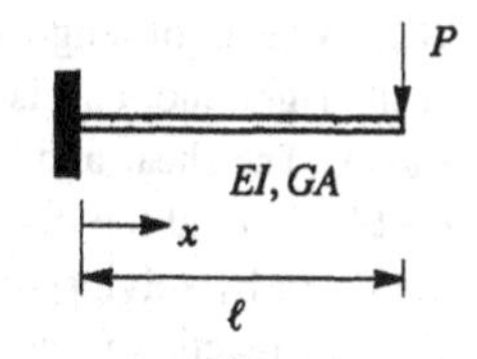

$$w(\xi) = a_2\ell\xi + \tfrac{1}{2}a_0\ell\xi^2 + \tfrac{1}{3}a_1\ell\xi^3, \qquad \theta(\xi) = a_0\xi + a_1\xi^2$$

198. A beam of length ℓ and modulus EI rests on two *linearly* elastic springs, each of modulus k. The beam is subjected to point loads P at the ends. Axial and shear deformations of the beam can be neglected. What is the virtual-work form of the equilibrium equations? What are the essential and natural boundary conditions? Solve the discrete virtual-work equations $\mathbf{Ka} = \mathbf{f}$ for this system using a three-term polynomial Ritz approximation.

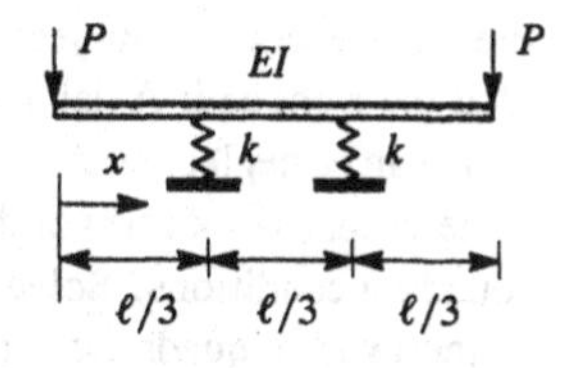

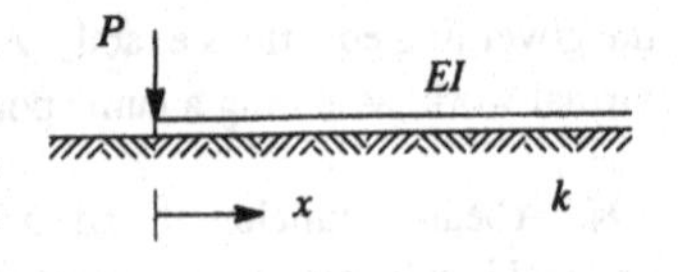

199. A semi-infinite (i.e., x extends to infinity) beam of modulus EI rests on an elastic foundation of modulus k and is subjected to a concentrated force P at $x = 0$. The classical differential equation for the beam is given by $EIw^{iv} + kw = 0$, where w^{iv} means fourth derivative. The shear in the beam is given by $Q(x) = -EIw'''$. Use the Ritz method with the principle of virtual work to find an approximation of the displacement field using the single term approximation $w(x) = ae^{-\beta x}$, where β is a given constant and a is the unknown Ritz displacement parameter. Find an expression for the error in equilibrium at each point x in the beam. For what value of β is the shear boundary condition at $x = 0$ satisfied exactly?

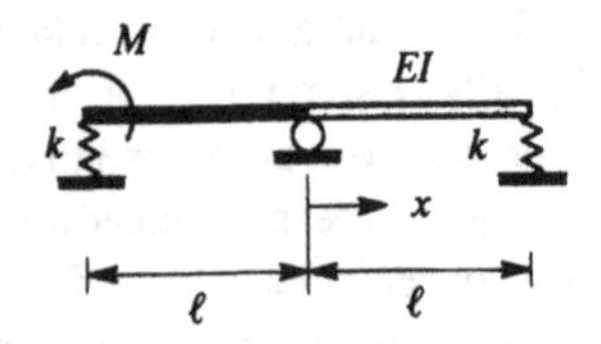

200. A flexible beam of length ℓ and modulus EI is connected to a rigid beam of length ℓ at a point that rests on a roller support. The left end of the rigid part of the beam is restrained by a linear elastic spring of modulus k as is the right end of the flexible part of the beam. The beam is subjected to an end moment M. Axial and shear deformations of the flexible beam can be neglected. What is the virtual-work form of the equilibrium equations? What are the essential and natural boundary conditions for the flexible segment of the beam? Solve the discrete virtual-work equations $\mathbf{Ka} = \mathbf{f}$ for this system using a three-term polynomial Ritz approximation.

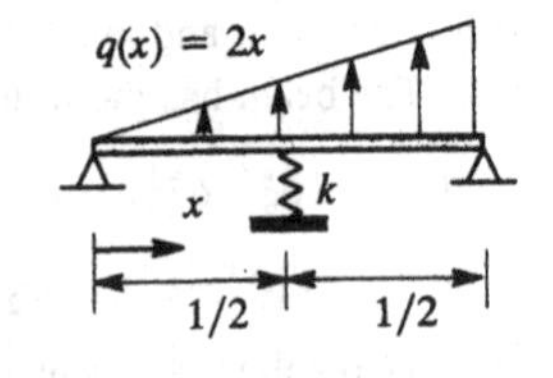

201. Consider the simply supported beam of length 1 and constant modulus $EI = 1$, subjected to a linearly varying force $q(x) = 2x$, as shown. The beam is supported by a spring at midspan that has modulus $k = 64$. Shear and axial deformations can be neglected. What are the natural boundary conditions? What are the essential boundary conditions? Find an approximate solution with the Ritz method. Use a cubic polynomial.

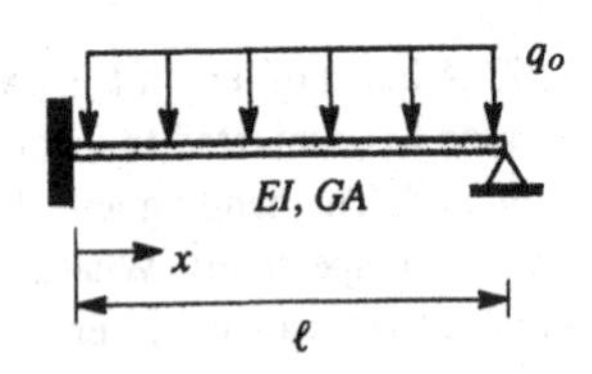

202. A beam of length ℓ is fixed at the left end, pinned at the right end, and is subjected to a uniform load, as shown. The shear and bending moduli are related as $EI/GA\ell^2 = 1$. Find the displacement and rotation fields for the beam by solving the classical governing equations. Find the reaction forces at the supports.

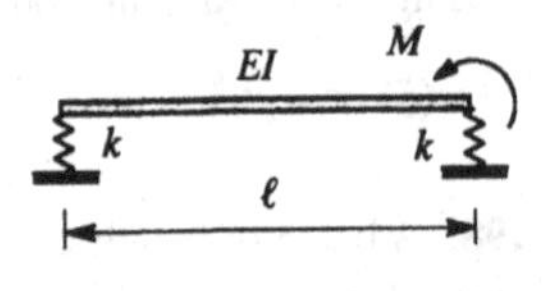

203. A beam of length ℓ and modulus EI rests on two linearly elastic springs, each of modulus k. The springs accrue force in proportion to the amount by which they stretch. The beam is subjected to a concentrated moment M at the right end. Axial and shear deformations of the beam can be neglected. Assume that motion along the axis of the beam is restrained. What is the virtual-work form of the equilibrium equations? What are the essential and natural boundary conditions? Solve the discrete equations of equilibrium $\mathbf{Ka} = \mathbf{f}$ using the Ritz method with a quadratic approximation of the displacement field. Describe the error in approximation.

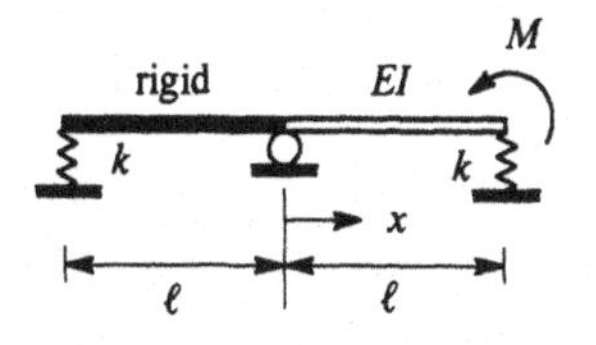

204. A flexible beam of length ℓ and modulus EI is connected to a rigid beam of length ℓ at a point that rests on a roller support. The left end of the rigid beam is restrained by a linear elastic spring of modulus k. The beam is subjected to an end moment M. Axial and shear deformations of the flexible beam can be neglected. What is the virtual-work form of the equilibrium equations? What are the essential and natural boundary conditions for the flexible segment of the beam? Solve the discrete equations of equilibrium $\mathbf{Ka} = \mathbf{f}$ using the Ritz method with a quadratic approximation of the displacement field. Describe the error in approximation.

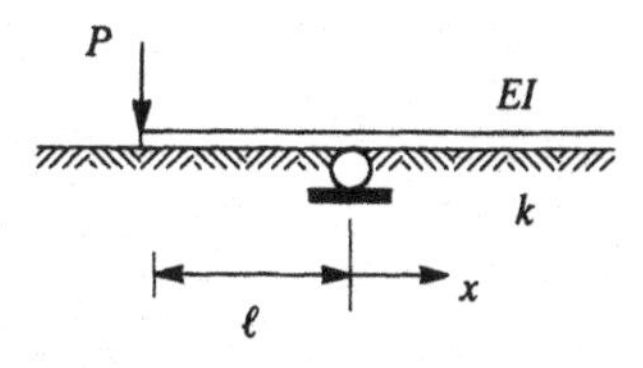

205. A semi-infinite beam (i.e., the beam extends to infinity in the positive x direction) of modulus EI rests on an elastic foundation of modulus k. The beam is supported at x=0 and is subjected to a concentrated force P at a distance ℓ from the support. Discuss how you would solve this problem. Include in your discussion comments on both classical and variational approaches. Assume that Bernoulli-Euler beam theory is adequate to describe the response of this system. The classical equations of a Bernoulli-Euler beam on an elastic foundation are

$$EIw^{iv} + kw = 0 \qquad M(x) = EIw'' \quad Q(x) = -EIw'''$$

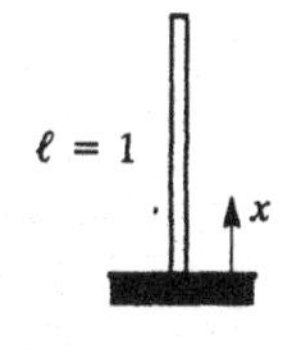

206. Consider a beam of *unit length* and circular cross section, fixed at $x = 0$ and free at $x = 1$. The axis of the beam ($x_3 = x$) points along the $\mathbf{e}_3$ direction. The origin of the cross sectional coordinates x_1 and x_2 are at the center of the centroid of the section. The internal resultant force is given by the explicit expression $\mathbf{Q}(x) = (x^2-1)\mathbf{e}_1 + (x^3-1)\mathbf{e}_2$. Find the applied force $\mathbf{q}(x)$ that must be present. Assume that the applied moment $\mathbf{m}(x) = \mathbf{0}$. Find the internal moment field $\mathbf{M}(x)$. Find the rotation field $\boldsymbol{\theta}(x)$

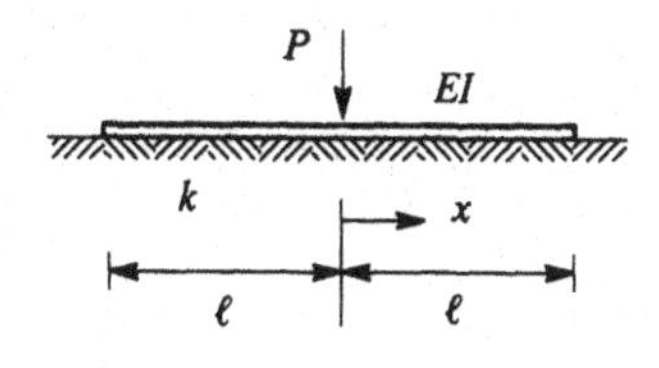

207. A flexible beam of length 2ℓ and modulus EI rests on an elastic foundation of modulus k. The properties have values such that the dimensionless ratio $k\ell^4/EI = 15$. The beam is subjected to a load P at its midpoint. Axial and shear deformations of the flexible beam can be neglected. Find the deflection at the middle and ends of the beam using virtual work and the Ritz method with a polynomial approximation. (Note: due to symmetry odd functions—i.e., linear, cubic, etc.—need not be included.)

8
The Linear Theory of Plates

A plate is a body with one geometric dimension that is significantly smaller than the other two. We call this dimension the *thickness* of the plate. We shall treat the thickness dimension of a plate much the same as we did the cross section of the beam. Like the beam, we shall characterize the behavior of the plate with a particular kinematic assumption, and we shall consider the average response of the body through the thickness. These assumptions allow a reduction of the governing differential equations from spatial dimension three to two. Unlike the beam, the governing equations of a plate are still partial differential equations. In some sense, a plate is simply a two-dimensional beam, and there are many analogies between the two theories.

One of the most valuable aspects of the approach to plate theory taken here is the clear display of the striking similarities between beam theory and plate theory. In fact, one can compare the derivation of the two theories almost equation for equation. The reader would be well advised to reconsider the previous chapter on linear beam theory while reading through the derivation of plate theory. In many ways, the plate and the beam are exact complements of each other. Some of the classical treatments obscure this complementarity.

Our order of tasks is analogous to those in the chapter on linear beam theory. First, we derive the equilibrium equations for the plate by defining resultants of the traction vector over the thickness of the plate and seeing how the three-dimensional equilibrium equations relate to the rate of change of the resultants. We then introduce a kinematic hypothesis that describes the motion of the plate. Finally, we introduce the three-dimensional elastic constitutive equations into the theory and deduce definitions of strain resultants that are conjugate to the stress resultants as well as constitutive equations for the resultants.

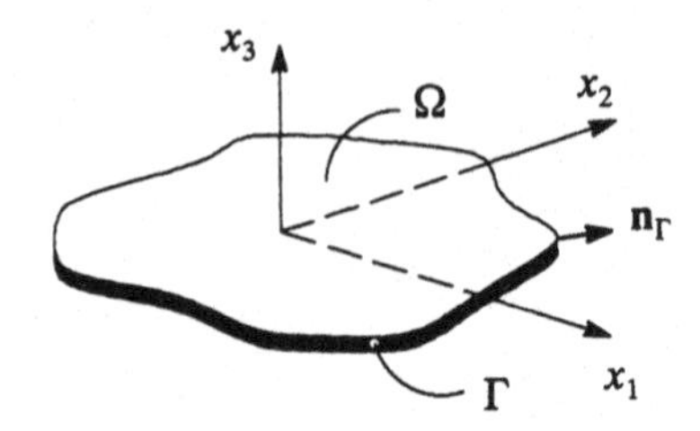

Figure 100 A plate occupying three-dimensional space

Once the classical form of the boundary value problem is laid out, we will proceed to develop a virtual-work form of the plate equations.

Notation. We will describe our plate in accord with the convention shown in Fig. 100. Accordingly, we choose the x_3 coordinate axis to be perpendicular to the *reference surface* of the plate, a flat two-dimensional surface in the x_1 - x_2 plane from which any point in the plate can be described by its elevation. Often, the reference surface will be taken as the surface midway between the two faces. The domain of the reference surface, and, hence, the plate itself, is a closed geometric figure Ω having a boundary Γ that can be parameterized by its arc length s. The boundary has a normal $\mathbf{n}_\Gamma$ at every point, and this normal is unique, except possibly at a finite number of corners. The normal to the boundary is orthogonal to the unit base vector pointing in the x_3 direction, that is, $\mathbf{n}_\Gamma \cdot \mathbf{e}_3 = 0$. The origin of the x_3 axis is at the reference surface, the top surface of the plate is positioned at $x_3 = \overline{h}$ above the reference surface, and the bottom surface of the plate at $x_3 = \underline{h}$. The total plate thickness is, therefore, $h \equiv \overline{h} - \underline{h}$. The plate need not be of constant thickness. Thus, the quantities $\overline{h}$ and $\underline{h}$ can depend upon x_1 and x_2. The plate is subjected to a body force of density $\mathbf{b}$ and tractions on the top, bottom, and lateral surfaces of $\mathbf{t}^+$, $\mathbf{t}^-$, and $\mathbf{t}_\Gamma$, respectively, as shown in Fig. 101.

The notation is largely the same as that used for the three-dimensional theory. Unless otherwise indicated, Latin subscripts $(i, j, k, \ldots)$ are assumed to range from 1 to 3, while Greek subscripts $(\alpha, \beta, \gamma, \ldots)$ range from 1 to 2 only. Summation over repeated indices is implied, unless otherwise indicated. We

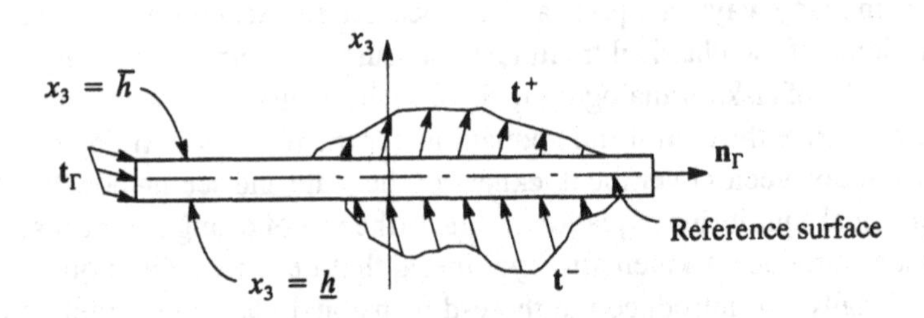

Figure 101 View through the thickness of a plate

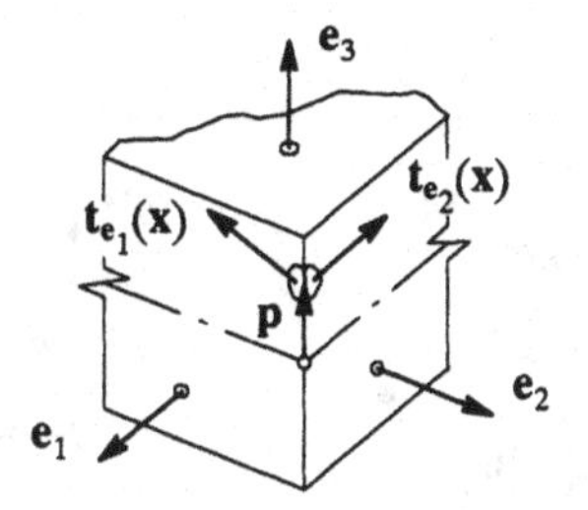

Figure 102 Resultants of the traction vector over the thickness

shall use the "comma" notation $(\cdot)_{,a} = \partial(\cdot)/\partial x_a$ for the partial derivative. Multiple subscripts following the comma indicate higher-order differentiation, e.g., $(\cdot)_{,a\beta} = \partial^2(\cdot)/\partial x_a \partial x_\beta$.

Equations of Equilibrium

A plate is subject to the same requirements of equilibrium as every other body. The concept that distinguishes a beam or a plate from a continuum is the *stress resultant*. The stress resultants for plates will represent the aggregate effect of all of the traction forces acting over the thickness of the plate. Like the beam, we shall find that a simple net resultant is not adequate to describe those tractions, so we shall also compute the first moment of these tractions about the reference surface. By considering the rate of change of these resultants, we can deduce equations of equilibrium for the resultants from the three-dimensional theory.

The main difference between the stress resultants of a plate and those of a beam is that, for the plate, we consider the resultants per unit length, integrating only through the thickness. For the beam, the resultant was a resultant over the entire cross section. The resultant force and moment can be computed by integrating the tractions through the thickness, as shown in Fig. 102. The traction vector on a plane with normal $\mathbf{e}_a$ is

$$\mathbf{t}_{\mathbf{e}_a}(\mathbf{x}) = \mathbf{S}(\mathbf{x})\mathbf{e}_a$$

where the free index a takes values 1 and 2. The resultant forces $\mathbf{Q}_a(x_1, x_2)$ and the resultant moments $\mathbf{M}_a(x_1, x_2)$ are computed as the integral of the tractions and the first moment of the tractions through the thickness

$$\boxed{\begin{aligned} \mathbf{Q}_a(x_1, x_2) &\equiv \int_{\underline{h}}^{\bar{h}} \mathbf{t}_{\mathbf{e}_a}(\mathbf{x})\, dx_3 \\ \mathbf{M}_a(x_1, x_2) &\equiv \int_{\underline{h}}^{\bar{h}} \mathbf{p}(x_3) \times \mathbf{t}_{\mathbf{e}_a}(\mathbf{x})\, dx_3 \end{aligned}} \qquad (437)$$

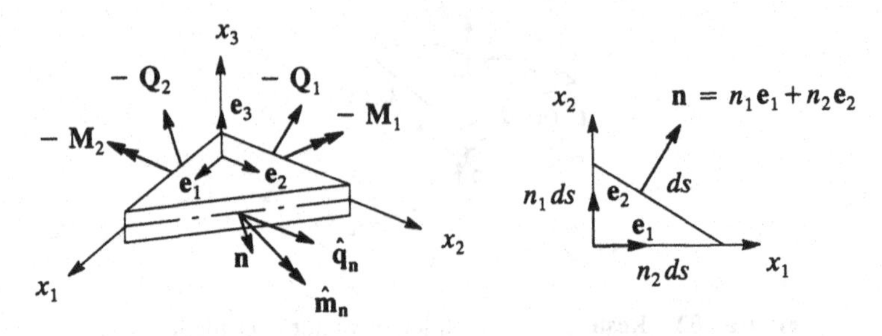

Figure 103 Cauchy triangle construction for plates

Where $\mathbf{p} \equiv x_3\mathbf{e}_3$. It is evident that $\mathbf{M}_a \cdot \mathbf{e}_3 = 0$. Physically, this condition means that there is no resultant moment about the x_3 axis, a consequence of our decision to characterize the body as a plate and thus to define the resultant relative to the normal to the reference surface. As a result, the resultant moment vectors have the form $\mathbf{M}_a = M_{a1}\mathbf{e}_1 + M_{a2}\mathbf{e}_2$, with no component in the $\mathbf{e}_3$ direction. One main difference between beam theory and plate theory is that in the former we have only one resultant force vector and one resultant moment vector, while in the latter we have two of each.

As was the case for beams, the stress resultants $\mathbf{Q}_a$ and $\mathbf{M}_a$ appear to be vectors, and they will behave like vectors in almost every regard. However, these vectors characterize the state of stress in the beam and, therefore, we must examine how the Cauchy relations $\mathbf{t}_{\mathbf{n}} = \mathbf{S}\mathbf{n}$ manifest for the plate. For plates, the appropriate analogy is the infinitesimal Cauchy triangle shown in Fig. 103. By geometry, the values n_1 and n_2 are the ratios of the lengths of the sides with normals $-\mathbf{e}_1$ and $-\mathbf{e}_2$, respectively, to the length of the oblique side. They are also the components of the normal vector $\mathbf{n}$ to the oblique side. Let us compute the resultant traction force $\hat{\mathbf{q}}_{\mathbf{n}}$ on the face with normal vector $\mathbf{n} = n_a\mathbf{e}_a$

$$\hat{\mathbf{q}}_{\mathbf{n}} = \int_{\underline{h}}^{\bar{h}} \mathbf{S}(\mathbf{x})\big(n_a\mathbf{e}_a\big)\,dx_3 = n_a\int_{\underline{h}}^{\bar{h}} \mathbf{S}(\mathbf{x})\mathbf{e}_a\,dx_3 = n_a\mathbf{Q}_a$$

Since n_a does not depend upon the cross-sectional coordinates x_1 and x_2, it can be factored out of the integral to give the one-dimensional version of the Cauchy formula relating stress to tractions

$$\hat{\mathbf{q}}_{\mathbf{n}} = n_a\mathbf{Q}_a \tag{438}$$

An identical argument produces an equivalent result for the moments. The resultant traction moment $\hat{\mathbf{m}}_{\mathbf{n}}$ acting on the face with normal vector $\mathbf{n} = n_a\mathbf{e}_a$ is related to the resultant moments as

$$\hat{\mathbf{m}}_{\mathbf{n}} = n_a\mathbf{M}_a \tag{439}$$

As was the case for three-dimensional solids, the Cauchy relationship shows us how to take freebody diagrams in the sense that it tells us what force to place at an exposed section to represent the state of stress resultants there.

Equilibrium of forces. We can derive equilibrium equations that correspond with the three-dimensional equations of equilibrium of a continuum. Let us integrate $\operatorname{div}\mathbf{S} + \mathbf{b}$ through the thickness of the plate

$$\begin{aligned}\int_{\underline{h}}^{\bar{h}} \big(\operatorname{div}\mathbf{S} + \mathbf{b}\big)\, dx_3 &= \int_{\underline{h}}^{\bar{h}} \left(\frac{\partial \mathbf{S}\mathbf{e}_\alpha}{\partial x_\alpha} + \frac{\partial \mathbf{S}\mathbf{e}_3}{\partial x_3} + \mathbf{b} \right) dx_3 \\ &= \frac{\partial \mathbf{Q}_\alpha}{\partial x_\alpha} + \int_{\underline{h}}^{\bar{h}} \left(\frac{\partial \mathbf{S}\mathbf{e}_3}{\partial x_3} + \mathbf{b} \right) dx_3\end{aligned} \tag{440}$$

where summation is implied on α from 1 to 2. Further note that

$$\int_{\underline{h}}^{\bar{h}} \frac{\partial \mathbf{S}}{\partial x_3} \mathbf{e}_3 \, dx_3 = \big[\mathbf{S}(x_1, x_2, \bar{h}) - \mathbf{S}(x_1, x_2, \underline{h})\big]\mathbf{e}_3$$

One can observe that Cauchy's relation implies that $\mathbf{S}(x_1, x_2, \bar{h})\mathbf{e}_3 = \mathbf{t}^+$, the applied traction on the top surface, and $-\mathbf{S}(x_1, x_2, \underline{h})\mathbf{e}_3 = \mathbf{t}^-$, the applied traction on the bottom face. The applied tractions $\mathbf{t}^+$ and $\mathbf{t}^-$ are vectors, independent of the orientation of the vector normal to planes in the body. These tractions are, of course, the known prescribed loads on the surfaces of the plate. Since the body forces $\mathbf{b}$ are also known, we are led to define the applied loading per unit of area as

$$\boxed{\mathbf{q}(x_1, x_2) \equiv \big(\mathbf{t}^+ + \mathbf{t}^-\big) + \int_{\underline{h}}^{\bar{h}} \mathbf{b}\, dx_3} \tag{441}$$

Thus, from Eqn. (440), we have

$$\int_{\underline{h}}^{\bar{h}} \big(\operatorname{div}\mathbf{S} + \mathbf{b}\big)\, dx_3 = \mathbf{Q}_{\alpha,\alpha} + \mathbf{q} \tag{442}$$

Observe that if $\mathbf{Q}_{\alpha,\alpha} + \mathbf{q} = \mathbf{0}$ then $\operatorname{div}\mathbf{S} + \mathbf{b} = \mathbf{0}$ is satisfied on the average through the thickness of the plate. Contrast this result with the beam result that the three-dimensional equilibrium equations are satisfied on the average over the cross section.

Equilibrium of moments. We can follow the same approach to the equilibrium of moments. Let us integrate $\mathbf{p} \times \big(\operatorname{div}\mathbf{S} + \mathbf{b}\big)$ through the thickness of the plate

$$\int_{\underline{h}}^{\overline{h}} \mathbf{p} \times (\operatorname{div}\mathbf{S}+\mathbf{b})\, dx_3 = \int_{\underline{h}}^{\overline{h}} \mathbf{p} \times \left(\frac{\partial \mathbf{S}\mathbf{e}_\alpha}{\partial x_\alpha} + \frac{\partial \mathbf{S}\mathbf{e}_3}{\partial x_3} + \mathbf{b} \right) dx_3$$
$$= \frac{\partial}{\partial x_\alpha} \int_{\underline{h}}^{\overline{h}} \mathbf{p} \times \mathbf{S}\mathbf{e}_\alpha \, dx_3 + \int_{\underline{h}}^{\overline{h}} \mathbf{p} \times \left(\frac{\partial \mathbf{S}\mathbf{e}_3}{\partial x_3} + \mathbf{b} \right) dx_3 \tag{443}$$

The first term is simply the divergence of the moments so that

$$\int_{\underline{h}}^{\overline{h}} \mathbf{p} \times (\operatorname{div}\mathbf{S}+\mathbf{b})\, dx_3 = \frac{\partial \mathbf{M}_\alpha}{\partial x_\alpha} + \int_{\underline{h}}^{\overline{h}} \mathbf{p} \times \left(\frac{\partial \mathbf{S}\mathbf{e}_3}{\partial x_3} + \mathbf{b} \right) dx_3 \tag{444}$$

To make further progress let us note the following identity

$$\frac{\partial}{\partial x_3}\left(\mathbf{p} \times \mathbf{S}\mathbf{e}_3\right) = \frac{\partial \mathbf{p}}{\partial x_3} \times \mathbf{S}\mathbf{e}_3 + \mathbf{p} \times \frac{\partial \mathbf{S}}{\partial x_3}\mathbf{e}_3$$

By the definition of $\mathbf{p}$, we know that $\partial \mathbf{p}/\partial x_3 = \mathbf{e}_3$. We also know that balance of angular momentum of the continuum implies that

$$\mathbf{e}_i \times \mathbf{S}\mathbf{e}_i = \mathbf{e}_3 \times \mathbf{S}\mathbf{e}_3 + \mathbf{e}_\alpha \times \mathbf{S}\mathbf{e}_\alpha = \mathbf{0}$$

Using these relationships we find that

$$\mathbf{p} \times \frac{\partial \mathbf{S}}{\partial x_3}\mathbf{e}_3 = \frac{\partial}{\partial x_3}\left(\mathbf{p} \times \mathbf{S}\mathbf{e}_3\right) + \mathbf{e}_\alpha \times \mathbf{S}\mathbf{e}_\alpha$$

Integrating this equation through the thickness yields the following result

$$\int_{\underline{h}}^{\overline{h}} \left(\mathbf{p} \times \frac{\partial \mathbf{S}}{\partial x_3}\mathbf{e}_3 \right) dx_3 = \int_{\underline{h}}^{\overline{h}} \frac{\partial}{\partial x_3}\left(\mathbf{p} \times \mathbf{S}\mathbf{e}_3\right) dx_3 + \int_{\underline{h}}^{\overline{h}} \mathbf{e}_\alpha \times \mathbf{S}\mathbf{e}_\alpha \, dx_3$$

Explicitly evaluating the first integral on the right side (the integral of an exact differential), and recognizing the definition of the resultants $\mathbf{Q}_\alpha$ in the second term, we obtain

$$\int_{\underline{h}}^{\overline{h}} \left(\mathbf{p} \times \frac{\partial \mathbf{S}}{\partial x_3}\mathbf{e}_3 \right) dx_3 = \mathbf{p}(\overline{h}) \times \mathbf{t}^+ + \mathbf{p}(\underline{h}) \times \mathbf{t}^- + \mathbf{e}_\alpha \times \mathbf{Q}_\alpha$$

where $\mathbf{p}(\overline{h}) = \overline{h}\mathbf{e}_3$ and $\mathbf{p}(\underline{h}) = \underline{h}\mathbf{e}_3$. Since the body force $\mathbf{b}$ and the tractions $\mathbf{t}^+$ and $\mathbf{t}^-$ are given as data, let us define the applied moment per unit area as

$$\boxed{\mathbf{m}(x_1,x_2) \equiv \mathbf{e}_3 \times \left(\overline{h}\mathbf{t}^+ + \underline{h}\mathbf{t}^-\right) + \mathbf{e}_3 \times \int_{\underline{h}}^{\overline{h}} x_3\, \mathbf{b}\, dx_3} \tag{445}$$

From this expression, it is easy to see that $\mathbf{m} \cdot \mathbf{e}_3 = 0$, that is, the loading causes no resultant about $\mathbf{e}_3$. As a result, when we describe the applied moment

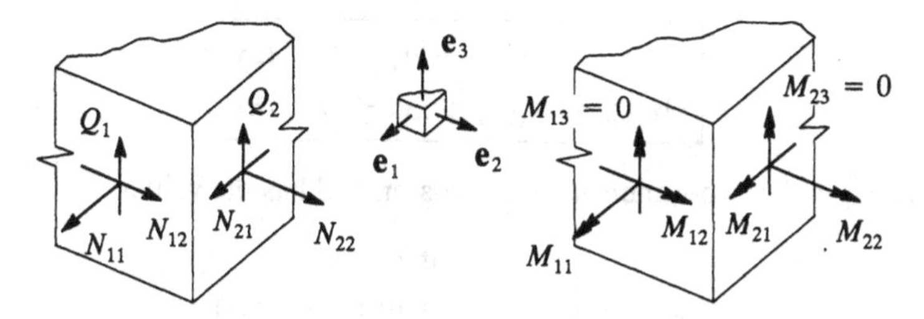

Figure 104 Definition of the components of the resultants (shown positive)

vector in terms of its components, it will have the form $\mathbf{m} = m_1\mathbf{e}_1 + m_2\mathbf{e}_2$. With these definitions we get the following equation

$$\int_{\underline{h}}^{\bar{h}} \mathbf{p} \times (\operatorname{div}\mathbf{S} + \mathbf{b})\, dx_3 = \mathbf{M}_{\alpha,\alpha} + \mathbf{e}_\alpha \times \mathbf{Q}_\alpha + \mathbf{m} \tag{446}$$

From these equations we can see that if $\mathbf{M}_{\alpha,\alpha} + \mathbf{e}_\alpha \times \mathbf{Q}_\alpha + \mathbf{m} = \mathbf{0}$, then the first moment of $\operatorname{div}\mathbf{S} + \mathbf{b}$ over the thickness of the plate is equal to zero. Contrast this result with the analogous one for beams. The equilibrium equations for the stress resultants can be summarized as follows (note that summation over the repeated index is implied)

$$\boxed{\begin{aligned} \mathbf{Q}_{\alpha,\alpha} + \mathbf{q} &= \mathbf{0} \\ \mathbf{M}_{\alpha,\alpha} + \mathbf{e}_\alpha \times \mathbf{Q}_\alpha + \mathbf{m} &= \mathbf{0} \end{aligned}} \tag{447}$$

Some interesting features of these equations will become evident if we examine them in component form. Let us define some nomenclature to help us with the component description. First, we shall call the components of the resultant force

$$\begin{aligned} \mathbf{Q}_\alpha &= N_{\alpha 1}\mathbf{e}_1 + N_{\alpha 2}\mathbf{e}_2 + Q_\alpha \mathbf{e}_3 \\ \mathbf{M}_\alpha &= M_{\alpha 1}\mathbf{e}_1 + M_{\alpha 2}\mathbf{e}_2 \end{aligned}$$

The components of the resultants are shown, with positive sign convention, in Fig. 104. Remember that the first subscript of $N_{\alpha\beta}$ and $M_{\alpha\beta}$ keeps track of the plane on which the resultant vector acts, while the second subscript keeps track of the component of the vector. If we write out Eqns. (447) in extenso, we get a system of equations describing the equilibrium of the in-plane forces

$$\boxed{\begin{aligned} N_{\alpha\beta,\alpha} + q_\beta &= 0 \quad \text{(a)} \\ N_{12} - N_{21} &= 0 \quad \text{(b)} \end{aligned}} \tag{448}$$

and a system of equations describing the bending of the plate

$$\boxed{\begin{aligned} Q_{\alpha,\alpha} + q_3 &= 0 \quad \text{(a)} \\ M_{\alpha\beta,\alpha} + \epsilon_{\beta\alpha} Q_\alpha + m_\beta &= 0 \quad \text{(b)} \end{aligned}} \tag{449}$$

where $\epsilon_{\alpha\beta}$ is the two-dimensional alternator, and has the values

$$\epsilon_{\alpha\beta} = \begin{cases} 0 & \text{if } \alpha = \beta \\ 1 & \text{if } (\alpha,\beta) = (1,2) \\ -1 & \text{if } (\alpha,\beta) = (2,1) \end{cases} \tag{450}$$

Equations $(448)_a$ and $(449)_a$ both come from the resultant force vector equation, while Eqns. $(448)_b$ and $(449)_b$ come from the resultant moment vector equation. Just as balance of angular momentum told us that the stress tensor **S** is symmetric, we get the symmetry condition $(448)_b$ from balance of moments about the x_3 axis. Equations (448) concern the equilibrium of the in-plane resultants. These equations are nothing more than the *plane stress* equations, and are uncoupled from the bending equations in the linear theory if the reference surface lies exactly halfway between the faces of the plate. The symmetry condition $(448)_b$ is the familiar symmetry of conjugate shears. Equations (449) concern the bending of the plate. In the linear theory of plates, only the transverse shears (not in-plane forces) are coupled with the bending moments.

We can see the similarity between linear plate theory and linear beam theory. In linear beam theory, the equilibrium of axial forces was uncoupled from the bending equations. The beam has one axial equation and two shear equations. The plate has one shear equation and two axial equations. For both the plate and the beam, there are two bending equations. For the beam, the axial moment equation gave rise to torsional equilibrium; for the plate, we get symmetry of the in-plane shears. The complementarity of these two theories is evident.

The Kinematic Hypothesis

As with the beam, we need a kinematic hypothesis to complete plate theory. A kinematic hypothesis is nothing more than a restriction placed on the deformation map. We assume that the body moves in a very specific manner, an assumption that must be verified either by observation of nature or by examining the consequences of imposing the constraints with a theory that does not make those assumptions (i.e., the general three-dimensional theory).

The basic idea behind beam theory was the hypothesis that cross sections that are plane and normal to the beam axis before deformation remain plane and unstretched after deformation. The cross-sectional area was the primary geometric object used in the description of the constrained deformation map. In fact, we could completely describe the map by tracking the motion of a typical cross section, parameterized by the axial coordinate x_3. The reduction in

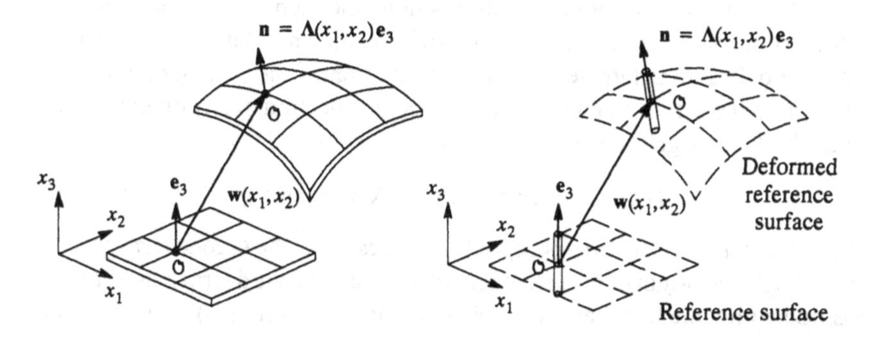

Figure 105 Displacement map for a plate

the dimensionality of the problem from three dimensions to one dimension was accomplished by introducing a two-dimensional geometric constraint: the rigid cross section.

For the plate, we shall introduce a kinematic constraint analogous to the one introduced for the beam. We shall assume that a straight line that is normal to the reference surface of the plate before deformation remains straight and unstretched after deformation. Let us consider the motion of a typical line normal to the undeformed reference surface of the plate. The initial orientation of the line is along the $\mathbf{e}_3$ direction. The kinematic hypothesis suggests that the line will remain straight, but will translate and rotate rigidly from its original position, as shown in Fig. 105. (The line is shown in the figure with exaggerated thickness for the purpose of visualization; it is, in reality, a one-dimensional geometric object.) Let us consider the point of intersection of the line and the reference surface $\mathcal{O}$. As described in the figure, the point displaces by an amount $\mathbf{w}$. It takes three quantities to keep track of the motion of the point, the three components of $\mathbf{w}$.

Keeping track of the motion of a single point is not sufficient to describe the motion of the line because the line also rotates. If, in addition to $\mathbf{w}$, we keep track of the orientation of the vector $\mathbf{n}$ that lies along the axis of the line, then our task of tracking the motion of the plane will be complete. It takes three quantities to describe the rotation of a vector in three-dimensional space: the three parameters of the rotation tensor $\mathbf{\Lambda}$. As we did for the rotation of the cross section of a beam, let the rotation tensor be precisely that needed to orient the normal vector as $\mathbf{n} = \mathbf{\Lambda}(x_1, x_2)\mathbf{e}_3$. Thus, it appears that we must keep track of six quantities in order to uniquely track the motion of the line in space. From those six quantities, we can find the location of any other particle in the body.

There is an important geometric consideration here, however, that affects the count of the number of parameters required to characterize our map of plate motion. Since we need only track the position of the line, we do not need all three components of the rotation of the normal vector. The rotation of the line about its own axis does not affect its position, and, hence, will not affect the deformation map of the plate. Thus, we arrive at the important conclusion that we need only five parameters to track the motion of the plate (see Problem 208). From those five quantities, we can find the location of any other particle on the body through the map

$$\boldsymbol{\phi}(\mathbf{x}) = x_\alpha \mathbf{e}_\alpha + \mathbf{w}(x_1,x_2) + \boldsymbol{\Lambda}(x_1,x_2)\mathbf{p}(x_3) \tag{451}$$

The first term gets us to the appropriate thickness line, the second term gets us to the displaced origin of the reference surface, and the third term gets us to the position within the line that was originally at the position $\mathbf{p}(x_3) = x_3\mathbf{e}_3$ in the undeformed configuration, i.e., $\mathbf{p}$ locates the position of points along the line relative to the point $\mathcal{O}$.

If the rotation of the normal vector is small, the above map can be simplified. In particular, for small rotations we have $\boldsymbol{\Lambda}\mathbf{p} \approx \mathbf{p} + \boldsymbol{\theta} \times \mathbf{p}$, where $\boldsymbol{\theta} = \theta_\alpha \mathbf{e}_\alpha$ is called the rotation vector. We can now describe our deformation through a displacement map. Let $\mathbf{u}(\mathbf{x})$ be the displacement of a point originally located at position $\mathbf{x}$ in our undeformed plate. The displacement of that point caused by the deformation is

$$\boxed{\mathbf{u}(\mathbf{x}) = \mathbf{w}(x_1,x_2) + \boldsymbol{\theta}(x_1,x_2) \times \mathbf{p}(x_3)} \tag{452}$$

We can clearly see the explicit dependence of the map on the transverse coordinate x_3 and the plate surface coordinates x_1 and x_2. The components of the displacement vector $\mathbf{w} = \{w_1, w_2, w_3\}$ and the components of the rotation vector $\boldsymbol{\theta} = \{\theta_1, \theta_2, 0\}$ are collectively called the *generalized displacements* and are functions of the plate surface coordinates x_1 and x_2. The displacement map can be written out in terms of its components as follows

$$\begin{aligned}
u_1(x_1,x_2,x_3) &= w_1(x_1,x_2) + x_3\theta_2(x_1,x_2) \\
u_2(x_1,x_2,x_3) &= w_2(x_1,x_2) - x_3\theta_1(x_1,x_2) \\
u_3(x_1,x_2,x_3) &= w_3(x_1,x_2)
\end{aligned}$$

From the explicit expression for the map, we can see that there is no dependence on a rotation about the x_3 axis. This rotation is often referred to as the *drilling degree of freedom*, because the motion it describes is reminiscent of a drill making a straight bore into the plate along the deformed normal direction.

The physical significance of the generalized displacements can be seen by examining the individual terms of the map. Figure 106 shows the displaced plate projected onto the $x_2 - x_3$ plane. Clearly, the component w_2 measures the displacement along the x_2 axis, while the component w_3 measures displacement transverse to the middle surface in the x_3 direction. The component θ_1 measures rotation about the x_1 axis and has a positive sense according to the right-hand rule. Displacements are, of course, positive if they are in the direc-

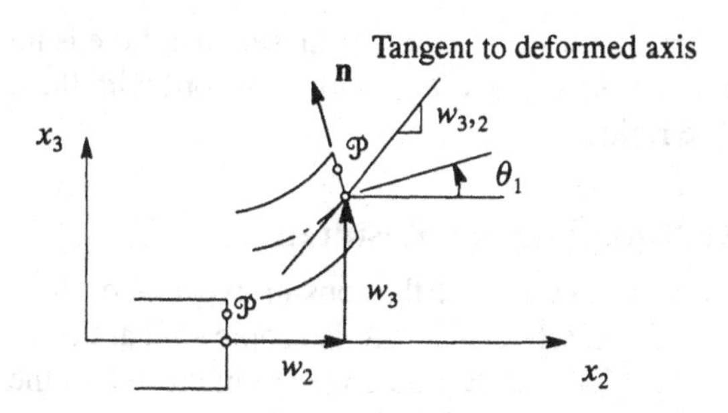

Figure 106 Components of displacement in a plane

tion of the coordinate axes. Consider the displacement of the point $\mathcal{P}$ a distance x_3 from the reference surface of the plate. For the purpose of illustration, suppose that we have a state of cylindrical bending, i.e., $w_2 = 0$, $\theta_1 = 0$, and $\theta_3 = 0$. Relative to the point $\mathcal{O}$, point $\mathcal{P}$ moves in the negative x_2 direction by an amount equal to $x_3 \sin\theta_1 \approx x_3\theta_1$, and in the negative x_3 direction by an amount $x_3(1-\cos\theta_1) \approx 0$. Because the motion is cylindrical, there is no motion in the x_1 direction (out of the plane of the page). Clearly, this is the motion that our deformation map captures.

If we have an explicit expression for the deformation map, it is simple to compute the strains implied by that map. Here we shall confine our attention to the linearized strain tensor $\mathbf{E} = \frac{1}{2}[\nabla\mathbf{u} + \nabla\mathbf{u}^T]$. Recall that the gradient of $\mathbf{u}$ can be expressed in the form $\nabla\mathbf{u}(\mathbf{x}) = \mathbf{u}_{,i} \otimes \mathbf{e}_i$. We can thus compute the gradient of the displacement map for our plate as

$$\nabla\mathbf{u}(\mathbf{x}) = \left(\mathbf{w}_{,\alpha} + \boldsymbol{\theta}_{,\alpha} \times \mathbf{p}\right) \otimes \mathbf{e}_\alpha - \left(\mathbf{e}_3 \times \boldsymbol{\theta}\right) \otimes \mathbf{e}_3$$

(We got the sign change on the last term by reversing the order of the cross product.) Before we use $\nabla\mathbf{u}$ to compute the linearized strain tensor, let us make a valuable simplification of the above expression. Recall from Chapter 7 that the tensor $\boldsymbol{\Theta} \equiv \left(\mathbf{e}_i \times \boldsymbol{\theta}\right) \otimes \mathbf{e}_i = -[\boldsymbol{\theta}\times]$ is skew-symmetric. We can rewrite the gradient of displacement as

$$\nabla\mathbf{u}(\mathbf{x}) = \left(\boldsymbol{\epsilon}_\alpha + \boldsymbol{\varkappa}_\alpha \times \mathbf{p}\right) \otimes \mathbf{e}_\alpha - \boldsymbol{\Theta}$$

where the strain resultants are defined as

$$\boxed{\begin{aligned} \boldsymbol{\epsilon}_\alpha &\equiv \mathbf{w}_{,\alpha} + \mathbf{e}_\alpha \times \boldsymbol{\theta} \\ \boldsymbol{\varkappa}_\alpha &\equiv \boldsymbol{\theta}_{,\alpha} \end{aligned}} \tag{453}$$

It is important to note that, because the generalized displacements depend only upon x_1 and x_2, $\boldsymbol{\epsilon}_\alpha$ and $\boldsymbol{\varkappa}_\alpha$ also depend only on x_1 and x_2. We are now in a position to compute the linearized strain tensor for the plate kinematic description

$$\mathbf{E} = \frac{1}{2}\left[\left(\boldsymbol{\epsilon}_\alpha + \boldsymbol{\varkappa}_\alpha \times \mathbf{p}\right) \otimes \mathbf{e}_\alpha + \mathbf{e}_\alpha \otimes \left(\boldsymbol{\epsilon}_\alpha + \boldsymbol{\varkappa}_\alpha \times \mathbf{p}\right)\right] \tag{454}$$

From this expression for the strain, we can see that there is no strain through the thickness of the plate, since $E_{33} = 0$, in accord with the assumption that normal lines are rigid.

Constitutive Equations for Resultants

The strains imply stresses through the constitutive relationships. Let us assume that the material is linearly elastic and isotropic so that the stress-strain relationship is $\mathbf{S} = \lambda \operatorname{tr}(\mathbf{E})\mathbf{I} + 2\mu\mathbf{E}$. Since we are interested in the resultant tractions only on planes with normal $\mathbf{e}_\alpha$, it is sufficient to compute only the values of $\mathbf{t}_{\mathbf{e}_\alpha} = \mathbf{S}\mathbf{e}_\alpha = \lambda \operatorname{tr}(\mathbf{E})\mathbf{e}_\alpha + 2\mu\mathbf{E}\mathbf{e}_\alpha$. From Eqn. (454) one can compute

$$2\mathbf{E}\mathbf{e}_\alpha = \left[\delta_{\alpha\beta}\mathbf{I} + \mathbf{e}_\beta \otimes \mathbf{e}_\alpha\right]\left(\boldsymbol{\epsilon}_\beta + \boldsymbol{\varkappa}_\beta \times \mathbf{p}\right)$$
$$\operatorname{tr}(\mathbf{E}) = \mathbf{e}_\beta \cdot \mathbf{E}\mathbf{e}_\beta = \mathbf{e}_\beta \cdot \left(\boldsymbol{\epsilon}_\beta + \boldsymbol{\varkappa}_\beta \times \mathbf{p}\right)$$

The expression for the traction on a plane with normal $\mathbf{e}_\alpha$ is

$$\mathbf{S}\mathbf{e}_\alpha = \boldsymbol{\Xi}_{\alpha\beta}\left(\boldsymbol{\epsilon}_\beta + \boldsymbol{\varkappa}_\beta \times \mathbf{p}\right) \tag{455}$$

where $\boldsymbol{\Xi}_{\alpha\beta} \equiv \lambda\mathbf{e}_\alpha \otimes \mathbf{e}_\beta + \mu\left(\delta_{\alpha\beta}\mathbf{I} + \mathbf{e}_\beta \otimes \mathbf{e}_\alpha\right)$. We will use this expression in our definitions for the stress resultants to derive constitutive equations relating the stress resultants to the strain resultants that we have defined above. It will then become clear that $\boldsymbol{\epsilon}_\alpha$ and $\boldsymbol{\varkappa}_\alpha$ are indeed appropriate measures of strain for a plate. Recall that our definitions of stress resultant and moment resultant were given by Eqn. (437). Substituting Eqn. (455) into these expressions yields the following results

$$\mathbf{Q}_\alpha = \int_{\underline{h}}^{\bar{h}} \boldsymbol{\Xi}_{\alpha\beta}\left(\boldsymbol{\epsilon}_\beta + \boldsymbol{\varkappa}_\beta \times \mathbf{p}\right) dx_3$$
$$\mathbf{M}_\alpha = \int_{\underline{h}}^{\bar{h}} [\mathbf{p}\times]\boldsymbol{\Xi}_{\alpha\beta}\left(\boldsymbol{\epsilon}_\beta + \boldsymbol{\varkappa}_\beta \times \mathbf{p}\right) dx_3 \tag{456}$$

We can integrate out the x_3 dependence of these expressions at a typical point in the $x_1 - x_2$ plane. If we recognize that only the vector $\mathbf{p}$ depends upon x_3, we realize that these integrations are rather simple. Equations (456) can be integrated to give the constitutive equations for the plate

$$\boxed{\begin{aligned}\mathbf{Q}_\alpha &= \mathsf{A}_{\alpha\beta}\boldsymbol{\epsilon}_\beta + \mathsf{S}_{\alpha\beta}\boldsymbol{\varkappa}_\beta \\ \mathbf{M}_\alpha &= \mathsf{S}^T_{\alpha\beta}\boldsymbol{\epsilon}_\beta + \mathsf{I}_{\alpha\beta}\boldsymbol{\varkappa}_\beta\end{aligned}} \tag{457}$$

where

$$\mathsf{A}_{\alpha\beta} = h\boldsymbol{\Xi}_{\alpha\beta} \qquad \mathsf{S}_{\alpha\beta} = S\boldsymbol{\Xi}_{\alpha\beta}[\mathbf{e}_3\times]^T$$
$$\mathsf{S}^T_{\alpha\beta} = S[\mathbf{e}_3\times]\boldsymbol{\Xi}_{\alpha\beta} \qquad \mathsf{I}_{\alpha\beta} = I[\mathbf{e}_3\times]\boldsymbol{\Xi}_{\alpha\beta}[\mathbf{e}_3\times]^T$$

where h, S, and I are geometric properties related to the thickness of the plate and the definition of the location of the reference surface

$$h \equiv \int_{\underline{h}}^{\overline{h}} dx_3, \quad S \equiv \int_{\underline{h}}^{\overline{h}} x_3\, dx_3, \quad I \equiv \int_{\underline{h}}^{\overline{h}} x_3^2\, dx_3 \tag{458}$$

One usually refers to h as the thickness of the plate, S as the first moment of the normal line about the reference surface, and I as the second moment of the normal line about the reference surface.

The constitutive tensors look somewhat ominous, but upon computing their components explicitly, we can see that they are actually quite simple. The various tensor parts of these expressions are three by three tensors with simple features. The various constitutive tensors have the explicit expressions

$$\mathbf{A}_{11} = h\begin{bmatrix} E & 0 & 0 \\ 0 & \mu & 0 \\ 0 & 0 & \mu \end{bmatrix}, \quad \mathbf{A}_{22} = h\begin{bmatrix} \mu & 0 & 0 \\ 0 & E & 0 \\ 0 & 0 & \mu \end{bmatrix}, \quad \mathbf{A}_{12} = h\begin{bmatrix} 0 & \lambda & 0 \\ \mu & 0 & 0 \\ 0 & 0 & 0 \end{bmatrix}$$

$$\mathbf{S}_{12} = S\begin{bmatrix} -\lambda & 0 & 0 \\ 0 & \mu & 0 \\ 0 & 0 & 0 \end{bmatrix}, \quad \mathbf{S}_{21} = S\begin{bmatrix} -\mu & 0 & 0 \\ 0 & \lambda & 0 \\ 0 & 0 & 0 \end{bmatrix}, \quad \mathbf{S}_{11} = S\begin{bmatrix} 0 & E & 0 \\ -\mu & 0 & 0 \\ 0 & 0 & 0 \end{bmatrix}$$

$$\mathbf{I}_{11} = I\begin{bmatrix} \mu & 0 & 0 \\ 0 & E & 0 \\ 0 & 0 & 0 \end{bmatrix}, \quad \mathbf{I}_{22} = I\begin{bmatrix} E & 0 & 0 \\ 0 & \mu & 0 \\ 0 & 0 & 0 \end{bmatrix}, \quad \mathbf{I}_{12} = I\begin{bmatrix} 0 & -\mu & 0 \\ -\lambda & 0 & 0 \\ 0 & 0 & 0 \end{bmatrix}$$

where $E \equiv \lambda + 2\mu$ has been introduced to economize the notation. The remaining tensors can be obtained by symmetry

$$\mathbf{A}_{21} = (\mathbf{A}_{12})^T, \quad \mathbf{S}_{22} = (\mathbf{S}_{11})^T, \quad \mathbf{I}_{21} = (\mathbf{I}_{12})^T$$

The entire set of tensors $\mathbf{S}_{\alpha\beta}^T$ can be obtained from $\mathbf{S}_{\alpha\beta}$ as follows

$$\mathbf{S}_{11}^T = -\mathbf{S}_{22}, \quad \mathbf{S}_{22}^T = -\mathbf{S}_{11}, \quad \mathbf{S}_{12}^T = \mathbf{S}_{12}, \quad \mathbf{S}_{21}^T = \mathbf{S}_{21}$$

Clearly, if the reference surface is taken as the one exactly halfway between the two faces, then $S = 0$ and $I = h^3/12$. For this canonical choice of reference surface, the constitutive equations are uncoupled in the sense that the resultant forces $\mathbf{Q}_\alpha$ do not depend upon the curvatures $\boldsymbol{\kappa}_\alpha$, and the resultant moments $\mathbf{M}_\alpha$ do not depend upon the stretches and shears $\boldsymbol{\epsilon}_\alpha$. This situation is much like the case for beams when centroidal axes are chosen. Unlike the beam, there is no further simplification in the constitutive equations, i.e., there is no concept analogous to principal axes.

The constitutive equations are of interest not only because they relate the generalized displacements to the stress resultants, but also because they help us identify the concept of ***strain resultant***. The strain resultant is the counterpart of the strain tensor in the three-dimensional theory. As its name indicates, it is the net result of all of the local straining across the normal line. With the canonical choice of reference surface, i.e., the middle surface, the stress resultants $\mathbf{Q}_a$ are linearly related to the deformation measures $\boldsymbol{\epsilon}_a = \mathbf{w}_{,a} + \mathbf{e}_a \times \boldsymbol{\theta}$. Accordingly, we shall view these quantities as the associated strain resultant. Similarly, the moments $\mathbf{M}_a$ are linearly related to the measures of deformation $\boldsymbol{\varkappa}_a = \boldsymbol{\theta}_{,a}$. We shall consider $\boldsymbol{\varkappa}_a$ to be the strain resultant associated with $\mathbf{M}_a$.

These resultants have a clear physical interpretation. Let us write out the components of each of the strain resultants

$$\begin{aligned} \boldsymbol{\epsilon}_1 &= w_{1,1}\,\mathbf{e}_1 + w_{2,1}\,\mathbf{e}_2 + (w_{3,1} + \theta_2)\mathbf{e}_3 \\ \boldsymbol{\epsilon}_2 &= w_{1,2}\,\mathbf{e}_1 + w_{2,2}\,\mathbf{e}_2 + (w_{3,2} - \theta_1)\mathbf{e}_3 \end{aligned} \tag{459}$$

and the curvatures have component expressions

$$\begin{aligned} \boldsymbol{\varkappa}_1 &= \theta_{1,1}\,\mathbf{e}_1 + \theta_{2,1}\,\mathbf{e}_2 \\ \boldsymbol{\varkappa}_2 &= \theta_{1,2}\,\mathbf{e}_1 + \theta_{2,2}\,\mathbf{e}_2 \end{aligned} \tag{460}$$

Consider again the case of planar deformation in the $x_2 - x_3$ plane, shown in Fig. 106. For cylindrical motion we have null displacements and rotations (and their derivatives) for all quantities that give rise to motion out of the plane. Accordingly, $w_1 = 0$, $\theta_2 = 0$, and $\theta_3 = 0$. One can see that $w_{2,2}$ measures the rate of stretch in the x_2 direction, i.e., the net axial strain of the plate. It is associated with the in-plane force N_{22}. The quantity $w_{3,2}$ is the slope of the deformed middle surface of the plate. As we can clearly see in the figure, the tangent to the deformed middle surface does not coincide with the direction perpendicular to the normal vector $\mathbf{n}$. The angle between these two lines is due to shearing of the plate. The strain resultant $w_{3,2} - \theta_1$ directly measures this component of deformation, and is associated with the shear force Q_2. The rate of change of the rotation of the normal vector $\mathbf{n}$ is $\theta_{1,2}$, the curvature of the plate flexing about the x_1 axis. The curvature is associated with the bending moment M_{21} about the x_1 axis on the face with normal pointing in the x_2 direction. By extension, the meaning of the other terms in the three-dimensional case is evident.

The shear strain resultant $w_{3,1} + \theta_2$ has a sign for the rotation term different than the shear strain in the other direction. This difference is due to the right-hand-rule convention for the rotations. The resultant shearing angle is always measured as the angle between the tangent to the deformed axis and the normal to the section. The rotation angle is always measured relative to the undeformed middle surface of the plate. Figure 107 shows positive values for the displacements, displacement gradients, and rotations for two cases: (a) cylindrical deformation in the $x_2 - x_3$ plane and (b) cylindrical deformation in the $x_1 - x_3$ plane. The shearing angle is shown shaded. Note that for the first case the x_1 axis is directed out of the page, while in the second case the x_2 axis is directed into the page. In both cases, a positive transverse displacement is up-

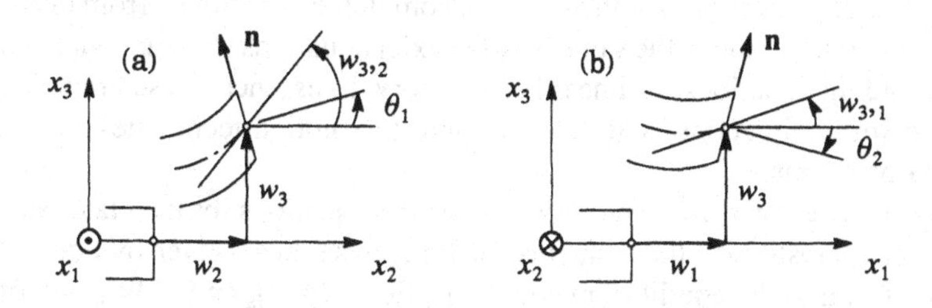

Figure 107 Why is there a sign difference in the two resultant shear strains?

ward, in the direction of the associated coordinate direction. The rate of change of the transverse displacement, or the slope of the deformed axis, is positive if it points in the direction up and to the right. On the other hand, according to the right-hand rule, the rotation θ_1 is positive if it is anticlockwise, while the rotation θ_2 is positive if it is clockwise. Thus, in the first case the shear angle is the difference between these two positive quantities, while in the second case the shear angle is the sum of these two positive quantities.

Constitutive equations in terms of displacements. We can write out the canonical constitutive equations for the resultants in terms of displacements. Using Eqn. (453) in Eqn. (457), we arrive at the following explicit relationships for the components of the stress resultants, for the in-plane forces

$$N_{\alpha\beta} = h\left[\lambda w_{\gamma,\gamma}\delta_{\alpha\beta} + \mu\left(w_{\beta,\alpha} + w_{\alpha,\beta}\right)\right] \tag{461}$$

where $\delta_{\alpha\beta}$ is the Kronecker delta, for the transverse shears

$$Q_\alpha = h\mu\left[w_{3,\alpha} + \epsilon_{\alpha\beta}\theta_\beta\right] \tag{462}$$

where $\epsilon_{\alpha\beta}$ is the alternator, and for the bending moments

$$M_{\alpha\beta} = \frac{h^3}{12}\left[\lambda\epsilon_{\alpha\beta}\epsilon_{\gamma\eta}\theta_{\eta,\gamma} + \mu\left(\epsilon_{\gamma\beta}\epsilon_{\alpha\eta}\theta_{\eta,\gamma} + \theta_{\beta,\alpha}\right)\right] \tag{463}$$

To get a clearer idea of how these constitutive equations look, it is instructive to write out the explicit expressions for Eqns. (461), (462), and (463) as suggested in Problem 209.

Navier equations for plates. Substituting the expressions for the in-plane forces into the equilibrium equations, Eqns. $(448)_a$ and $(448)_b$, we find a system of second-order differential equations for the in-plane displacements w_1 and w_2, often referred to as the *Navier equations*. These equations are

$$h\left[(\lambda+\mu)w_{\beta,\beta\alpha} + \mu w_{\alpha,\beta\beta}\right] + q_\alpha = 0 \tag{464}$$

The in-plane equations for the plate are completely uncoupled from the bending equations, in much the same way the axial equations were uncoupled from the bending equations for linear beam theory. In essence, these are averaged *plane-stress* equations, and are amenable to solution techniques for plane-stress problems.

We can perform the same operation on the equations for the plate-bending problem. Substituting the moments and transverse shears given by Eqns. (462) and (463) into the equilibrium equations, Eqns. $(449)_a$ and $(449)_b$, we find a system of second-order differential equations governing the transverse displacement w_3 and the rotations θ_1 and θ_2, often referred to as the *Mindlin plate equations.* The equations are given by

$$\frac{h^3}{12}\Big[\lambda\epsilon_{\alpha\beta}\epsilon_{\gamma\nu}\theta_{\nu,\gamma\alpha}+\mu\big(\epsilon_{\gamma\beta}\epsilon_{\alpha\nu}\theta_{\gamma,\nu\alpha}+\theta_{\beta,\alpha\alpha}\big)\Big] + h\mu\epsilon_{\alpha\beta}\big[w_{3,\alpha}+\epsilon_{\alpha\gamma}\theta_{\gamma}\big]+m_\beta = 0 \tag{465}$$

$$h\mu\big[w_{3,\alpha\alpha}+\epsilon_{\alpha\beta}\theta_{\beta,\alpha}\big]+q_3 = 0 \tag{466}$$

We can observe that when the equations are expressed in this form, they lose much of their physical appeal. It is difficult to see the simplicity of the concept of the stress resultant, the simplicity of the equations of equilibrium, and the simplicity of the kinematic hypothesis in these equations. Furthermore, classical solutions to these equations are few and far between. It is instructive to write out the explicit expressions for Eqns. (465) and (466), as suggested in Problem 210.

Boundary Conditions

We have already addressed the subject of prescribing tractions on the top and bottom surfaces of the plate. In fact, the definitions of applied force and moment include the resultants of these prescribed tractions. We must also consider the tractions that exist at the edges of the plate. Let $\mathbf{t}_\Gamma$ be the applied traction field on the edge Γ, which is parameterized by its arc length s. Computing the net force and moment acting along the edge leads to the definition of the edge resultants as

$$\mathbf{q}_\Gamma(s) \equiv \int_{\underline{h}}^{\bar{h}} \mathbf{t}_\Gamma(s,x_3)\,dx_3, \quad \mathbf{m}_\Gamma(s) \equiv \int_{\underline{h}}^{\bar{h}} \mathbf{p}(x_3)\times\mathbf{t}_\Gamma(s,x_3)\,dx_3 \tag{467}$$

These forces represent the net resultants of the applied tractions at the edge of the plate.

The plate suffers from the same problem on the boundary as the beam does. Technically, we should consider the edge of the plate as having either applied tractions or prescribed displacements. Our theory is expressed in terms of resultants, and, as such, is not equipped to differentiate between a region on a line through the thickness with prescribed tractions and a region on that same line with prescribed displacements. However, it should be clear that this point of view would force us to admit only completely fixed or completely free edges. We would not be able to model a simply supported plate!

The kinematic hypothesis again comes to our rescue. Because the kinematic hypothesis implies that each line through the thickness is rigid along its own length, and because the displacement map is expressed in terms of motions of a fixed point (usually the middle surface) on that line, we can imagine prescribing the displacement at a single point. In the three-dimensional theory, such a prescription would not be admissible because a point force is a finite force applied over a vanishingly small area and, thus, leads to infinite tractions and stresses. The assumption of rigidity, while not really justifying the concept of a point load, certainly allows the theory to accommodate it. In view of this special feature of plate theory, we can now imagine a thickness line where the net displacement $\mathbf{w}$ is prescribed, but the net rotation is not. This condition, known as the simple support, plays an important role in structural engineering. If the displacement $\mathbf{w}(s) = 0$ is known, then some corresponding force must be unknown. We can demonstrate through a virtual-work argument that the unknown force is $\mathbf{q}_\Gamma$. Further, since the cross section is free to rotate, there must be some force that is prescribed. Again, we can demonstrate through a virtual-work argument that that prescribed force is $\mathbf{m}_\Gamma$.

Such a condition of mixed boundary conditions can only be realized through a condition of constraint. Imagine simply that our plate is attached to a rigid band at the edge. The tractions that the plate feels are those transmitted to it from the band. Now we can imagine that the rigid plate is attached to a piano hinge that is free to rotate about the axis of the hinge, but is not free to translate. This device constitutes our version of the simple support. Since plate theory actually provides the rigid band, we need not worry about its physical implementation to carry out calculations.

Since we are in the business of concocting support devices for our plate, why not imagine a whole collection of such devices. We have five generalized displacements (including rotations) at each point. We can imagine a device that independently prescribes the associated force or displacement for each one. Hence, each component of the edge resultant vectors can exist as either a prescribed force or a reaction force if the corresponding displacement is prescribed. We must prescribe either the force or the displacement at each point on the edge. Thus, we always have exactly five boundary conditions. These conditions are always exactly enough to determine the constants of integration that we get when we solve the governing differential equations.

At each point s on the boundary of the plate, which has normal $\mathbf{n}_\Gamma(s)$, either the force $\mathbf{Q}_{\mathbf{n}_\Gamma}(s) = \mathbf{q}_\Gamma(s)$ will be prescribed, or the displacement $\mathbf{w}(s)$ will be prescribed, but not both; either the moment on the edge of the plate

$\mathbf{M}_{\mathbf{n}_\Gamma}(s) = \mathbf{m}_\Gamma(s)$ will be prescribed, or the rotation $\boldsymbol{\theta}(s)$ will be prescribed, but not both. If both the displacement and the rotation are prescribed to be zero, then the condition is called a *fixed edge*. If both the force and the moment are prescribed to be zero, then the condition is called a *free edge*. Various mixed conditions can also be realized, as mentioned above.

The Limitations of Plate Theory

Plate theory, like beam theory, suffers from an inconsistency brought on by the constraint implicit in the kinematic hypothesis. The ramification of the Kirchhoff assumption is that the normal strain through the thickness of the plate vanishes. In our coordinate system, this means that the strain component $E_{33} = 0$. Because normal strains are coupled in the isotropic elastic constitutive equations, this constraint implies that $S_{33} \neq 0$. However, from physical observations, we know that $S_{33} = 0$ comes closer to representing the actual stress state.

What would have happened if we had made the assumption of vanishing stress and not the assumption of vanishing strain? We have done it before. We made exactly that assumption for the condition of plane stress (see Problem 97 in Chapter 4). If we made the assumption that $S_{33} = 0$ we can write the constitutive relationship in the form

$$S_{\alpha\beta} = \lambda^* E_{\gamma\gamma}\delta_{\alpha\beta} + 2\mu E_{\alpha\beta}, \quad S_{\alpha 3} = 2\mu E_{\alpha 3} \tag{468}$$

where the new constant λ^* is given by

$$\lambda^* \equiv \frac{2\mu\lambda}{\lambda + 2\mu} = \frac{\nu C}{1-\nu^2} \tag{469}$$

where ν is Poisson's ratio and C is Young's modulus. This gives us a way to partially recover from our difficulties. In the constitutive relationships for plates, if we simply substitute the value λ^* each time λ appears, then the results of plate theory are remarkably good. In most classical treatments, $2\mu = C/(1+\nu)$ is used so that the elastic constants are Young's modulus and Poisson's ratio. Thus, in the constitutive equations, we will replace $E = \lambda + 2\mu$ with $E^* = \lambda^* + 2\mu = C/(1-\nu^2)$.

The Kirchhoff assumption also requires that normal lines remain straight after deformation. As the restraint of out-of-plane warping does in beams, the restraint of curvature of the normal line in plate theory compromises the representation of the shear stress distribution. In fact, this constraint implies that the shear stresses have the form $S_{13} = \mu(w_{3,1} + \theta_2)$ and $S_{23} = \mu(w_{3,2} - \theta_1)$, and, hence, are constant with respect to x_3. The stress components S_{11} and S_{22} are both linear in x_3. The constant shear stresses are inconsistent with local equilibrium of the linear normal stresses because, for example, $S_{11,1} + S_{12,2} + S_{13,3}$ is a nonzero linear function of x_3, and, therefore, cannot vanish at every point along the normal line. The extra stiffness induced by the constraint can be ameliorated by replacing h in Eqn. (462) with $5h/6$. This adjustment is entirely analogous to the shear coefficient in Timoshenko beam theory.

The Principle of Virtual Work for Plates

The principle of virtual work for a three-dimensional continuum can be used to develop an equivalent principle for plate theory. We shall compute the appropriate external work from the three-dimensional theory. The advantage of starting with the three-dimensional theory is that we need only to know that work is the product of force and displacement. Straightforward operations will yield the result that work for a plate includes terms computed as the product of moment and rotation.

The displacement map is constrained by the kinematic hypothesis. We must construct our virtual displacement field in accord with the same hypothesis. Hence, our (three-dimensional) virtual displacement field is

$$\bar{\mathbf{u}}(x_1, x_2, x_3) = \bar{\mathbf{w}}(x_1, x_2) + \bar{\boldsymbol{\theta}}(x_1, x_2) \times \mathbf{p}(x_3) \tag{470}$$

where $\bar{\mathbf{w}}$ and $\bar{\boldsymbol{\theta}}$ represent the generalized virtual displacements of the plate. The external virtual work is simply the product of the applied body forces and tractions with their respective virtual displacements, integrated over the volume of the structure

$$\begin{aligned} \overline{W}_E = & \int_\Omega \left[\mathbf{t}^+ \cdot \bar{\mathbf{u}}(x_1, x_2, \bar{h}) + \mathbf{t}^- \cdot \bar{\mathbf{u}}(x_1, x_2, \underline{h})\right] dA \\ & + \int_\Omega \int_{\underline{h}}^{\bar{h}} \mathbf{b} \cdot \bar{\mathbf{u}} \, dx_3 dA + \int_\Gamma \int_{\underline{h}}^{\bar{h}} \mathbf{t}_\Gamma \cdot \bar{\mathbf{u}} \, dx_3 ds \end{aligned} \tag{471}$$

where $\mathbf{b}$ is the body force, $\mathbf{t}^+$ and $\mathbf{t}^-$are the applied tractions on the surfaces of the plate, $\mathbf{t}_\Gamma$ is the traction field along the edge of the plate. Substituting Eqn. (470) into (471) and carrying out the appropriate integrals, we obtain a two-dimensional expression for the external virtual work. The first two terms are

$$\int_\Omega \bar{\mathbf{w}} \cdot \left(\mathbf{t}^+ + \mathbf{t}^- + \int_{\underline{h}}^{\bar{h}} \mathbf{b} \, dx_3\right) dA + \int_\Omega \bar{\boldsymbol{\theta}} \cdot \left(\mathbf{p} \times \left(\mathbf{t}^+ + \mathbf{t}^-\right) + \int_{h^-}^{\bar{h}} \mathbf{p} \times \mathbf{b} \, dx_3\right) dA$$

The terms in parentheses are precisely our definitions of the resultant of the applied loads $\mathbf{q}$ and $\mathbf{m}$, respectively. We can use a similar argument for the third term in Eqn. (471). This term can be rearranged to read

$$\int_\Gamma \left(\bar{\mathbf{w}} \cdot \int_{\underline{h}}^{\bar{h}} \mathbf{t}_\Gamma(s, x_3) \, dx_3 + \bar{\boldsymbol{\theta}} \cdot \int_{\underline{h}}^{\bar{h}} \mathbf{p}(x_3) \times \mathbf{t}_\Gamma(s, x_3) \, dx_3\right) ds$$

Clearly, the two integrals over the thickness are precisely our definition of the resultants of the tractions on these lines—$\mathbf{q}_\Gamma$ and $\mathbf{m}_\Gamma$. Thus, in the context of the kinematic hypothesis, these two terms exactly account for all of the virtual work done by the traction forces on the edges of the plate. We can also see that $\mathbf{q}_\Gamma$ and $\mathbf{m}_\Gamma$ are the natural forces conjugate to $\overline{\mathbf{w}}$ and $\overline{\boldsymbol{\theta}}$ on the boundary, respectively, in the sense that they completely characterize the virtual work done.

Combining these results, we can see that the external virtual work done by the forces acting on our plate in going through a virtual displacement is

$$\overline{W}_E = \int_\Omega \left(\mathbf{q}\cdot\overline{\mathbf{w}} + \mathbf{m}\cdot\overline{\boldsymbol{\theta}}\right) dA + \int_\Gamma \left(\mathbf{q}_\Gamma\cdot\overline{\mathbf{w}} + \mathbf{m}_\Gamma\cdot\overline{\boldsymbol{\theta}}\right) ds \tag{472}$$

Observe that, within the context of the kinematic hypothesis, this expression for the external virtual work is exactly consistent with the three-dimensional theory. As with the beam, the moment and rotation are duals.

The principle of virtual work is a valuable tool with which to consider the conjugateness of stress and strain resultants. We saw in the derivation of the principle of virtual work that a measure of internal virtual work involving the product of stress and virtual displacement gradient appeared naturally. It had the form

$$\overline{W}_I = \int_{\mathfrak{B}} \mathbf{S}\cdot\nabla\overline{\mathbf{u}}\, dV$$

You might expect that, if a reduced theory is truly compatible with the three-dimensional theory, then an analogous expression for internal virtual work in terms of the resultant quantities should result. In fact, we could use this equivalence to define which resultant strain measures are appropriately conjugate to the defined stress resultant measures. This equivalence is particularly important since we defined stress resultants without regard to the specific kinematic hypothesis, and the kinematic hypothesis had nothing to do with the definition of stress resultants.

Let us substitute the virtual strain implied by our kinematic map into the expression for the internal virtual of a three-dimensional continuum

$$\begin{aligned}
\overline{W}_I &= \int_\Omega \int_{\underline{h}}^{\overline{h}} \mathbf{S}\cdot\left(\left(\overline{\boldsymbol{\epsilon}}_\alpha + \overline{\boldsymbol{\kappa}}_\alpha \times \mathbf{p}\right)\otimes\mathbf{e}_\alpha - \overline{\boldsymbol{\Theta}}\right) dx_3 dA \\
&= \int_\Omega \int_{\underline{h}}^{\overline{h}} \left(\overline{\boldsymbol{\epsilon}}_\alpha + \overline{\boldsymbol{\kappa}}_\alpha \times \mathbf{p}\right)\cdot\mathbf{S}\mathbf{e}_\alpha\, dx_3 dA \\
&= \int_\Omega \left(\overline{\boldsymbol{\epsilon}}_\alpha \cdot \int_{\underline{h}}^{\overline{h}} \mathbf{S}\mathbf{e}_\alpha dx_3 + \overline{\boldsymbol{\kappa}}_\alpha \cdot \int_{\underline{h}}^{\overline{h}} \mathbf{p}\times\mathbf{S}\mathbf{e}_\alpha dx_3\right) dA
\end{aligned}$$

where $\overline{\boldsymbol{\epsilon}}_\alpha = \overline{\mathbf{w}}_{,\alpha} + \mathbf{e}_\alpha \times \overline{\boldsymbol{\theta}}_\alpha$ and $\overline{\boldsymbol{\kappa}}_\alpha = \overline{\boldsymbol{\theta}}_{,\alpha}$ are the virtual strains associated with the virtual displacements and we have noted that $\mathbf{S}\cdot\overline{\boldsymbol{\Theta}} = S_{ij}\overline{\Theta}_{ij} = 0$ because $\mathbf{S}$ is symmetric and $\overline{\mathbf{\Theta}}$ is antisymmetric. Recognizing the definitions of resultant force and resultant moment, the expression for internal virtual work takes the following form

$$\boxed{\overline{W}_I = \int_{\Omega} \left(\mathbf{Q}_a \cdot \bar{\boldsymbol{\varepsilon}}_a + \mathbf{M}_a \cdot \bar{\boldsymbol{\kappa}}_a\right) dA} \tag{473}$$

The final form of the internal virtual work is interesting and important. Each term in the expression is analogous to stress times virtual strain. In the present case, this analogy translates to stress resultant times strain resultant. Thus, we can see that the resultant strains are conjugate to the resultant stresses in the sense of virtual work. Notice that the demonstration of conjugateness did not involve the constitutive equations.

We are now in a position to state the principle of virtual work for plates. If the external work is equal to the internal work for all virtual displacements satisfying the strain displacement relationships, then the equations of equilibrium are automatically satisfied. Let us adopt our usual convention of defining the functional $G(\mathcal{M}, \overline{\mathcal{W}}) \equiv \overline{W}_I - \overline{W}_E$ to be the difference between the internal virtual work and the external virtual work. The notation $\mathcal{M} \equiv \{\mathbf{Q}_a, \mathbf{M}_a\}$ stands for the stress resultants of the plate and $\overline{\mathcal{W}} \equiv \{\overline{\mathbf{w}}, \overline{\boldsymbol{\theta}}\}$ the virtual displacements. The principle of virtual work for plates is

$$\boxed{\begin{array}{c} \text{If } \; G(\mathcal{M}, \overline{\mathcal{W}}) = 0 \quad \forall \overline{\mathcal{W}} \in \mathcal{P}(\Omega) \; \text{ then} \\ \left.\begin{aligned} \mathbf{Q}_{a,a} + \mathbf{q} &= \mathbf{0} \\ \mathbf{M}_{a,a} + \mathbf{e}_a \times \mathbf{Q}_a + \mathbf{m} &= \mathbf{0} \end{aligned}\right\} \text{ in } \Omega \\ \left.\begin{aligned} \mathbf{Q}_a n_a &= \mathbf{q}_\Gamma \\ \mathbf{M}_a n_a &= \mathbf{m}_\Gamma \end{aligned}\right\} \text{ on } \Gamma \end{array}} \tag{474}$$

The virtual-work functional $G(\mathcal{M}, \overline{\mathcal{W}})$ can be expressed explicitly by collecting the internal and external virtual work definitions as follows

$$\boxed{\begin{aligned} G(\mathcal{M}, \overline{\mathcal{W}}) \equiv & \int_{\Omega} \left(\mathbf{Q}_a \cdot \bar{\boldsymbol{\varepsilon}}_a + \mathbf{M}_a \cdot \bar{\boldsymbol{\kappa}}_a - \mathbf{q} \cdot \overline{\mathbf{w}} - \mathbf{m} \cdot \overline{\boldsymbol{\theta}}\right) dA \\ & - \int_{\Gamma} \left(\mathbf{q}_\Gamma \cdot \overline{\mathbf{w}} + \mathbf{m}_\Gamma \cdot \overline{\boldsymbol{\theta}}\right) ds \end{aligned}} \tag{475}$$

In the principle of virtual work, we have designated $\mathcal{P}(\Omega)$ as the collection of admissible functions defined over the domain of the plate. Since there are no derivatives of order greater than one in the functional, the functions must have square-integrable first derivatives. The principle of virtual work can be modified to exclude the unknown boundary reaction forces by suitably restricting the space of admissible functions. Let Γ_q be that portion of the boundary where the resultant force is prescribed, and Γ_m that portion of the boundary where the resultant moment is prescribed. Define a new set of admissible variations to be the collection of functions in $\mathcal{P}_e(\Omega) = \{\bar{\mathbf{w}}, \bar{\boldsymbol{\theta}}\}$ with $\bar{\mathbf{w}} = \mathbf{0}$ on Γ_w, the portion of the boundary where the displacements are prescribed and $\bar{\boldsymbol{\theta}} = \mathbf{0}$ on Γ_θ, the portion of the boundary where the rotations are prescribed. Clearly, we must account for the conditions on the entire boundary. Therefore, we must have $\Gamma_w \bigcup \Gamma_q = \Gamma$ and $\Gamma_\theta \bigcup \Gamma_m = \Gamma$. The regions cannot overlap. Therefore, the intersection of the regions is empty: $\Gamma_w \bigcap \Gamma_q = \emptyset$ and $\Gamma_\theta \bigcap \Gamma_m = \emptyset$.

The Kirchhoff-Love Plate Equations

A much simpler set of equations results if we neglect transverse shearing deformations. To do so leads to the famous Kirchhoff-Love plate equations. As with the beam, neglecting the transverse shearing deformations has two important effects. First, it allows us to express the rotations in terms of the transverse deflection, thereby reducing the above equations to a higher-order equation in a single variable. Second, the constraint uses up the two constitutive equations for the transverse shear forces, and, hence, those forces must be determined from equilibrium equations. Let us call the transverse displacement $w_3 \equiv w$, and implement the constraints $w_{,1} + \theta_2 = 0$ and $w_{,2} - \theta_1 = 0$. Let us assume that there are no applied moments, $m_1 = m_2 = 0$, and let us designate the transverse force simply as $q_3 \equiv q$.

We can eliminate the shear forces Q_α from Eqn. (449) by differentiating $(449)_b$ with respect to x_γ, multiplying the result by $\epsilon_{\eta\gamma}$, and substituting Eqn. $(449)_a$. Upon doing so, we arrive at the equilibrium equation in terms of moments alone. To wit

$$\epsilon_{\eta\gamma} M_{\alpha\eta,\alpha\gamma} - q = 0 \tag{476}$$

With the constraint on the shear deformations we have $\theta_\alpha = \epsilon_{\alpha\beta} w_{,\beta}$. Substituting these relationships into the constitutive relationships for the moments, Eqn. (463), we have the following constitutive equation for the moments

$$\boxed{M_{\alpha\beta} = D\big(\nu\epsilon_{\beta\alpha} w_{,\gamma\gamma} + (1-\nu)\epsilon_{\beta\gamma} w_{,\alpha\gamma}\big)} \tag{477}$$

These equations can be written out as

$$\begin{aligned} M_{11} &= D(1-\nu)w_{,12} = -M_{22} \\ M_{12} &= -D\big(w_{,11} + \nu w_{,22}\big) \\ M_{21} &= D\big(w_{,22} + \nu w_{,11}\big) \end{aligned}$$

where we have defined the *plate modulus* D as

$$D \equiv \frac{1}{12}h^3(\lambda^* + 2\mu) = \frac{Ch^3}{12(1-\nu^2)} \tag{478}$$

where C is Young's modulus and ν is Poisson's ratio. Recall also the relationships $\lambda^*/(\lambda^* + 2\mu) = \nu$ and $2\mu = C/(1+\nu) = C(1-\nu)/(1-\nu^2)$.

The shear forces must be computed from an equilibrium equation because of the vanishing shear deformation constraint. From equation (449), with no **m**, we get

$$\begin{aligned} Q_1 &= -D(w_{,111} + w_{,221}) \\ Q_2 &= -D(w_{,222} + w_{,112}) \end{aligned} \tag{479}$$

Differentiating Eqns. (477) in the appropriate manner and substituting the results into Eqn. (476), we arrive at the governing equations of the Kirchhoff-Love plate theory

$$\boxed{D\nabla^4 w = q} \tag{480}$$

where D is the plate modulus and the differential operator ∇^4 is given in terms of the mixed fourth partial derivatives as

$$\nabla^4 w \equiv \frac{\partial^4 w}{\partial x_1^4} + 2\frac{\partial^4 w}{\partial x_1^2 \partial x_2^2} + \frac{\partial^4 w}{\partial x_2^4} \tag{481}$$

The analogy to Bernoulli-Euler beam theory is quite evident here. In fact, if it were not for the cross-derivative term, these equations would be exactly equivalent to a grid of orthogonally placed beams interacting through frictionless contact acting to resist the load q. Early attempts at plate theory actually used this approximation, known as the *Grashoff approximation*, to simplify design calculations. The mixed term comes about from the presence of twisting moments in the plate and contributes a great deal to the stiffness of the plate.

Since the constitutive equations for shear have been sold for the price of a reduction in the number of independent descriptors of the kinematic field, you might expect that the boundary conditions of the plate will also require some attention. In particular, the shear-free boundary conditions must be obtained from the equilibrium equations.

Constraining the shearing deformations to be zero allows us to state the kinematic hypothesis in a slightly different form. Now the vector normal to the middle surface remains orthogonal to the tangent plane of the surface (that is, the collection of all vectors that are tangent to the middle surface). As a consequence, we often hear of the *Kirchhoff hypothesis* that lines normal to the middle surface in the undeformed configuration remain normal to the middle surface after deformation. The orthogonality of lines with the tangent plane is analogous to coincidence of normals to the tangent line (deformed axis) of a beam, both representing the zero-shear deformation constraint. One must be careful not to be seduced into seeing the similarity between the Kirchhoff hypothesis and the statement of the fundamental beam hypothesis that lines normal to the cross section remain normal after deformation. Both speak of lines remaining normal to something after deformation, but in one case shear is constrained out, while in the other it is not.

Virtual work for the Kirchhoff-Love plate. We can derive the appropriate expression for the virtual-work functional for the Kirchhoff-Love plate by implementing the vanishing shear deformation constraint directly into the virtual-work functional of the ordinary plate theory. To simplify the discussion, let us ignore the in-plane aspects of the plate problem in favor of the plate-bending problem. Let us also assume, for simplicity, that there is no applied bending moment **m**. It should be obvious how to implement a nonzero moment into the theory. With these assumptions, the internal virtual work is given by

$$\overline{W}_I = \int_{\Omega} \mathbf{M}_\alpha \cdot \overline{\boldsymbol{\varkappa}}_\alpha \, dA \tag{482}$$

Noting that $\mathbf{M}_\alpha \cdot \overline{\boldsymbol{\varkappa}}_\alpha = M_{\alpha\beta}\overline{\theta}_{\beta,\alpha}$ and $\overline{\theta}_\beta = \epsilon_{\beta\gamma}\overline{w}_{,\gamma}$ along with the constitutive equations for moments given in Eqn. (477), we get

$$\boxed{\overline{W}_I = \int_{\Omega} D\left(\nu w_{,\alpha\alpha}\overline{w}_{,\beta\beta} + (1-\nu)w_{,\alpha\beta}\overline{w}_{,\alpha\beta}\right) dA} \tag{483}$$

The external work can also be computed from the complete form of the virtual work (again assuming **m** = **0**)

$$\overline{W}_E = \int_{\Omega} q\overline{w}\, dA + \int_{\Gamma} \left(\mathbf{m}_\Gamma \cdot \overline{\boldsymbol{\theta}} + n_\alpha Q_\alpha \overline{w}\right) ds \tag{484}$$

where $Q_\alpha n_\alpha$ is that part of $\mathbf{q}_\Gamma \cdot \overline{\mathbf{w}}$ exclusive of the in-plane components along the edge with normal **n**.

The boundary term is the source of some difficulty because of the vanishing shear constraint. To appreciate the problem, consider the simply supported boundary along an edge of length ℓ with normal $\mathbf{n} = \mathbf{e}_1$, as shown in Fig. 108. Along this edge we have the following contribution to the external virtual work

$$\int_0^{\ell} \left(M_{11}\overline{w}_{,2} - M_{12}\overline{w}_{,1} + Q_1\overline{w}\right) ds \tag{485}$$

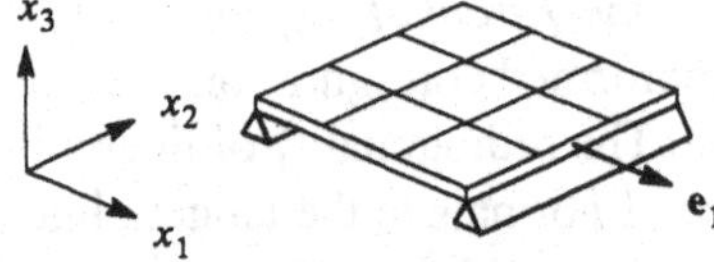

Figure 108 Simple support conditions on the boundary of a plate

Along this edge the displacement is constrained to be zero. Rotation can occur about the x_2 axis, but rotation is restrained about the x_1 axis. The key observation is that the restraint of the edge displacement $\overline{w} = 0$ implies vanishing of the rotation $\overline{w}_{,2}$ about the x_1 axis. Thus, the rotation $\overline{w}_{,2}$ is not independent from $\overline{w}$, while $\overline{w}_{,1}$ is independent. With this lack of independence, the fundamental theorem of the calculus of variations cannot be applied.

To get the expression to a form where the fundamental theorem can be applied, we must integrate the term containing $\overline{w}_{,2}$ by parts to get

$$\int_0^{\ell} \left[(Q_1 - M_{11,2})\overline{w} - M_{12}\overline{w}_{,1} \right] ds + M_{11}\overline{w}\Big|_0^{\ell} \tag{486}$$

This expression affords some interesting observations about the boundary conditions of the Kirchhoff-Love plate that appear because of the constraint of vanishing shear deformation. Like the Bernoulli-Euler beam, the Kirchhoff-Love plate must have exactly two boundary conditions at each point along the edge. These two conditions can be any of a variety of possible conditions. The fixed edge has vanishing transverse displacement w and vanishing rotation ($w_{,1}$ for an edge with normal $\mathbf{e}_1$, and $w_{,2}$ for an edge with normal $\mathbf{e}_2$). The simple support has vanishing transverse displacement w and vanishing tangential moment (M_{12} for an edge with normal $\mathbf{e}_1$, and M_{21} for an edge with normal $\mathbf{e}_2$).

The free edge boundary condition is the mysterious one. It has vanishing tangential moment and vanishing effective shear ($Q_1 - M_{11,2}$ for an edge with normal $\mathbf{e}_1$ and $Q_2 + M_{22,1}$ for an edge with normal $\mathbf{e}_2$). It would appear, from purely statical considerations, that the shear force Q_α should vanish on a free edge, as it does for the beam. However, because of the kinematic constraint, we find that the effective shear must vanish instead. Kirchhoff was the first to recognize this peculiar feature of the constrained plate theory so the effective shear is often called the Kirchhoff shear (he called it the *erzatzkräfte*). The virtual-work argument is the clearest way of seeing the need for this boundary condition because the conjugate conditions always appear in the virtual-work statement of the boundary value problem. In equation (486), we can see that the effective shear multiplies the virtual displacement, while the tangential moment multiplies the tangential virtual rotation. On any edge, either the displacement or the effective shear can be prescribed, but not both; either the tangential rotation or the tangential moment can be prescribed, but not both. The four possible boundary conditions for an edge with normal $\mathbf{e}_1$ are as follows

$$\begin{aligned} w &= 0 && \text{vanishing displacement} \\ w_{,1} &= 0 && \text{vanishing tangential rotation} \\ w_{,11} + \nu w_{,22} &= 0 && \text{vanishing tangential moment} \\ w_{,111} + (2-\nu)w_{,122} &= 0 && \text{vanishing effective shear} \end{aligned}$$

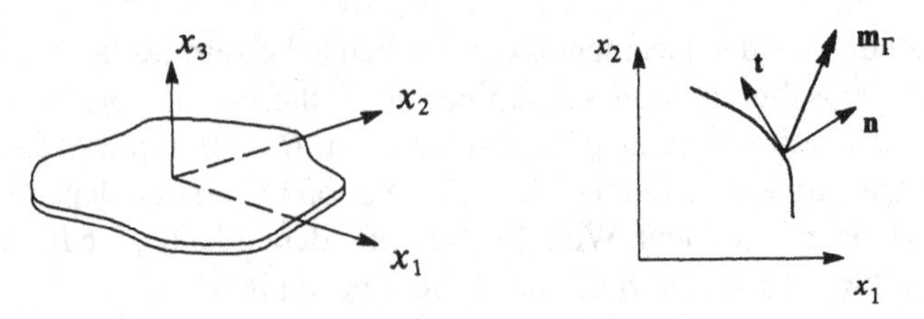

Figure 109 General edge conditions on the boundary of a plate

The second peculiar feature of the boundary conditions of the Kirchhoff-Love plate is the last term in Eqn. (486). This term, $M_{11}(\ell)\overline{w}(\ell) - M_{11}(0)\overline{w}(0)$, resulted from the integration by parts that got rid of the $\overline{w}_{,2}$ rotation in Eqn. (485). Each edge will generate terms of this variety. For two edges that meet in a corner, the terms may not cancel, giving rise to the so-called *corner forces* of Kirchhoff-Love plate theory. Again, these concentrated forces are an artifact of the constrained theory. However, they do approximate a phenomenon that is observed in plates: a tendency for the corners to curl if not restrained. If there are no corners on the boundary of the plate (as in a circular plate), these terms vanish.

The treatment of the boundary conditions for the general curved boundary is of interest. Consider the plate shown in Fig. 109. The boundary of the plate can be described by the unit normal vector $\mathbf{n} = n_\alpha \mathbf{e}_\alpha$. The vector tangent to the edge (and in the plane) of the plate is $\mathbf{t} = t_\alpha \mathbf{e}_\alpha$. The components of $\mathbf{n}$ and $\mathbf{t}$ are related by

$$t_\alpha = \epsilon_{\beta\alpha} n_\beta \quad \Leftrightarrow \quad n_\alpha = \epsilon_{\alpha\beta} t_\beta \tag{487}$$

where $\epsilon_{\alpha\beta}$ is the two-dimensional alternator. The moment $\mathbf{m}_\Gamma$ acting on the boundary can be decomposed along the normal and tangential directions as

$$\mathbf{m}_\Gamma = (\mathbf{m}_\Gamma \cdot \mathbf{n})\mathbf{n} + (\mathbf{m}_\Gamma \cdot \mathbf{t})\mathbf{t}$$

Therefore, the dot product of the moment with the virtual rotation is

$$\mathbf{m}_\Gamma \cdot \overline{\boldsymbol{\theta}} = (\mathbf{m}_\Gamma \cdot \mathbf{n})(\overline{\boldsymbol{\theta}} \cdot \mathbf{n}) + (\mathbf{m}_\Gamma \cdot \mathbf{t})(\overline{\boldsymbol{\theta}} \cdot \mathbf{t})$$

This key term will appear in the computation of the external virtual work. The vanishing shear deformation constraint relates the components of $\overline{\boldsymbol{\theta}}$ and $\overline{\mathbf{w}}$ as follows

$$\overline{\theta}_\alpha = \epsilon_{\alpha\beta} \overline{w}_{,\beta} \quad \Leftrightarrow \quad \overline{w}_{,\alpha} = \epsilon_{\beta\alpha} \overline{\theta}_\beta$$

With this observation, we can easily show that

$$\overline{\boldsymbol{\theta}} \cdot \mathbf{n} = \nabla\overline{w} \cdot \mathbf{t}, \quad \overline{\boldsymbol{\theta}} \cdot \mathbf{t} = -\nabla\overline{w} \cdot \mathbf{n}$$

where $\nabla\overline{w} \cdot \mathbf{n}$ gives the rate of change of $\overline{\mathbf{w}}$ in the direction $\mathbf{n}$. The external virtual work done by the moments on the boundary can be expressed as

$$\int_{\Gamma}(\mathbf{m}_{\Gamma}\cdot\bar{\boldsymbol{\theta}})\,ds = \int_{\Gamma}(\mathbf{m}_{\Gamma}\cdot\mathbf{n})(\nabla\bar{w}\cdot\mathbf{t})\,ds - \int_{\Gamma}(\mathbf{m}_{\Gamma}\cdot\mathbf{t})(\nabla\bar{w}\cdot\mathbf{n})\,ds \tag{488}$$

We need one more result concerning the integration of a function around a closed curve before we can compute the external work.

Lemma. Let $v(s)$ and $w(s)$ be two scalar functions defined along a closed curve Γ, parameterized by its arc length s, in three-dimensional space. Let $\mathbf{t}(s)$ be a unit vector field tangent to the curve at every point. Then

$$\int_{\Gamma} v\nabla w\cdot\mathbf{t}\,ds = -\int_{\Gamma} w\nabla v\cdot\mathbf{t}\,ds \tag{489}$$

Proof. Note first that $\nabla(vw) = v\nabla w + w\nabla v$. We can integrate this expression over the closed curve as follows

$$\int_{\Gamma} v\nabla w\cdot\mathbf{t}\,ds = \int_{\Gamma}\nabla(wv)\cdot\mathbf{t}\,ds - \int_{\Gamma} w\nabla v\cdot\mathbf{t}\,ds$$

Finally, observe that the first integral on the right side is the integral of an exact differential along the curve, and can be evaluated as

$$\int_{\Gamma}\nabla(wv)\cdot\mathbf{t}\,ds = wv\Big|_a^b$$

If the curve is closed, then the initial point $s = a$ and the end point $s = b$ occupy the same position on the curve. Since the curve is continuous we have $w(b)v(b) - w(a)v(a) = 0$. □

Identify $v = \mathbf{m}_{\Gamma}\cdot\mathbf{n}$ in the above lemma, and the first term on the right side of Eqn. (488) can be integrated to give the work of the edge moments as

$$\int_{\Gamma}(\mathbf{m}_{\Gamma}\cdot\bar{\boldsymbol{\theta}})\,ds = -\int_{\Gamma}\bar{w}\nabla(\mathbf{m}_{\Gamma}\cdot\mathbf{n})\cdot\mathbf{t}\,ds - \int_{\Gamma}(\mathbf{m}_{\Gamma}\cdot\mathbf{t})(\nabla\bar{w}\cdot\mathbf{n})\,ds \tag{490}$$

Let us define the *tangential moment* $\hat{M}_s$ and the *effective shear* $\hat{Q}_s$ as

$$\boxed{\begin{aligned}\hat{M}_s &\equiv \mathbf{m}_{\Gamma}\cdot\mathbf{t}\\ \hat{Q}_s &\equiv Q_\alpha n_\alpha - \nabla(\mathbf{m}_{\Gamma}\cdot\mathbf{n})\cdot\mathbf{t}\end{aligned}} \tag{491}$$

These quantities can be computed in terms of the components of the moment, shear, and normal vector by noting that $\mathbf{m}_\Gamma = n_\alpha \mathbf{M}_\alpha = n_\alpha M_{\alpha\beta}\mathbf{e}_\beta$ and that $\mathbf{t}$ is related to $\mathbf{n}$ by Eqn. (487). Substituting these results into Eqns. (491), we get the component expressions

$$\begin{aligned} \hat{M}_s &= n_\alpha M_{\alpha\beta} t_\beta = n_\alpha M_{\alpha\beta} \epsilon_{\gamma\beta} n_\gamma \\ \hat{Q}_s &= Q_\alpha n_\alpha - \left(n_\alpha M_{\alpha\beta} n_\beta\right)_{,\gamma} \epsilon_{\mu\gamma} n_\mu \end{aligned}$$

For the edge with $\mathbf{n} = \mathbf{e}_1$, we have $n_1 = 1$ and $n_2 = 0$. It is straightforward to show that the general boundary conditions reduce to the special ones derived previously for this case, that is, $\hat{M}_s = M_{12}$ and $\hat{Q}_s = Q_1 - M_{11,2}$.

Finally, adding the loading term and the shear term to Eqn. (490), which expresses the external work done by the boundary moments, we find a more convenient form for the external virtual work. Substituting our newly defined terms into Eqn. (484), we have the following expression for external virtual work

$$\boxed{\overline{W}_E = \int_\Omega q\overline{w}\, dA + \int_\Gamma \left(\hat{Q}_s \overline{w} - \hat{M}_s \overline{w}_{,\gamma} n_\gamma\right) ds} \tag{492}$$

As we saw in the case of beams, the expression for the external virtual work can be quite valuable in determining the appropriate boundary conditions because the external virtual work always involves the product of a force quantity with the appropriate conjugate displacement quantity. For the Kirchhoff-Love plate, the transverse displacement w is conjugate to the effective shear $\hat{Q}_s$, and the tangential rotation $w_{,\gamma} n_\gamma$ is conjugate to the tangential moment $\hat{M}_s$ on the boundary. Therefore, we can deduce the possible boundary conditions for each point along an edge to be two conditions taken from the following four

$$\boxed{\begin{aligned} w &= 0 && \text{vanishing displacement} \\ w_{,\gamma} n_\gamma &= 0 && \text{vanishing tangential rotation} \\ \hat{M}_s &= 0 && \text{vanishing tangential moment} \\ \hat{Q}_s &= 0 && \text{vanishing effective shear} \end{aligned}} \tag{493}$$

We can define the virtual-work functional as the difference between the internal and the external virtual work. As usual, we will want to restrict our functions in a way that forces the boundary terms to be zero. We can accomplish this goal by selecting virtual displacements that satisfy the essential boundary conditions, that is, boundary conditions on the displacement and the tangential rotation. Boundary conditions on the tangential moment and effective shear are natural boundary conditions, and need not be restricted in order for the boundary term to drop out of the virtual-work functional. As before, let us call Γ_w that portion of the boundary where displacements are prescribed, and Γ_θ

that portion of the boundary where tangential rotations are prescribed. Define the admissible virtual displacements to be those functions $\overline{w}(x_1, x_2)$ that satisfy $\overline{w} = 0$ on Γ_w and $\overline{w}_{,\gamma} n_\gamma = 0$ on Γ_θ. Let us call this restricted collection of functions $\mathcal{P}_e(\Omega)$. With this restricted set of functions, the virtual-work functional can be expressed as

$$G(w, \overline{w}) \equiv \int_\Omega \left[D\left(\nu w_{,\alpha\alpha} \overline{w}_{,\beta\beta} + (1-\nu) w_{,\alpha\beta} \overline{w}_{,\alpha\beta}\right) - q\overline{w} \right] dA \qquad (494)$$

and the principle of virtual work can be stated as

$$\begin{array}{l} \text{If } \; G(w, \overline{w}) = 0 \quad \forall \overline{w} \in \mathcal{P}_e(\Omega) \\ \text{then } \; \epsilon_{\eta\gamma} M_{\alpha\eta,\alpha\gamma} - q = 0 \; \text{ in } \Omega \end{array}$$

As usual, the principle of virtual work simply states that if internal work is equal to external work for all test functions, then equilibrium is automatically satisfied. Since we restricted the test functions to be zero on the portion of the boundary where displacements or rotations are prescribed, the principle of virtual work no longer tries to satisfy the reaction boundary conditions automatically. However, as we observed for the little boundary value problem, since the edge is adjacent to a portion of the domain, and the principle of virtual work is satisfying force equilibrium inside the domain, then the natural edge conditions should also be satisfied. Therefore, the reactions we compute at those points from the equations of equilibrium should be appropriate to the given problem.

The Ritz method. Having the virtual-work functional at our disposal, we can proceed to develop approximations based on the Ritz method. The method follows exactly the same idea as all of its previous incarnations. We must approximate the transverse displacement from basis functions that belong to $\mathcal{P}_e(\Omega)$. Let $h_n(x_1, x_2)$ be among those functions. We can then approximate the real and virtual displacements as

$$w(x_1, x_2) = \sum_{n=1}^{N} a_n h_n(x_1, x_2), \quad \overline{w}(x_1, x_2) = \sum_{n=1}^{N} \overline{a}_n h_n(x_1, x_2) \qquad (495)$$

where a_n and $\overline{a}_n$ are constants to be determined, and N is the number of basis functions included in the approximation. If we define the matrices $\mathbf{K}$ and $\mathbf{f}$ to have components

$$K_{mn} \equiv \int_\Omega D\left[\nu h_{m,\alpha\alpha} h_{n,\beta\beta} + (1-\nu) h_{m,\alpha\beta} h_{n,\alpha\beta}\right] dA \qquad (496)$$

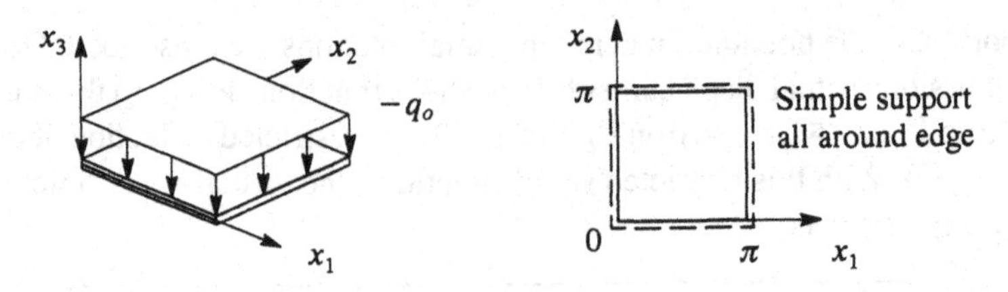

Figure 110 Simply supported square plate example

$$f_n \equiv \int_\Omega q h_n \, dA \tag{497}$$

then the virtual-work functional reduces to $G(\mathbf{a}, \bar{\mathbf{a}}) = \bar{\mathbf{a}}^T(\mathbf{K}\mathbf{a} - \mathbf{f})$, which is zero for all $\bar{\mathbf{a}}$ only if $\mathbf{K}\mathbf{a} = \mathbf{f}$. As with beams, $\mathbf{K}$ is often called the stiffness matrix, and $\mathbf{f}$ the load vector.

Example 51. Consider a square plate of depth h and sides of length π, subjected to a downward uniform load of intensity $-q_o$, shown in Fig. 110. Let us consider a single-term approximation for the real and virtual displacement

$$w(x_1, x_2) = a_1 \sin x_1 \sin x_2, \qquad \bar{w}(x_1, x_2) = \bar{a}_1 \sin x_1 \sin x_2$$

The second partial derivatives of the displacement needed to compute the stiffness matrix are

$$\begin{aligned} w_{,11} &= -\, a_1 \sin x_1 \sin x_2 \\ w_{,12} &= a_1 \cos x_1 \cos x_2 \\ w_{,22} &= -\, a_1 \sin x_1 \sin x_2 \end{aligned}$$

The virtual displacements can be treated in the same manner. Eqn. (494) suggests that we need

$$\begin{aligned} w_{,\alpha\alpha} \bar{w}_{,\gamma\gamma} &= 4 a_1 \bar{a}_1 \sin^2 x_1 \sin^2 x_2 \\ w_{,\alpha\eta} \bar{w}_{,\alpha\eta} &= 2 a_1 \bar{a}_1 \left(\sin^2 x_1 \sin^2 x_2 + \cos^2 x_1 \cos^2 x_2 \right) \end{aligned}$$

Substituting these results into Eqn. (494) and carrying out the integrals, we get

$$G(a_1, \bar{a}_1) = \left(\pi^2 D a_1 + 4 q_o \right) \bar{a}_1 = 0$$

where D is the plate modulus. Clearly, G can be zero only if $a_1 = -4q_o/\pi^2 D$. Thus, the approximate solution is

$$w(x_1, x_2) = -\frac{4 q_o}{\pi^2 D} \sin x_1 \sin x_2$$

We generally would not expect a one-term approximation to be very accurate; however, the present approximate solution happens to be quite good. In fact, if we take the displacement to be a sine series of the form

$$w(x_1, x_2) = \sum_{m=1}^{\infty} \sum_{n=1}^{\infty} a_{mn} \sin nx_1 \sin mx_2$$

then we can use the Ritz method to find the coefficients (Problem 212) to be

$$a_{mn} = -\frac{4q_o}{\pi^2 D}\left[\frac{4}{mn(m^2+n^2)^2}\right], \qquad m, n = 1, 3, 5, \ldots$$

The even coefficients are zero, i.e., $a_{mn} = 0$ for $m, n = 2, 4, \ldots$. The series converges quite rapidly, with most of the contribution coming from the first term (i.e., the one we used originally). Part of the reason the series converges so rapidly is that the base functions satisfy not only the essential boundary condition $w = 0$, but also the natural boundary condition $\hat{M}_s = 0$. While the principle of virtual work does not require that the basis functions satisfy the natural boundary conditions, to do so will generally improve the approximation.

The preceding example illustrates the essentially two-dimensional character of the principle of virtual work for plates, and shows some distinct differences from beam theory. Plate theory is governed by partial differential equations, while beam theory is governed by ordinary differential equations. While we can often find classical solutions to beam problems, we can seldom find classical solutions to plate problems. The principle of virtual work can be very valuable in the solution of these problems.

Additional Reading

Y. C. Fung, *Foundations of solid mechanics*, Prentice Hall, Englewood Cliffs, N.J., 1965.

H. L. Langhaar, *Energy methods in applied mechanics*, Wiley, New York, 1962.

S. P. Timoshenko and S. Woinowsky-Krieger, *Theory of plates and shells*, 2nd ed. McGraw-Hill, New York, 1959.

R. Szilard, *Theory and analysis of plates: Classical and numerical methods*, Prentice Hall, Englewood Cliffs, N.J., 1974.

Problems

208. The three-dimensional rotation tensor $\boldsymbol{\Lambda}$ without drilling rotation can be obtained from two successive rotations, first ψ about the x_1 axis, and then ϕ about the new x_2 axis

$$\boldsymbol{\Lambda}(\psi,\phi) = \begin{bmatrix} \cos\phi & 0 & \sin\phi \\ 0 & 1 & 0 \\ -\sin\phi & 0 & \cos\phi \end{bmatrix} \begin{bmatrix} 1 & 0 & 0 \\ 0 & \cos\psi & \sin\psi \\ 0 & -\sin\psi & \cos\psi \end{bmatrix}$$

Compute the product of the two tensors to find $\boldsymbol{\Lambda}$. Demonstrate that the tensor $\boldsymbol{\Lambda}$ is orthogonal by showing that $\boldsymbol{\Lambda}^{-1} = \boldsymbol{\Lambda}^T$. Show that for small values of the parameters ψ and ϕ, the tensor can be expressed in the form $\boldsymbol{\Lambda} \approx \mathbf{I} + \mathbf{W}$, where $\mathbf{I}$ is the identity and $\mathbf{W}$ is a skew-symmetric tensor. Show, therefore, that when the parameters are small, they can be viewed as the components of the rotation vector $\boldsymbol{\theta}$ with $\psi = \theta_1$, $\phi = -\theta_2$, and $\theta_3 = 0$ such that $\mathbf{W} = \boldsymbol{\theta}\times$.

209. Write out the explicit constitutive expressions for the stress resultants

$$N_{\alpha\beta} = h\Big[\lambda w_{\gamma,\gamma}\delta_{\alpha\beta} + \mu\Big(w_{\beta,\alpha} + w_{\alpha,\beta}\Big)\Big]$$

$$Q_\alpha = h\mu\Big[w_{3,\alpha} + \epsilon_{\alpha\beta}\theta_\beta\Big]$$

$$M_{\alpha\beta} = \frac{h^3}{12}\Big[\lambda\epsilon_{\alpha\beta}\epsilon_{\gamma\eta}\theta_{\eta,\gamma} + \mu\Big(\epsilon_{\gamma\beta}\epsilon_{\alpha\eta}\theta_{\eta,\gamma} + \theta_{\beta,\alpha}\Big)\Big]$$

210. Write out the explicit equilibrium expressions for the Mindlin plate equations

$$\frac{h^3}{12}\Big[\lambda\epsilon_{\alpha\beta}\epsilon_{\gamma\nu}\theta_{\nu,\gamma\alpha} + \mu\Big(\epsilon_{\gamma\beta}\epsilon_{\alpha\nu}\theta_{\gamma,\nu\alpha} + \theta_{\beta,\alpha\alpha}\Big)\Big] + h\mu\,\epsilon_{\alpha\beta}\Big(w_{3,\alpha} + \epsilon_{\alpha\gamma}\theta_\gamma\Big) + m_\beta = 0$$

$$h\mu\Big(w_{3,\alpha\alpha} + \epsilon_{\alpha\beta}\theta_{\beta,\alpha}\Big) + q_3 = 0$$

211. For a smooth boundary, the expression for the external virtual work for the Kirchhoff-Love plate is

$$\overline{W}_E = \int_\Omega q\overline{w}\,dA + \int_\Gamma \Big(\hat{Q}_s\overline{w} - \hat{M}_s\overline{w}_{,\gamma}n_\gamma\Big)\,ds$$

Modify the equation to account for point loads in the domain and corners on the boundary. What terms need to be added for a plate that has the shape of a regular polygon with n sides?

212. Consider the simply supported square plate of depth h, sides of length π, moduli λ^* and μ, subjected to a downward uniform load of intensity $-q_o$ (shown in Fig. 110). Assume an approximate transverse displacement of the form

$$w(x_1, x_2) = \sum_{m=1}^{\infty}\sum_{n=1}^{\infty} a_{mn}\sin mx_1 \sin nx_2$$

with a similar approximation for the virtual displacement. Assume that shear deformations are negligible and compute the coefficients a_{mn} using the principle of virtual work. Is it possible to consider an infinite number of terms in the displacement function? How should the solution be modified to solve the problem of a rectangular plate of dimensions $\ell_1 \times \ell_2$?

213. Reconsider the plate in Problem 212. We can compute an approximate solution using the Ritz method with a polynomial basis. Note that, in order to satisfy the boundary conditions, the polynomial must have the form

$$w(x_1,x_2) = x_1x_2(x_1-\pi)(x_2-\pi)\left[a_{00}+a_{10}x_1+a_{01}x_2+a_{11}x_1x_2+\cdots\right]$$

(a) Compute the approximate displacement considering only the first term a_{00}.

(b) Unlike the beam, where we can add one term at a time with good results, the next term we might want to add to improve the solution is more complicated for the plate. Since the displacement is a function of two variables, it is possible to introduce an asymmetry if we are not careful in the introduction of new terms. One strategy is to select all terms with the same exponents, as shown below by the dashed lines.

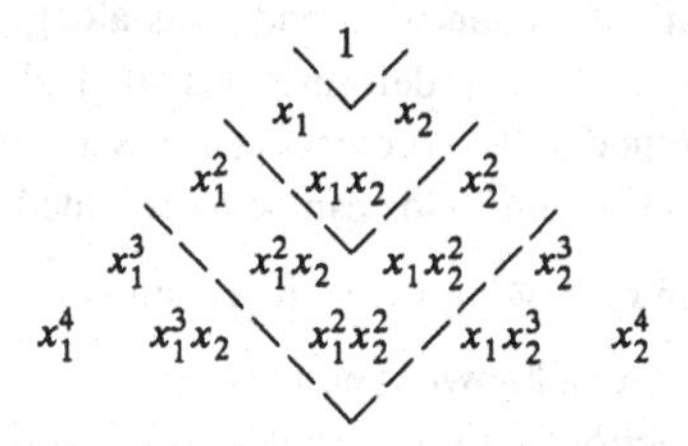

Resolve the problem using a four-term and a nine-term approximation.

(c) A mixed strategy for selecting the basis functions might also be fruitfully employed. Solve the problem with the four-term approximation

$$w(x_1,x_2) = \sin x_1 \sin x_2\left[a_{00} + a_{10}x_1 + a_{01}x_2 + a_{11}x_1x_2\right]$$

214. Consider a square plate of depth h, sides of length ℓ, moduli λ^* and μ, subjected to a downward uniform load of intensity $-q_o$. Assume that shear deformations are negligible (i.e., Kirchhoff-Love plate theory is applicable). Find an approximate displacement for the following boundary conditions

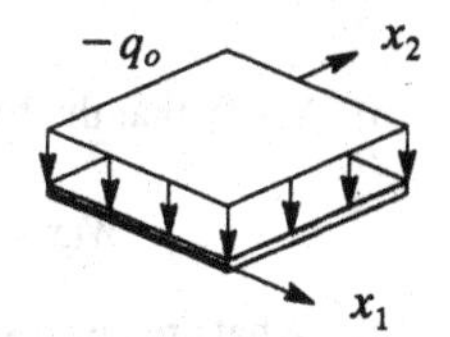

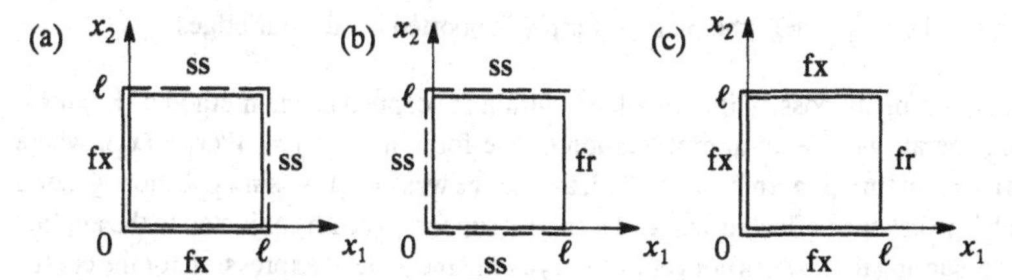

where the designation ss indicates a simple support, fx indicates a fixed edge, and fr indicates a free edge.

215. Solve any of the above variations on the square-plate problem with any combination of the following differences

(a) Consider a rectangular plate of dimension $\ell_1 \times \ell_2$.

(b) Consider a load form different from a uniform load.

216. Solve the square ($\pi \times \pi$), simply supported plate under uniform load (shown in Fig. 110) considering shearing deformations. Note: You can still neglect the in-plane problem because it is uncoupled from the bending problem. Use the Ritz method. Note that $w(x_1, x_2)$ must vanish on the boundary, but that $\theta_1(x_1, x_2)$ and $\theta_2(x_1, x_2)$ are not restricted. Let $w(x_1, x_2) = a_1 \sin x_1 \sin x_2$. Select the functions for θ_1 and θ_2 so that the shearing deformation is linear, that is

$$\theta_1 = w_{,2} + b_1(2x_2 - \pi), \quad \theta_2 = -w_{,1} + b_2(2x_1 - \pi)$$

What is the result of making such an assumption? Noting that the tangential moment must vanish on the boundary, is there a means of finding a better displacement function? Can we use statical considerations to improve the approximation of the rotation field?

217. Consider a square ($\pi \times \pi$) plate subjected to a uniform transverse load q. The plate is simply supported along the edges $x_2 = 0$ and $x_2 = \pi$ with any boundary conditions along the other two edges. Assume that shear deformations are negligible. According to the method of Kantorovich let us assume that the real and virtual displacement fields can be represented as

$$w(x_1, x_2) = W(x_1)\sin x_2, \quad \overline{w}(x_1, x_2) = \overline{W}(x_1)\sin x_2$$

where $W(x_1)$ and $\overline{W}(x_1)$ are unknown functions.

(a) Substitute these functions into the virtual-work functional

$$G(w, \overline{w}) \equiv \int_\Omega \Big[D\big(\nu w_{,\alpha\alpha} \overline{w}_{,\beta\beta} + (1-\nu) w_{,\alpha\beta} \overline{w}_{,\alpha\beta}\big) - q\overline{w} \Big] dA$$

and show that $W(x_1)$ must satisfy the ordinary differential equation

$$W^{IV} - 2W'' + W = \frac{4q}{\pi D}$$

(b) Verify that the following function satisfies the above equation

$$W(x_1) = \frac{4q}{\pi D} + (a_1 + a_2 x_1)\cosh x_1 + (a_3 + a_4 x_1)\sinh x_1$$

(c) What are the possible boundary conditions for $W(x_1)$? Find the constants of integration for the case where the plate is simply supported on all four edges.

218. There are many possibilities available within the context of the method of Kantorovich. In general, we use an approximation of the form $w(x_1, x_2) = W(x_1)\phi(x_2)$, where $\phi(x_2)$ is a known function. In Problem 217, the choice was $\phi(x_2) = \sin x_2$. A more general approach is to use the deflected shape of a simply supported beam subjected to the applied loading (assuming that it does not vary with x_2). Find the general expression for the coefficients a, b, c, and $\hat{q}$ of the differential equation

$$a W^{IV} - b W'' + c W = \hat{q}$$

that results from applying the method of Kantorovich with the known function $\phi(x_2)$. Express the coefficients as integrals of $\phi(x_2)$. Can Kantorovich's method be applied to boundary conditions other than simple supports along parallel edges? How would the approach change to accommodate more general boundary conditions?

9
Energy Principles and Static Stability

We have presented the principles of the mechanics of deformable bodies first in the context of vector mechanics, and second in the context of what we have called the principle of virtual work. Clearly, these two approaches are opposite sides of the same coin. They are duals of one another. Armed with the fundamental theorem of the calculus of variations, we can argue circularly that the classical form of the governing differential equations (those obtained from vector mechanics) implies the validity of the virtual-work form of those equations, and that the virtual-work form of the equations implies the validity of the classical form. Of course, two principles that are truly equivalent must travel in this same circle.

Under certain conditions, the principles of virtual work can be recast as extremum principles. We call such an alternative form of the principle of virtual work an *energy principle*. The name "energy" derives from the physical principle relating real work to energy, i.e., energy is the capacity to do work. In physics, we identify several forms of energy. Those most important in mechanics are potential energy, the energy of position (the relative height above datum in a gravitational field, for example), and kinetic energy, the energy of motion (proportional to the product of the mass and the square of the velocity). A system can have thermal energy, electrical energy, atomic energy, or energy associated with a chemical reaction. The law of conservation of energy states that energy can be neither created nor destroyed. Historically, energy has played a fairly significant role in physics as well as in engineering. In structural mechanics we have the so-called energy principles like Castigliano's theorems, Maxwell's law, Betti's law, and the principle of least work. In dynamics, we have Hamilton's principle.

Figure 111 A column can be in equilibrium under the same loading system in both straight and bent positions

Not all differential equations have an associated energy principle. On the other hand, a principle of virtual work always exists. (Recall that we can create a virtual-work functional simply from a weighted residual expression. The associated principle of virtual work is then a direct consequence of the fundamental theorem of the calculus of variations). We shall demonstrate, using a theorem of Vainberg, that if an energy principle exists for a certain differential equation then we can find the energy functional from the associated virtual-work functional.

The motivation for studying energy principles is that it provides a means to assess the stability of equilibrium of structures. Recall that the governing equations of solid mechanics developed in Chapters 2 through 4, extended to variational principles in Chapters 5 and 6 via the principle of virtual work, and applied to beams and plates in Chapters 7 and 8, allow us only to determine equilibrium configurations of a structural system.† Those equations tell us nothing about the stability of equilibrium.

To illustrate this point, consider an initially straight column subjected to a compressive axial force P, shown in Fig. 111. The axial force in the straight column $N(x) = P$ is constant and satisfies the equations of equilibrium at each point in the column. It is our common experience that if the column has flexural flexibility (i.e., if $EI < \infty$), it cannot sustain arbitrarily large compressive loads and remain straight (try it with your plastic straw the next time you eat at McDonalds). Euler was the first to actually compute a value for the maximum load that the column could sustain. If the column is elastic, and if shear and axial deformations are negligible, then the critical load (for a pinned column) is $P_{cr} = \pi^2 EI/\ell^2$, where EI is the flexural modulus of the column and ℓ is its length. This value often goes by the name *Euler load*, in honor of its discoverer. A column that is asked to sustain a load less than the Euler load will remain straight, while a column that is asked to sustain a load greater than the Euler load will buckle into a bent configuration.

---

† We will think of a *configuration* $\mathbf{u}(\mathbf{x})$ as a vector field which may or may not satisfy the governing differential equations (either in classical or variational form). Any configuration that satisfies the equilibrium equations is called an ***equilibrium configuration***. This concept is clearest if we think of a given displacement map $\mathbf{u}(\mathbf{x})$, which can be differentiated to give the strain field, which can be substituted into the constitutive equations to give a stress field, which can be tested (by substituting it into the equilibrium equations) to see if it satisfies the equilibrium equations. If it does then it is an equilibrium configuration.

Clearly, the concept of equilibrium alone is not sufficient for describing the behavior of mechanical systems. We also must examine the stability of that equilibrium. In the example above, the straight position is an equilibrium position for any value of the load P (let us leave aside, for the moment, the question of whether the material can sustain the stress without failure). Equilibrium holds for all values of P. However, the straight position is in *stable* equilibrium only for certain values of P. Euler found the critical load by considering the possibility that the column might also be in equilibrium in a bent position and he rewrote the equilibrium equations in the deformed configuration.† What he found was that there exists another equilibrium configuration. However, again the stability of this equilibrium configuration cannot be established from the equilibrium equations.

The principal goal of this chapter is to develop the appropriate machinery to examine the stability of certain mechanical systems. We shall restrict our considerations to static stability and to conservative systems. By a system, we mean the body in conjunction with its loads. A conservative system is simply one for which an energy functional exists. This class of systems includes all hyperelastic systems that are subjected to conservative loads, such as forces induced by a gravitational field. We shall develop a criterion for assessing the stability of the system called the *energy criterion*.

The order of tasks in this chapter is as follows. First we shall introduce the directional derivative of a functional as necessary background to establish Vainberg's theorem. Vainberg's theorem will provide the connection between energy functionals and virtual-work functionals by giving a criterion to establish when an energy functional exists and a formula to compute it from the virtual-work functional. We will apply Vainberg's theorem to derive energy functionals for all of the important theories we have already seen (i.e., the little boundary value problem, the Bernoulli-Euler beam, the Kirchhoff-Love plate, and the linearly elastic three-dimensional solid). Finally, we shall introduce the energy criterion to ascertain the stability of equilibrium.

We can generalize the idea of the equilibrium configuration to include more than just the displacement map (with the implication that stress is a function of displacement via the constitutive and strain-displacement relationships). In fact, we can think of the configuration of the system as comprising the displacement, strain, and stress fields. An "equilibrium" configuration in this more general setting is any combination of these three fields that satisfies all of the governing equations. This viewpoint gives rise to some other energy principles. We shall take a brief look at the Hu-Washizu and Hellinger-Reissner (left as an exercise) energy principles.

---

† Herein lies the great departure from the analysis of linear systems, where we assume that we can write the equations of equilibrium in the undeformed position. In general, we must establish equilibrium equations in the deformed configuration in order to assess the stability of a system.

Virtual Work and Energy Functionals

When an energy functional exists, its derivative is a virtual-work functional. In order to explore this connection we must first learn how to differentiate a functional. In this section we consider methods of differentiating and integrating functionals.

The directional derivative. As we discussed in Chapter 5, a functional is an operator that assigns a scalar value to a function. A functional is a function of a function. Consider a functional $J(\mathbf{u})$, which takes as its argument the function $\mathbf{u}(\mathbf{x})$. The output of $J(\mathbf{u})$ is a number—the value of J. We can compute the value of J for any input function (at least among those having the proper character and defined on the proper region).

Let us compute the rate of change of the functional J. It may be sufficient to observe that because the input to a functional is a function we cannot think about rate of change in the same way we do ordinary functions (i.e., the limit of the ratio of the difference between the values of the function at different points in space to the distance between those points). In other words, we cannot compute rates of change like "div" and "grad" like we can for ordinary functions because the inputs to a functional do not enjoy the same spatial organization that ordinary functions of position do. However, we will find that the concept of directional derivative is still useful (recall that we used that concept to develop our ideas for rates of change of vector and tensor fields).

Consider a fixed function $\mathbf{u}(\mathbf{x})$.† We can evaluate the functional for the function $\mathbf{u}$ to get $J(\mathbf{u})$, the value of J at the function $\mathbf{u}$. Now consider a second fixed function $\mathbf{v}(\mathbf{x})$ defined on the same region as $\mathbf{u}(\mathbf{x})$. We can define a new function $\mathbf{w}(\varepsilon) \equiv \mathbf{u}+\varepsilon\mathbf{v}$ as a one-parameter family of functions generated from the fixed functions $\mathbf{u}$ and $\mathbf{v}$. We can, of course, evaluate the functional at the function $\mathbf{w}(\varepsilon)$ to get

$$J(\varepsilon) \equiv J(\mathbf{w}(\varepsilon)) = J(\mathbf{u}+\varepsilon\,\mathbf{v}) \tag{498}$$

Observe that, because $\mathbf{u}$ and $\mathbf{v}$ are fixed, $J(\varepsilon)$ is an ordinary function of ε. Note that the function $\mathbf{w}(\varepsilon)$ has the properties $\mathbf{w}(0) \equiv \mathbf{u}$ and $\mathbf{w}'(0) \equiv \mathbf{v}$. In essence, we can think of $\mathbf{w}(\varepsilon)$ as being initially at the function $\mathbf{u}$ and moving in the direction of the function $\mathbf{v}$. Because $J(\varepsilon)$ is an ordinary function (a scalar function of a scalar variable, in fact) we can compute the rate of change with respect to ε by ordinary differentiation as $J'(\varepsilon)$.

We are most interested in the rate of change of the functional just as we begin to move in the direction of $\mathbf{v}$, that is, at $\varepsilon = 0$. The rate of change is the slope of the curve $J(\varepsilon)$ versus ε at $\varepsilon = 0$, as shown in Fig. 112. We call this rate of change the *directional derivative* of the functional and designate it with the

† The term *fixed function* means that the function remains the same throughout the discussion. The fixed function is to the functional what a fixed point in space is to an ordinary function. For a given fixed function there is just one value of the functional.

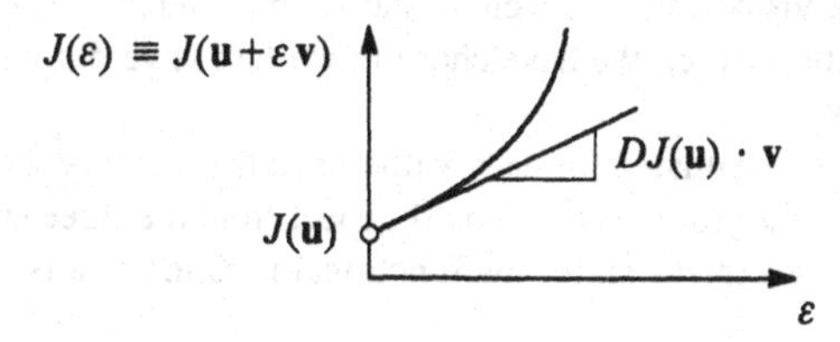

Figure 112 Variation of $J(\varepsilon)$ for fixed functions $\mathbf{u}(\mathbf{x})$ and $\mathbf{v}(\mathbf{x})$

notation $DJ(\mathbf{u}) \cdot \mathbf{v}$, which reads: the rate of change of J evaluated at $\mathbf{u}$ and moving in the direction of $\mathbf{v}$. Before we formalize our definition, let us consider a specific example.

Example 52. *The directional derivative.* Consider the functional $J(u)$ of a scalar function $u(x)$, defined on $x \in [0,1]$ that has the expression

$$J(u) = u(0) + \int_0^1 u^2(x)\,dx \tag{499}$$

This functional takes functions $u(x)$ and adds the value at $x = 0$ to the integral of the square of the function over the region. Let us consider the functions

$$u(x) = \sin(5\pi x), \quad v(x) = \tfrac{25}{4}x^2(1-x)$$

These functions are shown in Figs. 113(a,b). The first function oscillates over two and a half complete cycles between 0 and 1, while the second is simply a polynomial with zero values at the end points. Both functions have a maximum value of one. Figures 113(d,e,f) show the function that is created by adding a multiple of v to the original function u. In other words, they plot the function

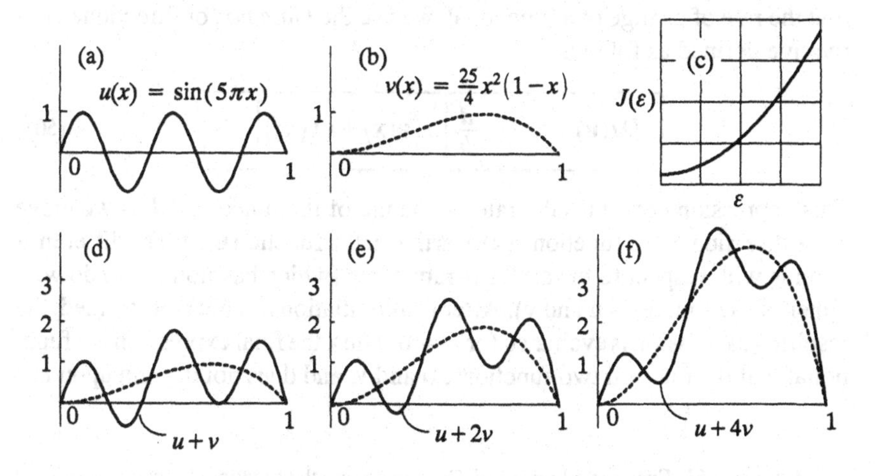

Figure 113 Variation of $u(x)$ in the direction of $v(x)$

$u(x) + \varepsilon\, v(x)$. To aid the visualization, the curve shown in dotted lines in each of these plots is the function $\varepsilon\, v(x)$. We have chosen for illustrative purposes the values $\varepsilon = 1, 2$, and 4.

The functional can be evaluated for each value of ε; the result is shown in Fig. 113(c). The value of J grows as one would expect from the three specific cases plotted. In fact, the expression for the functional in terms of ε is

$$J(\varepsilon) = \frac{1}{2} + \frac{1}{5\pi^3}\varepsilon + \frac{125}{336}\varepsilon^2 \tag{500}$$

The slope of the curve at $\varepsilon = 0$ is $J'(0) = 1/5\pi^3$, as can be seen by differentiating Eqn. (500) with respect to ε and evaluating the result at $\varepsilon = 0$. We can conclude that the initial rate of change of J as we move in the direction of v is $DJ(u) \cdot v = 1/5\pi^3$. Clearly, we cannot compute a nice, tidy function of ε for each possible choice of the function $v(x)$, because we generally are interested in finding this rate of change for a whole family of such functions. Thus, we must establish a more convenient method of computing this rate of change. Let us try to formalize these notions.

Let us consider functions $\mathbf{u}(\mathbf{x}) \mapsto \mathcal{B} \subset \mathbb{R}^3 \rightarrow \mathbb{R}^3$. Read this notation as, "The function $\mathbf{u}$ takes arguments $\mathbf{x}$ from the collection $\mathcal{B}$, which is a subset of the entire three-dimensional space (the domain of $\mathbf{u}$), and maps them to vectors $\mathbf{u}$ that live in three-dimensional space $\mathbb{R}^3$ (the range of $\mathbf{u}$)." For example, $\mathbf{u}$ might be the displacement vector in a three-dimensional body, in which case $\mathcal{B}$ would contain any position vector that extended from the origin to a point in the body. Let us also consider the functional $J(\mathbf{u}) \mapsto \mathcal{F} \rightarrow \mathbb{R}$ that takes the functions $\mathbf{u}$ from the collection $\mathcal{F}$ and produces a real number. We wish to find the rate of change of the functional as we move in the direction of another function $\mathbf{v}(\mathbf{x}) \mapsto \mathcal{B} \subset \mathbb{R}^3 \rightarrow \mathbb{R}^3$ taken from the same class of functions as $\mathbf{u}(\mathbf{x})$. To find the rate of change of a functional, we use the *Gateaux*, or directional, derivative defined as follows

$$\boxed{DJ(\mathbf{u}) \cdot \mathbf{v} \equiv \frac{d}{d\varepsilon}\Big[J\big(\mathbf{u}(\mathbf{x}) + \varepsilon\, \mathbf{v}(\mathbf{x})\big)\Big]_{\varepsilon=0}} \tag{501}$$

This expression computes the rate of change of the functional J as we move in the direction of the function $\mathbf{v}(\mathbf{x})$, starting at the function $\mathbf{u}(\mathbf{x})$. The differentiation is with respect to the scalar parameter ε (which has nothing to do with either of the functions $\mathbf{u}$ and $\mathbf{v}$). After differentiation has been performed, the resulting expression is evaluated for $\varepsilon = 0$. Thus, the final expression is a functional that operates on two functions, $\mathbf{u}$ and $\mathbf{v}$, and does not depend upon ε.

Example 53. *Example 52 revisited.* Consider again the functional given in Eqn. (499). For this functional, the directional derivative can be computed as

$$DJ(u) \cdot v = \frac{d}{d\varepsilon}\left[\big(u(0) + \varepsilon\, v(0)\big) + \int_0^1 \big(u(x) + \varepsilon\, v(x)\big)^2 \, dx \right]_{\varepsilon = 0}$$

Differentiating with respect to ε and evaluating the result at $\varepsilon = 0$, we get

$$DJ(u) \cdot v = v(0) + \int_0^1 2u(x)v(x)\, dx \tag{502}$$

For the previous example, we used $u(x) = \sin(5\pi x)$ (which gave a value of the functional of 0.5) and $v(x) = \frac{25}{4}x^2(1-x)$. The rate at which the value of J changes as we move away from $u(x)$ in the direction $v(x)$ is

$$DJ(u) \cdot v = \int_0^1 \frac{25}{2}x^2(1-x)\sin(5\pi x)\, dx = \frac{1}{5\pi^3} \tag{503}$$

As we shall soon see, the merit of Eqn. (501) is not that the directional derivative can be computed prior to knowing the functions **u** and **v**, but rather that it will provide variational formulas that must be true, say, for all functions **v**. The significance of this observation will be evident when we develop energy principles.

Directional derivatives of ordinary functions. It should be clear that the aforementioned formula for the directional derivative is a natural extension of the ordinary derivative of a function. In fact, the formula amounts to a simple application of the chain rule of differentiation for an ordinary function. Often, an analogy with ordinary functions that have geometric significance can help shed light on the geometric significance of the derivative of a functional.

Consider the ordinary scalar function $g(x)$ of the scalar variable x. Applying the above formula for the directional derivative, we obtain

$$Dg(x) \cdot \bar{x} = \frac{d}{d\varepsilon}\Big[g(x + \varepsilon \bar{x})\Big]_{\varepsilon = 0} = \frac{dg}{dx}\bar{x}$$

which is nothing more that the product of the ordinary derivative and the number $\bar{x}$. The concept of directional derivative is rather degenerate here because there is only one direction to go, and we lose no generality if we simply make the specific choice $\bar{x} = 1$.

If $g(\mathbf{x})$ is a scalar-valued function of a vector-valued variable $\mathbf{x}$, then the directional derivative in the direction $\bar{\mathbf{x}}$ is

$$Dg(\mathbf{x}) \cdot \bar{\mathbf{x}} = \frac{d}{d\varepsilon}\Big[g(\mathbf{x} + \varepsilon \bar{\mathbf{x}})\Big]_{\varepsilon = 0} = \nabla g \cdot \bar{\mathbf{x}}$$

where the gradient ∇g is a vector that points in the direction of greatest change of the function $g(\mathbf{x})$, that is, normal to the contour lines $g(\mathbf{x}) = \text{constant}$.

If we have an ordinary vector-valued function of a vector-valued argument $\mathbf{g}(\mathbf{x})$, then the directional derivative in the direction $\bar{\mathbf{x}}$ is

$$D\mathbf{g}(\mathbf{x}) \cdot \bar{\mathbf{x}} = \frac{d}{d\varepsilon}\Big[\mathbf{g}(\mathbf{x}+\varepsilon\bar{\mathbf{x}})\Big]_{\varepsilon=0} = [\nabla\mathbf{g}]\bar{\mathbf{x}}$$

where $\nabla\mathbf{g}$ is the gradient of the function, i.e., the matrix of partial derivatives $[\nabla\mathbf{g}]_{ij} = \partial g_i/\partial x_j$. In this case, the rate of change of the function depends upon which direction you are headed in.

Extremizing a functional. Like an ordinary function, a functional $J(\mathbf{u})$ might have certain inputs $\mathbf{u}$ that yield the greatest (or least) value of the functional. Hence, we can think of maximizing (or minimizing) the functional. The process for finding the extremum of a functional is very much like finding the extremum of an ordinary function. The location of the extremum is the point where the rate of change in any direction causes no change in the value of the functional. As shown in Fig. 114, the maximum and minimum of a function $g(x)$ occur at points where the first derivative is equal to zero. The character of an extremum can be deduced from the second-derivative test. If the second derivative of the function is positive at the extremum, then the point is a minimum; if the second derivative of the function is negative at the extremum, then the point is a maximum. The second derivative of the function $g(x)$ has meaning for all values of x, of course, but we use it in the second-derivative test only at a candidate extremum, that is, for points where $g'(x) = 0$. We shall construct a second-derivative test for functionals analogous to the one for ordinary functions. Finding extrema and testing the character of those extrema will be our primary use of the directional derivative of functionals.

Of course, we must be aware of the anomalous cases. A function may have a saddle point, an extremum that is neither maximum or minimum. An extremum may occur at a cusp in the curve where the derivatives of the function fail to exist. An extremum may occur at an endpoint of a curve, or in general on the boundary of the domain. The curve may be so flat at an extremum that the second-derivative test is insufficient to determine its character, in which case we must consider higher derivatives. All of these special cases of extrema for ordinary functions should be considered possible for functionals.

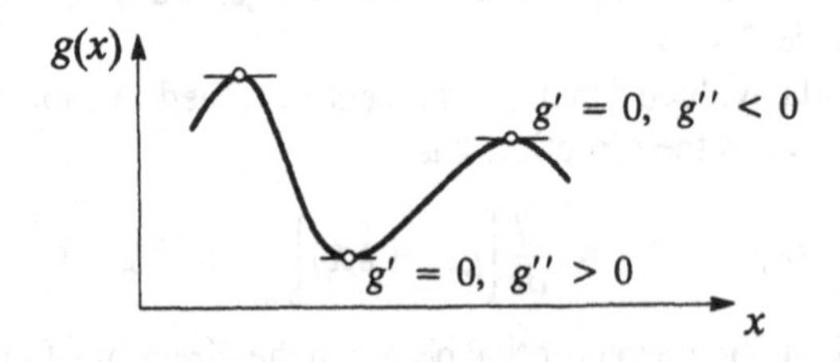

Figure 114 Maxima and minima of an ordinary function

Vainberg's theorem. Recall that the principle of virtual work guarantees that the function $\mathbf{u}(\mathbf{x})$ is an equilibrium configuration if

$$G(\mathbf{u}, \bar{\mathbf{u}}) = 0 \quad \forall \bar{\mathbf{u}} \in \mathcal{F} \tag{504}$$

where $\mathcal{F}$ is the collection of admissible functions.† The virtual-work functional $G(\mathbf{u}, \bar{\mathbf{u}}) \equiv \overline{W}_I - \overline{W}_E$ is the difference between the internal virtual work and the external virtual work for an arbitrary virtual displacement $\bar{\mathbf{u}}(\mathbf{x})$.

The virtual-work functional involves both the real displacement $\mathbf{u}(\mathbf{x})$ and the virtual displacement $\bar{\mathbf{u}}(\mathbf{x})$. It seems reasonable to ask whether or not there is a functional that depends only on the real displacement, from which we could deduce equilibrium. The answer to this question is: sometimes. When it does exist, we call the functional the *energy functional* $\mathcal{E}(\mathbf{u})$. The relationship between the energy and equilibrium is that equilibrium can be shown to hold for configurations $\mathbf{u}$ that make the energy an extremum. Since the virtual-work functional also describes equilibrium, we shall define the energy functional in terms of the virtual-work functional because when an energy functional exists, it can be obtained from the virtual-work functional. A virtual-work functional always exists.

The virtual-work functional is a functional $G(\mathbf{u}, \mathbf{v})$ having two arguments. It is important to note that, by construction, this functional is linear in the second slot (i.e., the one occupied by the virtual displacement). This linearity allows us to write the expression

$$G(\mathbf{u}, \alpha\mathbf{v} + \beta\mathbf{w}) = \alpha G(\mathbf{u}, \mathbf{v}) + \beta G(\mathbf{u}, \mathbf{w}) \tag{505}$$

Let us define a directional derivative of the functional $G(\mathbf{u}, \mathbf{v})$ in the following manner

$$DG(\mathbf{u}, \mathbf{v}) \cdot \mathbf{w} \equiv \frac{d}{d\varepsilon}\Big[G(\mathbf{u} + \varepsilon\mathbf{w}, \mathbf{v})\Big]_{\varepsilon=0} \tag{506}$$

This directional derivative is like a partial derivative in the sense that the argument in the second slot (i.e., $\mathbf{v}$) is not the subject of differentiation. In other words, it can be viewed as a constant for the purposes of directional differentiation. Observe that, by construction, $DG(\mathbf{u}, \mathbf{v}) \cdot \mathbf{w}$ is linear in both $\mathbf{v}$ and $\mathbf{w}$.

Lemma. Consider a functional $G(\mathbf{u}, \mathbf{v})$. The directional derivative of this functional satisfies the equation

$$\int_a^b DG(\mathbf{u} + s\mathbf{w}, \mathbf{v}) \cdot \mathbf{w}\, ds = G(\mathbf{u} + b\mathbf{w}, \mathbf{v}) - G(\mathbf{u} + a\mathbf{w}, \mathbf{v})$$

---

† To keep the notation simple, let us refer to the parameters associated with the deformation map generally as $\mathbf{u}(\mathbf{x})$. This notation will allow us to refer to a beam theory in the same manner as a three-dimensional continuum, and we will simply reinterpret the meaning of $\mathbf{u}$ and $\mathbf{x}$ for the special cases. For the little boundary value problem the function will be $u(x)$. For beam theory the function will be $\{u(x), w(x), \theta(x)\}$.

Proof. Observe that the directional derivative can be written as

$$\frac{d}{ds}G(\mathbf{u}+s\mathbf{w},\mathbf{v}) = \left[\frac{d}{d\varepsilon}G(\mathbf{u}+s\mathbf{w}+\varepsilon\mathbf{w},\mathbf{v})\right]_{\varepsilon=0}$$

$$= DG(\mathbf{u}+s\mathbf{w},\mathbf{v}) \cdot \mathbf{w}$$

for scalars s and ε. We can see the first equality by introducing a new variable $z(\varepsilon) \equiv s+\varepsilon$. Note that $dz/d\varepsilon = 1$ and that $z(0) = s$. By the chain rule we can compute

$$\left[\frac{d}{d\varepsilon}G(\mathbf{u}+s\mathbf{w}+\varepsilon\mathbf{w},\mathbf{v})\right]_{\varepsilon=0} = \left[\frac{d}{dz}G(\mathbf{v}+z\mathbf{w},\mathbf{v})\frac{dz}{d\varepsilon}\right]_{\varepsilon=0}$$

Now we can integrate the exact differential to get

$$\int_a^b \frac{d}{ds}G(\mathbf{u}+s\mathbf{w},\mathbf{v})\,ds = G(\mathbf{u}+b\mathbf{w},\mathbf{v})-G(\mathbf{u}+a\mathbf{w},\mathbf{v})$$

thereby proving the lemma. ❑

The following theorem due to Vainberg will allow us to make the connection between the virtual-work functional and the energy functional. It tells us when an energy functional exists and it tells us how to compute the energy functional.

Theorem (Vainberg). Consider a functional $G(\mathbf{u},\mathbf{v})$ that is linear in $\mathbf{v}$. If the functional is symmetric in the sense that

$$DG(\mathbf{u},\mathbf{v}) \cdot \mathbf{w} = DG(\mathbf{u},\mathbf{w}) \cdot \mathbf{v} \tag{507}$$

then there exists a functional $\mathcal{E}(\mathbf{u})$ that satisfies

$$D\mathcal{E}(\mathbf{u}) \cdot \mathbf{v} = G(\mathbf{u},\mathbf{v}) \tag{508}$$

Furthermore, $\mathcal{E}(\mathbf{u})$, if it exists, can be obtained from $G(\mathbf{u},\mathbf{v})$ as

$$\mathcal{E}(\mathbf{u}) = \int_0^1 G\big(t\mathbf{u},\mathbf{u}\big)\,dt + c \tag{509}$$

where c is an arbitrary constant.

The proof of Vainberg's theorem depends upon three things: (1) the lemma given above, (2) the linearity of $G(\mathbf{u},\mathbf{v})$ with respect to $\mathbf{v}$, and (3) a change in the order of integration of an iterated integral.

Proof. Let us start the proof by asserting Eqn. (509). It is sufficient then to show that the directional derivative of $\mathcal{E}(\mathbf{u})$ in the direction $\mathbf{v}$ is $G(\mathbf{u},\mathbf{v})$. By definition of $\mathcal{E}(\mathbf{u})$ we can compute

$$\mathcal{E}(\mathbf{u}+\mathbf{h})-\mathcal{E}(\mathbf{u}) = \int_0^1 G(t\mathbf{u}+t\mathbf{h},\mathbf{u}+\mathbf{h})\,dt - \int_0^1 G(t\mathbf{u},\mathbf{u})\,dt$$
$$= \int_0^1 \big[G(t\mathbf{u}+t\mathbf{h},\mathbf{u})-G(t\mathbf{u},\mathbf{u})\big]\,dt + \int_0^1 G(t\mathbf{u}+t\mathbf{h},\mathbf{h})\,dt$$

Note that linearity of $G(\mathbf{u},\mathbf{v})$ in the second slot enabled the second equality. Let us call the first integral on the second line I and compute

$$I = \int_0^1 \big[G(t\mathbf{u}+t\mathbf{h},\mathbf{u})-G(t\mathbf{u},\mathbf{u})\big]\,dt$$
$$= \int_0^1 \left[\int_0^t DG(t\mathbf{u}+s\mathbf{h},\mathbf{u})\cdot\mathbf{h}\,ds\right]dt$$
$$= \int_0^1 \left[\int_s^1 DG(t\mathbf{u}+s\mathbf{h},\mathbf{u})\cdot\mathbf{h}\,dt\right]ds$$
$$= \int_0^1 \left[\int_s^1 DG(t\mathbf{u}+s\mathbf{h},\mathbf{h})\cdot\mathbf{u}\,dt\right]ds$$
$$= \int_0^1 \big[G(\mathbf{u}+s\mathbf{h},\mathbf{h})-G(s\mathbf{u}+s\mathbf{h},\mathbf{h})\big]\,ds$$

The first step is the result of the lemma, with $\{\mathbf{u}, \mathbf{v}, \mathbf{w}, s, a, b\}$ in the lemma replaced by $\{t\mathbf{u}, \mathbf{u}, \mathbf{h}, s, 0, t\}$, repectively above. The second step is the result of changing the order of integration over the triangular region shown in Fig. 115, which holds for any integrand. The third step relies on linearity in the second slot of G. The last step is an application of the lemma again, with $\{\mathbf{u}, \mathbf{v}, \mathbf{w}, s, a, b\}$ in the lemma replaced by $\{s\mathbf{h}, \mathbf{h}, \mathbf{u}, t, s, 1\}$, repectively above.

Therefore, we can write

$$\mathcal{E}(\mathbf{u}+\mathbf{h})-\mathcal{E}(\mathbf{u}) = \int_0^1 G(\mathbf{u}+s\mathbf{h},\mathbf{h})\,ds$$

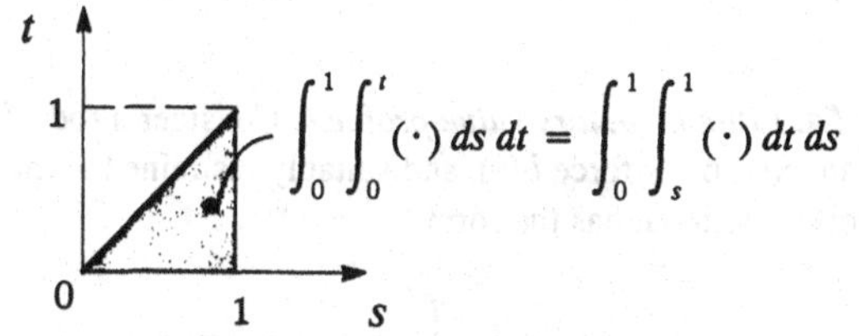

t, 1, 0, 1, s

$$\int_0^1\int_0^t (\cdot)\,ds\,dt = \int_0^1\int_s^1 (\cdot)\,dt\,ds$$

Figure 115 Change in order of integration over the triangle

Now substitute $\mathbf{h} = \varepsilon \mathbf{v}$ in the above expressions and compute

$$\mathcal{E}(\mathbf{u}+\varepsilon\mathbf{v}) - \mathcal{E}(\mathbf{u}) = \int_0^1 G(\mathbf{u}+s\varepsilon\mathbf{v}, \varepsilon\mathbf{v})\, ds = \varepsilon \int_0^1 G(\mathbf{u}+s\varepsilon\mathbf{v}, \mathbf{v})\, ds$$

where, again, we have used linearity in the second slot of G. Differentiating this result with respect to ε we get

$$\frac{d}{d\varepsilon}\mathcal{E}(\mathbf{u}+\varepsilon\mathbf{v}) = \int_0^1 G(\mathbf{u}+s\varepsilon\mathbf{v}, \mathbf{v})\, ds + \varepsilon \int_0^1 \frac{d}{d\varepsilon} G(\mathbf{u}+s\varepsilon\mathbf{v}, \mathbf{v})\, ds$$

Finally, setting $\varepsilon = 0$ gives

$$\left[\frac{d}{d\varepsilon}\mathcal{E}(\mathbf{u}+\varepsilon\mathbf{v})\right]_{\varepsilon=0} = \int_0^1 G(\mathbf{u}, \mathbf{v})\, ds = G(\mathbf{u}, \mathbf{v})$$

which is what we set out to prove. ☐

Corollary (energy principle). Assume that an energy functional $\mathcal{E}(\mathbf{u})$ exists. The configuration $\mathbf{u}$ is an equilibrium configuration if the energy is an extremum, that is, if the directional derivative vanishes for all functions $\bar{\mathbf{u}}$

$$D\mathcal{E}(\mathbf{u}) \cdot \bar{\mathbf{u}} = 0 \quad \forall \bar{\mathbf{u}} \in \mathfrak{B} \tag{510}$$

Proof. The corollary is proved by combining Eqn. (508) and the principle of virtual work. ☐

The integration to get the energy expression in Eqn. (509) is a line integral, and it is important to note that in the first slot of G one replaces $\mathbf{u}$ with $t\mathbf{u}$, where t is a dummy variable of integration; in the second slot one replaces $\bar{\mathbf{u}}$ with $\mathbf{u}$.

It should be evident that since $G(\mathbf{u}, \bar{\mathbf{u}})$ is the derivative of the energy $\mathcal{E}(\mathbf{u})$, we can add a constant to the energy without changing this relationship. This constant would represent the energy at zero deformation. Since we are never interested in the value of the energy itself, we can take this constant to be zero without loss of generality. You can see from the following examples that finding the energy for a linear theory is always quite simple.

Example 54. *Little boundary value problem.* Consider a rod of length ℓ, subjected to an axial body force $b(x)$, and suitably restrained at the boundary. The virtual-work functional has the form

$$G(u, \bar{u}) \equiv \int_0^\ell (Cu'\bar{u}' - b\bar{u})\, dx$$

and $u(x)$ represents an equilibrium configuration if $G(u,\bar{u}) = 0$, for all $\bar{u} \in \mathcal{F}_e(0,\ell)$, where $\mathcal{F}_e(0,\ell)$ is the collection of all functions, defined on the segment $x \in [0,\ell]$, whose first derivatives are square-integrable and satisfy the homogenous essential boundary conditions. The directional derivative of G in the direction $\hat{u}(x)$ is given by

$$DG(u,\bar{u}) \cdot \hat{u} = \int_0^\ell C\hat{u}'\bar{u}'\,dx = D_u G(u,\hat{u}) \cdot \bar{u}$$

The symmetry of the derivative of G is evident either by repeating the computation changing the roles of $\bar{u}$ and $\hat{u}$ or simply by reversing their roles in the expression for the derivative and noting that the reversal has no effect on the result. Since symmetry of the weak form holds, there must be an energy functional. The energy functional is given by Eqn. (509) as

$$\mathcal{E}(u) = \int_0^1 \int_0^\ell \left[C(tu')(u') - bu\right] dx\,dt$$

Carrying out the integration with respect to t and evaluating the result at the limits 0 and 1, we obtain the expression for the potential energy of the rod as

$$\mathcal{E}(u) = \int_0^\ell \left(\tfrac{1}{2}C(u')^2 - bu\right) dx \tag{511}$$

The first term in the energy expression of the rod is the potential energy stored in the rod owing to elastic extension. This term is typical of elastic systems, and represents the energy stored in going from a strain- (and stress-) free condition to the configuration $u(x)$. The second term is the potential energy of the applied load $b(x)$ with datum taken to be the undeformed position of the rod.

As shown in Fig. 116, the internal energy per unit length is simply the area under the stress-strain curve. For the little boundary value problem, the stress is $\sigma = Cu'$, and the strain is u'. Analogous results hold for all linearly elastic systems. For example, a translational spring of modulus k that develops force according to the relationship $F = k\Delta$, where Δ is the amount the spring has stretched, has energy $\mathcal{E} = \frac{1}{2}k\Delta^2$. A rotational spring of modulus k that develops moment according to the relationship $M = k\theta$, where θ is the amount of relative rotation experienced by the spring, has energy $\mathcal{E} = \frac{1}{2}k\theta^2$.

Example 55. *Bernoulli-Euler beam.* Consider an inextensible (i.e., no axial deformation) Bernoulli-Euler beam of length ℓ, subjected to a transverse force q, and suitably restrained at the boundary. The virtual-work functional is

$$G(w,\bar{w}) \equiv \int_0^\ell (EI\,w''\bar{w}'' - q\bar{w})\,dx$$

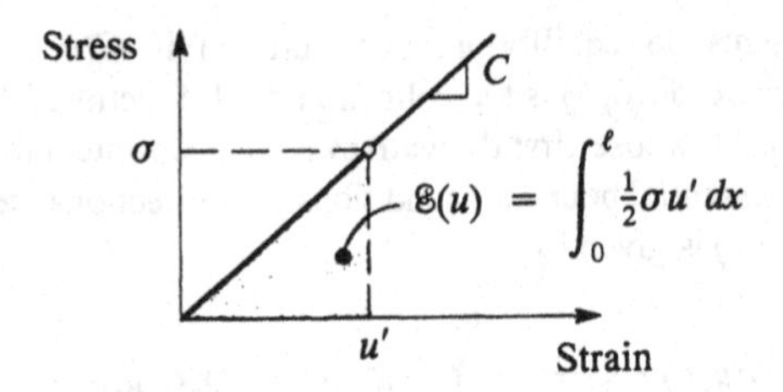

Figure 116 Internal energy as area under the stress-strain curve

The necessary symmetry of the derivative of G can be demonstrated in the same manner as for the little boundary value problem. Thus, an energy functional exists. The energy functional can be computed from Eqn. (509). Substituting tw each time w appears in the virtual-work functional and w each time $\overline{w}$ appears, and carrying out the integration with respect to t, we obtain

$$\mathcal{E}(w) = \int_0^\ell \left(\frac{1}{2} EI (w'')^2 - qw \right) dx \tag{512}$$

One can observe the similarities among the energy functionals for the Bernoulli-Euler beam and the little boundary value problem. Finding the energy functional for the Timoshenko beam is straightforward and is left as an exercise (Problem 220).

Example 56. *Kirchhoff-Love plate.* Consider a Kirchhoff-Love plate with planar domain Ω, subjected to a transverse force q, and suitably restrained at the boundary. The virtual-work functional has the form

$$G(w, \overline{w}) \equiv \int_\Omega \left[D\left(\nu w_{,\alpha\alpha} \overline{w}_{,\beta\beta} + (1-\nu) w_{,\alpha\beta} \overline{w}_{,\alpha\beta} \right) - q\overline{w} \right] dA$$

where D is the plate modulus and ν is Poisson's ratio. The necessary symmetry of the derivative of G can be demonstrated in the same manner as for the previous two cases. Thus, an energy functional exists. The energy functional can be computed from Eqn. (509). Substituting tw each time w appears in the virtual-work functional and w each time $\overline{w}$ appears, and carrying out the integration with respect to t, we obtain

$$\mathcal{E}(w) = \int_\Omega \left[\frac{1}{2} D\left(\nu (w_{,\alpha\alpha})^2 + (1-\nu) w_{,\alpha\beta} w_{,\alpha\beta} \right) - qw \right] dA \tag{513}$$

It should be clear from the foregoing examples that there is a close relationship between the virtual-work functionals and the energy functionals. Furthermore, going from one to the other can be accomplished simply through Vainberg's theorem.

Energy Principles

The energy functionals given by Vainberg's theorem allow the statement of equilibrium as a problem of finding the extrema of the energy. According to the corollary to Vainberg's theorem, the *energy extremum principle* is

$$\boxed{\underset{\mathbf{u}\in\mathcal{F}_e(\mathcal{B})}{\text{extremum}}\ \mathcal{E}(\mathbf{u}) \ \Rightarrow\ \text{Equilibrium}} \tag{514}$$

where the search is over all functions in $\mathcal{F}_e(\mathcal{B})$ that satisfy the essential boundary conditions. For ordinary functions, the necessary condition for an extremum is that the directional derivative of the function vanish in all directions. Since we have a suitable extension of the concept of the directional derivative for functionals, the same condition can be applied here, that is, $D\mathcal{E}(\mathbf{u}) \cdot \bar{\mathbf{u}} = 0$ for all $\bar{\mathbf{u}} \in \mathcal{F}_e(\mathcal{B})$. This necessary condition is obviously identical to the principle of virtual work. As such, the energy extremum principle is simply another way of looking at the principle of virtual work, for those cases where an energy functional exists.

Hu-Washizu energy functional. Some interesting energy principles can be formed with functionals that consider not only the displacement field as the independent function to be varied in the minimization process, but also the stress and the strain. Consider the *Hu-Washizu energy functional* given by

$$\begin{aligned}\mathcal{H}(\mathbf{u},\mathbf{S},\mathbf{E}) \equiv &\int_{\mathcal{B}} \Big(W(\mathbf{E}) - \mathbf{S}\cdot\big(\mathbf{E}-\nabla\mathbf{u}\big) - \mathbf{b}\cdot\mathbf{u}\Big)\,dV \\ &- \int_{\Omega_t} \hat{\mathbf{t}}\cdot\mathbf{u}\,dA - \int_{\Omega_u} \mathbf{Sn}\cdot\big(\mathbf{u}-\hat{\mathbf{u}}\big)\,dA\end{aligned} \tag{515}$$

where $\mathbf{u}$ is the displacement field, $\mathbf{S}$ is the stress field, $\mathbf{E}$ is the strain field, $\hat{\mathbf{t}}$ is the prescribed traction over Ω_t (the portion of the boundary where tractions are prescribed), $\hat{\mathbf{u}}$ is the prescribed displacement over Ω_u (the portion of the boundary where displacements are prescribed), $\mathbf{b}$ is the body force, and $W(\mathbf{E})$ is the strain-energy function. By taking the directional derivative of the Hu-Washizu functional in the directions of $\bar{\mathbf{u}}$, $\bar{\mathbf{E}}$, and $\bar{\mathbf{S}}$ (varying $\mathbf{u}$, $\mathbf{E}$, and $\mathbf{S}$, respectively) we can deduce the necessary conditions for an extremum. Let us examine these conditions in order.

Compute first the derivative of $\mathcal{H}(\mathbf{u},\mathbf{S},\mathbf{E})$ in the direction $\bar{\mathbf{u}}$, regarding the other field variables as fixed, to get

$$D_{\mathbf{u}}\mathcal{H}\cdot\bar{\mathbf{u}} = \int_{\mathcal{B}} \big(\mathbf{S}\cdot\nabla\bar{\mathbf{u}} - \mathbf{b}\cdot\bar{\mathbf{u}}\big)\,dV - \int_{\Omega_t} \hat{\mathbf{t}}\cdot\bar{\mathbf{u}}\,dA - \int_{\Omega_u} \mathbf{Sn}\cdot\bar{\mathbf{u}}\,dA$$

Noting that $\mathbf{S} \cdot \nabla\bar{\mathbf{u}} = \operatorname{div}(\mathbf{S}\bar{\mathbf{u}}) - \operatorname{div}\mathbf{S} \cdot \bar{\mathbf{u}}$, and transforming the first term on the right side with the divergence theorem, we get

$$D_{\mathbf{u}}\mathcal{H} \cdot \bar{\mathbf{u}} = -\int_{\mathcal{B}} \left(\operatorname{div}\mathbf{S} + \mathbf{b}\right) \cdot \bar{\mathbf{u}}\, dV - \int_{\Omega_t} \left(\hat{\boldsymbol{\tau}} - \mathbf{S}\mathbf{n}\right) \cdot \bar{\mathbf{u}}\, dA$$

From this expression we can see that the necessary condition for an extremum with respect to $\mathbf{u}$ (i.e., $D_{\mathbf{u}}\mathcal{H} \cdot \bar{\mathbf{u}} = 0$) gives, by the fundamental theorem of the calculus of variations, the classical equilibrium equations and traction boundary conditions

$$D_{\mathbf{u}}\mathcal{H} \cdot \bar{\mathbf{u}} = 0 \quad \forall \bar{\mathbf{u}} \quad \Rightarrow \begin{cases} \operatorname{div}\mathbf{S} + \mathbf{b} = \mathbf{0} & \text{in } \mathcal{B} \\ \mathbf{S}\mathbf{n} = \hat{\boldsymbol{\tau}} & \text{on } \Omega_t \end{cases}$$

Next, let us consider the derivative of $\mathcal{H}(\mathbf{u}, \mathbf{S}, \mathbf{E})$ in the direction $\bar{\mathbf{S}}$

$$D_{\mathbf{S}}\mathcal{H} \cdot \bar{\mathbf{S}} = \int_{\mathcal{B}} \bar{\mathbf{S}} \cdot \left(\nabla\mathbf{u} - \mathbf{E}\right) dV - \int_{\Omega_u} \bar{\mathbf{S}}\mathbf{n} \cdot \left(\mathbf{u} - \hat{\mathbf{u}}\right) dA$$

Again we can apply the fundamental theorem of the calculus of variations. Note that since $\bar{\mathbf{S}}$ is symmetric, $2\bar{\mathbf{S}} \cdot \nabla\mathbf{u} = \bar{\mathbf{S}} \cdot \left[\nabla\mathbf{u} + \nabla\mathbf{u}^T\right]$. The necessary conditions for an extremum with respect to $\mathbf{S}$ gives the strain-displacement conditions and the displacement boundary conditions

$$D_{\mathbf{S}}\mathcal{H} \cdot \bar{\mathbf{S}} = 0 \quad \forall \bar{\mathbf{S}} \quad \Rightarrow \begin{cases} \mathbf{E} = \frac{1}{2}\left[\nabla\mathbf{u} + \nabla\mathbf{u}^T\right] & \text{in } \mathcal{B} \\ \mathbf{u} = \hat{\mathbf{u}} & \text{on } \Omega_u \end{cases}$$

Finally, let us take the derivative of $\mathcal{H}(\mathbf{u}, \mathbf{S}, \mathbf{E})$ in the direction $\bar{\mathbf{E}}$

$$D_{\mathbf{E}}\mathcal{H} \cdot \bar{\mathbf{E}} = \int_{\mathcal{B}} \left(\frac{\partial W(\mathbf{E})}{\partial \mathbf{E}} - \mathbf{S}\right) \cdot \bar{\mathbf{E}}\, dV$$

Applying the fundamental theorem of the calculus of variations to this expression, we obtain the equivalence between the extremum of the energy with respect to $\mathbf{E}$ and the constitutive equations

$$D_{\mathbf{E}}\mathcal{H} \cdot \bar{\mathbf{E}} = 0 \quad \forall \bar{\mathbf{E}} \quad \Rightarrow \quad \mathbf{S} = \frac{\partial W(\mathbf{E})}{\partial \mathbf{E}} \quad \text{in } \mathcal{B}$$

The key idea behind the Hu-Washizu energy functional is that one can construct energy principles with some rather interesting properties built into them. By considering more of the field variables as independent fields, we can make the energy functional responsible for enforcing certain conditions among them, rather than requiring that those relations be enforced in a classical sense. In our usual implementations of the principles of virtual work, we satisfy the constitutive equations and the strain-displacement equations exactly and let the principle of virtual work do its best to satisfy equilibrium. We eliminate the boundary terms by suitably restricting the class of admissible functions. With the Hu-Washizu energy functional, we need not enforce conditions on the field variables (not even boundary conditions).

The Hellinger-Reissner variational principle is another multi-field energy principle, but it uses only the stress and displacement fields as independent fields. This variational principle is the subject of Problem 222.

The Euler equation of a functional. Some of the earliest work on the calculus of variations was done by the mathematician L. Euler. One of the classical results is the so-called Euler equation associated with a functional in integral form. Consider, as an example, a functional of the form

$$J(u) = \int_a^b F(u, u')\, dx \tag{516}$$

where the function $F(u, u')$ can be any function in which those arguments appear. We consider the functional to depend only upon the argument u because the derivative u' is not really independent of u. We will, however, find it useful also to think of F as having two arguments when it comes time to take derivatives of F. Let us compute the directional derivative of $J(u)$ in the direction $\bar{u}$ and set the result equal to zero to find a stationary point of the functional

$$DJ(u) \cdot \bar{u} = \frac{d}{d\varepsilon}\left[\int_a^b F(u+\varepsilon\bar{u}, u'+\varepsilon\bar{u}')\, dx\right]_{\varepsilon=0} = 0$$

Carrying out the derivatives, we obtain the result

$$\int_a^b \left(\frac{\partial F}{\partial u}\bar{u} + \frac{\partial F}{\partial u'}\bar{u}'\right) dx = 0$$

To put this equation in a form suitable for application of the fundamental theorem of the calculus of variations, we must integrate by parts any term with derivatives on $\bar{u}$. Thus, the second term must be integrated once with the result

$$\int_a^b \left(\frac{\partial F}{\partial u} - \frac{d}{dx}\left(\frac{\partial F}{\partial u'}\right)\right)\bar{u}\, dx + \frac{\partial F}{\partial u'}\bar{u}\Big|_a^b = 0$$

From the fundamental theorem of the calculus of variations, we conclude that the function F must satisfy the differential equation

$$\frac{\partial F}{\partial u} - \frac{d}{dx}\left(\frac{\partial F}{\partial u'}\right) = 0 \tag{517}$$

in the region $x \in [a,b]$, and that either $\partial F/\partial u' = 0$ (a natural boundary condition) or we can select $\overline{u}$ such that $\overline{u} = 0$ (homogeneous essential boundary conditions) at the boundary points $x = a$ and $x = b$. Equation (517) is called the Euler equation of the functional $J(u)$ of Eqn. (516).

Example 57. *Euler equations for the little boundary value problem.* Consider the energy associated with the little boundary value problem given in Eqn. (511). The function F can be identified as

$$F(u,u') \equiv -\tfrac{1}{2}C(u')^2 + bu$$

with partial derivatives

$$\frac{\partial F}{\partial u} = b, \quad \frac{\partial F}{\partial u'} = -Cu'$$

The Euler equation for F is the equilibrium equation of the little boundary value problem

$$\frac{\partial F}{\partial u} - \frac{d}{dx}\left(\frac{\partial F}{\partial u'}\right) = (Cu')' + b = 0$$

The natural boundary condition is $\partial F/\partial u' = -Cu' = 0$ at the traction-free ends, while the essential boundary conditions would require $\overline{u} = 0$ at the ends where displacements are prescribed.

Energy principles and the Ritz method. Because the energy functional is so closely related to the virtual-work functional, you might expect that the Ritz method for finding approximate solutions to boundary value problems might have application in energy methods. We shall find that a Ritz approximation reduces the energy functional to an algebraic function of the unknown parameters of the Ritz expansion. The tools of minimization of this functional are those of the ordinary calculus of several variables.

To make the ideas concrete, let us examine the energy functional for the Bernoulli-Euler beam, given in Eqn. (512). Assume that we can approximate the transverse displacement $w(x)$ in terms of known base functions in the usual manner $\mathbf{h}(x) = [h_1(x), \ldots, h_N(x)]^T$ as

$$w(x) = \sum_{n=1}^{N} a_n h_n(x) = \mathbf{a} \cdot \mathbf{h}(x)$$

where N is the number of basis functions included in the approximation. Differentiating the approximate expression and substituting the result into the energy functional gives the result

$$\mathcal{E}(\mathbf{a}) = \tfrac{1}{2}\mathbf{a}^T\mathbf{K}\mathbf{a} - \mathbf{a}^T\mathbf{f} \tag{518}$$

where $[\mathbf{a} = [a_1,\ldots, a_N]^T$ represents the unknown coefficients of the approximation. The matrix $\mathbf{K}$ and the matrix $\mathbf{f}$ are given respectively by the definitions

$$\mathbf{K} \equiv \int_0^\ell EI[\mathbf{h}''][\mathbf{h}'']^T dx, \quad \mathbf{f} \equiv \int_0^\ell q\mathbf{h}\, dx$$

just as they were for the Ritz method for virtual work.

The discretization of the energy functional is remarkably similar to the discretization of the virtual-work functional, but there are some key differences. The most important difference between the two is that, for the energy functional, only the real displacement field need be approximated, while in virtual work both the real and the virtual displacement fields must be approximated. We have generally advocated the Galerkin approach in virtual work, wherein we approximate the virtual displacements with exactly the same base functions used for the approximation of the real displacement field. This choice is what makes the $\mathbf{K}$ and $\mathbf{f}$ matrices identical in the two cases. If we were to approximate the virtual displacements with different base functions, then the resulting coefficient matrices would turn out to be different.

Equation (518), defining the discrete energy, is called a *quadratic form* because it is a quadratic function of the individual parameters a_n. The terms in $\mathbf{a}^T\mathbf{K}\mathbf{a}$ are purely quadratic (i.e., only products $a_i a_j$ appear). The terms in $\mathbf{a}^T\mathbf{f}$ are purely linear in the a_i. Since $\mathcal{E}(\mathbf{a})$ is an ordinary function, its minimization is straightforward. Much is known about the minimization of quadratic functions. The necessary conditions for a minimum is that the directional derivative vanish in all directions. Thus

$$D\mathcal{E}(\mathbf{a}) \cdot \bar{\mathbf{a}} = \bar{\mathbf{a}}^T(\mathbf{K}\mathbf{a} - \mathbf{f}) = 0 \quad \forall \bar{\mathbf{a}}$$

This condition is one that we have seen before. It is the same one that results for the Ritz approach to virtual work. Since there are no restrictions on the $\bar{\mathbf{a}}$, the necessary conditions for an extremum are simply $\mathbf{K}\mathbf{a} = \mathbf{f}$. Since the function is quadratic, we know that there is only one extremum. Thus, the solution to the problem is unique. The discrete energy is quadratic because the underlying beam theory is linear. When we get into nonlinear theories, the energy functional will not necessarily be quadratic, and, hence, uniqueness of solution will not necessarily hold.

Static Stability and the Energy Criterion

Let us consider a system for which an energy functional exists, and examine the concept of the stability of equilibrium of that system. The system has energy $\mathcal{E}(\mathbf{u})$, and is in equilibrium for functions $\mathbf{u}$ that are extrema of the functional $\mathcal{E}$. We can locate these extrema by taking the derivative of $\mathcal{E}$ and setting the result equal to zero. Accordingly, we can make a statement of equilibrium in the following form

$$D\mathcal{E}(\mathbf{u})\cdot\bar{\mathbf{u}} = 0 \quad \forall\bar{\mathbf{u}}\in\mathcal{F}$$

The directional derivative of the energy is identical to G

$$D\mathcal{E}(\mathbf{u})\cdot\bar{\mathbf{u}} = G(\mathbf{u},\bar{\mathbf{u}})$$

The stability of equilibrium can be deduced from the second derivative of the energy. Like an ordinary function, if the second derivative is positive, then the energy is a minimum; if it is negative, then the energy is a maximum. A configuration of minimum energy is a point of *stable equilibrium*, whereas a point of maximum energy is a point of *unstable equilibrium*. Configurations that are not in equilibrium are not classifiable as either stable or unstable. The second derivative test does not make sense for these configurations. Let us define the second derivative of the energy as a functional

$$\boxed{A(\mathbf{u},\bar{\mathbf{u}}) \equiv \frac{d^2}{d\varepsilon^2}\Big[\mathcal{E}(\mathbf{u}+\varepsilon\bar{\mathbf{u}})\Big]_{\varepsilon=0}} \tag{519}$$

The energy criterion for static stability can be stated as follows.

Theorem (the energy criterion for static stability). Consider an elastic body $\mathcal{B}$ with energy functional $\mathcal{E}(\mathbf{u})$. Let the configuration $\mathbf{u}(\mathbf{x})$ be an equilibrium configuration, i.e., $\mathbf{u}$ satisfies

$$D\mathcal{E}(\mathbf{u})\cdot\bar{\mathbf{u}} = 0 \quad \forall\bar{\mathbf{u}}\in\mathcal{F}$$

This equilibrium configuration is stable if and only if

$$A(\mathbf{u},\bar{\mathbf{u}}) > 0 \quad \forall\bar{\mathbf{u}}\in\mathcal{F} \tag{520}$$

If the energy functional $A(\mathbf{u},\bar{\mathbf{u}})$ fails to be positive for any test function $\bar{\mathbf{u}}$, then the system is unstable.

Proof. The elements of a proof are contained in the following discussion. □

Consider an equilibrium configuration $\mathbf{u}$. Let us first show that the energy is a minimum at $\mathbf{u}$. The energy at a neighboring configuration $\mathbf{u}+\varepsilon\bar{\mathbf{u}}$ can be expressed by expanding the energy functional in a Taylor series about $\mathbf{u}$ as

$$\mathcal{E}(\mathbf{u}+\varepsilon\bar{\mathbf{u}}) = \mathcal{E}(\mathbf{u}) + \varepsilon G(\mathbf{u},\bar{\mathbf{u}}) + \tfrac{1}{2}\varepsilon^2 A(\mathbf{u},\bar{\mathbf{u}}) + O(\varepsilon^3) \tag{521}$$

by definition of the functionals $G(\mathbf{u},\bar{\mathbf{u}})$ and $A(\mathbf{u},\bar{\mathbf{u}})$. If the third derivative of the energy is finite, then for sufficiently small values of ε, the third term dominates the $O(\varepsilon^3)$ term. Since $G(\mathbf{u},\bar{\mathbf{u}}) = 0$ for an equilibrium configuration, we conclude that $\mathcal{E}(\mathbf{u}+\varepsilon\bar{\mathbf{u}}) > \mathcal{E}(\mathbf{u})$ if and only if $A(\mathbf{u},\bar{\mathbf{u}}) > 0$.

The energy criterion for static stability depends upon the law of conservation of total (kinetic plus potential) energy. Let us write the total energy as

$$\Pi(\mathbf{u}, \dot{\mathbf{u}}) = \mathcal{T}(\dot{\mathbf{u}}) + \mathcal{E}(\mathbf{u})$$

where $\mathcal{T}(\dot{\mathbf{u}})$ is the kinetic energy and is always positive (the smallest value of kinetic energy is zero, when the system is at rest). For a solid body, the kinetic energy is given by the expression

$$\mathcal{T}(\dot{\mathbf{u}}) = \int_{\mathfrak{B}} \tfrac{1}{2}\varrho \, \| \dot{\mathbf{u}} \|^2 \, dV$$

Consider a system initially at rest, $\dot{\mathbf{u}}(0) = \mathbf{0}$, with its energy totally invested in potential energy, $\Pi(\mathbf{u}, \mathbf{0}) = \mathcal{E}(\mathbf{u})$. Let us perturb the system by imparting a velocity $\dot{\mathbf{v}}$, which displaces the system to the position $\mathbf{u}(t) = \mathbf{u} + t\dot{\mathbf{v}} + O(t^2)$ for small time t. The potential energy in the perturbed position is, from Eqn. (521)

$$\mathcal{E}(\mathbf{u} + t\dot{\mathbf{v}}) = \mathcal{E}(\mathbf{u}) + tG(\mathbf{u}, \dot{\mathbf{v}}) + \tfrac{1}{2}\, t^2 A(\mathbf{u}, \dot{\mathbf{v}}) + O(t^3)$$

Let $\Delta\mathcal{T}$ be the change in kinetic energy that results from the subsequent motion caused by the perturbation. The kinetic energy in the perturbed state is then equal to the kinetic energy of the perturbation plus the change in kinetic energy with time

$$\mathcal{T} = \mathcal{T}(\dot{\mathbf{v}}) + \Delta\mathcal{T}$$

Since the total energy must be constant

$$\mathcal{E}(\mathbf{u} + t\dot{\mathbf{v}}) + \mathcal{T}(\dot{\mathbf{v}}) + \Delta\mathcal{T} = \mathcal{E}(\mathbf{u}) + \mathcal{T}(\dot{\mathbf{v}})$$

that is, equal to the energy just after the perturbation. Since $G(\mathbf{u}, \dot{\mathbf{v}}) = 0$ by definition of an equilibrium configuration $\mathbf{u}$, conservation of energy gives

$$\Delta\mathcal{T} = -\tfrac{1}{2}\, t^2 A(\mathbf{u}, \dot{\mathbf{v}}) + O(t^3) \tag{522}$$

From Eqn. (522) we can see that if $A(\mathbf{u}, \dot{\mathbf{v}}) < 0$, then the kinetic energy grows with time (at least for small values of time) because the system experiences a decrease in potential energy. The increase in kinetic energy implies a nonzero value of velocity, which, over time, will cause the system to experience further displacement. The additional displacement will decrease the potential energy further implying even more increase in kinetic energy and hence velocity. Therefore, if $A(\mathbf{u}, \dot{\mathbf{v}}) < 0$, a small perturbation leads to increasing motion, and the configuration moves away from the equilibrium position. This sequence of events describes our understanding of instability.

If, on the other hand, $A(\mathbf{u}, \dot{\mathbf{v}}) > 0$, then the kinetic energy decreases with time. Since the kinetic energy associated with the perturbation is small, and since the minimum value of the kinetic energy is zero, the system returns to a state of rest. As the kinetic energy decreases, the potential energy increases. According to Eqn. (521), the increase in potential energy implies that the system moves back to the equilibrium configuration.

Perturbations and virtual velocities. The preceding discussion suggests that there is a close correlation between the perturbation velocity $\dot{\mathbf{v}}$ used to prove the energy criterion and the virtual displacement $\bar{\mathbf{u}}$ that we have been using throughout our discussion of the principle of virtual work. Recall from Chapter 5 that, under the assumptions made to introduce the notion of virtual work, we can think of the virtual displacement as a velocity if the arc length parameter s is interpreted as time. In fact, the early work in the principles of virtual work used the term *virtual velocity* to describe the arbitrary motion. Although we will not explore dynamical systems in this text, we can come to appreciate the connection between the perturbation velocity and the virtual displacement.

Let us reexamine the energy criterion for static stability by interpreting the process as an exchange in virtual work done in an arbitrary virtual displacement. Recall that the virtual-work functional is the difference between the internal and external virtual work, $G(\mathbf{u},\bar{\mathbf{u}}) = W_I(\mathbf{u},\bar{\mathbf{u}}) - W_E(\mathbf{u},\bar{\mathbf{u}})$. The rate of change of G, then, is the difference between the rate of change of the internal work and the rate of change of the external work. Also, we can show that

$$A(\mathbf{u},\bar{\mathbf{u}}) = \frac{d}{d\varepsilon}\Big[\,\overline{W}_I(\mathbf{u}+\varepsilon\bar{\mathbf{u}},\bar{\mathbf{u}}) - \overline{W}_E(\mathbf{u}+\varepsilon\bar{\mathbf{u}},\bar{\mathbf{u}})\Big]_{\varepsilon=0}$$

Therefore, if the rate at which the system accrues internal virtual work is greater than the rate at which the external loads remove it in undergoing a virtual displacement $\bar{\mathbf{u}}$, then the system is stable. On the other hand, if the loads remove work faster than the system stores it, then the system is unstable.

We can illustrate the preceding discussion of the energy criterion with balls resting on two different surfaces, as shown in Fig. 117. In each case, the ball is in static equilibrium because the normal force of contact between the two surfaces is oriented exactly to counterbalance the downward force caused by the weight of the ball. The arrows show admissible perturbations (we will not allow the ball to lift off of the surface or to penetrate into it). For the ball on the

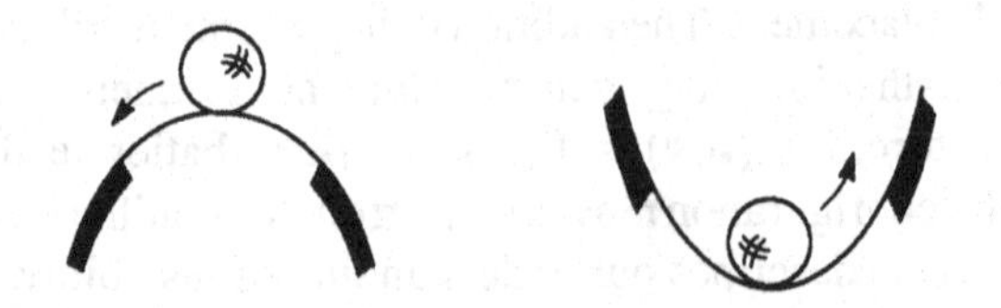

Figure 117 Simple illustration of the energy criterion

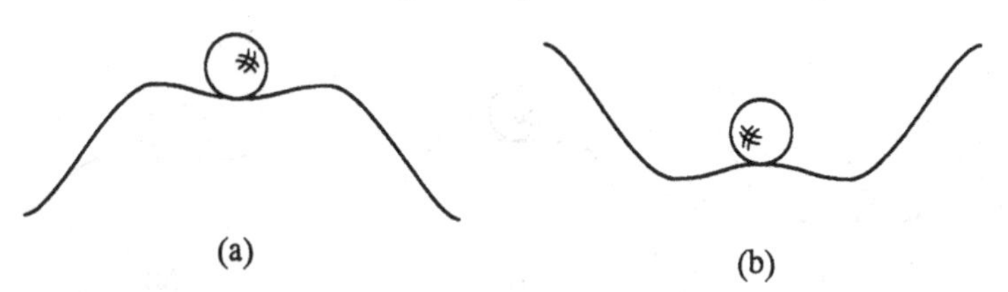

Figure 118 Limitations of a local stability criterion

left, the potential energy of the gravitational field is a maximum. For the ball on the right, the potential energy is a minimum. Assume that each of the balls is at rest, and let us impart a small velocity. For the ball on the left, the motion causes a decrease in the potential energy because the potential energy is maximum to begin with. The law of conservation of energy suggests that this decrease in potential energy will result in an increase in the kinetic energy. An increase in the kinetic energy implies further motion and the ball rolls even further from its initial position. For the ball on the right, the motion causes an increase in the potential energy with a corresponding loss in kinetic energy, thereby slowing the motion, eventually stopping it and returning the ball to its original position. The first case is clearly unstable because the motions are ever increasing. The second case is stable because the motions are arrested.

The stability criterion based on the second derivative test is a local criterion. The proof clearly depends upon the perturbation being small because it uses a Taylor series expansion in the neighborhood of the equilibrium configuration. This limitation of the local energy criterion is illustrated in Fig. 118. In case (a) we have $A(\mathbf{u}, \overline{\mathbf{u}}) > 0$ at the position of the ball, implying that the configuration is stable, but it is intuitively clear that a somewhat larger perturbation would send the ball over the small humps, never to return. Thus, case (a) shows stability in the small with instability in the large. Case (b) is the opposite of case (a). Instability is implied by $A(\mathbf{u}, \overline{\mathbf{u}}) < 0$, but a perturbation would lead only to a small motion. Thus, case (b) shows instability in the small with stability in the large. We will encounter situations in structural stability that have the features of this simple example. In those cases, we will simply find all of the equilibrium configurations and classify each one as stable or unstable. Equilibrium points in close proximity will be suspected of this type of behavior, but our static analysis will not allow us to examine it any further.

The stability of cases shown in Fig. 119 cannot be determined from the second derivative test because $A(\mathbf{u}, \overline{\mathbf{u}}) = 0$. If all of the derivatives higher than the second derivative are also zero, then the potential energy functional is perfectly flat at the equilibrium point. We call the stability of such a configuration *neutral stability*. For a neutral equilibrium configuration, a perturbation will not lead to a change in potential energy. The potential energy functional may be very flat at the equilibrium point, but not perfectly flat. In such a case, $A(\mathbf{u}, \overline{\mathbf{u}}) = 0$, but one of the higher derivatives of the energy functional may be

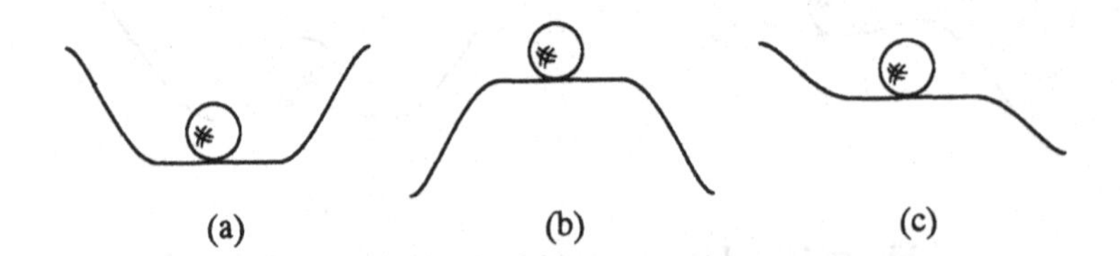

Figure 119 Stability with a locally flat potential energy functional

nonzero. Stability must then be determined from the sign of the lowest-order nonzero derivative of the energy. Case (a) has a higher derivative of even order with positive value, while case (b) has a higher derivative of even order with a negative value. Case (c) has a nonzero higher derivative of odd order. In such a case, the stability depends upon the direction of the perturbation. For this example, if the perturbation is to the left the potential energy increases, but to the right it decreases.

The general form of the stability requirement can be obtained from the Taylor series expansion of the energy functional. To wit,

$$\mathcal{E}(\mathbf{u}+\varepsilon\overline{\mathbf{u}}) = \mathcal{E}(\mathbf{u}) + \sum_{n=1}^{N}\frac{1}{n!}\varepsilon^n B^n(\mathbf{u},\overline{\mathbf{u}}) + O(\varepsilon^{N+1}) \tag{523}$$

where $n! = n(n-1)\cdots(2)(1)$ and the nth order directional derivative of the functional is given by the formula

$$B^n(\mathbf{u},\overline{\mathbf{u}}) \equiv \frac{d^n}{d\varepsilon^n}\big[\mathcal{E}(\mathbf{u}+\varepsilon\overline{\mathbf{u}})\big]_{\varepsilon=0} \tag{524}$$

Observe that $B^1(\mathbf{u},\overline{\mathbf{u}}) = G(\mathbf{u},\overline{\mathbf{u}})$ and $B^2(\mathbf{u},\overline{\mathbf{u}}) = A(\mathbf{u},\overline{\mathbf{u}})$. If $B^n(\mathbf{u},\overline{\mathbf{u}}) = 0$ for all values of $n = 2,\ldots, N-1$, and $B^N(\mathbf{u},\overline{\mathbf{u}}) \neq 0$, then the stability of the system is determined by the algebraic sign of B^N. If N is even then the stability criterion is the same as before, i.e., $B^N(\mathbf{u},\overline{\mathbf{u}}) > 0$ for all $\overline{\mathbf{u}}$ implies stability and $B^N(\mathbf{u},\overline{\mathbf{u}}) < 0$ for any $\overline{\mathbf{u}}$ implies instability. If N is odd then stability also depends upon the direction of the perturbing motion because the coefficient ε^N will be positive if $\varepsilon > 0$ and negative if $\varepsilon < 0$.

The second-derivative functional for discrete systems. For discrete systems, governed by algebraic equations, we can develop a useful form of the energy criterion if we recognize that the second derivative of the energy will always have the form

$$A(\mathbf{u},\overline{\mathbf{u}}) = \overline{\mathbf{a}}^T\mathbf{A}(\lambda,\mathbf{a})\overline{\mathbf{a}}$$

where $\mathbf{A}(\lambda,\mathbf{a})$ is the Hessian matrix of second derivatives for an ordinary function of the energy, λ represents a loading parameter, and $\mathbf{a}$ represents the parameters describing the motion. Application of the Ritz method to an energy functional always results in a discrete system. We include the load parameter λ as an argument of **A** to remind us that **A** generally depends upon the loads (if it does not depend upon the loads, we generally do not have a stability problem). Such details are most clear in particular applications and are deferred to the next chapter.

Since the second derivative must be positive for all arbitrary variations $\bar{\mathbf{a}}$, the second-derivative test amounts to testing the matrix **A** for positive definiteness. The eigenvalues of a matrix provide the most direct means of assessing positive definiteness.

Definition (positive definiteness). An $N \times N$ matrix **A** is positive definite if either of the following criteria are met

$$\bar{\mathbf{a}}^T\mathbf{A}(\lambda, \mathbf{a})\bar{\mathbf{a}} > 0 \quad \forall \bar{\mathbf{a}}$$

or if all of the eigenvalues of **A** are greater than zero. ❑

Recall that the eigenvalues and eigenvectors of a matrix **A** are the scalars γ and the vectors **u**, respectively, that satisfy the eigenvalue problem

$$\mathbf{A}\mathbf{u} = \gamma \mathbf{u} \tag{525}$$

If **A** is an $N \times N$ matrix, then there are exactly N pairs $(\gamma_i, \mathbf{u}_i)$ of associated eigenvalues and eigenvectors that satisfy Eqn. (525). If the matrix **A** is symmetric (as it usually is for structural mechanics problems), then all of the eigenvalues and eigenvectors have purely real values.

Stability of linear systems. Before we go on to the discussion of more general stability problems, let us examine the stability of the linear systems we have discussed in this chapter. Let us perform the second-derivative test on the previously defined energy functionals.

The second-derivative functional for the little boundary value problem is

$$A(u, \bar{u}) = \int_0^\ell C(\bar{u}')^2 \, dx \tag{526}$$

The value of the energy functional is the integral of the square of a function multiplied by Young's modulus. The square of a function is never negative. Therefore, the second derivative will always be positive if Young's modulus is positive, $C > 0$. If we interpolate the virtual displacement as $\bar{u} = \bar{\mathbf{a}} \cdot \mathbf{h}(x)$ then we have

$$\mathbf{A}(\lambda, \mathbf{a}) = \int_0^\ell C[\mathbf{h}'][\mathbf{h}']^T \, dx \tag{527}$$

Note that this matrix does not depend upon any loading parameter. The loading b does not appear in the second derivative functional and hence we can conclude that it does not contribute to the stability of the system. Also note that the matrix **A** is identical to the stiffness matrix **K**.

The second-derivative functional for the Bernoulli-Euler beam is

$$A(w,\overline{w}) = \int_0^\ell EI(\overline{w}'')^2\,dx \tag{528}$$

Again, we have the condition that the second-derivative functional will always be positive if the bending modulus EI is positive. Again, if we interpolate w we find that the stability matrix is identical to the stiffness matrix **K**.

The second-derivative functional for the Kirchhoff-Love plate is

$$A(w,\overline{w}) = \int_\Omega D\big(\nu\,\overline{w}_{,\alpha\alpha}\,\overline{w}_{,\beta\beta} + (1-\nu)\overline{w}_{,\alpha\beta}\,\overline{w}_{,\alpha\beta}\big)\,dA \tag{529}$$

In order for the second derivative to be positive, then D must be positive (there are also some restrictions on Poisson's ratio).

We can conclude that these linear systems are guaranteed to be stable if their elastic constants meet certain criteria that are commonly met by all materials. Therefore, we arrive at the conclusion that all of our linear theories give rise to stable equilibria. Furthermore, we should expect the **K** matrix that comes from a Ritz approximation with these theories always to be positive definite.

As mentioned in the introductory comments, buckling is a well-known phenomenon for structural systems. Since our linear theories always predict stable behavior, we should not look to these theories to explain the phenomenon of buckling. We shall discuss the basic issue of structural stability in the next chapter. We shall demonstrate that the assumption of small deformations, which led us to our linear theories, precludes the modeling of buckling. If we remove this assumption and express equilibrium in the deformed position of the system, then we can model buckling phenomena.

Additional Reading

H. L. Langhaar, *Energy methods in applied mechanics*, Wiley, New York, 1962.

I. M. Gelfand and S. V. Fomin, *Calculus of variations*, Prentice Hall, Englewood Cliffs, N.J., 1963.

Problems

219. Consider a one-dimensional rod of length ℓ and modulus C subjected to a body force $b(x)$ and a traction τ_o at the right end. The left end has a prescribed displacement of $u(0) = u_o$. The Hu-Washizu energy functional for the rod is given in terms of the independent variables $\sigma(x)$ (stress), $u(x)$ (displacement), and $\epsilon(x)$ (strain) as

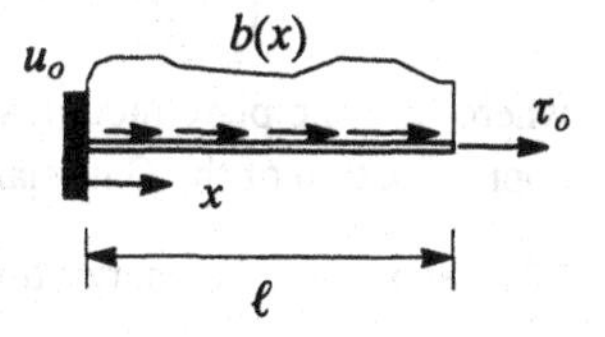

$$\mathcal{H}(u,\sigma,\epsilon) \equiv \int_0^\ell \left(\tfrac{1}{2}C\epsilon^2 - bu - \sigma(\epsilon - u')\right) dx - \tau_o u(\ell) + \sigma(0)\left(u(0) - u_o\right)$$

Show that by taking the directional derivative of $\mathcal{H}$ in the direction of variations of each of the variables, i.e.,

$$D\mathcal{H}(u,\sigma,\epsilon)\cdot(\bar{u},\bar{\sigma},\bar{\epsilon}) = \frac{d}{da}\left[\mathcal{H}(u+a\bar{u},\ \sigma+a\bar{\sigma},\ \epsilon+a\bar{\epsilon})\right]_{a=0}$$

and setting the result equal to zero (to find the extremum), all of the classical governing equations for the one-dimensional bar result from applying the fundamental theorem of the calculus of variations

$$\left.\begin{aligned} \sigma' + b &= 0 \\ \epsilon - u' &= 0 \\ \sigma - C\epsilon &= 0 \end{aligned}\right] \quad x \in (0,\ell) \qquad \begin{aligned} \sigma(\ell) - \tau_o &= 0 \\ u(0) - u_o &= 0 \end{aligned}$$

Note that the fields that appear in the functional are each functions of x themselves.

220. Find the energy functional $\mathcal{E}(u,w,\theta)$ for a Timoshenko beam.

221. Find a Hu-Washizu energy functional for a simply supported Bernoulli-Euler beam of length ℓ and modulus EI subject to a transverse load $q(x)$. The appropriate field variables are the transverse displacement $w(x)$, the moment $M(x)$, and the curvature $\varkappa(x)$. Show that the extremum of the energy functional with respect to the three field variables gives the classical equations of Bernoulli-Euler beam theory. How must the functional change if the beam is fixed at $x = 0$ and pinned at $x = \ell$?

222. The Hellinger-Reissner energy functional for a three-dimensional hyperelastic solid body $\mathfrak{B}$ with boundary Ω is given by

$$\mathcal{R}(\mathbf{u},\mathbf{S}) = \int_{\mathfrak{B}} \left(\mathbf{S}\cdot\nabla\mathbf{u} - \mathbf{b}\cdot\mathbf{u} - U(\mathbf{S})\right) dV - \int_{\Omega_t} \hat{\boldsymbol{\tau}}\cdot\mathbf{u}\, dA - \int_{\Omega_u} \mathbf{S}\mathbf{n}\cdot\left(\mathbf{u} - \hat{\mathbf{u}}\right) dA$$

where $\mathbf{u}$ is the displacement field, $\mathbf{S}$ is the stress field, $\hat{\boldsymbol{\tau}}$ is the prescribed traction over Ω_t (the portion of the boundary where tractions are prescribed), $\hat{\mathbf{u}}$ is the prescribed displacement over Ω_u (the portion of the boundary where displacements are prescribed), $\mathbf{b}$ is the body force, and $U(\mathbf{S})$ in the stress-energy function. What do the necessary conditions for an extremum imply? (Hint: take the directional derivative of the functional in the directions of $\bar{\mathbf{u}}$ and $\bar{\mathbf{S}}$, and apply the fundamental theorem of the calculus of variations.)

223. Show that the energy functional for the Kirchhoff-Love plate, given in Eqn. (513), can be expressed in the equivalent form

$$\mathcal{E}(w) = \int_{\Omega} \left[\tfrac{1}{2} D\Big((w_{,11} + w_{,22})^2 - 2(1-\nu)\big(w_{,11} w_{,22} - w_{,12}^2\big)\Big) - qw \right] dA$$

where D is the plate modulus and ν is Poisson's ratio. The term $w_{,11}\, w_{,22} - w_{,12}^2$ is an approximation of the Gaussian curvature of the deformed reference surface of the plate.

224. Show that the energy functional for a three-dimensional linear elastic solid is

$$\mathcal{E}(\mathbf{u}) = \int_{\mathcal{B}} \left(\tfrac{1}{2}\lambda (\operatorname{div} \mathbf{u})^2 + \tfrac{1}{2}\mu \left[\nabla \mathbf{u}^T + \nabla \mathbf{u} \right] \cdot \nabla \mathbf{u} - \mathbf{b} \cdot \mathbf{u} \right) dV - \int_{\Omega_t} \mathbf{t} \cdot \mathbf{u}\, dA$$

Show also that the extremum of the energy gives the same equations as the principle of virtual work.

225. Show that the energy functional for a Bernoulli-Euler beam on an elastic foundation can be expressed in the form

$$\mathcal{E}(w) = \int_0^{\ell} \left(\tfrac{1}{2} EI (w'')^2 + \tfrac{1}{2} k w^2 - qw \right) dx$$

where EI is the flexural modulus of the beam, k is the modulus of the foundation, and q is the transverse load.

226. Find the Euler equation and boundary conditions for the functional

$$J(w) = \int_a^b F(w, w', w'')\, dx$$

Use the Euler equation on the energy functional for a Bernoulli-Euler beam on an elastic foundation to find the classical differential equation governing the beam.

227. Using the fundamental theorem of the calculus of variations, find the classical form of the governing differential equation for $w(x)$ implied by the minimum of the energy functional

$$\mathcal{E}(w) = \int_0^{\ell} \left(\tfrac{1}{2} EI\, w^{iv} w - qw \right) dx - \tfrac{1}{2} EI\, w''' w \Big|_0^{\ell} + \tfrac{1}{2} EI\, w'' w' \Big|_0^{\ell}$$

228. The potential energy of a simply supported, symmetrically loaded circular plate of radius R is

$$\mathcal{E}(w) = \pi \int_0^R \left[D\left(r (w'')^2 + \tfrac{1}{r} (w')^2 + 2\nu w' w'' \right) - 2rqw \right] dr$$

where the function $w(r)$ is the transverse deflection of the plate, D and ν are constants, and q is a known function of r. Find the variational (virtual work) form of the governing differential equation. Find the classical (strong) form of the governing differential equation. What can you say, if anything, about the boundary conditions for the problem?

229. Find an approximate solution to the problem of a simply supported, circular plate of radius R and modulus D, subjected to a uniform load of q. Use the Ritz method with the energy functional given in Problem 228. Assume that the displacement is of the form

$$w(r) = a_0 \cos\left(\frac{\pi r}{2R}\right)$$

where a_0 is the, as yet, undetermined constant. If you had to pick additional terms in the approximation, what would you choose? Why is the cosine function a good choice?

230. A beam of length ℓ and modulus EI rests on a nonlinearly elastic foundation that accrues transverse force in proportion to the cube of the transverse displacement, i.e., $f(x) = k_o w^3$. The beam is subjected to downward transverse loading $q(x)$. Axial and shear deformations are negligible. Take $w(x)$ as positive when it is upward. Find the virtual-work form of the equilibrium equations. Find the energy functional $\mathcal{E}(w)$ for the system.

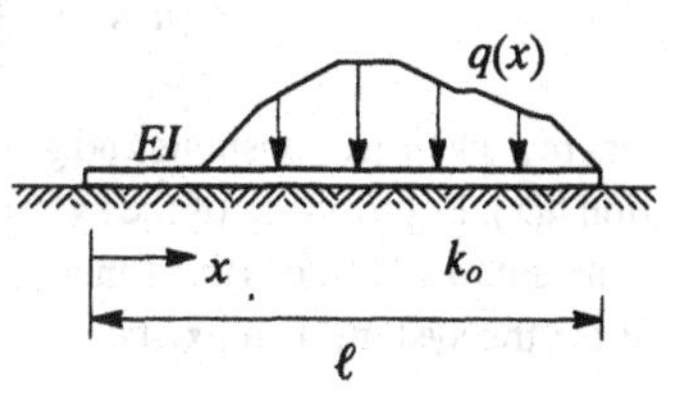

231. Consider a functional that takes scalar functions $u(x)$ as input. The independent scalar variable is defined on the range $x \in [0, 1]$. The functional has the explicit form:

$$\mathcal{E}(u) = \int_0^1 \left\{ \frac{1}{1+x} \left(u'(x)\right)^2 - 2u(x) \right\} dx$$

where a prime indicates differentiation with respect to x. The functions are constrained at the boundary to satisfy the conditions $u(0) = 0$ and $u'(1) = 0$. Find the classical differential equation implied by stationarity (i.e., max, min, or saddle point) of the functional. Solve the classical differential equation. Compute the second derivative functional associated with the given functional.

232. The virtual-work functional for a system is given by the following expression

$$G(w, \overline{w}) = \int_0^\ell \left(a\left(w''\overline{w} + \overline{w}''w\right) + bw^3\overline{w} \right) dx$$

where a, b, and ℓ are constants and $w(x)$ and $\overline{w}(x)$ are functions of the independent variable x. Does this functional have an associated energy? Find the energy functional for the system, if it exists.

233. Resolve Problem 232 with the functional

$$G(w, \overline{w}) = \int_0^\ell \left(-aw''\overline{w} + bw\overline{w}\right) dx - a\,w(\ell)\overline{w}'(\ell) + a\,w(0)\overline{w}'(0)$$

234. The deformation state of a particular system is characterized by the scalar function $\theta(x)$, where the scalar variable $x \in [0, \ell]$. The virtual-work functional for the system is given by the following expression

$$G(\theta, \overline{\theta}) = \int_0^\ell \left[a\theta'\overline{\theta}' + b\overline{\theta}\sin\theta \right] dx$$

where a, b, and ℓ are constants. Equilibrium of the system holds if $G(\theta, \overline{\theta}) = 0$ for all $\overline{\theta}$. Does this functional have an associated energy? Find the energy functional for the system, if it exists.

235. The deformation state of a particular system is characterized by the scalar function $u(x)$, where the scalar variable $x \in [0, 1]$. The virtual-work functional for the system is given by the following expression

$$G(u, \bar{u}) = \int_0^1 \left[a\, u' \bar{u}' + b u' \bar{u} + g(u) \bar{u} \right] dx$$

where a and b are constants and $g(u)$ is a given nonlinear function of the displacement function $u(x)$. Equilibrium of the system holds if $G(u, \bar{u}) = 0$ for all $\bar{u}$. For what values of the constants a and b does this functional have an associated energy? Find the energy functional for the system, if it exists.

236. A beam of length ℓ rests on an elastic foundation of modulus k (per unit length). It is pinned at the left end and is subjected to a point load P at the right end. The elastic foundation accrues a transverse force in proportion to the transverse displacement w. The energy of the system can be expressed as:

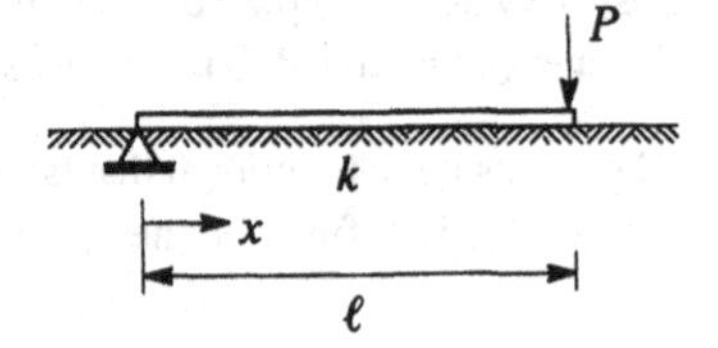

$$\mathscr{E}(w) = \int_0^{\ell} \tfrac{1}{2}\left(EI(w'')^2 + k w^2 \right) dx - P w(\ell)$$

Find the virtual-work form of the equilibrium equations. What are the essential boundary conditions? Find the classical form of the equilibrium equations and the boundary conditions. Which of the three functions given below are suitable for approximating the solution with the Ritz method? Explain why or why not in each case.

$$w(x) = x(\ell - x)(a_1 + a_2 x), \quad w(x) = x^2(a_1 + a_2 x), \quad w(x) = x(a_1 + a_2 x)$$

237. Consider a rectangular (rigid) block of height h and width ℓ and weight W. The block is prevented from sliding by a small obstruction at the lower right corner and is pushed by a force P at the upper left corner. Write the potential energy of the system in terms of the angle of rotation θ of the block. Find the force P as a function of h, ℓ, W, and the angle θ needed to have static equilibrium. Find an expression for the angle at which equilibrium goes from being stable to being unstable.

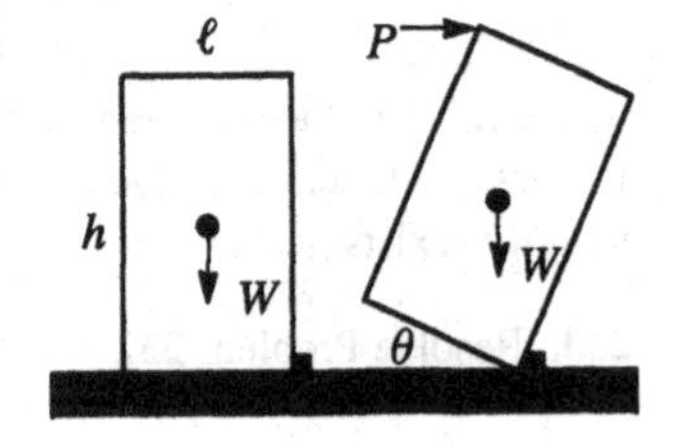

238. Consider the solid spherical region shown in the sketch. Assume that there exists a scalar field $w(\mathbf{x})$ for which we can define the functional

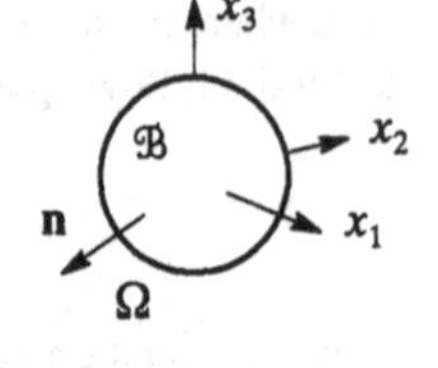

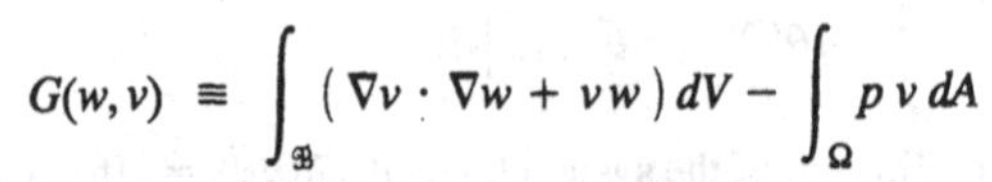

$$G(w, v) \equiv \int_{\mathscr{B}} \left(\nabla v \cdot \nabla w + v w \right) dV - \int_{\Omega} p\, v\, dA$$

that has the property that if $G(w, v) = 0$ for all (virtual) scalar functions $v(\mathbf{x})$ then the classical differential equations governing the real field $w(\mathbf{x})$ are satisfied (i.e., $G(w, v)$ is a "virtual-work" functional). Note that the scalar field p is defined on the surface of the solid region. Show that an "energy" functional exists for this theory if the function p depends only upon the position vector $\mathbf{x}$ and not the function $w(\mathbf{x})$, i.e., $p = p(\mathbf{x})$. Determine the energy functional in terms of the field w.

239. Reconsider Problem 238 for the case where the function p depends upon the field w and the position $\mathbf{x}$. Under what conditions would an "energy" functional exist in this case? (Hint: would an energy exist if p depends upon w itself? What if it depends upon derivatives of w, i.e., $p = p(\mathbf{x}, w(\mathbf{x}), \nabla w(\mathbf{x}), \ldots)$?

240. A beam of length ℓ rests on an elastic foundation of modulus k (per unit length). It is fixed at the left end, pinned at the right end, and is subjected to a point load P at midspan. The energy of the system can be expressed in terms of the transverse displacement $w(x)$ as:

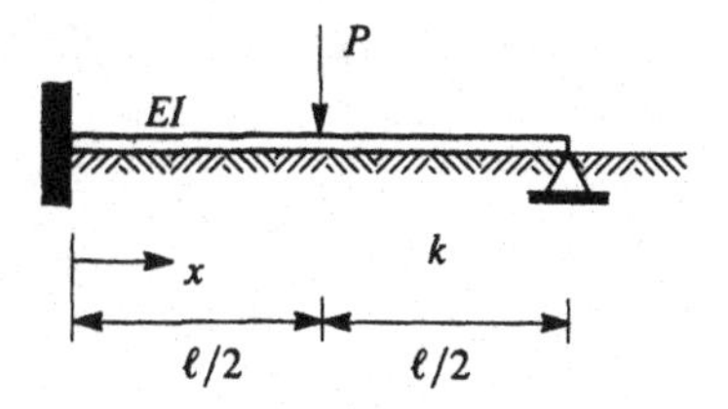

$$\mathcal{E}(w) = \int_0^\ell \tfrac{1}{2}\left(EI(w'')^2 + kw^2\right) dx + Pw(\ell/2).$$

Find the virtual-work form of the equilibrium equations. What are the essential and natural boundary conditions? Use the Ritz method to find a one-term approximation of the displacement field (use a polynomial approximation).

10
Fundamental Concepts in Static Stability

The limit to the load carrying capacity of many structures is buckling resulting from the loss of stability of equilibrium. Any structure that carries load through compression is a candidate for loss of stability. (Some tensile structures are candidates for instability, but that is less common, and it remains a good rule of thumb to think of tensile forces as essentially stabilizing). The primary function of many structures is to elevate space and the fight with gravitational forces can induce compression in many members of a structure. Hence, structures subjected to gravity forces can suffer stability problems. Loss of stability must be well understood and accounted for in the design of structural systems.

In the previous chapter we developed the energy criterion to assess the stability of static equilibrium of certain systems (i.e., those systems for which an energy functional exists). One of the observations that we made, based upon the energy criterion, is that an elastic system whose governing equations are linear† will lose stability only if certain of the elastic moduli (i.e., Young's modulus) are less than zero. Furthermore, the loss of stability of a linear system is independent of the motion of the system. In fact, Vainberg's theorem shows that, if $G(u, \bar{u})$ is linear in u, then the energy $\mathcal{E}(u)$ must be quadratic in u and $A(u, \bar{u})$ is, therefore, independent of u. Linear theories of structural mechanics are not very interesting from the standpoint of stability. Many more interesting possibilities arise when the governing equations are nonlinear.

Nonlinear theories of mechanics can arise from three basic sources: (1) nonlinearity in equations of equilibrium, which generally arise because equilibri-

† In a linear theory all of the equations—kinematics, equilibrium, and constitution—must be linear. If nonlinearity is present in any of these three aspects of the theory then the governing equations are nonlinear.

um in the deformed body generally depends upon the motion, (2) nonlinearity in the strain-displacement equations, and (3) nonlinearity in the constitutive equations. The first two sources of nonlinearity are coupled in a consistently formulated theory. Constitutive nonlinearities can take any number of forms from nonlinear elasticity to inelasticity. In this chapter we focus primarily on systems for which the first two sources are active, but the constitutive equations are linear.

The analysis of nonlinear systems is considerably more difficult than the analysis of linear systems, but the rewards are much higher. There is a beauty in a fully revealed solution to a nonlinear problem that cannot be matched in the realm of predictable, positive-definite linear systems where the principle of superposition applies and doubling the load means simply doubling the displacements and stresses. The stakes can be much higher for nonlinear systems, too. Failure of a system owing to instability is often dramatic, and often takes place without much warning.

The primary purpose of the present chapter is to motivate the ideas and concepts of static stability theory. Many new ideas and a lot of new terminology beyond those needed for the linear theory must be introduced. There is no better way to introduce these concepts than to explore some simple examples that contain them. While we do not aim for complete coverage of all of the issues of static stability, the problems examined here should provide a good starting point for the novice in stability theory.

We shall explore the ideas associated with static stability using the three simple systems shown in Fig. 120. All three of the systems consist of a rigid bar of length ℓ subjected to a vertical force P (taken positive downward). Each of these systems has elastic resistance provided by a discrete spring of modulus k. In each case the elastic resistance manifests in a different way and, as a result, the system displays a different response and thereby shows a different feature of nonlinear response. All three of the systems have the characteristic that, in the nominal configuration, they carry the load P purely through axial force in the bar. The elasticity in each case is provided by the springs, but that elasticity is not mobilized until the bar rotates from the vertical position. Because of the rigidity of the member and the support conditions, the system has only one

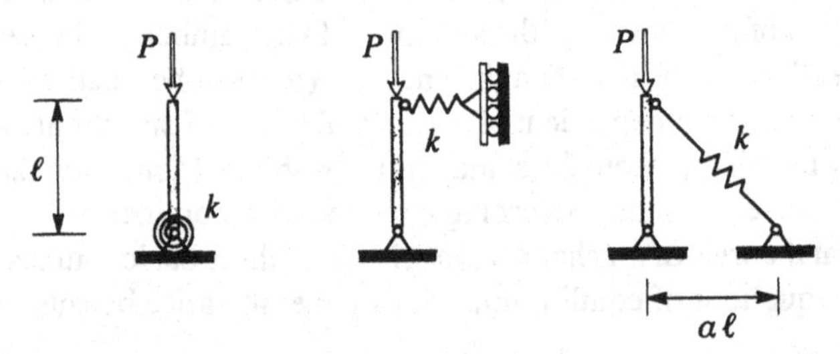

Figure 120 Example problems that will be used to illustrate various features of nonlinear response in this chapter

degree of freedom. The deformation of the system can be completely characterized by the parameter θ, measuring the rotation of the column from its original vertical position.

Bifurcation of Geometrically Perfect Systems

Bifurcation is the name given to structural response associated with a branching of the solution to a nonlinear equation (or system of equations) at a point. Figure 121 illustrates the features of lack of uniqueness and bifurcation for a nonlinear equation of two variables, i.e., $g(\lambda, \theta) = 0$. There are several issues that are important to recall. First, there is no guarantee that there will be only one solution to a nonlinear equation (actually there might not be any). Second, a single equation relating two variables can be represented as a curve in two-dimensional space (as in the figure). In our problems we will generally refer to these lines as *equilibrium paths* because they will come from equilibrium equations and they will relate the load λ to the deformation θ. Each branch is a continuous sequence of points $\{\lambda, \theta\}$ that satisfy the equation.

A bifurcation point is a point where two branches intersect. Imagine a loading sequence that generates a sequence of equilibrium points along a certain branch, say Branch 1 in the figure. At some stage of loading the system will encounter the bifurcation point. At that point there will be four choices to advance the solution (one of which is returning along the path just traversed). If the system switches to the other branch (say Branch 2 in the figure) then there will be a change in the mode of behavior. A bifurcation point is associated with a zero value of the second derivative functional and, hence, represents a point where a branch can change from stable behavior to unstable behavior.

W. T. Koiter wrote a dissertation entitled *Over de Stabiliteit van het Elastisch Evenwicht* (in Dutch) in 1945 to earn his Ph.D. from the Delft University of Technology. In this work he laid the modern foundations of structural stability. He determined that there were only three possible types of bifurcation in structural systems: Stable symmetric bifurcation, unstable symmetric bifurcation, and asymmetric bifurcation. In this section we shall study those three types of bifurcation through a series of examples. The main purpose of these examples is to see, in a familiar structural setting, the meaning and implications of nonlinear structural response that includes bifurcation.

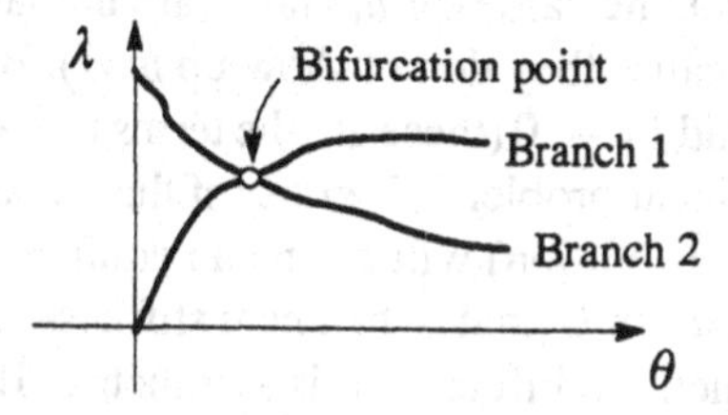

Figure 121 Lack of uniqueness and bifurcation of the nonlinear equation $g(\lambda,\theta)=0$.

To study bifurcation we shall make use of energy principles, virtual work (to establish equilibrium), and the second derivative test (to establish the stability of equilibrium). For the single-degree-of-freedom systems that we will examine the energy will have the form

$$\mathcal{E}(\theta) = \mathcal{A}(\theta) + \lambda\mathcal{B}(\theta) \tag{530}$$

where θ will be the measure of deformation of the system and λ will be the loading parameter. All of our systems will have an energy that is linear in the loading parameter and nonlinear in the displacement parameter. The nature of the functions $\mathcal{A}(\theta)$ and $\mathcal{B}(\theta)$ will determine the character of the response and distinguish the different types of bifurcations.

The virtual-work functional and the second derivative functional can be easily computed from the energy as

$$\begin{aligned} G(\theta,\bar{\theta}) &= \left[\mathcal{A}'(\theta)-\lambda\mathcal{B}'(\theta)\right]\bar{\theta} \\ A(\theta,\bar{\theta}) &= \left[\mathcal{A}''(\theta)-\lambda\mathcal{B}''(\theta)\right]\bar{\theta}^2 \end{aligned} \tag{531}$$

where $(\cdot)' = d(\cdot)/d\theta$. Equilibrium holds if $G(\theta,\bar{\theta}) = 0$ for all $\bar{\theta}$. This equation allows the determination of the load parameter as a function of θ as

$$\lambda = \frac{\mathcal{A}'(\theta)}{\mathcal{B}'(\theta)} \tag{532}$$

For states that satisfy this relationship, the stability is determined from the second derivative test. Since $\bar{\theta}^2$ is always positive, the term in brackets in Eqn. (531) determines the algebraic sign of the second derivative functional. The system is stable if

$$\mathcal{A}''(\theta) - \frac{\mathcal{A}'(\theta)}{\mathcal{B}'(\theta)}\mathcal{B}''(\theta) > 0 \tag{533}$$

We can draw some simple, but general, conclusions if we expand the functions $\mathcal{A}(\theta)$ and $\mathcal{B}(\theta)$ in Taylor series to give the energy expression

$$\begin{aligned} \mathcal{E}(\theta) = {} & a_0 + a_1\theta + a_2\theta^2 + a_3\theta^3 + a_4\theta^4 \\ & -\lambda\left(b_0 + b_1\theta + b_2\theta^2 + b_3\theta^3 + b_4\theta^4\right) \end{aligned} \tag{534}$$

It should be evident that the values of a_0 and b_0 are immaterial to the equilibrium of the system (because they do not show up in G). Bifurcation is not possible unless $a_1 = 0$ and $b_1 = 0$ (these are the terms that are usually associated with the loading in a linear problem). If either of these coefficients are nonzero then $\theta = 0$ (the trivial solution) will not be an equilibrium configuration. In general, we will have $a_2 > 0$ (that is the linear stiffness of the system). We will find that if $a_3 = 0$ then the bifurcation is symmetric. If $a_3 \neq 0$ then the bifurcation is asymmetric.

The following examples illustrate these general ideas.

Stable Symmetric Bifurcation. A symmetric bifurcation gives the same response regardless of the algebraic sign of θ (i.e., it does not matter which direction the system moves). The functions $\mathcal{A}(\theta)$ and $\mathcal{B}(\theta)$ must be even functions (e.g., the odd power terms will not appear in Eqn. (534)). The trivial configuration $\theta = 0$ will be an equilibrium configuration for all values of the loading parameter and the nontrivial solution will be stable and ascending. The following example illustrates this type of response.

Example 58. *Stable symmetric bifurcation.* Consider the rigid column with a rotational spring, shown in Fig. 122. The structure is composed of a rigid bar hinged at the base and restrained from rotation by a rotational spring at that point. The column is free to move at the point of loading. The force on the system is the vertical force P. The rotational spring accrues moment M in proportion to the relative rotation θ it experiences, i.e., $M = k\theta$, and constitutes the elasticity of the system. Recall that the potential energy of the spring is $\frac{1}{2}k\theta^2$.

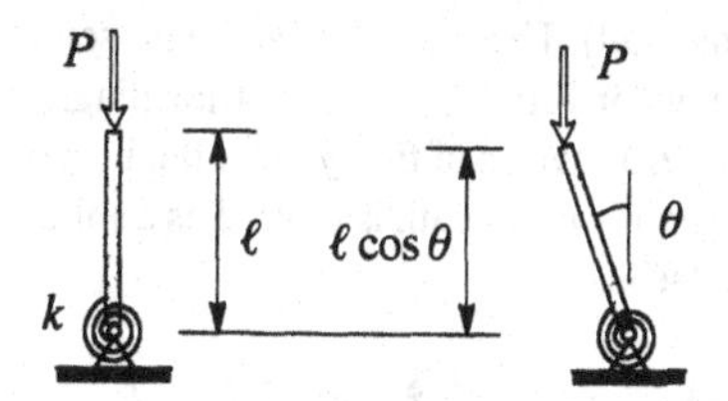

Figure 122 A rigid column with a rotational spring

The energy, virtual-work functional, and second-derivative functional are†

$$\begin{aligned}
\mathcal{E}(\theta) &= \tfrac{1}{2}k\theta^2 + P\ell\cos\theta \\
G(\theta,\bar{\theta}) &= \left[k\theta - P\ell\sin\theta\right]\bar{\theta} \\
A(\theta,\bar{\theta}) &= \left[k - P\ell\cos\theta\right]\bar{\theta}^2
\end{aligned}$$

Note that datum for the potential energy of the load is at the base of the column.

As usual, θ represents an equilibrium configuration if $G(\theta,\bar{\theta}) = 0$ for all values of the arbitrary virtual displacement $\bar{\theta}$. Since the expression must hold for all values of the arbitrary constant $\bar{\theta}$, the term in brackets must be identically zero (again, our old friend the fundamental theorem of the calculus of variations). To wit, the equation governing the equilibrium of the column is

$$k\theta - P\ell\sin\theta = 0$$

† For these examples we will simply write down the energy expression because the energy associated with springs is quite simple to derive. The skeptical reader can take a Newtonian approach and write the equations of equilibrium in the deformed configuration, use a weighted residual to create the virtual-work functional, and then determine the energy functional using Vainberg's theorem. In this case the moment in the spring, $M = k\theta$, must balance the moment created by the force, $M = P\ell\sin\theta$, to give the equilibrium equation $k\theta = P\ell\sin\theta$.

This equation is interesting in that it admits more than one equilibrium path (load-deflection curve) and possesses a bifurcation point or branching point. We must examine the stability of each of these equilibrium paths with the energy criterion.

Clearly, $\theta = 0$ is a solution to this problem. This solution corresponds to the straight position of the column. The equation is satisfied for *any* value of the load parameter P. This observation is in accord with our expectations from taking a freebody diagram of the structure in the straight position. Not all values of the load constitute stable equilibrium. For $\theta = 0$, the energy criterion reduces to

$$A(\theta,\bar{\theta}) = [k - P\ell]\bar{\theta}^2 = \begin{cases} > 0 & \text{for } P < k/\ell \quad \text{(stable)} \\ < 0 & \text{for } P > k/\ell \quad \text{(unstable)} \end{cases}$$

The energy criterion tells us that equilibrium is stable for all values of the load P less than the value k/ℓ (even negative values, which represent tension on the column) and is unstable for all values of the load exceeding this critical value. We call this load the critical load (note that we do not yet have enough evidence to call it the buckling load). These results can be presented in a *bifurcation diagram* like the one shown in Fig. 123. A bifurcation diagram is nothing more than a plot of load versus deformation of the system, but it shows all possible equilibrium paths. A stable equilibrium path is plotted as a solid line, while an unstable path is plotted as a dashed line.

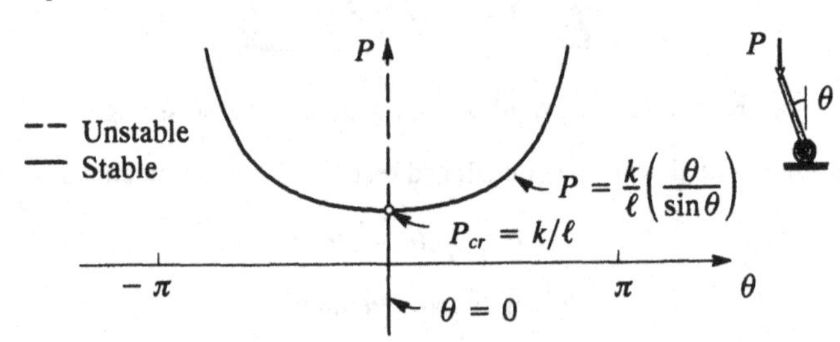

Figure 123 Bifurcation diagram for a rigid column with a rotational spring

A second equilibrium path can be found for values of θ not equal to zero, i.e., the bent position. In the bent position, the value of the load parameter depends upon the state of deformation, so we will say that P is a function of θ. Equilibrium is satisfied if the load is given by

$$P = \frac{k}{\ell}\left(\frac{\theta}{\sin\theta}\right)$$

Interestingly, this equilibrium path does not emanate from the origin, but rather branches from the load axis at the load value $P = k/\ell$, the critical load (use l'Hospital's rule to show that this is true). A point, such as this one, where two or more equilibrium paths branch from a single point, is called a *bifurcation point*. The path corresponding to the bent position is such that the load increases as the angle increases. We can also see that the second branch is symmetric with respect to θ, meaning that it has zero slope at $\theta = 0$. The symmetry tells us that if the structure is initially perfectly straight, then it has no preference in its buckling direction.

The stability of the second equilibrium path can, again, be established from the second-derivative test. In this case, we have

$$A(\theta,\bar{\theta}) = k\left(1-\frac{\theta}{\tan\theta}\right)\bar{\theta}^2 > 0 \quad \text{for all } \theta \quad \text{(stable)}$$

The expression comes from substituting $P = k\theta/\ell \sin\theta$, required for equilibrium, into the expression for the second derivative of the energy. Since

$$0 < \theta/\tan\theta < 1 \qquad \theta \in [-\pi/2, \pi/2]$$
$$\theta/\tan\theta < 0 \qquad \theta \in [-\pi, -\pi/2]$$

this equilibrium path is stable everywhere (and, thus, is plotted as a solid line).

Unstable Symmetric Bifurcation. An unstable symmetric bifurcation also gives the same response regardless of the algebraic sign of θ (i.e., it does not matter which direction the system moves), and hence the functions $\mathcal{A}(\theta)$ and $\mathcal{B}(\theta)$ must be even functions. The trivial configuration $\theta = 0$ will be an equilibrium configuration for all values of the loading parameter and the nontrivial solution will be unstable and descending. The following example illustrates this type of response.

Example 59. *Unstable symmetric bifurcation.* Consider the structure shown in Fig. 124. The structure is identical to the previous one, except that instead of a rotational spring at the base, the top is restrained from lateral motion by a translational spring. The column is free at the top and hinged at the base. Again, the force on the system is the vertical force P. The translational spring accrues force F in proportion to its extension Δ, i.e., $F = k\Delta$, and constitutes the elasticity of the system. Recall that the potential energy of the spring is $\frac{1}{2}k\Delta^2$.

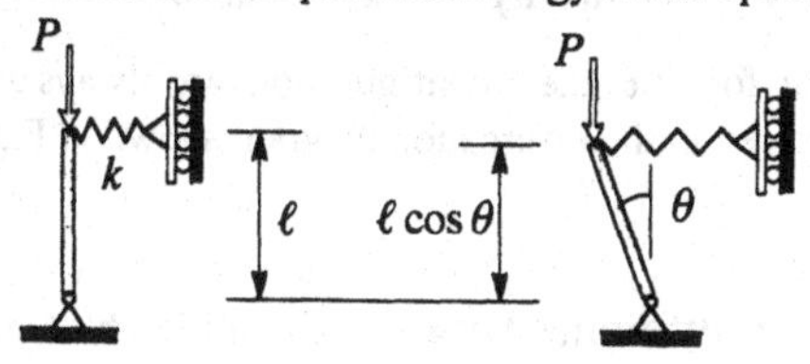

Figure 124 A rigid column with a translational spring

The energy, virtual-work functional, and second-derivative functional are

$$\mathcal{E}(\theta) = \tfrac{1}{2}k(\ell\sin\theta)^2 + P\ell\cos\theta$$
$$G(\theta,\bar{\theta}) = \ell\sin\theta\big[k\ell\cos\theta - P\big]\bar{\theta}$$
$$A(\theta,\bar{\theta}) = \ell\big[k\ell(\cos^2\theta - \sin^2\theta) - P\cos\theta\big]\bar{\theta}^2$$

As usual, θ represents an equilibrium configuration if $G(\theta,\bar{\theta}) = 0$ for all values of the arbitrary virtual displacement $\bar{\theta}$. Since the expression must hold for all values of the arbitrary constant $\bar{\theta}$, the remaining terms must be identically zero. To wit, the equation governing the equilibrium of the column is

$$\sin\theta\left[k\ell\cos\theta - P\right] = 0$$

As was the case for the last problem, this equation is interesting in that it has multiple equilibrium paths. One way to satisfy the equation is to have $\sin\theta = 0$, which is true for $\theta = 0, \pm\pi, \pm 2\pi, \ldots$, that is, all positive and negative integer multiples of π. These values of θ all correspond to straight configurations of the column. Let us ignore all solutions that require that θ make a full circle (although such a mechanism is quite possible). Hence, we will consider only the solutions $\theta = 0, \pi, -\pi$. For these values of the rotation, any value of the load P is possible. For values of θ that do not correspond to straight configurations, the system can be in equilibrium only if

$$P = k\ell\cos\theta$$

Notice again that the nontrivial equilibrium path at $\theta = 0$ does not emanate from the origin, but rather at the critical load $P_{cr} = k\ell$. The nontrivial equilibrium path also branches at the critical load $P_{cr} = -k\ell$ from the two other two straight configurations, i.e., $\theta = \pm\pi$. For the straight configurations, the second derivative test gives, for $\theta = 0$

$$A(0,\bar{\theta}) = \ell(k\ell - P)\bar{\theta}^2 = \begin{cases} > 0 & \text{for } P < k\ell \quad \text{(stable)} \\ < 0 & \text{for } P > k\ell \quad \text{(unstable)} \end{cases}$$

and for $\theta = \pm\pi$

$$A(\pm\pi,\bar{\theta}) = \ell(k\ell + P)\bar{\theta}^2 = \begin{cases} > 0 & \text{for } P > -k\ell \quad \text{(stable)} \\ < 0 & \text{for } P < -k\ell \quad \text{(unstable)} \end{cases}$$

For the configurations where $P = k\ell\cos\theta$, the second-derivative test gives

$$A(\theta,\bar{\theta}) = -k\ell^2\bar{\theta}^2\sin^2\theta$$

Since $A(\theta,\bar{\theta}) < 0$ for all θ, these configurations are always unstable. These results are summarized in the bifurcation diagram shown in Fig. 125.

There is one major difference between Examples 58 and 59. At the bifurcation points in Example 59, three of the four branches are unstable, and only one is stable. We can imagine that, if the system is loaded from zero in the undeformed configuration, catastrophe awaits at the bifurcation point. This is indeed the case. The system has no choice but to snap through to a stable configuration, either at $\theta = \pi$ or at $\theta = -\pi$, because it cannot remain on an unstable equilibrium path. This phenomenon is often called *snap-through buckling*. Snap-through buckling is, of course, a dynamic phenomenon, and our static model is not able to predict the path the structure will take in getting to a stable

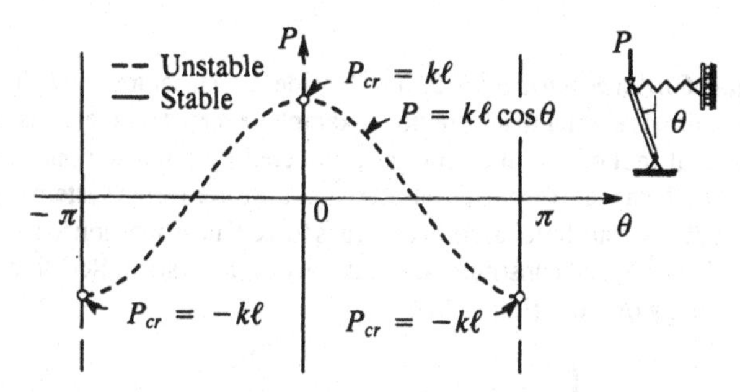

Figure 125 Bifurcation diagram for a rigid column with a translational spring

configuration, but if there is little inertia, we can imagine that the process will be quite fast. Clearly, there is danger associated with this type of bifurcation diagram, and this danger is the primary concern in the design of such a system (unlike the previous system, where the buckling caused some fairly significant cosmetic disturbance, but did not shed load in the buckling process).

Asymmetric Bifurcation. In both of the previous examples there was no preferential direction for buckling from the trivial state. Asymmetric buckling is possible when the function $\mathcal{A}(\theta)$ contains an odd function of θ, at least one that gives rise to a θ^3 term in a Taylor series expansion of $\mathcal{A}(\theta)$. In the simplest case, we can would have

$$\mathcal{E}(\theta) = a_2\theta^2 + a_3\theta^3 - \lambda\, b_2\theta^2$$

The virtual-work and second-derivative functional for this case are

$$G(\theta,\bar{\theta}) = \left(2a_2\theta + 3a_3\theta^2 - \lambda\, 2b_2\theta\right)\bar{\theta}$$
$$A(\theta,\bar{\theta}) = \left(2a_2 + 6a_3\theta - \lambda\, 2b_2\right)\bar{\theta}^2$$

Setting $G(\theta,\bar{\theta}) = 0$ for all $\bar{\theta}$ gives the load in terms of θ and the second derivative functional for that load tests the stability of the configuration. For this case we have

$$\lambda = \frac{2a_2 + 3a_3\theta}{2b_2}, \qquad A(\theta,\bar{\theta}) = \left(3a_3\theta\right)\bar{\theta}^2$$

We can, without loss of generality, assume that $a_2 > 0$ and $b_2 > 0$. This assumption gives a positive value of the bifurcation load $\lambda_{cr} = a_2/b_2$ for a bifurcation at $\theta = 0$. If we also assume that $a_3 > 0$ then it is evident from the presence of the linear term in the expression for the load that the load will increase for positive values of θ and will decrease for negative values of θ. The second derivative indicates that the branch for positive values of θ is stable and the branch for negative values of θ is unstable.

The following example illustrates the important features of the asymmetric bifurcation.

Example 60. *Asymmetric bifurcation.* Consider the structure shown in Fig. 126. The structure is similar to the previous examples except that the top is restrained from lateral motion by an elastic guy, modeled by a translational spring anchored at a distance $a\ell$ from the base. Again, the force on the system is the vertical force P. The translational spring accrues force F in proportion to its extension Δ, i.e., $F = k\Delta$, and constitutes the elasticity of the system. Recall that the potential energy of the spring is $\frac{1}{2}k\Delta^2$.

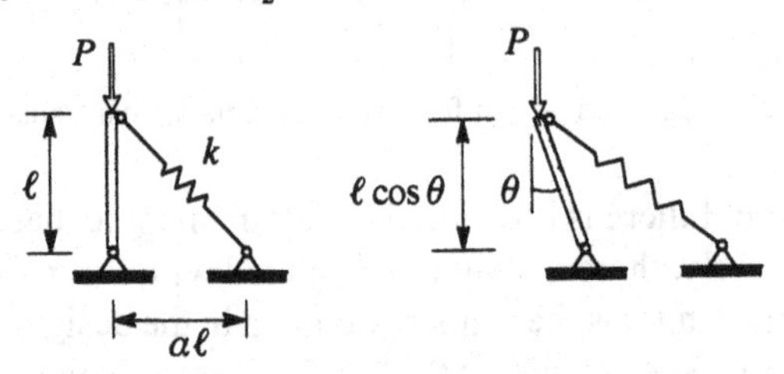

Figure 126 A rigid column with an elastic guy

The energy, virtual-work functional, and second-derivative functional are

$$\mathcal{E}(\theta) = \tfrac{1}{2}k\big(L(\theta)-L_o\big)^2 + P\ell\cos\theta$$

$$G(\theta,\bar{\theta}) = \left[ak\ell^2\big(1-\lambda(\theta)\big)\cos\theta - P\ell\sin\theta\right]\bar{\theta}$$

$$A(\theta,\bar{\theta}) = \left[ak\ell^2\big(\gamma\lambda^3(\theta)\cos^2\theta - (1-\lambda(\theta))\sin\theta\big) - P\ell\cos\theta\right]\bar{\theta}^2$$

where the current length $L(\theta)$ of the spring is given by the expression

$$L(\theta) \equiv \ell\sqrt{1+a^2+2a\sin\theta}$$

$L_o = L(0) = \ell\sqrt{1+a^2}$ is the initial length of the spring, $\lambda(\theta) \equiv L_o/L(\theta)$ is the inverse of the stretch of the spring, and $\gamma \equiv a/(1+a^2)$. Note that the rate of change of L is $L'(\theta) = a\ell^2\cos\theta/L(\theta)$. As usual, if $G(\theta,\bar{\theta}) = 0$ for all $\bar{\theta}$ then equilibrium holds. Therefore, the guyed column is in equilibrium if

$$ak\ell^2\big(1-\lambda(\theta)\big)\cos\theta - P\ell\sin\theta = 0 \tag{535}$$

The straight configurations $\theta = 0$ and $\theta = \pm\pi$ are equilibrium configurations for all values of the load since $L(0) = L(\pm\pi) = L_o$ for those cases. The bent configuration is in equilibrium only for the loads

$$P = ak\ell\left(\frac{1-\lambda(\theta)}{\tan\theta}\right)$$

For the straight configuration we have $\theta = 0$ and $\lambda = 1$, from which we can observe that the critical load is

$$P_{cr} = a\gamma k\ell = \frac{a^2k\ell}{1+a^2}$$

At $\pm\pi$ we also have bifurcation points with critical loads equal to $-P_{cr}$. We can observe that, in the limit as $a \to \infty$, the critical load approaches that of the column with the horizontal spring. As $a \to 0$, the critical load approaches zero.

The bifurcation diagram for the geometrically perfect column is shown in Fig. 127 for the specific case of $a = 1.2$.

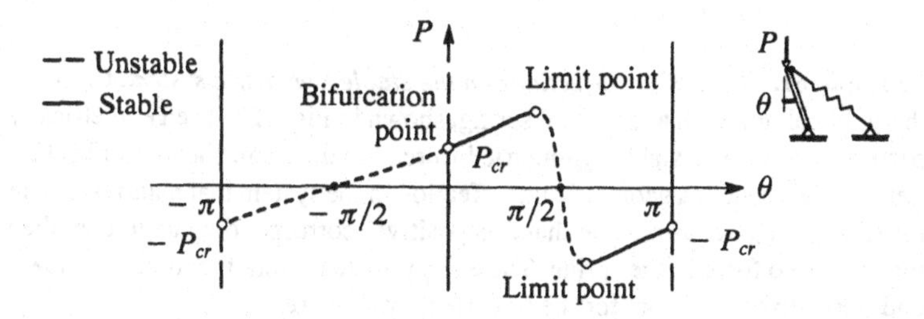

Figure 127 Bifurcation diagram for rigid column with an elastic guy (perfect case)

The bifurcation diagram in Example 60 shows some very interesting features. We can observe that the bifurcation diagram is not symmetric, unlike the previous two examples. Clearly, the behavior of the system is different if the column moves to the right, as opposed to the left. This lack of symmetry manifests in a nonzero slope of the equilibrium path in the neighborhood of the bifurcation point. The straight configuration is stable for all values of the load less than the critical load P_{cr} and unstable for all values of the load greater than P_{cr}, like the symmetric systems. The branch to the right (positive θ) of the bifurcation point is stable (at least for a while), while the branch to the left (negative θ) is unstable. The system exhibits two limit points on the equilibrium path for positive values of θ, and crosses the axis (zero load) at the positions $\pi/2$ and $-\pi/2$.

The Effect of Imperfections

The preceding example assumed that the initial position of the column was perfectly straight. In reality, there is no such thing as a geometrically perfect system. Imperfections can manifest in many ways. The geometry may be imperfect, the load may be imperfectly placed or directed, the material properties may be imperfectly distributed, and the boundary conditions may be imperfectly implemented. Thus, the study of imperfections is complicated for even the simplest system. Throughout this chapter, we will focus on the imperfection in the initial geometry of the system to get an idea of the effects of imperfections.

We shall see that some systems are sensitive to imperfections. A relatively small perturbation in the geometry leads to a relatively large change in the response. Linear systems are not generally sensitive to imperfections and hence, this sort of analysis is not commonly done for linear systems. Not all nonlinear systems are sensitive to imperfections.

Let us reconsider our three example systems to see how geometric imperfections affect the response of the system.

Example 61. *Effect of imperfections on the stable symmetric system.* Consider the rigid column with a rotational spring, shown in Fig. 122. The geometrically perfect system has a stable symmetric bifurcation diagram, shown in Fig. 123. Let us now consider a geometric imperfection in the system that manifests as an initial angle θ_o (let us assume that it is positive) corresponding to zero applied load and zero force in the spring. The energy, virtual-work functional, and second-derivative functional for the imperfect system are

$$\mathcal{E}(\theta) = \tfrac{1}{2}k(\theta-\theta_o)^2 + P\ell\cos\theta$$

$$G(\theta,\bar{\theta}) = \left[k(\theta-\theta_o)-P\ell\sin\theta\right]\bar{\theta}$$

$$A(\theta,\bar{\theta}) = \left[k-P\ell\cos\theta\right]\bar{\theta}^2$$

We can see that $\theta = 0$ is no longer a solution. The system is in equilibrium at the deformation θ only if the load has the value

$$P = \frac{k}{\ell}\left(\frac{\theta-\theta_o}{\sin\theta}\right)$$

There are two equilibrium paths that satisfy this expression. These equilibrium paths are shown in Fig. 128 along with those for the perfect system.

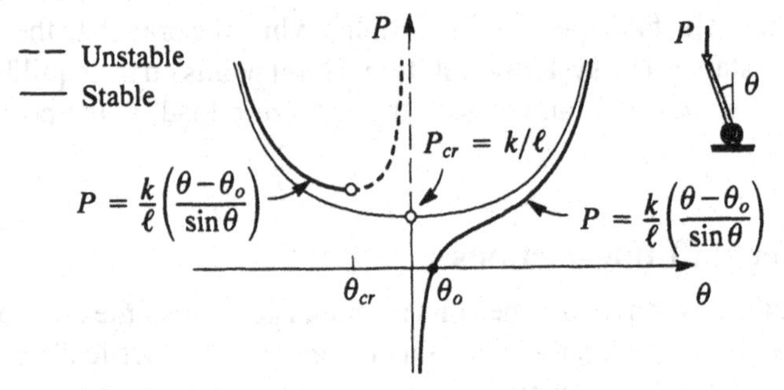

Figure 128 The effect of an imperfection for a rigid column with rotational spring

The second-derivative te . gives

$$A(\theta,\bar{\theta}) = k\left(1+\frac{\theta_o-\theta}{\tan\theta}\right)\bar{\theta}^2 = \begin{cases} > 0 & \text{for } \theta > 0 & \text{(stable)} \\ < 0 & \text{for } \theta_{cr} < \theta < 0 & \text{(unstable)} \\ > 0 & \text{for } \theta < \theta_{cr} & \text{(stable)} \end{cases}$$

where θ_{cr} is the solution to the equation $\theta-\theta_o = \tan\theta$, as illustrated in Fig. 129.

Thus, the equilibrium path that passes through the point of zero load (with positive values of θ) is stable. The path above the secondary path of the perfect system shows a *limit load* (a point of transition from a stable branch to an unsta-

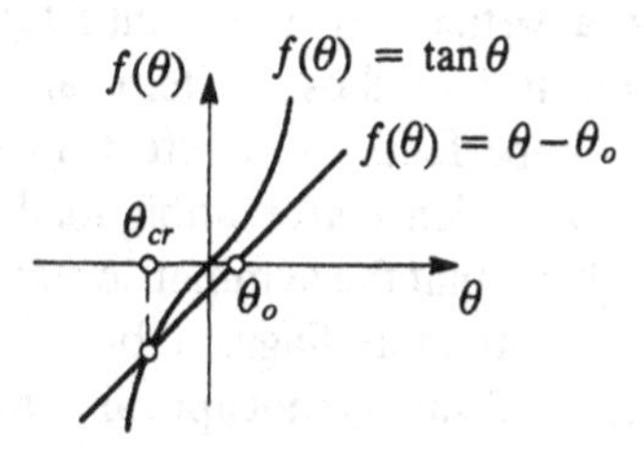

Figure 129 Graphical solution of $\theta - \theta_o = \tan\theta$

ble one without a bifurcation of paths) at θ_{cr}, with the path being stable to the left of the limit load and unstable to the right. Because of the nature of the imperfection, the straight configuration $\theta = 0$ cannot be reached at finite values of the load. As $\theta \to 0$ from the right, the force P goes to large negative (tensile) values, indicating that a tensile axial force cannot completely straighten the initial imperfection.

As $\theta \to 0$ from the left, the load P takes large positive values, indicating that if we could get to a configuration with negative values of θ (we could force it over with a lateral force and then remove the lateral force), it would take an infinitely large compressive load to keep it from snapping through to the other side if it got close enough to the straight configuration. On the other hand, if the compressive load was large enough and the system was bent enough, equilibrium in the bent position would be quite stable. In either case, as θ increases, the equilibrium path is asymptotic to the perfect path.

There is no bifurcation in the imperfect system in this example. Note the presence of the term $a_1 = -k\theta_o\theta$ (the linear term) in the energy. Hence, there is no meaning to the concept of a critical load in the sense that we have been using it. However, we can clearly see that the equilibrium paths for the perfect system provide a backbone to the imperfect system. The smaller the imperfections are, the closer the imperfect paths hug the perfect ones. The critical load roughly represents the point in the imperfect curve where the system transitions from a relatively stiff to a relatively flexible system. The critical load is an indicator of the value at which buckling starts to progress rapidly.

The previous example demonstrates many of the features that are important to problems of stability of equilibrium. It illustrates the juxtaposition of the perfect system and the imperfect one. It illustrates that, even for this simple system, there are solutions you might never imagine. In fact, we have not found all of the possible solutions here. There are other solutions that correspond to complete windings of the rotational spring. We have defined the important concepts of bifurcation point and limit point. We have illustrated the role of the bifurcation diagram. All of these concepts will carry over to the case where equilibrium is governed by differential equations rather than algebraic ones. We can use these simple systems as sounding boards for the more complicated cases where we might not be able to make as much analytical headway.

Imperfections have a particularly important effect on systems with unstable post-buckling behavior. In fact, these systems are called *imperfection sensitive.* The sensitivity to imperfections manifests in a limit point having a limit load that is lower than the critical (bifurcation) load of the associated geometrically perfect system. (Note that the symmetric stable system had a limit point on the left side of the bifurcation diagram, but it was associated with a load greater than the critical load and did not appear to be reachable from the initial unloaded state.) The reduction in load carrying capacity in imperfection sensitive structures can be substantial.

Example 62. *Effect of imperfections on the unstable symmetric system.* Consider the rigid column with a translational spring, shown in Fig. 124. The geometrically perfect system has an unstable symmetric bifurcation diagram, shown in Fig. 125. Consider a geometric imperfection in the system that manifests as an initial angle θ_o corresponding to zero applied load and zero force in the spring. The energy, virtual-work functional, and second-derivative functional are

$$\begin{aligned}
\mathcal{E}(\theta) &= \tfrac{1}{2}k\ell^2(\sin\theta - \sin\theta_o)^2 + P\ell\cos\theta \\
G(\theta,\bar{\theta}) &= \left[k\ell^2(\sin\theta - \sin\theta_o)\cos\theta - P\ell\sin\theta\right]\bar{\theta} \\
A(\theta,\bar{\theta}) &= \left[k\ell^2\left(\cos^2\theta - \sin^2\theta + \sin\theta\sin\theta_o\right) - P\ell\cos\theta\right]\bar{\theta}^2
\end{aligned}$$

We can see that $\theta = 0$ is no longer a solution. The system is in equilibrium at the deformation θ only if the load has the value

$$P = k\ell\left(1 - \frac{\sin\theta_o}{\sin\theta}\right)\cos\theta$$

There are two equilibrium paths that satisfy this expression. These equilibrium paths are shown in Fig. 130 along with those for the perfect system.

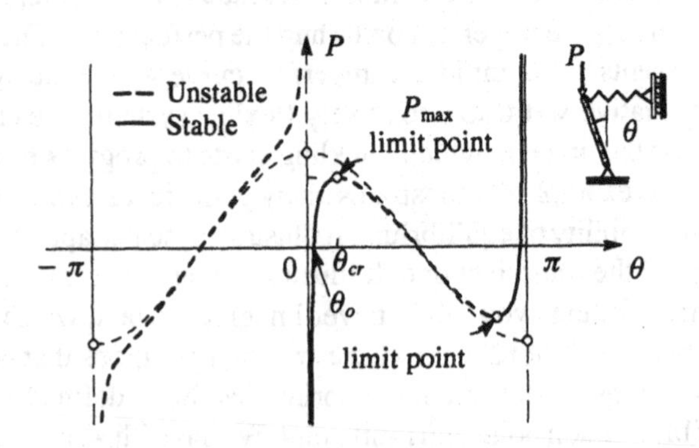

Figure 130 The effect of an imperfection on the rigid column with translational spring

The second-derivative test gives

$$A(\theta,\bar{\theta}) = k\ell^2\left(\frac{\sin\theta_o}{\sin\theta} - \sin^2\theta\right)\bar{\theta}^2 = \begin{cases} < 0 \text{ for } \theta < 0 & \text{(unstable)} \\ > 0 \text{ for } 0 < \theta < \theta_{cr} & \text{(stable)} \\ < 0 \text{ for } \theta_{cr} < \theta < \pi - \theta_{cr} & \text{(unstable)} \\ > 0 \text{ for } \pi - \theta_{cr} < \theta < \pi & \text{(stable)} \end{cases}$$

where θ_{cr} is the solution to the equation $\sin\theta = \sin^{1/3}\theta_o$ (which gives the value of θ that makes the second derivative equal to zero). The equilibrium path reaches a maximum load-carrying capacity at the limit point. The maximum load can be computed by substituting θ_{cr} into the expression for the load. We shall call the maximum load, or limit load, $P_{max} = P(\theta_{cr})$. It depends upon the initial imperfection in the following way

$$P_{max} = P_{cr}\left[1 - \sin^{2/3}\theta_o\right]^{3/2}$$

where $P_{cr} = k\ell$ is the critical load of the perfect system.

We can clearly see that the imperfection tends to reduce the limit load for this type of system. The greater the imperfection, the greater the reduction. In the present example, if the imperfection is only 1°, the maximum load is reduced by 10% from the perfect critical load. If the imperfection is 5°, the maximum load is reduced by 30% from the perfect critical load. The exponent of 2/3 is significant, giving the two-thirds power law of Koiter. The result that we have here came from a straightforward computation with this specific system, but it has a much greater significance. According to Koiter, any system that experiences a symmetric bifurcation with unstable post-buckling behavior for the perfect system will be sensitive to imperfections, and the reduction in the limit capacity will vary according to the value of the imperfection raised to the two-thirds power.

Upon loading from zero, the initial equilibrium path is stable. If the loading is tensile, the forces act to straighten the bar. As in the previous example, a straight configuration can never be realized by this system of forces. When loaded in compression, the equilibrium path is stable up to the limit point. The system loses stability at that point only to regain it at the second limit point at $\theta = \pi - \theta_{cr}$. Because the equilibrium path between the limit points is unstable, the system will snap to a stable configuration upon reaching a limit load.

Example 63. ***Effect of imperfections on the asymmetric system.*** Consider the rigid column with a translational spring, shown in Fig. 126. The geometrically perfect system has an asymmetric bifurcation diagram, shown in Fig. 127. Let us now consider a geometric imperfection in the system that manifests as an initial angle θ_o corresponding to zero applied load and zero force in the spring. All of the equations of Example 60 are valid for the imperfect case if we redefine the original length as

$$L_o \equiv \ell\sqrt{1+a^2+2a\sin\theta_o}$$

The bifurcation diagram for the geometrically imperfect column is shown in Fig. 131 for the value of the imperfection of $\theta_o = 0.2$. The perfect case is also plotted in a lighter line weight to illustrate the connection between the perfect and imperfect cases. This diagram shows features typical of the unstable symmetric case to the right and features of the stable symmetric case to the left. As is typical of these systems, the bifurcation point does not manifest for the imperfect case, but the geometrically perfect case provides a backbone curve to which the imperfect case is asymptotic. The imperfect case exhibits six limit points, and shows a peculiar departure from the perfect system in the neighborhood of $\theta = \pm\pi$.

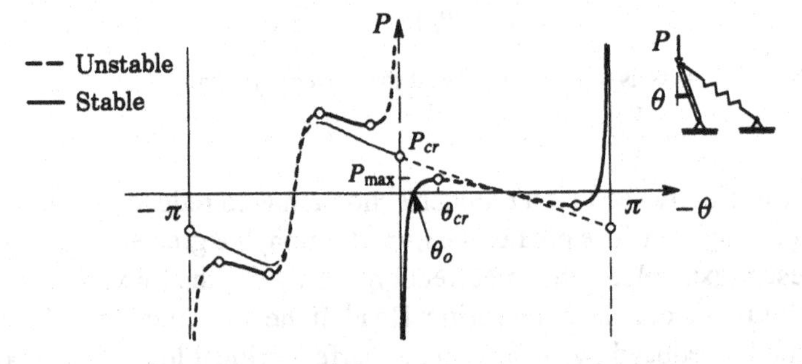

Figure 131 Bifurcation diagram for rigid column with an elastic guy (imperfect case)

In a typical circumstance, we might have an imperfect system with the load level initially at zero and increased to positive values. In such a circumstance, the value of the maximum load P_{max}, which occurs at the first limit point, is of singular importance. The limit point occurs at the critical angle θ_{cr} and is associated with the load P_{max}. To find this state note that this point has both $G(\theta,\bar{\theta}) = 0$ and $A(\theta,\bar{\theta}) = 0$. For this system that implies

$$ak\ell^2\big(1-\lambda(\theta_{cr})\big)\cos\theta_{cr} - P_{max}\ell\sin\theta_{cr} = 0$$

$$ak\ell^2\big(\gamma\lambda^3(\theta_{cr})\cos^2\theta_{cr} - (1-\lambda(\theta_{cr}))\sin\theta_{cr}\big) - P_{max}\ell\cos\theta_{cr} = 0$$

The solution of these two equations yields $\{\theta_{cr}, P_{max}\}$. To find a closed-form solution for the maximum load is not practical even for this simple system. However, we can develop an approximate formula for the maximum load by expanding the terms in the equation in a Taylor series for θ_{cr}, keeping only the first few terms. Carrying out these operations for the present example, we get a critical angle of

$$\theta_{cr} \approx \sqrt{\frac{2\theta_o}{3\gamma}}, \qquad P_{max} \approx a\gamma\lambda^3(\theta_{cr})\cos^3\theta_{cr}$$

We can substitute this value of the critical angle into Eqn. (535) to compute the maximum load with good accuracy. For the present case, we get $\theta_{cr} = 0.52$ and $P_{max} = 0.369 P_{cr}$. (Note the significant reduction from the bifurcation load!)

We can also get an approximate expression for the maximum load by substituting the above value of the critical load into a Taylor series expansion for P_{max}. The first few terms give

$$P_{max} \approx P_{cr}\left(1 - \sqrt{6\gamma\theta_o} + \frac{1}{3\gamma}\left(8\gamma^2 - 1\right)\theta_o\right)$$

for the maximum load. These expressions are good for any value of α, but only for relatively small values of θ_o (in fact, for $\theta_o = 0.2$ we get $0.359 P_{cr}$). The expression for the maximum serves to illustrate Koiter's half-power law. The maximum load of an imperfect system is a reduction from the critical load of the perfect system. According to Koiter, the dominant term in that reduction is proportional to the value of the imperfection raised to the one-half power for asymmetric systems. Contrast this result with the two-thirds power law of Koiter for symmetric systems.

The Role of Linearized Buckling Analysis

For the simple, single-degree-of-freedom systems we have just analyzed, the equations were amenable to algebraic manipulation, and we were able to get closed-form solutions for the nonlinear equilibrium paths. For more complicated systems, a closed-form solution is rarely possible, and we must resort to numerical computations. There will be few cases where we cannot trace the equilibrium paths of a system by taking small increments along the path and solving the nonlinear equations with Newton's method, but these numerical solutions do not always give the same crisp insight as an analytical solution.

There is a parcel of middle ground on this issue that has been exploited for centuries in the solution of buckling problems: *linearized buckling analysis*. Euler's analysis of column buckling was, in fact, an example of linearized buckling analysis. In this section we shall take a look at what happens to our two symmetric systems when we subject them to a linearized analysis. Our aim is to find out what we retain and what we have given away in the linearization process.

A linearized buckling analysis is one in which the equations of equilibrium are linear in the deformation variable. For a stability problem, the equilibrium equations will invariably involve the product of the load parameter and the deformation variable. In order to have linear equilibrium equations, the potential energy must be quadratic (a linear energy does not have an extremum, and, hence, would not give rise to equilibrium configurations of any kind). For the examples discussed above, we can approximate the trigonometric functions with second-order polynomials (using Taylor series expansions) to get quadratic energies.

Example 64. *Linearized buckling analysis of the stable symmetric bifurcation.* Consider the rigid column with a rotational spring, shown in Fig. 122. For a linearized buckling analysis we express the energy functional only up to quadratic terms. Let us truncate the Taylor series approximation at quadratic terms

$$\cos\theta = 1-\tfrac{1}{2}\theta^2 + \tfrac{1}{24}\theta^4 - O(\theta^6) \approx 1-\tfrac{1}{2}\theta^2$$

The energy, virtual-work functional, and second-derivative functional are

$$\begin{aligned} \mathcal{E}(\theta) &= \tfrac{1}{2}k\theta^2 + P\ell\left(1-\tfrac{1}{2}\theta^2\right) \\ G(\theta,\bar{\theta}) &= \left[k-P\ell\right]\theta\,\bar{\theta} \\ A(\theta,\bar{\theta}) &= \left[k-P\ell\right]\bar{\theta}^2 \end{aligned}$$

The equilibrium equation for the system is, therefore, $[k-P\ell]\theta = 0$. This equation is satisfied for any load P if $\theta = 0$ (i.e., the straight configuration). The stability of this configuration can be assessed from the second derivative to show that the straight configuration is stable for loads $P < k/\ell$ and is unstable if $P > k/\ell$.

The equilibrium equations are also satisfied for any value of θ if $P = k/\ell$. The second derivative is exactly zero for $P = k/\ell$ so the stability of this branch cannot be determined.

We can see from the previous example that the linearized analysis gives a complete picture of the stability of the straight configuration and it tells us that the system will buckle into a bent configuration at the critical load $P_{cr} = k/\ell$. However, the linearized buckling analysis yields no information on the post buckling response. In particular, it is unable to distinguish stable from unstable post buckling response.

Example 65. *Linearized buckling analysis of the unstable symmetric bifurcation.* Consider the rigid column with a translational spring, shown in Fig. 124. For a linearized buckling analysis we express the energy functional up to quadratic terms. Let us truncate the Taylor series approximations at quadratic terms

$$\cos\theta \approx 1-\tfrac{1}{2}\theta^2, \qquad \sin\theta \approx \theta + O(\theta^3)$$

The energy, virtual-work functional, and second-derivative functional are

$$\begin{aligned} \mathcal{E}(\theta) &= \tfrac{1}{2}k\ell^2\theta^2 + P\ell\left(1-\tfrac{1}{2}\theta^2\right) \\ G(\theta,\bar{\theta}) &= \left[k\ell^2-P\ell\right]\theta\,\bar{\theta} \\ A(\theta,\bar{\theta}) &= \left[k\ell^2-P\ell\right]\bar{\theta}^2 \end{aligned}$$

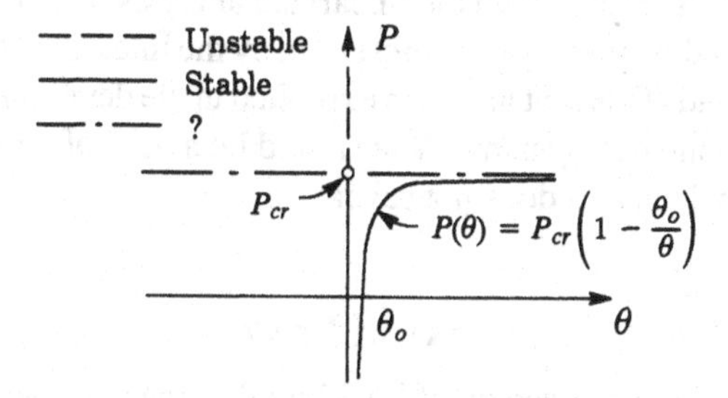

Figure 132 The effect of an imperfection in a linearized analysis

The equilibrium equation for the system is, therefore, $\left[k\ell^2 - P\ell\right]\theta = 0$. This system yields results identical to Example 64 except that the critical load for the present case is $P_{cr} = k\ell$. All of the same conclusions apply.

The second-derivative test is able to tell us everything we knew before about the straight configurations. In particular, loads below the critical load are stable for both systems, and loads above the critical load are unstable for both systems (the critical loads are different for the two systems). What we lose in linearizing the analysis is all of the information about the behavior in the bent configuration. The solution of the equilibrium equations suggests that for $P = P_{cr}$, the equations are satisfied for any value of θ. We plot this equilibrium branch as a horizontal line on the bifurcation diagram, as shown in Fig. 132. This equilibrium path is dubious because our assumption made in linearizing ceases to hold as we get further from the P axis. The second-derivative test is unable to tell us whether this branch is stable or unstable. It does, however, point in the right direction initially (that is, it predicts a symmetric bifurcation).

The effect of imperfections can be examined through a linearized buckling analysis also. The equilibrium equation for both cases turns out to be (with the critical load suitably interpreted for the two cases)

$$P(\theta) = P_{cr}\left(1 - \frac{\theta_o}{\theta}\right)$$

where $P_{cr} = k\ell$ for the translational spring, and $P_{cr} = k/\ell$ for the rotational spring. All of these results are summarized in Fig. 132. The imperfect system is asymptotic to the perfect one, but because the perfect system does not give adequate information on the post-buckling behavior, the response of the imperfect system is actually asymptotic to the wrong response. The linearization limits the applicability of the analysis to small values of the angle θ. It is important to observe that the linearized buckling analysis does not yield information about the imperfection sensitivity of the system and cannot predict the maximum load of an imperfection sensitive structure like the one in Example 65.

Clearly, we lose a great deal in the linearized analysis. However, one of the key features of a stability analysis is preserved by the linearized buckling analysis: the critical load. This fact has been exploited in the development of design formulas for complicated systems. You should be aware of what a linearized buckling analysis does and does not reveal.

Systems with Multiple Degrees of Freedom

There is another important aspect of buckling that the single-degree-of-freedom systems do not exhibit. Because the system has only one degree of freedom, it has little choice as to how it will deform; the only issue is whether or not it will deform. For systems with more than one degree of freedom, additional possibilities arise. These systems give rise to multiple bifurcation points and buckling modes. They also make the second-derivative test a little more interesting.

To see some of the aspects of the stability of systems with more than one degree of freedom, we shall consider the following example of a discrete two-degree-of-freedom system.

Example 66. *System with two degrees of freedom.* Consider the system shown in Fig. 133. The structure is composed of two rigid links of length ℓ, like the previous two examples, cantilevered from the base. The elasticity of the system is manifested in two rotational springs both with modulus k. The motion of the system is completely characterized by the two independent rotations θ_1 and θ_2. We often will refer to the deformation with the vector $\boldsymbol{\theta}^T \equiv \{\theta_1, \theta_2\}$. Let us take datum for the potential energy of the load P to be the ground. The potential energy in the undeformed state is then $2P\ell$.

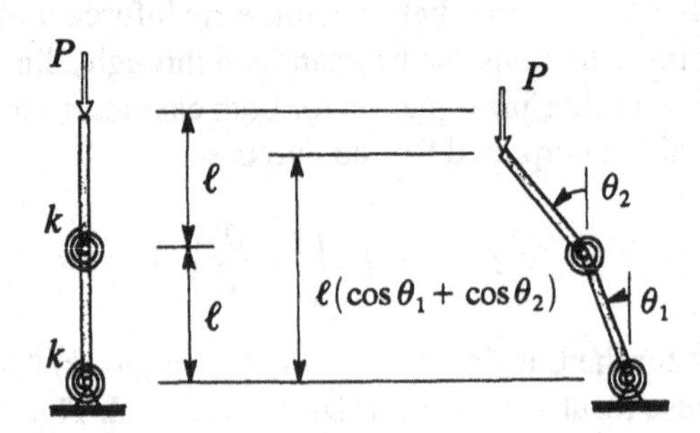

Figure 133 A two-degree-of-freedom example

The potential energy of the system, in a deformed position, can be written as

$$\mathcal{E}(\boldsymbol{\theta}) = \tfrac{1}{2}k(\theta_1)^2 + \tfrac{1}{2}k(\theta_2 - \theta_1)^2 + P\ell(\cos\theta_1 + \cos\theta_2)$$

The directional derivative of the energy in the direction $\bar{\boldsymbol{\theta}}^T = \{\bar{\theta}_1, \bar{\theta}_2\}$ gives the virtual-work function

$$G(\boldsymbol{\theta},\bar{\boldsymbol{\theta}}) = \left[2k\theta_1 - k\theta_2 - P\ell\sin\theta_1\right]\bar{\theta}_1 + \left[-k\theta_1 + k\theta_2 - P\ell\sin\theta_2\right]\bar{\theta}_2$$

As usual, $\boldsymbol{\theta}$ is an equilibrium configuration if $G(\boldsymbol{\theta},\bar{\boldsymbol{\theta}}) = 0$ for all $\bar{\boldsymbol{\theta}}$. Thus, equilibrium can hold only if the two terms in brackets vanish independently. The equations of equilibrium are, therefore

$$\begin{aligned} 2k\theta_1 - k\theta_2 - P\ell\sin\theta_1 &= 0 \\ -k\theta_1 + k\theta_2 - P\ell\sin\theta_2 &= 0 \end{aligned}$$

The equilibrium equations are a system of two nonlinear algebraic equations in the unknowns θ_1 and θ_2 (assuming that the load level P is given). There are two of them because the system has two degrees of freedom. Like the previous examples, this system has multiple equilibrium paths. We shall find those paths and investigate their stability presently.

The second derivative of the energy is $A(\boldsymbol{\theta},\bar{\boldsymbol{\theta}}) = k\bar{\boldsymbol{\theta}}^T\mathbf{A}(\boldsymbol{\theta},p)\bar{\boldsymbol{\theta}}$, where the load parameter $p \equiv P\ell/k$ is a normalized version of the applied load P, and the Hessian matrix is given by

$$\mathbf{A}(\boldsymbol{\theta},p) = \begin{bmatrix} 2-p\cos\theta_1 & -1 \\ -1 & 1-p\cos\theta_2 \end{bmatrix}$$

As discussed Chapter 9, the stability of the discrete system is judged by the signs of the eigenvalues of the matrix $\mathbf{A}$ at an equilibrium configuration. The eigenvalues of $\mathbf{A}$ are the values λ that satisfy the eigenvalue problem $\mathbf{A}\boldsymbol{\phi} = \lambda\boldsymbol{\phi}$. Since the dimension of the matrix is two, there are two eigenvalues. Corresponding to each of these eigenvalues is an eigenvector $\boldsymbol{\phi}$. We shall see the significance of the eigenvectors soon.

The bifurcation diagram for this problem is shown in Fig. 134. Note that the equilibrium paths in three-dimensional space are accentuated by showing their projections on the θ_1 - θ_2 plane and connecting those two curves with a vertical curtain. This curtain is only for help in visualizing the three-dimensional curve. The equilibrium path is always a line in space. The bifurcation diagram shows several features that the one-dimensional case did not possess. In particular, it has two bifurcation points with symmetric branches emanating from each. Further, all deformations have a shape, dictated by the relative proportions of θ_1 and θ_2. The shape of the deformation changes as we move along an equilibrium path.

Clearly, the straight configuration $\boldsymbol{\theta}^T = \{0,0\}$ is a solution for all values of the load parameter p. The stability can be judged by the eigenvalues of the matrix $\mathbf{A}(\mathbf{0},p)$. A straightforward computation shows that these eigenvalues are

$$\lambda_1 = \tfrac{1}{2}\left(3-\sqrt{5}\right) - p, \quad \lambda_2 = \tfrac{1}{2}\left(3+\sqrt{5}\right) - p$$

The system is stable for values of the load parameter $p < \frac{1}{2}(3-\sqrt{5})$, and unstable for $p > \frac{1}{2}(3-\sqrt{5})$. These results are summarized as Branch 0 on the bifurcation diagram shown in Fig. 134. It would appear that λ_2 has no significance to the question of stability of the system, since a single negative eigenvalue is sufficient to conclude that the system is unstable, and $\lambda_2 > \lambda_1$ for all values of the load p. However, we shall see that both eigenvalues are important to the question of the bifurcation of equilibrium at critical points.

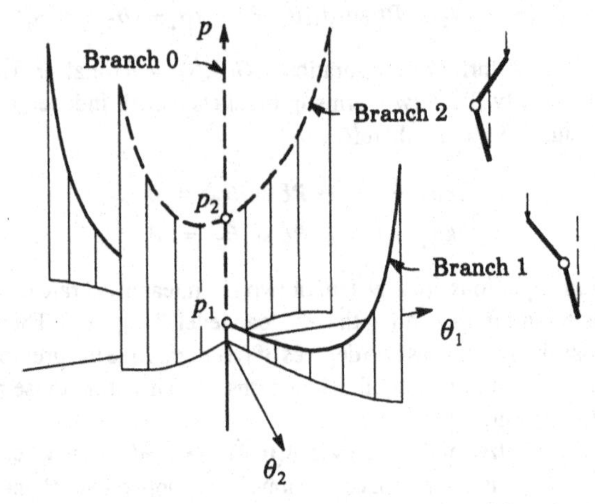

Figure 134 Bifurcation diagram for the two-degree-of-freedom example

Consider the bifurcation diagram shown in Fig. 134. At $p = 0$, both λ_1 and λ_2 are positive, as they should be for stability. These two eigenvalues get more positive for negative (tensile) values of p, indicating more robust stability.† As the load is increased from zero in the positive (compressive) direction, the values of λ_1 and λ_2 begin to decrease. We reach a critical point when the first eigenvalue goes to zero. The critical point defined by $\lambda_1 = 0$, corresponding to a load value of $p_1 = (3-\sqrt{5})/2$, is a bifurcation point at the boundary between stable and unstable behavior in the straight position. If we continue to increase p along the straight configuration, λ_1 becomes negative and gets increasingly negative. The second eigenvalue λ_2 is still positive, but continues to decrease. The condition $\lambda_2 = 0$, corresponding to a load value of $p_2 = (3+\sqrt{5})/2$, defines another critical point where bifurcation can occur. For load values of $p > p_2$, both of the eigenvalues are negative and get increasingly negative as p increases. So much for the straight configuration. What about bifurcations to bent configurations?

As we compute Branch 1 and Branch 2, we can evaluate the eigenvalues of $\mathbf{A}(\boldsymbol{\theta}, p)$ to monitor the stability of those branches. On Branch 1, which emanates from the critical point where $\lambda_1 = 0$ and $\lambda_2 > 0$, both eigenvalues become increasingly positive the further out on the branch you go. Thus, Branch 1 is stable and gets increasingly so. On Branch 2, which emanates from the critical point where $\lambda_1 < 0$ and $\lambda_2 = 0$, λ_2 regains its positivity and becomes increasingly positive the further out on the branch one goes. However, λ_1 remains negative. Thus Branch 2 is unstable.

---

† According to our criterion, stability is like being pregnant: Either you are, or you are not. In reality, we can consider stability (and pregnancy) to be a matter of degree. Some configurations are more stable than others. The problem lies not with the systems, but with our definition of stability. The degree of instability can best be understood in a dynamic setting.

The effect of imperfections. The analysis of systems with multiple degrees of freedom is essentially the same as for system with a single degree of freedom except that there is more variety to the way in which the geometric imperfections can manifest. In a system with multiple degrees of freedom there can be an imperfection parameter associated with each degree of freedom. Often the imperfection is simply taken as a nonzero initial value of the response parameter itself, as the following example illustrates.

Example 67. *Effects of imperfections for the system with two degrees of freedom.* Reconsider the system shown in Fig. 133. Let us examine the effects of imperfections in the system. Here we find an interesting feature that was not present in the one-dimensional problems. The imperfections must be specified as a pair of values $\boldsymbol{\theta}_o^T = \{\theta_{o1}, \theta_{o2}\}$. The key question now is: What is the behavior of the equilibrium path for the imperfect system? The energy for the imperfect system can be written as

$$\mathcal{E}(\boldsymbol{\theta}) = \tfrac{1}{2}k\big(\theta_1 - \theta_{o1}\big)^2 + \tfrac{1}{2}k\big[\big(\theta_2 - \theta_{o2}\big) - \big(\theta_1 - \theta_{o1}\big)\big]^2 + P\ell\big(\cos\theta_1 + \cos\theta_2\big)$$

where we continue to measure the angles of rotation from the vertical position, and have adjusted the expression for the energy of the springs to be zero at the point where the rotations are exactly equal to the initial values.

The virtual-work functional is $G(\boldsymbol{\theta}, \bar{\boldsymbol{\theta}}) = D\mathcal{E}(\boldsymbol{\theta}) \cdot \bar{\boldsymbol{\theta}}$. The equilibrium equations can be obtained from setting $G(\boldsymbol{\theta}, \bar{\boldsymbol{\theta}}) = 0$ for all $\bar{\boldsymbol{\theta}}$. This process results in the equations

$$\begin{aligned} 2k\big(\theta_1 - \theta_{o1}\big) - k\big(\theta_2 - \theta_{o2}\big) - P\ell\sin\theta_1 &= 0 \\ -k\big(\theta_1 - \theta_{o1}\big) + k\big(\theta_2 - \theta_{o2}\big) - P\ell\sin\theta_2 &= 0 \end{aligned}$$

As in the previous examples, $\boldsymbol{\theta}^T = \{0,\ 0\}$ is no longer a solution. In fact, as the load is increased in the tensile (negative p) direction, the system asymptotically approaches the straight configuration. At the configuration $\boldsymbol{\theta} = \boldsymbol{\theta}_o$, the load p must be zero. As the load increases in the compressive direction, our intuition tells us that the equilibrium path should eventually grow close to the equilibrium path of the perfect system. However, now we have two such paths. If the initial imperfection is in the direction of the first eigenvector, we would expect the equilibrium path to approach Branch 1 asymptotically, as indeed happens.

What happens if the initial imperfection is in the direction of the second mode? The result of one case for the previous example is shown in Fig. 135. The equilibrium path shows a very brief inclination to follow the second path, but soon unwinds to a mode that resembles the first mode and proceeds to try to follow Branch 1. We can speculate that the urge to follow Branch 2 initially will depend upon how close the second critical point is to the first. If it is too far away, the attraction of the higher path is small. If it is close, its attraction is greater. In the present case, the two critical points are rather far from each other, and, thus, the urge to follow Branch 2 is almost imperceptible. (Examination of some imperfections closer to mode 2 reveals this phenomenon under magnification.) Although it is impossible for this structure because of its geometry, we might

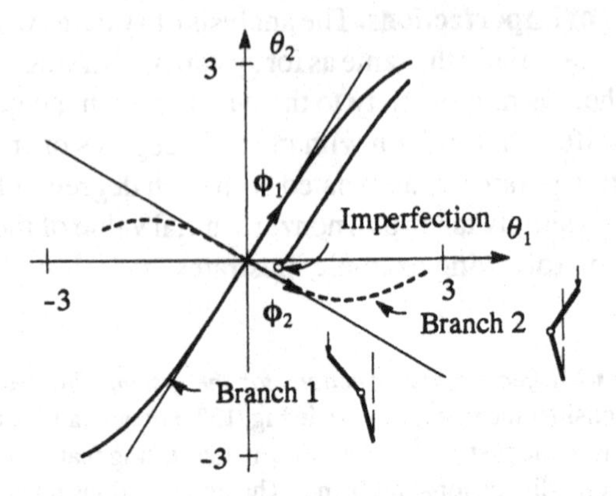

Figure 135 The effect of an initial imperfection for the two-degree-of-freedom example

also ask what would happen if the two critical points were very close together (or, in the limit, on top of each other).

Linearized buckling of MDOF systems. From our earlier examples, we suspect that we will have a bifurcation of equilibrium at the critical points, and this is indeed true. For the previous examples, we found a closed-form expression for the load as a function of θ, and took the limit as $\theta \to 0$ to see that equilibrium bifurcates from the straight configuration at the critical load. For the present case, we cannot proceed in the same manner because we cannot write those closed-form expressions. However, we can appeal to the linearized buckling problem to see if the solution branches at those points. For small values of θ_1 and θ_2, the equations of equilibrium reduce to the eigenvalue problem

$$\begin{bmatrix} 2 & -1 \\ -1 & 1 \end{bmatrix} \begin{Bmatrix} \theta_1 \\ \theta_2 \end{Bmatrix} = p \begin{Bmatrix} \theta_1 \\ \theta_2 \end{Bmatrix}$$

As usual, we can still see that $\boldsymbol{\theta}^T = \{0, 0\}$ is a solution for all values of the load parameter p, Branch 0 on the bifurcation diagram. However, this equation is a linear eigenvalue problem, and thus suggests that there may be other solutions for certain values of p. It should be quite clear that this eigenvalue problem is very closely related to the one we solved to determine the stability of equilibrium of the straight configuration. The eigenvalues and eigenvectors of this system are easily found to be

$$p_1 = \tfrac{1}{2}\left(3 - \sqrt{5}\right), \quad \boldsymbol{\Phi}_1 = c_1 \begin{Bmatrix} 2 \\ 1 + \sqrt{5} \end{Bmatrix}$$

$$p_2 = \frac{1}{2}\left(3+\sqrt{5}\right), \quad \boldsymbol{\phi}_2 = c_2\left\{\begin{array}{c} 2 \\ 1-\sqrt{5} \end{array}\right\}$$

where the constants c_1 and c_2 appear in the eigenvectors to remind us that the magnitude of these vectors cannot be established from the eigenvalue problem, only the direction. The eigenvalues are just the values of the load p at the critical points found earlier. The eigenvectors represent the direction that displacement must occur in order to satisfy equilibrium at these loads. As such, these eigenvectors represent the tangents to the equilibrium branches at the critical points. These vectors are plotted in Fig. 135, which shows the projection of the equilibrium paths on the $\theta_1 - \theta_2$ plane (load p is normal to the page). The equilibrium equations are really nonlinear, so the paths do not remain straight, but the eigenvectors initially point in exactly the right direction.

The most interesting feature of the eigenvectors $\boldsymbol{\phi}_1$ and $\boldsymbol{\phi}_2$ is that they give a specific shape into which the structure must buckle at the critical load. No other shape is possible. This feature of the multidimensional problem clearly is not captured by the one-dimensional problem because the latter problem has no freedom in the shape of deformation, while the former problem does. The shapes are qualitatively sketched in the figures, and are often called *buckling modes*. The first mode has $\theta_1 > 0$ and $\theta_2 > 0$ (or, by symmetry, both negative), whereas the second mode has $\theta_1 > 0$ and $\theta_2 < 0$ (or, by symmetry, the reverse signs). Furthermore, since they correspond to distinct eigenvalues, these eigenvectors are orthogonal. While the shape of deformation changes along a branch, it retains its original character.

The relationship between the linearized buckling eigenvalue problem and the second derivative test for the trivial configuration can be seen by noting that

$$\begin{aligned} G(\boldsymbol{\theta},\bar{\boldsymbol{\theta}}) &= \bar{\boldsymbol{\theta}}^T\mathbf{g}(\boldsymbol{\theta},\lambda) = \bar{\boldsymbol{\theta}}^T\mathbf{A}(\mathbf{0},\lambda)\,\boldsymbol{\theta} \\ A(\mathbf{0},\bar{\boldsymbol{\theta}}) &= \bar{\boldsymbol{\theta}}^T\mathbf{A}(\mathbf{0},\lambda)\,\bar{\boldsymbol{\theta}} \end{aligned} \tag{536}$$

If we set $G(\boldsymbol{\theta},\bar{\boldsymbol{\theta}}) = 0$ for all $\bar{\boldsymbol{\theta}}$ then we get the equation

$$\mathbf{A}(\mathbf{0},\lambda)\,\boldsymbol{\theta} = \mathbf{0} \tag{537}$$

which is a linear eigenvalue problem. The trivial solution $\boldsymbol{\theta} = \mathbf{0}$ is obviously a solution. There are solutions with $\boldsymbol{\theta} \neq \mathbf{0}$ only if $\det\mathbf{A}(\mathbf{0},\lambda) = 0$, which gives an equation (the characteristic equation) for the load parameter λ for which such nontrivial solutions are possible. The eigenvectors $\boldsymbol{\theta}_i$ that correspond to the values λ_i that are roots of the characteristic equation give the directions in which buckling is possible. These are the *buckling mode shapes*.

The second derivative test says that the system is stable if the eigenvalues of $\mathbf{A}(\mathbf{0},\lambda)$ are all positive. Hence, we set up the eigenvalue problem

$$\mathbf{A}(\mathbf{0},\lambda)\,\boldsymbol{\phi} = \mu\boldsymbol{\phi} \tag{538}$$

We can solve this eigenvalue problem by observing that it is equivalent to the equation

$$\left[\mathbf{A}(0,\lambda)-\mu\,\mathbf{I}\right]\boldsymbol{\phi} = \mathbf{0} \tag{539}$$

This equation has a solution with $\boldsymbol{\phi} \neq \mathbf{0}$ only if $\det\left[\mathbf{A}(0,\lambda)-\mu\,\mathbf{I}\right] = 0$. This characteristic equation can be viewed as an equation for $\mu(\lambda)$. In other words, the eigenvalues of the second derivative matrix are a function of the loading parameter λ. In fact, $\mu(\lambda)$ can be evaluated for any value of λ (that is how we determine the stability of the trivial branch). If we are seeking the bifurcation points then we need to find those places where $\mu(\lambda) = 0$. If $\mu = 0$ then we can observe that Eqn. (538) is identical to Eqn. (537), and the eigenvectors have the same meaning in both cases.

Additional Reading

Z. P. Bazant and L. Cedolin, *Stability of structures: Elastic, inelastic, fracture and damage theories*, Oxford University Press, New York, 1991.

H. L. Langhaar, *Energy methods in applied mechanics*, Wiley, New York, 1962.

Problems

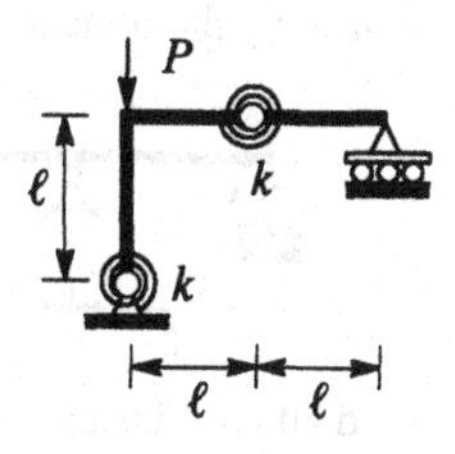

241. The frame shown is composed of two rigid members connected by rotational springs. The moment developed by a spring is related to the rotation by $M = k\phi$, where ϕ is the rotation experienced by the spring. Both springs have modulus k. The frame is subjected to a load P acting downward. The motion of the structure can be completely characterized by the rotation of the vertical member from its original position. Examine the stability of the system. In particular, find the critical load and plot the bifurcation diagram. Note that the bifurcation diagram is not symmetric. Can you explain, in physical terms, why it is not?

242. Examine the effect of an imperfection in the system of Problem 241. Let the imperfection be an initial value of the angle of rotation used to describe the motion, and assume that the springs are such that they have no moment at this initial position. Plot the maximum load versus the size of the initial imperfection.

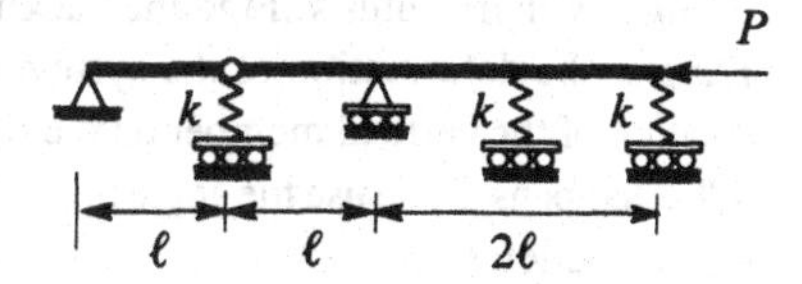

243. Consider the two rigid bars hinged together and subjected to axial load P, as shown. The bars have length ℓ and 3ℓ, and are restrained by three elastic springs, with modulus k, that resist vertical motion. Find all equilibrium paths for the system. Find the bifurcation loads of the system. Assess the stability of the straight and bent configurations.

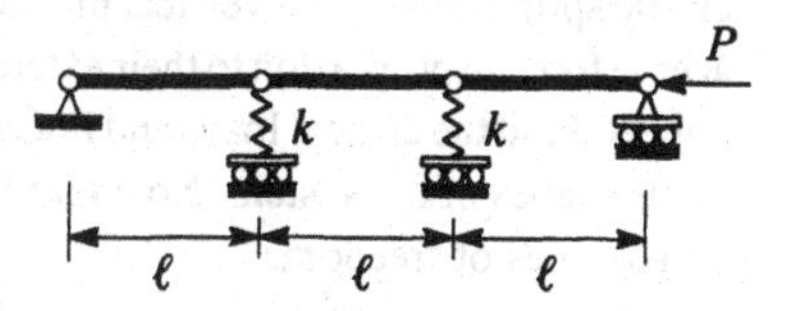

244. Consider the three-bar rigid linkage shown. The bars are hinged together and are restrained by elastic springs that resist vertical motion. The springs accrue force in proportion to their extension, with modulus k. The system is subjected to an axial force P. Write an expression for the potential energy of the system. What are the equations of equilibrium governing the response of the system? Find the critical loads and the buckling mode shapes of the system. Feel free to linearize the geometry of deformation as you see fit.

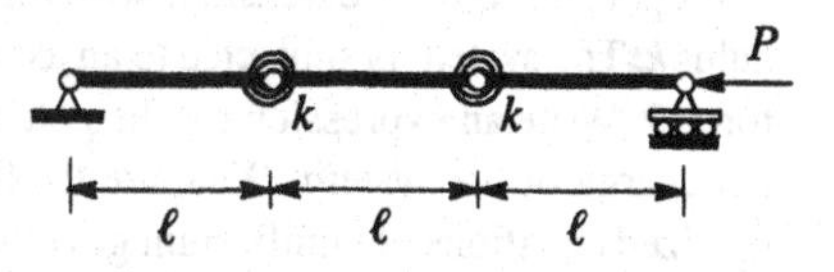

245. Consider the three-bar rigid linkage shown. The bars are hinged together and are restrained by elastic rotational springs. The springs accrue force in proportion to the relative angle of distortion, with modulus k. The system is subjected to an axial force P. Write an expression for the potential energy of the system. What are the equations of equilibrium governing the response of the system? Find the critical loads and the buckling mode shapes of the system. Feel free to linearize the geometry of deformation as you see fit.

246. Consider the rigid bar subjected to axial load P and transverse load εP as shown. The bar has length 2ℓ, is restrained against horizontal and vertical motion at the midpoint, and is supported by two elastic springs that resist vertical motion at the ends. The springs accrue force in proportion to their extension, with modulus k. The deformation can be characterized by the rotation of the bar relative to the horizontal position.

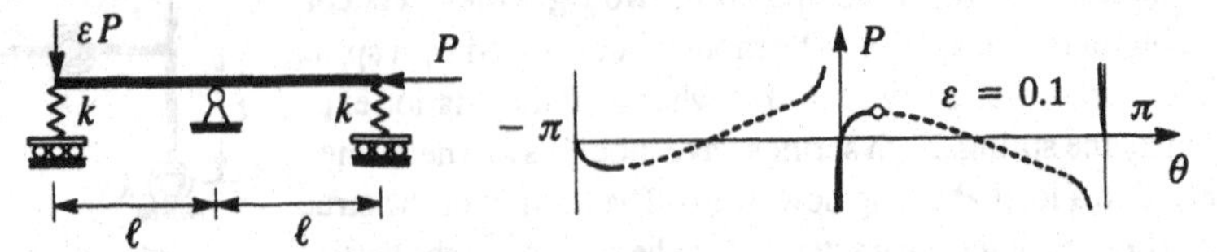

Find all equilibrium paths $P(\theta)$ for the system ($-\pi < \theta < \pi$). Determine the stability of these branches. Find the critical load of the system when $\varepsilon = 0$. Locate the *limit point* on the bifurcation diagram plotted for $\varepsilon = 0.1$. Is the limit load at this point greater or less than the critical load?

247. Consider the frame composed of rigid bars subjected to the load P as shown. The rigid members are hinged at the top right corner, with an elastic spring that resists relative rotation. The rotational spring accrues force in proportion to its relative angle change, with modulus k. Find the buckling load for this system. Express the deformation of the system in terms of the angle of rotation of the vertical member on the right side of the structure. What happens if you use the angle of rotation of the vertical member on the left?

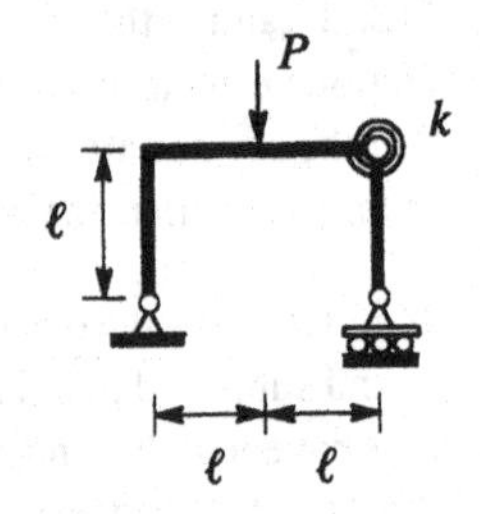

248. Consider the rigid bar subjected to axial load P as shown. The bar has length 2ℓ and is supported by elastic springs that resist vertical motion. The springs accrue force in proportion to their extension, with modulus k. Find the critical loads and linearized buckling mode shapes of the system. Note that this system has two degrees of freedom.

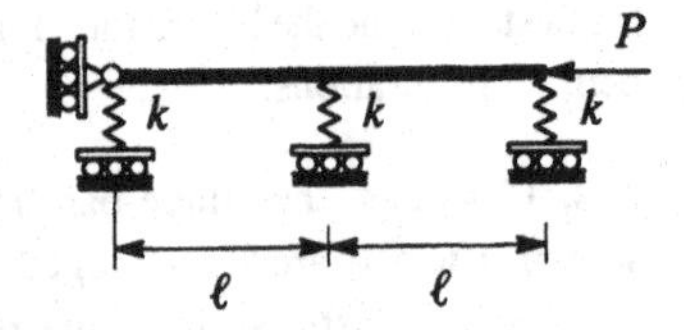

249. Consider the three-bar rigid linkage shown. The bars are hinged together and are restrained by elastic springs that resist vertical motion. The springs accrue force in proportion to their extension, with modulus k. The system is subjected to an axial force P. Write an expression for the potential energy of the system. What are the linearized equations of equilibrium governing the response of the system? Find the critical loads of the system. A convenient choice of degrees of freedom is shown in the diagram.

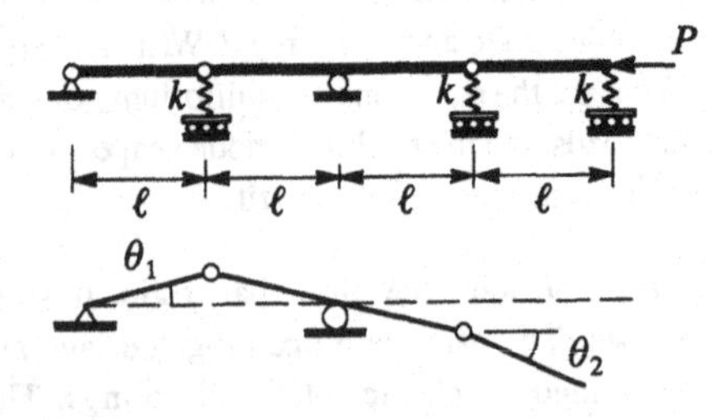

250. Two rigid bars are hinged together and rest on a linearly elastic foundation. The foundation accrues a force per unit length proportional to the transverse displacement, i.e., $f(x) = kw(x)$. The system is subjected to an axial load P as shown. Find an expression for the energy functional for the system. Find an expression for the virtual-work functional for the system. Find the buckling loads of the system by solving the *linearized* buckling eigenvalue problem.

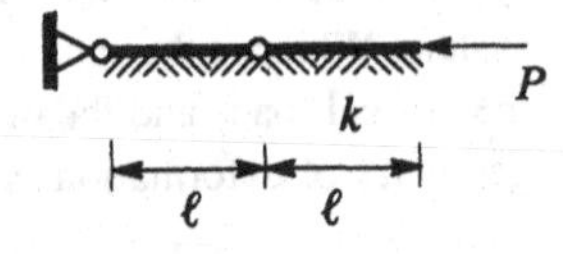

251. The vertical rigid bar is subjected to axial load P and is hinged to the horizontal rigid bar which has length 2ℓ. A rotational spring restrains the change in angle between the two bars. The horizontal bar is restrained against horizontal and vertical motion at the midpoint, and is supported by two elastic springs that resist vertical motion at the ends. The springs accrue force in proportion to their extension, with modulus k. Find an expression for the energy $\mathscr{E}$ of the system. Find the (nonlinear) equations of equilibrium of the system. Find the critical loads of the system.

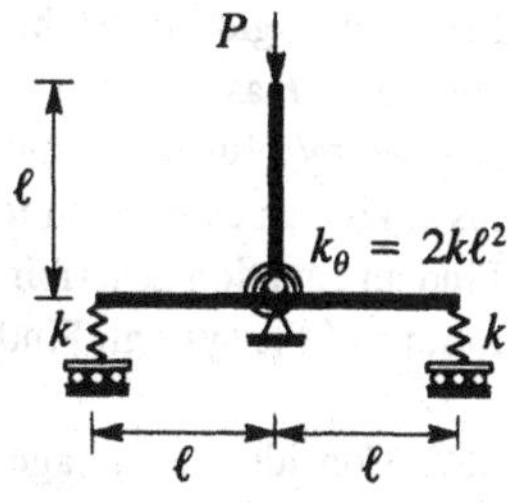

252. A rigid bent of height 2ℓ and length 3ℓ rests on three elastic springs, each with modulus k. The springs accrue force in proportion to the amount by which they stretch. The bent is pinned at the corner end and is subjected to a load P at the top and a load of P at the right end. Find the virtual-work form of the equilibrium equations. Find the second-derivative functional A for the system. Find all equilibrium configurations of the system and assess their stability. Sketch the result on a bifurcation diagram.

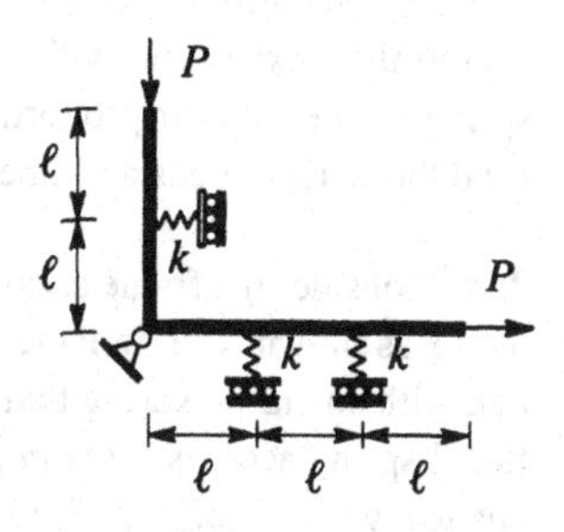

253. A rigid bar of length ℓ is pinned and restrained by a rotational spring of modulus k at the bottom. It is subjected to a force P at the top. The force changes its direction with the motion of the bar. If the bar rotates by an angle θ then the load rotates an angle $\alpha\theta$ in the opposite sense (α is a known constant). Find a suitable virtual-work function for the system? (Hint: start with a classical equilibrium equation from a freebody diagram of the bar). Does an energy function exist? If so, then find it. Estimate the buckling load of the system.

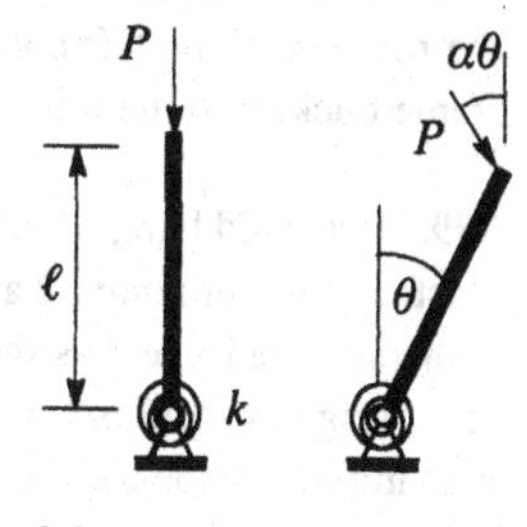

254. Two rigid bars, each of length ℓ are hinged together and attached to two *linearly* elastic springs of modulus k. The bottom end of the vertical member is on a roller that rolls on a horizontal plane. The right end of the horizontal member is on a roller that rolls on a slope. The column is subjected to a vertical force P. Find an expression for the energy of the system. Find the equilibrium configurations of the system. Find the critical loads of the structure.

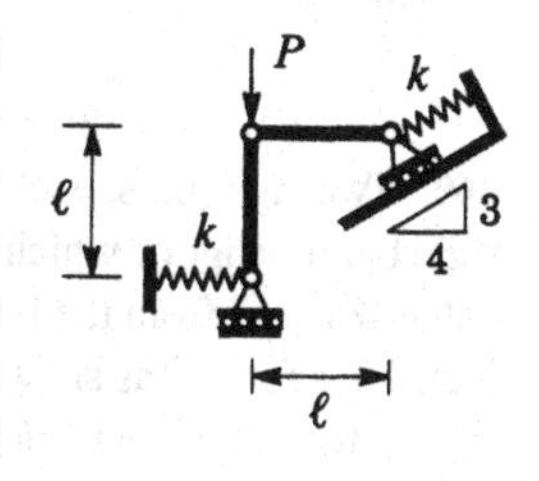

255. A ladder of length $\ell = 20$ ft leans against a wall with the base 4 ft from the wall. Both ends are frictionless and the bottom end is restrained by an elastic spring of modulus $k = 10$ lb/ft. What is the maximum height x (measured along the ladder as shown) that a person of weight $W = 200$ lb can climb? The ladder can be assumed rigid, the rollers are very small relative to the length of the ladder, and the person climbs slowly enough to neglect dynamic effects.

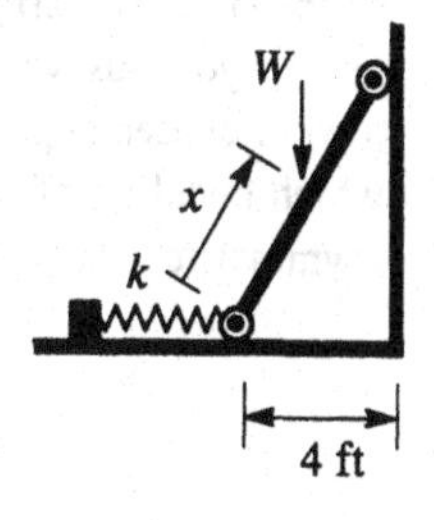

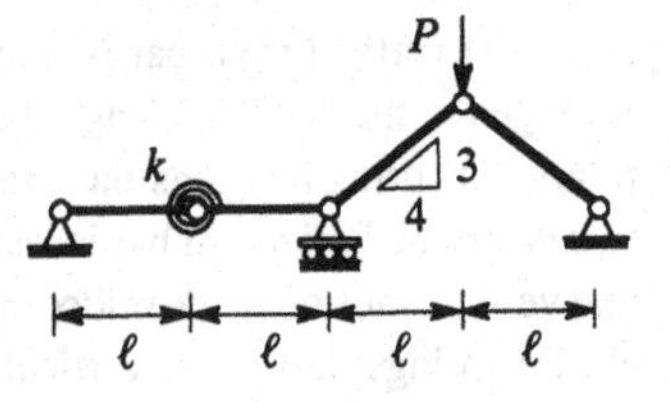

256. Four rigid bars are hinged together and subjected to the load P as shown. The two horizontal bars are restrained by a linear, elastic rotational spring of modulus k. Find an expression for the energy of the system. Find an equation describing the equilibrium configurations of the system. Find the bifurcation load.

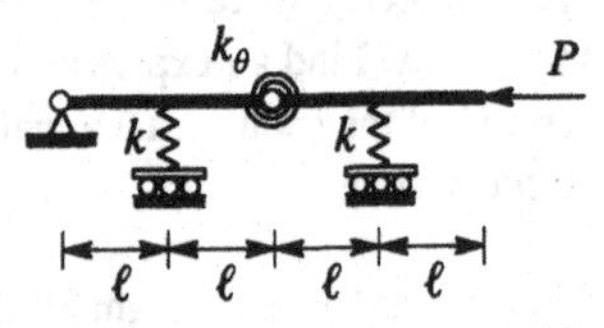

257. Consider the linkage of two rigid bars subjected to axial load P as shown. The linkage has length 4ℓ, is pinned at the left end, and has elastic springs that resist motion. The translational springs accrue force in proportion to their extension, with modulus k. The rotational spring accrues force in proportion to the its relative angle change, with modulus $k_\theta = k\ell^2$. Find the critical loads and linearized buckling mode shapes of the system.

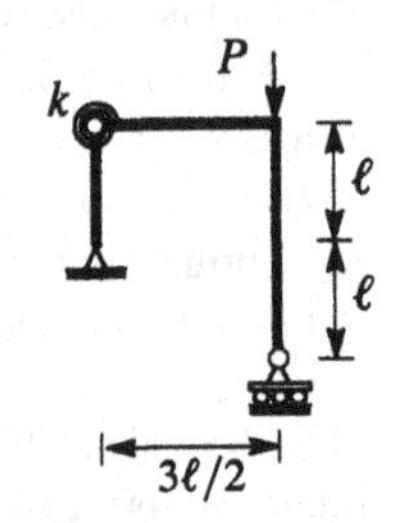

258. Consider the frame composed of rigid bars subjected to the load P as shown. The rigid members are hinged at the top left corner, with an elastic spring that resists relative rotation. The rotational spring accrues force in proportion to the its relative angle change, with modulus k. Find the buckling load for this system. Determine the post-buckling response of the system in terms of the rotation of the left column. Does it make a difference if the frame buckles to the left or to the right?

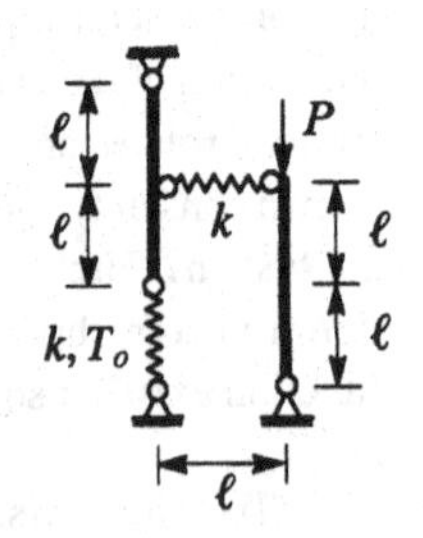

259. Two rigid bars, each of length 2ℓ are connected by a linear elastic spring of length ℓ and modulus k. The right vertical bar is subjected to a force P as shown. The left bar is attached to a vertical spring of modulus k that has been stretched into place, giving it an initial tension force of T_o (i.e., when $P=0$). Write the energy functional for the system. Find the lowest buckling load P_{cr} of the system.

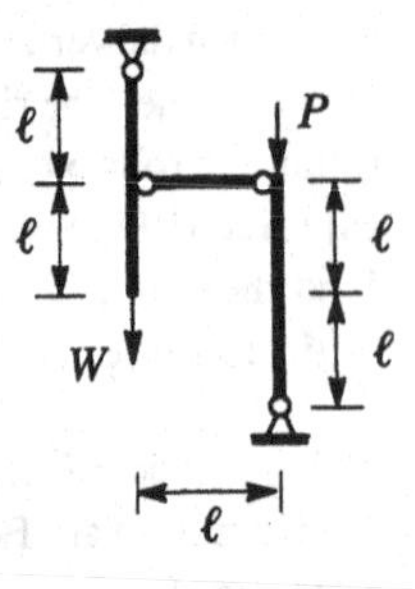

260. Two rigid bars, each of length 2ℓ are connected by a single rigid bar of length ℓ which is hinged at the ends. A weight of fixed value W hangs from the left bar while the right vertical bar is subjected to a force P as shown. Note: there are no elastic elements in this system! How many degrees of freedom does the system have? Write the exact energy functional for the system. (Hint: You can describe the deformation in terms of the rotation angles of each member, but you must write equations of constraint relating those angles to your chosen degrees of freedom). Find the critical value of P at which buckling of the system takes place. Is the post-buckling behavior symmetric or asymmetric? Do you expect the post-buckling behavior to be stable or unstable?

11

The Planar Buckling of Beams

Armed with some understanding of the stability of discrete systems, we now move on to the stability of continuous systems. The equations that govern continuous systems are differential equations, and, hence, are considerably more complicated to solve than discrete systems. However, most of the issues of stability are the same. As mentioned previously, in order to investigate the stability of a system, we must work with the nonlinear equations that govern the behavior of that system. For mechanical systems, this nonlinearity can accrue from a variety of causes, as we discussed in Chapter 10, but we shall focus here on nonlinearity in the equilibrium and strain-displacement equations (and not constitutive nonlinearities). The description of a body in a deformed configuration requires that we work with nonlinear equations of the geometry of deformation and, thus, nonlinear equations of equilibrium. Without even considering the effects of nonlinear constitutive behavior, we are led to the interesting and important phenomenon of elastic buckling of structures, first discovered by the great mathematician Leonhard Euler centuries ago.

In order to make some headway in the understanding of the buckling of continuous systems, we shall consider the case of the planar beam. Our study begins with a simple derivation of a geometrically exact planar beam theory† . The kinematic hypothesis that plane sections remain plane after deformation will again play the key role, but the simple derivation here will disguise the importance of that hypothesis somewhat. Unlike the derivation of the linear beam theory, we shall start immediately with stress resultants and establish the equa-

† A geometrically exact theory is one in which we make no approximations of the type $\sin\theta \approx \theta$, nor do we neglect any terms that arise naturally in the derivation of the theory.

tions of equilibrium. This derivation requires a leap of faith in the definition of stress resultants as the resultant of stress over a cross section, but this leap is easier to make since we have seen the rigorous development of the linear beam theory. The advantage of this approach is that it will gain us access to an important nonlinear theory without much difficulty.

Once our simple nonlinear theory has been derived, we can make some common approximations and identify some of the classical theories, such as Euler's elastica and the linearized buckling theory. We will take a close look at the linearized buckling theory, considering classical solutions to the resulting boundary value problem as well as approximating techniques based upon the virtual-work form of the equations. In particular, we will find a method of accurately approximating the critical loads of an axially compressed beam. In accord with the analyses from the previous chapter, we shall consider the effect of imperfections and transverse loads on the linearized buckling of an axially compressed beam.

Consider the planar cantilever beam shown in Fig. 136. This beam has length ℓ and cross sections that are symmetric with respect to the plane of the page. Often we shall consider prismatic beams, that is, beams with cross sections that do not vary along the length of the beam. We consider the line of centroids to be the axis of the beam, and we consider only beams that are initially straight (unless explicitly characterized otherwise, e.g., as an initial imperfection). The forces that act on the beam include the distributed transverse and axial forces, $q(x)$ and $p(x)$, and the distributed moment $m(x)$. For the present discussion, we assume that these forces do not change direction as the deformation progresses. The beam has boundary conditions at its ends that complete the specification of the problem. There are two boundary conditions on each end, on either the displacement or the force (mixed conditions can also be implemented), as was the case for the linear theory. We take this model problem as our point of departure.

Derivation of the Nonlinear Planar Beam Theory

There are many ways to derive a beam theory. The approach used in Chapter 7 to derive the equations of linear beam theory showed the relationship between the one-dimensional beam equations and the equations governing the mechanics of a three-dimensional continuum. In that derivation, we assumed that the deformations were infinitesimally small so that the equations could be

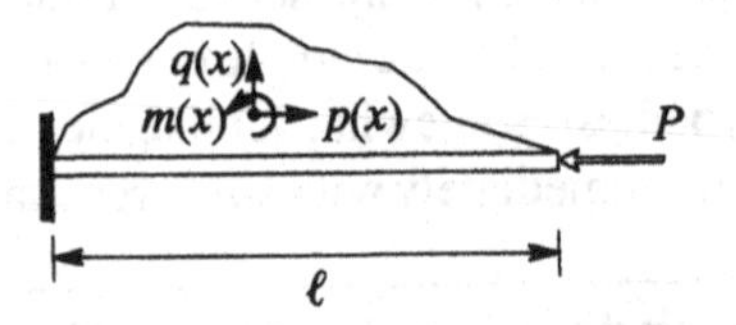

Figure 136 A planar beam subjected to axial thrust

characterized in the undeformed geometry of the body. In this chapter, we shall derive a beam theory that does not assume small deformations. As such, this theory will be useful for examining the stability of beams. The approach taken to the derivation of the theory is distinctly different from that of Chapter 7, and it assumes that you already understand the meaning of resultant force and moment. In particular, we assume that we know how to add forces to moments in equilibrium equations. This type of derivation is often found in the literature and can be quite enlightening, particularly when viewed in light of the approach of Chapter 7.

Our derivation of the nonlinear beam theory will proceed as follows. First, we shall establish the equations of equilibrium of stress resultants. Next, we cast those equations in a weighted residual (virtual-work) form by multiplying them by arbitrary virtual displacement functions and integrating them over the length of the beam. Integrating the resulting expression by parts to unload the differentiation from the stress resultants to the virtual displacements, we define the virtual strains that must, by construction, be associated with the stress resultants. Using Vainberg's theorem, these virtual strains can be integrated to give the real strains. Finally, we hypothesize constitutive equations in accord with the linear theory.

Equilibrium. Consider the segment of beam bounded by the cross sections located at distance x and $x+\Delta x$ from the left end. The displacement field is characterized by the displacement of the centroid in the axial direction $u(x)$, the displacement of the centroid in the transverse direction $w(x)$, and the rotation of the normal to the cross-sectional plane $\theta(x)$, as shown in Fig. 137.

The resultant force $\mathbf{R}$ acting on a cross section can be expressed in components, either relative to the rotated cross section, axial force $N(x)$ and shear force $Q(x)$, or relative to the axial and transverse direction of the undeformed beam, horizontal force $H(x)$ and vertical force $V(x)$, as shown in Fig. 138. Hence, we have the equivalence

$$\mathbf{R} = H\mathbf{e}_1 + V\mathbf{e}_2 = N\mathbf{g}_1 + Q\mathbf{g}_2$$

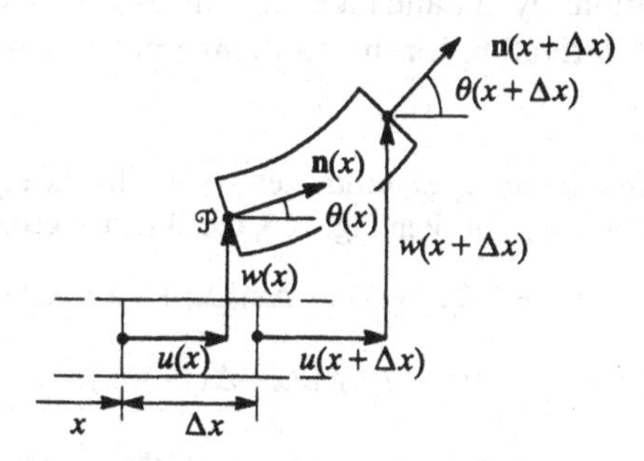

Figure 137 The geometry of deformation

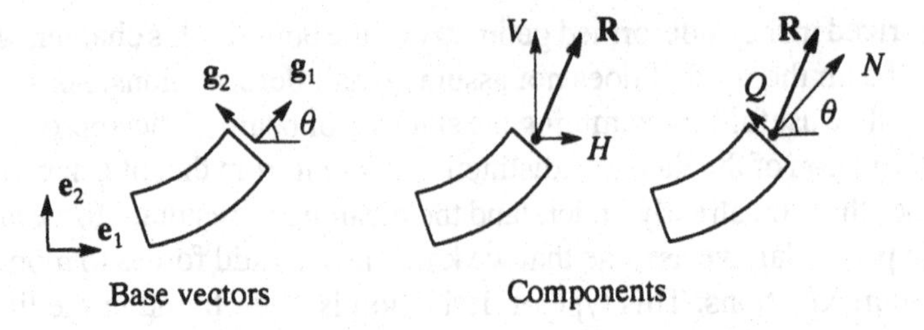

Figure 138 The components of the resultant force at a cross section

where $\mathbf{e}_1$ and $\mathbf{e}_2$ are the base vectors along the coordinate axes, and $\mathbf{g}_1$ and $\mathbf{g}_2$ are base vectors normal and transverse to the cross section. We can change from one set of components to the other through the relationships

$$\begin{aligned} H &= N\cos\theta - Q\sin\theta \\ V &= N\sin\theta + Q\cos\theta \end{aligned} \tag{540}$$

With this notation at hand, we can proceed to establish equilibrium of the segment by summing forces and moments. The forces acting on the segment $[x, x+\Delta x]$ are shown in Fig. 139.

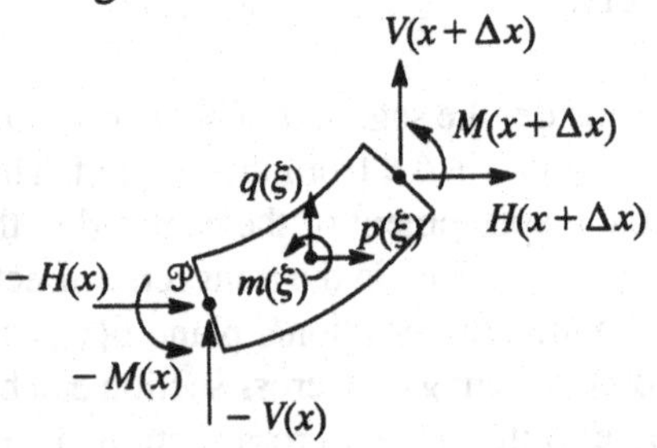

Figure 139 The equilibrium of a beam segment

Let us set the sum of forces in the horizontal and vertical direction and the sum of the moments about the point $\mathcal{P}$ (the centroid at the left end of the segment in the deformed position) equal to zero to establish equilibrium of the segment. Divide these equations by Δx and take the limit as $\Delta x \to 0$. The following examaple shows the derivation for the moment equilibrium equation.

Example 68. *Equilibrium of moments.* Let us consider the equation of balance of moments about the point $\mathcal{P}$ in Fig. 139, recalling the geometry of Fig. 137.

$$\begin{aligned} M(x+\Delta x) - M(x) + V(x+\Delta x)\big(\Delta x + u(x+\Delta x) - u(x)\big) \\ - H(x+\Delta x)\big(w(x+\Delta x) - w(x)\big) \\ + \int_0^{\Delta x} \big[\,(w(\xi)-w(x))p(\xi) - (\xi + u(\xi))q(\xi) + m(\xi)\,\big]\,d\xi = 0 \end{aligned}$$

where ξ measures distance (on the undeformed configuration) from the left end of the segment. Dividing the equation by Δx gives

$$\frac{M(x+\Delta x) - M(x)}{\Delta x} + \frac{V(x+\Delta x)\left[\Delta x + u(x+\Delta x) - u(x)\right]}{\Delta x}$$

$$- \frac{H(x+\Delta x)\left[w(x+\Delta x) - w(x)\right]}{\Delta x}$$

$$+ \frac{1}{\Delta x}\int_0^{\Delta x} \left[\left(w(\xi) - w(x)\right)p(\xi) - \left(\xi + u(\xi)\right)q(\xi) + m(\xi)\right] d\xi = 0$$

Taking the limit as $\Delta x \rightarrow 0$, recognizing the definition of the derivative of a function, gives

$$M' + V(1+u') - Hw' + m = 0$$

The first two loading terms in the integral vanish because the moment arm in the integrand goes to zero as the length of the segment goes to zero.

The governing differential equations of equilibrium for the planar beam are

$$\boxed{\begin{array}{cc} H' + p = 0 & V' + q = 0 \\ \multicolumn{2}{c}{M' + V(1+u') - Hw' + m = 0} \end{array}} \qquad (541)$$

These equations look remarkably like the linear equations of equilibrium. The main difference is the deformation measures $1+u'$ and w' in the equation of moment equilibrium. These terms give rise to buckling phenomena in beams. These equations are exact within the context of beam theory and the assumption of planar behavior.

Virtual-work functional. We have seen previously that the virtual-work form of the equations can be obtained simply by multiplying the equilibrium equations by an arbitrary virtual displacement and integrating the result over the domain of the body. We will use this approach here to show how it helps us define the appropriate strain resultants to go along with the stress resultants.

We know that we want the final result to have a well-defined expression for external virtual work. Therefore, we shall multiply the first equation by $\bar{u}(x)$, a horizontal virtual displacement; the second by $\bar{w}(x)$, a vertical virtual displacement; and the third by $\bar{\theta}(x)$, a virtual rotation. These selections are motivated by the presence of the loading terms $p(x)$, $q(x)$, and $m(x)$ in the equations and our desire to compute the virtual work done by these forces. The other terms in the equation are of much less help in figuring out what the character of the arbitrary function should be. As we shall soon see, simple manipulation of these equations will tell us what the virtual strains must be in order for the internal work to make sense.

Let us denote the three displacement functions as $\mathbf{u} = \{u, w, \theta\}$ and the virtual displacement functions as $\overline{\mathbf{u}} = \{\overline{u}, \overline{w}, \overline{\theta}\}$. In accord with the specification of the virtual displacements above, we can define the following functional

$$G(\mathbf{u}, \overline{\mathbf{u}}) \equiv -\int_0^\ell \Big[(H' + p)\overline{u} + (V' + q)\overline{w} + \big(M' + V(1+u') - Hw' + m\big)\overline{\theta}\Big]\, dx$$

Note that, in accord with the fundamental theorem of the calculus of variations, $\mathbf{u}$ is an equilibrium configuration if $G(\mathbf{u}, \overline{\mathbf{u}}) = 0$ for all $\overline{\mathbf{u}} \in \mathcal{T}(0, \ell)$, where $\mathcal{T}$ is the space of admissible functions. (Note that $\mathcal{T}$ is the same space we used for the linear Timoshenko beam.) This form of the equations is not very interesting in itself, but it will be upon some simple manipulation. Integrating all of the terms involving derivatives of the stress resultants by parts, we arrive at a suitable definition of the internal and external virtual work

$$\overline{W}_I \equiv \int_0^\ell \left(H\overline{u}' + V\overline{w}' + M\overline{\theta}' - V(1+u')\overline{\theta} + Hw'\overline{\theta}\right) dx$$
$$\overline{W}_E \equiv \int_0^\ell \left(p\overline{u} + q\overline{w} + m\overline{\theta}\right) dx + M\overline{\theta}\Big|_0^\ell + V\overline{w}\Big|_0^\ell + H\overline{u}\Big|_0^\ell \tag{542}$$

If the virtual-work functional is to have the form $G = \overline{W}_I - \overline{W}_E$, then the integral in Eqn. (542) must be the internal virtual work. Let us substitute Eqns. (540) to eliminate H and V in favor of Q and N. The internal virtual work can be written in the form

$$\overline{W}_I \equiv \int_0^\ell \Big[(N\cos\theta - Q\sin\theta)\overline{u}' + (N\sin\theta + Q\cos\theta)\overline{w}' + M\overline{\theta}' - (N\sin\theta + Q\cos\theta)(1+u')\overline{\theta} + (N\cos\theta - Q\sin\theta)w'\overline{\theta}\Big]\, dx \tag{543}$$

Regrouping terms, this expression can be written in the form

$$\overline{W}_I = \int_0^\ell \Big[N\big[\cos\theta\,\overline{u}' + \sin\theta\,\overline{w}' - ((1+u')\sin\theta - w'\cos\theta)\overline{\theta}\big] + Q\big[\cos\theta\,\overline{w}' - \sin\theta\,\overline{u}' - ((1+u')\cos\theta + w'\sin\theta)\overline{\theta}\big] + M\overline{\theta}'\Big]\, dx \tag{544}$$

Let us finally summarize the internal work as

$$\overline{W}_I \equiv \int_0^\ell \left(M\overline{\kappa}_o + Q\overline{\beta}_o + N\overline{\epsilon}_o\right) dx$$

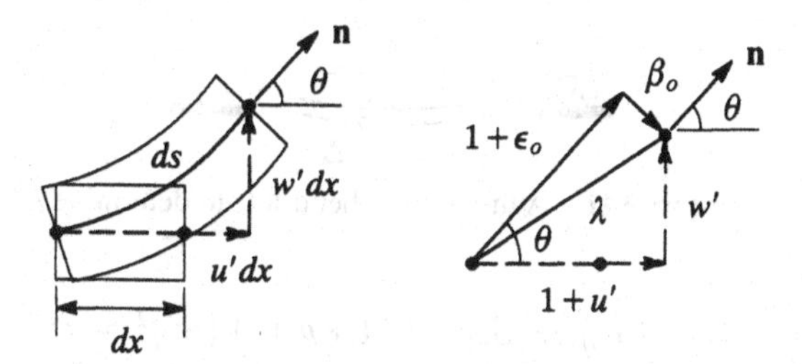

Figure 140 The components of the stretch of the line of centroids

This final form of the expression for the internal virtual work suggests that the quantities $\overline{\varkappa}_o$, $\overline{\beta}_o$, and $\overline{\epsilon}_o$ have the character of virtual strains associated with M, Q, and N, respectively. Accordingly, the virtual curvature, shear strain, and axial strain are given, respectively, by

$$\begin{aligned} \overline{\varkappa}_o &\equiv \overline{\theta}' \\ \overline{\beta}_o &\equiv \overline{w}'\cos\theta - \overline{u}'\sin\theta - \big(w'\sin\theta + (1+u')\cos\theta\big)\overline{\theta} \\ \overline{\epsilon}_o &\equiv \overline{w}'\sin\theta + \overline{u}'\cos\theta + \big(w'\cos\theta - (1+u')\sin\theta\big)\overline{\theta} \end{aligned} \tag{545}$$

Strain resultants. We can find the real strains by applying Vainberg's theorem to the question of the integrability of the virtual strains (in the same spirit as we integrate the virtual-work functional to get an energy functional). To be integrable, the directional derivatives of the virtual strains must be symmetric in the sense of Vainberg. This symmetry is easy to verify and is left as an exercise (Problem 261). We shall call the real strains $\varkappa_o$, β_o, and ϵ_o. These strains are the functions that, when differentiated in the direction of the virtual displacements $\overline{\mathbf{u}} = \{\overline{u}, \overline{w}, \overline{\theta}\}$, give the virtual strains. The real strains are

$$\boxed{\begin{aligned} \varkappa_o &= \theta' \\ \beta_o &= w'\cos\theta - (1+u')\sin\theta \\ \epsilon_o &= w'\sin\theta + (1+u')\cos\theta - 1 \end{aligned}} \tag{546}$$

An interpretation of the strain resultants β_o and ϵ_o can be obtained by examining Fig. 140, which shows a segment of beam originally of length dx stretched to a deformed configuration with length ds. The stretch ratio is called $\lambda \equiv ds/dx$. If we translate the deformed configuration back on top of the undeformed configuration, we can see that the right end of the deformed segment has moved horizontally by $u'dx$ and vertically by $w'dx$ relative to the undeformed configuration. The stretch is, thus, given as $\lambda^2 = (1+u')^2 + (w')^2$. Using the relationships for strain resultants in terms of displacements in Eqn. (546), we can show that

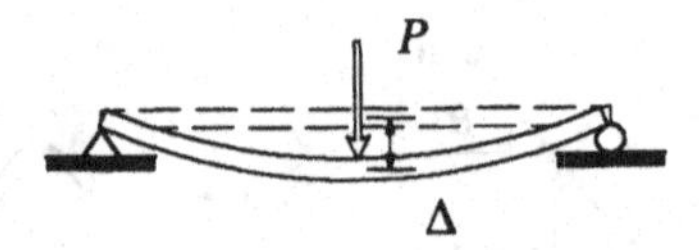

Figure 141 A three-point bend test to determine EI

$$\left(1+\epsilon_o\right)^2+\left(\beta_o\right)^2 = \left(1+u'\right)^2+\left(w'\right)^2 = \lambda^2$$

Furthermore, the strains ϵ_o and β_o are related to the quantities w' and $1+u'$ through a rotation of magnitude θ, indicating that they are the components of the same vector in two different coordinate systems, the latter in the undeformed coordinate system and the former in the coordinate system attached to the cross section. Thus, we can interpret ϵ_o and β_o as the components of the stretch of the line of centroids in a coordinate system that moves with the cross section through the deformation. In a sense, then, ϵ_o measures axial stretch and β_o measures transverse stretch. The transverse stretch must be caused by shearing strain.

Constitutive equations. The only component of the theory that remains to be established is the constitutive equations. In the linear theory, we were able to derive the constitutive equations from the constitutive equations of three-dimensional elasticity. We found that beam theory was not entirely consistent with the three-dimensional theory, but it could be fixed by modifying the constitutive equations. Since the constitutive parameters for any theory (including the general three-dimensional theory) must be determined empirically, we can consider the constitutive models of the resultant theory as relationships between resultant stresses and resultant strains that need to be established through laboratory testing.

We could, for example, use a three-point bend test to determine the value of EI for the linear theory, rather than looking at EI as the product of Young's modulus E and the second moment of the cross-sectional area I. The test is illustrated in Fig. 141. From the linear theory, we know that $\Delta = P\ell^3/48EI$. We can measure the length ℓ and plot P versus Δ for the test. Then EI is the slope of the $P-\Delta$ curve multiplied by $\ell^3/48$. Thus, it is possible to evaluate EI without evaluating E and I separately. Similar tests for GA and EA can be devised.

We can extend these ideas to the nonlinear theory. We can also use the linear theory as a guide in the sense that the nonlinear theory should reduce to the linear theory when the strains and displacements are small. From the linear theory, we found that the constitutive equations were uncoupled if centroidal, principal axes were used to describe the axis of the beam. Thus, the moment is a function of curvature, $M = \mathcal{M}(\varkappa_o)$, the shear force is a function of shear strain, $Q = \mathcal{Q}(\beta_o)$, and the axial force is a function of axial strain, $N = \mathcal{N}(\epsilon_o)$. The specific functions could be determined experimentally.

As a first approximation, we shall adopt linear constitutive equations for the nonlinear theory. These constitutive equations should remain accurate if the strains are relatively small, no matter how large are the displacements and rotations. We postulate the following constitutive model for our nonlinear beam

$$M = EI\varkappa_o, \quad Q = GA\beta_o, \quad N = EA\epsilon_o \tag{547}$$

where *EI, GA,* and *EA* can be interpreted in the same manner as the linear theory. The specification of constitutive equations completes the field equations for our nonlinear planar beam theory. In order to have a properly posed boundary value problem, we must augment the field equations with boundary conditions. These conditions are identical to the linear theory and, therefore, will not be discussed here.

A Model Problem: Euler's Elastica

The fully nonlinear theory can be constrained to produce some interesting classical results. Clearly, the present theory includes shear and axial strains. For long, slender beams, the influence of the shear and axial strains on the deflections of the beam are typically small. We can constrain these strains to be zero a priori to generate a model generally attributed to Euler. In the linear theory, the shear and axial strains were uncoupled, and it made no difference whether we constrained one or the other. In the nonlinear theory they are coupled, and we must constrain both in order to realize a simplification in the governing equations. In the nonlinear theory, β_o and ϵ_o are components of the stretch of the axis. Therefore, setting $\epsilon_o = 0$ and $\beta_o = 0$ is tantamount to saying that the length of the beam cannot change. We call such a beam *inextensible*. From Eqn. (546) we can determine that the inextensibility constraints imply

$$w' = \sin\theta, \quad 1+u' = \cos\theta \tag{548}$$

These constraints allow us to recast the theory purely in terms of the rotation of the cross section θ. Substituting these expressions into the moment equation, we obtain

$$M' + V\cos\theta - H\sin\theta + m = 0 \tag{549}$$

in addition to the original horizontal and vertical equilibrium equations. The force equilibrium equations can be integrated to give

$$V(x) = V(0) - \int_0^x q(\xi)d\xi, \quad H(x) = H(0) - \int_0^x p(\xi)d\xi \tag{550}$$

For statically determinate problems, *H* and *V* can be expressed in terms of the applied forces without appealing to the other equations governing the behavior

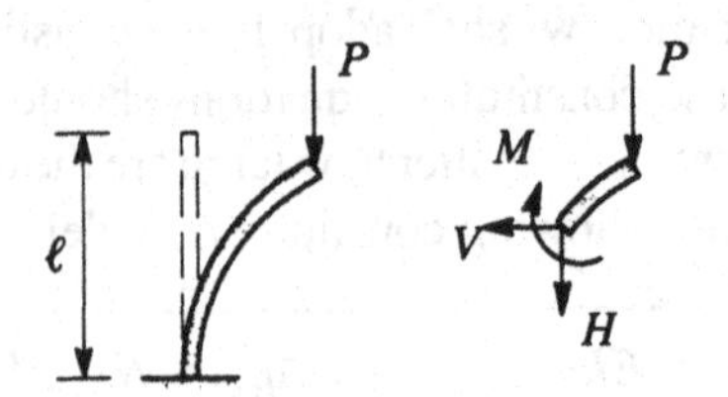

Figure 142 The model problem for Euler's elastica

of the system. When there are no distributed loads, these problems are known as *Euler's elastica* because they were studied by the famous mathematician Euler centuries ago.

Classical differential equation. One particular case of Euler's elastica is the cantilever beam under a compressive tip load P, shown in Fig. 142. We shall adopt this case as our model problem and use it to demonstrate some features of the nonlinear theory. There are no transverse loads, so $p = q = m = 0$. Furthermore, from overall equilibrium we see that $H = -P$ and $V = 0$. Consider a prismatic column, i.e., EI is constant. Using the constitutive equation for moment, and substituting the above relations for H and V into Eqn. (549) we find the classical equations and boundary conditions governing the elastica

$$\boxed{\begin{array}{c} EI\theta'' + P\sin\theta = 0 \\ \theta(0) = 0 \quad \theta'(\ell) = 0 \end{array}} \tag{551}$$

This classical problem has been studied extensively. Solutions to this problem are given by elliptic integrals. The classical solution is rather involved and will not be pursued here. You may wish to consult a classic text, e.g., Love (1944), for the solution to this problem.

When we speak of the elastica, we generally mean the whole class of problems subject to the inextensibility constraint. Accordingly, other boundary conditions are also possible. The interesting observation about the elastica is that the equations govern only the rotation field of the beam, not the displacement field. As such, a solution can be obtained for which the actual position of the beam in space is not determined. Once the rotation field is known, however, Eqns. (548) can be integrated to find the axial and transverse displacements. The integration of these two first-order differential equations would introduce two more constants that can be used to establish a unique position of the beam in space. These solutions also give the shape of the beam by locating the position of the centroidal axis.

Virtual-work and energy functionals. The problem of the elastica can be cast in weak form. For the present example, Eqn. (542) reduces to

$$G(\theta,\bar{\theta}) \equiv \int_0^{\ell} \left(EI\theta'\bar{\theta}' - P\bar{\theta}\sin\theta\right) dx \tag{552}$$

The essential boundary condition is vanishing rotation at $x = 0$, that is, $\theta(0) = 0$, while the natural boundary condition is vanishing moment at $x = \ell$, that is, $\theta'(\ell) = 0$. If we select our virtual rotations $\bar{\theta}$ such that they satisfy the essential boundary condition, then the variational equation $G(\theta,\bar{\theta}) = 0$ for all $\bar{\theta} \in \mathfrak{B}_e(0,\ell)$ is equivalent to the classical governing differential equation, and the functions θ that satisfy this equation represent equilibrium configurations.

Symmetry of the virtual-work functional given in Eqn. (552) can be verified by taking the directional derivative of G as follows

$$DG(\theta,\bar{\theta}) \cdot \hat{\theta} \equiv \int_0^{\ell} \left[EI\hat{\theta}'\bar{\theta}' - P\hat{\theta}\bar{\theta}\cos\theta\right] dx = DG(\theta,\hat{\theta}) \cdot \bar{\theta}$$

Vainberg's theorem guarantees the existence of an energy functional and tells us how to compute it. Carrying out the computations, we get

$$\mathscr{E}(\theta) \equiv \int_0^{\ell} \left(\tfrac{1}{2}EI(\theta')^2 + P\cos\theta\right) dx \tag{553}$$

The energy and virtual-work functionals for the model problem are remarkably similar to the energy and virtual-work expressions for the column of rigid links with a rotational spring between them. The main difference, of course, is that the present problem involves continuous functions and derivatives. You might expect some similarities in the stability of these two problems, and, indeed, this is the case. The elastica has a stable ascending equilibrium branch that bifurcates from the critical load (i.e., the Euler load $P_1 = \pi^2 EI/4\ell^2$).

Remark. There is another way to construct the virtual-work functional that lends insight to the boundary conditions. Let us take the residual of the classical differential equation, multiply it by an arbitrary function $\bar{\theta}$, and integrate the product over the length of the beam. The resulting weighted residual is

$$G(\theta,\bar{\theta}) \equiv -\int_0^{\ell} \left(EI\theta'' + P\sin\theta\right)\bar{\theta}\, dx$$

Obviously, if θ satisfies the classical differential equation, then $G(\theta,\bar{\theta}) = 0$ for all $\bar{\theta}$. The fundamental theorem of the calculus of variations also suggests that if $G(\theta,\bar{\theta}) = 0$ for all $\bar{\theta}$, then $EI\theta'' + P\sin\theta = 0$. This virtual-work functional is perfectly suitable, but we generally prefer to balance the derivatives between the real displacement variable and the virtual displacement. We can do so by integrating the first term by parts to get

$$G(\theta,\bar{\theta}) = \int_0^\ell \left(EI\theta'\bar{\theta}' - P\bar{\theta}\sin\theta\right)dx - EI\theta'\bar{\theta}\Big|_0^\ell$$

The boundary term is the result of integration by parts. The boundary condition at either end must be either vanishing moment M or vanishing rotation θ (but not both at the same point). For the present problem, the rotation vanishes at $x = 0$ and the moment vanishes at $x = \ell$. In order to have the boundary term vanish completely from our functional, we must restrict the class of virtual rotations that we allow in our function space $\mathfrak{B}_e(0, \ell)$ to functions that satisfy $\bar{\theta}(0) = 0$. With this understanding, the boundary term vanishes and the virtual-work expression reduces to Eqn. (552). ❑

Example 69. *Solution to the linearized Elastica.* We can make some analytical headway with the classical differential equation if we linearize it. Let us make the approximation that the angle of rotation is small. Thus, $\sin\theta \approx \theta$, and the governing equation takes the linearized form $EI\theta'' + P\theta = 0$. This equation has the general solution

$$\theta(x) = a_1 \sin\mu x + a_2 \cos\mu x \qquad (554)$$

where $\mu^2 \equiv P/EI$ is the ratio of axial load to bending modulus. Verify this solution by substituting it back into the linearized differential equation. For the problem at hand, the load P is positive if it is compressive. The modulus EI is always positive. Therefore, μ^2 is positive and, hence, μ is real for the model problem.

The general solution to the linearized model problem, given by Eqn. (554), has two arbitrary constants. These constants can be determined from the boundary conditions. For the present case, we have $\theta(0) = 0$, which gives $a_2 = 0$, and $\theta'(\ell) = 0$, which gives

$$\mu a_1 \cos\mu\ell = 0$$

This equation has the solution $a_1 = 0$, and that solution corresponds to the straight configuration. Thus, we arrive at the conclusion that the straight configuration is an equilibrium configuration for all values of the load P. Like the discrete problems in the preceding chapter, there are other solutions for certain values of the load P. These solutions are given by

$$\mu\ell = \pm\frac{\pi}{2}, \pm\frac{3\pi}{2}, \pm\frac{5\pi}{2}, \ldots$$

These values of μ are the bifurcation points from which nontrivial equilibrium paths branch from the straight configuration. Let us designate these critical values of μ as $\mu_n \equiv (2n-1)\pi/2\ell$ for all positive integer values n. There are an infinite number of such points. Note that it is sufficient to consider only the positive values of the solutions $\mu\ell$ given above because cosine is an even function. The nth critical load is given by the definition of μ as $P_n = \mu_n^2 EI$. The lowest critical load, $n = 1$, is called the ***Euler load***. We will see later that the straight configuration is stable for all loads below the Euler load, and unstable for all loads above the Euler load.

At these critical loads, it is possible for the column to bend into certain buckled shapes. These shapes are proportional to the eigenfunctions

$$\theta_n(x) = \sin \mu_n x$$

obtained by substituting the eigenvalue μ_n back into Eqn. (554). These are the solutions for which $a_1 \neq 0$. Clearly, for each eigenvalue there exists an eigenfunction, much like the discrete case. Also like the discrete case, we can observe that the buckled configurations are determined only up to an undetermined magnitude. The linearized equations are not sufficient to determine the value of this constant. For this reason, we often refer to the eigenfunctions as the ***buckling mode shapes***. We can determine only the shape that the column must have initially on each equilibrium branch. These functions tell us the direction that the equilibrium path takes upon buckling, and we can use this information to develop approximate solutions in the neighborhood of the bifurcation points. We will also find that these functions provide an excellent basis for Ritz approximations for solving the complete nonlinear problem.

Remark. If the load is tensile, then $\mu^2 < 0$ and μ is imaginary. Trigonometric functions with imaginary arguments can be readily converted to hyperbolic trigonometric functions with real arguments with Euler's formulas

$$\sinh x = -i \sin ix, \quad \cosh x = \cos ix$$

where $i \equiv \sqrt{-1}$ is the imaginary unit. Thus, for tension problems, the solution can be taken in the more suitable form $\theta(x) = b_1 \sinh \mu x + b_2 \cosh \mu x$, to avoid complex numbers. The character of the trigonometric functions is oscillatory, while the hyperbolic trigonometric functions are exponentially decaying. There is a great difference in these two types of behavior, and oneshould keep this observation in mind when solving problems with tensile loads.

Example 70. *An approximate solution to the nonlinear Elastica.* The linearized buckling theory gives us a good start on a nonlinear analysis. In particular, we know that the solution bifurcates at the load $\mu_1 = \pi/2\ell$, and that the configuration of the column on the equilibrium path is initially proportional to the first eigenfunction $\theta_1(x) = \sin \mu_1 x$. Thus, we know how the column must deflect in order to get onto the first nontrivial equilibrium branch. We can use this information to launch an approximate analysis of the stability of Euler's elastica. Let us assume that the solution $\theta(x)$ is approximately proportional to the first eigenfunction

$$\theta(x) \approx \gamma \theta_1(x)$$

where γ is a scalar parameter measuring the amplitude of deformation. What we are really assuming is that, over a limited range of deformation, the shape of the column will not change, but the amplitude will increase in accord with the load level. Thus, we shall characterize our equilibrium path by finding the load P as a function of γ.

We can use this assumption in the virtual-work form of the equation governing the deformation of the elastica, Eqn. (552). Let us also assume that the virtual rotation has the same form as the real rotation, that is, $\bar{\theta}(x) = \bar{\gamma}\theta_1(x)$, where $\bar{\gamma}$ is an arbitrary scalar parameter (virtual rotation). Substituting the approximate rotation function into the weak form, we get an ordinary function in place of the functional

$$G(\gamma,\bar{\gamma}) = \int_0^\ell \bar{\gamma}\left[\gamma EI(\theta_1')^2 - P\theta_1 \sin(\gamma\theta_1)\right] dx$$

For small angles, the sine function can be expanded as a Taylor series as

$$\sin(\gamma\theta_1) = \gamma\theta_1 - \tfrac{1}{6}\gamma^3\theta_1^3 + \tfrac{1}{120}\gamma^5\theta_1^5 - \cdots$$

Substituting the first two terms of the expansion into the virtual-work functional, we obtain the result

$$G(\gamma,\bar{\gamma}) = \bar{\gamma}\int_0^\ell \left[\gamma EI(\theta_1')^2 - P\gamma\left(\theta_1^2 - \tfrac{1}{6}\gamma^2\theta_1^4\right)\right] dx$$

Since the function $\theta_1(x) = \sin \pi x/2\ell$ is explicitly known, the integrals can be carried out. These integrals have the values

$$\int_0^\ell EI(\theta_1')^2\, dx = \int_0^\ell P_1 \cos^2\left(\frac{\pi x}{2\ell}\right) dx = \tfrac{1}{2}P_1\ell$$

$$\int_0^\ell \theta_1^2\, dx = \int_0^\ell \sin^2\left(\frac{\pi x}{2\ell}\right) dx = \tfrac{1}{2}\ell$$

$$\int_0^\ell \theta_1^4\, dx = \int_0^\ell \sin^4\left(\frac{\pi x}{2\ell}\right) dx = \tfrac{3}{8}\ell$$

where $P_1 = \pi^2 EI/4\ell^2$ is the first critical load. With these results, we finally arrive at the discrete form of the virtual-work functional

$$G(\gamma,\bar{\gamma}) = \tfrac{1}{2}\ell\left[\gamma P_1 - \gamma P\left(1 - \tfrac{1}{8}\gamma^2\right)\right]\bar{\gamma} \tag{555}$$

We are now in the same position as we were in the analysis of the discrete systems. The virtual-work form of the equilibrium equations is algebraic rather than integral. The parameters are scalars rather than functions. The analysis proceeds along the same lines as the discrete system. Since $G(\gamma,\bar{\gamma}) = 0$ for all $\bar{\gamma}$ implies an equilibrium configuration, we have

$$\gamma\left[P_1 - P\left(1 - \tfrac{1}{8}\gamma^2\right)\right] = 0 \tag{556}$$

This algebraic equation has two solutions. The solution $\gamma = 0$ is the straight configuration, for which all values of the load P satisfy equilibrium. The second solution is

$$P(\gamma) = \frac{P_1}{\left(1 - \tfrac{1}{8}\gamma^2\right)} \tag{557}$$

This equation tells us that the load P increases along Branch 1 in a *pitchfork* bifurcation. Branch 1 is symmetric with respect to the deformation parameter γ. In the limit as $\gamma \to 0$, we can see that Branch 1 emanates from the critical load P_1. The bifurcation diagram is shown in Fig. 143.

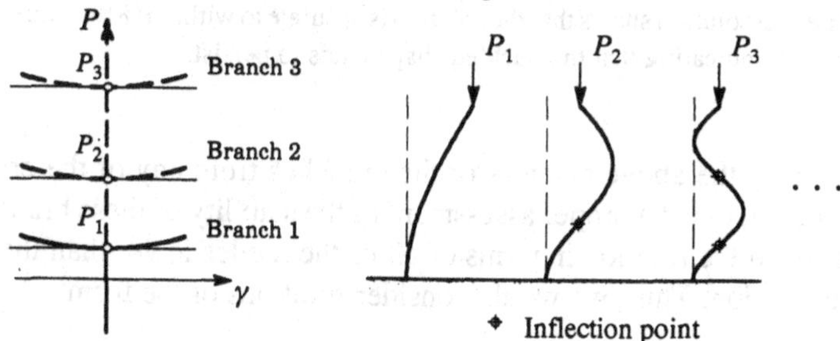

Figure 143 The bifurcation diagram and buckling mode shapes of the cantilever column

Note that, in order to plot the buckled shapes, we needed the transverse deflection $w(x)$ rather than the slope $\theta(x)$. These can be computed approximately from the constraint $w' = \sin\theta \approx \theta$. Upon integrating this equation and substituting the boundary condition $w(0) = 0$, we find that

$$w_n(x) \approx 1 - \cos\mu_n x \tag{558}$$

Since the approximation is not linearized, we can examine the stability of the non-trivial equilibrium path. The second derivative of the energy can be computed from G, Eqn. (555). It has the following expression

$$A(\gamma,\bar{\gamma}) = \tfrac{1}{2}\ell\left[P_1 - P\left(1 - \tfrac{3}{8}\gamma^2\right)\right]\bar{\gamma}^2 \tag{559}$$

The second-derivative test is now a test of an ordinary function. For the solution $\gamma = 0$, the straight configuration, the energy criterion suggests that

$$A(0,\bar{\gamma}) = \tfrac{1}{2}\ell\left[P_1 - P\right]\bar{\gamma}^2 = \begin{cases} > 0 & \text{for } P < P_1 \quad \text{(stable)} \\ < 0 & \text{for } P > P_1 \quad \text{(unstable)} \end{cases}$$

The second-derivative test on the trivial equilibrium path tells us that loads below the first critical load are stable, and loads above the first critical load are unstable. The second-derivative test on the first branch gives

$$A(\gamma,\bar{\gamma}) = \tfrac{1}{2}P_1\ell\left(\frac{2\gamma^2}{8-\gamma^2}\right)\bar{\gamma}^2$$

which is greater than zero for all values of $\gamma < \sqrt{8}$. Therefore, the first branch is stable.

Clearly, the analysis holds only in a small neighborhood of the critical point. How small is small? Since we took a cubic approximation for the sine function, that approximation should hold out for relatively large values of $\theta(x) = \gamma\theta_1(x)$. Since the eigenfunction θ_1 is never greater than 1, this limitation applies essentially to γ. The source of error that we really cannot assess is the desire of the system to change its shape as it moves along a branch. We saw in the exact analysis of the two-degree-of-freedom system that such changes are natural in the evolution of the response. Since we have insisted that the shape remains the same, we are adding a constraint that we cannot evaluate. However, comparison with the exact solution shows that the solution is accurate to within 1% for values of $\gamma = 60°$, indicating that the buckled shape tends to persist.

We can apply the above analysis of the branches from any of the critical points. However, to get a proper assessment of the stability of these branches, we must expand the rotation in terms of all of the modes lower than the one under investigation. Thus, we would consider solutions of the form

$$\theta(x) = \sum_{n=1}^{N} \gamma_n \theta_n(x)$$

Clearly, upon substituting this series into the cubic term of the expansion for the sine function, we obtain a system of N equations that contains the full, coupled cubic combinations of the deformation parameters. These algebraic equations are, of course, amenable to iterative solution by Newton's method, but even for the case $n = 2$, a closed-form solution is not feasible. If we assume that the solution emanates in a pure mode, i.e., $\theta(x) = \gamma\theta_n(x)$, we lose the influence of the lower modes (particularly the first) on the second-derivative test, and the conclusions on the stability of equilibrium are erroneous. If we were to carry out the solution for $n = 2$, we would see that the shape of the higher branches is similar to the first branch, and that they are all unstable.

It should be evident that looking along an eigenfunction to assess the stability of a nontrivial branch near a critical point is a method that is generally applicable. If any nonlinear function that appears in the virtual-work functional is expanded as a Taylor series, then the resulting equations will be polynomial in the deformation parameter, making equilibrium and stability easy to assess. Thus, we have seen how the linearized buckling eigenvalue problem can be used as a preprocessor for a stability analysis of a system. Even though the linearized buckling problem tells us nothing about the stability of the branching solutions, it does tell us the points that those branches emanate from and the directions that they follow initially. Eigenfunction expansions are also useful for investigating problems with imperfections and transverse loads.

Example 71. *The effect of geometric imperfections.* One can modify the above analysis to include the effect of imperfections on the behavior of the cantilever column. We will continue to measure the deformation of the column from the straight position, but we will consider a column that is not initially straight. We characterize the imperfection as the angle of rotation of the cross section that exists without load on the column or flexural strain in the beam. Accordingly, let $\psi(x)$ be the initial rotation field of the beam.

We saw in the preceding analysis that we could make some progress by recognizing that the column buckles into the first linearized buckling mode $\theta_1(x) = \sin(\pi x/2\ell)$ at the critical load $\mu_1 = \pi/2\ell$, and that we could look in that direction to examine the behavior of the bifurcation diagram in the neighborhood of the critical point. We can extend this analysis to include the imperfect column. For the imperfect column, the potential energy of the system is

$$\mathcal{E}(\theta) = \int_0^{\ell} \left(\tfrac{1}{2}EI(\theta' - \psi')^2 + P\cos\theta\right)dx$$

We can see that at the configuration $\theta = \psi$, there is no flexural energy stored in the column. The variational form of the equilibrium equations can be obtained by taking the directional derivative of the energy. Doing so, we obtain

$$G(\theta, \bar{\theta}) = \int_0^{\ell} \left(EI\theta'\bar{\theta}' - P\bar{\theta}\sin\theta - EI\psi'\bar{\theta}'\right)dx$$

The first two terms of this expression are exactly the same as the perfectly straight column. The third term reflects the effect of the geometric imperfection.

Let us examine the particular case where the imperfection is in exactly the same shape as the initial buckling mode, that is, $\psi(x) = \gamma_o\theta_1(x)$, where γ_o represents the amplitude of the initial imperfection and is a fixed positive scalar value. In this case, it is reasonable to again assume that the deformations of the column will be proportional to the first buckling mode $\theta(x) = \gamma\theta_1(x)$, where, as before, γ is the total amplitude of the rotation field, measured relative to the straight configuration. The virtual rotation can also be expressed as a multiple of the first eigenfunction $\bar{\theta}(x) = \bar{\gamma}\theta_1(x)$. The expression $\sin(\gamma\theta_1)$ is again expressed as a cubic Taylor series approximation. If we substitute these expressions into the expression for G and carry out the requisite integrals of the eigenfunction, we obtain

$$G(\gamma, \bar{\gamma}) = \tfrac{1}{2}\ell\left[(\gamma - \gamma_o)P_1 - \gamma P\left(1 - \tfrac{1}{8}\gamma^2\right)\right]\bar{\gamma} \tag{560}$$

thereby reducing the functional to an ordinary function of the real deformation γ and the virtual displacement parameter $\bar{\gamma}$. The analysis proceeds along the same lines as it did for the perfect system. Since $G(\gamma, \bar{\gamma}) = 0$ for all $\bar{\gamma}$, we have

$$P(\gamma) = P_1\left(\frac{\gamma - \gamma_o}{\gamma - \frac{1}{8}\gamma^3}\right) \tag{561}$$

Clearly, $\gamma = 0$ (the straight configuration) is no longer an equilibrium configuration. For the value $\gamma = \gamma_o$ we have $P(\gamma_o) = 0$, indicating that loading starts from zero rather than branching from a critical point. As γ gets large ($\gamma \gg \gamma_o$), the imperfect curve becomes asymptotic to the perfect curve. The second derivative of the energy is identical to the perfect case and is given by Eqn. (559). Substituting Eqn. (561) into Eqn. (559) we arrive at the result that

$$A(\gamma,\bar{\gamma}) = \tfrac{1}{2}P_1\ell\left(\frac{2\gamma^3+8\gamma_o-3\gamma_o\gamma^2}{8\gamma-\gamma^3}\right)\bar{\gamma}^2 = \begin{cases} > 0 \text{ for } \gamma > 0 & \text{(stable)} \\ < 0 \text{ for } \gamma_{cr} < \gamma < 0 & \text{(unstable)} \\ > 0 \text{ for } \gamma < \gamma_{cr} & \text{(stable)} \end{cases}$$

This result holds only for values of the deformation that are appropriate to the approximation of the sine function. We can see a remarkable similarity between this problem and the discrete one-degree-of-freedom problem with a rotational spring. The bifurcation diagram for that problem is shown in Fig. 128. For that problem, negative values of the deformation gave rise to a secondary branch above the branch for the perfect system. That branch had a limit point and was unstable for values of the deformation closer to zero (on the negative side) and stable for values more negative than the limit deformation. The value of the limit deformation was related to the magnitude of the initial imperfection.

For the present problem, the same phenomenon exists and the same interpretations hold. The value of the critical deformation γ_{cr} is given by the solution to the cubic equation

$$2\gamma^3 - 3\gamma_o\gamma^2 + 8\gamma_o = 0$$

It is straightforward to show that this cubic equation has one real root that can be computed for specific values of the initial imperfection. Numerical computations show that this root is indeed negative. Some values of the critical deformation, relative to the size of the imperfection, are given in Table 8. For small values of the imperfection, we can observe that the critical deformation is given approximately by

$$\gamma_{cr} \approx -\left(4\gamma_o\right)^{1/3}$$

Using this approximate value in the equation for the load, Eqn. (561), we find that the limit load is given approximately by the expression

$$P_{max} \approx P_1\left(1+0.95\gamma_o^{2/3}\right)$$

in accord with the two-thirds power law of Koiter for the symmetric bifurcation. Hence, we see that the two-thirds power law applies to limit points above the perfect bifurcation curve (limit points associated with stable post-buckling behavior) as well as limit points below (limit points associated with unstable post-buckling behavior).

Table 8 Solutions to equation for critical deformation
$2\gamma^3 - 3\gamma_o\gamma^2 + 8\gamma_o = 0$

γ_o	γ_{cr}
0.00001	-0.0342
0.0001	-0.0736
0.001	-0.1582
0.01	-0.3371
0.1	-0.6900

The bifurcation diagram for the model problem is shown in Fig. 144 (the imperfection diagram was computed with $\gamma_o = 0.04$).

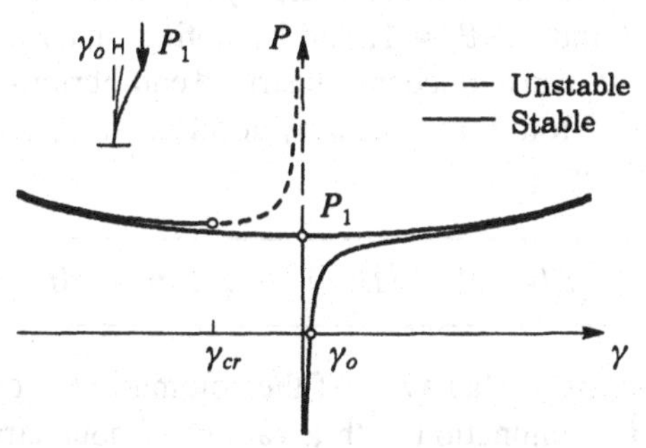

Figure 144 The effect of an imperfection for Euler's elastica

We can repeat the foregoing analysis and show that similar results hold for the elastica with simple supports. Although the values of the critical loads are different in the two cases (the fundamental critical load of the pinned column is $\mu_1 = \pi/\ell$ rather than $\mu_1 = \pi/2\ell$ for the cantilever column), the post-critical load has the same expression. As such, one might consider the stability of the first branch as a property of the differential operator, independent of the boundary conditions, while the value of the buckling load is very much a property of the boundary conditions, too. This observation applies only to cases in which the internal forces can be related directly to the applied load, i.e., statically determinate columns. Assemblages of elastic members can display remarkably different behavior; the asymmetric bifurcation of frames is a case in point.

The remarkable similarity between the behavior of the elastica and the simple systems with rigid links and rotational springs suggests that these simple systems are actually very good models of the behavior of their continuous counterparts. Thus, there is great value in studying these discrete systems, as the analytical overhead is considerably smaller. We must be extremely careful, however, in extrapolating results from simple systems to more complicated systems. We must also recognize that, in effect, we have reduced the continuous system to a single-degree-of-freedom system with a Ritz approximation using the critical mode as the base function. The Ritz approach always generates a discrete system (i.e., algebraic equations rather that differential or integral) and provides a rigorous connection between discrete and continuous systems. The discretization process must be done with extreme caution, however, because the Ritz method will discretize the system in exactly the manner that you ask it to, hidden constraints and all. Clearly, we are on fairly solid ground choosing an eigenfunction of the linearized problem as our base function since this function, at least initially, satisfies all boundary conditions, essential and natural.

The General Linearized Buckling Theory

Let us linearize the inextensible beam theory by making small angle approximations $\sin\theta \approx \theta$ and $\cos\theta \approx 1$, and by noting that $w' \approx \theta$ and $u' \approx 0$. From Eqns. (541), we can deduce the linearized equilibrium equations. We can eliminate the force V from Eqn. $(541)_c$ by differentiating it once and substituting Eqn. $(541)_b$ to get

$$\boxed{(EIw'')'' - (Hw')' - q + m' = 0} \tag{562}$$

This equation governs the behavior of the column in the domain $x \in [0, \ell]$ and can be used in conjunction with a variety of boundary conditions. The boundary conditions are, for the most part, the same as for the linear theory, and are illustrated in Fig. 145. The fixed support has displacement $w = 0$ and rotation $w' = 0$ with unknown moment M and transverse reaction V, the simple support has displacement $w = 0$ and moment $M = 0$ with unknown rotation and transverse reaction, the free end has moment $M = 0$ and transverse reaction $V = 0$. This last condition deserves special attention.

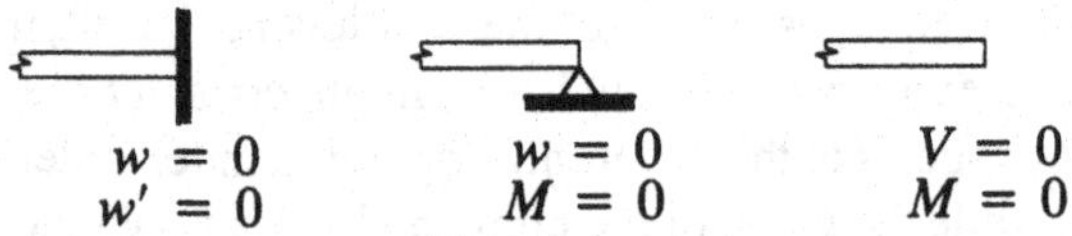

$w = 0$, $w' = 0$ $w = 0$, $M = 0$ $V = 0$, $M = 0$

Figure 145 Boundary conditions for a beam

Because we have used the constraints that axial and shear deformations are negligible, we have given away our ability to determine shear and axial force from a constitutive equation. Any force associated with a kinematic constraint must be determined from an equilibrium equation. Just as we did for the Bernoulli-Euler beam, we shall appeal to the moment equilibrium equation to determine how to translate the condition $V = 0$ to terms involving the displacement field $w(x)$. From Eqn. (541), with $u' = 0$ and $M = EIw''$, we find that

$$V = -EIw''' + Hw' - m \tag{563}$$

Thus, we see that transverse force is no longer related simply to the third derivative of displacement because of the effect of axial force. This boundary condition is very important, as we will see in one of the following examples.

The force H is also associated with a constraint, and, therefore, cannot be determined from a constitutive equation. All of the problems we consider here will be statically determinate with respect to the axial force, and, hence, we will be able to find H from purely statical considerations. Remember that H is positive if it produces a net tension. For the model problem, we have the relationship $H(x) = -P$.

The virtual-work and energy functionals. In accord with the conditions of linearization the general virtual-work functional, Eqn. (542), takes the form

$$G(w,\overline{w}) = \int_0^\ell \left(EI w''\overline{w}'' + Hw'\overline{w}' - q\overline{w} - m\overline{w}'\right) dx \tag{564}$$

In accord with the principle of virtual work, $w(x)$ is an equilibrium configuration if $G(w,\overline{w}) = 0$ for all $\overline{w} \in \mathfrak{B}_e(0,\ell)$. The admissible functions in the collection $\mathfrak{B}_e(0,\ell)$ must have square-integrable second derivatives and must satisfy the essential boundary conditions of the particular problem. The energy functional for the linearized problem is

$$\mathscr{E}(w) = \int_0^\ell \left(\tfrac{1}{2}EI(w'')^2 + \tfrac{1}{2}H(w')^2 - qw - mw'\right) dx \tag{565}$$

The astute reader might be wondering what happened to the external work done by the distributed axial force $p(x)$. It is, indeed, contained in the H term as the following example shows.

Example 72. *Accounting for the energy of the applied axial load.* Consider the model problem with an axial force P applied at $x = \ell$ and a distributed axial load $p(x)$. The axial displacement must conform to the inextensibility constraint $1+u' = \cos\theta$. Since $\theta \approx w'$, we have that $u' \approx -\frac{1}{2}(w')^2$, in accord with the Taylor expansion of the cosine function. Thus, the axial displacement is

$$u(x) = u(0) - \int_0^x \tfrac{1}{2}(w')^2\, dx \tag{566}$$

The potential energy possessed by the end load P and the distributed load p, relative to the undeformed position of the beam can be expressed as

$$\mathscr{E}_E = Pu(\ell) - \int_0^\ell p(x)u(x)\, dx \tag{567}$$

Substituting Eqn. (566) into Eqn. (567), noting that $u(0) = 0$, we get

$$\mathscr{E}_E = -P\int_0^\ell \tfrac{1}{2}(w')^2 dx + \int_0^\ell p(x)\int_0^x \tfrac{1}{2}(w'(\xi))^2 d\xi\, dx \tag{568}$$

Integrating the double integral on the right by parts, we obtain

$$\mathscr{E}_E = -P\int_0^\ell \tfrac{1}{2}(w')^2\, dx + \int_0^\ell \tfrac{1}{2}(w')^2\, dx \int_0^\ell p\, dx - \int_0^\ell \tfrac{1}{2}(w')^2 \int_0^x p(\xi)d\xi\, dx$$

Finally, noting from Eqn. (550) that the axial force $H(x)$ is given by

$$H(x) = -P + \int_x^\ell p(\xi)\, d\xi$$

we arrive at the result that the potential energy possessed by the end load and the distributed axial load can be computed as

$$\mathcal{E}_E = \int_0^\ell \tfrac{1}{2} H(w')^2 \, dx$$

The resultant axial force $H(x)$ contains the contribution of both the end load and the distributed load. The derivation shows that Eqn. (564) rigorously accounts for all of the potential energy for any distribution of axial forces.

Classical solution to the model problem. Consider again the model problem shown in Fig. 142. The column has no distributed loads applied along its length. Hence, $q = p = m = 0$. In addition, the axial force in the column is given by $H = -P$. For constant *EI*, Eqn. (562) takes the special form

$$EIw^{iv} + Pw'' = 0 \tag{569}$$

where $(\cdot)^{iv}$ indicates the fourth derivative of $(\cdot)$. This equation is, in essence, a linear eigenvalue problem for a differential operator. It asks whether there are functions that satisfy the condition that their fourth derivative is equal to the negative of their second derivative multiplied by a constant. The trigonometric functions $\sin\mu x$ and $\cos\mu x$ are just such functions. Actually, linear combinations of these functions are the only functions that satisfy this relationship. Such functions are called ***harmonic***. Clearly, since the lowest-order derivative in the equation is a second derivative, any constant and linear expression will also satisfy the equation. Therefore, we can express the general solution to Eqn. (562) as

$$w(x) = a_0 + a_1 x + a_2 \sin\mu x + a_3 \cos\mu x \tag{570}$$

where, again, the notation $\mu^2 \equiv P/EI$ is introduced for convenience. The fourth-order equation gave rise to four constants of integration, as they always do. These constants can be determined from the boundary conditions appropriate to a specific problem.

Example 73. *Classical solution to the linearized problem.* The boundary conditions for the model problem are vanishing displacement at the base, $w(0) = 0$, which gives the equation

$$a_0 + a_3 = 0 \tag{a}$$

vanishing rotation at the base, $w'(0) = 0$, which gives the equation

$$a_1 + \mu a_2 = 0 \tag{b}$$

vanishing moment at the tip, $w''(\ell) = 0$, which gives the equation

$$\mu^2[a_2 \sin\mu\ell + a_3 \cos\mu\ell] = 0 \tag{c}$$

and vanishing shear at the tip, $w'''(\ell) + \mu^2 w'(\ell) = 0$, which gives the equation

$$\mu^3[a_2 \cos\mu\ell - a_3 \sin\mu\ell] - \mu^2[a_1 + \mu(a_2 \cos\mu\ell - a_3 \sin\mu\ell)] = 0 \tag{d}$$

These equations can be solved to give $a_1 = a_2 = 0$ and $a_0 = -a_3$ along with the important condition

$$a_3 \cos\mu\ell = 0 \tag{571}$$

This equation is identical to the one obtained for the model problem previously. It has the trivial solution $a_3 = 0$, corresponding to the straight configuration, as well as an infinite number of nontrivial solutions for load values

$$\mu_n = \frac{(2n-1)\pi}{2\ell}$$

which satisfy $\cos\mu\ell = 0$. We can extract the critical loads as $P_n = \mu_n^2 EI$. The eigenfunction corresponding to μ_n is given by Eqn. (570) with $a_1 = a_2 = 0$ and $a_0 = -a_3$. To wit

$$w_n(x) = a_3(1 - \cos\mu_n x)$$

Again, the buckled configuration can be determined up to an arbitrary amplitude. The constant a_3 cannot be determined from the governing equations. The bifurcation diagram and the first three buckling mode shapes are shown in Fig. 146. As we shall soon see, for loads $P < P_1$ the straight configuration is stable, but for $P > P_1$ it is unstable. Because we have performed a linearized buckling analysis, we are unable to determine the stability of the bent configurations.

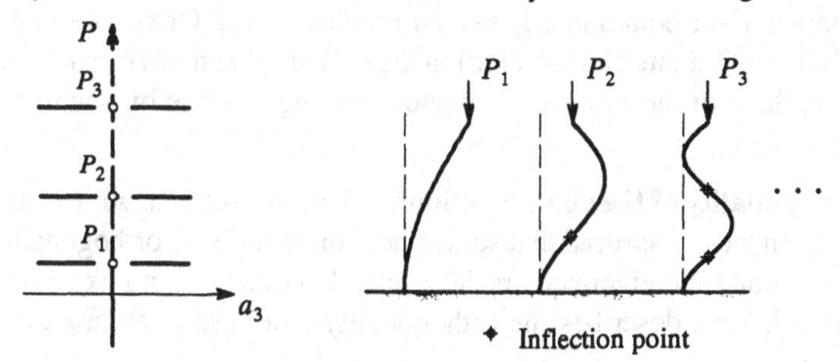

Figure 146 The bifurcation diagram and buckling mode shapes of the cantilever column

The buckling eigenfunctions have some interesting properties. First, the eigenfunctions become increasingly tortuous the higher the mode number n. Tortuosity is the measure of how much the axis curves around in space. For a fixed amplitude of motion, the higher the tortuosity is, the more flexural potential energy the beam has stored. From the point of view of energy, then, it is easy to see why the first mode has the lowest critical buckling load. It is the mode that requires the least energy to deform into. A second measure of tortuosity is the number of inflection points (an *inflection point* is a point where $w'' = 0$). We can see that mode number n has $n-1$ inflection points between its ends. This observation will hold for all statically determinate boundary conditions. Each degree of indeterminacy adds another inflection point (i.e., a fixed-fixed beam has two inflection points in the first mode; a fixed-pinned beam has one).

The four equations that come from the boundary conditions that relate the four arbitrary constants $\mathbf{a} = \{a_1, a_2, a_3, a_4\}$ and the load parameter μ will not always reduce to such a simple equation, as they did in this example. These four equations will always be linear in $\mathbf{a}$, but usually nonlinear in μ. The equations will always be homogeneous. Thus, we can write the general form of the equations as

$$\mathbf{B}(\mu)\mathbf{a} = \mathbf{0} \tag{572}$$

where the exact character of the four by four matrix $\mathbf{B}(\mu)$ will depend upon the specific boundary conditions of the problem. Eqn. (572) is solvable for nonzero $\mathbf{a}$ only if the determinant of the coefficient matrix vanishes. Thus, the characteristic equation for the bifurcation load parameter μ is

$$\boxed{\det \mathbf{B}(\mu) = 0} \tag{573}$$

The characteristic equation is generally nonlinear in μ. Often, one or more of the a_i can be eliminated by substitution to give a smaller matrix. In the previous example, three of the a_i were eliminated, leading to a one by one matrix.

Orthogonality of the eigenfunctions. Whenever we deal with an eigenvalue problem we are assured that somewhere there lurks an orthogonality relationship among the eigenvectors. The present problem is no exception. The following lemma describes the orthogonality relationship among the eigenfunctions.

Lemma (*Orthogonality of the eigenfunctions*). Let μ_n and $w_n(x)$ be the nth eigenvalue and eigenfunction that satisfy the eigenvalue problem $w^{iv} + \mu^2 w'' = 0$. For distinct eigenvalues, i.e., $\mu_n^2 \neq \mu_m^2$, the eigenfunctions $w_n(x)$ and $w_m(x)$ satisfy the following orthogonality relationships

$$\int_0^\ell w_n' w_m' \, dx = 0 \tag{574}$$

$$\int_0^\ell w_n{}''w_m{}''\,dx = 0 \tag{575}$$

$$\int_0^\ell (w_n{}'')^2\,dx = \mu_n^2\int_0^\ell (w_n{}')^2\,dx \tag{576}$$

Proof. The proof of orthogonality of the first derivatives is straightforward. Let us start with an expression we know to be zero

$$\int_0^\ell \left[(w_n^{iv} + \mu_n^2 w_n{}'')w_m - (w_m^{iv} + \mu_m^2 w_m{}'')w_n\right]dx = 0$$

This integral is zero because each of the terms in parentheses is zero. We shall proceed to integrate each term by parts until all of the derivatives balance. The result of these integrations is

$$\begin{aligned}(\mu_m^2 - \mu_n^2)\int_0^\ell w_n{}'w_m{}'\,dx &+ \int_0^\ell \left[w_n{}''w_m{}'' - w_m{}''w_n{}''\right]dx \\ &+ (w_n{}''' + \mu_n^2 w_n{}')w_m\Big|_0^\ell - (w_m{}''' + \mu_m^2 w_m{}')w_n\Big|_0^\ell \\ &- w_n{}''w_m{}'\Big|_0^\ell + w_m{}''w_n{}'\Big|_0^\ell = 0\end{aligned}$$

Clearly, the second integral vanishes identically. The boundary terms all vanish because, in order for the boundary value problem to be properly posed, we must have either zero displacement or zero shear, and either zero slope or zero moment at an end point. All of the boundary terms have products of both pairs of items, one of which must be zero. We are left with the condition

$$(\mu_m^2 - \mu_n^2)\int_0^\ell w_n{}'w_m{}'\,dx = 0 \tag{577}$$

thus completing the proof of Eqn. (574). One can prove orthogonality of the second derivatives by considering the virtual-work equation with $w = w_n$ and $\overline{w} = w_m$. With these choices of functions, we have

$$G(w_n, w_m) = \int_0^\ell \left(w_n{}''w_m{}'' - \mu_n^2 w_n{}'w_m{}'\right)dx = 0$$

Since orthogonality of the first derivatives has already been established, this result proves orthogonality of the second derivatives. $G(w_n, w_n) = 0$ gives Eqn. (576) directly. ❑

If the eigenvalues are distinct, then the orthogonality relationship holds. If $\mu_n^2 = \mu_m^2$, then Eqn. (577) is satisfied without the eigenfunctions being orthogonal. In this case, like the discrete case, we have the result that the eigenfunctions associated with the repeated eigenvalue form a subspace. Any function that is a linear combination of eigenfunctions from this subspace is also an eigenfunction. Thus, orthogonality is not necessary in the subspace. We can, as usual, create orthogonal functions from any set of functions in this subspace by Gram-Schmidt orthogonalization. You should be aware that the orthogonality condition applies to the first and second derivatives of the eigenfunctions, not to the functions themselves.

The eigenfunctions provide a convenient basis for computations of problems that are almost like the eigenvalue problem, such as problems with transverse loading in addition to the axial thrust and problems with initial imperfections. Any function can be expressed as an infinite sum of eigenfunctions. These functions are particularly convenient because they have all of the boundary conditions satisfied at the outset. An eigenfunction expansion is also convenient for establishing the stability criterion.

The stability of equilibrium. Although we are working with a linearized buckling theory, we can still expect the second-derivative test to give insight into the stability of the straight configuration. The second derivative of $\mathcal{E}(w)$ for the present problem is

$$A(w, \overline{w}) = \int_0^\ell \left[EI(\overline{w}'')^2 - P(\overline{w}')^2\right] dx$$

Testing the sign of the second derivative for a continuous problem is not quite the same as for the discrete problem. We must establish the algebraic sign of A for all functions $\overline{w} \in \mathfrak{B}_e(0, \ell)$. The easiest way to implement this criterion is to use an eigenfunction expansion. Let us assume that our test function is a linear sum of eigenfunctions, as follows

$$\overline{w}(x) = \sum_{n=1}^{\infty} \overline{a}_n w_n(x)$$

where w_n is the nth eigenfunction and the constants $\overline{a}_n$ are arbitrary. The second-derivative functional is now a function of the arbitrary constants $\overline{a}_n$ and has the form

$$A(\mathbf{a}, \overline{\mathbf{a}}) = \sum_{n=1}^{\infty} \sum_{m=1}^{\infty} \overline{a}_n \overline{a}_m \int_0^\ell \left(EIw_n''w_m'' - Pw_n'w_m'\right) dx$$

Noting the orthogonality of the eigenfunctions, this expression reduces to

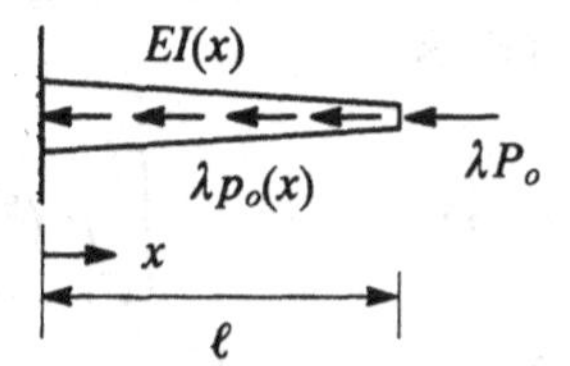

Figure 147 Beam of varying modulus subjected to proportional loads

$$A(\mathbf{a},\bar{\mathbf{a}}) = \sum_{n=1}^{\infty} \bar{a}_n^2 (P_n - P) \int_0^{\ell} (w_n')^2 \, dx$$

Since the integral of the square of the slope of an eigenfunction is never negative (in fact, the eigenfunctions can be normalized so that this integral is unity), the second-derivative test reduces to

$$A(\mathbf{a},\bar{\mathbf{a}}) = \sum_{n=1}^{\infty} \bar{a}_n^2 (P_n - P) = \begin{cases} > 0 & \text{for } P < P_1 \quad \text{(stable)} \\ < 0 & \text{for } P > P_1 \quad \text{(unstable)} \end{cases}$$

because if $P > P_1$, one need only choose $\bar{a}_1 \neq 0$ with all others equal to zero to show that the second derivative is less than zero for some choice of $\bar{\mathbf{a}}$. Clearly, the second-derivative test tells us nothing about the stability of the nontrivial equilibrium branches.

Ritz and the Linearized Eigenvalue Problem

The virtual-work functional for the linearized buckling problem is given by Eqn. (564). The variational principle suggests that if $G = 0$ for all suitable virtual displacements, then the system is in equilibrium. As we have seen previously, the virtual displacement functions need only satisfy the essential boundary conditions. The real displacements can be expressed in terms of functions that also satisfy only the essential boundary conditions. As we saw in the example for the little boundary value problem with a sinusoidal load, the natural boundary conditions are recovered through the principle of virtual work as the size of the approximating basis increases. We shall see that the principle of virtual work gives rise to the buckling eigenvalue problem, which provides us with a tool for estimating the critical points and buckling modes of our continuous column.

To set up the discussion of applying the Ritz method to the buckling problem, consider the beam shown in Fig. 147. A requirement of the buckling analysis is that the loads be proportional, that is, the spatial distribution of the loads is fixed and the magnitude of each load varies in accord with the load factor λ. In the figure, P_o is an applied end load of fixed magnitude and $p_o(x)$ is an applied distributed load of fixed amplitude. The total load is given by the aggregate of the loads λP_o and $\lambda p_o(x)$. Let us designate the axial force under the nominal loads P_o and $p_o(x)$ as

$$H_o(x) = -P_o - \int_x^\ell p_o(\xi)\, d\xi$$

so that $H(x) = \lambda H_o(x)$. Note that the algebraic sign of all loading terms is in accord with the figure, that is, all of the loads are oriented to induce compression in the beam. According to our convention, $H(x)$ is positive for tensile loads and negative for compressive loads.

We are now prepared to apply the Ritz method. Consider the set of base functions $\mathbf{h}(x) = [h_1(x), h_2(x), \ldots, h_n(x)]^T$. The real and virtual displacements can be expressed as linear combinations of these base functions, to wit

$$w(x) = \sum_{i=1}^{n} a_i h_i(x) = \mathbf{a} \cdot \mathbf{h}(x), \quad \overline{w}(x) = \sum_{i=1}^{n} \overline{a}_i h_i(x) = \overline{\mathbf{a}} \cdot \mathbf{h}(x)$$

Let $\mathbf{a} = [a_1, \ldots, a_n]^T$ and $\overline{\mathbf{a}} = [\overline{a}_1, \ldots, \overline{a}_n]^T$ be vectors containing the coefficients of the base functions used in the approximations. We can substitute these approximations into Eqn. (564) to get the discrete functional

$$\boxed{G(\mathbf{a}, \overline{\mathbf{a}}) = \overline{\mathbf{a}}^T \left[\mathbf{K}\mathbf{a} - \lambda \mathbf{G}\mathbf{a}\right]} \tag{578}$$

where the matrices $\mathbf{K}$ and $\mathbf{G}$ have components given by integrals of derivatives of the base functions

$$\mathbf{K} \equiv \int_0^\ell EI[\mathbf{h}''][\mathbf{h}'']^T\, dx, \quad \mathbf{G} \equiv -\int_0^\ell H_o[\mathbf{h}'][\mathbf{h}']^T\, dx \tag{579}$$

The matrix $\mathbf{K}$ is generally referred to as the stiffness matrix while the matrix $\mathbf{G}$ is generally referred to as the geometric (stiffness) matrix. As usual, the variational statement that $G(\mathbf{a}, \overline{\mathbf{a}}) = 0$ for all $\overline{\mathbf{a}}$ implies equilibrium of the system. In this case, those equilibrium equations give the classical buckling eigenvalue problem

$$\boxed{\mathbf{K}\mathbf{a} = \lambda \mathbf{G}\mathbf{a}} \tag{580}$$

Clearly, $\mathbf{a} = \mathbf{0}$ is a solution to Eqn. (580) for any value of λ, and, thus, represents an equilibrium configuration. Like any eigenvalue problem, we can expect Eqn. (580) to have a nontrivial solution, $\mathbf{a} \neq \mathbf{0}$, only for certain values of the load parameter λ. These values are the bifurcation loads. The two coefficient matrices $\mathbf{K}$ and $\mathbf{G}$ are n by n. Therefore, the matrix eigenvalue problem gives rise to n pairs of eigenvalues and eigenvectors: $\{\lambda_i, \mathbf{a}_i\}$, for $i = 1, \ldots, n$. The eigenvalues are estimates of the actual eigenvalues of the continuous system. The eigenvectors, when used as coefficients of the base functions $h_i(x)$, are estimates of the eigenfunctions $w_i(x)$. As such, the Ritz method gives us a tool for approximating the eigenvalues and eigenfunctions from any set of base functions that satisfy the essential boundary conditions.

If the eigenfunctions themselves are used as the base functions, both **K** and **G** are diagonal. The ratio of the diagonal element K_{ii}/G_{ii} (no sum implied) is the eigenvalue μ_i^2, and the eigenvectors are given by the standard base vectors in $\mathcal{R}^n$ (i.e., the ith standard base vector has a one in the ith slot and zeros in all of the other slots). This result is a direct consequence of the orthogonality property of the eigenfunctions.

There are many techniques for solving the algebraic eigenvalue problem, Eqn. (580), numerically. Certainly, we can endeavor to find the roots of the determinantal characteristic equation

$$\mathcal{P}(\lambda) \equiv \det\left[\mathbf{K}-\lambda\mathbf{G}\right] = 0 \tag{581}$$

These roots are the critical values we seek. This approach is the one we used to find the principal values of the stress and strain tensors. The key difference in the present problem is that the characteristic polynomial $\mathcal{P}(\lambda)$ is of nth order if **K** and **G** are n by n matrices, compared with order three for the principal-values problem. Clearly, the mechanics of finding the n roots gets increasingly difficult as n gets large. Most methods for large systems use either a matrix iteration technique or a matrix diagonalization technique. For the small problems we tackle here, n usually will not be too large, and we can continue to view the algebraic eigenvalue problem as one of finding the roots of $\mathcal{P}(\lambda)$.

Example 74. *Column buckling by the Ritz method.* Let us reexamine the linearized buckling problem for the cantilever column with constant modulus EI, length ℓ, and compressive end load λP. We know that the exact eigenfunctions are given by cosine functions. Can we get reasonable results using a polynomial basis? Consider the polynomial approximation of the real and the virtual displacement given by the base functions

$$h_1(x) = \frac{x^2}{\ell}, \quad h_2(x) = \frac{x^3}{\ell^2}, \quad \cdots, \quad h_n(x) = \frac{x^{n+1}}{\ell^n}$$

We have discarded the constant and linear base functions because the boundary conditions insist that displacement and slope vanish at $x = 0$. The stiffness and geometric matrices are easily computed to have the components

$$K_{ij} = \frac{ij(i+1)(j+1)}{i+j-1}\frac{EI}{\ell}, \quad G_{ij} = \frac{(i+1)(j+1)}{i+j+1}P\ell \tag{582}$$

Let us carry out the computation for a two-term basis, i.e., $n = 2$. For this case, the stiffness and geometric stiffness have the specific values

Table 9 Approximations of the buckling eigenvalues with polynomial basis

n	λ_1	λ_2	λ_3	λ_4	λ_5	λ_6
1	3.0000					
2	2.4860	32.181				
3	2.4678	23.391	109.14			
4	2.4674	22.322	69.404	265.81		
5	2.4674	22.214	63.028	148.21	545.75	
6	2.4674	22.207	61.863	127.21	271.61	1002.7
Exact	2.4674	22.207	61.685	120.90	199.9	298.6

$$\mathbf{K} = \frac{EI}{\ell}\begin{bmatrix} 4 & 6 \\ 6 & 12 \end{bmatrix}, \qquad \mathbf{G} = \frac{P\ell}{60}\begin{bmatrix} 80 & 90 \\ 90 & 108 \end{bmatrix}$$

In order for there to be a nontrivial solution we must have $\det[\mathbf{K}-\lambda\mathbf{G}] = 0$. Let us define $\bar{\lambda} \equiv \lambda P\ell^2/EI$. Multiplying $\mathbf{K}-\lambda\mathbf{G}$ by ℓ/EI and taking the determinant, we get

$$\det\begin{bmatrix} 4-\frac{4}{3}\bar{\lambda} & 6-\frac{3}{2}\bar{\lambda} \\ 6-\frac{3}{2}\bar{\lambda} & 12-\frac{9}{5}\bar{\lambda} \end{bmatrix} = \frac{3}{20}\bar{\lambda}^2 - \frac{26}{5}\bar{\lambda} + 12 = 0$$

The roots to this quadratic equation are $\bar{\lambda}_1 = 2.4860$ and $\bar{\lambda}_2 = 32.1807$. These eigenvalues are approximations to the first two critical loads of the column. The eigenvectors can be obtained by substituting the eigenvalues back into

$$[\mathbf{K}-\lambda_i\mathbf{G}]\mathbf{a}_i = \mathbf{0}$$

assuming that one of the components of $\mathbf{a}_i$ is known, and solving for the remaining components. For example, for $\bar{\lambda}_1 = 2.4860$ we have

$$\begin{bmatrix} 41.12 & 136.26 \\ 136.26 & 451.51 \end{bmatrix}\begin{bmatrix} 1 \\ a \end{bmatrix} = \begin{bmatrix} 0 \\ 0 \end{bmatrix}$$

from which we get $a = 0.3018$ (from either of the two equations). Therefore, the eigenvector associated with $\bar{\lambda}_1 = 2.4860$ is $\mathbf{a}_1 = (1.0, 0.3018)$. One can follow the same procedure for $\bar{\lambda}_2 = 32.1807$ to get $\mathbf{a}_2 = (1.0, -0.9204)$.

Table 9 shows the results of increasing the number of terms n in expansion for $w(x)$ and $\bar{w}(x)$ in the previous example along with the exact results obtained previously. The one-term expansion gives a surprisingly good result, indicating that the quadratic function is a reasonably good approximation of the eigenfunction. Of course, the one-term expansion gives rise to only one eigenva-

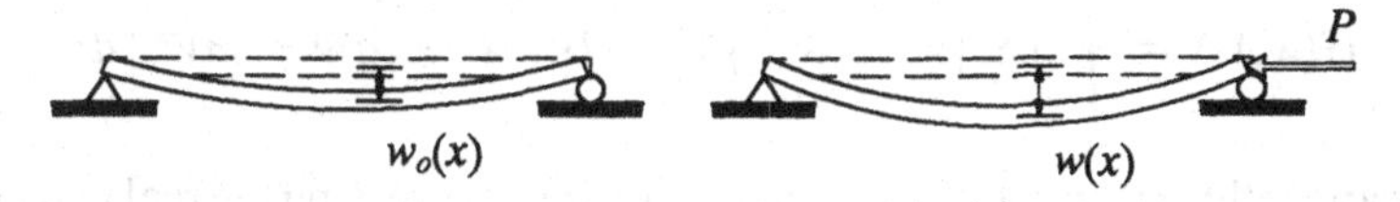

Figure 148 Beam with an initial geometric imperfection

lue estimate. The two-term expansion gives rise to two estimates. The lower value is a remarkably accurate estimate of the first critical load. Presumably, the second value is an estimate of the second critical load, but the accuracy is not very good. Three terms in the expansion improves on the existing estimates and introduces an estimate of the third critical load. Clearly, as the higher modes come in, they are increasingly inaccurate, a consequence of the higher-order polynomials being less and less suitable approximations of the higher eigenfunctions. None of them has an inflection point. Thus, none of them, alone, is a good approximation of any mode other than the first. However, in combination, they are able to capture the shapes with inflection points. When a new base function is introduced, its shape is used by the functional mostly to improve the representation of the lower modes, and very little to represent the new mode that has appeared owing to the increase in the order of the discretization. Therefore, the mode shapes converge much more slowly than the estimates of the critical loads, by an order of magnitude, in fact.

We can observe that the value of the critical loads converges from above, a hallmark of displacement-based approximations. It is tempting to talk about rules of thumb regarding how many terms we need to get an acceptable approximation of the nth critical load. For this problem, we might be tempted to say that about $2n$ terms are needed to get the nth critical load approximately correct. Such a rule of thumb depends a great deal on the specific base functions we use. The best possible base functions are the eigenfunctions, and even they require n terms to get the nth critical load (but it is exact as soon as it comes into the picture).

Ritz analysis of imperfections. When studying the discrete systems in the previous chapter, we saw that systems with imperfections displayed behavior different than those without imperfections if the system without imperfections had bifurcation points. We can apply the Ritz method to the analysis of continuous systems with imperfections using the general linearized buckling theory. Let us continue to measure the transverse displacement $w(x)$ from the straight position, but assume that the deflection is known to be $w_o(x)$ when the system is unloaded (i.e., when $P = 0$), as shown in Fig. 148. With this convention, the moment at any section is $M = EI(w'' - w_o'')$. Note that for the configuration $w(x) = w_o(x)$, there is no moment in the beam. The virtual-work functional given in Eqn. (564) can be revised to reflect the imperfection as

$$G(w,\overline{w}) = \int_0^\ell \left[EI\left(w''-w_o''\right)\overline{w}''+Hw'\overline{w}' - q\overline{w} - m\overline{w}'\right]dx \quad (583)$$

Let us apply the Ritz method as in the last section, expressing the real and virtual displacement fields as

$$w(x) = \sum_{i=1}^{n} a_i h_i(x) = \mathbf{a}\cdot\mathbf{h}(x), \quad \overline{w}(x) = \sum_{i=1}^{n} \overline{a}_i h_i(x) = \overline{\mathbf{a}}\cdot\mathbf{h}(x)$$

Let the axial forces be $H(x) = \lambda H_o(x)$where λ is a loading parameter and the force $H_o(x)$ is the internal axial force for the nominal pattern of applied axial loads. A straightforward computation shows that the discrete principle of virtual work takes the form

$$\boxed{G(\mathbf{a},\overline{\mathbf{a}}) = \overline{\mathbf{a}}^T\left[\mathbf{Ka}-\lambda\,\mathbf{Ga}-\mathbf{f}\right]} \quad (584)$$

where the matrices **K** and **G** are exactly the same as those given in Eqn. (579). The constant term **f** is due to the transverse forces and initial imperfections. Its ith component is given by

$$\mathbf{f} = \int_0^\ell \left(EIw_o''\mathbf{h}_i''+q\mathbf{h}_i+m\mathbf{h}_i'\right)dx \quad (585)$$

Equilibrium holds if $G(\mathbf{a},\overline{\mathbf{a}}) = 0$ for all $\overline{\mathbf{a}}$. From the discrete fundamental theorem of the calculus of variations, we must have

$$\left[\mathbf{K}-\lambda\mathbf{G}\right]\mathbf{a} = \mathbf{f} \quad (586)$$

which is a linear system of equations in the unknowns **a**. Note that for $\lambda = 0$ (no axial forces), the equations are exactly the same as those obtained for linear Bernoulli-Euler beam theory. For most values of λ, the equations yield a unique value of **a** for each **f**. However, it is evident that when $\lambda = \lambda_{cr}$ (the buckling eigenvalues from the preceding section), the coefficient matrix $\mathbf{K}-\lambda\mathbf{G}$ is singular, and, in general, no solution exists. It is apparent from this formulation that geometric imperfections affect the behavior of the beam in the same way that transverse loads do.

Example 75. ***Column with imperfection by Ritz method.*** **Reconsider the cantilever beam example from the previous section, except let the beam have an initial imperfection of $w_o(x) = \gamma_o x^2/\ell$, that is, a quadratic initial displacement with value $\gamma_o \ell$ at the end. Let us further assume that there are no transverse loads, $q = m = 0$. Using an n-term polynomial Ritz approximation with base functions $h_i(x) = x^{i+1}/\ell^i$, we get the same K and G of Eqn. (582). The components of the matrix f are given by**

$$f_i = \frac{2(i+1)EI\gamma_o}{\ell}$$

Let us again examine the case of a two-term expansion. For this case, Eqn. (586) takes the explicit form

$$\frac{EI}{\ell}\begin{bmatrix} 4-\frac{4}{3}\bar{\lambda} & 6-\frac{3}{2}\bar{\lambda} \\ 6-\frac{3}{2}\bar{\lambda} & 12-\frac{9}{5}\bar{\lambda} \end{bmatrix}\begin{bmatrix} a_1 \\ a_2 \end{bmatrix} = \frac{EI\gamma_o}{\ell}\begin{bmatrix} 4 \\ 6 \end{bmatrix}$$

where $\bar{\lambda} \equiv \lambda P\ell^2/EI$. These equations can be solved to give a_1 and a_2

$$a_1(\bar{\lambda}) = \frac{\gamma_o\left(80+12\bar{\lambda}\right)}{\left(\bar{\lambda}-\bar{\lambda}_1\right)\left(\bar{\lambda}-\bar{\lambda}_2\right)}, \quad a_2(\bar{\lambda}) = \frac{-80\gamma_o\bar{\lambda}}{6\left(\bar{\lambda}-\bar{\lambda}_1\right)\left(\bar{\lambda}-\bar{\lambda}_2\right)}$$

where $\bar{\lambda}_1 = 2.4860$ and $\bar{\lambda}_2 = 32.1807$ are the estimates of the critical load factors for the problem without imperfection. The approximate deflection of the beam under load is given by the expression

$$w(x) = \frac{80\gamma_o}{\left(\bar{\lambda}-\bar{\lambda}_1\right)\left(\bar{\lambda}-\bar{\lambda}_2\right)}\left[\frac{x^2}{\ell} + \bar{\lambda}\left(\frac{3x^2}{20\ell} - \frac{x^3}{6\ell^2}\right)\right]$$

Observe that $\bar{\lambda}_1\bar{\lambda}_2 = 80$. Therefore, $a_1(0) = \gamma_o$ and $a_2(0) = 0$, indicating that the displaced configuration at zero load is simply the initial imperfection. For small values of the load ($\bar{\lambda} \ll \bar{\lambda}_1 < \bar{\lambda}_2$), the displacement increases approximately linearly with load in the direction $\Delta w = w - w_o$ as

$$\Delta w(x) = \gamma_o\bar{\lambda}\left(\frac{3x^2}{20\ell} - \frac{x^3}{6\ell^2}\right)$$

As the load approaches the first critical value, i.e., $\bar{\lambda} \to \bar{\lambda}_1$, the displacements tend toward infinity, as we expect from the linearized analysis.

It is interesting to note that the expression for the displacement for an imperfection problem will generally have the form

$$w(x) = \frac{g(x,\lambda)}{\mathcal{P}(\lambda)}$$

where $\mathcal{P}(\lambda) = (\lambda-\lambda_1)(\lambda-\lambda_2)\cdots(\lambda-\lambda_n)$, with λ_i being the ith critical load factor, is the characteristic polynomial for the buckling load of the associated problem without imperfections, and $g(x)$ is a function of x that depends upon the particular characteristics of the problem. Because of the nature of the denominator, the displacements will always increase without bound as the load approaches the first critical load.

Additional Reading

Z. P. Bazant and L. Cedolin, *Stability of structures: Elastic, inelastic, fracture and damage theories*, Oxford University Press, New York, 1991.

H. L. Langhaar, *Energy methods in applied mechanics*, Wiley, New York, 1962.

A. E. H. Love, *A treatise on the mathematical theory of elasticity*, Dover, New York, 1944.

Problems

261. Vainberg's theorem is simply a statement of integrability. This theorem can be applied to the strain variations that we derive through a virtual work argument for a nonlinear planar beam. Let $\mathbf{u} \equiv \{u, w, \theta\}$ and $\bar{\mathbf{u}} \equiv \{\bar{u}, \bar{w}, \bar{\theta}\}$ be the real and virtual displacements and rotation. From the principle of virtual work, we have found that the virtual curvature is given by $\bar{\kappa}_o(\mathbf{u}, \bar{\mathbf{u}}) = \bar{\theta}'$. Show that the symmetry condition holds for the virtual curvature, i.e., $D\bar{\kappa}_o(\mathbf{u}, \bar{\mathbf{u}}) \cdot \hat{\mathbf{u}} = D\bar{\kappa}_o(\mathbf{u}, \hat{\mathbf{u}}) \cdot \bar{\mathbf{u}}$ and, hence, that it is integrable. Show that the real curvature is given by $\kappa_o = \theta'$. Note that the directional derivative of $\bar{\kappa}_o$ is

$$D\bar{\kappa}_o(\mathbf{u}, \bar{\mathbf{u}}) \cdot \hat{\mathbf{u}} = \frac{d}{d\varepsilon}\left[\bar{\kappa}_o(\mathbf{u} + \varepsilon\,\hat{\mathbf{u}}, \bar{\mathbf{u}})\right]_{\varepsilon=0}$$

and the integral of the virtual curvature can be computed by Vainberg's formula as

$$\kappa_o = \int_0^1 \bar{\kappa}_o(t\mathbf{u}, \mathbf{u})\, dt$$

where $t\mathbf{u} = \{tu, tw, t\theta\}$. Repeat the calculation for the virtual shear and axial strains

$$\bar{\beta}_o(\mathbf{u}, \bar{\mathbf{u}}) = \bar{w}' \cos\theta - \bar{u}' \sin\theta - [w' \sin\theta + (1+u')\cos\theta]\bar{\theta}$$
$$\bar{\epsilon}_o(\mathbf{u}, \bar{\mathbf{u}}) = \bar{u}' \cos\theta + \bar{w}' \sin\theta + [w' \cos\theta - (1+u')\sin\theta]\bar{\theta}$$

to get the real shear and axial strains

$$\beta_o = w' \cos\theta - (1+u')\sin\theta$$
$$\epsilon_o = w' \sin\theta + (1+u')\cos\theta - 1$$

Take the directional derivatives of the real strains to verify that these results are correct.

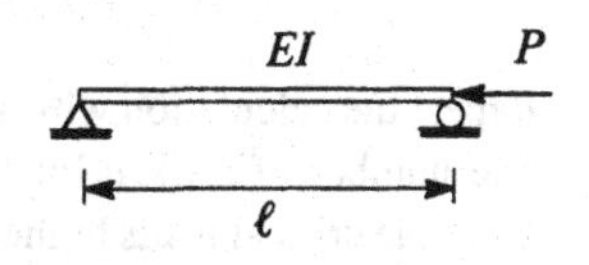

262. Consider the simply supported column of length ℓ and flexural modulus EI. Assume that shear and axial deformations are negligible, so that the constraints of Euler's elastica are appropriate. Compute the critical loads for this column by solving the equation $EIw^{iv} + Pw'' = 0$ with the appropriate boundary conditions. Carry out the stability analysis parallel to the analysis done for the cantilever model problem in the text.

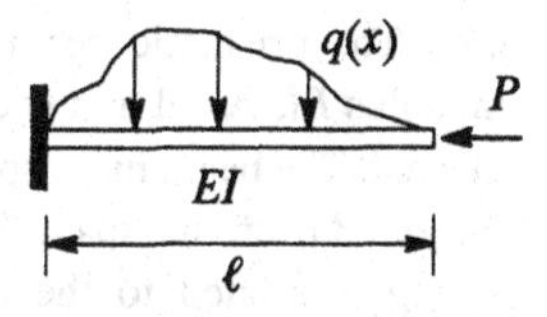

263. Consider the simply supported column of length ℓ and flexural modulus EI. Assume that shear and axial deformations are negligible so that constraints of Euler's elastica are appropriate. Can the classical elastica theory be extended to accommodate the transverse load $q(x)$? What difficulties do you encounter when you attempt to do so? Can the virtual-work principle for the elastica be modified to account for the transverse load?

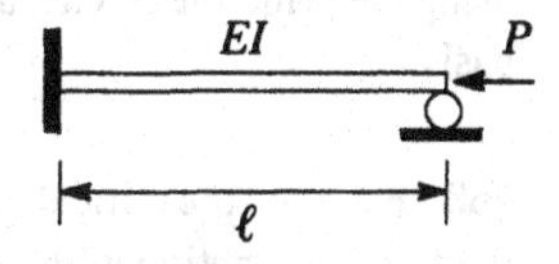

264. Consider the bar of length ℓ with bending modulus EI, fixed at the left end, propped at the right end, and subjected to axial load P as shown. Assume that shear and axial deformations are negligible. Compute the critical loads for this column by solving the classical differential equation. Estimate the critical loads using the principle of virtual work in conjunction with the Ritz method. Use a polynomial basis.

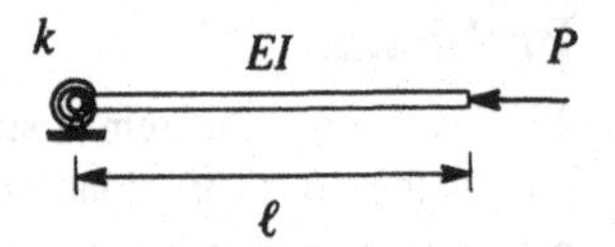

265. The prismatic beam shown below has a cross section that is symmetric with respect to the plane of the page. The cross section has flexural modulus EI. Axial and shear deformations can be neglected. The beam has a deformable spring support at the left end that elastically restrains rotations. The moment developed by the spring is related to the rotation at that point by $M_s = k\theta_s$, where $\theta_s = w'(0)$ is the rotation experienced by the spring. The right end of the beam is free to translate and to rotate. Solve the linearized buckling problem by the classical method, i.e., by integrating the differential equation. What are the appropriate boundary conditions for this problem? Solve the problem by integrating the differential equations and using the boundary conditions to find the constants of integration. What are the critical loads of the system? What is the smallest critical load as $k \to 0$? What is the smallest critical load as $k \to \infty$? Into what shapes does the beam deform at the critical loads? What are the shapes as $k \to 0$? What are the shapes as $k \to \infty$?

266. For Problem 265, the virtual-work functional that accounts for the work done by the springs and by the axial force is given by the expression

$$G(w, \bar{w}) = \int_0^\ell \left[EIw''\bar{w}'' - Pw'\bar{w}'\right] dx + kw'(0)\bar{w}'(0)$$

Estimate the buckling loads of the beam using a two-term polynomial expansion for the transverse deflection. That is, assume the real and virtual transverse deflections to be

$$w(x) = a_1 x + a_2 \frac{x^2}{\ell}, \quad \bar{w}(x) = \bar{a}_1 x + \bar{a}_2 \frac{x^2}{\ell}$$

Repeat the calculation with a three-term polynomial. The classical solution gives an infinite number of critical loads. How many did the two-term approximation give? Why? Were the critical loads higher or lower than the exact values? Why? Discuss what is good and bad about the assumed shapes. Could the approximating functions be improved easily? Suggest a better approximation.

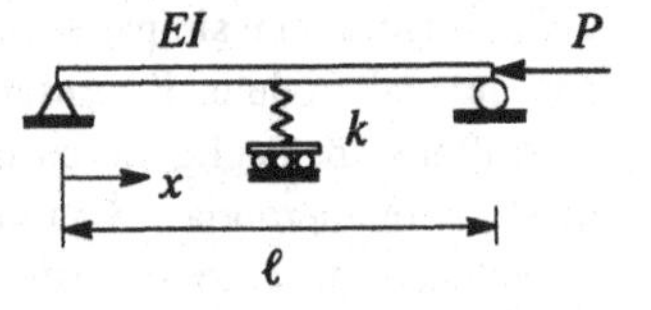

267. The prismatic beam shown below has flexural modulus EI. Axial and shear deformations can be neglected. The beam has a spring of modulus k located at the middle of the span. The force developed by the spring is related to the deflection at that point by $f(\ell/2) = kw(\ell/2)$. Find a suitable expression for the virtual-work functional that accounts for the virtual work done by the spring. Estimate the critical loads of the column using the principle of virtual work in conjunction with the Ritz method. Use a polynomial basis.

268. Estimate the critical loads of the column in Problem 267 using the principle of virtual work in conjunction with the Ritz method. Use the eigenbasis of the problem *without the spring*, that is $w_n(x) = \sin\mu_n x$, where $\mu_n = n\pi/\ell$.

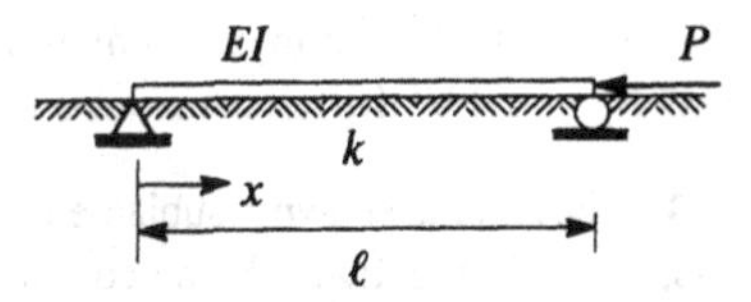

269. The prismatic beam shown below has flexural modulus EI. Axial and shear deformations can be neglected. The beam is supported on an elastic foundation of modulus k. The force developed, per unit length, by the foundation is related to the deflection at that point by $f(x) = kw(x)$. A beam on an elastic foundation with axial thrust is governed by the following (linearized) differential equation and boundary conditions

$$EIw^{iv} + Pw'' + kw = 0$$

$$w(0) = 0, \quad w''(0) = 0, \quad w(\ell) = 0, \quad w''(\ell) = 0$$

The eigenfunctions of the beam without the elastic foundation are $w_n = \sin n\pi x/\ell$. Verify that the virtual-work functional, accounting for the elastic foundation, is

$$G(w, \bar{w}) = \int_0^\ell \left(EI w''\bar{w}'' - Pw'\bar{w}' + kw\bar{w}\right) dx$$

Does the presence of the elastic foundation affect the boundary conditions? Find the buckling loads of the system using the Ritz method, assuming that the real and virtual displacements have the shape of the nth eigenfunction

$$w(x) = a \sin\frac{n\pi x}{\ell}, \quad \bar{w}(x) = \bar{a} \sin\frac{n\pi x}{\ell}$$

How does the buckling load vary with the elastic properties of the system, namely EI and k? Express your result in terms of $P_1 \equiv \pi^2 EI/\ell^2$ and the ratio of foundation stiffness to beam stiffness, given by the dimensionless parameter $\beta \equiv k\ell^4/\pi^4 EI$. (Hint: the critical buckling mode depends upon β.) Is your answer exact?

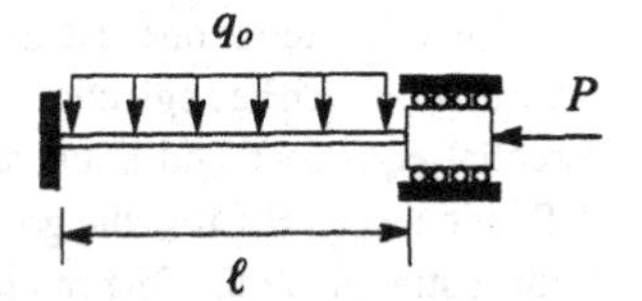

270. The prismatic beam shown has a cross section that is symmetric with respect to the plane of the page. The cross section has flexural modulus EI. Axial and shear deformations can be neglected. The beam is fixed against transverse deflection and rotation at both ends, but the supports provide no resistance to the axial force P. The beam is also subjected to a uniform transverse load of magnitude q_o. The linearized buckling theory for a beam with transverse load and axial thrust gives rise to the following differential equations and boundary conditions for the present configuration

$$EIw^{iv} + Pw'' = q(x)$$

$$w(0) = 0, \quad w'(0) = 0, \quad w(\ell) = 0, \quad w'(\ell) = 0$$

Solve the governing differential equations by the classical method to find an expression for the transverse deflection $w(x)$ and the bending moment $M(x)$.

In the design of beams subjected to transverse load and thrust, sometimes called beam-columns, the concept of the magnification factor is often used. The idea behind the magnification factor is that the influence of the axial thrust is to magnify the values of displacement and moment that would be present if the axial thrust were not (i.e., the solution if $P = 0$). Show that the maximum deflection and moment can be expressed as

$$w(\ell/2) = w_o\left(\frac{1}{1-P/P_1}\right), \quad M(0) = M_o\left(\frac{1-0.4P/P_1}{1-P/P_1}\right)$$

where w_o and M_o are the maximum deflection and moment that would occur if $P = 0$, and P_1 would be the fundamental critical load of the column if $q_o = 0$.

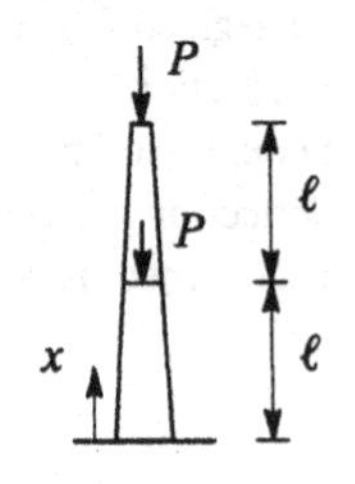

271. The column shown is subjected to axial forces at the midpoint and top, both of magnitude P. The column has variable flexural modulus given by the expression $EI(x) = EI_o(2 - x/2\ell)$. Estimate the buckling load of the system using the Ritz method with a one-term polynomial basis. Is your estimate higher or lower than the actual buckling load? Explain your answer. Propose a function for a one-term Ritz approximation that will give better results than you got in the first part. Why do you think it is better? Estimate the buckling load of the system using the Ritz method with a two-term polynomial basis. Repeat with a three-term polynomial basis.

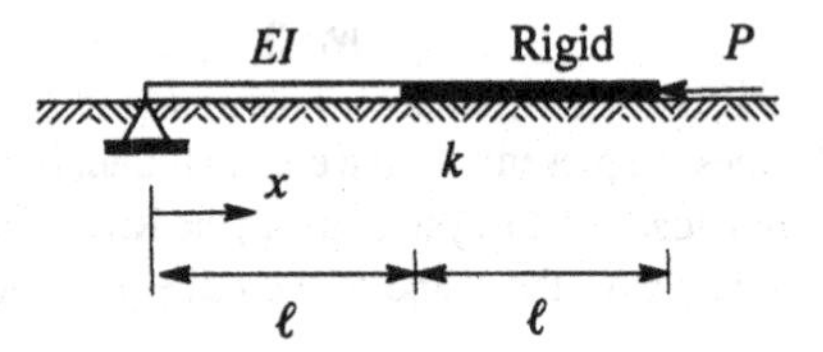

272. A flexible beam of length ℓ and modulus EI is welded to a rigid beam of length ℓ and rests on an elastic foundation of modulus k (per unit length). It is pinned at the left end and is subjected to a compressive axial load P at the right end. The elastic foundation accrues a transverse force in proportion to the transverse displacement w. Shear and axial deformations in the beam are negligible. Write the expression for the energy of the system. What are the essential and natural boundary conditions for the flexible beam? Find an approximate solution for the buckling loads and mode shapes using a two-term polynomial Ritz basis.

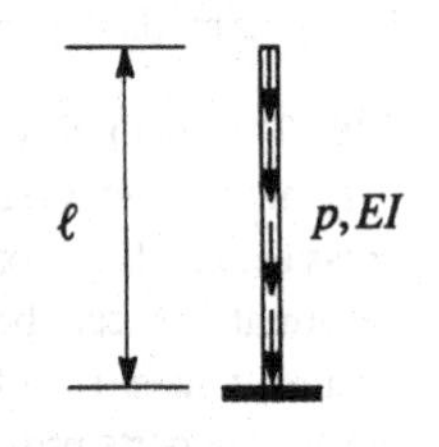

273. The column shown has modulus EI and weight per unit length p. It is fixed at one end and free at the other. Shear and axial deformations can be neglected. Find the (classical) governing differential equations and boundary conditions for the transverse deflection $w(x)$. Express the governing equations in virtual-work form. Estimate by the Ritz method the maximum length the column can have before it buckles under its own weight.

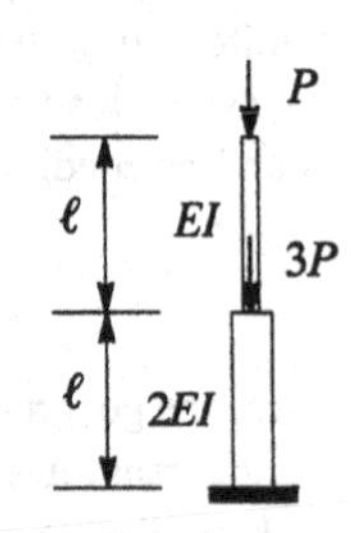

274. The stepped column shown has a variable modulus and is subjected to vertical forces at two points. It is fixed at one end and free at the other. Shear and axial deformations can be neglected. Find the (classical) governing differential equations and boundary conditions for the transverse deflection $w(x)$. Express the governing equations in virtual-work form. Using a two-term polynomial approximation for w, estimate the critical load using the Ritz method.

275. Consider the bar of length ℓ with bending modulus EI and shear modulus GA, subjected to axial load P as shown. Show that the linearized virtual-work functional for the buckling of a beam with shear deformation is given by

EI, GA   P

ℓ

$$G(w,\theta,\bar{w},\bar{\theta}) = \int_0^\ell \left[EI\theta'\bar{\theta}' + GA(w-\theta)(\bar{w}'-\bar{\theta}) - P\left(w'\bar{\theta}+\bar{w}'\theta-\theta\bar{\theta}\right)\right] dx$$

Make the assumption that the (generalized) shear strain in the beam is constant. Estimate the critical loads of the beam using a polynomial approximation with the Ritz method. For example, a three-parameter approximation would have the expression

$$\theta = a_0\frac{x}{\ell} + a_1\frac{x^2}{\ell^2}, \quad w = a_1 x + \tfrac{1}{2}a_0\frac{x^2}{\ell} + \tfrac{1}{3}a_1\frac{x^3}{\ell^2}$$

Describe at least two ways of improving the approximation, and rank them according to which is likely to give the most improvement (no calculations necessary). If EI is very large in comparison to $GA\ell^2$, what will the buckled shape of the beam look like?

276. Consider the beam of length ℓ, fixed at both ends, with constant modulus EI shown in the sketch. The beam is subjected to a compressive axial load P. When the beam is not loaded, the initial shape can be described as

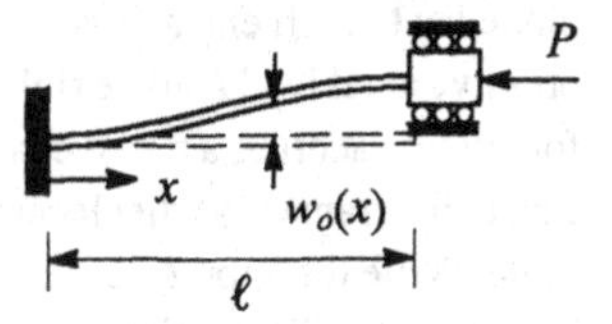

$$w_o(x) = c_o\left(3\frac{x^2}{\ell^2} - 2\frac{x^3}{\ell^3}\right)$$

where $c_o \ll 1$ is known as given data. Assume that shear and axial deformations are negligible. Find an expression for the energy functional $\mathcal{E}$ and the virtual-work functional G for this problem. Estimate the deflection of the beam as a function of load P using the Ritz method and a one-term approximation as follows

$$w(x) = w_o(x) + a\left(\frac{x^2}{\ell^2} - 2\frac{x^3}{\ell^3} + \frac{x^4}{\ell^4}\right)$$

assuming that the displacements are small enough to use the linearized buckling theory.

277. A flexible beam of length 2ℓ and modulus EI rests on an elastic foundation of modulus k. The beam is compressed by a known fixed force $2P$ and is subjected to a transverse load P at its midpoint. The properties have values such that $k\ell^4 = EI$. Axial and shear deformations of the flexible beam can be neglected. Estimate the deflection at the middle and ends of the beam using virtual work and the Ritz method. (Note: due to symmetry a odd base function need not be included.)

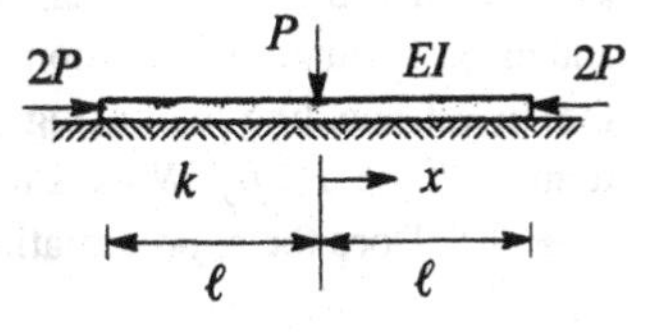

278. A square frame of dimension ℓ is composed of two columns connected together by a beam (the beam can be considered rigid). The left column, which is rigid and pinned at both ends, is subject to a force $2P$. The right column, which has flexural modulus EI, is subjected to a force P. What are the essential and natural boundary conditions on the flexible beam. Express all boundary conditions in terms of the transverse displacement $w(x)$ of the flexible beam. Find the classical characteristic equation that determines the buckling load of the system. Find the exact value of the primary buckling load from the characteristic equation. Recall that the classical differential equation $w^{iv} + \mu^2 w'' = 0$ has the general solution in the form

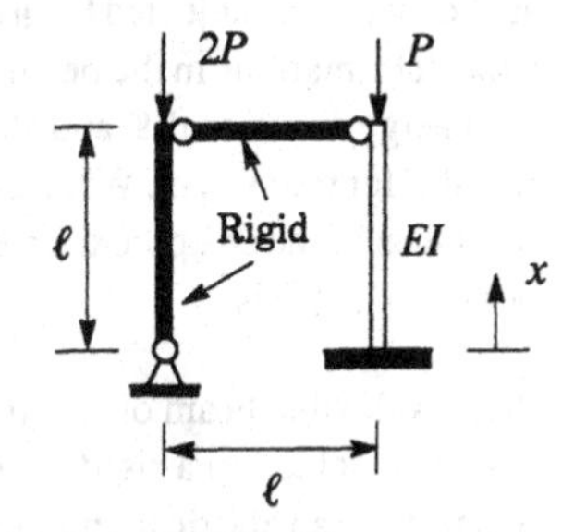

$$w(x) = a_0 + a_1 x + a_2 \sin\mu x + a_3 \cos\mu x.$$

Estimate the buckling capacity of the structure using the Ritz method in conjunction with the principle of virtual work. Compare the classical and variational solutions. [Note: The left column is often referred to as a "leaner" because it leans on the right column to find resistance to sway. By itself the left column has no lateral stiffness, but it carries a destabilizing force.]

279. A flexible beam of length 2ℓ and modulus EI is stuck on an elastic foundation of modulus k (per unit length) over half of its length. It is pinned at the left end and is subjected to a compressive axial load P at the right end. The elastic foundation accrues a transverse force in proportion to the transverse displacement w. Shear and axial deformations in the beam are negligible. Write the expression for the energy of the system. What are the essential and natural boundary conditions for the beam? Find an approximate solution for the buckling loads using a polynomial Ritz basis.

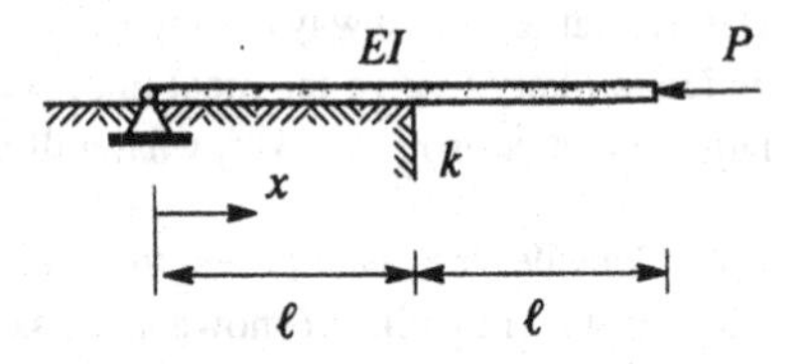

280. A flexible beam of length ℓ and modulus EI is connected to rigid beams of length ℓ at both ends. The beams are supported by two linear springs with modulus $k = \beta EI\ell^3$, where β is a give constant. The beam is supported as shown and is subjected to an end load P. Shear and axial deformations in the beam are negligible. Write the (quadratic) energy functional $\mathscr{E}$ and the virtual-work functional G for the system. What are the essential and natural boundary conditions for the flexible beam? Find an approximate solution for the lowest buckling load using a two-term polynomial Ritz basis. Express the result in terms of β, i.e. $P_{cr}(\beta)$. What is the buckling load for very large spring stiffnesses (i.e., as $\beta \to \infty$)? Does the approximation appear to make sense in the limit? Explain.

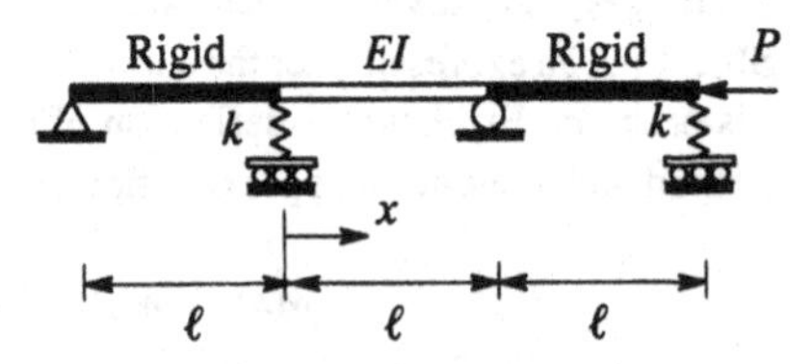

281. A flexible beam of length ℓ and modulus EI is welded to a rigid beam of length ℓ which rests on a spring of modulus $k = 2EI/\ell^3$. The beam is supported as shown and is subjected to an end load P. Shear and axial deformations in the beam are negligible. Write the energy functional $\mathscr{E}$ and the virtual-work functional G for the system. What are the essential and natural boundary conditions for the flexible beam? Find an approximate solution for the displacement $w(x)$ using a two-term polynomial Ritz basis.

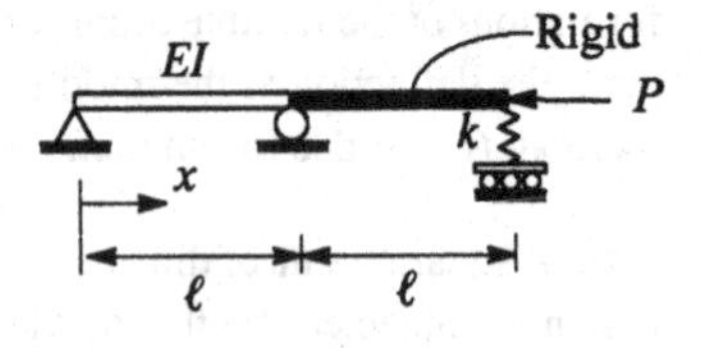

282. A flexible beam of length ℓ and modulus EI is welded to a rigid beam of length ℓ and rests on an elastic foundation of modulus $k = 20EI/\ell^4$. The beam is simply supported and is subjected to an end load P. The elastic foundation accrues a transverse force in proportion to the transverse displacement w. Shear and axial deformations in the beam are

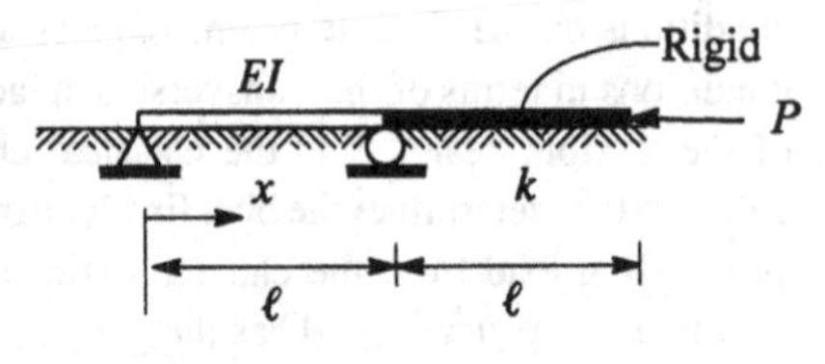

negligible. Write the energy functional $\mathcal{E}$ and the virtual-work functional G for the system. What are the essential and natural boundary conditions for the flexible beam? Find an approximate solution for the displacement $w(x)$ using a polynomial Ritz basis.

283. A flexible bar of length ℓ and bending modulus EI is welded to a rigid bar of length ℓ. The structure is fixed at the bottom and subjected to a compressive axial load P at the top as shown. What are the appropriate essential and natural boundary conditions for this problem? Find an appropriate energy functional $\mathcal{E}(w)$ for the system, where $w(x)$ is the transverse deflection of the flexible bar. Compute an approximation of the critical load of the system using the Ritz method and a polynomial basis function.

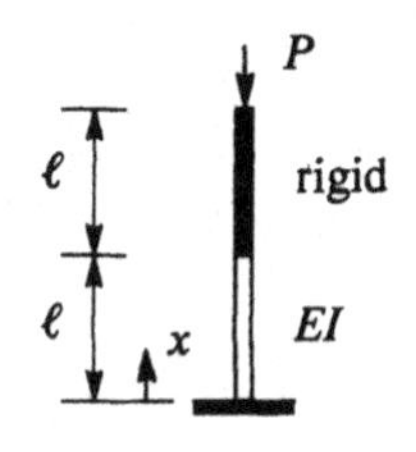

284. A beam of flexural modulus $4EI$ carries the load P to the frame as shown in the figure. The frame is made of two columns pinned together by a beam at midheight. The frame members all have length 2ℓ and flexural modulus EI as shown. The force P is applied directly above the left column. The members have axial modulus $EA \gg EI/\ell^2$. Estimate the smallest buckling capacity P_{cr} of the structure using the Ritz method. Resolve the problem assuming that the load can be placed anywhere along the top beam.

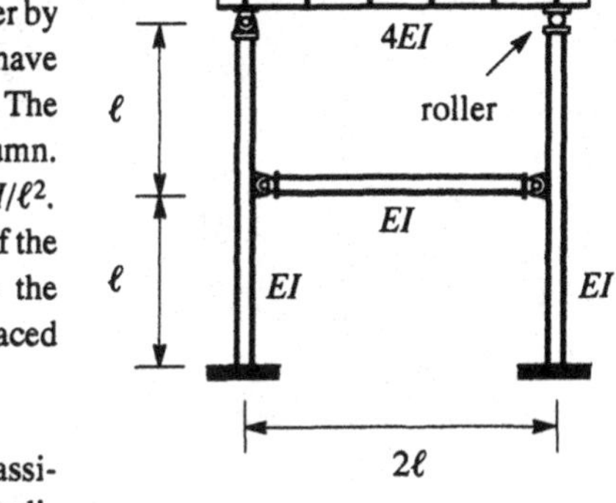

285. Resolve Problem 284 by solving the classical differential equations and boundary conditions.

12

Numerical Computation for Nonlinear Problems

The overwhelming feeling you get from the preceding chapter is that the only thing you can really hope to do with a complex nonlinear system is to compute its critical loads and the corresponding modes with the linearized buckling theory. The examples we have seen have clearly demonstrated that nonlinear systems do not have to be very complicated before we find ourselves unable to find a closed-form solution to the problem of finding the equilibrium paths. Even for some rather modest one-dimensional problems, the possibility of finding a closed-form solution is a dismal prospect. Often, even if we do find a closed-form solution, it is so complicated that the only way to appreciate it is to evaluate the expression at a number of discrete points and plot the bifurcation diagram by connecting those points. There is little motivation for executing monumental feats of algebra if there is an alternative means of generating the discrete points along the path. An incremental numerical solution method provides such a tool.

We shall exploit some simple observations on nonlinear equations to develop an approach to tracing the equilibrium paths of a structural system. First, let us observe that it is always easy to tell whether a certain deformation state (e.g., the displacement field $w(x)$ for a beam or $\mathbf{u}(\mathbf{x})$ for three-dimensional elasticity) represents an equilibrium configuration. With displacements we can compute strains; with strains we can compute stresses; and with stresses we can check to see if equilibrium is satisfied†. If the equilibrium equations are satisfied, then the state is an equilibrium configuration. If it is an equilibrium configuration,

† If we are working with a displacement-based variational statement of the boundary value problem with virtual-work functional G, then all we need to do is to substitute the displacement field directly into G to see if equilibrium is satisfied in a weak sense.

we can plug it into the second-derivative functional to test its stability. Second, let us observe that, for most problems, we start out with a known point on the equilibrium path. Usually, this point is the one with no applied load and no deformation. Finally, let us observe that, in a small enough neighborhood of any point on a curve, the curve is essentially linear. The direction in which this line points (i.e., the tangent to the curve) gives us a good indication of where the curve is headed. We can move along this line to a new trial state of deformation. The trial state can be tested to see if it satisfies equilibrium. If it does not, the estimate of the state can be modified to improve it. What we need is an orderly way of making the improvements to the linear guesses.

One of the most ingenious and popular methods for iteratively improving linear estimates of the equilibrium state is Newton's method. Other methods are available, but many of them are slight variations of Newton's method or have a spirit similar to it. Here, we shall adopt Newton's method as the prototypical algorithm for iteratively computing an equilibrium path. Fletcher (1987) and Luenberger (1984) give detailed accounts of some of the other methods.

The notion of the equilibrium path is illustrated in Fig. 149 for a system with two kinematic degrees of freedom, θ_1 and θ_2, and a single load parameter λ, which is the multiplier of some fixed nominal pattern of loads. The equilibrium path is the curve described by a system of two algebraic equations in three variables: $g_1(\theta_1, \theta_2, \lambda) = 0$ and $g_2(\theta_1, \theta_2, \lambda) = 0$. As such, it describes a curve in three-dimensional space. All of the systems that we discuss in this chapter will conform to this model. For discrete systems, we will typically have one load parameter and N kinematic degrees of freedom. There will be N equations of equilibrium, and, hence, the equilibrium path will be a curve in $N+1$ dimensional space. It is impossible to graphically represent the equilibrium path in dimensions higher than three, but we can extrapolate our understanding of the geometry of the path from three-dimensional space.

Our approach to solving nonlinear problems can be summarized as follows. We start our computation at some point in configuration space where we know everything about the solution (like the origin or a bifurcation point), and we inch our way along the curve from one point on the equilibrium path to the next, iterating to convergence with Newton's method at each step. At converged

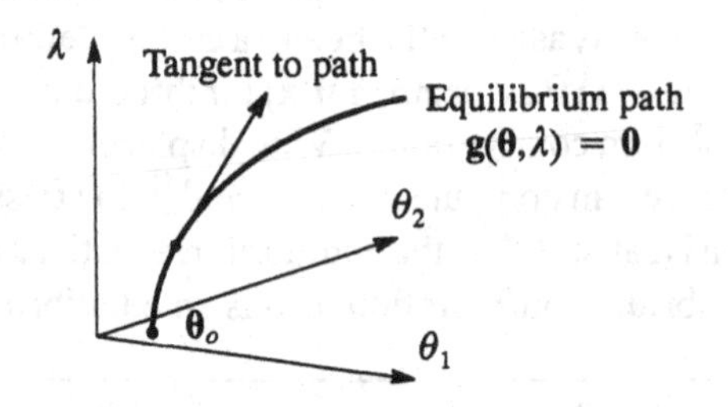

Figure 149 Example of an equilibrium path

states (i.e., states that are actually on the equilibrium path), we can evaluate things like the eigenvalues of the second derivative of the energy to assess the stability of the path we are on and to look for bifurcation points.

We shall initiate our discussion of nonlinear computations with the simple problem of finding the roots to the nonlinear equation $g(x) = 0$, where g is a scalar function of a scalar variable x. We will use this problem to illustrate Newton's method as he actually conceived it (the first application was to the problem of finding the roots to a cubic polynomial). We will then extend Newton's method to the analysis of the equilibrium paths of discrete systems of several variables. During the discussion of discrete systems, we shall introduce the notion of the arc-length constraint that will help us move along the curve in configuration space. Finally, we consider the computations associated with the fully nonlinear planar beam theory introduced in the last chapter.

Newton's Method

Newton's method provides the basic building block for the more general numerical algorithms in this chapter. This section gives an introduction to Newton's method in the context of solving nonlinear algebraic equations.

Finding roots of univariate functions. Let us first attempt to establish Newton's method for finding the roots of a nonlinear, univariate algebraic equation. The problem is illustrated in Fig. 150. We want to find the solutions to the problem

$$g(x) = 0 \tag{587}$$

where $g(x)$ is some known function (e.g., $g(x) = x^3 - 2x + 1$). We can compute the value of $g(x_o)$ and the first derivative $g'(x_o)$ for a known value x_o. Thus, we can find an approximate linear function that is tangent to the curve at the point x_o

$$\hat{g}(x) = g(x_o) + (x - x_o)g'(x_o) \tag{588}$$

We can see from the figure that the linear function $\hat{g}(x)$ is quite close to the nonlinear function $g(x)$ in the neighborhood of x_o, but deviates from it at remote

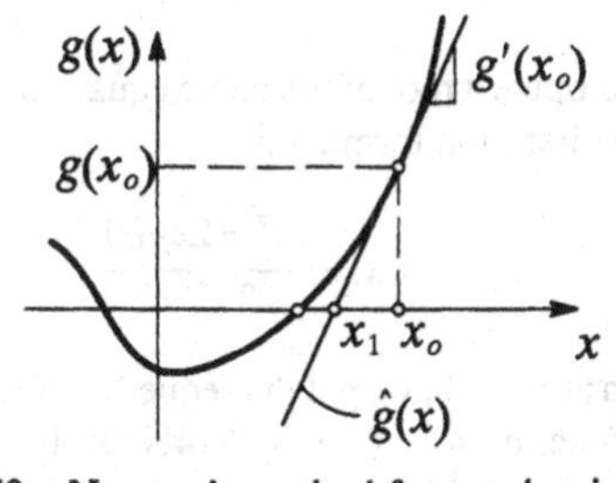

Figure 150 Newton's method for a univariate function

points. What Newton suggested was that the solution to the linear equation $\hat{g}(x) = 0$ would yield a value of x that approximated the solution to the original nonlinear equation. Setting Eqn. (588) equal to zero and solving for x, we get

$$x = x_o - \frac{g(x_o)}{g'(x_o)} \tag{589}$$

This point is labeled x_1 in the figure. Clearly, this point does not satisfy the original nonlinear equation, that is, $g(x_1) \neq 0$, but it is apparently closer. Indeed, x_1 is a point at which we can evaluate g and its first derivative. Thus, we could repeat the calculation starting at x_1 rather than x_o. Newton's method is the iterative scheme that starts with some known point x_o and computes successive iterates, as follows

$$\boxed{x_{i+1} = x_i - \frac{g(x_i)}{g'(x_i)}} \tag{590}$$

We can terminate the iteration when the solution is close enough to the exact solution to the problem. How do we know when we are close enough? If $|g(x_n)| < tol$, where *tol* is a tolerance established a priori, then the solution x_n is close enough.

A few comments about Newton's method are in order. First, like any iterative method, we must specify the starting value x_o and the termination tolerance *tol*. The starting point and solution tolerance will generally require a good understanding of the problem at hand. Some experimentation may be required to establish these values. Second, Newton's method is guaranteed to converge only if we start within the basin of attraction of the solution. If there are other solutions to the nonlinear problem, a starting value may converge to one of the other solutions. Third, the rate of convergence is quite fast for Newton's method in the vicinity of the solution. Finally, Newton's method will fail if it encounters any point x_i that has $g'(x_i) = 0$, because the algorithm would require division by zero at such a point. Newton's method is not particularly attracted to such points, but we shall see that these points can present problems in tracing equilibrium paths past limit and bifurcation points when solving stability problems.

Example 76. Let us compute a root of the cubic equation $x^3 - 2x + 1 = 0$ using Newton's method. The iteration formula is

$$x_{i+1} = x_i - \frac{x_i^3 - 2x_i + 1}{3x_i^2 - 2}$$

If we start the iteration at $x_o = 0$, we get the sequence of iterates shown in Table 10. The value of the function for $g(x_6) = 0.1480 \times 10^{-11}$ is very nearly zero compared to the error of the initial estimate x_o of $g(0) = 1$. The important ob-

Table 10 Iteration history for the univariate example problem

i	x_i	$g(x_i)$
1	0.000000	0.1000E+01
2	0.500000	0.1250E+00
3	0.600000	0.1600E-01
4	0.617391	0.5496E-03
5	0.618033	0.7631E-06
6	0.618034	0.1480E-11

servation from this example is that we obtained a numerical solution to the nonlinear problem by executing a sequence of arithmetic operations involving the evaluation of the function and its first derivative. This feature is the hallmark of iterative methods.

Equations with several variables. Newton's method is based upon iteratively solving a linearized version of the nonlinear equations. Let us assume that we wish to solve a system of nonlinear equations

$$\mathbf{g}(\mathbf{x}) = \mathbf{0} \tag{591}$$

where $\mathbf{g}(\mathbf{x}) \in \mathbb{R}^N$ is a vector valued function of the unknowns $\mathbf{x} \in \mathbb{R}^N$. It is important that we have the same number of equations as unknowns. Again, we can linearize the equations about some configuration $\mathbf{x}_o$ (a point that does not necessarily satisfy the equations). Let us define the linear function $\hat{\mathbf{g}}(\mathbf{x})$ to be the first-order Taylor series expansion of the function $\mathbf{g}(\mathbf{x})$. To wit

$$\hat{\mathbf{g}}(\mathbf{x}) \equiv \mathbf{g}(\mathbf{x}_o) + \nabla\mathbf{g}(\mathbf{x}_o)\left(\mathbf{x} - \mathbf{x}_o\right) \tag{592}$$

where $\nabla\mathbf{g}(\mathbf{x}_o)$ is the matrix of first derivatives of $\mathbf{g}$ with respect to $\mathbf{x}$, evaluated at the point $\mathbf{x}_o$, $\left[\nabla\mathbf{g}\right]_{ij} = \partial g_i/\partial x_j$. The matrix $\nabla\mathbf{g}(\mathbf{x}_o)$ is N by N if the original system has N equations in N unknowns.

We can extend Newton's idea to multiple dimensions and suggest that the solution to the linear equation $\hat{\mathbf{g}}(\mathbf{x}) = \mathbf{0}$ will yield an estimate of the solution to $\mathbf{g}(\mathbf{x}) = \mathbf{0}$ that is closer than $\mathbf{x}_o$. The Newton estimate of the solution is

$$\mathbf{x} = \mathbf{x}_o - \left[\nabla\mathbf{g}(\mathbf{x}_o)\right]^{-1}\mathbf{g}(\mathbf{x}_o) \tag{593}$$

Of course, the new point generally will not satisfy the nonlinear equation, but it should be better than $\mathbf{x}_o$. We could replace the old estimate with the new one, $\mathbf{x}_o \leftarrow \mathbf{x}$, and repeat the calculation. Therefore, the Newton iteration takes the form

$$\boxed{\mathbf{x}_{i+1} = \mathbf{x}_i - \left[\nabla\mathbf{g}(\mathbf{x}_i)\right]^{-1}\mathbf{g}(\mathbf{x}_i)} \tag{594}$$

with $\mathbf{x}_o$ specified as given data. Eventually, the new estimate will look very much like the previous one, and we call this estimate a *converged state*. Again, there is no need to continue the iteration if the solution is, within some tolerance, acceptable. In analogy with the univariate problem, a good termination criterion for the multivariate problem is $\| \mathbf{g}(\mathbf{x}_n) \| < tol$, where $\| \cdot \|$ is some suitable norm, for example, the Euclidean norm. We need to test the norm of the residual because we are trying to satisfy several equations at the same time. The norm measures the aggregate satisfaction of the equations rather than the satisfaction of any one of them.

Again, there are two things we must specify in Newton's method. First, like any iterative method, we must select the starting point $\mathbf{x}_o$. Second, we must select a suitable tolerance for judging convergence. Newton's method is not guaranteed to converge from any arbitrary starting point, but if a point is close enough to the solution, the rate of convergence is quadratic. The exact features of the basin of attraction of Newton's method depend upon the problem, so it is difficult to make any sweeping statements beyond "good luck with your initial choice." (All joking aside, in our problems we generally have very good choices for starting points.) There are many methods available to improve on these weaknesses of Newton's method, but we shall stick with the basic version here. Newton's method will fail if the matrix $\nabla\mathbf{g}(\mathbf{x}_i)$ is singular (i.e., not invertible) at some point $\mathbf{x}_i$.

The algorithm is rarely implemented with the matrix inversion indicated in Eqn. (594). Rather, we would solve the system of equations $\mathbf{A}_i \Delta\mathbf{x}_i = \mathbf{b}_i$, where $\mathbf{A}_i \equiv \nabla\mathbf{g}(\mathbf{x}_i)$ and $\mathbf{b}_i \equiv -\mathbf{g}(\mathbf{x}_i)$, for the increment $\Delta\mathbf{x}_i$. The new estimate can then be found by adding this increment to the previous value to give the update $\mathbf{x}_{i+1} = \mathbf{x}_i + \Delta\mathbf{x}_i$. There are many algorithms available for solving a linear system of equations.

The basic algorithm. The organization of the Newton iteration algorithm is straightforward. It includes an initialization step, an iteration loop, and a termination criterion. At each step in the iteration loop, we establish and solve the linearized version of the nonlinear equations for an estimate of the increment $\Delta\mathbf{x} \equiv \mathbf{x} - \mathbf{x}_i$ in the unknown $\mathbf{x}$. This increment is added to the previous estimate to give a new estimate by the update $\mathbf{x}_{i+1} = \mathbf{x}_i + \Delta\mathbf{x}$. The following pseudocode illustrates the organization of the algorithm.

Algorithm 1 (Newton's method)

1. Select $\mathbf{x}_o$. Initialize counter, $i = 0$.
2. Compute residual and gradient, $\mathbf{b}_i \equiv -\mathbf{g}(\mathbf{x}_i)$ and $\mathbf{A}_i \equiv \nabla\mathbf{g}(\mathbf{x}_i)$.
3. Test for convergence. If $\| \mathbf{g}(\mathbf{x}_i) \| < tol$, then Stop.
4. Solve linear system of equations $\mathbf{A}_i \Delta\mathbf{x}_i = \mathbf{b}_i$.
5. Update the estimate, $\mathbf{x}_{i+1} = \mathbf{x}_i + \Delta\mathbf{x}_i$.
6. Increment counter $i \leftarrow i+1$, Go to 2.

Note, in particular, that we need only evaluate the functions and their gradients. The solution of equations in step 4 is generally carried out by Gaussian elimination, but any equation-solving method is suitable. This format for iterative nonlinear solution of algebraic equations is suitable for problems of any dimension N.

Example 77. Let us employ Newton's method to solve the nonlinear, two by two system of algebraic equations $\mathbf{g(x)} = \mathbf{0}$, where the functions $\mathbf{g}$ are given by

$$g_1(\mathbf{x}) \equiv x_1^3 - 2x_1x_2^2 + x_1 - 3$$
$$g_2(\mathbf{x}) \equiv x_2^3 - 2x_1^2x_2 + x_2 - 2$$

The gradient is simple to compute, and has the explicit form

$$\nabla \mathbf{g(x)} = \begin{bmatrix} 3x_1^2 - 2x_2^2 + 1 & -4x_1x_2 \\ -4x_1x_2 & 3x_2^2 - 2x_1^2 + 1 \end{bmatrix}$$

The MATLAB code to compute the solution is

```
%-- Chapter 12, Example 77
  clear; tol = 1.e-8; maxit = 30; x = [0; 0]; test = 1.; i = 0;
     while (test > tol) & (i < maxit)
     y = x(1); z = x(2);
     b = [ y^3 - 2*y*z^2 + y - 3 ;
           z^3 - 2*z*y^2 + z - 2];
     A = [ 3*y^2 - 2*z^2 + 1 ,         -4*y*z          ;
           - 4*y*z           , 3*z^2 - 2*y^2 + 1];
     test = norm(b);
     dx = -A\b;    x = x + dx;
     fprintf('%5i%9.5f%9.5f%13.2e%13.2e%13.2e\n',i,x',b',test)
     i = i + 1;
   end
```

The result of the Newton iteration on the example problem is

i	x1	x2	g1	g2	|| g ||
0	3.00000	2.00000	-3.00e+000	-2.00e+000	3.61e+000
1	1.98373	1.27811	3.00e+000	-2.80e+001	2.82e+001
2	1.25394	0.62220	3.09e-001	-8.69e+000	8.70e+000
3	0.64287	-0.58445	-7.45e-001	-3.09e+000	3.18e+000
4	1.72547	-0.02197	-2.53e+000	-2.30e+000	3.42e+000
5	1.34269	-0.41551	3.86e+000	-1.89e+000	4.30e+000
6	1.43236	-0.79342	3.00e-001	-9.89e-001	1.03e+000
7	1.45732	-0.73069	-4.32e-001	-3.73e-002	4.34e-001
8	1.46010	-0.73392	-3.81e-003	-1.72e-002	1.76e-002
9	1.46010	-0.73390	-2.28e-005	4.10e-005	4.69e-005
10	1.46010	-0.73390	-5.21e-010	4.29e-011	5.23e-010

The first column of the output table gives the iteration number i. The second and third columns give the ith estimate of the solution $\mathbf{x}_i$. The fourth and fifth columns give the values of the function $\mathbf{g}(\mathbf{x}_i)$, and the last column gives the Euclidean norm of the residual $\| \mathbf{g}(\mathbf{x}_i) \|$. The starting point is $\mathbf{x}_o = (0,0)$, and it took 10 iterations to get a solution within a tolerance of $tol = 10^{-8}$ on the Euclidean norm of the residual.

There are some interesting features of the previous example that give an indication of what sort of behavior to expect from a Newton iteration. First, because of the choice of the initial estimate $\mathbf{x}_o$, the algorithm initially heads in the wrong direction, that is, into the positive quadrant. As a result, the norm of the residual actually increases in going from $\mathbf{x}_o$ to $\mathbf{x}_1$. Clearly, a Newton step does not always give a better estimate of the solution. However, these misdirections generally occur when we are far from the solution, where the actual functions are not well represented by the linear function $\hat{\mathbf{g}}(\mathbf{x})$. Satisfaction of the two functions occurs at different rates, but they both wind up satisfied to within the tolerance. Because convergence is tested with the Euclidean norm of the residual, one of the equations will contribute more than the other. In this case, convergence was controlled by g_1. As the iteration closes in on the solution to the problem, Newton's method has quadratic convergence. One can see the speed of quadratic convergence by examining the exponent of the norm of the residual in the last few iterations. We can observe that, with Newton's method, we really need not worry very much about the exact value of the tolerance because, if the solution is close, one more iteration will generally nail it.

We must be careful not to specify a tolerance smaller than the machine precision of the computer can tolerate. There is a point in each calculation that limits the accuracy of the computation. If the residual gets stuck at some value (usually small) then one might expect that the tolerance is tighter than the calculation will allow.

Tracing the Equilibrium Path of a Discrete System

We have found in the previous two chapters that the governing equations of equilibrium of a discrete system can be expressed in the form

$$\mathbf{g}(\boldsymbol{\theta},\lambda) = \mathbf{0} \tag{595}$$

where $\boldsymbol{\theta}$ is the vector of displacement parameters and λ is the load-level parameter. For the single-degree-of-freedom problems solved in Chapter 10, $\boldsymbol{\theta}$ had one component, which measured the rotation of the rigid bar from the vertical position. For the two-degree-of-freedom example of Chapter 10, $\boldsymbol{\theta} = (\theta_1,\theta_2)$ had two components, the first of which measured the rotation of the lower bar from vertical, and the second of which measured the rotation of the upper bar from vertical. When we applied the Ritz discretization to the continuous systems of Chapter 11, we got $\boldsymbol{\theta} = (a_1, a_2, \ldots, a_N)$, where a_i is the coefficient of the ith base function $h_i(x)$ in an N-term expansion for the displacement field, e.g., $w(x)$ for the beam. For most of the examples we have considered, there has been only a single load P. The load parameter λ can be viewed as a multiplier of some fixed pattern of applied loads. In any case, the load level will always be controlled by a scalar parameter.

It should be evident that Eqn. (595) can be solved by Newton's method. At the simplest level, we can consider that the load level λ is prescribed at some

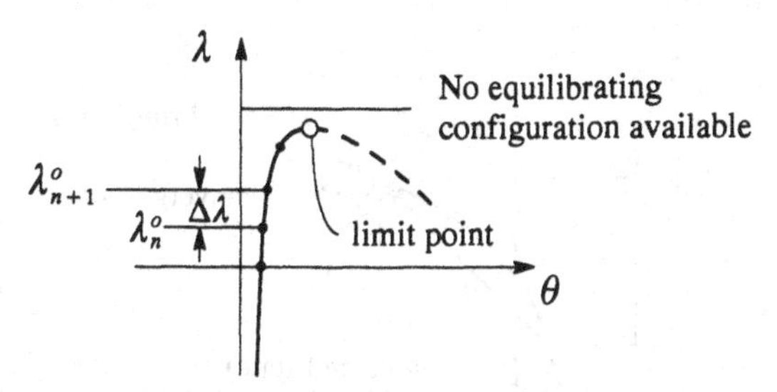

Figure 151 Fixed load incrementation with Newton's method

value λ^o, and Eqn. (595) represents N nonlinear of equations for the N unknown displacement parameters $\boldsymbol{\theta}$ (recall that we always have N equations if we have N displacement parameters). The problem reduces to finding the state $\boldsymbol{\theta}$ that equilibrates the loads at precisely the level λ^o. This problem can be solved by applying the Newton algorithm exactly as it was outlined in the previous section.

You have probably already figured out that the gradient of the function $\mathbf{g}(\boldsymbol{\theta},\lambda^o)$, with respect to $\boldsymbol{\theta}$, is $\nabla\mathbf{g}(\boldsymbol{\theta},\lambda^o) = \mathbf{A}(\boldsymbol{\theta},\lambda^o)$, the Hessian of the discrete energy function. Clearly, at bifurcation points and limit points, this matrix becomes singular. Newton's method is destined to fail at these points. While bifurcation points are somewhat more delicate, there is a simple remedy for limit points.

To see the difficulty in prescribing the load level, consider Fig. 151, which shows an equilibrium path with a limit point (indicated by an open circle). To compute this path, we would start at the configuration with zero load and the deformation equal to its imperfection values. The initial configuration happens to be an equilibrium configuration, so no iteration is needed to establish equilibrium. Assume that we have successfully located the equilibrium state $\{\lambda^o_n, \boldsymbol{\theta}_n\}$. We locate the next point on our curve by incrementing the load λ^o_n by a fixed amount $\Delta\lambda$. At this new fixed load level λ^o_{n+1}, we iterate to find the equilibrium configuration $\boldsymbol{\theta}_{n+1}$. Clearly, if the load increments are small enough, we can usually guarantee convergence to the next point on the curve. This process, called *load-control incrementation*, continues by incrementing the load and iterating to find the associated configuration $\boldsymbol{\theta}$. The algorithm fails when the load increment takes the total load above the limit point. There is no configuration that will satisfy equilibrium at this load level. The algorithm fails because force control eventually asks the impossible of an equilibrium path with a limit point. You could argue that if the increments were made small enough, then we could approach the limit point slowly by trial and error. This is indeed true, but as we approach the limit point, the condition of the Hessian matrix $\mathbf{A}(\boldsymbol{\theta},\lambda^o)$ gets worse and worse because it has an eigenvalue that is approaching zero at the limit point. Thus, the numerical computations break down at the limit point. We could even conceive of decrementing the load if

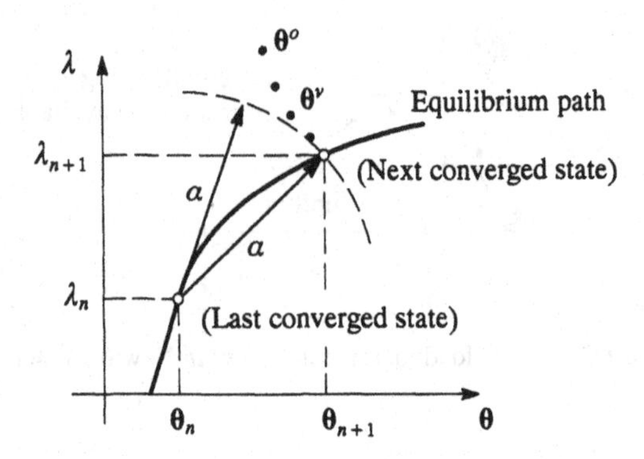

Figure 152 The arc-length constraint

we could get past the limit point, but as we concoct these remedies, we must assume more and more knowledge of the path that we are trying to compute. For most problems, we are navigating in the dark. Every ad hoc algorithm has its Achilles heel.

One popular alternative to a load-control Newton method is the so-called *arc-length method.* As in the previous method, we inch along the curve, but rather than incrementing the load, we introduce the constraint that the distance between the next estimate of the solution $\{\boldsymbol{\theta}, \lambda\}$ and some fixed state $\{\boldsymbol{\theta}_n, \lambda_n\}$ will be constant, as shown in Fig. 152. Let us introduce the scalar equation of constraint

$$c(\boldsymbol{\theta},\lambda) \equiv \| \boldsymbol{\theta} - \boldsymbol{\theta}_n \|^2 + (\lambda - \lambda_n)^2 - a^2 = 0 \tag{596}$$

Now we can view the load λ as an independent variable in exactly the same way we do $\boldsymbol{\theta}$. Generally, a wise choice for the fixed state $\{\boldsymbol{\theta}_n, \lambda_n\}$ is the last converged state. This point is good because we know that it lies on the equilibrium path, and, thus, our constraint will allow us to find a new equilibrium configuration for arbitrarily small values of the arc-length a. The constraint equation is really a ball of radius a centered at $\{\boldsymbol{\theta}_n, \lambda_n\}$ in the state space. The constraint insists that any new solution be found on the surface of the ball. It should be clear that, if the ball is small enough, the equilibrium path will pierce it at least at two points. We specify the distance a in the same spirit as $\Delta\lambda$ in the load incrementation scheme. It is possible to specify a so large that Newton's method cannot converge, but the algorithm with this constraint has no trouble with either vertical or horizontal tangents on the equilibrium path. There are now $N+1$ equations in $N+1$ unknowns (N displacement unknowns and 1 load level unknown). This nonlinear system of equations can be summarized as

$$\mathbf{g}(\boldsymbol{\theta},\lambda) = \mathbf{0}, \quad c(\boldsymbol{\theta},\lambda) = 0 \tag{597}$$

There are many variants of the constraint equation $c(\boldsymbol{\theta},\lambda) = 0$, so we will use the general form throughout our subsequent discussions. We can vary the constraint to suit the particular application (see Problem 300).

Newton's method can be applied to the augmented system by recognizing that the equations can be linearized about the state $\{\boldsymbol{\theta}^\nu,\lambda^\nu\}$ to give the linear functions $\hat{\mathbf{g}}$ and $\hat{c}$

$$\begin{aligned}\hat{\mathbf{g}}(\boldsymbol{\theta},\lambda) &\equiv \mathbf{g}(\boldsymbol{\theta}^\nu,\lambda^\nu) + \nabla_\theta \mathbf{g}(\boldsymbol{\theta}^\nu,\lambda^\nu)(\boldsymbol{\theta}-\boldsymbol{\theta}^\nu) + \nabla_\lambda \mathbf{g}(\boldsymbol{\theta}^\nu,\lambda^\nu)(\lambda-\lambda^\nu)\\ \hat{c}(\boldsymbol{\theta},\lambda) &\equiv c(\boldsymbol{\theta}^\nu,\lambda^\nu) + \nabla_\theta c(\boldsymbol{\theta}^\nu,\lambda^\nu)(\boldsymbol{\theta}-\boldsymbol{\theta}^\nu) + \nabla_\lambda c(\boldsymbol{\theta}^\nu,\lambda^\nu)(\lambda-\lambda^\nu)\end{aligned}$$

The notation $\nabla_\theta \mathbf{g}(\boldsymbol{\theta},\lambda)$ means the derivative of $\mathbf{g}$ with respect to $\boldsymbol{\theta}$, holding λ constant. Similarly, $\nabla_\lambda \mathbf{g}(\boldsymbol{\theta},\lambda)$ means the derivative of $\mathbf{g}$ with respect to λ, holding $\boldsymbol{\theta}$ constant. The same notation holds for the function c. We have taken great pains to distinguish the state $\{\boldsymbol{\theta}^\nu,\lambda^\nu\}$ from the converged state $\{\boldsymbol{\theta}_n,\lambda_n\}$ here. The states with superscripts will be intermediate results of our Newton iteration, while states with subscripts will represent converged load steps. If we set the linearized functions equal to zero, $\hat{\mathbf{g}}(\boldsymbol{\theta},\lambda) = \mathbf{0}$ and $\hat{c}(\boldsymbol{\theta},\lambda) = 0$, we can compute a (presumably) better estimate of the equilibrium state. Let us call this new state $\{\boldsymbol{\theta}^{\nu+1},\lambda^{\nu+1}\}$. For convenience, let us define the increments to the configuration and the load parameter as

$$\Delta\boldsymbol{\theta}^\nu \equiv \boldsymbol{\theta}^{\nu+1}-\boldsymbol{\theta}^\nu, \quad \Delta\lambda^\nu \equiv \lambda^{\nu+1}-\lambda^\nu$$

Let us gather the configuration and the load parameter into a single matrix $\mathbf{x}$ as $\mathbf{x} \equiv [\boldsymbol{\theta},\lambda]$, and let us define the matrices

$$\mathbf{A}^\nu \equiv \begin{bmatrix}\nabla_\theta \mathbf{g}(\boldsymbol{\theta}^\nu,\lambda^\nu) & \nabla_\lambda \mathbf{g}(\boldsymbol{\theta}^\nu,\lambda^\nu)\\ \nabla_\theta c(\boldsymbol{\theta}^\nu,\lambda^\nu) & \nabla_\lambda c(\boldsymbol{\theta}^\nu,\lambda^\nu)\end{bmatrix} \qquad \mathbf{b}^\nu \equiv \begin{Bmatrix}-\mathbf{g}(\boldsymbol{\theta}^\nu,\lambda^\nu)\\ -c(\boldsymbol{\theta}^\nu,\lambda^\nu)\end{Bmatrix}$$

Now, in order for the linearized functions to be equal to zero at $\mathbf{x}^{\nu+1} = \{\boldsymbol{\theta}^{\nu+1},\lambda^{\nu+1}\}$, the increments $\Delta\mathbf{x}^\nu$ must satisfy the equations

$$\mathbf{A}^\nu\Delta\mathbf{x}^\nu = \mathbf{b}^\nu \tag{598}$$

The increment can be found as $\Delta\mathbf{x}^\nu = [\mathbf{A}^\nu]^{-1}\mathbf{b}^\nu$. With the increment known, the new state can be found by the simple update formula

$$\mathbf{x}^{\nu+1} = \mathbf{x}^\nu + \Delta\mathbf{x}^\nu \tag{599}$$

All we need to start this iteration is the initial value of the state $\mathbf{x}^o = \{\boldsymbol{\theta}^o,\lambda^o\}$ and an initial point on the equilibrium path $\mathbf{x}_o = \{\boldsymbol{\theta}_o,\lambda_o\}$ with which we can reckon our arc-length constraint. Iteration can be terminated when

$$\|\mathbf{b}^\nu\| = \sqrt{\|\mathbf{g}(\boldsymbol{\theta}^\nu,\lambda^\nu)\|^2 + |c(\boldsymbol{\theta}^\nu,\lambda^\nu)|^2} < tol \tag{600}$$

where *tol* is the preset convergence tolerance†. As before, we can identify our old friend $\nabla_\theta \mathbf{g}(\boldsymbol{\theta},\lambda) = \mathbf{A}(\boldsymbol{\theta},\lambda)$, the second derivative of the energy, but now the matrix we must invert for Newton's method has been augmented and no longer has an eigenvalue that goes to zero at a limit point.

The curve-tracing algorithm. The organization of the Newton iteration algorithm is straightforward. To economize the notation in the Newton iteration, we will combine the unknowns into the single vector $\mathbf{x} = (\boldsymbol{\theta}, \lambda)$. The following pseudocode illustrates the organization of the algorithm

Algorithm 2

(Newton's method with arc-length constraint)

1. Select tolerance *tol*, step size a, appropriate limits to variables, maximum number of iterations, etc. Initialize load step counter, $n = 0$.
2. Select starting values $\mathbf{x}_o$ and initial direction $\mathbf{d}_o$ in which to move, such that $\mathbf{x}_o$ is an equilibrium configuration and $\| \mathbf{d}_o \| = a$.
3. Initialize iteration state $\mathbf{x}^o = \mathbf{x}_n + \mathbf{d}_n$ to be last converged state plus a move in the desired direction that satisfies the arc-length constraint. Initialize the iteration counter, $\nu = 0$.
4. At state $\mathbf{x}^\nu$ do the following:
   (a) Compute residual and gradient

$$\mathbf{b}^\nu = \begin{Bmatrix} -\mathbf{g}(\boldsymbol{\theta}^\nu, \lambda^\nu) \\ -c(\boldsymbol{\theta}^\nu, \lambda^\nu) \end{Bmatrix} \qquad \mathbf{A}^\nu = \begin{bmatrix} \nabla_{\boldsymbol{\theta}} \mathbf{g}(\boldsymbol{\theta}^\nu, \lambda^\nu) & \nabla_\lambda \mathbf{g}(\boldsymbol{\theta}^\nu, \lambda^\nu) \\ \nabla_{\boldsymbol{\theta}} c(\boldsymbol{\theta}^\nu, \lambda^\nu) & \nabla_\lambda c(\boldsymbol{\theta}^\nu, \lambda^\nu) \end{bmatrix}$$

   (b) Test for convergence. If $\| \mathbf{b}^\nu \| < tol$, then Go to 5.
   (c) Solve linear system of equations $\mathbf{A}^\nu \Delta \mathbf{x}^\nu = \mathbf{b}^\nu$.
   (d) Update the estimate, $\mathbf{x}^{\nu+1} = \mathbf{x}^\nu + \Delta \mathbf{x}^\nu$.
   (e) Increment counter $\nu \leftarrow \nu + 1$, Go to 4(a).
5. Update converged state $\mathbf{x}_{n+1} \leftarrow \mathbf{x}^\nu$, estimate the direction for the next step as $\mathbf{d}_{n+1} = \mathbf{x}_{n+1} - \mathbf{x}_n$.
6. Increment counter $n \leftarrow n + 1$. If n is equal to the maximum number of steps, then Stop, else Go to 3.

There are many variants of this algorithm, but the one presented here gives the basic flavor. This algorithm will experience difficulties at a bifurcation point because it is unable to choose among branches. It will choose one, but it may not be the one that you want. This algorithm was the one used to compute the bifurcation diagrams shown in Figs. 134 and 135. To compute the branches

† The load and displacement parameters can have vastly different magnitudes, particularly if the system is stiff. Care must be exercised in setting up the arc-length constraint and iteration equations to make sure that none of the solution parameters get swamped out by the others. Often, the unknowns can be scaled to avoid such problems.

that emanate from the bifurcation points, the algorithm was started with $\mathbf{x}_o = (\mathbf{0}, \lambda_{cr})$, the trivial state at either the first or the second critical load. To get the algorithm to follow the nontrivial branch, we need only give an initial estimate of the direction to go of $\mathbf{d}_o = (\alpha\boldsymbol{\psi}, 0)$, where $\boldsymbol{\psi}$ is a unit vector that points in the direction of the linearized eigenvector. After the first step, we can encourage continuation along the same path by setting $\mathbf{d}_n = \mathbf{x}_n - \mathbf{x}_{n-1}$. Since $\mathbf{x}_n$ was found by iterating with the arc-length constraint, this choice has $\| \mathbf{d}_n \| = \alpha$ automatically. Also, with this choice, the starting iterate is $\mathbf{x}^o = 2\mathbf{x}_n - \mathbf{x}_{n-1}$. In essence, we are suggesting, for starters, another step just like the last one.

We can augment the equilibrium equations with a condition that will lock onto the bifurcation point. Let $\mathbf{A}(\boldsymbol{\theta}, \lambda)$ be the Hessian of the energy function, that is, the second derivative function is $A(\lambda, \boldsymbol{\theta}, \bar{\boldsymbol{\theta}}) = \bar{\boldsymbol{\theta}}^T \mathbf{A}(\boldsymbol{\theta}, \lambda)\bar{\boldsymbol{\theta}}$. The constraint equation

$$c(\boldsymbol{\theta}, \lambda) = \det \mathbf{A}(\boldsymbol{\theta}, \lambda) = 0$$

then describes a point on the equilibrium path at which one of the eigenvalues of $\mathbf{A}(\boldsymbol{\theta}, \lambda)$ goes to zero, i.e., a bifurcation point or a limit point. If we add this extra equation, we must discard the arc-length constraint to retain the feature of having $N+1$ equations in $N+1$ unknowns. Algorithms that locate bifurcation points exactly are very useful because they allow us to execute switches from one branch to another at the critical points. If we have converged on a bifurcation point, we can compute the eigenvectors of $\mathbf{A}(\boldsymbol{\theta}, \lambda)$. The eigenvector associated with the zero eigenvalue corresponds to the direction of the branching solution. You can suggest to the Newton algorithm that the next step should be in that direction, and, in doing so, switch to that branch. Many systems bifurcate from nontrivial branches. Clearly, we would not want to include this additional equation for points remote from a bifurcation point because the Newton algorithm would attempt to iterate directly to that load level. One strategy is to start with a normal arc-length constraint and monitor the eigenvalues of $\mathbf{A}(\boldsymbol{\theta}, \lambda)$. When an eigenvalue of the second-derivative functional gets small enough, switch from the arc-length constraint to the determinant constraint to lock onto the critical point. Once converged on the critical point, select the eigenvector as the direction vector $\mathbf{d}_n$, and return to the arc-length constraint.

Example 78. Let us reconsider the two-degree-of-freedom example problem from Chapter 10, described in Fig. 133. The system consists of two rigid links connected by rotational springs and connected to a fixed base with a rotational spring. Both links have the same length, and both springs have the same stiffness. The cantilevered links are subjected to a compressive axial force P at the free end. The nonlinear equations, $\mathbf{g}(\mathbf{x}) = \mathbf{0}$, for the present case, are

$$\begin{aligned} g_1(\boldsymbol{\theta}, \lambda) &= 2\theta_1 - \theta_2 - \lambda \sin\theta_1 = 0 \\ g_2(\boldsymbol{\theta}, \lambda) &= -\theta_1 + \theta_2 - \lambda \sin\theta_2 = 0 \end{aligned} \tag{601}$$

where $\lambda \equiv P\ell/k$. The arc-length constraint equation has the form

$$c(\boldsymbol{\theta},\lambda) = \| \boldsymbol{\theta} - \boldsymbol{\theta}_n \|^2 + (\lambda - \lambda_n)^2 - a^2 = 0 \qquad (602)$$

where $\boldsymbol{\theta}_n$ and λ_n represent the previous converged state, a distance a from which we want to find a new point on the equilibrium path. The matrix of gradients $\mathbf{A}^\nu$ and the residual $\mathbf{b}^\nu$ at the state $\{\boldsymbol{\theta}^\nu, \lambda^\nu\}$ are given by

$$\mathbf{A}^\nu = \begin{bmatrix} 2-\lambda^\nu \cos\theta_1^\nu & -1 & -\sin\theta_1^\nu \\ -1 & 1-\lambda^\nu\cos\theta_2^\nu & -\sin\theta_2^\nu \\ 2(\theta_1^\nu - \theta_{1n}) & 2(\theta_2^\nu - \theta_{2n}) & 2(\lambda^\nu - \lambda_n) \end{bmatrix} \qquad \mathbf{b}^\nu = -\begin{bmatrix} g_1(\boldsymbol{\theta}^\nu, \lambda^\nu) \\ g_2(\boldsymbol{\theta}^\nu, \lambda^\nu) \\ c(\boldsymbol{\theta}^\nu, \lambda^\nu) \end{bmatrix}$$

The portion of the gradient within the shaded and dotted box is $\mathbf{A}(\boldsymbol{\theta}^\nu, \lambda^\nu)$, the Hessian of the energy function. The eigenvalues of this two by two matrix can be readily computed so that the stability of the equilibrium branch can be monitored as we compute our way along the path. An implementation of the algorithm for this problem is given by the MATLAB program, called NEWTON, that follows. The program should clarify some of the details about the implementation of Newton's method for computing an equilibrium path. This code was used to compute the equilibrium paths shown in Figs. 134 and 135.

The Program NEWTON

```
%  *-----------------------------------------------------------------------*
%  |                          Program NEWTON                                |
%  |      Fundamentals of Structural Mechanics, 2nd Edition                 |
%  |                  K. D. Hjelmstad, July 1, 2004                         |
%  *-----------------------------------------------------------------------*

%.. Set problem parameters, Chapter 12, Example 79,
    clear; tol = 1.e-8; alpha = 0.5; maxsteps = 10; maxit = 20;
    xo = [0; 0; 0.3820]; x = [0.8510; 05260; 0.3820];

%.. Initialize values for load step zero,set next trial state
    b = x - xo;   x = xo + alpha*b/norm(b);

%.. Compute MAXSTEPS points along the Equilibrium Path
    for n = 1:maxsteps

%.... Perform Newton iteration at each load step
      nu = 0; test = 1.0;
      while (test > tol) & (nu < maxit)
        nu = nu + 1;

%...... Compute residual and Hessian at current state
        b = [2*x(1) - x(2) - x(3)*sin(x(1)) ;
              -x(1) + x(2) - x(3)*sin(x(2)) ;
                 dot(x-xo,x-xo) - alpha^2] ;
        K = [ 2 - x(3)*cos(x(1)),        -1              ;
                       -1        , 1 - x(3)*cos(x(2))];
        c = [ -sin(x(1)); -sin(x(2))];
        A = [ K , c; 2*(x-xo)'];
```

```
%...... Compute residual norm, eigenvalues of tangent stiffness
        test = norm(b);   e = eig(K);

%...... Compute increment and update state vector
        dx = -A\b;   x = x + dx;

      end % while

%.... Output results, set converged state, guess at next state
      fprintf('%5i%9.5f%9.5f%9.5f%13.4e%13.4e%4i%12.2e\n',
              n,x',e',nu,test)
      temp = xo;   xo = x; x = 2*x - temp;

    end % loop on n
%.. End of program Newton
```

Program notes. The MATLAB code contains the input in the first few lines. The values set are for the following example. The variable names are basically the same as the notation used in the text with a few exceptions. Note that the name `alpha` stands for the arc length a, `maxsteps` is the number of load steps, and `maxit` is the maximum number of Newton iterations allowed at each load step. The iteration counter ν is called `nu` and the norm of the residual is called `test`. In MATLAB the backslash indicate solution of a linear system of equations so that `dx = A\b` means solve the equations $\mathbf{A}\Delta\mathbf{x} = \mathbf{b}$. The name `xo` stands for the reference value $\mathbf{x}_n$ that serves as the anchor point for the arc length constraint. Many of these same naming conventions will be used in the programs later in the chapter.

Example 79. The program *Newton* was run for the case of the two-degree-of-freedom linkage example from Chapter 10 for the equilibrium path that branches from the first bifurcation point located at $\lambda_{cr} = 0.3820$. The solution tolerance was set at $tol = 10^{-8}$, and the step length was set at $a = 0.5$. The iteration was started at the bifurcation point (0, 0, 0.382) with a suggestion to move to the state (0.851, 0.526, 0.382), that is, in the direction of the first buckling mode. The results of 10 load steps are given here

n	θ1	θ2	Pl/k	EV 1	EV 2	NU	$\|\|\mathbf{b}\|\|$
1	0.26343	0.42487	0.39166	1.9198e-02	2.2458e+00	5	1.85e-09
2	0.53016	0.84665	0.42252	7.8012e-02	2.2775e+00	4	4.98e-14
3	0.80357	1.26124	0.48051	1.8060e-01	2.3395e+00	4	4.97e-13
4	1.08655	1.66184	0.57768	3.3581e-01	2.4478e+00	4	1.20e-11
5	1.37871	2.03580	0.73515	5.6014e-01	2.6292e+00	4	4.78e-10
6	1.66858	2.36008	0.98176	8.7667e-01	2.9161e+00	5	6.66e-16
7	1.92720	2.60639	1.33169	1.2924e+00	3.3177e+00	5	3.87e-14
8	2.13007	2.76947	1.75859	1.7748e+00	3.7965e+00	5	1.18e-14
9	2.28006	2.87216	2.22438	2.2849e+00	4.3079e+00	4	4.94e-09
10	2.39171	2.93830	2.70725	2.8028e+00	4.8298e+00	4	9.81e-11

We can observe the changing of the eigenvalues of the second derivative of the energy in the columns labeled *EV 1* and *EV 2*. The number of Newton iterations required to converge to the specified tolerance are listed in the column labeled *NU*. The norm of the residual is listed in the column marked $\|\mathbf{b}\|$. To get the symmetric counterpart of the equilibrium path given here, we need only suggest negative values for the next state in the input (-0.851, -0.526, 0.382). To get the second branch, we would specify the initial value of (0, 0, 2.618) and the next value as (0.851, -0.526, 2.618).

Newton's Method and Virtual Work

The Newton algorithm, and its variants, provide a numerical tool that can be applied to virtually any nonlinear computation problem. The algorithm for algebraic systems can be extended to functionals without much difficulty. As we saw in the discrete problem, the main idea of Newton was to replace the nonlinear equation with a linear approximation, and solve the linear problem to give a better estimate of the equilibrium state. The process can be iterated to give a solution that is arbitrarily close to the exact solution of the problem. To extend the method to functionals, we need only find an analogy to the approximating linear function $\hat{\mathbf{g}}(\mathbf{x})$. The directional derivative of a functional provides the mathematical machinery we need to make this definition.

Let us consider a virtual-work functional $G(\lambda, \mathbf{u}, \bar{\mathbf{u}})$, where λ is the load parameter, $\mathbf{u}(\mathbf{x})$ is the displacement field, and $\bar{\mathbf{u}}(\mathbf{x})$ is the arbitrary virtual displacement field. The family of configurations $\{\lambda, \mathbf{u}\}$ represents an equilibrium path if $G(\lambda, \mathbf{u}, \bar{\mathbf{u}}) = 0$ for all virtual displacements $\bar{\mathbf{u}} \in \mathfrak{F}_e(\mathfrak{B})$, where $\mathfrak{F}_e$ is the collection of suitable functions, satisfying the essential boundary conditions, defined on the domain $\mathfrak{B}$. The functional G is, by definition, linear in the virtual displacement $\bar{\mathbf{u}}$, but may be nonlinear in the state $\{\lambda, \mathbf{u}\}$.†

The linear functional $\hat{G}(\lambda, \mathbf{u}, \bar{\mathbf{u}})$ will be our analog to the linear function $\hat{\mathbf{g}}(\mathbf{x})$ used in the solution of nonlinear algebraic equations. The linear functional can be obtained as the Taylor series approximation of the functional about the known state $\{\lambda^o, \mathbf{u}^o\}$ as

$$\hat{G}(\lambda, \mathbf{u}, \bar{\mathbf{u}}) \equiv G(\lambda^o, \mathbf{u}^o, \bar{\mathbf{u}}) + DG(\lambda^o, \mathbf{u}^o, \bar{\mathbf{u}}) \tag{603}$$

We can compute the directional derivatives of the functional to be

$$DG(\lambda^o, \mathbf{u}^o, \bar{\mathbf{u}}) \equiv \left[\frac{d}{d\varepsilon} \left[G(\lambda + \varepsilon\,\Delta\lambda, \mathbf{u} + \varepsilon\,\Delta\mathbf{u}, \bar{\mathbf{u}}) \right]_{\varepsilon=0} \right]_{\mathbf{u}=\mathbf{u}^o, \lambda=\lambda^o} \tag{604}$$

where $\Delta\lambda \equiv \lambda - \lambda^o$ and $\Delta\mathbf{u} \equiv \mathbf{u} - \mathbf{u}^o$ are the increments in the state. The principle of virtual work suggests that if $\hat{G}(\lambda, \mathbf{u}, \bar{\mathbf{u}}) = 0$ for all virtual displacements $\bar{\mathbf{u}} \in \mathfrak{F}_e(\mathfrak{B})$, then equilibrium holds for the linearized problem at the

† When we say that a functional is nonlinear in $\mathbf{u}(\mathbf{x})$, we are not referring to the nonlinear variation of $\mathbf{u}$ as a function of $\mathbf{x}$, but rather nonlinearity in the sense that terms like $\|\mathbf{u}\|^2$ appear in the functional. For example, in linear beam theory, the virtual-work functional is linear in the displacement $w(x)$, and yet $w(x)$ generally turns out to be a nonlinear function of x.

load level λ. In accord with Newton's method, we will endeavor to find the state that satisfies $\hat{G}(\lambda, \mathbf{u}, \bar{\mathbf{u}}) = 0$ for all $\bar{\mathbf{u}} \in \mathcal{F}_e(\mathcal{B})$. The linearized virtual-work equation is linear in the incremental state $\Delta\mathbf{u}$ and the incremental load level $\Delta\lambda$. We shall endeavor to solve the linearized problem to determine this incremental state. The state $\mathbf{u}^o + \Delta\mathbf{u}$ should then come closer to satisfying the nonlinear equilibrium equation at the load level $\lambda^o + \Delta\lambda$ than did the configuration $\mathbf{u}^o$ at the load level λ^o. If the new configuration does not satisfy equilibrium well enough, the process can be repeated until it does. This procedure leads to the Newton iteration for functionals

$$\boxed{\begin{aligned} G(\lambda^\nu, \mathbf{u}^\nu, \bar{\mathbf{u}}) + DG(\lambda^\nu, \mathbf{u}^\nu, \bar{\mathbf{u}}) &= 0 \\ \mathbf{u}^{\nu+1} = \mathbf{u}^\nu + \Delta\mathbf{u}^\nu, \quad \lambda^{\nu+1} &= \lambda^\nu + \Delta\lambda^\nu \end{aligned}} \tag{605}$$

where ν counts the iterations. Iteration should cease when $|G(\lambda^\nu, \mathbf{u}^\nu, \bar{\mathbf{u}})| < tol$ (for all virtual displacements).

Remark. We can solve the linear problem approximately using the Ritz method. The Ritz method transforms a continuous problem into a discrete problem expressed in terms of the coefficients of the Ritz expansion. Thus, we can look at the Newton process in another way. Let us apply the Ritz approximation first to get a nonlinear function $G(\lambda, \mathbf{a}, \bar{\mathbf{a}})$ in place of the functional $G(\lambda, \mathbf{u}, \bar{\mathbf{u}})$. The functional will always be linear in the virtual constants $\bar{\mathbf{a}}$, so the discrete form of the virtual-work functional can always be written

$$G(\lambda, \mathbf{a}, \bar{\mathbf{a}}) = \bar{\mathbf{a}}^T\mathbf{g}(\lambda, \mathbf{a})$$

The variational condition that $\bar{\mathbf{a}}^T\mathbf{g}(\lambda, \mathbf{a}) = 0$ for all $\bar{\mathbf{a}}$ implies the equation $\mathbf{g}(\lambda, \mathbf{a}) = \mathbf{0}$, which is simply a nonlinear algebraic equation. The machinery of Newton's method for algebraic functions can, quite obviously, be used to solve this system of equations. This equivalence should become more evident through the following model problem.

A model problem (Euler's elastica). Let us consider the model problem of Euler's cantilever subject to a compressive tip load of P, which we studied in Chapter 11, to illustrate the solution of nonlinear continuous problems. This problem has beauty because it depends on a single displacement function $\theta(x)$. At the same time, the elastica is a bona fide geometrically exact theory in mechanics. We do not need to worry about when the approximations give out as we compute our way along an equilibrium path. Consequently, this classic example will allow us to see the details of the nonlinear computation procedure sketched in the previous section.

The virtual-work functional for the cantilever elastica is given by

$$G(\lambda,\theta,\bar{\theta}) = \int_0^\ell \left(\theta'\bar{\theta}' - \lambda\bar{\theta}\sin\theta\right) dx \tag{606}$$

where $\lambda \equiv P/EI$ is the load parameter. Note that we have factored out the constant EI from the virtual-work functional. This functional is clearly nonlinear in the rotation field $\theta(x)$. The derivatives of the functional in the direction of the increments $\Delta\theta(x)$ and $\Delta\lambda$ can be computed from Eqn. (604) to be

$$DG(\lambda^o,\theta^o,\bar{\theta}) = \int_0^\ell \left(\Delta\theta'\bar{\theta}' - \lambda^o\bar{\theta}\Delta\theta\,\cos\theta^o - \Delta\lambda\bar{\theta}\sin\theta^o\right) dx$$

where $\Delta\lambda \equiv \lambda - \lambda^o$ and $\Delta\theta(x) \equiv \theta(x) - \theta^o(x)$. The state $\{\lambda^o,\theta^o\}$ is known, but does not necessarily represent an equilibrium configuration of the system. Using this result in the definition of the linearized functional, Eqn. (603), we obtain a linear functional for the present case of

$$\hat{G}(\lambda,\theta,\bar{\theta}) = \int_0^\ell \Big[(\theta^o)'\bar{\theta}' - \lambda^o\bar{\theta}\sin\theta^o + \Delta\theta'\bar{\theta}' - \Delta\lambda\bar{\theta}\sin\theta^o - \lambda^o\bar{\theta}\Delta\theta\,\cos\theta^o\Big]\,dx \tag{607}$$

This functional is linear in the increments $\Delta\theta(x)$ and $\Delta\lambda$, and, hence, is amenable to solution methods for linear problems. A solution to the linearized principle of virtual work can be stated as follows: If $\hat{G}(\lambda,\theta,\bar{\theta}) = 0$ for all virtual displacements $\bar{\theta} \in \mathfrak{B}_e(0,\ell)$, then the state $\{\lambda,\theta\}$ is an equilibrium configuration of the linearized problem (but generally not of the nonlinear problem). We can construct the Newton iteration by observing that once we have solved the linear problem, we can take the new estimated state as the point about which we linearize, set up a new linear problem, and solve the new problem for a new state. The iteration can be repeated until convergence obtains.

To solve this problem we can to use the Ritz method. Let us approximate the real and virtual rotation fields with base functions as

$$\theta(x) = \mathbf{a}\cdot\mathbf{h}(x), \quad \bar{\theta}(x) = \bar{\mathbf{a}}\cdot\mathbf{h}(x)$$

where $\mathbf{a} \equiv [a_1, a_2, \ldots, a_N]^T$ and $\mathbf{h} \equiv [h_1, h_2, \ldots, h_N]^T$ are the coefficients and base functions, respectively. The known state $\theta^o(x)$ of the rotation field can be interpolated in the same manner as the state $\theta(x)$. If we do so, then the incremental state also has the same interpolation. To wit, we have

$$\theta^o(x) = \mathbf{a}^o\cdot\mathbf{h}(x), \quad \Delta\theta(x) = \Delta\mathbf{a}\cdot\mathbf{h}(x)$$

where the $\mathbf{a}^o$ are known constants and $\Delta\mathbf{a} = \mathbf{a} - \mathbf{a}^o$. According to our definitions, we have $\mathbf{a} = \mathbf{a}^o + \Delta\mathbf{a}$. Substituting these approximations into the linear functional of Eqn. (607), we arrive at the function

$$\hat{G}(\lambda,\mathbf{a},\bar{\mathbf{a}}) = \bar{\mathbf{a}}^T\left(\mathbf{K}^o\Delta\mathbf{a} + \mathbf{k}^o\Delta\lambda + \mathbf{g}^o\right)$$

where the N by N matrix $\mathbf{K}^o$ is

$$\mathbf{K}^o \equiv \int_0^\ell \left([\mathbf{h}'][\mathbf{h}']^T - \lambda^o \cos\theta^o [\mathbf{h}][\mathbf{h}]^T\right) dx \tag{608}$$

the N by one matrix $\mathbf{k}^o$ is

$$\mathbf{k}^o \equiv -\int_0^\ell \sin\theta^o \mathbf{h}\, dx \tag{609}$$

and the N by one matrix $\mathbf{g}^o$ is

$$\mathbf{g}^o \equiv \int_0^\ell \left((\theta^o)'[\mathbf{h}'] - \lambda^o \sin\theta^o \mathbf{h}\right) dx \tag{610}$$

The statement of equilibrium, i.e., $\hat{G}(\lambda, \mathbf{a}, \bar{\mathbf{a}}) = 0$ for all $\bar{\mathbf{a}}$, implies that the following equation must hold for the increment in the state $\{\Delta\lambda, \Delta\mathbf{a}\}$

$$\mathbf{K}^o \Delta\mathbf{a} + \mathbf{k}^o \Delta\lambda + \mathbf{g}^o = \mathbf{0} \tag{611}$$

To implement the arc-length constraint, we must recognize that the state is now parameterized by the load λ and the displacement parameters $\mathbf{a}$. An appropriate constraint is given by

$$c(\lambda, \mathbf{a}) \equiv \| \mathbf{a} - \mathbf{a}_n \|^2 + (\lambda - \lambda_n)^2 - \alpha^2 = 0$$

where $\{\lambda_n, \mathbf{a}_n\}$ represents a known point, generally on the equilibrium path. It is important to note the distinction between the states $\{\lambda_n, \mathbf{a}_n\}$ and $\{\lambda^o, \mathbf{a}^o\}$ here. The former is the previously found converged equilibrium state, while the latter is simply the latest best guess at the next equilibrium state. We can linearize the arc-length constraint at the state $\{\lambda^o, \mathbf{a}^o\}$ to give

$$\hat{c}(\lambda, \mathbf{a}) \equiv c(\lambda^o, \mathbf{a}^o) + 2(\mathbf{a}^o - \mathbf{a}_n)^T \Delta\mathbf{a} + 2(\lambda^o - \lambda_n)\Delta\lambda = 0 \tag{612}$$

We can use Eqns. (611) and (612) to set up a Newton algorithm. First, let us interpret the state $\{\lambda^o, \mathbf{a}^o\}$ as being the result of the previous Newton iteration, and designate this state as $\{\lambda^\nu, \mathbf{a}^\nu\}$. The starting value of this iteration sequence is $\{\lambda^o, \mathbf{a}^o\}$ and will be taken as the converged state of the previous load step $\{\lambda_n, \mathbf{a}_n\}$. Let us define the $N+1$ by $N+1$ matrix $\mathbf{A}^\nu$ and the N by one matrix $\mathbf{b}^\nu$ in the following manner

$$\mathbf{A}^\nu \equiv \begin{bmatrix} \mathbf{K}^\nu & \mathbf{k}^\nu \\ 2(\mathbf{a}^\nu - \mathbf{a}_n)^T & 2(\lambda^\nu - \lambda_n) \end{bmatrix} \qquad \mathbf{b}^\nu \equiv \begin{bmatrix} -\mathbf{g}^\nu \\ -c(\lambda^\nu, \mathbf{a}^\nu) \end{bmatrix}$$

where $\mathbf{K}^\nu$, $\mathbf{k}^\nu$, and $\mathbf{g}^\nu$ are given by Eqns. (608), (609), and (610), respectively, with the state $\{\lambda^\nu, \mathbf{a}^\nu\}$ substituted in the place of $\{\lambda^o, \mathbf{a}^o\}$. With these matrices identified, Algorithm 2 can be employed without much modification. In order

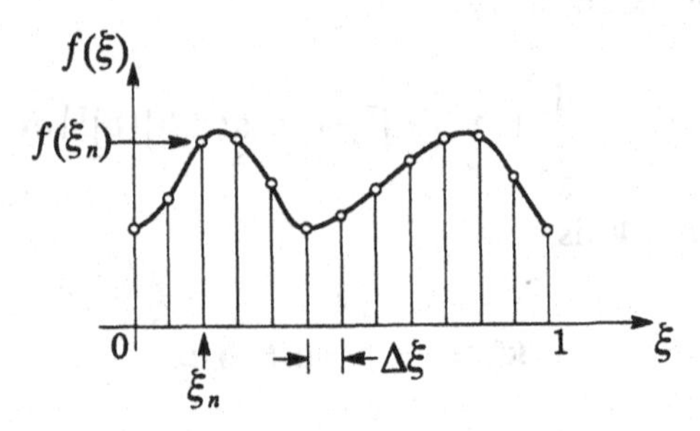

Figure 153 The idea behind numerical integration

to see how the algorithm must be modified, we must consider the problem of integration of the system matrices.

Numerical integration of the system matrices. There is one key difference between the discrete system and the continuous system that we have casually brushed over. The matrices $\mathbf{K}^\nu$, $\mathbf{k}^\nu$, and $\mathbf{g}^\nu$ involve integrals over the domain of the body. In contrast with the solution of linear problems, these integrals involve terms beyond simply the base functions and their derivatives. Each of these integrals has terms like $\cos\theta^\nu(x)$ or $\sin\theta^\nu(x)$ in the integrand. Even if we use polynomial base functions $\{1, x, x^2, \ldots\}$, we would, at each step, need to integrate terms like $x^2 \cos\left(a_o^\nu + a_1^\nu x + a_2^\nu x^2\right)$, where the $\mathbf{a}^\nu$ are constants determined from the previous iteration. Even for the best expert in the integral calculus, the evaluation of such integrals would be a terrible chore. There is an alternative that can be easily implemented into the existing algorithm. That alternative is numerical integration.

Numerical integration procedures can be used to compute any definite integral, provided that the integral exists, to any desired degree of accuracy. Let $f(\xi)$ be any function defined on the interval $\xi \in [0,1]$, for example, the curve shown in Fig. 153. The integral of $f(\xi)$ is simply the shaded area under the curve. Let ξ_m be the location of the mth quadrature point, and ω_m the weight associated with that quadrature point. Then the integral can be computed numerically with the formula

$$\int_0^1 f(\xi)d\xi \approx \sum_{m=0}^{M} \omega_m f(\xi_m) \tag{613}$$

where $M+1$ is the number of quadrature points in the interval. Clearly, numerical quadrature requires only that the function in question be evaluated at selected points. Function evaluation is always a simple computation! Different integration formulas like Gaussian quadrature, Simpson's rule, and the trapezoidal rule are distinguished by their specific weights and integration stations. Integrals on different intervals can always be converted to integrals on the unit interval with a suitable change of variable. For example, if x is defined on the interval $x \in [a,b]$, then the change of variable $x = a(1-\xi) + b\xi$ converts it to the interval $\xi \in [0,1]$. Since $dx = (b-a)d\xi$, the integral can be written as

$$\int_a^b f(x)\,dx = (b-a)\int_0^1 \hat{f}(\xi)\,d\xi$$

where $\hat{f}(\xi) \equiv f\big(a(1-\xi)+b\xi\big)$. Two of the simplest numerical integration formulas are the trapezoidal rule and Simpson's rule.

For the *trapezoidal rule*, the interval is subdivided into M equal segments of length $\Delta\xi$. The $M+1$ points (including the endpoints) that distinguish those segments are the quadrature points. The weights associated with these quadrature points are given by

$$\omega_m = \begin{cases} \dfrac{1}{2M} & \text{if } m = 0 \text{ or } m = M \\[2ex] \dfrac{1}{M} & \text{otherwise} \end{cases}$$

that is, all interior points have weight $\omega_m = 1/M$, while the endpoints have the weight $\omega_0 = \omega_M = 1/2M$. The physical interpretation of the trapezoidal rule is simple. The area is approximated by the sum of the trapezoids formed by connecting two adjacent points with a straight line.

Simpson's rule requires that the domain be divided into an even number of M equal segments of length $\Delta\xi$. The $M+1$ points that distinguish those segments are the quadrature points. The weights associated with these quadrature points are given by

$$\omega_m = \begin{cases} \dfrac{1}{3M} & \text{if } m = 0 \text{ or } m = M \\[2ex] \dfrac{4}{3M} & \text{if } m \text{ is an odd interior point} \\[2ex] \dfrac{2}{3M} & \text{if } m \text{ is an even interior point} \end{cases}$$

For Simpson's rule, three adjacent points are fit with a parabola, and the area under the parabola is computed exactly. The use of three adjacent points to define the parabola is the reason behind needing an even number of segments. For either rule, greater accuracy can be obtained by taking finer and finer subdivisions.

We can incorporate a numerical integration scheme into Algorithm 2. The matrices $\mathbf{K}^\nu$, $\mathbf{k}^\nu$, and $\mathbf{g}^\nu$ are computed by numerical integration at Step 4.a. before incorporation into the matrices $\mathbf{A}^\nu$ and $\mathbf{b}^\nu$. One can clearly see how Algorithm 2 must be amended to incorporate the numerical integration through the specific implementation of the nonlinear analysis procedure for the cantilever elastica example below. You should spend some time comparing and contrasting the program NEWTON with ELASTICA in order to clearly see the connections between the Ritz method and the discrete examples used to motivate stability theory.

The Program ELASTICA

```
%    *---------------------------------------------------------------*
%    |                     Program ELASTICA                          |
%    |    Fundamentals of Structural Mechanics, 2nd Edition          |
%    |           K. D. Hjelmstad, July 1, 2004                       |
%    *---------------------------------------------------------------*

%.. Set problem parameters, Chapter 12, Example 80
    clear; tol = 1.e-8; alpha = 0.5; maxsteps = 10; maxit = 20;
    xo = [0; 0; 2.467];           % Starting value for load path
    x  = [1.0; 0.0; 2.467];       % Guess at initial direction
    xlength = 1.0;                % Length of beam

%.. Initialize values for load step zero,set next trial state
    b = x - xo;  x = xo + alpha*b/norm(b);

%.. Compute MAXSTEPS points along the Equilibrium Path
    for n = 1:maxsteps

%.... Perform Newton iteration at each load step
      nu = 0; test = 1.0;
      while (test > tol) & (nu < maxit)
        nu = nu + 1;

%...... Compute Hessian and residual
        [A,b] = fcn(x,xo,xlength,alpha);

%...... Compute residual norm, eigenvalues of tangent stiffness
        test = norm(b);  e = eig(A(1:2,1:2));

%...... Compute increment and update state vector
        dx = -A\b;  x = x + dx;

      end % while for Newton loop

%.... Output results, set converged state, guess at next state
      fprintf('%5i%9.5f%9.5f%9.5f%10.5f%10.5f%4i%12.2e\n',
                  n,x',e',nu,test)
      temp = xo;  xo = x; x = 2*x - temp;

    end % loop on n Load Steps

%.. End of program Elastica

%---Compute A and b matrices by Simpson integration-----FCN
    function [A,b] = fcn(x,xo,xlength,alpha)

%.. Set weights for Simpson integration
    wt = [1,4,2,4,2,4,2,4,2,4,2,4,2,4,2,4,2,4,2,4,1];
    intpts = length(wt); dz = 1/(intpts-1);

%.. Initialize A and b to zero
    b = zeros(3,1); A = zeros(3,3); z = 0;
```

```
%.. Loop on integration points
    for n = 1:intpts
      factor = wt(n)*dz*xlength/3;

%.... Compute base functions and their derivatives
      ev1 = 1.570796327;                  ev2 = 4.712388981;
      h1  = sin(ev1*z);                   h2  = sin(ev2*z);
      dh1 = ev1*cos(ev1*z)/xlength; dh2 = ev2*cos(ev2*z)/xlength;

%.... Compute rotation and first derivative at current point
      theta  = x(1)*h1  + x(2)*h2;
      dtheta = x(1)*dh1 + x(2)*dh2;
      c1 = x(3)*cos(theta);
      c2 = x(3)*sin(theta);

%.... Compute integral part of residual vector
      b(1) = b(1) - (c2*h1 - dtheta*dh1)*factor;
      b(2) = b(2) - (c2*h2 - dtheta*dh2)*factor;

%.... Compute integral part of Hessian matrix
      A(1,1) =  A(1,1) + (dh1*dh1 - c1*h1*h1)*factor;
      A(1,2) =  A(1,2) + (dh1*dh2 - c1*h1*h2)*factor;
      A(1,3) =  A(1,3) - (sin(theta)*h1)*factor;
      A(2,1) =  A(2,1) + (dh2*dh1 - c1*h2*h1)*factor;
      A(2,2) =  A(2,2) + (dh2*dh2 - c1*h2*h2)*factor;
      A(2,3) =  A(2,3) - (sin(theta)*h2)*factor;

      z = z + dz;
    end % Loop on n

%.. Add part associated with arc length constraint
    b(3) = b(3) + dot(x-xo,x-xo) - alpha^2; A(3,:) = 2*(x-xo)';
    return

%.. End of function FCN
```

Program notes. The program *Elastica* uses many of the same naming conventions as the program *Newton*. This program uses Simpson integration and the weights are the product of the values stored in the array `wt` and scalar `factor`. The function *fcn* carries out the integration of the Hessian and residual and, therefore, contains all of the information about the base functions $h_i(x)$. It should be clear that the addition of the numerical integration procedure changed the program very little.

The program considers a two-term approximation of the rotation field with the two base functions taken to be

$$h_1(x) = \sin\left(\frac{\pi x}{2\ell}\right), \quad h_2(x) = \sin\left(\frac{3\pi x}{2\ell}\right)$$

These base functions are the linearized eigenfunctions for the elastica and, hence, should be excellent choices for this problem.

The program is set up to move along any branch of the bifurcation diagram. The input requires the specification of a direction to head at the start of the computation. This allows the user to start at a bifurcation point and move along a nontrivial path. Without the hint of direction, the iteration will always quickly converge to the trivial branch. The initialization of the first trial state simply moves in the specified direction with a scaling to make sure the initial step has length a.

Example 80. The program *Elastica* was run for the case of the cantilever elastica example from Chapter 11 for the equilibrium path that branches from the first bifurcation point located at $\lambda_{cr} = 2.467$. The solution tolerance was set at $tol = 10^{-8}$, and the step length was set at $a = 0.5$. The iteration was started at the bifurcation point with a suggestion to move to the state (1.0, 0.0, 2.467), that is, in the direction of the first buckling mode. The results of 10 load steps are given as follows

n	*a1*	*a2*	*P/EI*	*EV 1*	*EV 2*	*nu*	‖ **b** ‖
1	0.49401	0.00064	2.54413	0.07586	9.90782	4	1.60e-09
2	0.94382	0.00457	2.76245	0.28290	10.01509	5	1.08e-13
3	1.32196	0.01310	3.08946	0.57255	10.17231	5	9.60e-15
4	1.62598	0.02553	3.48622	0.89812	10.35867	5	4.10e-16
5	1.86824	0.04052	3.92335	1.23187	10.55978	4	1.02e-09
6	2.06338	0.05698	4.38340	1.56176	10.76792	4	6.75e-11
7	2.22332	0.07418	4.85682	1.88378	10.97936	4	4.84e-12
8	2.35673	0.09165	5.33837	2.19727	11.19238	4	4.16e-13
9	2.46983	0.10914	5.82510	2.50281	11.40619	4	4.47e-14
10	2.56705	0.12645	6.31525	2.80133	11.62045	4	5.70e-15

We can see from the above results that the first branch is dominated by the first mode shape long after bifurcation occurs, but that the second mode contributes more and more to the actual shape of the column for states remote from the bifurcation point. If we carry out the computations for the path branching from the second critical point, we find that the participation of the first mode along the second path is negligible. Clearly, there is much in common between the discrete and continuous problems examined in this chapter.

Armed with the Ritz method and numerical integration, we find that the solution of nonlinear problems is quite accessible. The crucial observation that nonlinear problems can be solved by stepping from point to point along an equilibrium path and iterating to convergence at each point is quite powerful. Although there are many difficulties in computational mechanics that will require modification of the algorithms presented here, this basic framework should facilitate the general understanding of why these modifications are needed, and how they come about.

The Fully Nonlinear Planar Beam

The primary purpose of this last section in the chapter is to dispel the notion that somehow all of the preceding computational methods are ad hoc. We can, indeed, extend the basic approach to virtually any computational mechanics problem that we face. The main change is that the bookkeeping for the more

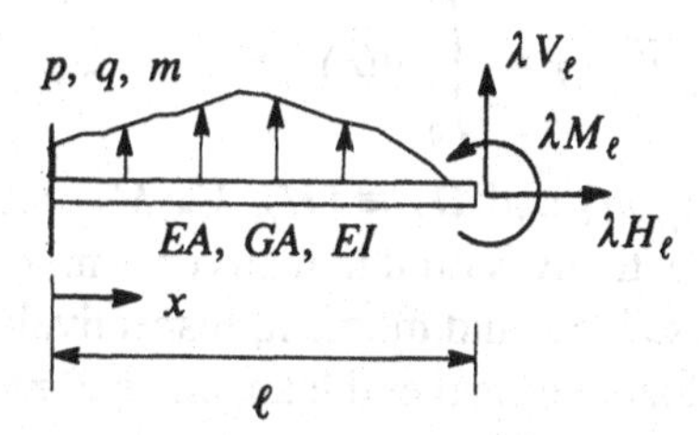

Figure 154 Example configuration for the fully nonlinear beam problem

general theories is a little more tedious. In view of the observation that we will generally write a computer program to carry out the details of our computations, we can afford a high tolerance for the tedium.

Let us carry out the steps of the previous section for the general nonlinear planar beam theory presented in the previous chapter. Specifically, let us consider our model problem of the cantilever column of length ℓ and moduli EA, GA, and EI. The beam is subjected to distributed loads $p(\lambda,x)$, $q(\lambda,x)$, and $m(\lambda,x)$, as well as concentrated end loads λH_ℓ, λV_ℓ, and λM_ℓ, as shown in Fig. 154. It should be clear from the discussion how to accommodate different boundary conditions.

Let us assume that the beam is elastic with the standard linear elastic constitutive equations given by Eqn. (547). Accordingly, the moment M is proportional to the curvature $\varkappa_o$, the shear Q is proportional to the shear strain β_o, and the axial force N is proportional to the axial strain ϵ_o. The constants of proportionality are the moduli EI, GA, and EA, respectively. The virtual-work functional then has the form

$$G(\lambda, \mathbf{u}, \bar{\mathbf{u}}) = \int_0^\ell \left(EI\varkappa_o\bar{\varkappa}_o + GA\beta_o\bar{\beta}_o + EA\epsilon_o\bar{\epsilon}_o\right) dx - \overline{W}_E(\lambda) \qquad (614)$$

where $\overline{W}_E(\lambda)$ is the external-work functional for the applied forces. In our expression for the virtual work, we use the notation $\mathbf{u} \equiv \{u(x), w(x), \theta(x)\}$, a set containing our three displacement functions, to describe the real configuration of the beam. The curvature, shear, and axial strains, $\varkappa_o$, β_o, and ϵ_o, are given in terms of the displacement functions in accord with Eqn. (546). The variations of these strains, $\bar{\varkappa}_o$, $\bar{\beta}_o$, and $\bar{\epsilon}_o$, are given by the directional derivative of the real strains in the direction of the variations of the displacement functions, as described in Chapter 11. The real strains are functions of the real displacements, which are, in turn, functions of the independent variable x. Similarly, the virtual strains are functions of the real and virtual displacements, which are, in turn, functions of the variable x.

The precise expression for the external work depends upon the forces present, and we can customize the expression to each problem. For our example problem, the expression has the form

$$\overline{W}_E(\lambda) = \int_0^\ell \mathbf{q}(\lambda) \cdot \overline{\mathbf{u}}\, dx + \mathbf{Q}_\ell \cdot \overline{\mathbf{u}}(\ell)$$

where $\mathbf{q}(\lambda) \equiv \{p, q, m\}$ and $\mathbf{Q}_\ell \equiv \{H_\ell, V_\ell, M_\ell\}$ are the applied distributed and end loads, respectively, stored in matrix form., with p, q, and m being the axial load, transverse load, and moment, respectively, and H_ℓ, V_ℓ, M_ℓ, being the horizontal end load, vertical end load, and end moment, respectively, and λ is the loading parameter. The designation $q(\lambda, x)$ suggests that the transverse force is not simply proportional to λ, but is a function of it. The other distributed loads are treated similarly. With this designation, we can, for example, specify a loading $q(\lambda, x) = q_1(x) + \lambda q_2(x)$ that has a fixed part and a proportional part. The work of the forces is positive if the forces act in the positive direction of the virtual displacement.

As usual, the equation we wish to solve is the variational equation

$$G(\lambda, \mathbf{u}, \overline{\mathbf{u}}) = 0, \quad \forall \overline{\mathbf{u}} \in \mathcal{T}_e(0, \ell) \tag{615}$$

where $\mathcal{T}_e(0, \ell)$ is a collection of functions $\overline{\mathbf{u}} = \{\overline{u}, \overline{w}, \overline{\theta}\}$, that satisfy the homogeneous essential boundary conditions $\overline{u}(0) = 0$, $\overline{w}(0) = 0$, and $\overline{\theta}(0) = 0$, and whose first derivatives are square-integrable. This collection of functions is the same one required for the linear Timoshenko beam.

Let us consider the known configuration $\mathbf{u}^o = \{u^o, w^o, \theta^o\}$ and load level λ^o. This configuration is not necessarily an equilibrium configuration, i.e., it may be true that $G(\lambda^o, \mathbf{u}^o, \overline{\mathbf{u}}) \neq 0$ for all virtual displacements. We must try to improve our configuration to one that does satisfy equilibrium. A linear incremental expression for the equilibrium equation can be obtained by linearizing the functional G about the known configuration. The linear part of G is

$$\hat{G}(\lambda, \mathbf{u}, \overline{\mathbf{u}}) \equiv G(\lambda^o, \mathbf{u}^o, \overline{\mathbf{u}}) + DG(\lambda^o, \mathbf{u}^o, \overline{\mathbf{u}}) \tag{616}$$

The directional derivatives are computed in accord with Eqn. (604) with the appropriate interpretation of $\mathbf{u}$.

The principle of virtual work suggests that if $\hat{G}(\lambda, \mathbf{u}, \overline{\mathbf{u}}) = 0$ for all virtual displacements $\overline{\mathbf{u}} \in \mathcal{T}_e(0, \ell)$, then equilibrium holds for the linearized problem at the load level λ. The linearized virtual-work equation is linear in the incremental state $\Delta\mathbf{u} \equiv \{\Delta u, \Delta w, \Delta\theta\}$ and the incremental load level $\Delta\lambda$. We shall endeavor to solve the linearized problem to determine this incremental state. The state $\mathbf{u}^o + \Delta\mathbf{u}$ should then come closer to satisfying the nonlinear equilibrium equations at the load level $\lambda^o + \Delta\lambda$ than did the configuration $\mathbf{u}^o$ at the load level λ^o. If the new configuration does not satisfy equilibrium well enough, the process can be repeated until it does. This procedure leads to the Newton iteration for functionals as described in Eqn. (605).

Let us compute the derivatives of the functional G in the direction of an increment in displacement $\Delta\mathbf{u} = \{\Delta u, \Delta w, \Delta\theta\}$ and the increment in load level $\Delta\lambda$. The first directional derivative is given by the expression

$$
\begin{aligned}
DG(\lambda, \mathbf{u}, \overline{\mathbf{u}}) = \int_0^{\ell} & \left[EI\left(D\varkappa_o \cdot \Delta\mathbf{u}\right)\overline{\varkappa}_o + M\left(D\overline{\varkappa}_o \cdot \Delta\mathbf{u}\right) \right. \\
& + GA\left(D\beta_o \cdot \Delta\mathbf{u}\right)\overline{\beta}_o + Q\left(D\overline{\beta}_o \cdot \Delta\mathbf{u}\right) \\
& \left. + EA\left(D\epsilon_o \cdot \Delta\mathbf{u}\right)\overline{\epsilon}_o + N\left(D\overline{\epsilon}_o \cdot \Delta\mathbf{u}\right) \right] dx
\end{aligned}
$$

where $M = EI\varkappa_o$, $Q = GA\beta_o$, and $N = EA\epsilon_o$. The derivatives of the real strains in the direction of the increment in the displacement fields have the explicit form

$$
\begin{aligned}
D\varkappa_o \cdot \Delta\mathbf{u} &= \Delta\theta' \\
D\beta_o \cdot \Delta\mathbf{u} &= \Delta w' \cos\theta - \Delta u' \sin\theta - \left(1+\epsilon_o\right)\Delta\theta \\
D\epsilon_o \cdot \Delta\mathbf{u} &= \Delta w' \sin\theta + \Delta u' \cos\theta + \beta_o \Delta\theta
\end{aligned}
$$

where $1+\epsilon_o = w' \sin\theta + \left(1+u'\right)\cos\theta$ and $\beta_o = w' \cos\theta - \left(1+u'\right)\sin\theta$ are the real axial and shear strains. Of course, the incremental displacements are, as yet, unknown functions of x. The directional derivatives of the virtual strains in the direction of the displacement increments are given as follows

$$
D\overline{\varkappa}_o \cdot \Delta\mathbf{u} = 0
$$

$$
\begin{aligned}
D\overline{\beta}_o \cdot \Delta\mathbf{u} = & - \overline{w}' \sin\theta\Delta\theta - \overline{u}' \cos\theta\Delta\theta \\
& - [\Delta w' \sin\theta + w' \cos\theta\Delta\theta + \Delta u' \cos\theta - (1+u')\sin\theta\Delta\theta]\overline{\theta}
\end{aligned}
$$

$$
\begin{aligned}
D\overline{\epsilon}_o \cdot \Delta\mathbf{u} = & \ \overline{w}' \cos\theta\Delta\theta - \overline{u}' \sin\theta\Delta\theta \\
& + [\Delta w' \cos\theta - w' \sin\theta\Delta\theta - \Delta u' \sin\theta - (1+u')\cos\theta\Delta\theta]\overline{\theta}
\end{aligned}
$$

A consolidated notation will help to keep the formulation clear. Let us introduce the differential operator

$$
\mathbf{F}(\mathbf{u}) \equiv \left[u', w', \theta, \theta'\right]^T
$$

that takes the functions $u(x)$, $w(x)$, and $\theta(x)$ and produces a four by one matrix with the first derivative of u in the first slot, the first derivative of w in the second slot, and θ and its first derivative in the third and fourth slots, respectively. Note that we can use the operator on the virtual displacements to compute the array $\mathbf{F}(\overline{\mathbf{u}}) \equiv [\overline{u}', \overline{w}', \overline{\theta}, \overline{\theta}']^T$ and on the incremental displacements to compute the array $\mathbf{F}(\Delta\mathbf{u}) \equiv [\Delta u', \Delta w', \Delta\theta, \Delta\theta']^T$. Let us store the strains in a matrix as $\mathbf{e}_o = [\varkappa_o, \beta_o, \epsilon_o]^T$ and the resultant forces in a matrix $\mathbf{s} = [M, Q, N]^T$. Let us also introduce a matrix of constitutive properties, which is diagonal for the present constitutive model

$$
\mathbf{D} \equiv \begin{bmatrix} EI & 0 & 0 \\ 0 & GA & 0 \\ 0 & 0 & EA \end{bmatrix}
$$

With this notation at hand, the derivative of the functional G in the direction of the increment in displacement can be expressed as

$$DG \cdot \Delta\mathbf{u} = \int_0^{\ell} \mathbf{F}^T(\overline{\mathbf{u}})\left[\mathbf{E}^T(\mathbf{u})\mathbf{D}\,\mathbf{E}(\mathbf{u}) + \mathbf{G}(\mathbf{u},\mathbf{s})\right]\mathbf{F}(\Delta\mathbf{u})\,dx \tag{617}$$

where the matrix $\mathbf{E}(\mathbf{u})$ is defined to be

$$\mathbf{E}(\mathbf{u}) \equiv \begin{bmatrix} 0 & 0 & 0 & 1 \\ -\sin\theta & \cos\theta & -(1+\epsilon_o) & 0 \\ \cos\theta & \sin\theta & \beta_o & 0 \end{bmatrix} \tag{618}$$

so that $De_o \cdot \overline{\mathbf{u}} = \mathbf{E}(\mathbf{u})\mathbf{F}(\overline{\mathbf{u}})$ and $De_o \cdot \Delta\mathbf{u} = \mathbf{E}(\mathbf{u})\mathbf{F}(\Delta\mathbf{u})$. These identities can be verified from the above equations. The matrix $\mathbf{G}(\mathbf{u},\mathbf{s})$ in Eqn. (617) is defined to be

$$\mathbf{G}(\mathbf{u},\mathbf{s}) \equiv \begin{bmatrix} 0 & 0 & -V & 0 \\ 0 & 0 & H & 0 \\ -V & H & \Xi & 0 \\ 0 & 0 & 0 & 0 \end{bmatrix} \tag{619}$$

where $\Xi \equiv -Vw' - H(1+u')$. Recall that the forces H and V are related to the axial and shear forces N and Q through the relations $H = N\cos\theta - Q\sin\theta$ and $V = N\sin\theta + Q\cos\theta$. This notation allows us to express

$$\mathbf{s} \cdot \left(D\overline{\mathbf{e}}_o \cdot \Delta\mathbf{u}\right) = \mathbf{F}^T(\overline{\mathbf{u}})\mathbf{G}(\mathbf{u},\mathbf{s})\mathbf{F}(\Delta\mathbf{u})$$

The first term in Eqn. (617) gives rise to what we shall call the tangent stiffness matrix when we discretize the problem by the Ritz method. The second term gives rise to the so-called geometric (stiffness) matrix, and is stress-dependent, as it involves the forces H and V. If the tangent stiffness is evaluated at the configuration $\mathbf{u} = \{0,0,0\}$, it reduces to the stiffness from the linear theory.

The derivative of the functional G in the direction of the increment in load level can be expressed as

$$DG \cdot \Delta\lambda = -\int_0^{\ell} \overline{\mathbf{u}} \cdot \frac{\partial \mathbf{q}(\lambda)}{\partial\lambda}\Delta\lambda\,dx - \overline{\mathbf{u}}(\ell) \cdot \mathbf{Q}_{\ell}\Delta\lambda \tag{620}$$

Note that only the external virtual work contributes to this derivative.

Before we discretize Eqn. (617) with the Ritz method, it is worth a brief digression on the stability of equilibrium. We will, at each converged state, need to answer the question of whether or not the equilibrium configuration found is stable. For this we need the second derivative of the energy functional. It is interesting to note that, in essence, we already computed the second derivative of the energy functional $\mathfrak{G}(\mathbf{u})$ when we computed the linearized form of $G(\lambda, \mathbf{u}, \overline{\mathbf{u}})$ for the purpose of the Newton iteration. The energy functional has the form

$$\mathcal{E}(\lambda, \mathbf{u}) = \int_0^{\ell} \tfrac{1}{2}\left(EI\varkappa_o^2 + GA\beta_o^2 + EA\epsilon_o^2\right) dx - \mathcal{E}_E(\lambda, \mathbf{u}) \tag{621}$$

where the energy of the external loads $\mathcal{E}_E(\mathbf{u})$ is given by

$$\mathcal{E}_E(\lambda, \mathbf{u}) = \int_0^{\ell} \mathbf{q}(\lambda) \cdot \mathbf{u}\, dx + \lambda \mathbf{Q}_{\ell} \cdot \mathbf{u}(\ell)$$

Since the external energy is linear in the displacements, the second derivative of $\mathcal{E}_E(\mathbf{u})$ vanishes. Hence, the second derivative of the energy functional comes entirely from the strain energy and has the expression

$$A(\lambda, \mathbf{u}, \bar{\mathbf{u}}) = \int_0^{\ell} \mathbf{F}^T(\bar{\mathbf{u}})\left[\mathbf{E}^T(\mathbf{u})\mathbf{D}\mathbf{E}(\mathbf{u}) + \mathbf{G}(\mathbf{u}, \mathbf{s})\right]\mathbf{F}(\bar{\mathbf{u}})\, dx \tag{622}$$

For any state $\mathbf{u} \equiv \{u(x), w(x), \theta(x)\}$ that satisfies equilibrium at the load level λ, stability is assured if A is positive for all virtual displacements $\bar{\mathbf{u}}$. When we use the Ritz method, this stability condition reduces to the condition that all of the eigenvalues of the tangent stiffness matrix be positive.

Discretization by the Ritz method. The Ritz method is straightforward to apply to the present problem. The main change to what we have already seen is that, in addition to expressing the displacements and their variations in terms of base functions, we must also express the incremental displacements in terms of those base functions. To wit, we have

$$\mathbf{u}(x) = \mathbf{H}(x)\mathbf{a}, \quad \Delta\mathbf{u}(x) = \mathbf{H}(x)\Delta\mathbf{a}, \quad \bar{\mathbf{u}}(x) = \mathbf{H}(x)\bar{\mathbf{a}}$$

where we define the three by $3N$ matrix $\mathbf{H} \equiv \left[\mathbf{H}_1, \mathbf{H}_2, \ldots, \mathbf{H}_N\right]$ with the three by three submatrices $\mathbf{H}_i(x) \equiv h_i(x)\mathbf{I}$, $i = 1, \ldots, N$. Note that the interpolation has N terms. The ordering of $\mathbf{H}$ implies that the unknowns $\mathbf{a}$ are stored in a $3N$-vector $\mathbf{a} \equiv [\mathbf{a}_1^T, \ldots, \mathbf{a}_N^T]^T$, where each three by one vector $\mathbf{a}_i$ is associated with the u, w, θ functions, respectively.

For simplicity, we have chosen to interpolate all of the functions exactly alike. Clearly, there is no reason that we must do so. Expanding all of the various versions of the same function with the same base functions leads to symmetry of the matrices that are generated by the Ritz method. The different displacement types can be interpolated differently, however. For example, one might choose to interpolate the displacements with a polynomial of order N and the rotations with a polynomial of order $N-1$. It is quite clear from the above expression for the Ritz expansions that we must keep track of many things. Fortunately, the computer does this task quite well. You should not really think in terms of carrying out these computations by hand, even for a very small number of base functions.

We need to define some more notation to help us with the bookkeeping. Let us define the four by three matrix $\mathbf{B}_i(x)$ as

$$\mathbf{B}_i(x) \equiv \begin{bmatrix} h_i' & 0 & 0 \\ 0 & h_i' & 0 \\ 0 & 0 & h_i \\ 0 & 0 & h_i' \end{bmatrix}$$

and let us concatenate these matrices into a four by $3N$ matrix of base function derivatives as $\mathbf{B} \equiv [\mathbf{B}_1, \mathbf{B}_2, \ldots, \mathbf{B}_N]$. With this notation, we can express the differential expressions $\mathbf{F}(\Delta\mathbf{u})$ and $\mathbf{F}(\bar{\mathbf{u}})$ in terms of the coefficients of the base functions as

$$\mathbf{F}(\bar{\mathbf{u}}) = \mathbf{B}(x)\bar{\mathbf{a}}, \quad \mathbf{F}(\Delta\mathbf{u}) = \mathbf{B}(x)\Delta\mathbf{a}$$

With these definitions, the linearized virtual-work functional, defined in Eqn. (616), takes the discrete form

$$\hat{G}(\lambda, \mathbf{a}, \bar{\mathbf{a}}) = \bar{\mathbf{a}}^T\big(\mathbf{K}(\mathbf{a})\Delta\mathbf{a} + \mathbf{k}(\lambda)\Delta\lambda + \mathbf{g}(\lambda, \mathbf{a})\big) \tag{623}$$

where $\mathbf{K}(\mathbf{a})$ is the $3N$ by $3N$ tangent stiffness matrix defined as

$$\boxed{\mathbf{K}(\mathbf{a}) \equiv \int_0^{\ell} \mathbf{B}^T\big[\mathbf{E}^T\mathbf{D}\mathbf{E} + \mathbf{G}\big]\mathbf{B}\,dx} \tag{624}$$

where $\mathbf{k}(\lambda)$ is the $3N$ by one matrix defined as

$$\boxed{\mathbf{k}(\lambda) \equiv -\int_0^{\ell} \mathbf{H}^T \frac{\partial \mathbf{q}(\lambda)}{\partial \lambda}\,dx - \mathbf{H}^T(\ell)\mathbf{Q}_{\ell}} \tag{625}$$

and where the residual force $\mathbf{g}(\lambda, \mathbf{a})$, the difference between the internal resisting forces and the externally applied forces, is defined as

$$\boxed{\mathbf{g}(\lambda, \mathbf{a}) \equiv \int_0^{\ell} \big(\mathbf{B}^T\mathbf{E}^T\mathbf{s} - \mathbf{H}^T\mathbf{q}(\lambda)\big)\,dx - \lambda\mathbf{H}^T(\ell)\mathbf{Q}_{\ell}} \tag{626}$$

The dependence upon x of each matrix in the integrand of these expressions is implicit. It should be clear that, for any fixed set of values of the coefficients $\mathbf{a}$, we can compute $\mathbf{K}(\mathbf{a})$, $\mathbf{k}(\lambda)$, and $\mathbf{g}(\lambda, \mathbf{a})$ by carrying out the integration as indicated. It is worth noting that the stiffness matrix at the initial configuration $\mathbf{K}(\mathbf{0})$ is the stiffness matrix of linear analysis. The tangent stiffness matrix $\mathbf{K}(\mathbf{a})$ and the matrix $\mathbf{k}(\lambda)$ can be evaluated by numerical integration. Of course, the integral part of $\mathbf{g}(\lambda, \mathbf{a})$ must also be computed by numerical quadrature. A straightforward computation shows that the integrand can be computed in three by one blocks explicitly as follows

$$\mathbf{B}_i^T\mathbf{E}^T\mathbf{s} - \mathbf{H}_i^T\mathbf{q}(\lambda) = \begin{bmatrix} h_i'H - h_i p \\ h_i'V - h_i q \\ h_i'M + h_i\big(w'H - (1+u')V - m\big) \end{bmatrix}$$

Newton's method with arc-length control. Applying the discrete version of the calculus of variation to the linearized functional in Eqn. (623) gives a linearized principle of virtual work. Since $\hat{G}(\lambda, \mathbf{a}, \bar{\mathbf{a}}) = 0$ for all virtual constants $\bar{\mathbf{a}}$, the term inside the brackets in Eqn. (623) must be identically equal to zero. Therefore, we arrive at an equation to estimate $\Delta\mathbf{a}^\nu$ and $\Delta\lambda^\nu$

$$\mathbf{K}(\mathbf{a}^\nu)\Delta\mathbf{a}^\nu + \mathbf{k}(\lambda^\nu)\Delta\lambda^\nu + \mathbf{g}(\lambda^\nu, \mathbf{a}^\nu) = \mathbf{0} \tag{627}$$

To uniquely determine these increments, we must augment Eqn. (627) with a constraint on how far along the equilibrium path we wish to move. As in the previous formulation, let us take the arc-length constraint as

$$c(\lambda, \mathbf{a}) \equiv \| \mathbf{a} - \mathbf{a}_n \|^2 + (\lambda - \lambda_n)^2 - \alpha^2 = 0 \tag{628}$$

where $\mathbf{a}_n$ and λ_n represent a known configuration on the equilibrium path. Generally, we take this configuration to be the most recent converged state. Linearizing this constraint at the configuration $(\lambda^\nu, \mathbf{a}^\nu)$ allows us to define the linear function

$$\hat{c}(\lambda, \mathbf{a}) = c(\lambda^\nu, \mathbf{a}^\nu) + 2(\mathbf{a}^\nu - \mathbf{a}_n)\Delta\mathbf{a} + 2(\lambda^\nu - \lambda_n)\Delta\lambda$$

Insisting that $\hat{c}(\lambda, \mathbf{a}) = 0$ gives the additional equation needed to determine the incremental state

$$2(\mathbf{a}^\nu - \mathbf{a}_n)\Delta\mathbf{a} + 2(\lambda^\nu - \lambda_n)\Delta\lambda = -c(\lambda^\nu, \mathbf{a}^\nu)$$

Let us define the matrix $\mathbf{A}(\lambda, \mathbf{a})$ and the matrix $\mathbf{b}(\lambda, \mathbf{a})$ as follows

$$\mathbf{A}(\lambda, \mathbf{a}) \equiv \begin{bmatrix} \mathbf{K}(\mathbf{a}) & \mathbf{k}(\lambda) \\ 2(\mathbf{a} - \mathbf{a}_n) & 2(\lambda - \lambda_n) \end{bmatrix}$$

$$\mathbf{b}(\lambda, \mathbf{a}) \equiv \begin{bmatrix} -\mathbf{g}(\lambda, \mathbf{a}) \\ -c(\lambda, \mathbf{a}) \end{bmatrix} \tag{629}$$

The matrix $\mathbf{b}(\lambda, \mathbf{a})$ precisely records the amount by which equilibrium and the arc-length constraint fail to be satisfied at the state $(\lambda, \mathbf{a})$. Thus, any state that is not a fixed distance from the previous point and is not an equilibrium configuration can be improved by solving

$$\mathbf{A}(\lambda^\nu, \mathbf{a}^\nu)\Delta\mathbf{x}^\nu = \mathbf{b}(\lambda^\nu, \mathbf{a}^\nu) \tag{630}$$

where $\mathbf{x} \equiv (\mathbf{a}, \lambda)$ is a matrix that stores both the coefficients of the displacement state and the load factor. Clearly, $\mathbf{x}^\nu = (\mathbf{a}^\nu, \lambda^\nu)$ and $\Delta\mathbf{x}^\nu = (\Delta\mathbf{a}^\nu, \Delta\lambda^\nu)$. Hence, the updated configuration can be obtained as

$$\mathbf{x}^{\nu+1} = \mathbf{x}^{\nu} + \Delta\mathbf{x}^{\nu} \tag{631}$$

Iteration on Eqns. (630) and (631) can be continued until the solution converges to within a specified tolerance, i.e., $\| \mathbf{b}(\lambda^{\nu}, \mathbf{a}^{\nu}) \| < tol$. Clearly, the success of the method depends upon the radius of convergence of the system. Convergence can generally be guaranteed by selecting suitably small values of the arc-length parameter α. The matrix $\mathbf{A}(\lambda^{\nu}, \mathbf{a}^{\nu})$ should be invertible at limit points. Bifurcation points cause the algorithm difficulty because there is no way to specify which branch the next solution point must be on. With the definitions of the state and the appropriate matrices, Algorithm 2 applies to this problem, as modified with numerical integration of the coefficient matrices.

The discretization also gives us a discrete version of the second-derivative test for stability of equilibrium. The second derivative of the energy can now be expressed in terms of the tangent stiffness matrix as

$$A(\mathbf{a}, \bar{\mathbf{a}}) = \bar{\mathbf{a}}^T \mathbf{K}(\mathbf{a}) \bar{\mathbf{a}} \tag{632}$$

As we saw previously, the requirement that A be positive is tantamount to all of the eigenvalues of the tangent stiffness matrix $\mathbf{K}(\mathbf{a})$ being positive. A critical point on the path is a point where one of the eigenvalues of the tangent stiffness matrix is identically equal to zero.

Remark. There are two kinds of errors we must be concerned with: (a) the error in the Ritz approximation, and (b) the error in equilibrium that we are trying to iterate away. For a fixed set of base functions, the error in the spatial distribution of the functions, like the transverse displacement, is fixed. The only way to reduce this error is to take more base functions, and thereby add unknown coefficients to the problem. Adding more base functions is called improving the spatial approximation. Newton's method, on the other hand, can find the solution to any level of accuracy (permitted by the finite precision arithmetic of the computer, of course). We must always specify the solution tolerance to tell Newton when the solution is adequate. As such, we will always view equilibrium as a condition that can be exactly satisfied. These issues may be clearer in the context of the following computer program, NONLINEARBEAM, which implements the foregoing derivations.

The Program NONLINEARBEAM

```
%  *-----------------------------------------------------------------*
%  |                     Program NonlinearBeam                       |
%  |         Fundamentals of Structural Mechanics, 2nd Edition       |
%  |                  K. D. Hjelmstad, July 1, 2004                  |
%  *-----------------------------------------------------------------*

%.. Set problem parameters, Chapter 12, Example 81,
    clear; tol = 1.e-8; alpha = 10; maxsteps = 10; maxit = 40;
    nbasis = 6; ndm = 3*nbasis + 1; nnstep = 1;     npts = 21;
```

```
%.. Applied loads
    loads = [  0.0,  0.0,  0.0   ;          % p1, p2, H(L)
               0.0,  0.0,  0.0   ;          % q1, q2, V(L)
               0.0,  0.0,  6.28 ];          % m1, m2, M(L)

%.. Beam moduli D = [EI, GA, EA], beam length is xlength
    D = [1e3, 1e6, 1e6]; xlength = 10.0;

%.. Initialize values for load step zero, set next trial state
    x = zeros(ndm,1); xo = zeros(ndm,1); x(ndm) = alpha;

%.. Compute MAXSTEPS points along the Equilibrium Path
    for n = 1:maxsteps

%.... Perform Newton iteration at each load step
      nu = 0; test = 1.0;
      while (test > tol) & (nu < maxit)
        nu = nu + 1;

%...... Compute Hessian and residual
        [A,b] = fcn(D,x,xo,loads,xlength,ndm,nbasis,alpha);

%...... Compute residual norm, min eigenvalue of tangent stiffness
        test = norm(b);  e = min(eig(A(1:ndm-1,1:ndm-1)));

%...... Compute increment and update state vector
        dx = -A\b;  x = x + dx;

      end % while for Newton loop

%.... Output results, set converged state, guess at next state
      results(x,xlength,n,nu,test,ndm,nbasis,nnstep,npts)
      temp = xo;  xo = x; x = 2*x - temp;

    end % loop on n Load Steps

%.. End of program NonlinearBeam

%--- Compute A and b for NonlinearBeam element -------------FCN
    function [A,b] = fcn(D,x,xo,loads,xlength,ndm,nbasis,alpha)

%.. Initialize A and b to zero
    b = zeros(ndm,1); A = zeros(ndm,ndm); z = 0;

%.. Set weights for Simpson integration
    wt = [1,4,2,4,2,4,2,4,2,4,2,4,2,4,2,4,2,4,2,4,1];
    intpts = length(wt); dz = 1/(intpts-1);

%.. Loop on integration points
    for n = 1:intpts
      factor = wt(n)*dz*xlength/3;

%.... Compute displacements and derivatives
      du = 0; dw = 0; dtheta = 0; theta = 0;
      for i = 1:nbasis
         [h,dh] =  basis(i,z,xlength);
         du = du + x(3*i-2)*dh;
         dw = dw + x(3*i-1)*dh;
         dtheta = dtheta + x(3*i)*dh;
         theta  = theta  + x(3*i)*h;
      end
```

```
      ct = cos(theta); st = sin(theta);

%.... Compute axial strain, shear strain, and curvature
      epsi = dw*st + (1+du)*ct - 1;
      beta = dw*ct - (1+du)*st;
      curv = dtheta;

%.... Compute axial force, shear force, bending moment, etc.
      bend  = D(1)*curv; shear = D(2)*beta; axial = D(3)*epsi;
      Hor = axial*ct - shear*st; Ver = axial*st + shear*ct;
      Xi = (1+du)*Hor + dw*Ver; Yi = dw*Hor - (1+du)*Ver;
      force = [Hor; Ver; bend]; fyi = [0; 0; Yi];
      c1 = loads(:,1) + loads(:,2)*x(ndm);
      c2 = fyi - c1;
      c3 = loads(:,2);
      c4 = loads(:,3);

%.... Compute components of [E(T)DE + G] store in matrix G
      G(1,1) = D(2)*st*st + D(3)*ct*ct;
      G(1,2) = ct*st*(D(3)-D(2));
      G(1,3) = D(2)*st*(1+epsi) + D(3)*ct*beta - Ver;
      G(1,4) = 0;
      G(2,1) = G(1,2);
      G(2,2) = D(2)*ct*ct + D(3)*st*st;
      G(2,3) = D(3)*st*beta - D(2)*ct*(1+epsi) + Hor;
      G(2,4) = 0;
      G(3,1) = G(1,3);
      G(3,2) = G(2,3);
      G(3,3) = D(2)*(1+epsi)^2 + D(3)*beta^2 - Xi;
      G(3,4) = 0;
      G(4,1) = G(1,4);
      G(4,2) = G(2,4);
      G(4,3) = G(3,4);
      G(4,4) = D(1);

%.... Form stiffness matrix and residual
      for i = 1:nbasis
        [hi,dhi] = basis(i,z,xlength); mm = 3*(i-1);
        for j = 1:nbasis
          [hj,dhj] = basis(j,z,xlength); nn = 3*(j-1);

%........ Compute B(T)[E(T)DE + G]B noting the sparse structure of B
          GB(1:4,1:2)  = dhj*G(1:4,1:2);
          GB(1:4,3)    =  hj*G(1:4,3) + dhj*G(1:4,4);
          BGB(1:2,1:3) = dhi*GB(1:2,1:3);
          BGB(3,1:3)   =  hi*GB(3,1:3) + dhi*GB(4,1:3);

%........ Assemble the result into the A matrix
          A(mm+1:mm+3,nn+1:nn+3) = A(mm+1:mm+3,nn+1:nn+3) + BGB*factor;
        end % loop on j

%...... Form integral part of residual force and assemble into matrix
        b(mm+1:mm+3) = b(mm+1:mm+3) + (dhi*force + hi*c2)*factor;
        A(mm+1:mm+3,ndm) = A(mm+1:mm+3,ndm) - hi*c3*factor;

      end % loop on i
      z = z + dz;
    end % Loop on n
```

```
%.. Add end load terms to the residual and coefficient matrix
    for i = 1:nbasis
      [hi,dhi] = basis(i,1,xlength); mm = 3*(i-1);
      b(mm+1:mm+3)      = b(mm+1:mm+3)      - hi*c4*x(ndm);
      A(mm+1:mm+3,ndm) = A(mm+1:mm+3,ndm) - hi*c4;
    end

%.. Add arc-length constraint terms to Hessian and residual
    b(ndm) = b(ndm) + dot(x-xo,x-xo) - alpha^2;
    A(ndm,1:ndm) = 2*(x-xo)';

    return

%.. End of function fcn

%--- Evaluate ith basis function h and derivative dh ----------BASIS
    function [h,dh] = basis(i,z,xlength)

%.. Compute sequence m=[1,1,2,2,3,3] for use in base functions
    n = mod(i,2); m = (i+n)/2; a = (2*m-1)*pi/2;
    if n == 0
      h = sin(a*z);       dh = a*cos(a*z)/xlength;
    else
      h = 1 - cos(a*z); dh = a*sin(a*z)/xlength;
    end
    return

%.. End of function basis

%--- Print and plot results of current step ------------------RESULTS
    function [] = results(x,xlength,n,nu,test,ndm,nbasis,nnstep,npts)

%.. Determine if current step is an output step
    if mod(n,nnstep) == 0

%.... Compute and print current geometry of beam
      axis('square'); hold on;
      z = 0; dz = 1/(npts-1);
      for ii = 1:npts
        u = zeros(3,1);
        for i = 1:nbasis
          [h,dh] = basis(i,z,xlength); mm = 3*(i-1);
          u(:) = u(:) + h*x(mm+1:mm+3);
        end
        y1(ii) = z*xlength + u(1); y2(ii) = u(2);
        z = z + dz;
      end
      fprintf('%5i%9.5f%14.4e%14.4e%14.4e%4i%12.2e\n',
                   n,x(ndm),u',nu,test)
      plot(y1,y2,'-');

    end
    return

%.. End of function results
```

Program notes. The program NonlinearBeam uses many of the same naming conventions as the previous programs. The initialization for Example 81 is to set the displacements to zero and the load factor to α. The function *results* produces output every `nnstep` load steps. The output consists of the usual summary table and a plot of the deformed shape of the structure, computed at `npts` stations along the length of the beam using the base functions to interpolate.

Again the formation of **A** and **b** is done in the function *fcn*, with the integrals being carried out by Simpson's rule. The matrix $\mathbf{E}^T\mathbf{D}\mathbf{E}+\mathbf{G}$ has been carried out by hand rather than forming the individual matrices. The multiplication by the matrices $\mathbf{B}_i$ has been done recognizing the sparse structure of those matrices. The information about the base functions $h_i(x)$ is now in the function *basis*. The number of base functions used in the analysis is controlled by the parameter `nbasis` in the main program. The base functions implemented in the program are

$$h_1(x) = \sin\left(\frac{\pi x}{2\ell}\right), \quad h_2(x) = 1-\cos\left(\frac{\pi x}{2\ell}\right)$$

$$h_3(x) = \sin\left(\frac{3\pi x}{2\ell}\right), \quad h_4(x) = 1-\cos\left(\frac{3\pi x}{2\ell}\right)$$

$$h_5(x) = \sin\left(\frac{5\pi x}{2\ell}\right), \quad h_6(x) = 1-\cos\left(\frac{5\pi x}{2\ell}\right)$$

The only "clever" part of the function *basis* is the method of computing the sequence $m = [1, 1, 2, 2, 3, 3, \ldots]$ to aid the pairwise definition of the base functions with indices $i = [1, 2, 3, 4, 5, 6, \ldots]$. An even number of base functions seems appropriate. The motivation for using these functions is the observation that the functions $\sin(\mu_i\xi)$ are the linearized eigenfunctions for the rotation field for the inextensible elastica, and the functions $1-\cos(\mu_i\xi)$ are the linearized eigenfunctions of the transverse displacement of the inextensible elastica, where the eigenvalues are $\mu_i = (2i-1)\pi/2\ell$. Since we are using the same functions for all three displacement fields, both base functions are needed for each displacement field.

The unknowns are stored in the matrix **x** with the ordering given earlier. The last element in the array `x`, i.e., `x(ndm)`, is the load factor λ. The array `xo` gives the values of `x` at the last converged state. The array `loads` contains the vectors $\mathbf{q}_1 \equiv [p_1, q_1, m_1]^T$ and $\mathbf{q}_2 \equiv [p_2, q_2, m_2]^T$ in its first two columns for the definition of the distributed load $\mathbf{q}(\lambda) \equiv \mathbf{q}_1+\lambda\mathbf{q}_2$. The third column contains the end loads $\mathbf{Q}_\ell = [H_\ell, V_\ell, M_\ell]^T$.

Example 81. The program NONLINEARBEAM was run for the case of a cantilever beam of length $\ell = 10$ and bending modulus $EI = 1000$ (the shear and axial moduli were taken to be $GA = EA = 10^6$). The beam was subjected to an end moment λM_o, with $M_o = 6.28$. The configuration for this example is shown in Fig. 155. The solution tolerance was set at $tol = 10^{-8}$, and the step length was set at $\alpha = 10$. The exact solution to this problem of pure bending is that the beam bends into a circular arc of radius $\varrho = M/EI$. The summary output results of 10 load steps are given below

n	Load Fact.	u(L)	w(L)	θ(L)	nu	|| b ||
1	9.39690	-5.6631e-01	2.8547e+00	5.8528e-01	8	3.96e-09
2	18.87009	-2.1281e+00	5.2136e+00	1.1605e+00	6	1.12e-09
3	28.35467	-4.3203e+00	6.7138e+00	1.7326e+00	7	1.74e-09
4	37.57238	-6.8925e+00	7.2049e+00	2.3570e+00	8	1.83e-09
5	47.06530	-9.3157e+00	6.6691e+00	2.9711e+00	10	2.19e-09
6	56.54820	-1.1177e+01	5.3321e+00	3.5940e+00	8	4.22e-09
7	66.12599	-1.2047e+01	3.6671e+00	4.1577e+00	9	3.12e-09
8	75.89923	-1.2012e+01	2.0095e+00	4.6987e+00	10	3.60e-09
9	85.71366	-1.1195e+01	5.9709e-01	5.3338e+00	8	4.42e-09
10	95.54433	-1.0069e+01	2.9521e-03	5.9969e+00	10	3.19e-09

Notice that the ratio of the load factor to the rotation is constant for all steps in accord with the exact solution. The displaced configuration for certain selected load levels is shown in Fig. 155(b). It is evident that the base functions are able to represent the circular arc of the exact solution. Problem 299 asks you to examine the polynomial basis for this problem. Can the polynomial basis accommodate the circular shape?

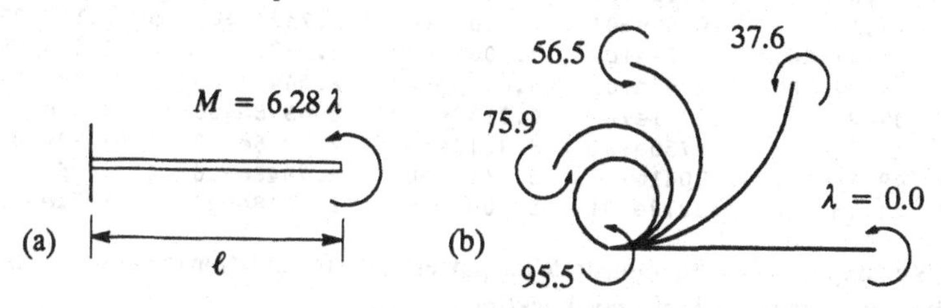

Figure 155 Example 81, (a) problem geometry and loading, (b) displaced shapes of beam (certain load steps only)

Example 82. The program NONLINEARBEAM was run for the case of a cantilever beam of length $\ell = 10$ and bending modulus $EI = 1000$ (the shear and axial moduli were taken to be $GA = EA = 10^6$). The beam was subjected to a compressive end load $-\lambda P_o$, with $P_o = 2.46$, and a nonproportional transverse load. Two cases of transverse load were considered, one with $q_1 = 0.2$ and the other with $q_1 = 0.05$. The configuration for this example is shown in Fig. 156(a). The solution tolerance was set at $tol = 10^{-8}$, and the step length was set at $a = 0.5$.

The bifurcation diagrams for this problem are given in Figure 156(c,d,e). Since six base functions were used and there are three independent displacement fields, it is not clear how to present the results in a bifurcation diagram. Figure 156(c,d,e) is one possible method of presentation wherein we plot the three displacement fields at the end of the beam against the load parameter. Several features of this example problem are of interest. First, it is clear that the buckling load is located in the vicinity of $\lambda = 10$ because both bifurcation diagrams take a sharp bend near that load level. This result is in accord with the value predicted by the linearized buckling theory for the beam without lateral load. Second, the difference between the two lateral load levels is evident early in the response, but the load-deformation curves coalesce at large deformations.

Displaced configurations of the beam are shown, to scale, for equal increments along the load path in Fig. 156(b). We can observe that the transverse displacement reaches a maximum and then decreases. An examination of Fig. 156(c,d,e) shows why. The transverse load q_1 remains fixed throughout the analysis. Hence, its magnitude diminishes relative to the axial load as the load factor increases. The effect of different transverse loads cannot be distinguished for load levels above the buckling load. In a sense, the fixed load acts the same as a geometric imperfection.

A sample of the output is given below. Note that the output has been edited to show only every tenth load step.

n	*Load*	*u(L)*	*w(L)*	*θ(L)*	*nu*	|| **b** ||
10	4.91611	-1.3609e-02	4.7803e-01	6.8900e-02	5	9.95e-10
20	9.27598	-3.7206e-01	2.4361e+00	3.8262e-01	7	4.62e-09
30	11.53529	-2.9260e+00	6.2347e+00	1.1105e+00	6	1.03e-09
40	15.12410	-6.1826e+00	7.8527e+00	1.6681e+00	6	1.62e-09
50	19.67508	-8.4840e+00	7.9860e+00	2.0050e+00	5	2.66e-09
60	24.45017	-1.0059e+01	7.6037e+00	2.2421e+00	5	1.12e-09
70	29.31286	-1.1129e+01	7.1069e+00	2.4177e+00	5	2.09e-09
80	34.24278	-1.1856e+01	6.6468e+00	2.5447e+00	5	3.68e-09
90	39.20700	-1.2371e+01	6.2533e+00	2.6368e+00	4	4.90e-09
100	44.18688	-1.2750e+01	5.9219e+00	2.7046e+00	4	2.87e-09
110	49.17399	-1.3041e+01	5.6427e+00	2.7548e+00	4	2.06e-09
120	54.16438	-1.3272e+01	5.4062e+00	2.7918e+00	4	1.71e-09

This example serves to demonstrate the power of the formulation to compute the bifurcation diagrams in the postbuckling region.

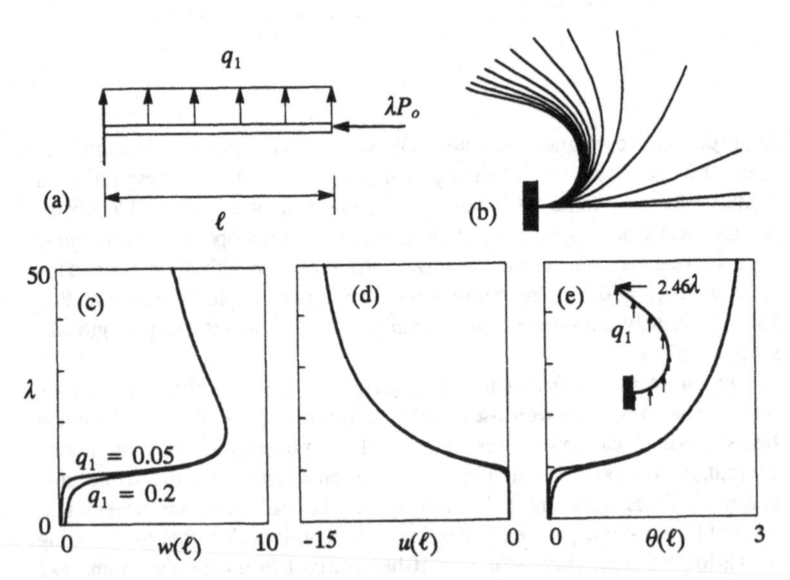

Figure 156 Bifurcation diagrams for a beam subjected to proportional axial load with constant transverse load (a) problem geometry, (b) deformed shapes at output points, (c) load versus vertical displacement at end, (d) load versus horizontal displacement at end, (e) load versus rotation at end

Summary

Many problems in mechanics are nonlinear. Some of the most interesting and important phenomena associated with the mechanical behavior of deformable bodies are artifacts of nonlinearity. This chapter has been a brief introduction to the possibility of numerically computing the response of a nonlinear system. Nonlinear analysis remains something of an art because every system exhibits nonlinearity in different degrees and manifests in different ways. An algorithm that converges well for one problem may not work as well for another.

The objective of the current chapter was simply to give a baseline understanding of nonlinear computations in mechanics and to arm you, the reader, with an algorithmic framework from which to launch further study. Treatment of the subject of nonlinear computations by exhaustively enumerating the many ad hoc methods that have appeared in the literature did not seem to have the same pedagogical merit as the approach of outlining a unified framework based upon Newton's method with arc-length constraints, and so we chose the latter over the former. The computer programs included in this chapter were meant more for study than for practical use. The limitations of these simple programs should be evident. Perhaps the simplicity will aid the understanding of nonlinear computations. Please pardon the biases of this presentation toward Newton's method, toward arc-length constraints, and toward the MATLAB programming language.

Good luck traversing the rocky terrain of nonlinear problems. May the convergence of all of your problems be quadratic.

Additional Reading

R. Fletcher, *Practical methods of optimization*, Wiley, New York, 1987.

D. G. Luenberger, *Linear and nonlinear programming*, Addison-Wesley, Reading, Mass., 1984.

R. Pratrap, *Getting started with MATLAB: A quick introduction for scientists and engineers*, Saunders College Publishing, New York, 1996.

M. A. Crisfield, *Non-linear finite element analysis of solids and structures, Volume* 1: *Essentials*, John Wiley & Sons, New York, 1991.

M. A. Crisfield, *Non-linear finite element analysis of solids and structures, Volume* 2: *Advanced Topics*, John Wiley & Sons, New York, 1997.

G. A. Holzapfel, *Nonlinear solid mechanics. A continuum approach for engineering*, John Wiley & Sons, 2000.

J. Bonet and R. D. Wood, *Nonlinear continuum mechanics for finite element analysis*, Cambridge University Press, 1997.

Problems

286. Modify the program NEWTON to account for initial geometric imperfections in the two-bar rigid linkage connected by rotational springs.

287. Modify the program NEWTON to analyze the three-bar rigid linkage shown. The bars are hinged together and are restrained by elastic springs that resist vertical motion. The springs accrue force in proportion to their extension, with modulus k. The system is subjected to an axial force P.

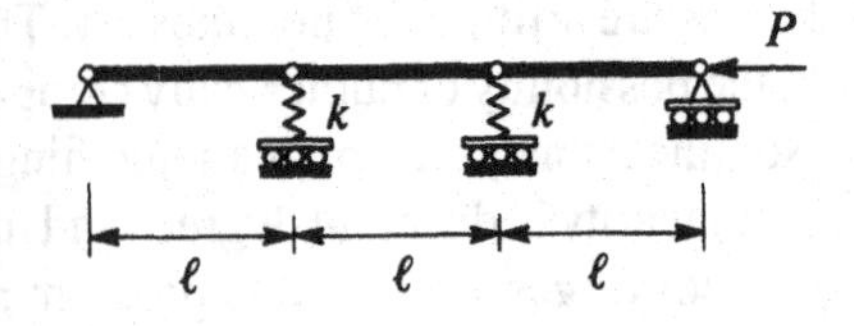

288. Implement the constraint $c(\boldsymbol{\theta},\lambda) = \det \mathbf{A}(\boldsymbol{\theta},\lambda) = 0$ into the program NEWTON to locate bifurcation points exactly. At the bifurcation point, compute the eigenvectors of the tangent stiffness matrix and switch to another equilibrium branch.

289. Modify the program NEWTON to analyze the three-bar rigid linkage shown below. This structure has three degrees of freedom. The bars are hinged together and are restrained by elastic springs that resist relative rotation. The springs accrue force in proportion to their extension, with modulus k. The system is subjected to an axial force P.

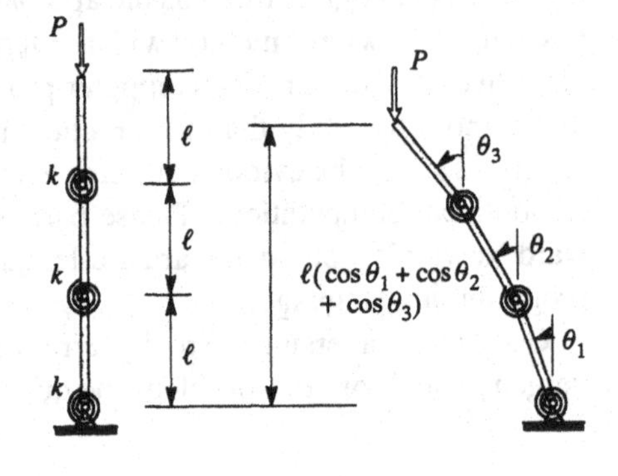

Lateral loads can be viewed as imperfections to a purely axial loading system. Modify the equations of equilibrium to allow the applications of the lateral loads $\epsilon_1 P$, $\epsilon_2 P$, and $\epsilon_3 P$ at locations ℓ, 2ℓ, and 3ℓ respectively, where ϵ_i is a fixed value recording the ratio of the lateral load to the axial load. Implement the load imperfections in the program.

290. Modify the program ELASTICA to incorporate a distributed transverse loading $q(x)$ on the cantilever column in addition to the load P. Examine the case where the transverse load is proportional to the axial load, as well as the case where the transverse load is fixed and the axial load is increased.

291. Modify the program ELASTICA to incorporate N base functions. Examine the performance of the system as the number of base functions is increased.

292. Modify the program ELASTICA to use polynomial base functions. Examine the performance of the system as the number of base functions is increased. Compare the performance of the polynomial functions with the eigenfunctions.

293. Modify the program ELASTICA to account for a nonlinear moment curvature relationship of the form

$$M(\varkappa_o) = \frac{EI_o \varkappa_o}{\sqrt{1+\mu_o \varkappa_o^2}}$$

where EI_o and μ_o are material constants. Note that for small values of μ_o, the constitutive model reduces to the linear model originally used. The material is hyperelastic because an energy function exists such that $M = \partial W(\varkappa_o)/\partial \varkappa_o$. What is the strain energy function W? Plot the bifurcation diagrams for various values of the material constants.

294. Modify the program ELASTICA to use piecewise linear finite element base functions. Examine the performance of the system as the number of base functions is increased. Compare the performance of the finite element base functions with the polynomial functions.

295. Describe a method for using the program NONLINEARBEAM to locate the bifurcation points of a system without imperfections.

296. Explore the features of the program NONLINEARBEAM by using it to solve the cantilever beam problem under a variety of loading scenarios.

297. Modify the program NONLINEARBEAM to account for initial geometric imperfections in the column. Is it sufficient to specify imperfections only in the field $w(x)$?

298. Modify the subroutine *basis* in the program NONLINEARBEAM to use piecewise linear finite-element base functions. Examine the performance of the system as the number of base functions is increased. Compare the performance of the finite element base functions with the sinusoids used in the original program.

299. Modify the *basis* subroutine in the program NONLINEARBEAM to use the polynomials $h_i(x) \in \{x, x^2, \ldots, x^N\}$. Examine the performance of the system as the number of base functions is increased using the pure bending problem. Are these functions able to capture the exact solution, which is a circular shape, as shown in the text example? Why is convergence so difficult with a large number of base functions? Implement the orthogonal base functions described in Chapter 6. Do these base functions work better than the original polynomials?

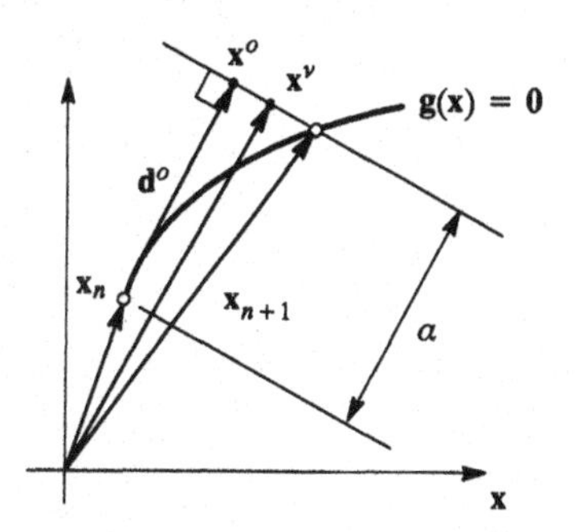

300. The arc-length constraint forces the next equilibrium configuration to be a fixed distance from the previous converged state. Therefore, all iterates must lie on a sphere of radius a centered on the converged state $\mathbf{x}_n$. One of the problems with this strategy is that the equilibrium path pierces the sphere at two points (at least). For a highly nonlinear equilibrium path the Newton iterations can converge to the other point on the sphere, which causes the loading direction to change. We can observe this phenomenon in the program NONLINEARBEAM if the step size is not judiciously chosen. Once the loading direction has turned around, it is not likely to change back. Consider another possible constraint pictured on the following page.

The difference between the estimate $\mathbf{x}^\nu$ and $\mathbf{x}^o$ is forced to lie in a hyperplane normal to the tangent direction $\mathbf{d}^o \equiv \mathbf{x}^o - \mathbf{x}_n$. The normal plane is set a distance a from $\mathbf{x}_n$.

Therefore, $\| \mathbf{d}^o \| = a$. How can one compute the tangent direction $\mathbf{d}^o$? Show that the normality condition is $\mathbf{d}^\nu \cdot \mathbf{d}^o = a^2$, where the νth increment in state is $\mathbf{d}^\nu \equiv \mathbf{x}^\nu - \mathbf{x}_n$. Develop a method based upon a secant direction where $\mathbf{d}^o \equiv \mathbf{x}_n - \mathbf{x}_{n-1}$ (the previous two converged states). How would you start a method based on this definition? Implement these constraints in the program NONLINEARBEAM and assess their performance. Are there advantages over the arc-length constraint? Are there disadvantages?

301. When using finite element base functions in the programs ELASTICA and NONLINEAR-BEAM, most of the integration points contribute nothing to the integrals because the base functions are zero over much of the region. Restructure the order of the programs to make them more efficient by putting the loop over integration points inside the loop over base functions.

302. Add a subroutine to find the eigenvalues and eigenvectors of $\mathbf{K}(\mathbf{a})$ at each point on the equilibrium path in the program NONLINEARBEAM to examine the stability of equilibrium. Implement a procedure for branch switching so that the program will trace the bifurcation diagram when there are no imperfections.

Index

Italic numbers refer to worked examples in the text. **Bold** numbers refer to problems at the back of a chapter.

C

D

E

F

Q

R

S

W

Y